Magmatic Systems

This is Volume 57 in the
INTERNATIONAL GEOPHYSICS SERIES
A series of monographs and textbooks
Edited by RENATA DMOWSKA AND JAMES R. HOLTON

A complete list of the books in this series appears at the end of this volume.

The eruption of the Krafla central volcano, northeast Iceland during September 4–18, 1984. (*Top*) 06:00 hours on September 5th. Silhouetted against dawn's light, the coalescing plumes from seven en-echelon-arranged eruptive fissures rise over the Gjastykki fissure system. Fountain heights during these first hours of eruption averaged 50–75 m and had flow rates that tended to be maximized over an individual fissure's center. Prior to the eruption, magma was stored within Krafla's reservoir located beneath the caldera floor (right margin of photograph). (*Bottom*) 21:30 hours on September 11th. Within 72 hours the eruption had localized at this northernmost vent, shown here in the process of building a large spatter cone. In this eastward view, the fountain heights ranged from 30–40 m above the cone top. Collectively, magma flow at Krafla illustrates a number of dynamic patterns and processes: lateral injection at depth along the horizon of neutral buoyancy, laminar flow during crack migration, turbulent flow above surface fissures, and thermally constricted flow beneath volcanic vents. U.S. Geological Survey photographs by Michael P. Ryan.

Magmatic Systems

Edited by

Michael P. Ryan
U.S. Geological Survey
Reston, Virginia

ACADEMIC PRESS
San Diego New York Boston
London Sydney Tokyo Toronto

Cover photo: Sakurajima volcano, Kyushu, Japan, in eruption at Minami-Dake (the southern crater) during the early morning hours (03:09) of February 17, 1988. This remarkable photograph was made possible through the use of an automated system that continuously records atmospheric pressure. The expanding atmospheric shock wave generated during the eruption outbreak triggered the camera at the Sakurajima Volcanological Observatory, 5.5 km from the summit. The integrated intensity of the resulting Strombolian bombardment has been recorded by the time-lapse photographic process. Incandescent volcanic bombs are seen to follow both steep and low-angle trajectories, blanketing the upper flanks and rupturing on impact. Bomb rupture produces secondary showers of fragments down paths revealed by their bright comet-shaped trails on the volcano flanks. High in the ash plume, lightning strokes discharge buildups of static electricity and usually emanate from, or terminate in, the higher conductivity incandescent plume core.

Sakurajima is a tightly coalesced pair of pyroxene andesite and dacite stratovolcanoes that have grown within the Aira caldera during the last 13,000 years. The formation of the Aira caldera itself was associated with the great eruption of the Ito pyroclastic flow about 22,000 years ago, which covered much of southern Kyushu with some 150 km^3 of rhyolitic ejecta. The base of Sakurajima today rests about 125 km above the Wadati-Benioff zone, and the dynamics, mechanics, thermal structure, and deep hydrology of the subduction zone influences the generation, the segregation, and the ascent of Sakurajima magma. Photograph by T. Takayama, and courtesy of the Sakurajima Volcanological Observatory of Kyoto University, Professor Kosuke Kamo, and Dr. Kazuhiro Ishihara.

This book is printed on acid-free paper. 

Academic Press, Inc.
A Division of Harcourt Brace & Company
525 B Street, Suite 1900, San Diego, California 92101-4495

United Kingdom Edition published by
Academic Press Limited

24-28 Oval Road, London NW1 7DX

Library of Congress Cataloging-in-Publication Data

Magmatic systems / edited by Michael P. Ryan.
p. cm. -- (International geophysics series : v. 57)
Includes index.
ISBN 0-12-605070-8
1. Magmatism. I. Ryan, Michael P., [DATE]. II. Series.
QE461.M237124 1994
552'.1--dc20 93-48074
CIP

PRINTED IN THE UNITED STATES OF AMERICA
94 95 96 97 98 99 QW 9 8 7 6 5 4 3 2 1

To

Herbert and Virginia Ryan
OF EATON RAPIDS, MICHIGAN

and

Tsuneo and Ayako Eguchi
OF KAMAKURA-SHI, KANAGAWA-KEN, JAPAN

Contents

Chapter 9 Lateral Water Transport across a Dynamic Mantle Wedge: A Model for Subduction Zone Magmatism

J. Huw Davies

Chapter 10 Buoyancy-Driven Fracture and Magma Transport through the Lithosphere: Models and Experiments

Moritz Heimpel and Peter Olson

Chapter 11 Accumulation of Magma in Space and Time by Crack Interaction

Akira Takada

Chapter 12 Generalized Upper Mantle Thermal Structure of the Western United States and Its Relationship to Seismic Attenuation, Heat Flow, Partial Melt, and Magma Ascent and Emplacement

Hiroki Sato and Michael P. Ryan

Chapter 13 Aspects of Magma Generation and Ascent in Continental Lithosphere

George W. Bergantz and Ralph Dawes

Chapter 14 Two-Component Magma Transport and the Origin of Composite Intrusions and Lava Flows

Charles R. Carrigan

Chapter 15 Fluid and Thermal Dissolution Instabilities in Magmatic Systems

J. A. Whitehead and Peter Kelemen

Contributors

Numbers in parentheses indicate the pages on which the authors' contributions begin.

George W. Bergantz (291), Department of Geological Sciences, University of Washington, Seattle, Washington 98195

F. Boudier (77), Laboratoire de Tectonophysique, Université Montpellier 2, URA 1370 CNRS, 34095 Montpellier, France

Charles R. Carrigan (319), Earth Science Division, Lawrence Livermore National Laboratory, Livermore, California 94550

Y. J. Chen (139), College of Oceanography, Oceanography Administration, Oregon State University, Corvallis, Oregon 97311

Prame N. Chopra (37), Australian Geological Survey Organization, Canberra, ACT 2601, Australia

Reid F. Cooper (19), Department of Materials Science and Engineering, University of Wisconsin—Madison, Madison, Wisconsin 53706

J. Huw Davies (197), Department of Earth Sciences, The Jane Herdman Laboratories, University of Liverpool, Liverpool L69 3BX, England

Ralph Dawes (291), Department of Geological Sciences, University of Washington, Seattle, Washington 98195

Tye T. Gribb (19), Department of Materials Science and Engineering, University of Wisconsin—Madison, Madison, Wisconsin 53706

A. Harding (139), Institute of Geophysics and Planetary Physics, Scripps Institution of Oceanography, University of California—San Diego, La Jolla, California 92093

Akira Hasegawa (179), Observation Center for Prediction of Earthquakes and Volcanic Eruptions, Faculty of Science, Tohoku University, Sendai 980, Japan

Moritz Heimpel (223), Department of Earth and Planetary Sciences, The Johns Hopkins University, Baltimore, Maryland 21218

B. Ildefonse (77), Laboratoire de Tectonophysique, Université Montpellier 2, URA 1370 CNRS, 34095 Montpellier, France

Peter Kelemen (355), Department of Physical Oceanography, Woods Hole Oceanographic Institution, Woods Hole, Massachusetts 02543

G. Kent (139), Institute of Geophysics and Planetary Physics, Scripps Institution of Oceanography, University of California—San Diego, La Jolla, California 92093

David L. Kohlstedt (37), Department of Geology and Geophysics, University of Minnesota, Minneapolis, Minnesota 55455

A. Nicolas (77), Laboratoire de Tectonophysique, Université Montpellier 2, URA 1370 CNRS, 34095 Montpellier, France

Peter Olson (1, 223), Department of Earth and Planetary Sciences, The Johns Hopkins University, Baltimore, Maryland 21218

J. Orcutt (139), Institute of Geophysics and Planetary Physics, Scripps Institution of Oceanography, University of California—San Diego, La Jolla, California 92093

E. M. Parmentier (55), Department of Geological Sciences, Brown University, Providence, Rhode Island 02912

J. Phipps Morgan (139), Institute of Geophysics and Planetary Physics, Scripps Institution of Oceanography, University of California—San Diego, La Jolla, California 92093

Michael P. Ryan (97, 259), U.S. Geological Survey, Reston, Virginia 22092

Hiroki Sato (259), Institute for the Study of the Earth's Interior, Okayama University, Tottori-Ken 682–01, Japan

David W. Sparks[1] (55), Department of Geological Sciences, Brown University, Providence, Rhode Island 02912

[1]Present address: Lamont-Doherty Earth Observatory, Columbia University, Palisades, New York 10964.

Akira Takada (241), Environmental Geology Department, Geological Survey of Japan, Ibaraki-Ken 305, Japan

J. A. Whitehead (355), Department of Physical Oceanography, Woods Hole Oceanographic Institution, Woods Hole, Massachusetts 02543

Shanyong Zhang (19), Department of Materials Science and Engineering, University of Wisconsin—Madison, Madison, Wisconsin 53706

Dapeng Zhao (179), Seismological Laboratory, California Institute of Technology, Pasadena, California 91125

Acknowledgments

The Magmatic Systems project has been a collective effort in several ways. Each manuscript has materially benefited from the critical commentary provided by two or more reviewers. These participants are given special acknowledgment for the expertise, thoroughness, energy, and time they have applied. They are:

Keiiti Aki Department of Geological Sciences, University of Southern California

Charles R. Bacon Branch of Volcanic and Geothermal Processes, U.S. Geological Survey

David Bercovici Department of Geology and Geophysics, University of Hawaii

David D. Blackwell Department of Geological Sciences, Southern Methodist University

C. Wayne Burnham Department of Geosciences, The Pennsylvania State University, and Department of Geology, Arizona State University

Ulrich Christensen Institut für Geophysik, Universität Göttingen

Reid F. Cooper Department of Materials Science and Engineering, University of Wisconsin-Madison

Paul T. Delaney Branch of Volcanic and Geothermal Processes, U.S. Geological Survey

Robert S. Detrick Department of Geology and Geophysics, Woods Hole Oceanographic Institution

John J. Dvorak Cascades Volcano Observatory, U.S. Geological Survey

Terence N. Edgar Branch of Atlantic Marine Geology, U.S. Geological Survey

Donald W. Forsyth Department of Geological Sciences, Brown University

Wes Hildreth Branch of Volcanic and Geothermal Processes, U.S. Geological Survey

H. Mahadeva Iyer Branch of Seismology, U.S. Geological Survey

Ian Jackson Research School of Earth Sciences, The Australian National University

Shun-Ichiro Karato Department of Geology and Geophysics, University of Minnesota

Ross C. Kerr Research School of Earth Sciences, The Australian National University

Christopher Kincaid Graduate School of Oceanography, University of Rhode Island

Arthur H. Lachenbruch Branch of Tectonophysics, U.S. Geological Survey

Brian R. Lawn Material Science and Engineering Laboratory, National Institute of Standards and Technology, U.S. Department of Commerce

John R. Lister Institute of Theoretical Geophysics and Department of Earth Sciences and Applied Mathematics and Theoretical Physics, Cambridge University

Nobuo Morita Research and Development Laboratories, Conoco Incorporated

Janet L. Morton Branch of Pacific Marine Geology, U.S. Geological Survey

Peter Olson Department of Earth and Planetary Sciences, The Johns Hopkins University

John S. Pallister Branch of Volcanic and Geothermal Processes, U.S. Geological Survey

Mervyn S. Paterson Research School of Earth Sciences, The Australian National University

Rishi Raj Department of Materials Science and Engineering, Cornell University

Neil M. Ribe Department of Geology and Geophysics, Yale University

Frank M. Richter Department of Geophysical Sciences, University of Chicago

Eugene C. Robertson 917 National Center, U.S. Geological Survey

Allan M. Rubin Department of Geological and Geophysical Sciences, Princeton University

John H. Sass Branch of Tectonophysics, U.S. Geological Survey

Wayne C. Shanks III Branch of Eastern Mineral Resources, U.S. Geological Survey

Norman H. Sleep Department of Geophysics, Stanford University

Yoshiyuki Tatsumi School of Earth Sciences, Kyoto University

Robert I. Tilling Branch of Volcanic and Geothermal Processes, U.S. Geological Survey

Rob van der Hilst Research School of Earth Sciences, The Australian National University

E. Bruce Watson Department of Earth and Environment Sciences, Rensselaer Polytechnic Institute

Sarah T. Watson Department of Earth Sciences, Oxford University

Stephen M. Wickham Department of the Geophysical Sciences, University of Chicago

Lionel Wilson Environmental Science Division, Lancaster University

The volume has been designed to address perceived needs that relate to both research and to graduate teaching in Earth Sciences. All chapters have been invited to fill specific niches within the overall physical processes scope. Aspects, however, of selected chapters have also been presented at the Symposium on the Generation, Segregation, Ascent and Storage of Magma at the 29th International Geological Congress, Kyoto, Japan. In this connection, I am indebted to Professor Ikuo Kushiro of the Geological Institute of the University of Tokyo for the invitation to convene the Symposium and for his kind support. Professor Kosuke Kamo and Dr. Kazuhiro Ishihara of Kyoto University provided logistical support and gracious hospitality as did professors Toshitsugu Fujii and Yoshiaki Ida of the Earthquake Research Institute of the University of Tokyo, and Yoshiyuki Tatsumi of the School of Earth Sciences of Kyoto University.

Administrative support within the U.S. Geological Survey has been provided by Bruce Hemingway, Chief, Branch of Lithospheric Processes and by Rob Wesson, Chief, Office of Earthquakes, Volcanoes and Engineering. Debbie Pasquale and Mary Woodruff have applied their wordprocessing skills toward typescript preparation. Shirley Brown, Fran Buchanan, Lendell Keaton, and Nancy Polend have applied their graphics skills in making several of the illustrations of this volume a reality. Lewis Thompson provided exceptional photographic support. Thanks to Scotty Livingston and Herb Wolford for additional administrative support.

The editorial staff of Academic Press has been a pleasure to work with and special appreciation is extended to Charles G. Arthur, Jacqueline Garrett, and Michael Remener.

Professor Kosuke Kamo, Dr. Kazuhiro Ishihara, and Mr. T. Takayama of the Sakurajima Volcanological Observatory of Kyoto University, have graciously provided the figure that is the cover of this volume. Permissions for additional figure usage have been kindly granted by:

Disaster Prevention Research Institute of Kyoto University
The Geologists Association of London
Professor C. D. Han, Polytechnic Institute of New York
Journal of Applied Polymer Science
Macmillan Magazines Ltd.
Professor Nobuhiko Minagawa, Niigata University
Nature
Polymer Engineering and Science
Sakurajima Volcanological Observatory, Kyoto University
Society of Plastics Engineers
Dr. H. W. Stockman, Sandia National Laboratories
U.S. Geological Survey
John Wiley and Sons, Ltd.

Prologue

The full title of this volume is *The Dynamics and Mechanics of Magmatic Systems,* but the contraction to *Magmatic Systems* serves as a useful shorthand. This book focuses on the core problems of igneous petrology and volcanology: magmatic heat and mass transport processes in the Earth's mantle and crust. In preparing this book, I have tried to combine viewpoints that represent the three principal scientific perspectives: the theoretical, the experimental, and the observational. These perspectives represent the three principal means of study, and together they blend into a powerful way of understanding Nature's plan for the regulatory mechanisms that control the movement of melts and thus the cooling and differentiation of our planet. The scope is devoted to physical processes and to the physics of magmatic systems in the Earth's interior and the subvolcanic environment.

The chapters present new research results within a surrounding framework of review material that summarizes the theoretical approach, the experimental technique, or the relevant physical environment. This provides a useful background for professional research workers, graduate students embarking on research in magma transport, and the general interest reader seeking a greater familiarity with the issues and concepts of this new field. The primary tools are continuum mechanics, analytic and computational fluid dynamics, and computational heat transfer. Summary results from high-temperature and high-pressure experimental geophysics and seismic tomography complement the theoretical studies.

The volume begins with studies of decompression melting in ascending mantle plumes and thermal diapirs and the rheology of basaltic partial melts. Next are studies of the kinematics of mantle flow beneath mid-ocean ridge spreading centers; the geologic field relations of intrusives within these centers; the three-dimensional magma buoyancy zonation structure of the magma reservoir; the sheeted dike complex; and the accreting oceanic crust. Island arc magmatism is studied from the perspectives of geophysical constraints on melting domains beneath the volcanic front, the kinematics and thermal structure of flow in the descending slab, and the melt generation—H_2O migration region of the recirculating mantle wedge. Studies of buoyancy-driven magma fracture and melt-filled crack interactions are presented next and have a general applicability that extends across specific tectonic settings. Magmatism in the continental interior is studied from the perspectives of generalized regions of partial melt within the subcontinental mantle and the mechanics of basaltic underplating; assessments of the relative roles of convective and conductive heat transfer and the rheology of partial melt in regions of silicic magma generation; two-component viscous segregation processes during concurrent silicic–basaltic magma flow; and the thermal and fluid instabilities in flow regimes that accompany surficial lava movements and may participate in deeper magma migration.

In Chapter 1, Olson discusses the fundamental process of decompression melting in ascending upper mantle thermal plumes and in rising thermal diapirs. Univariant melting that includes the effects of latent heat absorption and melt buoyancy is treated in a self-consistent approach. The chapter treats both solitary plumes and continuously driven plumes that rise beneath and then impact upon the base of thick (continental) lithosphere or, alternatively, the relatively thin lithosphere of the oceanic basins. The finite-difference solutions determined reveal a generally lens-shaped region of partial melt enrichment conformable with the basal topography of the lithospheric keel. The length of the terminal portion of the ascent pathway—as allowed by the thickness of the lithosphere itself—interacts strongly with the melting process and thus largely determines the overall volume of melt for a given class of plume. Therefore, solitary plumes are able to produce $\approx 10^6$ km^3 of magma over a few million years (Myr) beneath lithosphere of normal thickness, while thin lithospheric thicknesses promote correspondingly greater melting path lengths and are associated with $\approx 10^7$ km^3 of magma production over comparable time scales. These results accord with current estimates of the continental and oceanic plateau basalt volumes.

At the microscopic level, melting commences along the grain boundaries within the ascending lherzolite. Here, spreading melt may initiate the development of microporous networks and episodes of grain-scale fluid flow and thus assist in the attenuation of transiting seismic waves. In a highly original series of high-temperature, beam-bending experiments, Gribb, Zhang, and Cooper (Chapter 2) have induced alternating states of tension and compression in samples of synthetic olivine–basalt partial melts. Microporous flow of melt within the samples is thus experimentally induced via the production of transient gradients in the dilational stress. Application of the linear viscoelastic Burgers model to the rheology of this two-phase, liquid–solid system provides relationships for the aggregate shear viscosity and for the bulk viscosity.

Both hydrolytic weakening and melt phase enhanced grain boundary diffusion processes affect the rheology of rocks in regions of magma generation. Deformation experiments conducted by Kohlstedt and Chopra (Chapter 3) on synthetic basalt and olivine aggregates have thrown light on the important interactions between these two mechanisms of weakening. In unmelted mantle peridotites, distributed water-derived species within the framework of the crystalline silicates promote hydrolytic weakening and thus reductions in aggregate viscosity. The infrared spectroscopy portion of this study has revealed that ample amounts of a grain boundary melt may deplete the matrix olivine of hydrous species due to the relatively strong H_2O partitioning into the adjacent melt phase. Thus, while the presence of melt dramatically affects the rheology via creep strength and viscosity reductions (e.g., about a factor of 5 in strength reduction for $\approx 8\%$ basaltic melt), it tends to "dry out" the matrix olivines, reducing the *hydrolytic* component of aggregate flow regulation. As a result, the rheology of partial melts will depend on the H_2O partitioning into the grain boundary melt phase as well as the bulk permeability—both functions of the overall melt fraction.

The fluid mechanical richness of a mid-ocean ridge substructure is partially the result of the concurrent and differential flow of the deforming peridotite matrix and the intercrystalline melt phase. This flow is inherently three-dimensional, reflecting the offset or en-echelon arrangement of spreading center segments in plan view—an organizational pattern that further increases the complexity of the flow. Sparks and Parmentier (Chapter 4) have applied the mass balance and force balance equations for concurrent matrix and melt flow in a deforming mantle in three-dimensional simulations of the melt migration process. Combined with the energy equation, the determination of characteristic length scales for the solidification region has helped identify where in the system the permeability reduction due to solidification promotes enhanced melt pressures and a resulting matrix dilation. An intriguing result has been that melt focussing occurs in the decompacting (dilating) boundary layer that roofs the melting region. Melt enrichment and permeability enhancement in this *magmatic canopy* thus allows magma to migrate upslope toward the ridge axis, assisted by positive buoyancy forces. Geophysical evidence for this type of focussing in Iceland, where the magma-rich canopy lies at the top of the asthenosphere and dips symmetrically away from the neovolcanic zone axis, has been summarized by Ryan (1990).

The availability of relatively large volumes of magma near the top of the asthenosphere, combined with the steep thermal gradients and rapidly varying stress states associated with the divergent mantle flow below the Mohorovičić discontinuity, promotes diverse modes of magma fracture beneath an active ridge. Mapping campaigns in the Oman ophiolite conducted by Nicolas, Boudier, and Ildefonse (Chapter 5) combined with careful petrofabric analyses have defined the nature of these modes and their relationships to the flowing asthenosphere and the accreting lithosphere. A *first generation* of indigenous gabbro and pyroxenite dikes and veins crosses melt-impregnated peridotites while a *second generation* of microgabbro and diabase dikes has intruded into cooler crustal rocks and is regionally organized in patterns that mimic the strike of the reconstructed ridge itself. Diverse intrusive orientations that range from dikes to sills reflect the lithospheric and asthenospheric stress fields respectively, and sill-forming injections have been spatially linked with the strongly divergent asthenospheric flow fields at the very top of subridge diapirs. Thus, detailed studies of the Oman ophiolite open windows into the roots of mid-ocean ridges and illuminate their magmatic machinery.

Gravitational equilibrium and the quest for its periodic reinstatement are fundamental to the operation of magmatic systems. Until recently, the functional maintenance of this equilibrium state in active volcanic systems was thought to generally require the movement of magma completely through the lithosphere and its emplacement as lava on the Earth's surface. Lithospheric densities were generally regarded as always greater than magma density and *states of universal positive buoyancy* were believed to characterize the oceanic and continental lithosphere. These views have been widespread and pervade the geophysical and volcanological literature. The assumption of *complete fluid continuity* from the magma source region to the eruption site (now known to be generally false) complemented the universal positive buoyancy assumption and permitted the balancing of lithostatic and magmastatic pressures over great columns of virtually lithospheric dimensions. Existence criteria for mid-ocean ridge magma reservoirs were ascribed to be almost wholly the result of the initially assumed and then numerically modeled ridge thermal structures: robust development for high spreading rates, negligible development for low spreading rates. The detailed density states of mid-ocean ridges as the combined functions of the relevant range in magma compositions, suspended mineral phases, and the vertical succession of the nonlinear *in situ* density structure of the oceanic crust itself were not of general interest. These views, with few exceptions, characterized the period following the inception of plate tectonics to at least the mid-1980s, and to some extent remain today. That *states of positive magma buoyancy are, in fact, routinely complemented by states of neutral and negative buoyancy in active basaltic systems* was first demonstrated by Ryan (1987a) for Hawaii, and then again for Hawaii and Iceland with hypothesized extensions to the East Pacific Rise by Ryan (1987b). The recognition of the widespread role of neutral buoyancy states in the shallow subvolcanic environment has, I think, profound implications for how magmatic systems work. In these terms, the importance of the neutral buoyancy concept is three-fold: (a) an internally consistent existence criterion for shallow magma reservoirs; (b) a new tool for the quantitative understanding of the upward (and lateral) evolution of the reservoir; and (c) a powerful means of understanding magma dynamics in intrusion episodes within stratified reservoirs as well as in the sheeted-dike complex. In Chapter 6, I have reviewed several aspects of the neutral buoyancy structure of mid-ocean ridge magma reservoirs. The compositionally averaged horizon of neutral buoyancy (HNB) for the East Pacific Rise occurs at a depth of $\approx$1000 m to $\approx$3000 m beneath the rise axis (at 9°N latitude). Fractional crystallization and elastic crack stability criteria suggest that this generalized HNB should be subdivided into a deeper picritic horizon ($\approx$1400 to $\approx$3000 m depth) and a shallow tholeiitic horizon ($\approx$600 m to $\approx$1400 m depth). During the fractional crystallization that transforms a picritic melt and olivine mixture into tholeiitic melt, parcels of magma must elevate themselves to maintain gravitational equilibrium. In ophiolite complexes, the uppermost isotropic gabbros are correlated with the picritic HNB, whereas the sheeted-dike complex corresponds to the tholeiitic horizon of neutral buoyancy. These relations are consistent with the dominantly tholeiitic nature of the sheeted dikes as well as the paucity of ocean floor picrites. The combined tholeiitic and picritic HNB is, in addition, in virtually a 1 : 1 correspondence with the inferred sheeted-dike complex and the compressional wave velocity minima region (i.e., the magma-rich portion of the reservoir) for the East Pacific Rise as determined by seismic reflection surveys and by tomographic inversions of P-wave travel-time residuals. The dynamics of magma injections within the sheeted-dike complex are suggested to be analogous to their counterparts at Kilauea volcano, Hawaii, and to the Krafla central volcano, northeast Iceland, during, for example, the 1975–1984 intrusion-eruption episodes. [It is important to recognize that the present erosion levels in Hawaii and Iceland do not sample the heart of the tholeiitic HNB (centered at ≈3 km local depth) and thus contain components of vertical flow.] From an evolutionary perspective, as lithosphere is created at mid-ocean ridges and spreads laterally, the horizons of neutral buoyancy and negative buoyancy ride with it, and they are suggested to significantly influence the ascent of off-axis magma to about 30 Myr and provide the means for shallow off-axis storage as well.

Recent advances in the seismic resolution of mid-ocean ridge magma reservoirs have refined their images, considerably tightened their geometric extents, and have laid the basis for a new generation of numerical models that yield kinematic insights on how ridges work. Phipps Morgan, Harding, Orcutt, Kent and Chen (Chapter 7) have reviewed the results of seismic reflection experiments over the East Pacific Rise, suggesting that the magma reservoir comprises a relatively fluid-rich upper chamber (width: ≈1 km; thickness ≈50–200 m) that crowns a progressively crystal-rich mush where melt contents average 3–5%. Kinematic modeling of the two-dimensional structure of the flow field induced by steady-state lithospheric lid translation has used the conservation equations in conjunction with McKenzie's finite deformation formulation. The deformation results accord nicely with textural observations in ophiolite crustal sections: steeply dipping foliations high in the gabbros, with progessive increases in strain and considerable foliation flattening with depth.

About 439 of the world's 442 active andesitic volcanoes occur in regions of oceanic plate subduction (Gill, 1981). Frequently arranged in great sweeping island arcs, their magma generation systems, segregation mechanisms, and ascent networks remain shrouded behind a veil of complex rheology, often sluggish magma migration kinetics, and largely aseismic ascent pathways. Because so much of the inner workings of these systems remains unknown—even in outline—they comprise a great and relatively uncharted frontier for research in magma transport. Their petrologic diversity, violent eruptive behavior, and shear number underscore the great need for intensely focused and highly coordinated research in this area. A parting of this veil and an initial illumination of an arc system interior has made use of the velocity dependence of compressional seismic waves resulting from rock composition, temperature, regional structure, and the presence and distribution of included domains of fluid. Hasegawa and Zhao (Chapter 8) have employed a three-dimensional velocity model that explicitly incorporates the locally complex shapes of the Mohorovičić and Conrad discontinuities to refine the velocity structure of the northeast Japan arc based on least-squares inversions of P-wave travel-time residuals. The magma-impregnated low velocity zones revealed in their study are inclined, and they dip parallel to the dip of the Pacific Plate but are found at least some 30–60 km above the plate's upper surface. Produced by velocity contrasts of 2–6%, they are continuously distributed from the upper crust (shallow magma ascent regions) to a depth of 100–150 km within the mantle wedge (the magma generation and segregation regions).

The *refrigeration* of the mantle produced by the subduction of oceanic plates leads naturally to the paradox of prolific arc magmatism directly above a mantle wedge that is cooled by conductive heat transfer to a juxtaposed downgoing slab. Building on the theoretical work of McKenzie (1969) and Toksoz, Minear, and Julian (1971), the experimental studies from the laboratories of Burnham, Green, and Wyllie, and the geochemical studies of Tatsumi (1989), J. H. Davies (Chapter 9) addresses the *why* and *how* of this paradox by combining thermomechanical modeling with an incremental mechanism for water transfer from the slabs' upper surface well into the wedge interior. The model produces finite element solutions to the advection–conduction energy equation in response to the kinematically driven flow field of the subducting slab. The steady-state thermal structures of the wedge and slab as functions of the subduction velocity reflect the slow secondary wedge flow and set the stage for the H_2O migration mechanism. Water is released from amphibole at depths near 80 km, and in a cyclic series of dehydration (H_2O release) amphibole reformation stages, rides the downward streamlines of the wedge flow field in steps that, in inchworm fashion, carry it both laterally and downward into melt generation depths. These positions within the wedge are consistent with the roots of island arc magmas and the location of the volcanic front above.

The processes of linear elastic brittle fracture, creep rupture, and stress corrosion cracking occupy portions of a continuum of failure phenomena associated with the migration of magma by cracks. Within the oceanic and continental lithosphere and within the crust above subduction zones, the fracture mode of magma ascent, emplacement, and lateral intrusion is one of the great workhorses of magma migration. Without it the galaxy of veins, dikes, sills, sheets, fissure eruptions and shallow basaltic-to-andesitic magma reservoirs would not exist. Heimpel and Olson (Chapter 10) have used gelatine gel-based analogue experiments to study the mechanics of positive buoyancy-driven, fluid-filled fractures. In contrast with dry remotely loaded fracture processes which induce large amounts of elastic stored strain energy just prior to fracture and are associated with catastrophic crack growth, the *locally loaded* buoyancy-driven cracks suggest a set of dynamical constraints on the fracture process when the fluid-filled crack is imbedded in viscoelastic media. Thus, the coupled processes of time-dependent external stress relaxation and internal fluid flow that accompany increments of crack extension modulate the crack propagation velocities and, by analogy, may significantly influence the magma migration process.

The development and function of well-trodden magma migration conduits, as well as the nucleation of shallow magma reservoirs, all owe a great debt to the phenomena of magma fracture coalescence. Takada (Chapter 11) has used Westergaard stress functions and complex variables to compute the nature of the displacements and stress fields near two parallel offset fluid-pressurized cracks in a search for crack coalescence criteria. Complementary gelatine gel-based experiments on fluid buoyancy-driven cracks have revealed that magma-filled cracks *can* coalesce and undergo a resulting magma volume increase as long as there is an available *range* in crack sizes and ascent velocities. The coalescence process is further enhanced by small differential stress states in horizontal sections and by concomitant large magma supply rates.

The continents have been the birthplace of igneous petrology and most subdisciplines of geophysics, yet after the inception of plate tectonics, much of the energy and attention in these fields of research has been understandably directed offshore, toward the problems—and the promise—of the ocean basins. Perhaps the time is now ripe, however, for a rejuvenation of work within—and beneath—the continental interiors. Sato and Ryan (Chapter 12) have estimated generalized temperature profiles and degrees of partial melt in the upper mantle beneath the western United States. The study makes use of seismic velocity data for partially melted peridotites and published anelasticity data for the regional scale upper mantle. Experimental measurements of the compressional wave velocity and elastic wave attenuation in spinel lherzolites show a homologous temperature dependence: families of Vp, Qp^{-1} data for a variety of pressures plot as a single band in terms of the experimental temperature normalized by the sample melting temperature (T/T_m). Thus, a knowledge of the solidus as a function of pressure allows the estimation of temperature as a function of pressure—and hence depth in the mantle. Therefore, both Vp and Qp^{-1} are of use in estimating $T(Z)$ for various regions of interest. In addition, empirical plots of melt fraction versus (T/T_m) allow *in situ* melt fractions to be estimated once T_m and T have been determined. This procedure has produced generalized temperature-depth profiles for major regions of the western United States, and they cover the depth range 50–300 km. At shallow (crustal) depths they are comparable with conductive geotherms, while at depths greater than 200 km they are compatible with the adiabatic temperature gradient. Partial melt contents estimated for the intermountain and western margin regions cover the range $\approx 3\% \leq \varnothing_m \leq 10^+\%$, when regionally averaged. These melt contents occur over the 120–200 km-depth range, and transitions from porous media melt flow to vein and dike flow are expected over these depth intervals. A review of elastic fracture stability theory within the context of subcontinental magma ascent reveals the fundamentally disconnected nature of the ascent pathway: the finite strengths of fluid-weakened mantle peridotites at high temperatures mandate, in turn, finite height magma-filled fractures. Deep dike swarms are thus suggested to be the ascent mode for the deepest levels of basaltic underplating along the crust–mantle boundary. The crust–mantle interface itself, reflecting changes in both mineralogy and potential melt content, may be treated as a bi-elastic material boundary, and the morphology and mechanics of dikes that penetrate such an interface depend on the ratios of effective elastic moduli on either side, as well as the fluid pressure loading conditions within the fracture.

The shear viscosities of single-phase natural silicate melts span an extraordinary 13 orders of magnitude and the *melts* of the continental interiors must routinely encompass this entire spectrum. For regions of basaltic underplating and of the secondary generation of silicic crustal magmas, the single-phase contents of laboratory crucibles are but one end-member of a rheological continuum that includes melt-based suspensions, melt-weakened country rocks, and highly altered, but as yet unmelted, rock on all scales. The high-solids-fraction domains thus raise the aggregate effective shear viscosities to yet greater values. Accordingly, the migration kinetics of ascending magma in the continental crust cover a time range that extends far beyond the realm of human experience and renders much of the generation, ascent, emplacement, and replenishment process of silica-rich magmas aseismic. Like the pathways for magma above Wadati-Benioff zones, the continental *interior* regions of silicic melt generation remain heavily shrouded and thus deserving of carefully focused and well-coordinated research. Bergantz (Chapter 13) reviews aspects of continental magmatism around the general theme of basaltic underplating. Finite volume solutions of the conservation equations have employed the spatial and temperature dependence of viscosity, density, heat capacity, and thermal conductivity in a model of basaltic underplating that considers the progress of crystallization in the basaltic substrate as a function of partial melting in the overlying crust. Three questions are of interest: How does convection influence the timing of partial melting in the crust, and what is the style of convection? How does variable viscosity (as a function of melt composition and crystallinity) influence convection? What are the overall heat transfer rates from the system relative to those expected from conduction only? Nusselt (Nu) number computations during the evolution of the melting region compare the strength of the overall heat transfer components and suggest that for critical melt fractions over $0.3 \leq \varnothing_m \leq 0.5$, the amount of melt generated was indistinguishable from that produced during conduction only. This is broadly consistent with the notion that the rheological conditions associated with the onset of crystallization control the subsequent dynamic evolution of the body. Thus the *early* stages of underplating appear to be largely conductive, and multiple basaltic intrusions are required to thermally mature a deep crustal section and permit widespread regional melting episodes in the crust above.

Magma reservoirs may be viewed as great mechanical capacitors that, during magma influxes, slowly accumulate potential energy. This stored energy occurs by virtue of the vertical displacements of their roof rocks and caldera floors and by the compression of the immediate surroundings, including the magma. Conceptually, these are akin to compressed springs that are coupled in parallel. Rupture of the reservoir walls suddenly releases magma into the surrounding country rocks in dike-forming injections—a process that relaxes the conceptual springs but works against the environment by virtue of the crack wall displacements and the work-of-fracture (atomic bond breaking) at the advancing crack tip. Importantly, the energy dissipated in the high shear boundary layers of the near-wall dike interior represents a significant resistance term in the overall process of fluid flow. Inherently repetitious, the overall magma recharge—discharge cycle maximizes its efficiency by (1) maximizing the overall potential energy reductions, and (2) simultaneously minimizing all the associated work done in the dike formation episodes. Thus a globally integrated *least work* principle lies at the heart of the magma storage and fracture process. In regions of *mixed* magma storage, Nature also has a trick up her sleeve that helps in the effort to minimize the work done in moving fluids and thus enhance the efficiency of the magma injection process. Carrigan (Chapter 14) has derived the lubrication equations for the two-component flow of a power-law fluid in a dike-like geometry. The solutions describe the process of the hydrodynamic encapsulation of the high viscosity (interior) phase by the low viscosity (exterior) fluid. Importantly, the outer (low viscosity) phase occupies the high shear—and energy dissipating—boundary layers at the dike wall. The overall flow process is analogous with the *lubricated pipelining* process of industrial settings. It describes the means of providing a least work solution to the problem of mixed magma transport in settings as diverse as Long Valley, California, and southeast Iceland, for example. A substantially more realistic magma withdrawal model has also evolved from the work, wherein two or more layers of differing composition (and viscosity) can be *simultaneously* withdrawn without invoking the rather special set of circumstances inherent in the *draw-up-depth* parameter or in the *overtaking* requirement of other approaches. Overall, the model represents a significant step forward in advancing our understanding of this important class of problems.

The simultaneous attempts to achieve both thermal and mechanical equilibrium within regions of high temperature magma flow, lead to a number of thermomechanical feedback processes. Familiar examples relate to the potential for either dike wall solidification or meltback, depending on the flow rates, magma temperatures, dike widths and the ambient thermal environment (Bruce and Huppert, 1990). Other examples include the role of the evolving conduit geometric aspect ratio in flow localization during the course of an eruption (Delaney and Pollard, 1982). Whitehead and Kelemen (Chapter 15) explore, theoretically and experimentally, a new class of dynamic feedback phenomena. Gravity-driven laboratory flow experiments, for example, reveal the potential for interplay between fluid pressure, environmental and fluid temperatures, temperature-dependent viscosity fluctuations, and the resulting flow rates. These effects include flow choking, episodic pressure buildups, and transitions in flow rates that correlate with the radially inward growth of high viscosity 'walls' during environmental temperature drops. Additional effects have been modeled theoretically, and include oscillatory pressure–time (P–t) and fluid velocity–time histories that may have either sawtooth or sinusoidal P–t signatures, and are correlative with pulsatile fluid flow behavior. These types of relationships hold the potential for further application to the throttling of magmatic and volcanic flows through regions with high spatial gradients in environmental temperature and conduit geometry. For surficial volcanic process considerations, laboratory simulations by Whitehead and Kelemen (Chapter 15) of horizontally spreading flows have used paraffin as the working fluid. These show transitions from smooth perimeter radial flow regimes to finger flow regimes as a function of the solidification rate and temperature-dependent viscosity. Experiments with lateral spreading flows from line sources have charted a morphological continuum of surface textures depending on the ratio of the solidification to advection time scales. These forms range from pillows through surface ripples to flat sheeted flows.

In a direct way, this volume complements the book *Magma Transport and Storage* (M. P. Ryan, ed.), John Wiley and Sons, Ltd., 1990. While some authors have returned to write new chapters for this book, there is no overlap, and each chapter is a fresh, new and complementary effort, quite separate and distinct from the work cited above.

Increasingly, scientists with diverse backgrounds that have traditionally been spread out over several of the subdisciplines of geophysics, igneous petrology and volcanology have found that the subject area of magma transport offers an exciting and unifying process-oriented framework for study. Within this framework lies great cohesion yet remarkable room for individual and team research initiatives. It is my hope that the present book will further this process of unification by providing fertile points of departure for research as well as offering stimulating areas of discussion for the classroom and for the seminar hall.

Finally, detailed author, geographical and subject indices complete the volume and promote accessibility and cross-referencing.

Michael Ryan

References

Bruce, P. M., and Huppert, H. E. (1990). Solidification and melting along dykes by the laminar flow of basaltic magma, *in* "Magma Transport and Storage" (M. P. Ryan, ed.), Wiley, Chichester/Sussex, England.

Delaney, P. T., and Pollard, D. D. (1982). Solidification of basaltic magma during flow in a dike, *Am. J. Sci.* **282,** 856–885.

Gill, J. B. (1981). "Orogenic Andesites and Plate Tectonics," Springer-Verlag, New York, pp. 390.

McKenzie, D. P. (1969). Speculations on the consequences and causes of plate motions, *Geophys. J. R. Astron. Soc.* **18,** 1–32.

Ryan, M. P. (1987a). Elasticity and contractancy of Hawaiian olivine tholeiite and its role in the stability and structural evolution of sub-caldera magma reservoirs and volcanic rift systems, *in* "Volcanism in Hawaii" (R. W. Decker, T. L. Wright and P. H. Stauffer, eds.), U.S. Geol. Survey Prof. Paper 1350, v. 2, 1395–1448.

Ryan, M. P. (1987b). Neutral buoyancy and the mechanical evolution of magmatic systems, *in* "Magmatic Processes: Physicochemical Principles" (B. O. Mysen, ed.). The Geochemical Society. Special Publication No. 1. Yoder Symposium Volume. 259–288.

Ryan, M. P. (1990). The physical nature of the Icelandic magma transport system *in* Magma Transport and Storage, (M. P. Ryan, ed.), Wiley, Chichester/Sussex, England. p. 175–224.

Tatsumi, Y. (1989). Migration of fluid phases and genesis of basalt magmas in subduction zones, *J. Geophys. Res.* **94,** 4697–4707.

Toksoz, M. N., Minear, J. W., and Julian, B. R. (1971). Temperature field and geophysical effects of a downgoing slab. *J. Geophys. Res.* **76,** 1113–1138.

Chapter 1 Mechanics of Flood Basalt Magmatism

Peter Olson

Overview

The dynamics of magma generation in a mantle plume are investigated using time-dependent numerical calculations of pressure-release melting in axisymmetric thermal plumes and thermal diapirs in a viscous fluid with a rigid lid. The plumes ascend through the fluid and spread laterally beneath the lid, producing a broad surface uplift and a correlated geoid high. Decompression melting in the plume occurs in a lens-shaped region centered near 100 km in depth. The excess temperature in the melting lens is 100–300° C. The calculations demonstrate that the initial melting pulse from a single plume can produce 10^6 km^3 of magma in a few million years beneath normal lithosphere and more than 10^7 km^3 of magma beneath very thin lithosphere, in accord with recent estimates of continental flood basalt and ocean plateau volumes. After the initial melting event the plume constricts to a narrow thermal anomaly and the melt production rate decreases by nearly two orders of magnitude, consistent with the long-term history of many hotspots. The calculations also indicate that plume formation initiates an episode of enhanced heat flow in the source region. This behavior supports recent speculations that deep mantle plumes can thermally couple the core to the mantle.

Notation

		Units
C_p	specific heat at constant pressure	$kJ \cdot kg^{-1} \cdot °C^{-1}$
H	enthalpy	$kJ \cdot m^{-3}$
H_m	enthalpy at melting	$kJ \cdot m^{-3}$
H_s	enthalpy at saturation melt fraction	$kJ \cdot m^{-3}$
L	latent heat of melting	$kJ \cdot kg^{-1}$
R	cylinder radius	m, km
T	temperature	°C
T_I	interior mantle temperature	°C
T_m	melting temperature	°C
T_{m0}	surface melting temperature	°C
T_D	diapir temperature	°C
T_B	basal temperature	°C
g	gravitational acceleration	$m \cdot s^{-2}$
h_c	added crust thickness	m, km
k	thermal conductivity	$W \cdot m^{-1} \cdot °C^{-1}$
r	radial coordinate	m, km
t	time	s
u	radial velocity	$m \cdot s^{-1}$
w	vertical velocity	$m \cdot s^{-1}$
y	depth coordinate	m, km
z	vertical coordinate	m, km
z_L	lid thickness	m, km
z_0	cylinder height	m, km
α	thermal expansivity	$°C^{-1}$
β	melting expansivity	dimensionless
γ	melting point gradient	$°C \cdot km^{-1}$
ρ	mantle density	$mg \cdot m^{-3}$
ρ_c	crust density	$mg \cdot m^{-3}$
χ	melting progress variable	dimensionless
η	viscosity	pa · s
ϕ	melt fraction	dimensionless
ϕ_s	saturation melt fraction	dimensionless
ψ	streamfunction	$m^3 \cdot s^{-1}$
τ_0, τ_B	surface and basal boundary layer ages	s
∇	del, the gradient operator	m^{-1}
∇^2	nabla	m^{-2}
$\mathbf{u}$	velocity of matrix material	$m \cdot s^{-1}$
$\hat{\mathbf{r}}, \hat{\mathbf{z}}$	unit coordinate vectors	m

Magmatic Systems
Edited by M. P. Ryan

Table 1
Continental Flood Basalts[a]

Flood basalt	Age (Ma)	Hotspot	Basalt volume ($\times 10^6$ km^3)
Columbia River	15 ± 2	Yellowstone	0.3
North Atlantic	59 ± 2	Iceland	>1.5
Deccan	66 ± 1	Reunion	>1.7
Parana	120 ± 3	Tristan da Cunha	~ 1
Karoo	192 ± 3	Prince Edward	$\geqslant 1.0$
Siberia	$250 \pm ?$	?	~ 0.9

[a]From White and McKenzie (1989).

Introduction

In this chapter I describe the sequence and duration of magmatic events accompanying the development of a mantle plume, using calculations of decompression melting in axisymmetric thermal diapirs and in axisymmetric thermal plumes. These calculations are motivated by growing evidence in support of a mantle plume origin for many continental flood basalt provinces as well as some oceanic plateaus. The calculations are also intended to establish the temporal relationships between near-surface magmatic activity associated with mantle plumes and heat transport into the hot thermal boundary layer feeding the plume. This relationship is a key element in the current debate on the extent of the thermal coupling between the Earth's core and mantle.

Continental Flood Basalts, Oceanic Plateaus, and Mantle Plumes

Many of the largest flood basalt events preserved in (and on) the continental crust appear to be associated with the point of origin of linear volcanic hotspot tracks. Table 1 lists several prominent continental flood basalt provinces and the hotspot tracks that emerge from them. The association of flood basalt provinces with the beginning of a hotspot track, together with evidence that they are emplaced very rapidly and have a distinct isotopic signature (Duncan and Richards, 1991; Carlson, 1991), has led to the embellishment of a "starting plume" model of flood basalt formation originally proposed by Morgan (1981). In the starting plume model, the flood basalt province is interpreted to be the result of the partial melting of the head of a mantle thermal plume or diapir, and the hotspot track is interpreted to be the result of partial melting of the plume conduit (Richards *et al.*, 1989; Griffiths and Campbell, 1990).

Many large ocean plateaus are now thought to have a plume origin as well (Mahoney, 1987; White and McKenzie, 1989; Mahoney and Spencer, 1991; Larson, 1991; Tarduno *et al.*, 1991). There are, however, some important physical differences between ocean plateaus and the continental flood basalts. One significant difference is that ocean plateaus tend to be larger than continental flood basalts. As Table 1 indicates, continental flood basalt volumes are typically $1-2 \times 10^6$ km^3, whereas the data in Table 2 indicate ocean plateau volumes are typically 10 times that size and the largest, the Ontong–Java plateau, may exceed 50×10^6 km^3. It is unreasonable to explain this difference by postulating either a size

Table 2
Large Oceanic Plateaus[a]

Plateau	Age (Ma)	Basalt volume ($\times 10^6$ km^3)
Kerguelen	90–115	20
Manihiki	115–125	12
Mid-Pacific Mts.	75–130	21
Ontong–Java	117 ± 2	54
Shatsky	130–150	11

[a]From Larson (1991); Tarduno *et al.* (1991); Mahoney (1987); Schubert and Sandwell (1989).

difference or a temperature difference between plumes rising beneath continental and oceanic lithosphere. Also, it is unlikely that the difference in size is due to greater longevity of the sources of oceanic plateaus. For example, the mammoth Ontong–Java plateau appears to have been built in just a few million years, the same formation time as continental flood basalts (Tarduno *et al.*, 1991). If we suppose that both continental flood basalts and ocean plateaus are the result of mantle plumes, then it is reasonable to seek an explanation for the systematic difference in size by appealing to differences in the volumes of melt produced within plumes.

The amount of melt produced in a given plume is proportional to the amount of decompression the plume experiences in the region where the temperature is at or above the solidus. The calculations presented here show that temperatures above the solidus occur at 200 km depth or shallower, so the extent of melting in a mantle plume is controlled primarily by the thickness of the lithosphere. According to the calculations, melt production in a plume can vary by a factor of 40 or more because of differences in the thickness of the lithosphere beneath continental crust and young oceanic crust. The calculations also show that the rise of a leading diapir at the head of a continuous plume can produce a transient magmatic event with the magnitude and time scale of the largest ocean plateaus.

Constraints from Seismic Tomography and Geochemistry

Figure 1 shows vertical cross sections of the upper mantle at four hotspots from a model of shear wave velocity variations derived by Zhang and Tanimoto (1992). Low velocity regions between 100- and 200-km depths are found beneath the four hotspots, Hawaii, Iceland, Azores, and Tristan da

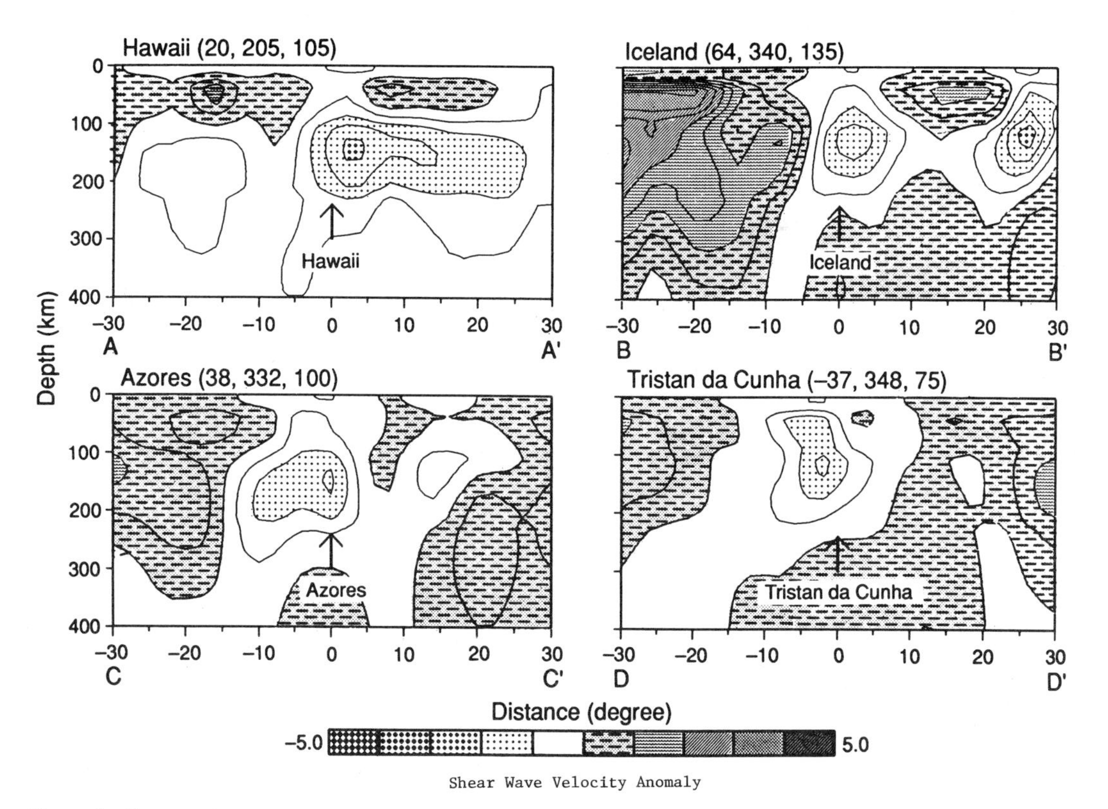

Figure 1 Cross sections of upper mantle shear wave velocity structure at four hotspot locations, as determined by Zhang and Tanimoto (1992). Abscissa scales are the great circle degrees from the hotspot. Latitude, longitude, and azimuth of the great circle cross section are indicated above each panel. The scale bar gives velocity perturbation in percent. Note the low velocity anomaly beneath each hotspot between 100 and 200 km in depth.

Cunha. The horizontal extent of each region is 500–1000 km, approximately. Note there is little indication of deep root structures beneath the low velocity regions; at least there is no evidence for broad roots on the scale of several hundred kilometers, the limit of resolution in the Zhang and Tanimoto study. However, regional-scale tomography indicates a deep, narrow root beneath Iceland to about 375 km in depth (Ryan, 1990), which is evidently not seen on the scale of Fig. 1.

Several interpretations of the broad structures beneath hotspots are plausible, but one based on the dynamics of a buoyant plume approaching the lithosphere is particularly simple and appealing. According to the plume model, the low seismic velocity regions represent portions of the upper mantle intruded by the hot and perhaps partially melted plume material as it spreads out beneath the lithosphere. Since each of these hotspots has a well-dated track and is known to have been active for at least 40 Ma, it is unlikely that these seismic anomalies represent newly formed plumes. A more likely interpretation is that the low velocity regions represent the structure of fully developed plumes stagnating beneath the lithosphere.

Seismic topographic images provide constraints on the pattern of mantle convection structures, but it is hazardous to use images such as those in Fig. 1 for quantitative estimates of mantle temperature or melt content. However, there are other ways to estimate the temperature of mantle hotspots. Schilling (1991) combined geochemical and topographic data from hotspot swells to infer the excess temperature ΔT in the mantle beneath near-ridge hotspots. The term excess temperature refers to the temperature in the plume material supporting the hotspot swell, relative to normal mantle at the same depth. For three of the hotspots in Fig. 1, Schilling finds ΔT = 162, 263, and 198 degrees Kelvin at Tristan, Iceland, and Azores, respectively.

Time Variations in Hotspot Activity

There appear to be large changes in hotspot activity with time. The evidence for variability in the strength of hotspot magmatism comes from comparisons between the present-day rate of hotspot volcanism and the rates inferred for the past from continental flood basalts and ocean plateaus. The present-day hotspot activity is dominated by Hawaii and is probably less than 0.3 $km^3\ yr^{-1}$ for all hotspots (Duncan and Richards, 1991). This activity amounts to only a few percent of the 13 $km^3\ yr^{-1}$ present-day rate of crust production estimated for the mid-ocean ridge system.

The present-day level of hotspot activity, however, is not representative of longer-term averages. Figure 2 shows rates of oceanic crust production over the past 150 Myr averaged in 5-Myr bins, for ocean plateaus and for the Earth as a whole, as determined by Larson (1991). Over the past 10 Myr, oceanic plateaus have contributed about 1.5 $km^3\ yr^{-1}$ of crust on average, or roughly 10% of the global rate for the same period. This fraction is generally consistent with the fraction of the total heat transport at hotspots, compared to the heat transport at mid-ocean ridges. Davies (1988) and Sleep (1990) have calculated that the present-day hotspots transport about 2–4 TW of heat to the lithosphere, while sea floor spreading transports roughly 10 times as much.

The data summarized in Fig. 2 also indicate that hotspot activity has been even greater at various times in the past. In particular, the crustal production rate at oceanic plateaus during the mid- to late Cretaceous (125–65 Ma) was three to five times larger than in the past 10 Myr. The pulse in hotspot activity accounts for a significant part of the increase in total oceanic crust production during that interval of time. Several particularly large oceanic events were responsible for the Cretaceous pulse, including the formation of the Ontong–Java plateau and Mid-Pacific Mountains on the Pacific plate and the Kerguelen plateau on the Antarctic plate. Larson's (1991) plate reconstruction for that time shown in Fig. 3 indicates that the largest events occurred in two clusters, one near the former Pacific–Farallon plate boundary and the other near the Indian–Antarctic plate boundary. In addition to defining two main centers of plume activity, Fig. 3 also suggests that the most massive magmatic events occur when hotspots form near spreading centers, where the oceanic lithosphere is thin.

Hotspot Activity and Geomagnetic Reversal Frequency

Evidence that thermal plumes originating from the deep mantle thermally couple the core to the mantle comes from the correlation between hot-

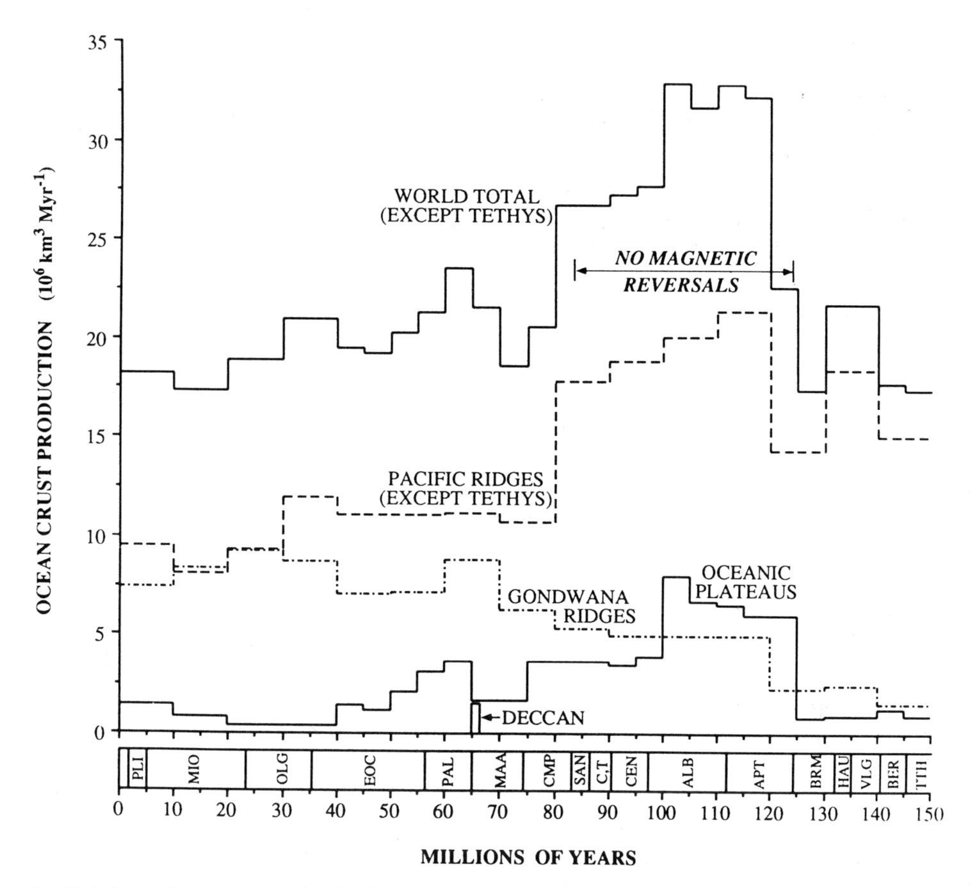

Figure 2 Global oceanic crust production for the past 150 Myr averaged in 5-Myr bins, from Larson (1991). Shown is the world total, plus the contributions from Pacific ridges, Gondwana (Atlantic and Indian) ridges, and oceanic plateaus. The interval marked "no magnetic reversals" is the Cretaceous normal geomagnetic polarity superchron.

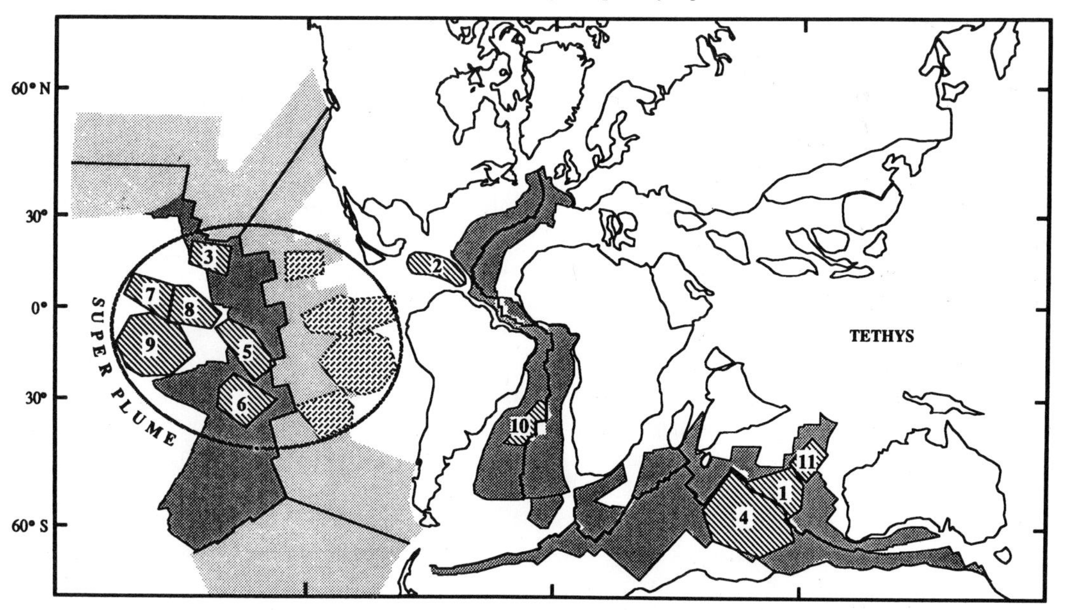

Figure 3 Plate boundary reconstruction near 83 Ma (magnetic anomaly 34) from Larson (1991) showing location of ocean plateau basalts formed during the mid-Cretaceous plume pulse. Striped polygons represent surviving ocean plateaus with 125–80 Ma age; stippled polygons indicate possible locations of ocean plateaus since subducted. The two clusters of plateaus suggest the sources were either two superplumes or two groups of smaller plumes. The absolute longitudes are unknown; tic marks on horizontal axes are 90° apart in longitude.

spot activity and the duration of geomagnetic polarity epochs. Larson and Olson (1991) have shown that the rate of crust production in flood basalt and oceanic plateau provinces is inversely correlated with the frequency of geomagnetic polarity reversals during the past 150 Myr. The Cretaceous plume pulse coincides very closely with the long Cretaceous normal polarity superchron, a 40-Myr interval without field reversals. Since the Cretaceous superchron, the length of an average polarity interval has decreased in step with hotspot activity. This correlation suggests that variations in mantle plume activity somehow affect the operation of the geodynamo.

One theoretically plausible mechanism for long-term coupling of the core and mantle is variations in heat flow at the core–mantle boundary. Thermal coupling is envisioned to act as follows. During periods of high plume activity, heat flow from the core to the mantle is high, and convection in the core is vigorous. Some models of the geodynamo indicate that vigorous convection tends to stabilize the dynamo against frequent reversals (Olson and Hagee, 1990) by increasing the strength and stability of the dipole field. These models predict an increase in polarity epoch length when core heat loss increases. During quiet periods when plume activity is low, heat transfer at the core–mantle boundary is low, and core convection is relatively weak. The same dynamo models predict that the magnetic field is then prone to frequent reversals.

The concept of thermal coupling between core and mantle entails the following sequence of events. During quiet times, when plume activity is low, the plume source region (assumed to be the hot thermal boundary layer in the D″ region at the base of the mantle) grows by conduction. The quiet interval ends when thermal instabilities, originating in the thermal boundary layer, develop into starting plumes. As the plumes rise through the mantle, the thickness of the thermal boundary layer is reduced in the region near the plume, and the heat flow from the core to the mantle is increased proportionally. The core responds by increased vigor of convection, changes in the convective pattern, or both, which alter the reversal frequency. Arrival of the plume at the base of the lithosphere results in massive amounts of decompression melting, leading to formation of continental flood basalts or ocean plateau basalts. The model is predicated on the assumptions that mantle convection is highly time dependent and that plumes are formed and destroyed repeatedly. In cases where the plume survives for a long time, a linear hotspot track is formed on the overriding plate. In other cases the plume is so short-lived that only an isolated thermal develops from the hot thermal boundary layer, and a flood basalt or ocean plateau is created, but no linear hotspot track.

One problem with using this model to explain the correlation between hotspot activity and geomagnetic reversal patterns is the uncertainty in the temporal relationship between the increased heat flow at the base of the mantle and the pulse in near-surface magmatism. The correlation discussed above indicates very little time difference between the two phenomena, with changes in reversal frequency synchronous with changes in hotspot activity to within 5–10 Myr (Larson and Olson, 1991). This picture is quite different from an earlier study by Courtillot and Besse (1987), who considered only continental flood basalts and concluded that hotspots and polarity epoch length variations were 90° out of phase.

Plume dynamics do not require that the temporal difference between basal heat flow and magmatic activity be small. It is even unclear which of the two phenomena should lead in time. Magmatic activity and basal heat flow occur at different ends of the plume and are governed by different physical processes. The onset time for magmatism is related to the transit time of the plume through the mantle. This time is difficult to estimate, because the lower mantle viscosity is poorly constrained, and also because the solid-state phase changes in the transition are expected to interrupt plume development (Liu *et al.*, 1991; Weinstein, 1993). The onset of increased core heat flow represents the time required to substantially thin the thermal boundary layer in the D″-layer source region. This heat flow increase may be related to the rise of the developing plume, or to a general increase in mantle flow velocities. Since the time scales of each of the processes may conceivably range from a few to several tens of millions of years, there is no theoretically based reason to suppose they will occur in close temporal proximity. This point was stressed recently by Loper (1992), who used a model of forced convection to argue that the D″ layer would require at least 100 Myr to respond to plume formation and would prevent core heat flow from varying phase with plume-related magmatism. In view of this uncertainty, it is important to establish the causal relationship between near-surface and deep-seated variations in thermal plumes.

Partial Melting in Thermal Plumes and Diapirs

An idealized but conceptually useful model of a developing mantle plume supposes a viscous fluid layer initially at rest and at a uniform temperature, which is heated from below and cooled from above. Conductive thermal boundary layers develop at the top and bottom of the fluid. At the upper boundary a rigid but thermally conducting lid, representing the mechanically strong portion of the lithosphere, caps the fluid layer. Instabilities first develop in the lower thermal boundary layer and grow into thermal plumes. At the leading edge of each plume is a large thermal diapir. If the plume is continuous in time, a narrow conduit is established behind the diapir (Olson *et al.,* 1987; Griffiths and Campbell, 1990). As the plume ascends, its highest temperature portions begin melting where the plume material intersects the melting curve. As the *in situ* melt fraction increases, the velocity of melt percolation through the mantle approaches and exceeds the transport velocity of the plume. The melt then separates from the plume and is eventually incorporated into the overlying lid as added crust. Because the rate of melt production is particularly large in the leading diapir, a starting plume produces a very large initial magmatic pulse.

A variation of this model consists of a solitary diapir thermal rising from depth in the mantle, without a trailing conduit. This model is intended to simulate a transient instability originating from the breakup of a lower mantle plume by transition zone phase changes (Liu *et al.,* 1991). The solitary diapir is a useful simplification for numerical experimentation because the only important source properties are the initial radius, depth, and temperature of the diapir.

Numerous techniques have been used for incorporating partial melting in models of mantle convection (Scott and Stevenson, 1989; Watson and McKenzie, 1991; Arndt and Christensen, 1992; Fowler, 1990a,b; Mutter *et al.,* 1988; Farnetani and Richards, 1992). All of these techniques involve approximations and rely on assumptions of how the melt is transported relative to the solid matrix. However, there is general agreement on several basic points. First, nearly all melting takes place by decompression, as rising material crosses the local solidus. Second, partially molten rock is permeable even for very low melt fractions (porosity), and therefore melt can migrate through the mantle even when the melt fraction is very low, on the order of 1% (McKenzie, 1985). At higher porosities, model studies indicate the melt percolation velocity is much greater than the solid matrix velocity associated with mantle convection (Scott and Stevenson, 1989), and therefore the processes of melt extraction and crustal growth can be regarded as instantaneous on the time scale of the larger-scale convective motions.

The specific problem of melt production in mantle plumes has been considered previously by Campbell and Griffiths (1990), Watson and McKenzie (1991), and Arndt and Christensen (1992). Watson and McKenzie applied relations for pressure-release melting of dry peridotite derived by McKenzie and Bickle (1988) to steady-state calculations of an axisymmetric plume beneath a conducting lid to model the Hawaiian hotspot. They determined the combination of temperature in the core of the plume, lid thickness, and asthenosphere viscosity that gave the best fit to the present-day crust production rate, geoid, and swell height for Hawaii. They also showed, in the context of their steady-state model, how lithosphere thickness limits the rate of melt production by limiting access to the zone where decompression melting can occur. The calculations by Arndt and Christensen (1992) focused on magma compositions and specifically sought to determine the fractions of magma produced by melting within the plume and by remelting of the lithosphere. They concluded that at most a few percent of the total magma is derived from remelted lithosphere; nearly all of the melt comes from the plume itself. The study by Campbell and Griffiths applied melting relations to a highly idealized analytical model of a rising thermal diapir to estimate the magma volumes produced in a plume head. Uniform viscosity plumes were assumed in the calculations by Campbell and Griffiths (1990) and Watson and McKenzie (1991), whereas temperature-dependent viscosity was used in the calculations by Arndt and Christensen (1992). In all of these studies, the melt production is calculated using parameterizations of experimentally derived melting relations for dry peridotites. These relations were obtained from batch melting experiments, whereas the condition assumed in the models, with melt continuously extracted from the matrix, is in fact closer to a fractional (Rayleigh) melting process. In addition, the effects of latent heat upon melting and the additional buoyancy due to melt within the plume were ignored.

The approach used here differs from these others in several important ways. First, a simple univariant melting formula with a constant pressure derivative is used. Second, the effects of latent heat in melting and melt buoyancy are included in the calculation in a self-consistent manner. I find that latent heat absorption is important in the energy balance of the plume and affects both the thermal structure and the extent of melting. Also, I assume in this model that the melt is transported with the matrix at low melt fractions and escapes from the mantle when it exceeds a critical value. In these calculations, percolation is neglected when the melt fraction ϕ is less than a saturation value ϕ_s, and is assumed to be instantaneous when $\phi > \phi_s$. Calculations of mantle structure and melt production based on this method are presented for both solitary thermal diapirs and continuous thermal plumes.

In all the numerical models of melting in thermal plumes, including the calculations presented here, the effects of plate motion on plume structure are not explicitly considered. Constant-velocity plate motion transforms an otherwise axisymmetric hotspot into a bilaterally symmetric hotspot track, elongated in the direction of plate motion (Olson, 1990). The resulting plume structure is then fully three-dimensional. It is beyond the scope of this study to incorporate fully three-dimensional effects, although this will soon be possible to do because numerical models of plumes with moving plates are now being developed (Ribe and Christensen, 1992).

The Numerical Model

The conservation equations for vorticity and energy for axially symmetric thermal convection in an infinite Prandtl number, uniform-viscosity Boussinesq fluid can be written in cylindrical (r, z) coordinates as

$$\left(\nabla^2 - \frac{2}{r}\frac{\partial}{\partial r}\right)^2 \psi = -\frac{r\rho g}{\eta}\left(\alpha\frac{\partial T}{\partial r} + \beta\frac{\partial \phi}{\partial r}\right) \qquad (1)$$

and

$$\frac{\partial H}{\partial t} + \mathbf{u} \cdot \nabla H = k\nabla^2 T. \qquad (2)$$

Here, ψ is the streamfunction of the motion; T is temperature; ϕ is the *in situ* melt fraction; $\mathbf{u}$ is the velocity of the matrix material; g is gravity; ρ is reference density; η is the viscosity; α and β are thermal and melting expansion coefficients, respectively; k is thermal conductivity; and ∇ is the gradient operator. The energy equation (2) is written using an enthalpy variable H, which in this context is the sum of latent and sensible, heat,

$$H = \rho C_p T + \rho L \phi, \qquad (3)$$

where C_p is specific heat at constant pressure, and L is the latent heat of melting. The streamfunction is related to the matrix velocity $\mathbf{u} = (\hat{\mathbf{r}}u, \hat{\mathbf{z}}w)$ by

$$(u, w) = \left(-\frac{1}{r}\frac{\partial \psi}{\partial z}, \frac{1}{r}\frac{\partial \psi}{\partial r}\right). \qquad (4)$$

Melting and solidification are specified as follows. I assume a simple, univariant melting law in which the melting temperature varies with depth according to

$$T_m = T_{m0} + \gamma y, \qquad (5)$$

where T_m is the melting temperature, γ is its depth derivative due to hydrostatic pressure, $y = z_0 - z$ is depth below the top surface $z = z_0$, and the subscript 0 denotes the melting temperature at surface pressure. The dry peridotite solidus of Takahashi and Kushiro (1983), corrected for the adiabatic gradient of the mantle, can be closely approximated using (5) with $T_{m0} \approx 1100°C$ and $\gamma \approx 4°C/km$.

It is convenient to introduce two reference enthalpy functions. The first corresponds to sensible heat at the melting temperature,

$$H_m = \rho C_p T_m, \qquad (6)$$

and the second corresponds to enthalpy at the saturation value of the *in situ* melt fraction, ϕ_s,

$$H_s = H_m + \rho L \phi_s. \qquad (7)$$

The subscript s refers to saturation conditions. In addition, I define a progress variable χ to denote the volume of melt extracted from a fixed-volume element of the mantle and added to the crust. Assuming the extracted melt moves vertically, the additional crustal thickness from melt extracted from the mantle is given in terms of χ by

$$h_c = \int_0^{z_0} \chi \, dy. \tag{8}$$

According to this prescription, three states are possible: subsolidus, partially molten without melt percolation, and partially molten with percolation. Melt fraction, temperature, and the progress variable can be determined from H in each of these three states using the following relations:

1. Subsolidus ($H < H_m$):

$$\phi = \frac{\partial \chi}{\partial t} = 0; \qquad T = \frac{H}{\rho C_p}. \tag{9}$$

2. Partially molten, no percolation ($H_m < H < H_s$):

$$\begin{aligned} \phi &= \frac{H - H_m}{\rho L}; \\ T &= T_m; \\ \frac{\partial \chi}{\partial t} &= 0. \end{aligned} \tag{10}$$

3. Partially molten, percolation ($H > H_s$):

$$\begin{aligned} \phi &= \phi_s; \\ T &= T_m; \\ \frac{\partial \chi}{\partial t} &= \frac{\partial}{\partial t}\frac{(H - H_s)}{\rho L}. \end{aligned} \tag{11}$$

We divide the cylinder into a uniformly viscous mantle $0 < z < z_L$ and a rigid mechanical lid $z_L < z < z_0$, where z_L is the mantle–lid boundary. Equations (1)–(11) are applied to the viscous mantle; only the heat conduction version of (2) is applied in the lid.

The boundary conditions are as follows. The bottom surface $z = 0$ is isothermal, impermeable, and free-slip (zero shear stress), conditions which permit the plume to draw material from the entire hot thermal boundary layer. The top surface $z = z_0$ is isothermal, and the base of the lid $z = z_L$ is no-slip (zero velocity). The rigid lid simulating the mechanically strong, elastic portion of the lithosphere has zero velocity and the same thermal conductivity as the underlying viscous mantle. The sidewall $r = R$ and the centerline $r = 0$ are reflecting boundaries (zero heat flow, impermeable, and zero shear stress).

Initial conditions in both sets of calculations include a cold thermal boundary layer at the surface and isothermal interior at temperature T_I. The upper thermal boundary layer, representing the thermal lithosphere, is specified by an error function diffusion profile with an age τ_0. In the solitary diapir calculations, an initially spherical thermal anomaly with temperature T_D is introduced near the base of the cylinder. The diapir position is fixed for a short time (typically a few million years), allowing it to develop a diffusive thermal halo, and is then released. In the continuous plume calculations, a hot basal thermal boundary layer is included as an initial condition, specified by a basal temperature T_B and an error function diffusion profile with an age τ_B. A small, long wavelength perturbation is added to the basal thermal boundary layer, in order to initiate a rising instability on the centerline.

Equations (1)–(11) have been solved for both solitary diapirs and for continuous plumes using finite difference methods. The enthalpy in Eq. (2) is advanced in time using an explicit method with the upstream differences for calculating the advection terms. Temperature and porosity are updated at each time step according to the criteria expressed in Eqs. (9)–(11), and the streamfunction is updated by solving (1) as two modified Poisson equations.

In these calculations the flow is restricted to a cylinder 750 km in depth, representing only the upper mantle. This restriction is not intended to mean that this model applies only to plumes originating in the upper mantle. I found that a grid size of 10 km or finer is necessary to resolve the thermal structure and the distribution of melt within the plume, and this requirement made calculations of the whole mantle depth prohibitively expensive. Nevertheless, the general behavior of these calculations provides some insight useful for interpreting the evolution of flood basalts, regardless of the depth of origin of the plumes.

In the solitary diapir calculations, the lid thickness and the initial excess temperature of the diapir were varied. In the continuous plume calculations, only the lid thickness was varied. Values of the physical parameters common to the calcula-

Table 3
Thermophysical Properties and Model Parameters

Parameter	Model value	Reference
Cylinder depth, z_0	750 km	
Cylinder radius, R	1000 km	
Grid spacing, Δz, Δr	10 km	
Interior temperature, T_1	1300°C	Jeanloz and Morris (1986)
Surface temperature, T_0	0°C	
Surface melting temperature, T_{m0}	1100°C	Takahashi and Kushiro (1983)
Melting point gradient, Υ	$4°C \cdot km^{-1}$	Takahashi and Kushiro (1983)
Latent heat of melting, L	$400\ kJ \cdot kg^{-1}$	Stebbins *et al.* (1983)
Saturation melt content, ϕ_s	0.01	McKenzie (1985)
Specific heat, C_p	$1\ kJ \cdot kg^{-1} \cdot °C^{-1}$	Turcotte and Schubert (1982)
Mantle density, ρ	$3.3\ mg \cdot m^{-3}$	Anderson (1989)
Crust density, ρ_c	$2.9\ mg \cdot m^{-3}$	Anderson (1989)
Thermal expansion, α	$3 \times 10^{-5} \cdot °C^{-1}$	Anderson (1989)
Melting expansion, β	0.12	Turcotte and Schubert (1982)
Gravity, g	$9.8\ m \cdot s^{-2}$	Anderson (1989)
Thermal conductivity, k	$3.3\ W \cdot m^{-1} \cdot °C^{-1}$	Horai and Simmons (1970)
Viscosity, η	$0.5 \times 10^{21}\ Pa \cdot s$	Nakada and Lambeck (1989)

tions are listed in Table 3. The diapir calculations were typically run for 15–20 Myr, and the plume calculations were typically run for 42 Myr. In addition to the temperature, velocity, and *in situ* melt fraction, the total crust production and production rate, surface geoid, dynamical surface uplift (due to normal stress acting on the lid), and isostatic uplift (due to crustal thickening) were calculated.

Melting in Solitary Diapirs

Figures 4–6 show a typical result of calculations on ascent and partial melting in an isolated thermal diapir with an initial 200-km radius.

Figure 4 shows the distribution of temperature and melt concentration, along with the distribution of surface observables at various times in the evolution of a diapir rising beneath a 40-km-thick lid and 50-Myr thermal lithosphere. The surface observables plotted are the geoid anomaly, the dynamical topography (the topography induced by vertical normal stress on the lid), and the .total topography (the sum of dynamical topography and the isostatic topography due to an increase in crustal thickness from melt extracted vertically from the diapir).

The first stage in the evolution of the diapir is represented in Fig. 4 by the image at $t = 5.2$ Myr (top left) and corresponds to the time when the rising diapir first undergoes decompression partial melting, but has not yet flattened out beneath the lid. This stage is characterized by partial melting to 200 km in depth, a large Gaussian-shaped dynamically supported uplift approximately 3 km in amplitude, and correlated with the topography, a positive geoid anomaly of nearly 70 m in amplitude. While the melt content of the diapir is large at this stage, very little melt has been extracted and added to the crust. The second stage in the evolution is represented in Fig. 4 at 9.1 Myr (top right). Here the diapir has begun to collapse beneath the lid, forming a lens of partial melt approximately 600 km in diameter and centered near 110 km in depth. Magma extraction and crustal formation are most rapid at this stage, as indicated by the large difference between the total topography and the dynamically supported topography. Due to the lateral spreading of the diapir, both the dynamic topography and the geoid subside rap-

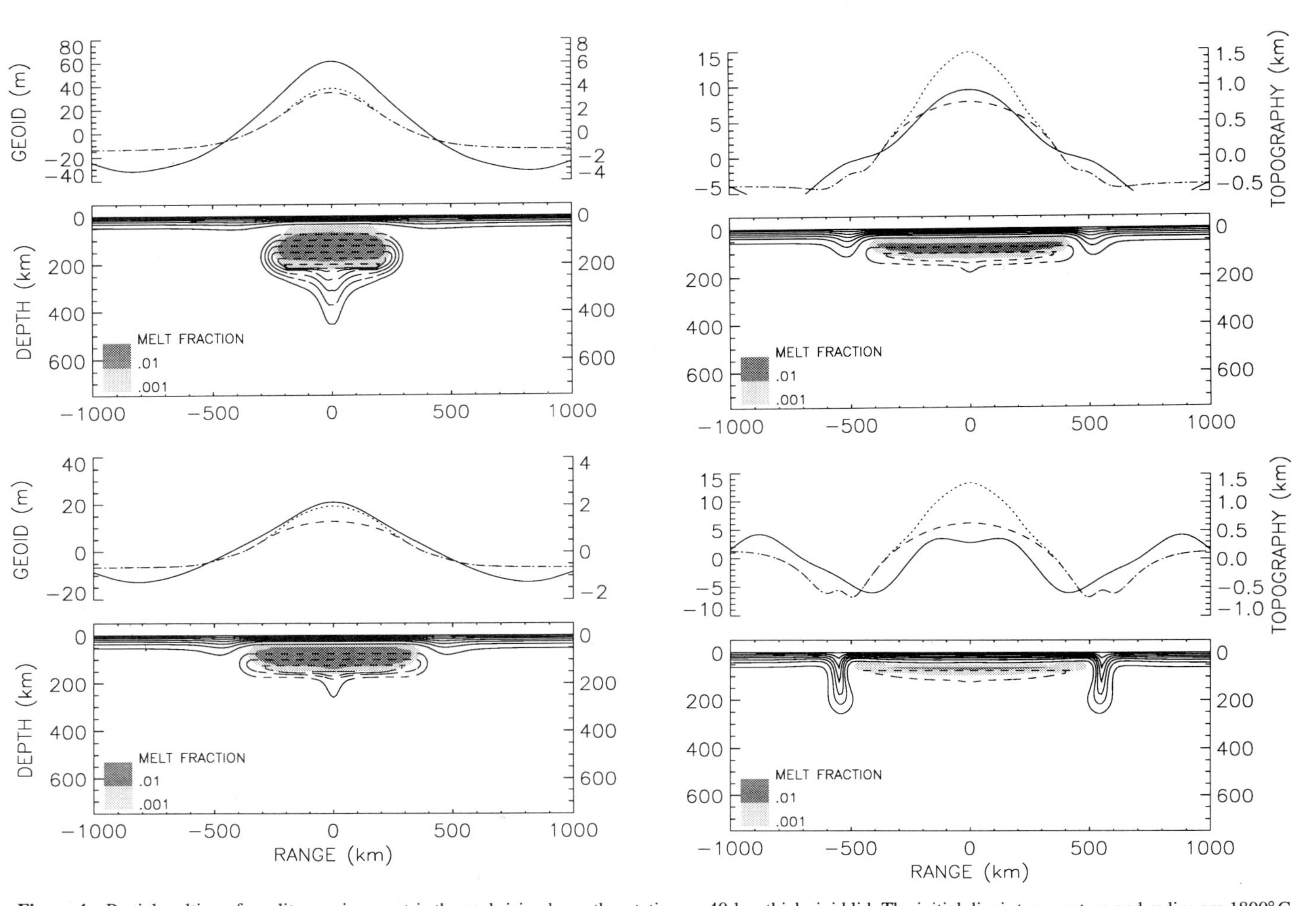

Figure 4 Partial melting of a solitary axisymmetric thermal rising beneath a stationary 40-km-thick rigid lid. The initial diapir temperature and radius are 1800° C and 200 km, respectively. Other parameters used in the calculation are given in Table 3. The times of each image in million years after release of the diapir are as follows: top left, 5.2; bottom left, 7.1; top right, 9.1; bottom right, 13.1. The upper plot in each image shows the distribution of geoid height (solid curve), dynamic topography (dashed curve), and total topography (dotted curve). The lower plot shows the distribution of temperature contours and *in situ* melt fractions (shaded). Lithosphere contours (solid) are 0–1200° C in 200° C intervals; diapir contours (dashed) increase from 1400° C in 100° C intervals. The images are reflected about the symmetry axis.

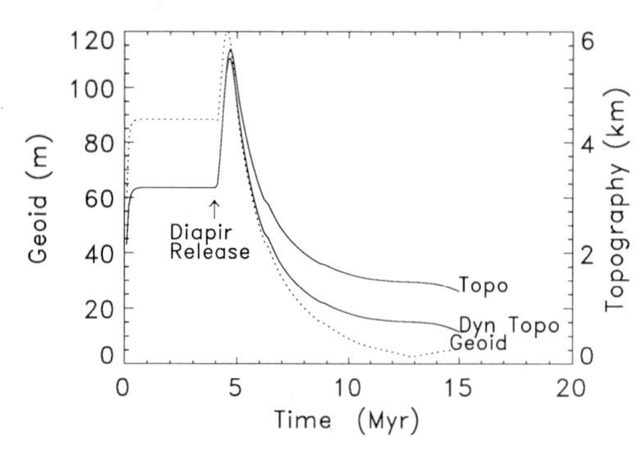

Figure 5 Time series of surface observables from the calculations in Figure 4. Shown are the time series of axial ($r = 0$) geoid height, axial topography, and axial dynamic topography.

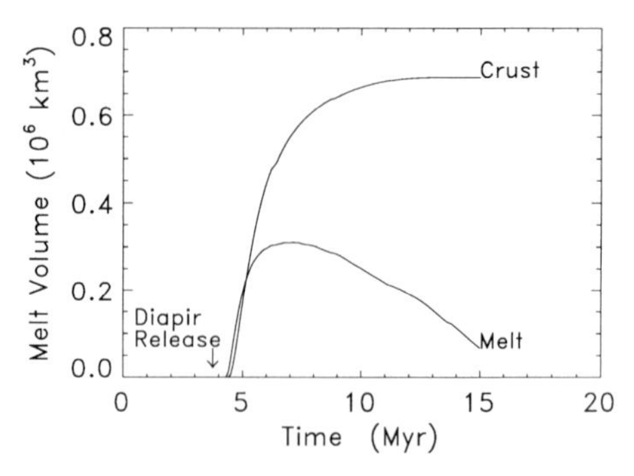

Figure 6 Time series of mantle melt volume and crust volume from the calculations in Figure 4.

idly, and the total topography rapidly decreases in elevation during this stage, despite the rapid addition of crustal material. Surprisingly, the period of most active crust building coincides with rapid subsidence, particularly in the central region above the diapir. Note that most of the melt is derived from portions of the diapir where the excess temperature is 200° C or less, and virtually all of the melt is derived from portions with less than 300° C excess temperature.

The third stage in the evolution is represented in Fig. 4 by times later than 13 Myr (bottom right column). This final stage is characterized by the cooling and resolidification of the partially molten diapir within the mantle, slow subsidence at the surface, and greatly diminished rates of crustal addition. The lateral spread of the diapir is checked by the formation of a ring-shaped instability in the cold thermal boundary layer, formed from the material displaced outward when the diapir first began to collapse. The motion induced by the ring instability destroys the thermal anomaly of the diapir on a faster time scale than that if the diapir simply stagnated beneath the lid and cooled only by conduction. Within 15 Myr the diapir is essentially gone, and the dynamics are dominated by the sinking of the ring instability. The three stages are also seen in the time series of axial topography, geoid, mantle melt content, and added crust volume and volcanic heat, as shown in Figs. 5 and 6 for the calculations in Fig. 4. The first stage is marked by rapid (1–2 Myr in duration) uplift and occurs as the diapir approaches the lid. The second stage is the main magmatic episode and is marked by rapid subsidence. The last stage includes slow subsidence, little crustal addition, and gradual refreezing of the melt remaining in the mantle.

Melting in Continuous Plumes

I have also made a series of calculations of partial melting in plumes initiated by the instability of a hot thermal boundary layer. The presence of a basal thermal boundary layer in axisymmetric geometry ensures the existence of a continuous plume and provides a simple way to model the long-term behavior of the hotspot, as well as its initial development. In the calculations summarized in Table 4, the basal temperature is fixed at either 2000 or 1800° C, and the thickness of the mechanical lid is varied between 0 and 100 km. The values of all the other parameters are given in Table 3.

Figure 7 shows the distribution of temperature and melt concentration, along with the distribu-

Table 4
Results of Plume Calculations

Lid (km)	Crustal volume at 20 Myr (10^6 km^3)	Crustal production peak (km^3/yr)	Crustal production at 40 Myr (km^3/yr)	Heat flow at 40 Myr (GW)
0	11.5	1.87	0.019	560
10	6.11	0.81	0.067	318
20	4.31	0.55	0.031	221
50	2.47	0.306	0.003	215
75	1.29	0.135	0.0046	201
100	0.45	0.048	0.004	200

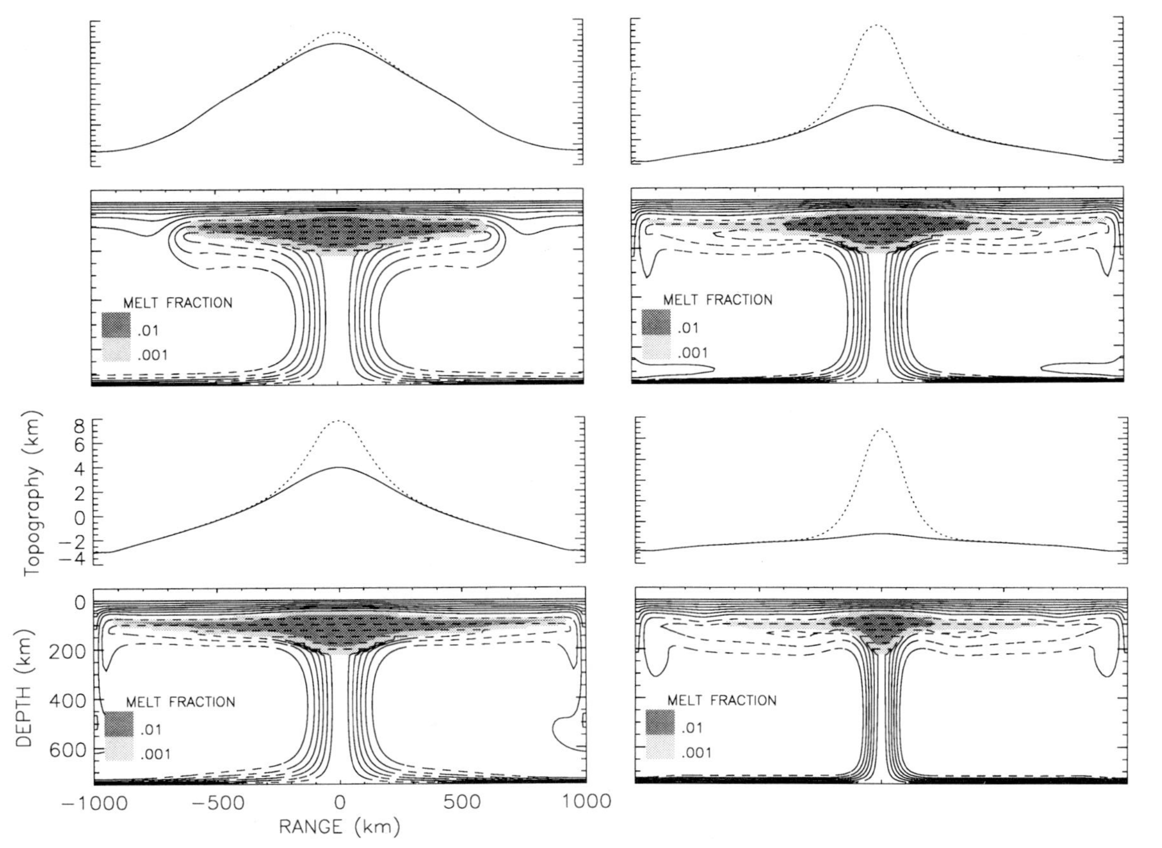

Figure 7 Partial melting of a thermal plume rising beneath a stationary 20-km-thick rigid lid. The basal temperature is 2000° C. Other parameters used in the calculation are given in Table 3. The times of each image in million years after plume release started by introducing a long-wavelength perturbation in the basal thermal boundary layer are as follows: top left, 6.0; bottom left, 12.1; top right, 18.0; bottom right, 36. The upper plot in each image shows the dynamically supported topography (solid curve) and the total topography, which includes the isostatic crustal component (dotted curve). The lower plot in each image shows the distribution of temperature contours and the *in situ* melt content (shaded). The lithosphere contours (solid) are 0–1200° C in 200° C intervals; plume contours (dashed) increase from 1400° C in 100° C intervals. The images are reflected about the symmetry axis.

tion of surface topography at various times in the evolution of a thermal plume originating from a 2000° C lower boundary, rising beneath a 20-km-thick mechanical lid. Both the dynamical topography and the total topography (including isostatic crustal topography) are shown above each cross section. The image at 6 Myr (top left) is equivalent to the first stage in the evolution of the solitary diapir shown in Fig. 4. The plume head has begun to spread out beneath the mechanical lid, and the melt content of the plume is high. The region occupied by partial melt is 1200 km in diameter and extends from 50 to 200 km in depth. The surface uplift has reached its maximum, approximately 4 km at the point directly above the plume centerline. Only a small amount of melt has been added to the crust at this time. The plume at 12 Myr (Fig. 7, lower left) is in the second stage, characterized by lateral spreading beneath the lid, peak melt concentrations near 100 km in depth, peak rates of crust production, and rapid subsidence due to loss of the dynamically supported topography. At this time, the partially molten zone is 1600 km in diameter, and the topography above the plume axis approaches 4 km. The images at 18 and 36 Myr (Fig. 7, right column) show the evolution of the plume from its initial structure, dominated by the head, to the final conduit structure, dominated by a thin cylinder-shaped thermal anomaly and a restricted zone of partial melting. The zone of partial melting shrinks to a 200-km-diameter lens centered near 100 km in depth. The dynamic topography is reduced to a long-wavelength Gaussian-shaped swell with only a 1.0-km elevation at 36 Myr (Fig. 7, lower right).

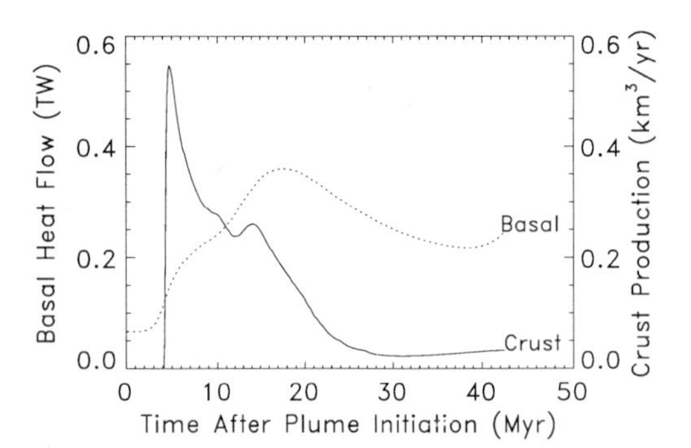

Figure 8 Time series from the calculation in Figure 7. Basal heat is the total heat flow into the base of the cylinder, supporting the plume. Crustal production is the rate of melt extracted from the plume.

Beyond 40 Myr the plume structure is essentially as shown in the last image of Fig. 7, although the calculation never reaches a true steady state. Small fluctuations in plume strength and melt production continue to occur, but the overall structure of the flow and the temperature fields do not change.

The images in Fig. 7 show the melt zone extends from 50 to 200 km in depth and is centered near 100 km in depth, where the excess temperature at 100 km is 300° C at all stages. This high-temperature core of the plume, with as much as 600° C excess temperature, extends upward from the hot boundary layer into the partially molten lens. However, most of the plume head and most of the melt lens have excess temperatures ranging from 100 to 350° C, generally consistent with excess temperatures deduced from hotspot magmas and topographic rises (Sleep, 1990; Schilling, 1991). The role of latent heat absorption is evident in Fig. 7, where the isotherms in the melt lens are horizontal and evenly spaced at the melting point gradient. Without latent heat absorption in the calculation, the high-temperature plume core would extend closer to the surface, so the "excess temperature" and the melt volume that would be inferred for such a plume would be much greater. Accordingly, latent heat is a necessary element in modeling mantle plume structure.

Figure 8 compares the time series of basal heat flow and crust production from the calculation in Fig. 7. The main pulse of crust production lasts about 15 Myr. At the peak of activity, the rate of production is 0.55 km^3/yr. In comparison, the production rate at 40 Myr, typical of the long-term rate, is only about 0.03 km^3/yr, nearly a factor of 20 lower. This reduction factor is typical for hotspot tracks originating from continental flood basalts (Richards *et al.,* 1989). The long-term average basal heat flow of 220 GW is comparable to Sleep's (1990) estimate for the present-day strength of the Hawaiian hotspot. During the period of highest activity, the basal heat flow is about 60% greater than the long-term average value.

Figure 8 demonstrates that the variations in basal heat flow occur in association with the variations in crustal production, although the two time series are not closely in phase over most of the calculation. Initially, the flow associated with the accelerating plume entrains a large part of the thermal boundary layer. This entrainment accounts

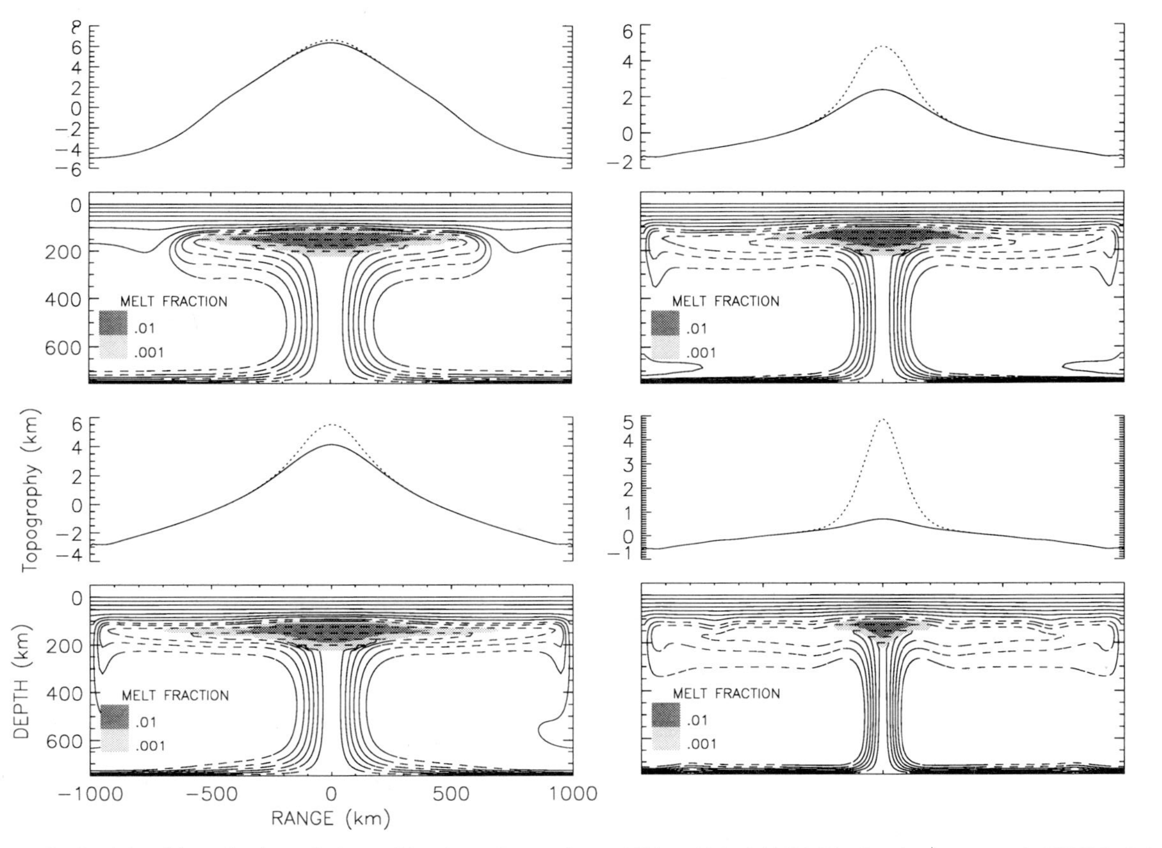

Figure 9 Partial melting of a thermal plume rising beneath a stationary 75-km-thick rigid lid. The basal temperature is 2000° C. Other parameters used in the calculation are given in Table 3. The times of each image in million years after plume release started by introducing a long-wavelength perturbation in the basal thermal boundary layer are as follows: top left, 6; bottom left, 12.1; top right, 18; bottom right, 36. The upper plot in each image shows the dynamically supported topography (solid curve) and the total topography, which includes the isostatic crustal component (dotted curve). The lower plot in each image shows the distribution of temperature contours and *in situ* melt content (shaded). The lithosphere contours (solid) are 0–1200° C in 200° C intervals; plume contours (dashed) increase from 1400° C in 100° C intervals. The images are reflected about the symmetry axis.

for the initial sharp increase in basal heat flow near $t = 5$ Myr, which is well correlated with the steep rise in magmatic activity. However, the correlation between the two signals degrades after the magmatic peak. As magmatic activity drops sharply after 7 Myr, basal heat flow continues to slowly increase, reaching a maximum near 18 Myr. Inspection of Fig. 7 reveals that the heat flow maximum occurs when cold sinking material displaced from the top thermal boundary layer approaches the basal thermal boundary layer. In this calculation the heat flow maximum lags behind the peak in magmatic activity by approximately the time needed for the material displaced by the plume head to sink to the hot thermal boundary layer. More generally, the peak heat flow corresponds to the time when the large-scale circulation is most developed.

In the mantle, the largest scale of the circulation is probably controlled more by subduction than by plumes. Consequently, periods of increased core heat loss are likely to be more causally related to increases in sea floor spreading rates than hotspot activity. The time series in Fig. 2, which show that sea floor spreading was also high during the Cretaceous geomagnetic polarity superchron, are consistent with this interpretation.

Figures 9 and 10 show the results of a calculation similar to Figs. 7 and 8, but with an 75-km-thick lid, representative of a plume rising beneath stable continental lithosphere. The same sequence of events found in the thin lid case is present in this case, with the major difference being slower rates of crustal generation when the lid is thick. This calculation produced 1.29×10^6 km^3 of crust in the main pulse (before 20 Myr), which is representative of the continental flood basalt volumes given in Table 1. The peak production rate is about 0.135 km^3/yr, the long-term average rate is about 0.005 km^3/yr, and the long-term average plume strength is about 200 GW, about 10% weaker than the plume in Fig. 7. As in Fig. 8, the time series in Fig. 10 show the basal heat flow and rate of crustal production increase nearly simultaneously, but the heat flow reaches a maximum about 12 Myr after the peak in magmatic activity, when the material displaced from the cold thermal boundary layer by the plume head reaches the basal boundary layer.

Table 4 and Fig. 11 show the total volume added to the crust from several thermal plume calculations for various lid thicknesses and basal temperatures. As expected, increasing the basal temperature increases the amount of melt produced. More importantly for the plume model, however, is the fact that the volume of melt produced is quite sensitive to the lid thickness. As indicated in Fig. 11, crustal volumes typical of continental flood basalts can easily be produced from a plume rising beneath a 40- to 80-km-thick mechanical lid. The same plume produces nearly 10 times as much melt when it ascends beneath very young lithosphere, where the mechanically strong lid is only 10–20 km thick. Lid thickness strongly controls the amount of melt produced in a plume by controlling the depth at which the leading diapir spreads out. Melting occurs only in that part of the plume which crosses the solidus before reaching the lid. The thicker the lid, the smaller is the portion of the diapir that enters the melting zone,

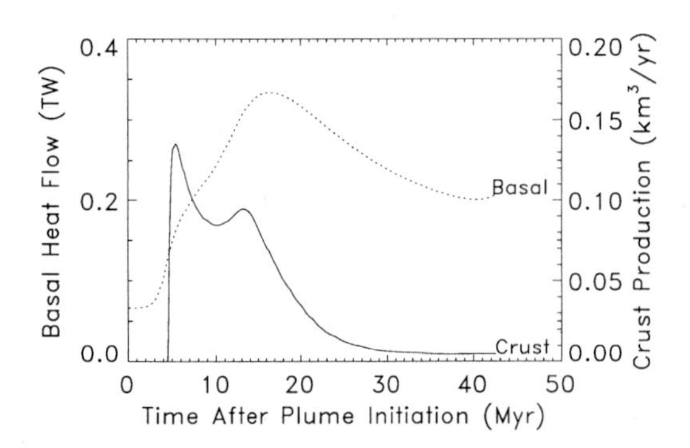

Figure 10 Time series from the calculations in Figure 9. Basal heat is the total heat flow into the base of the cylinder, supporting the plume. Crustal production is the rate of melt extracted from the plume.

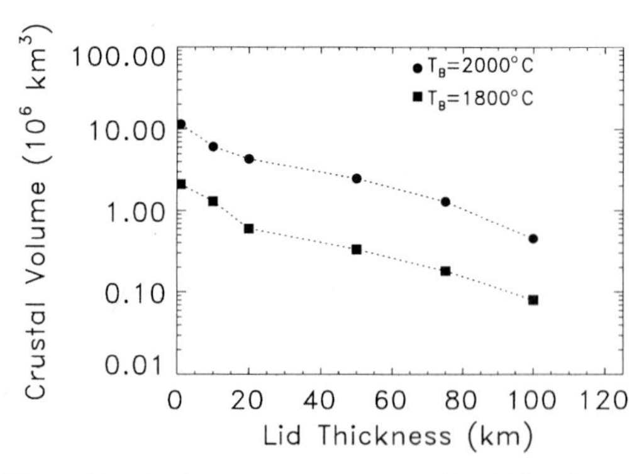

Figure 11 Total crustal volume produced in 40 Myr after initiation by partial melting in a plume rising beneath rigid lids of various thickness. The basal temperature is either 2000 or 1800° C, as indicated by symbols. Other parameters used in the calculations are given in Table 3.

and the smaller is the portion that melts. The large difference in the volume of the major oceanic plateaus versus the major continental flood basalts can be explained, in the context of the plume model, in terms of differences in the portions of plumes that undergo partial melting beneath thin versus thick lithosphere.

Conclusions

Calculations of partial melting in thermal plumes and diapirs indicate the following general properties for mantle plumes:

1. There are three stages in the magmatism of a mantle plume. The first stage occurs as the plume approaches the lithosphere from below and is characterized by rapid uplift, rapid melt generation, but relatively little crust production. The second stage occurs as the plume head spreads out beneath the lithosphere, and is marked by rapid subsidence and rapid crust production. The last stage corresponds to the contraction of the plume and the melt zone toward a steady state and is marked by slow subsidence and greatly diminished crustal production.

2. The amount of melt produced in a plume is very sensitive to the thickness of the mechanical lithosphere. The large difference between continental flood basalt and ocean plateau volumes is probably due to the differences in lithosphere thickness beneath continents and young ocean crust.

3. Basal heat flow increases during plume formation, supporting the suggestion that deep mantle plumes thermally couple the core to the mantle. However, the time of maximum heat flow is governed by the sinking of cold material displaced by the plume head, and generally lags behind the peak in magmatic activity.

Acknowledgments

The author thanks D. Bercovici, U. Christensen, S. Watson, M. Ryan, and J. Mahoney for helpful comments. This work was supported by NSF Grant EAR 8916152.

References

Anderson, D. L. (1989). "Theory of the Earth," Chap. 3, Blackwell, Oxford.

Arndt, N. T., and Christensen, U. R. (1992). The role of lithospheric mantle in continental flood basalt volcanism: Thermal and chemical constraints, *J. Geophys. Res.* **97,** 10967–10981.

Carlson, R. W. (1991). Physical and chemical evidence on the cause and source characteristics of flood basalt volcanism, *Austral. J. Earth Sci.* **38,** 525–544.

Campbell, I. H., and Griffiths, R. W. (1990). Implications of mantle plume structure for the evolution of flood basalts, *Earth Planet. Sci. Lett.* **99,** 79–93.

Courtillot, V., and Besse, J. (1987). Magnetic field reversals, polar wander, and core-mantle coupling, *Science* **237,** 1140–1147.

Davies, G. F. (1988). Ocean bathymetry and mantle convection. 1. Large-scale flow and hotspots, *J. Geophys. Res.* **93,** 10467–10480.

Duncan, R. A., and Richards, M. A. (1991). Hotspots, mantle plumes, flood basalts, and true polar wander, *Rev. Geophys.* **29,** 31–50.

Farnetani, C. G., and Richards, M. A. (1992). Numerical models of oceanic flood basalt events, *EOS Trans. Amer. Geophys. Union* **73,** 533.

Fowler, A. C. (1990a). A compaction model for melt transport in the Earth's asthenosphere. Part I. The basic model, *in* "Magma Transport and Storage" (M. P. Ryan, ed.), Wiley, Chichester/Sussex, England.

Fowler, A. C. (1990b). A compaction model for melt transport in the Earth's asthenosphere. Part II. Applications, *in* "Magma Transport and Storage" (M. P. Ryan, ed.), Wiley, Chichester/Sussex, England.

Griffiths, R. W., and Campbell, I. H. (1990). Stirring and structure in mantle starting plumes, *Earth Planet. Sci. Lett.* **99,** 66–78.

Horai, K., and Simmons, G. (1970). An empirical relationship between thermal conductivity and Debye temperature for silicates, *J. Geophys. Res.* **75,** 678–682.

Jeanloz, R., and Morris, S. (1986). Temperature distribution in the crust and mantle, *Annu. Rev. Earth Planet. Sci.* **14,** 377–415.

Larson, R. L. (1991). The latest pulse of the Earth: Evidence for a mid-Cretaceous superplume, *Geology* **19,** 547–550.

Larson, R. L., and Olson, P. (1991). Mantle plumes control magnetic reversal frequency, *Earth Planet. Sci. Lett.* **107,** 437–447.

Liu, M., Yuen, D. A., and Honda, S. (1991). Development of diapiric structures in the upper mantle due to phase transitions, *Science* **252,** 1836–1839.

Loper, D. C. (1992). On the correlation between mantle plume flux and the frequency of reversals of the geomagnetic field, *Geophys. Res. Lett.* **19,** 25–29.

Mahoney, J. J. (1987). An isotopic survey of Pacific oceanic plateaus: Implications for their nature and origin, *in* "Seamounts, Islands and Atolls" (B. Keating *et al.*, eds.), AGU Monograph 43, pp. 207–220. Washington D.C.

Mahoney, J. J., and Spencer, K. J. (1991). Isotopic evidence for the origin of the Manihiki and Ontong–Java oceanic plateaus, *Earth Planet. Sci. Lett.* **104,** 196–210.

McKenzie, D. P. (1985). The extraction of melt from the crust and mantle, *Earth Planet. Sci. Lett.* **74,** 81–91.

McKenzie, D. P., and Bickle, M. J. (1988). The volume and composition of melt generated by extension of the lithosphere, *J. Petrol.* **29,** 625–680.

Morgan, W. J. (1981). Hotspot tracks and the opening of the Atlantic and Indian oceans, *in* "The Sea," Vol. 7, pp. 443–487.

Mutter, J. C., Buck, W. R., and Zehnder, C. M. (1988). Convective partial melting. 1. A model for formation of thick basaltic sequences during the initiation of spreading, *J. Geophys. Res.* **93,** 1031–1048.

Nakada, M., and Lambeck, K. (1989). Late Pleistocene and Holocene sea-level change in the Australian region and mantle rheology, *Geophys. J. Intl.* **96,** 497–517.

Olson, P. (1990). Hotspots, swells and mantle plumes, *in* "Magma Transport and Storage" (M. P. Ryan, ed.), Wiley, Chichester/Sussex, England.

Olson, P., and Hagee, V. L. (1990). Geomagnetic polarity reversals transition field structure and convection in the outer core, *J. Geophys. Res.* **95,** 4609–4620.

Olson, P., Schubert, G., and Anderson, C. (1987). Plume formation in the D″-layer and the roughness of the core–mantle boundary, *Nature* **327,** 409–413.

Ribe, N., and Christensen, U. R. (1992). A 3D dynamical model of the Hawaiian plume, *EOS Trans. Amer. Geophys. Union,* 578.

Richards, M. A., Duncan, R. A., and Courtillot, V. E. (1989). Flood basalts and hotspot tracks: Plume heads and tails, *Science* **246,** 103–107.

Ryan, M. P. (1990). The physical nature of the Icelandic magma transport system, *in* "Magma Transport and Storage" (M. P. Ryan, ed.), Wiley, Chichester/Sussex, England.

Schilling, J.-G. (1991). Fluxes and excess temperatures of mantle plumes inferred from their interaction with migrating mid-ocean ridges, *Nature* **352,** 397–403.

Schubert, G., and Sandwell, D. (1989). Crustal volumes of the continents and of oceanic and continental submarine plateaus, *Earth Planet. Sci. Lett.* **92,** 234–246.

Scott, D. R., and Stevenson, D. J. (1989). A self-consistent model of melting, magma migration and buoyancy-driven circulation beneath mid-ocean ridges, *J. Geophys. Res.* **94,** 2973–2988.

Sleep, N. A. (1990). Hotspots and mantle plumes: Some phenomenology, *J. Geophys. Res.* **95,** 6715–6736.

Stebbins, J. F., Carmichael, I. S. E., and Weill, D. E. (1983). The high-temperature liquid and glass heat contents and heats of fusion of diopside, albite, sanidene and nepheline, *Amer. Mineral.* **68,** 717–730.

Takahashi, E., and Kushiro, I. (1983). Melting of a dry peridotite at high pressures and basalt magma genesis, *Amer. Mineral.* **68,** 859–879.

Tarduno, J. A., Sliter, W. V., Kroenke, L., Leckie, M., Mayer, H., Mahoney, J. J., Musgrave, R., Storey, M., and Winterer, E. L. (1991). Rapid formation of Ontong–Java plateau by aptian mantle plume volcanism, *Science* **254,** 399–403.

Turcotte, D. L., and Schubert, G. (1982). Applications of continuum physics to geological problems, *in* "Geodynamics," Wiley, New York.

Watson, S., and McKenzie, D. (1991). Melt generation by plumes: A study of Hawaiian volcanism, *J. Petrol.* **32,** 503–537.

Weinstein, S. A. (1993). Catastrophic overturn of the Earth's mantle driven by multiple phase changes and internal heat generation, *Geophys. Res. Lett.* **20,** 101–104.

White, R., and McKenzie, D. (1989). Magmatism at rift zones: The generation of volcanic continental margins and flood basalts, *J. Geophys. Res.* **94,** 7685–7729.

Zhang, Y.-S., and Tanimoto, T. (1992). Ridges, hotspots and their interaction as observed in seismic velocity maps, *Nature* **355,** 45–49.

Chapter 2 Melt Migration and Related Attenuation in Equilibrated Partial Melts

Tye T. Gribb, Shanyong Zhang, and Reid F. Cooper

Overview

Anelasticity associated with the migration of the melt phase in synthetic (Co–Mg) olivine–basalt partial melts is examined experimentally. Newtonian-viscous, texturally (quasi)equilibrated partial-melt aggregates are subjected to four-point flexural loading ($\sigma_{1MAX} = 3.45$ MPa) at elevated temperature (1070–1200° C); the creep response is characterized by a substantial, decelerating transient that would be absent but for the presence of the melt phase. The transient strain is created by the flow of liquid from the compression side of the flexure specimen to the tension side; the melt flow is driven by the gradient in dilatational stress caused by the flexural loading. The long-distance (relative to the grain size) migration of the melt phase is demonstrated via electron microscopy; the anelasticity of the melt-flow process is demonstrated by the strain recovery that occurs during stress-free annealing of the deformed specimens. Compression–compression attenuation measurements ($\Delta\sigma_1 = 20$ MPa; $T = 1070–1130°$C; $10^{-0.5} \geq f \geq 10^{-3.5}$ Hz) reveal an absorption band with $Q_E^{-1} \sim 1$ that is temperature and frequency independent. A numerical fit of the creep data to a discrete distribution of linear anelastic (Voigt/Kelvin) elements suggests that this attenuation behavior is effected by three distinct losses. A simple D'Arcy flow model for the melt migration in the flexed beams, however, demonstrates the same attenuation behavior. The discrete analyses are consistent with there being a continuous distribution of relaxation times associated with the melt migration in these high-strain-amplitude experiments. The combined results suggest that dilatational anelasticity may be the principal dynamic signature of texturally equilibrated partial melts.

Notation

		Units
E_a	activation enthalpy for anelastic flow	kJ · mol^{-1} or eV
E_{ss}	activation enthalpy for steady-state flow	kJ · mol^{-1} or eV
L	length between outer load points of flexure specimen	m
P	load	N
Q_E^{-1}	attenuation in Young's modulus mode	dimensionless
R_i	stiffness (modulus)	Pa
T	temperature	K or °C
$\mathbf{U}$	velocity of the solid phase in a partial melt (vector)	m · s^{-1}
a	length between inner load points of flexure specimen	m
b	specimen width	m
d	grain size	m
g	acceleration due to gravity	m · s^{-2}
h	specimen height	m
k	permeability	m^2
n	exponent relating stress and steady-state strain rate	dimensionless
p_{app}	applied pressure	Pa
p_e	effective pressure	Pa
p_0	pore (liquid-phase) pressure	Pa
r	radius of triple junction	m
t	time	s
$\mathbf{u}$	velocity of liquid phase (vector)	m · s^{-1}
(x, y, z)	Cartesian coordinate system variables for creep specimen	m
f	frequency	Hz
$\nabla\Pi$	pressure gradient (vector)	Pa · m^{-1}
$\Delta\rho$	density contrast between solid and liquid phases	kg · m^{-3}
Φ	compliance	Pa^{-1}
γ_{sl}	solid–liquid interfacial energy	J · m^{-2}
ξ	bulk viscosity of partial melt aggregate	Pa · s

Magmatic Systems
Edited by M. P. Ryan

		Units
η_i	viscosity	Pa · s
η	shear viscosity of partial melt aggregate	Pa · s
η_f	shear viscosity of the melt phase	Pa · s
ϕ	volumetric melt fraction	dimensionless
$\sigma_{1,2,3}$	principal stresses	Pa
σ_{1MAX}	maximum value of the maximum principal stress	Pa
ϵ	strain, inelastic strain	dimensionless
ϵ_a	anelastic strain	dimensionless
ϵ_0	maximum anelastic strain	dimensionless
$\dot{\epsilon}_{ss}$	steady-state strain rate	s^{-1}
τ	characteristic time for anelastic transient	s

Introduction and Theoretical Foundation

The physical processes by which steady-state viscosity, seismic wave attenuation, and melt migration and segregation occur in partial melts must be understood to develop testable hypotheses concerning the structure, chemistry, and dynamics of affected regions in the upper mantle. The steady-state mechanical behavior of partial melts (i.e., that unaffected by rupture of the crystalline residuum) is uniquely related to the (quasi)equilibrium distribution of the liquid phase on the scale of the grain size. In silicate systems this equilibrium state, dictated by the relative energies of solid–solid (i.e., grain boundaries and interphase boundaries) and solid–liquid interfaces, has been generally accepted as having the liquid phase confined to three-grain edge intersections ("triple junctions") and melt-free grain boundaries (e.g., Vaughan *et al.,* 1982; Waff and Bulau, 1982; Jurewicz and Watson, 1985; Daines and Richter, 1988). The degree of this restricted interconnectivity has been shown to be affected by volatiles and by a multiphase crystalline residuum (e.g., Fujii *et al.,* 1986), but wetted grain boundaries have not been demonstrated to be an equilibrium structure. A recent analysis suggesting that crystalline anisotropy, specifically of olivine in an ultramafic system, may allow for thermodynamically stable, wetted grain boundaries (Waff and Faul, 1992) is at odds with other experimental and observational studies (e.g., Cooper and Kohlstedt, 1982, 1984; Schwindinger and Anderson, 1989). The effect of partial melting on the steady-state viscosity of partially molten olivine–basalt aggregates supports the idea that grain boundaries in these systems remain melt free: the enhancement of deformation kinetics due to melting, in both Newtonian (Cooper *et al.,* 1989) and non-Newtonian (power-law) regimes (Beeman and Kohlstedt, 1993), is distinctly small (less than an order of magnitude change in viscosity compared to the subsolidus value) and fully accounted for by the presence of melt on triple junctions only.

The interconnected melt network on triple junctions allows for the porous-media permeation of melt through regions of partial melting in the Earth's mantle. Beyond this allowance, however, the physical extraction of melt from a partially molten zone requires that the crystalline residuum undergo compaction, specifically by a ductile process, and thus displace the liquid phase. Driven by gravity-induced positive buoyancy (i.e., the density contrast between the liquid and the residuum), this melt migration process has been described phenomenologically using D'Arcy's law: the relative velocities of the residuum and the liquid are linearly proportional to the inverse gradient in pore fluid pressure, with the constant of proportionality being the permeability of the residuum (e.g., Turcotte and Schubert, 1982, pp. 382, 413–416),

$$\mathbf{u} - \mathbf{U} = -[k(\phi, d)/\eta_f]\, \nabla\Pi, \tag{1}$$

where **u** and **U** are the velocities of the melt and the residuum, respectively, k is the permeability (which is a function of the volumetric melt fraction, ϕ, and the grain size of the residuum, d), η_f is the viscosity of the melt, and $\nabla\Pi$ is the pressure gradient. A pressure gradient in the melt phase is created by the momentum transferred to the liquid by the gravitational collapse of the residuum and it is resisted by (i) the buoyancy provided the residuum by the liquid phase and (ii) the capillarity force that seeks to preserve some amount of liquid in the triple junction network (McKenzie, 1984; Ribe, 1987; Riley *et al.,* 1990):

$$\begin{aligned}\nabla\Pi = {} & \eta\nabla^2\mathbf{U} + (\xi + \eta/3)\nabla(\nabla \cdot \mathbf{U}) \\ & - (1 - \phi)g\Delta\rho - H(\gamma_{sl}, \phi)\nabla\phi.\end{aligned} \tag{2}$$

In Eq. (2), the first two terms on the right-hand side represent pressure gradients in the melt due to shear deformation and dilatational deformation of the residuum, respectively (η is the shear viscosity and ξ the bulk viscosity of the partially molten aggregate), the third term is the buoyancy (g is the acceleration due to gravity and $\Delta\rho$ is the difference in density of the solid and liquid phases), and the final term is the force due to capillarity (H is a function of γ_{sl}, the solid–liquid interfacial energy, and of ϕ). Equations (1) and (2) have been evaluated numerically to characterize melt extraction in a gravity field (McKenzie, 1984; Richter and McKenzie, 1984; Scott and Stevenson, 1986; Ribe, 1987); application of the models to geologic conditions, however, requires knowledge of the physical parameters of the system, i.e., η_f, η, ξ, and the relationship between ϕ and the permeability for the environment of interest. These parameters must ultimately be characterized through experiments.

We have concentrated on flexural deformation experiments, using model partial-melt aggregates that are carefully prepared to be in microstructural equilibrium and working in the temperature-differential stress regime that allows characterization of steady-state behavior. (Phenomena involving fracture of the crystalline residuum have been carefully avoided.) The flexural creep approach so developed is predicated on a first-order analysis of the effect of hydrostatic pressure on the melt fraction in a microstructurally equilibrated partial melt (Cooper, 1990; Raj, 1982). The radius of curvature, r, of the melt-filled triple junctions (see Fig. 1) is related to the effective pressure in the partial melt system,

$$p_e = \gamma_{sl}/r, \tag{3}$$

where p_e, the effective pressure, is given by the difference in applied (e.g., overburden) pressure (p_{app}) and the pressure in the liquid phase (the pore fluid pressure, p_0):

$$p_e = p_{app} - p_0. \tag{4}$$

For a fixed value of the solid–liquid interfacial energy, one sees that an increase in p_e should lead to a reduction of r and therefore of ϕ (since the melt fraction can be described simply by a geometric argument for a known value of r [i.e., $\phi \propto r^2/d^2$; Waff, 1980]); of course, a decrease in p_e, assuming the system has access to a reservoir of liquid, should cause an increase in r (and ϕ). The potential mechanical behavior of this surface-tension phenomenon is easily illustrated by the removal

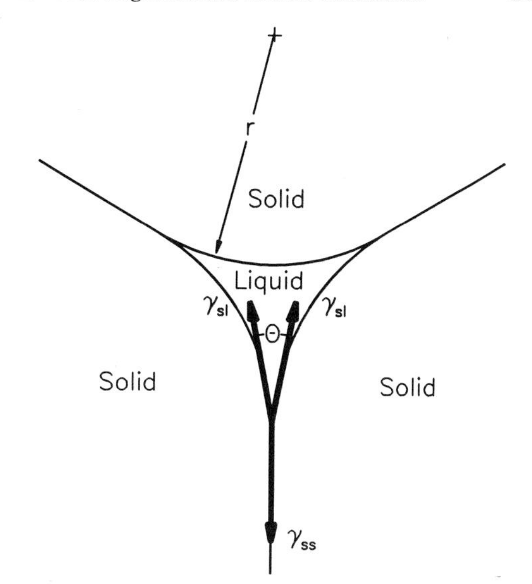

Figure 1 Schematic diagram of a melt-filled triple junction. The size of the junction is a function of the radius of curvature, r, and the dihedral angle, Θ. Θ is a function of the interfacial energies of the system. The melt fraction ϕ can be calculated from r, Θ, and the grain size of the crystalline phase, although one must make assumptions concerning grain shape to do so.

of water from a submerged sponge by squeezing it and monitoring the subsequent influx of the expulsed water by the release of the differential pressure while still submerged. The influx of the liquid into the sponge in this simple example illustrates the important fact that this effect of p_e on the liquid fraction is completely recoverable (given access to the liquid reservoir); the "compaction" process for the sponge is thus an *anelastic* effect, driven by the gradient in the *dilatational* component of the stress state, the kinetics of which are dictated by the variables in Eq. (2).

Placing a microstructurally equilibrated partial melt specimen in a four-point flexural stress condition maximizes the gradient in the dilatational component of the stress state relative to the deviatoric component, thus allowing the greatest resolution of the kinetics of melt migration in a creep experiment. In four-point flexure, the portion of the beam between the inner load points experiences "pure bending"; i.e., there are no applied shear forces in this region of the beam and, as such, the beam in this region experiences a maximum principal stress ($\sigma_1 = \sigma_{xx}$) that varies linearly from compressive to tensile across the specimen thickness (Crandall *et al.*, 1972, Chap. 7),

$$\sigma_1(y) = [6P(L - a)/bh^3]y, \tag{5}$$

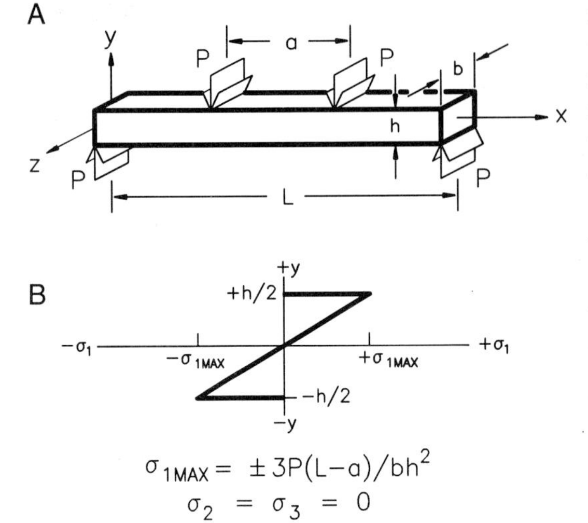

Figure 2 The configuration and stress distribution for four-point flexure (after Cooper (1990)). (**A**) Specimen dimensions (b and h), load point placement (L and a), loads (P), and coordinate system (x, y, z). (**B**) Elastic stress distribution for that portion of the beam between the center load points (i.e., from $x = \frac{1}{2}[L - a]$ to $x = \frac{1}{2}[L + a]$). This stress distribution holds for creep deformation if the beam has a Newtonian-viscous rheology.

where P is the load applied to each load point and $L, a, b, h,$ and y are specimen load and shape parameters as shown in Fig. 2. Equation (5) is an elastic flexure analysis that holds for Newtonian viscous behavior of the beam and is but slightly modified for non-Newtonian viscous flow (e.g., Hollenberg *et al.*, 1971). In the beam, σ_2 and σ_3 are both zero, and the maximum values of σ_1 (σ_{1MAX}) are determined by substituting $\pm h/2$ into Eq. (5). The state of loading thus creates a linear profile in p_{app} given by the mean-normal stress (cf. Dieter, 1986, pp. 46–48):

$$p_{app}(y) = (\sigma_1 + \sigma_2 + \sigma_3)/3 = [2P(L - a)/bh^3]y. \quad (6)$$

A comparison of Eq. (6) with Eq. (3) indicates that the flexural loading should force a reduction in ϕ on the compressive side of the specimen and an increase in ϕ on the tension side. Given that the forced motion of melt should cease when the local melt fraction is again in equilibrium with the applied flexural state of stress, the melt motion should produce a distinct, decelerating anelastic transient in the creep (strain versus time) curve. Of course, removal or reduction of the loads P requires the melt phase to subsequently flow back, thus recovering the transient strain of the specimen. One sees, then, that the flexural approach inherently avoids the problem of needing an external melt reservoir to study the recoverability of the melt-migration process.

We present within the results and interpretation of flexural creep experiments performed on a synthetic olivine–basalt aggregate, a simple model system (i.e., homogeneous melt phase in chemical equilibrium with a single-phase crystalline residuum) for a peridotitic upper mantle. The results are interpreted relative to the ideas discussed above, with an emphasis on the role of dilatation-stress-induced melt migration on the characteristic attenuation for an equilibrated zone of partial melt. Preliminary Young's modulus (compression–compression) attenuation experiments are compared with the predictions of the flexural creep data.

Experimental Approach

There are a number of practical requirements involved in the flexural creep approach to the study of melt migration. Portions of the specimen are subjected to tensile stress, and so, to avoid rupture of the specimen, the thermal and mechanical potentials—as well as the specimen grain size—must be selected so as to promote a Newtonian/superplastic rheology. Further, a Newtonian solid (i) has a negligibly small intrinsic transient strain, which allows the melt migration effect(s) to be isolated in the creep data, and (ii) allows the simple elastic stress distribution for four-point loading (i.e., Eqs. (5) and (6)) to be directly applicable to the experiments. The loads P used to produce a given maximum principal stress in a flexure specimen are very small compared to that required for the same stress in a tension or compression test; as such, small amounts of friction in the creep apparatus can produce substantial uncertainties in the actual mechanical potential to which the specimen is subjected. These factors, and others, influenced the design of both the specimens and the creep apparatus. For example, an O-ring seal or a bellows arrangement on the specimen chamber could easily produce a 20-to-30% uncertainty in σ_{1MAX} of a specimen. The ability to perform experiments in air (i.e., without any environmental seals on the apparatus) was thus crucial mechanically, although this approach has, of course, precluded the use of natural ferro-magnesian olivine–basalt specimens (which are unstable in air).

Experimental Specimens: Co–Mg Olivine–"Basalt"

The specimens employed in this study were an aggregate consisting of crystalline $(Co_{0.74}Mg_{0.26})_2SiO_4$ olivine in chemical and microstructural equilibrium with a liquid that is analogous to a basalt—at least in terms of its polymerization. Specimens were prepared by the sintering of previously fused, ground, and compacted powder of the appropriate bulk composition. Co–Mg olivine was used because of its stability in air and its ease of melting (the liquidus for the 0.74Co–0.26Mg composition is estimated as 1650° C). Having no constraints on the solid–liquid partitioning behavior in this system, the "basalt" chemistry was estimated using a direct analogy to Fe–Mg partitioning (Roeder and Emslie, 1970) (Co^{2+} and Fe^{2+} having identical ionic radii); the bulk chemistry was calculated assuming a melt fraction of approximately 10 vol%. Analytical-reagent-grade (or better) oxide, carbonate and nitrate powders ($CaCO_3$ was the calcium source; $NaNO_3$ the sodium source), precisely measured to provide the desired bulk olivine-basalt composition, were tumble-mixed and fused in a cobalt-soaked platinum crucible for 16 h at 1650° C in air. The melt thus produced was poured directly into a cold-water bath and the semicrystalline cullet that resulted was shattered into a powder using a fluid-energy particle mill. The mean particle size of the powder from which the beam specimens were produced was 2.0 μm, with 90% of the particles less than 5.3 μm in diameter.

Beam specimens for the flexural creep tests were prepared by sintering. The silicate powder was mixed with a polypropylene glycol binder (5 wt% binder in the mixture; the binder was first dissolved in dehydrated 2-propanol alcohol, sufficient to cause the powder/binder mixture to be slightly damp after mixing) and a pre-weighed amount of the mixture was placed in a rectangular die and compacted to a (nominally hydrostatic) pressure of 20 MPa. These beams were placed on a sheet of previously cobalt-soaked platinum and fired in air first at 400° C for 2 h to burn out the organic binder and then at 1200° C for 10 h to sinter the beam; the heating and cooling rates employed in the thermal schedule were $100° C \cdot h^{-1}$. The specimens thus prepared had final dimensions of approximately 40 × 10.8 × 3.2 mm. Microstructural analysis of the sintered beams revealed a mean grain size for the olivine of 4.9 μm, a melt fraction of ~12 vol% of the olivine-plus-basalt content, and a real porosity (isolated voids of a diameter similar to the grain size) of ~4 vol%. The microstructure of this material was typical of fine-grained olivine–basalt materials nearing microstructural equilibrium (cf. micrographs in Fig. 7): most of the melt phase was confined to well-formed triple junctions with diameters (i.e., of a circle scribed within the triple junction) of approximately 0.4 μm; pools of melt bordered by faceted olivine grains (i.e., the structures recently described by Waff and Faul (1992)) were additionally noted. The chemistry of the olivine and the residual glass phase, as determined by energy-dispersive x-ray spectroscopy (EDS) on a scanning transmission electron microscope, are presented in Table 1. The bulk composition calculated from these analyses is slightly enriched in SiO_2 and is depleted in CoO and MgO from the original batch that was fused; this fact can be explained by processing of the cullet in the fluid-energy mill: More of the less-dense, silica-rich glass phase was included in the "fine" powder used to fabricate the

Table 1

Compositions of Co–Mg Olivine–Basalt Material[a]

Oxide	Olivine (mol%)	Residual glass (mol%)
SiO_2	33.50 ± 1.42	77.83 ± 3.16
Al_2O_3	0.45 ± 0.50	13.35 ± 1.80
MgO	15.91 ± 0.95	1.17 ± 0.10
CoO	49.76 ± 2.45	2.02 ± 0.36
CaO	0.34 ± 0.09	5.46 ± 2.53
Na_2O	—	0.17 ± 0.10
	$(Co_{0.76}Mg_{0.24})_2SiO_4$	

[a]EDS data from a scanning transmission electron microscope. The "metallurgical thin film" data reduction algorithm (a standardless technique) was used to determine cation contents from the x-ray emission data; oxygen was calculated stoichiometrically. The errors indicated are ± 1 standard deviation of the measured compositions.

beams. We exploited this observation by reprocessing a small amount of this powder through the mill and then fabricating beams from the powder consequently placed by the mill in the "coarse" fraction; a few beams having a mean grain size of 6.0 μm and a melt fraction of ~7 vol% were thus produced (these, too, had ~4 vol% real porosity).

Creep Experiments and Specimen Analysis

Four-point flexural creep experiments were performed in a dead-weight creep apparatus; the apparatus and experimental protocol are described by Cooper (1990). The specimen cradle in this case, however, was fabricated from α-SiC; solid Al_2O_3 rods, 3.1 mm in diameter, were used to apply the load. The outer and inner span lengths used were $L = 30.5$ mm and $a = 15.2$ mm, respectively. Temperatures employed in these experiments ranged from 1050 to 1200° C; the temperature was both controlled and monitored by a Pt/Pt-13% Rh (R-type) thermocouple located ~7 mm from the specimen center and at the exact vertical location of the undeformed specimen; temperature control was accurate to ±0.5° C, while differences within the zone occupied by the specimen were ±2° C of the control temperature. The maximum principal stress applied to the specimens was $\sigma_{1MAX} = \pm 3.45$ MPa, corresponding to a vertical load on the specimen cradle of $2P \approx 10$ N (1.02 kgf). Strain (and strain rate) was determined by monitoring the vertical displacement of the top piston (i.e., the displacement of the inner load points) through the use of two direct-current displacement transducers whose output was summed, digitized, and fed into a microcomputer. Strain and strain rate determined by this method are accurate, provided the total inelastic strain does not exceed 0.02 (Hollenberg *et al.*, 1971); as such, experiments were terminated at this strain.

The shape of a beam that is plastically deformed in four-point flexure is very sensitive to the total strain as well as to the stress exponent that characterizes the relationship between the steady-state strain rate and the magnitude of differential stress. The curvature between the inner load points is constant (i.e., the beam has circular curvature in this region) with the maximum specimen strain being linearly proportional to the curvature ($\epsilon_{1MAX} = h/\rho$, where ϵ is inelastic strain and ρ is the radius of curvature). The natural logarithm of the curvature between the outer and inner load points should be linearly proportional to the natural logarithm of the horizontal distance along the beam, with the constant of proportionality being the stress exponent. As such, creep specimens were quenched from their deformation temperature while still under load. (The quenching of creep specimens was accomplished by disrupting the furnace power; the result was that specimens were quenched to 900° C at a rate of approximately 0.5° C · s^{-1}.) The shape of the crept specimen was then carefully measured by a device in which the specimen was dragged at a controlled rate under a gravity-fed displacement transducer (Jakus and Wiederhorn, 1988). The digitized data could then be numerically manipulated to determine the curvature of the beam, etc.

The anelastic nature of the measured transient strain was ascertained by a comparison of the specimen curvature (strain) in the as-deformed (and quenched) state and after stress-free annealing of the specimen at a temperature corresponding to that at which it was originally deformed. In these annealing experiments, specimens were placed such that if gravity were to have an effect, it would add to the creep strain of a specimen and not assist any strain recovery.

The microstructure of as-deformed-and-quenched specimens was analyzed using scanning electron microscopy in the backscattered electron imaging mode. Specimens from the pure-bending region of the specimen were sectioned sequentially along the y direction (Fig. 2) and the microstructure in the (x, z) plane recorded for each section on 12 to 15 random micrographs at a magnification corresponding to an image area of ~1200 μm^2 (approximately 100 olivine grains would be imaged in each micrograph). The areal percentage of the melt phase (approximately equal to ϕ) was determined by digital image analysis for each micrograph.

Attenuation Experiments

A modest number of Young's modulus attenuation experiments were performed on the Co–Mg olivine–basalt aggregates, to test ideas arising from the creep results. Small rectangular prismatic specimens (3.0 × 3.0 × 8.0 mm) of the $\phi = 0.12$ material were deformed in a compression–compression mode in a servomechanical-actuated creep apparatus employing a gravity-fed extensometer (the apparatus is fully described by Meyer *et al.* [1993]). Specimens were deformed in the temperature range 1100 to 1160°C (control to ±0.5°C; variation across specimen within ±1°C) with the maximum compressive stress varying sinusoidally from 5 to 25 MPa at controlled frequencies of $10^{-0.5} \geq f \geq 10^{-3.5}$ Hz. The strain-versus-time data thus produced consisted of a Newtonian steady-state component corresponding to the mean value of σ_1 (15 MPa) with the sinusoidal component of amplitude $\sim 10^{-3}$ superposed. The data analysis therefore consisted of subtracting the steady-state component from the data and subsequently determining the phase angle between the applied stress and the anelastic strain. The attenuation (Q_E^{-1}, where subscript E refers to Young's modulus) for this lossy material is given by the tangent of the phase angle (e.g., Green *et al.*, 1990).

Experimental Results

Flexural Creep Experiments

The flexural creep (inelastic strain versus time) behavior of the Co–Mg olivine–basalt specimens is illustrated by the data presented in Figs. 3A and 3B for the $\phi = 0.12$ and $\phi = 0.07$ aggregates, respectively. Qualitatively, one sees that each creep curve is characterized by a pronounced, decelerating strain rate that ultimately approaches a steady state. Using a regression analysis, the data have been described in the figures using the linear viscoelastic Burgers solid model (Fig. 10A; e.g., Findley *et al.*, 1976),

$$\epsilon(t) = \epsilon_0(1 - e^{-t/\tau}) + \dot{\epsilon}_{ss}t, \tag{7}$$

where ϵ denotes inelastic strain, ϵ_0 is the total anelastic strain, $\dot{\epsilon}_{ss}$ is the steady-state strain rate, and t is time. The variable τ is the characteristic time of the anelastic transient, that is, the time required to produce $1 - 1/e$ (~63.2%) of the total anelastic strain or, alternatively, the time at which the *anelastic* strain rate has decreased to $1/e$ (~36.8%) of its initial value. One can see that the Burgers solid-analysis predictions of ϵ_0 and $\dot{\epsilon}_{ss}$ match closely with values determined by a visual analysis of the data, a result confirmed by many measurements of flexural creep on similar, equilibrated glass-ceramic materials (Cooper *et al.*, 1989; Cooper, 1990). The highest temperature data (1130°C, $\phi = 0.12$ and 1200°C, $\phi = 0.07$) are an exception: these experiments did not reach steady state in the $\epsilon = 0.02$ limit of the experiment. For example, the *measured* transient strain (via the strain recovery experiments, see below) for the 1130°C, $\phi = 0.12$ experiment is 5.8×10^{-3}, whereas the Burgers solid analysis predicted $\epsilon_0 = 7.7 \times 10^{-3}$, a prediction consistent with the ϵ_0 results determined for the lower-temperature experiments that did achieve a steady-state strain rate.

Post-experiment measurements of the curvature of creep specimens revealed that all beams are characterized by a Newtonian (diffusion-dominated) rheology (i.e., $\dot{\epsilon}_{ss} \propto \sigma^n_{1MAX}$ and $n = 1$) for the temperature, stress, and grain-size regimes studied. The results of the measurements are shown in Fig. 4 and are compared to the line representing a power-law (dislocation-dominated) rheology with a stress exponent of $n = 3$. The ability of the curvature measurements to discriminate between the rheologies is plainly evident.

The anelastic nature of the transient strain is demonstrated in Fig. 5, which shows the effect of post-deformation, stress-free annealing experiments on the curvature of the tensile surface of the 1130°C, $\phi = 0.12$ beam. In the figure, one sees that, consistent with pure bending, the curvature of the beam is quite constant between the inner load points. (The curves themselves are the digitally analyzed displacement data; the markers on each curve have been added solely for curve identification.) Further, one sees that strain is recovered with annealing, with strain recovery reaching a limit of 5.7×10^{-3} after ~250 h. One should note that this recovered strain value matches well with the amount of transient strain measured during creep (5.8×10^{-3}, see text above). Equally noteworthy, though, is the fact that the relaxation time for this strain recovery is greater by a factor of $\sim 10^2$ than that characterized by the fit of the Burgers solid model to the creep curve. For each specimen, the recovered strain from stress-free annealing was within 10% of that measured for

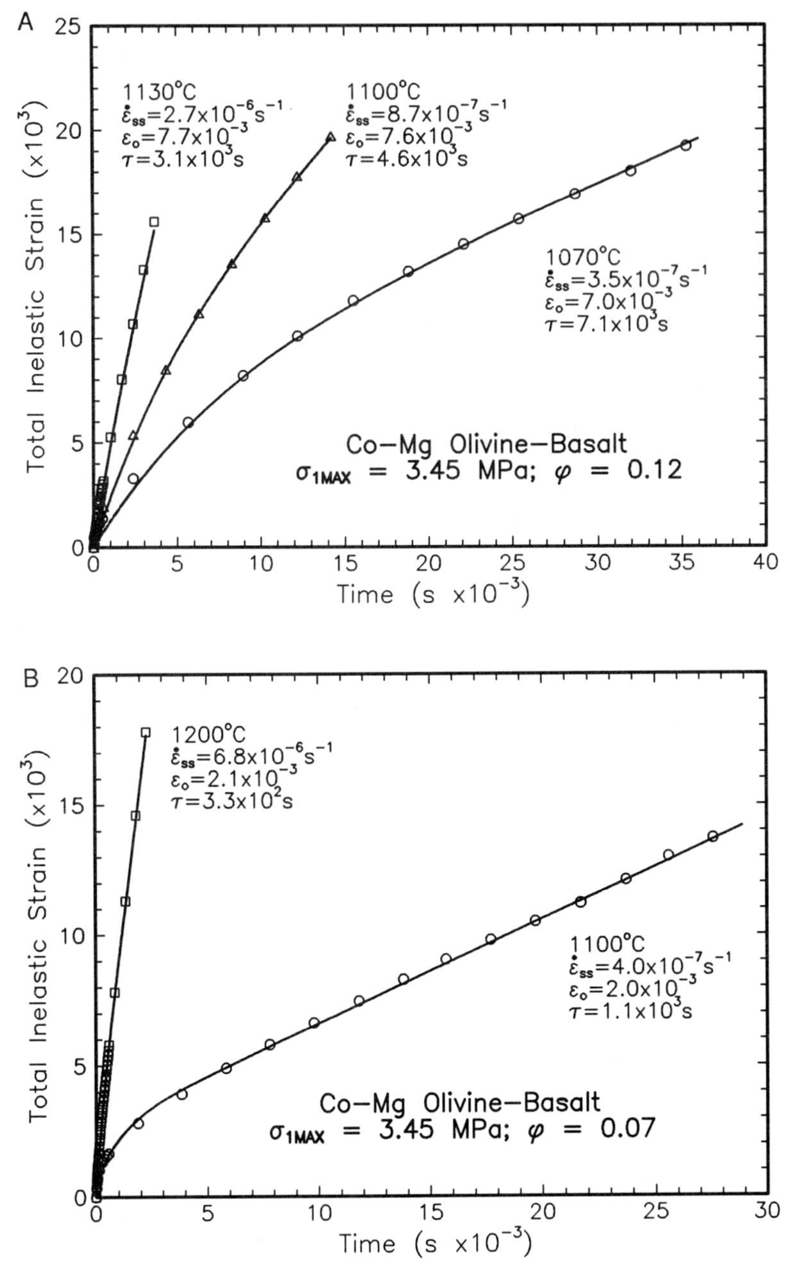

Figure 3 Four-point flexural creep curves for the Co–Mg olivine–basalt aggregates: (**A**) $\phi = 0.12$ specimens; (**B**) $\phi = 0.07$ specimens. The parameters listed (and the solid curves) are based on a Burgers solid-model regression of the data. The curves are characterized by a large transient strain, the anticipated effect of the migration of the melt phase across the specimen. Linear regression analyses of $\dot{\epsilon}_{ss}$ and of τ versus inverse temperature give activation enthalpies of $E_{ss} = 500 \pm 35$ kJ · mol^{-1} and $E_a = 210 \pm 10$ kJ · mol^{-1} for the steady-state and transient creep processes, respectively.

the transient on the creep curve. The errors were always negative (i.e., less strain recovered than expected).

The temperature dependencies of both the steady-state and the anelastic creep rates follow Boltzmann/Arrhenius behavior,

$$\dot{\epsilon} \propto \exp(-E/kT), \tag{8}$$

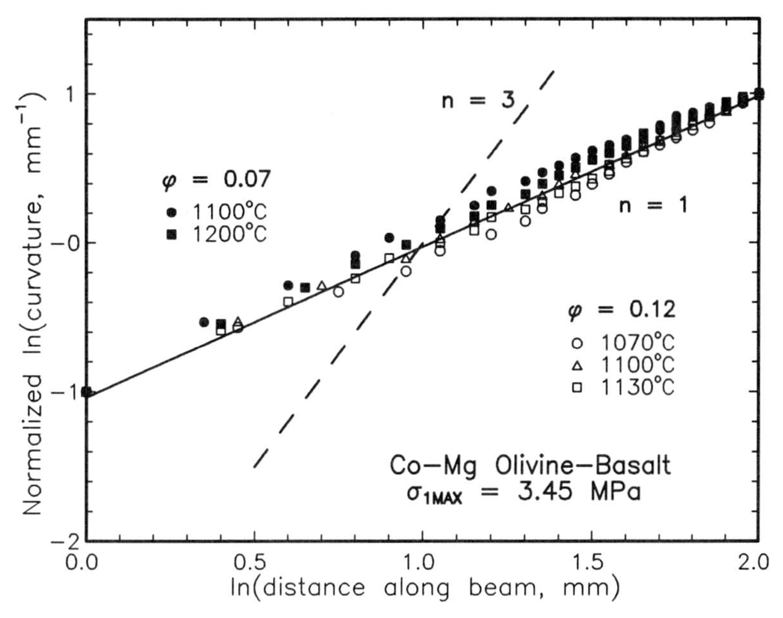

Figure 4 A ln–ln plot of specimen curvature versus distance between the outer and inner load points; the slope of the lines thus produced indicates the stress sensitivity of the steady-state creep response. The aggregates are characterized by a Newtonian-viscous, diffusional rheology (i.e., $n = 1$). A line for $n = 3$ (dislocation rheology) is shown for comparison.

where E is the activation enthalpy, T is the absolute temperature, and k is Boltzmann's constant. In the case of steady-state creep, a linear regression of $\ln(\dot{\epsilon}_{ss})$ data versus inverse temperature reveals an activation enthalpy of $E_{ss} = 500 \pm 35$ kJ $\cdot$ mol^{-1} (5.2 $\pm$ 0.4 eV). The activation enthalpy for the anelastic flow can be evaluated by determining the time necessary to complete a certain

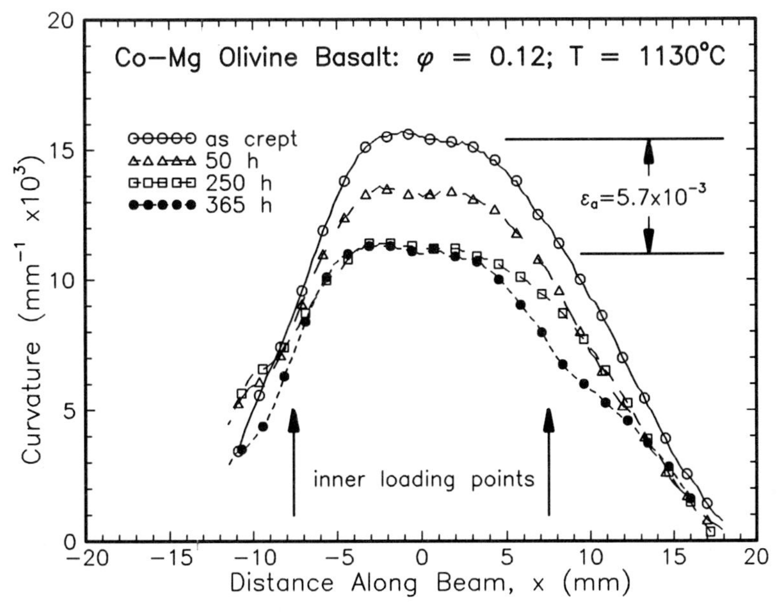

Figure 5 Relaxation of anelastic strain as revealed by specimen curvature change following stress-free annealing (1130° C, $\phi = 0.12$ specimen). A relaxation of 5.7×10^{-3} was achieved after annealing for ~250 h; this amount matched well the transient strain recorded in the creep curve (5.8×10^{-3}).

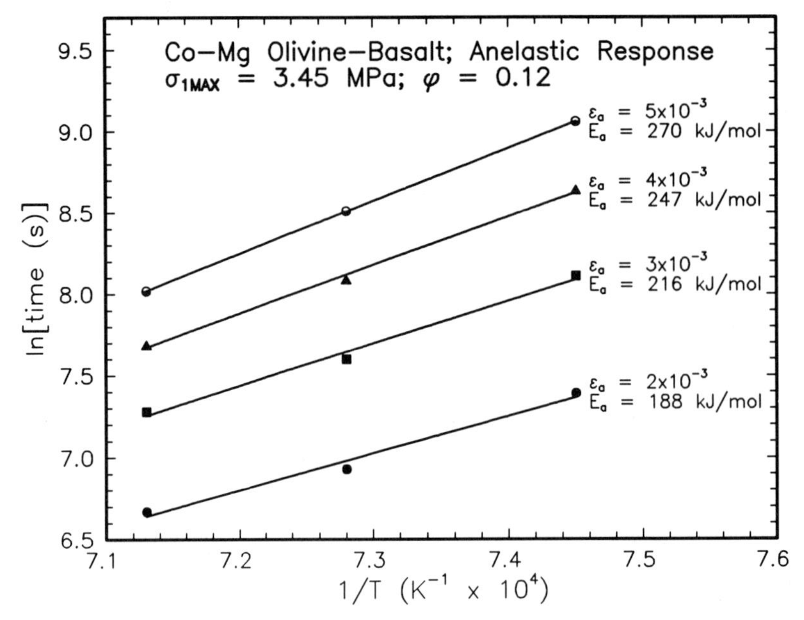

Figure 6 Activation enthalpy of the anelastic transient determined from the time required to achieve a specific anelastic strain, ϵ_a. The mean value of E_a is 230 ± 35 kJ · mol^{-1}.

amount of anelastic strain (the calculated steady-state strain is subtracted from the total strain, cf. Eq. (7)). Figure 6 displays the evaluations of the $\phi = 0.12$ data of Fig. 3A. Here one sees that the average of the many evaluations gives $E_a = 230 \pm 35$ kJ · mol^{-1} (2.4 ± 0.4 eV), although the value of E_a apparently increases modestly with increasing anelastic strain. A linear regression on the values of ln(τ) versus inverse temperature gives $E_a = 210 \pm 10$ kJ · mol^{-1} (2.2 ± 0.1 eV).

Microstructural Analysis

It was generally observed in the post-deformation sectional microstructural analysis that the melt fraction had decreased on the compression side of a crept specimen and increased on the tension side. Image analyses on the $\phi = 0.12$ specimens indicated that the mean value of ϕ measured across a specimen displayed such behavior, but as the sample standard deviations for ϕ at a given value of y was large, the analyses were inconclusive. Not surprisingly, this result prompted the development of the $\phi = 0.07$ specimens. Lowering the initial melt fraction allowed resolution of the melt migration effect, as demonstrated for the 1100°C specimen in the micrographs presented as Fig. 7 and the melt fraction profile in Fig. 8: The difference in ϕ across the specimen is clearly shown; the melt phase has indeed migrated across the specimen during flexural creep in a manner consistent with the stress-induced gradient in effective pressure (cf. Eqs. (3), (4), and (6)).

Attenuation Measurements

Results of the compression–compression attenuation measurements are shown in Fig. 9 (the curves marked "Exp."). The measured attenuation is distinctly insensitive to frequency and temperature: Q_E^{-1} is approximately unity (i.e., the phase angle measured between applied stress and strain response stayed near 45°) across the frequency range studied regardless of the temperature. (The errors in the measurements shown in the figure are approximately ±0.1 log unit in Q_E^{-1}.)

Discussion

Creep and Melt Migration

A general interpretation of the flexural creep behavior and the deformation-induced microstructure of these fine-grained, Newtonian-viscous partial-melt specimens is straightforward: Consistent with the thermodynamic argument made in

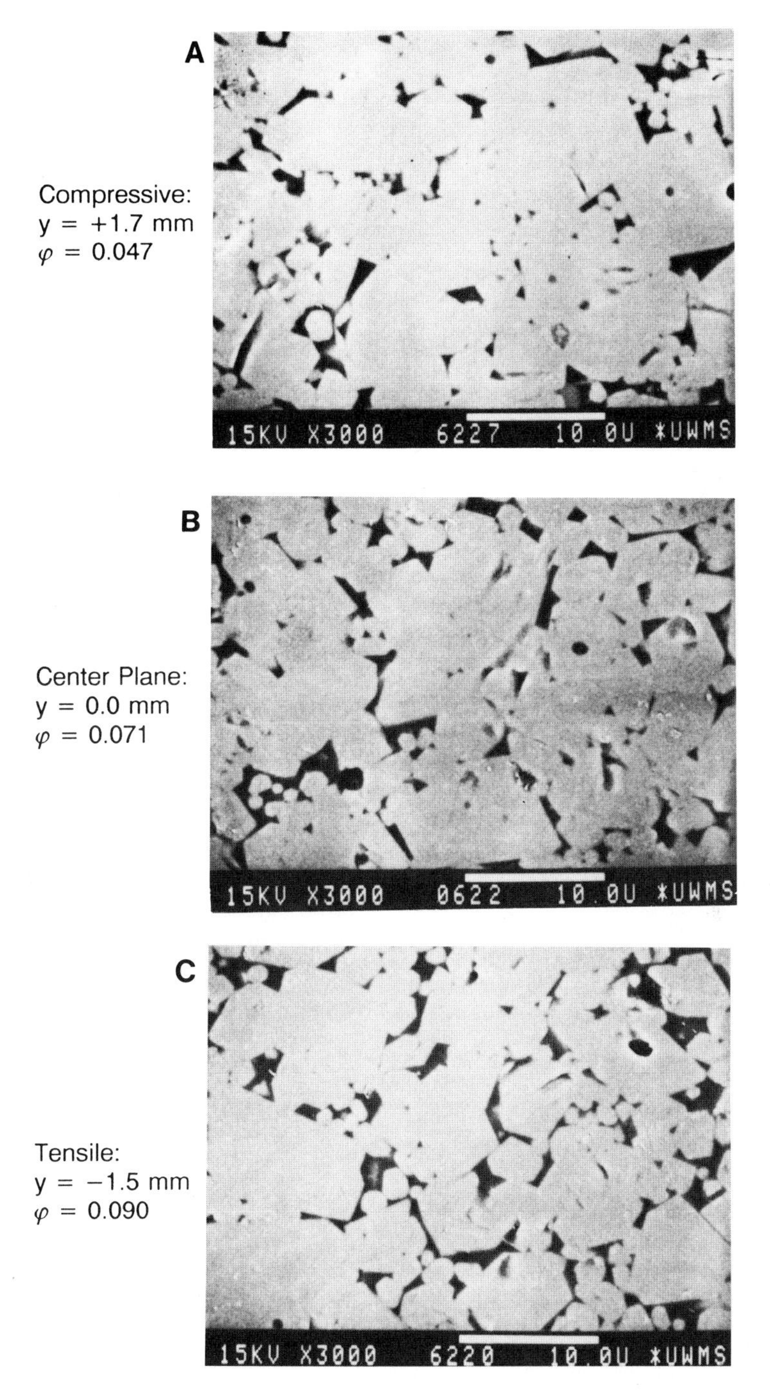

Figure 7 Backscattered electron SEM images from the $\phi = 0.07$ specimen deformed at 1100° C. These micrographs were taken in the (x, z) plane of the specimen (cf. Figure 2A). The values for ϕ indicated are areal percentages determined by digital image analysis. (**A**) $y = +1.7$ mm (at the surface of maximum compressive stress); (**B**) $y = 0$ (at the center plane); (**C**) $y = -1.5$ mm (near the surface of maximum tensile stress). In the micrographs, the light gray phase is olivine, the dark gray phase is glass, and black contrast is produced by porosity.

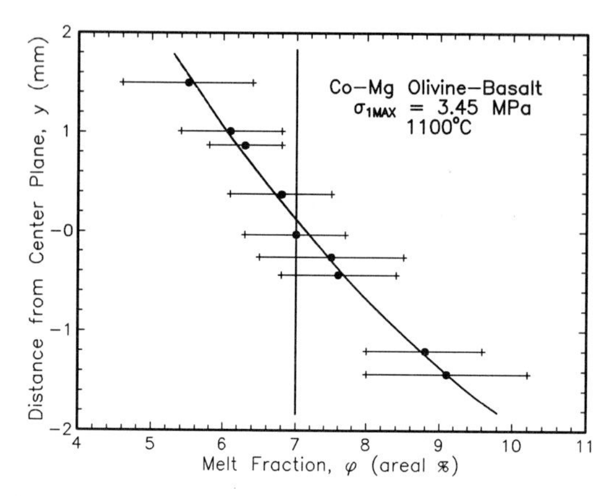

Figure 8 Melt distribution for the $\phi = 0.07$, 1100° C specimen after deformation and quenching. The data points represent the mean values of digital analyses of 12 to 15 micrographs at a given location; the error bars are ± 1 standard deviation of the mean. The heavy vertical line represents the initial (pre-deformation) condition of the specimen. While not a unique description, given the magnitude of uncertainty of the data, the curve compared to the final melt distribution is that expected from a straightforward application of Eqs. (3) and (6), i.e., $\phi \propto y^{-2}$.

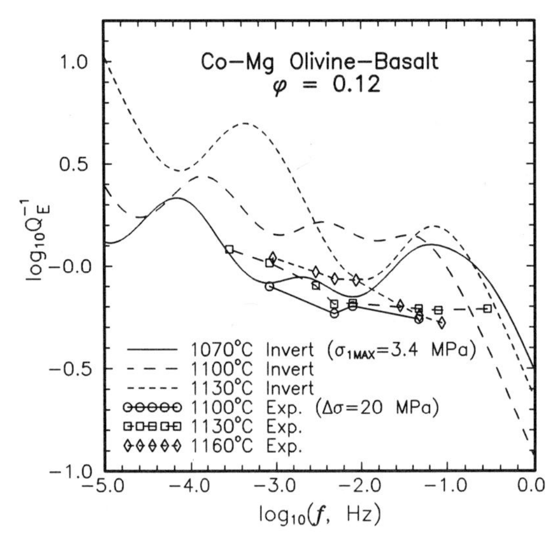

Figure 9 Young's modulus attenuation (Q_E^{-1}) of the $\phi = 0.12$ specimens in compression–compression experiments. The experimental data (marked "Exp." in the key) show a band of attenuation at $Q_E^{-1} \sim 1$ for the frequency range studied that is additionally temperature independent. (Errors on the experimental measurements are approximately ± 0.1 log units in Q_E^{-1}.) These data are compared with attenuation spectra that were calculated from the compliance behavior of the creep specimens (Figure 11), which indicated three discrete linear loss mechanisms account for the attenuation behavior of the partial melt.

the chapter introduction, the anelastic strain recorded in the data is directly and uniquely related to the migration of the melt phase, driven by the gradient in the dilatational component of the applied stress, from the compression side of the beam specimen (i.e., $y > 0$) to the tension side ($y < 0$). The final distribution of the melt phase in the specimens is consistent with the relationship between the linear gradient in effective pressure and the inverse relationship between the effective pressure and triple junction radius ($\phi \propto p_e^{-2} \propto y^{-2}$; Fig. 8). Further, the steady-state rheology of the two-phase, solid–liquid system is revealed in the steady-state creep rate of the beams. Application of the linear viscoelastic Burgers solid model to the data (Eq. (7); Fig. 10A) allows direct interpretation for the viscosity parameters employed in the melt-migration models: the shear viscosity is $\eta = \frac{1}{3}(\sigma_{1MAX}/\dot{\epsilon}_{ss})$; the bulk viscosity is $\xi = (\frac{1}{9})(\sigma_{1MAX}\tau/\epsilon_0)$ (Cooper, 1990; Green and Cooper, 1993).

The temperature dependencies of the bulk and shear viscosities provide clues as to the physical processes that rate limit the flow responses. The steady-state creep behavior of Newtonian-viscous olivine–basalt partial melts in this thermodynamic and microstructural regime has been demonstrated to consist of a series kinetic process involving solid-state grain boundary diffusion and solution–liquid-phase transport reprecipitation ("pressure solution") of the ionic components of olivine (Cooper *et al.*, 1989). The effect of the grain size on the steady-state creep rate isolates the rate-limiting step to either grain boundary or liquid-phase transport; the activation enthalpy measured for steady state indicates grain boundary diffusion as rate limiting. The activation enthalpy for liquid-phase transport of ions should match that for the liquid's viscosity. The E_{ss} determined for the flow of olivine–basalt was distinctly higher. This solid-state rate limitation occurs not because pressure solution kinetics are sluggish—on the contrary, they are very rapid: orders of magnitude faster than grain boundary diffusion in this system—but because the microstructural equilibrium, which is characterized by melt-free grain boundaries, does not allow access of the solid to the liquid phase except at triple junctions. The $\sim$500 kJ $\cdot$ mol^{-1} noted here for the activation enthalpy of the steady-state process in the Co–Mg

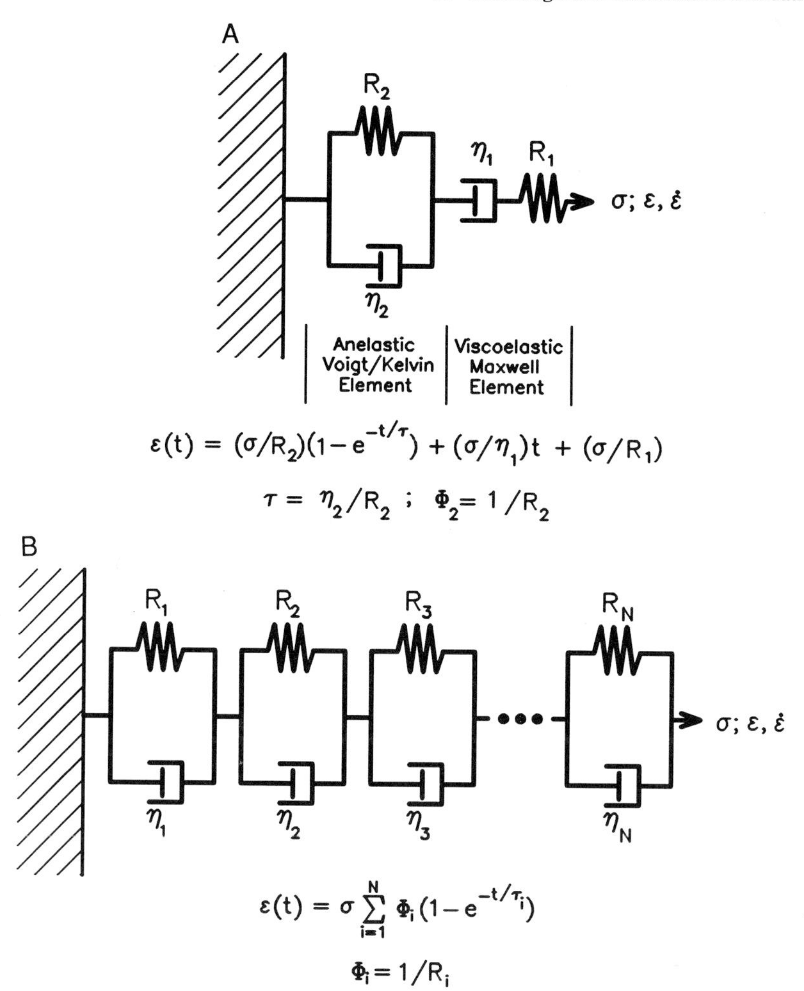

Figure 10 Linear viscoelastic models (after Findley *et al.* (1976)): (**A**) Burgers solid model; (**B**) generalized Voigt/Kelvin model. The creep function of the Burgers model (A) is given by the superposition of the elastic, steady-state plastic and anelastic responses. Systems containing a distribution of linear losses can be modeled by a number of Voigt/Kelvin elements placed in series (B). One should note that the magnitude of the stiffness or the viscosity of an element in the series representation allows for modeling of steady-state flow and elastic responses. If the stiffness of a single element goes to zero, the element then describes steady-state viscosity; obversely, if the viscosity of a single element goes to zero, the element then describes the elastic behavior. As such, the entire creep function—and not just the anelastic behavior—can be modeled by the series arrangement of Voigt/Kelvin elements.

olivine–basalt specimens is logically argued, then, as that characterizing grain boundary diffusion in the olivine residuum. The activation enthalpy for the anelastic transient, however, is distinctly different; the value for E_a, ~230 kJ · mol^{-1}, is approximately one-half that for the steady-state process and matches well that for the viscosity of a poorly polymerized silicate melt (e.g., an olivine-saturated basalt; cf. Richet, 1984).

The physical process of melt migration in the specimens is also a coupled, series kinetic process: in order for the melt to move, dilatation (tensile side) and contraction (compressive side) of the crystalline residuum must occur in addition to (i.e., simultaneously with) the physical flow of the liquid. One of these two processes will be rate limiting and thus be responsible for the magnitude of the activation enthalpy measured for anelastic flow. The dilation/compaction of the residuum must occur via the same grain-scale creep mechanism responsible for steady-state flow; as such, if it were rate limiting the anelastic flow, E_a would be equivalent to E_{ss}. Such is not the case for these Co–Mg olivine–basalt specimens: the clear difference between E_a and E_{ss} and the appropriate magnitude of E_a indicate that the melt migration process is rate limited by the ability of the liquid to flow.

Attenuation and Melt Migration

The adequacy of the linear viscoelastic Burgers solid model (Fig. 10A) to fully describe the dynamics of the partial-melt system is brought into question when one compares the prediction of dynamic properties via the Burgers model with the measured attenuation response. The Burgers model consists of a single, linear energy loss (the anelastic Voigt/Kelvin element of viscosity η_2, restorative stiffness R_2, and characteristic time $\tau = \eta_2/R_2$) plus the linear steady-state plastic and elastic responses (the viscoelastic Maxwell element of viscosity η_1 and stiffness R_1). The dynamic response for this model for the frequency region explored in the attenuation experiments here is an increasing Q_E^{-1} with decreasing frequency, caused by the steady-state creep response, punctuated by a superposed single anelastic peak (Green *et al.*, 1990). The model thus assigns the dilatational-stress-induced melt migration to an anelastic process having a single characteristic time. The attenuation "band" noted in the experiments suggests, instead, that there may be many energy losses that sum to produce the measured behavior, that is, a distribution of loss mechanisms (cf. Anderson and Given, 1982). To test this idea, the creep data were reanalyzed, allowing for an extremely large number of discrete linear loss elements (Fig. 10B) to be applied to the anelastic creep curve; in this case, the anelastic creep function becomes

$$\epsilon_a(t) = \sigma \sum_{n=1}^{N} \Phi_i(1 - e^{-t/\tau_i}), \qquad (9)$$

where Φ_i is the compliance ($\equiv R_i^{-1}$) of a single, given anelastic loss element that has a characteristic time of τ_i. One should note that the ϵ_0 of any anelastic loss element is equal to $\sigma\Phi$. The anelastic creep data can be regressed against Eq. (9) to create a retardation spectrum, $\Phi(\tau)$ versus τ (Gribb, 1992; cf. Thigpen *et al.*, 1983). Restrictions must be placed on the algorithm to ensure

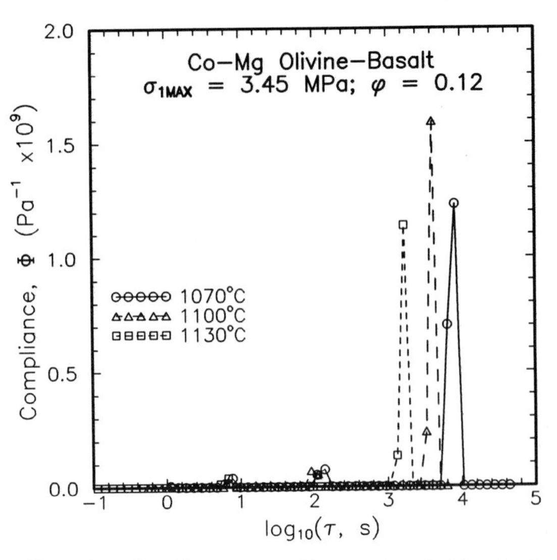

Figure 11 Compliance spectra (Φ versus log τ) determined by numerical regression to a large number of discrete linear anelastic elements (e.g., the model in Figure 10B). The result shown, that the compliance behavior represents three discrete anelastic loss mechanisms, was quite robust to the magnitude of the time step used in the regression analysis. These spectra have been Fourier transformed to produce the Q_E^{-1} spectra shown with the experimental attenuation data in Figure 9.

positive values of Φ_i as negative compliances have no physical meaning. The results of such an analysis for the $\phi = 0.12$ creep data are shown in Fig. 11: There are apparently three discrete linear losses that characterize the anelastic response of the material. This result is distinctly robust relative to the magnitude of the time element used in the numerical regression. Fourier transformation of these compliance spectra results in the attenuation spectra (the continuous curves) shown with the experimental attenuation data in Fig. 9. One sees that the three discrete linear losses sum to create an attenuation band near $Q_E^{-1} = 1$ that plots quite closely to the experimental data. Similar results have been noted for flexural attenuation experiments on metasilicate glass-ceramic specimens (Gribb and Cooper, 1991). It should be noted that the creep and attenuation experiments were performed on entirely different apparatus. As such, the close correlation noted in Fig. 9 is strong evidence that the attenuation measurements represent properties of the olivine–basalt specimens and not any dynamic property of the deformation apparatus.

Given that the material studied here is a Newtonian (i.e., linear) solid, it is tempting to ana-

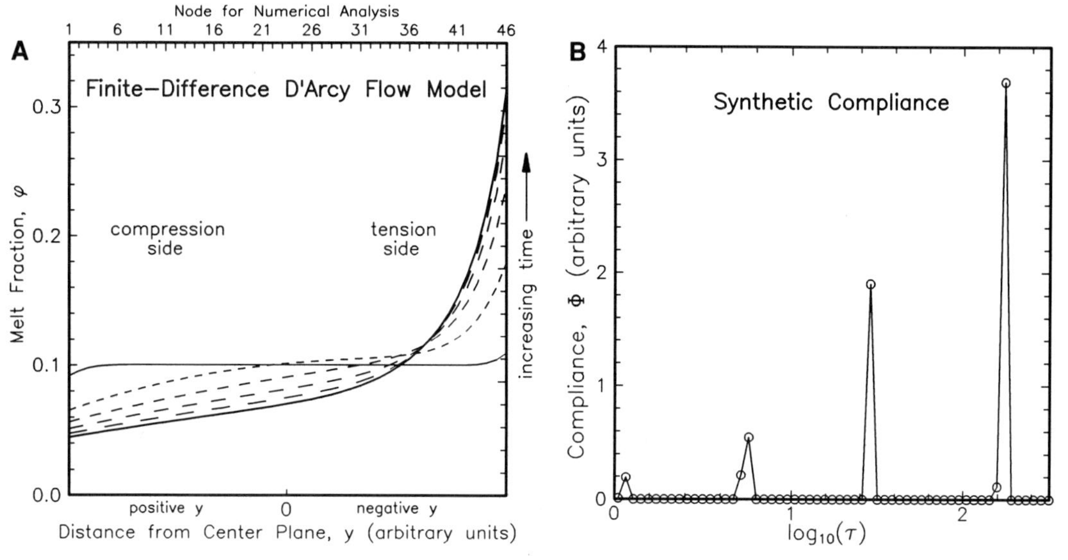

Figure 12 D'Arcy's law model for melt flow in pure bending. (**A**) ϕ versus y. One sees that the melt fraction is predicted to evolve to the $\phi \propto y^{-2}$ form characteristic of steady state. In this example, the magnitude of σ_{1MAX} was chosen sufficiently large to magnify the form of the results. The initial, uniform melt fraction is 0.1. The time interval between the curves is constant, with the earliest time shown as the fine solid line, the final time shown as the bold solid line, and each intermediate curve having a dash length proportional to its respective time step. (**B**) Compliance spectrum determined from the model. Application of the many-linear-element algorithm to the model indicates that the melt migration, a single-loss process as modeled here, is characterized by four discrete linear elements, a result that matches closely the behavior of the experimental specimens (cf. Figure 11). The analysis indicates that the melt-migration process in the flexure experiments cannot be characterized by a single relaxation time. This discrete spectrum is consistent with $\Phi(\tau)$ having a continuous form, one that is most likely related to the geometry of the flexure experiment.

lyze the three linear losses revealed for their physical significance. For example, while the microscopy results reported here provide compelling proof of a dilatational anelasticity in equilibrated partial melts, the Young's modulus attenuation experiments do not isolate dilatational from shear attenuation mechanisms. The measured attenuation response thus could contain also (i) an intrinsic, shear-stress-induced effect of melt-free grain boundaries with a very short τ (e.g., Raj, 1975) and (ii) a local (i.e., grain-scale) melt flow ("squirt"; Mavko, 1980), also shear-stress-induced, with a τ somewhat larger than the intrinsic grain boundary effect. While we cannot rule out these effects (they need to be tested for specifically in a driven-torsion test on an equilibrated partial melt), an alternative approach to the analysis suggests that the three discrete losses revealed uniquely relate to the dilatational-stress-induced melt migration.

We present in Fig. 12 the results of a simple finite-difference model (all units arbitrary) for flow of the liquid phase in a flexed, partially molten beam. D'Arcy's law is here applied to describe the motion of the liquid phase driven by the linearly varying pressure gradient and resisted by capillarity (i.e., as represented by Eqs. (3), (4), and (6)). This "zeroth-order" approach is appropriate for the case where D'Arcy flow of the liquid rate limits the behavior (Ribe, 1987)—such as the present case for Co–Mg olivine–basalt as evidenced by the magnitude of E_a. Figure 12A shows the evolution of ϕ with increasing time. While parameters in the model have been chosen to magnify the effect of bending on ϕ, one sees that the melt distribution assumes the $\phi \propto y^{-2}$ form expected at steady state. These analyses can be converted into strain versus time "data" by monitoring the volume change of the end elements and converting it to strain; these strain "data" can then be analyzed using the linear-element algorithm (Eq. (9)) to generate a synthetic compliance spectrum. The form of such a synthetic spectrum is shown as Fig. 12B: four discrete peaks are noted for what is obviously (in this exercise) a single physical loss process. Fourier transformation of these four peaks into an attenuation spectrum produces an attenuation band at $Q_E^{-1} \sim 1$, a result that is not influenced by parameters used in the model (e.g., original ϕ, σ_{1MAX}, η_f, etc.). The spacing of the discrete peaks in the compliance spectrum and their increasing magnitude with time match closely in form the compliance spectra determined from real creep data on the Co–Mg olivine–basalt aggregates (cf. Figs. 11 and 12B).

We can at least conclude from the model result and its comparison to the experimental data that the anelasticity created by dilatational-stress-driven melt migration in these Newtonian specimens cannot be described by a single characteristic time. Rather, the results are consistent with there being a *continuous compliance distribution,* $\Phi(\tau)$, for the melt-migration process in these experiments. Physically, one can argue that it is the evolution of $\phi(y)$ with increasing strain that produces a continuously changing anelastic compliance during creep. The model (Fig. 12A) indicates that the initial changes in $\phi(y)$ occur at the extremes of the specimen (i.e., $y = \pm h/2$), and with increasing time (that is, total specimen strain) the distribution effect migrates from these extremes toward the specimen center. However, the inverse relationship between p_e and r, combined with the fixed total volume of melt in the system, results in a greater portion of the melt migration path (that is, the y dimension) experiencing a decrease in ϕ. As a consequence the melt experiences an ever greater resistance to flow, throughout most of the specimen, with increasing anelastic strain.

There is likely a variety of continuous $\Phi(\tau)$ functions that can be reasonably fit to the creep and attenuation data collected on the Co–Mg olivine–basalt specimens. One of these, perhaps the simplest, is derived from a power law description of the anelastic flow, i.e.,

$$\epsilon_a \propto t^M, \tag{10}$$

where the exponent $M < 1$. Equating this power law expression with the anelastic creep function determined from a continuous distribution of Voigt/Kelvin elements (that is, the integral expression of Eq. (9)) produces (Findley *et al.*, 1976, pp. 69–70; cf. Anderson and Given, 1982)

$$\Phi(\tau) \propto \tau^{M-1}, \tag{11}$$

which on a log–log plot would produce a straight line with negative slope. One should note that, despite the gross differences in form between the continuous spectra represented by Eq. (11) and the experimental and synthetic spectra shown in Figs. 11 and 12B, respectively, they are consistent: the integral in τ of the continuous function would equal the sum of the compliances shown as discrete peaks in the figures.

The result, in both the physical data and the model, that the attenuation in the beams is nominally frequency and temperature independent and has a value of $Q_E^{-1} \sim 1$ suggests a strong geometry dependence for the loss mechanism in flexure. While the melt experiences a more difficult task of flow with increasing strain, as noted above, the rate capability of the solid residuum to accommodate the melt may be exceeded simultaneously (the dilatations required of the tensile side of the specimen can be quite large, cf. Fig. 12A). The phenomenon is such that the $Q_E^{-1} \sim 1$ result is expected: a phase angle of 45° can occur under conditions where the strain produced by anelastic flow is equivalent to that produced by the steady-state mechanism (each on the order of 10^{-3}); the fact that these two components of strain are intricately linked with increasing specimen deformation provides the result. We thus anticipate the discovery of a distinct attenuation peak associated with dilatational-stress-induced melt migration at very low strain amplitudes ($\leq 10^{-5}$) that has application to the dynamics in the geological setting. Whether or not this loss is represented by the shortest τ peak in the discrete compliance spectrum (Fig. 11) remains a question to be answered with additional experiments.

Application of the Experimental Observations to the Earth's Mantle

There exist a variety of thermally activated mechanisms for attenuation of seismic waves in the Earth's mantle. As carefully outlined in the reviews by Jackson (1986) and Anderson (1989, Chap. 14), these mechanisms include the motion of lattice dislocations, the motion of grain boundary dislocations and other diffusive motion on grain boundaries, the motion of lattice point defects, and the motion of liquid in partial melts. With this variety, a distribution of relaxation times is anticipated, and thus the absorption band thought to be characteristic of attenuation processes in the upper mantle is explained logically (e.g., Anderson and Given, 1982).

Experiments to determine internal friction response and mechanisms in earth materials and analogs place constraints on the geophysical arguments, however, and thus the capabilities to isolate individual mechanisms proves valuable in interpreting seismic data. For example, experiments indicate that the solid-state losses noted above are known primarily to affect shear waves. Further, in the case of dislocations, relaxation times may vary according to the lengths of dislocations and the magnitude of the thermal barrier to dislocation motion (e.g., the energy required for the formation of jogs and kinks on individual dislocations). In the case of grain boundary losses, the grain size is the distance over which a mechanical potential is dissipated (Raj, 1975); thus a grain-size distribution creates a spectrum of relaxation times. What can be said, then, of the attenuation effect of partial melting?

Many seismic studies have indicated that distinct absorption of S-waves occurs below volcanic zones. For example, Solomon (1973) interpreted observations of strong S-wave attenuation beneath the mid-Atlantic ridge as an indicator of high melt fraction ($\sim$10 vol%), citing the grain boundary melt film model of Walsh (1969). Anderson and Sammis (1970) applied this model as well to the low velocity zone to support the idea that its structure included $\sim$1 vol% melt. The experiments used to support these ideas, though, have often been dynamic melting experiments: measuring the attenuation response of a material both below and above the solidus temperature. As melting occurs initially on solid interphase (grain) boundaries, a substantial increase in shear attenuation measured in such experiments is not surprising. However, the knowledge of equilibrium microstructure formation in partial melts, as outlined in the introduction to this chapter, combined with the calculated efficiency of the compaction and melt extraction process that suggests that less than 1 vol% melt could be retained in an actively melting region at depth (Riley *et al.*, 1990), disavows any easy application of the results of dynamic melting experiments to the steady-state structure of the upper mantle. We have demonstrated clearly that seismic/subseismic-frequency anelasticity in an equilibrated silicate partial melt is primarily, if not uniquely, a dilatational phenomenon. A substantial shear attenuation effect due to melting may be limited, then, only to regions of incipient melting or regions where the melt fraction is sufficiently large to affect a significant loss of grain boundary area. The latter criterion may indeed represent the case in active volcanic zones at shallow depth.

Summary and Conclusions

Flexural creep experiments on microstructurally equilibrated olivine–basalt partial melts display a distinct anelastic transient that is related to the migration of the melt phase within the specimen. This anelasticity is argued thermodynamically to be dilatational. The fact that melt-free grain boundaries characterize the steady-state microstructure of partial melts of upper-mantle composition supports the idea that dilatational anelasticity may be the unique dynamic signature of partial melting under conditions that produce a low melt fraction and that disallow fracture of the crystalline residuum. The melt-migration attenuation effect noted in the experiments is sensitive to the magnitude of applied strain because of the coupling of the melt migration to the deformation kinetics of the crystalline residuum. Further experimentation into attenuation behavior of partial melts at seismic/subseismic frequencies thus requires working at low strain amplitudes.

Acknowledgments

We happily acknowledge many fruitful discussions with Drs. Douglas Green and Dallas Meyer on various aspects of this work. The fusion of the Co–Mg silicate melts used to fabricate the creep specimens was done by the Experimental Melting Group at Corning, Inc., Corning, New York; Elton Harris is thanked for his assistance in this regard. The electron microscopy work was performed in the Center for Materials Science at the University of Wisconsin-Madison; the instruments there are beautifully maintained by Dick Casper and Rick Noll. This manuscript benefited from careful reviews by Mervyn Paterson, Rishi Raj, and Michael Ryan. We gratefully acknowledge the National Science Foundation for financial support through Grant EAR-9005226.

References

Anderson, D. L. (1989). "Theory of the Earth," Blackwell Scientific, Boston.

Anderson, D. L., and Given, J. W. (1982). Absorption band Q model for the Earth, *J. Geophys. Res.* **87,** 3893–3904.

Anderson, D. L., and Sammis, C. G. (1970). Partial melting in the upper mantle, *Phys. Earth Planet. Inter.* **3,** 41–50.

Beeman, M. L., and Kohlstedt, D. L. (1993). Deformation of Fine-Grained Aggregates of Olivine Plus Melt at High Temperatures and Pressures, *J. Geophys. Res.* **98,** 6443–6452.

Cooper, R. F. (1990). Differential stress-induced melt migration: An experimental approach, *J. Geophys. Res.* **95,** 6979–6992.

Cooper, R. F., and Kohlstedt, D. L. (1982). Interfacial energies in the olivine–basalt system, *Adv. Earth Planet. Sci.* **12,** 217–228.

Cooper, R. F., and Kohlstedt, D. L. (1984). Solution-precipitation enhanced diffusional creep of partially molten olivine–basalt aggregates during hot-pressing, *Tectonophysics* **107,** 207–233.

Cooper, R. F., Kohlstedt, D. L., and Chyung, K. (1989). Solution-precipitation enhanced creep in solid–liquid aggregates which display a non-zero dihedral angle, *Acta Metall.* **37,** 1759–1771.

Crandall, S. H., Dahl, N. C., and Lardner, T. J. (1972). "An Introduction to the Mechanics of Solids," 2nd ed., McGraw–Hill, New York.

Daines, M. J., and Richter, F. M. (1988). An experimental method for directly determining the interconnectivity of melt in a partially molten system, *Geophys. Res. Lett.* **15,** 1459–1462.

Dieter, G. E. (1986). "Mechanical Metallurgy," 3rd ed., McGraw–Hill, New York.

Findley, W. N., Lai, J. S., and Onaran, K. (1976). "Creep and Relaxation of Nonlinear Viscoelastic Materials," North-Holland, Amsterdam.

Fujii, N., Osamura, K., and Takahashi, E.-I. (1986). The effect of water saturation on the distribution of partial melt in the olivine–pyroxene–plagioclase system, *J. Geophys. Res.* **91,** 9253–9259.

Green, D. H., and Cooper, R. F. (1993). Dilatational anelasticity in partial melts: Viscosity, attenuation and velocity dispersion, *J. Geophys. Res.* **98,** 19,807–19,817.

Green, D. H., Cooper, R. F., and Zhang, S. (1990). Attenuation spectra of olivine/basalt partial melts: Transformation of Newtonian creep response, *Geophys. Res. Lett.* **17,** 2097–2100.

Gribb, T. T. (1992). "Low-Frequency Attenuation in Microstructurally Equilibrated Silicate Partial Melts," M.Sc. Thesis, Univ. of Wisconsin-Madison.

Gribb, T. T., and Cooper, R. F. (1991). Attenuation in a model silicate partial melt: Comparison of direct measurements and inversions of creep data (abstract), *EOS Trans. Am. Geophys. Union* **72**(Supp.), 508.

Hollenberg, G. W., Terwilliger, G. R., and Gordon, R. S. (1971). Calculation of stresses and strains in four-point bending creep tests, *J. Am. Ceram. Soc.* **54,** 196–199.

Jackson, I. (1986). The laboratory study of seismic wave attenuation, *in* "Mineral and Rock Deformation: Laboratory Studies," Geophys. Monogr. Series, Vol. 36, pp. 11–23, American Geophysical Union, Washington, DC.

Jakus, K., and Wiederhorn, S. M. (1988). Creep deformation of ceramics in four-point bending, *J. Am. Ceram. Soc.* **71,** 832–836.

Jurewicz, S. R., and Watson, E. B. (1985). The distribution of partial melt in a granitic system: The application of liquid phase sintering theory, *Geochim. Cosmochim. Acta* **49,** 1109–1121.

Mavko, G. M. (1980). Velocity and attenuation in partially molten rocks, *J. Geophys. Res.* **85,** 5173–5189.

McKenzie, D. (1984). The generation and compaction of partially molten rock, *J. Petrol.* **25,** 713–765.

Meyer, D. W., Cooper, R. F., and Plesha, M. E. (1993). High-temperature creep and the interfacial mechanical response of a ceramic matrix composite, *Acta Metall.* **41,** 3157–3170.

Raj, R. (1975). Transient behavior of diffusion-induced creep and creep rupture, *Metall. Trans. A* **6A,** 1499–1509.

Raj, R. (1982). Creep in polycrystalline aggregates by matter transport through a liquid phase, *J. Geophys. Res.* **87,** 4731–4739.

Ribe, N. M. (1987). Theory of melt segregation—A review, *J. Volcan. Geotherm. Res.* **33,** 241–253.

Richet, P. (1984). Viscosity and configurational entropy of silicate melts, *Geochim. Cosmochim. Acta* **48,** 471–483.

Richter, F. M., and McKenzie, D. (1984). Dynamical models for melt segregation from a deformable matrix, *J. Geol.* **92,** 729–740.

Riley, G. N., Jr., Kohlstedt, D. L., and Richter, F. M. (1990). Melt infiltration in a silicate melt–olivine system: An experimental test of compaction theory, *Geophys. Res. Lett.* **17,** 2101–2104.

Roeder, P. L., and Emslie, R. F. (1970). Olivine–liquid equilibrium, *Contrib. Mineral. Petrol.* **29,** 275–289.

Schwindinger, K. R., and Anderson, A. T., Jr. (1989). Synneusis of Kilauea Iki olivines, *Contrib. Mineral. Petrol.* **103,** 187–198.

Scott, D. R., and Stevenson, D. J. (1986). Magma ascent by porous flow, *J. Geophys. Res.* **91,** 9283–9296.

Solomon, S. C. (1973). Shear wave attenuation and melting beneath the mid-Atlantic ridge, *J. Geophys. Res.* **78,** 6044–6059.

Thigpen, L., Hedstrom, G. W., and Bonner, B. P. (1983). Inversion of creep response for retardation spectra and dynamic viscoelastic functions, *J. Appl. Mech.* **105,** 361–366.

Turcotte, D. L., and Schubert, G. (1982). "Geodynamics: Applications of Continuum Physics to Geological Problems," Wiley, New York.

Vaughan, P. J., Kohlstedt, D. L., and Waff H. S. (1982). Distribution of the glass phase in hot-pressed, olivine–basalt aggregates: An electron microscopy study, *Contrib. Mineral. Petrol.* **81,** 253–261.

Waff, H. S. (1980). Effects of the gravitational field on liquid distribution in partial melts within the upper mantle, *J. Geophys. Res.* **85,** 1815–1825.

Waff, H. S., and Bulau, J. R. (1982). Experimental determination of near-equilibrium textures in partially molten silicates at high pressures, *Adv. Earth Planet. Sci.* **12,** 229–236.

Waff, H. S., and Faul, U. H. (1992). Effects of crystalline anisotropy on fluid distribution in ultramafic partial melts, *J. Geophys. Res.* **97,** 9003–9014.

Walsh, J. B. (1969). A new analysis of attenuation in partially melted rock, *J. Geophys. Res.* **74,** 4333–4337.

Chapter 3 Influence of Basaltic Melt on the Creep of Polycrystalline Olivine under Hydrous Conditions

David L. Kohlstedt and Prame N. Chopra

Overview

Constant displacement-rate and load-relaxation experiments were performed at high pressures and temperatures to investigate the rheological behavior of partially molten aggregates of fine-grained olivine with a small amount of included basaltic melt under hydrous conditions. Two-phase samples with melt fractions of either 0.9 or 8.6 vol% basalt and grain sizes of 12 or 15 μm were fabricated by hydrostatically hot-pressing powders of San Carlos olivine plus synthetic basalt. Single-phase samples with a grain size of 12 μm were prepared by hot-pressing powders of San Carlos olivine. Scanning and transmission electron microscopy observations demonstrated that the dihedral angle was ~30°, such that the melt formed an interconnected network along three-grain and through four-grain junctions but did not generally wet entire grain boundaries. Infrared spectra revealed that the olivine in the sample without basalt and in the sample with 0.9 vol% basalt was saturated with water-derived species, while the olivine in the sample with 8.6 vol% basalt was undersaturated because the water partitioned largely into the melt phase. The samples were deformed under hydrous conditions at temperatures of 1300 and/or 1200°C, a confining pressure of 300 MPa, and a strain rate of ~10^{-4} s^{-1}. Stress exponents of $1.0 \leq n \leq 1.4$ for both the two-phase and the single-phase samples combined with a comparison with published creep results indicated that the samples deformed predominantly by grain boundary diffusion creep. For a common grain size at 1300°C, the partially molten sample containing 8.6 vol% basalt was a factor of 5 weaker than the sample without melt. Both samples were substantially weaker than single-phase aggregates of anhydrous olivine. It is proposed that the difference in creep strength between the melt-free and the melt-rich samples was a result of two competing effects: First, the presence of melt caused the partially molten sample to be weaker than the single-phase olivine sample. Second, the presence of water resulted in hydrolytic weakening of both samples; however, because the partially molten sample was undersaturated with water while the single-phase sample was fully saturated, the hydrolytic weakening effect was larger in the latter than in the former. If both samples had been fully saturated with water, the difference in creep strength may have been larger. Hence, the rheology of partially molten upper mantle rocks in an ascending mantle plume or a mid-oceanic ridge environment will depend critically on the partitioning of water between the melt and solid phases and on the permeability of the rock (i.e., melt fraction).

Notation

		Units
A	Materials parameter in the strain-rate relationship	$s^{-1} \cdot m^{m} \cdot MPa^{-n}$
Q	Effective activation energy for creep	$kJ \cdot mol^{-1}$
R	Gas constant	$8.314(3) \pm 0.00008$ $J \cdot K^{-1} \cdot mol^{-1}$
T	Absolute temperature	K, kelvins
T	Temperature	°C
d, d_f	Grain size	m
Δd	Length of the grain boundary lost to the melt phase	m
m	Grain size exponent	dimensionless
n	Stress exponent	dimensionless
ϵ, ϵ_f	Strain	dimensionless, %
$\dot{\epsilon}, \dot{\epsilon}_f$	Strain rate	s^{-1}
$\dot{\epsilon}_1$	Single-phase strain rate	s^{-1}
$\dot{\epsilon}_2$	Two-phase strain rate	s^{-1}
ϕ	Melt fraction	dimensionless
σ_f	Stress	MPa
σ_1	Maximum principal stress	MPa
σ^n	Applied differential stress with stress exponent	MPa^n
θ	Equilibrium dihedral angle at melt crystal interfaces	°, degrees

Introduction

Pressure-release melting of ascending mantle material beneath mid-ocean ridges produces partially molten rocks at depths as great as 100 km. Trace element analyses of mid-ocean ridge basalts (MORBs) and of diopside grains in abyssal peridotites (Salters and Hart, 1989; Johnson *et al.*, 1990) as well as laboratory studies of the permeability of olivine-rich aggregates (Riley *et al.*, 1990; Riley and Kohlstedt, 1990, 1991) indicate that melt segregation is relatively rapid such that the amount of melt at any instant in the sub-ridge upwelling mantle is less than 1%. The presence of even this small amount of a melt phase can affect the physical properties of the rock, including not only the seismic velocity and seismic attenuation but also the electrical conductivity and viscosity of the aggregate.

To date, relatively few experiments have been carried out to determine the rheology of partially molten ultrabasic rocks. On the basis of published results from creep experiments on two-phase ceramic materials such as silicon nitride plus an oxynitride silicate glass, large decreases in flow strength might be anticipated. For fine-grained aggregates of silicon nitride, the flow strength decreases by over two orders of magnitude if a very small amount of an amorphous phase is added. For this system, the glass phase forms a thin (~1-nm-thick) film along the grain boundaries (for a review, see Kohlstedt, 1992). The melt distribution in two-phase aggregates of olivine plus basalt is quite different, however. In this case, the melt forms an interconnected network along triple junctions (three-grain edges) and through four-grain junctions (tetrahedral four-grain corners) but largely does not wet the grain boundaries (Waff and Bulau, 1979; von Bargen and Waff, 1988; Waff and Faul, 1992). Uniaxial hot-pressing experiments on fine-grained aggregates of olivine plus a few percent basalt under anhydrous conditions at 1 atm total pressure suggest that, at least in the diffusion creep regime, the presence of the melt phase results in a rather modest (factor of 2–5) increase in creep rate (Cooper and Kohlstedt, 1984b).

To extend this earlier work, the present study investigates the influence of a few percent basaltic melt on the flow properties of polycrystalline olivine under hydrous conditions at high temperature and pressure. Hence, compressive creep experiments have been carried out on fine-grained (~10-μm) aggregates synthesized from olivine powders plus a synthetic basalt. Fine-grain-size materials were used for two reasons: First, a primary role of the melt phase is to enhance the kinetics of grain boundary diffusion (Coble) creep (Cooper and Kohlstedt, 1984b; Cooper *et al.*, 1989); to observe diffusion creep at laboratory strain rates, small grain sizes are essential. The results can be extrapolated to the larger grain sizes expected for mantle rocks on the basis of constitutive equations for melt-enhanced diffusion creep. Second, because the distances over which the melt must redistribute are small, the solid–melt microstructure reaches, or at least closely approaches, textural equilibrium reasonably quickly in these samples (Cooper and Kohlstedt, 1984a, 1986). This point is important because in the mantle the rate of melt redistribution is rapid compared to the rate of deformation for a rock with a grain size of 3 mm. In contrast, in the laboratory the situation is reversed even for a rock with a grain size of only 50 μm (Kohlstedt, 1992). Consequently, unless laboratory samples are allowed to equilibrate for long time periods at elevated temperatures (~100 h for a grain size of 50 μm but only ~1 h for a grain size of 5 μm), the melt–solid distribution—and possibly also the associated flow behavior of the aggregate—is unlikely to be representative of that of a partially molten mantle rock.

Experimental Procedures

The olivine used in this study was obtained from San Carlos, Arizona. Small pieces of olivine, generally <7 mm in their longest dimension, were washed in acid and water to improve their clarity and to remove the bulk of other adhered materials. Fragments of this material that appeared clear and free from contamination by minerals other than trace amounts of spinel (i.e., notably free from pyroxene and layer silicates) were hand-picked over a light table. As illustrated in Fig. 1, after the selected grains were fractured with a steel jaw crusher, they were ground ~15 g at a time for 15 min in a small steel shatterbox to produce a fine powder. This powder was then processed through a cyclone air classifier set to separate out olivine grains smaller than 10 μm. This fine-grain-size

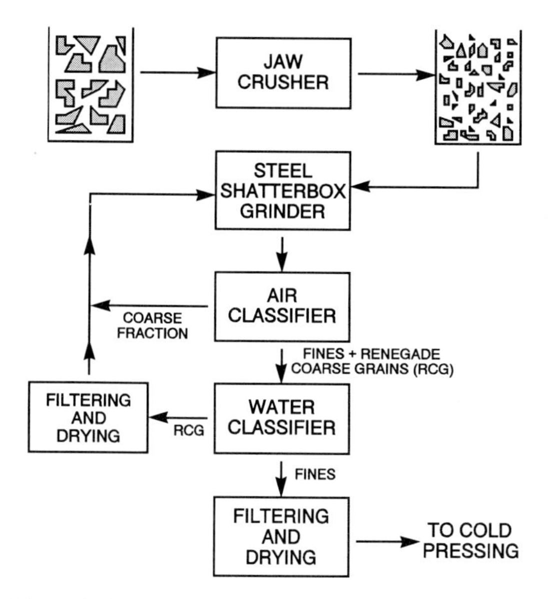

Figure 1 Schematic diagram illustrating the processing steps used to prepare fine-grained olivine powders.

Table 1
Chemical Compositions of Olivine and Basalts

	Hot-pressed San Carlos olivine (wt%)	Synthetic basalt (wt%)	Mid-ocean ridge basalt analogue (wt%)
SiO_2	40.82	51.54	51.0
Al_2O_3	—	14.05	17.6
Fe oxides	9.6	11.03	9.1
NiO	0.23	—	—
MgO	49.25	8.13	7.4
CaO	—	12.54	12.1
Na_2O	—	2.71	2.8
Total	99.89	100.00	100.0

fraction, however, always contained some large "renegade" olivine grains 40 to 60 μm in size as well as small quantities of small iron particles introduced during the crushing and grinding procedures. These contaminants were then separated from the fine-grained olivine by two processes. The iron was removed by vigorously stirring slurries of ~60 g of powder in 1 liter of water for ~1 h with a horseshoe magnet. Typically, the iron recovered on the magnet amounted to ~0.03 wt% of the powdered sample. The large renegade olivine grains were then removed by adding the slurries to plastic tanks filled with water. The large grains were allowed to settle, and the suspended fine-grained material ("fines") was filtered off and dried. These sample preparation procedures produced finely powdered olivine with angular, fragmented shapes as illustrated in the scanning electron microscope (SEM) micrograph in Fig. 1 of Chopra (1986). The average grain size was less than 10 μm, with the particle size distribution shown in Fig. 2 of Chopra (1986).

To introduce the basalt phase, the olivine powders were coated with a thin layer of a gel made from tetraethyl orthosilicate and compounds of magnesium, iron, calcium, aluminum and sodium by the method developed by Cooper and Kohlstedt (1984a). This gel coating, which after firing at 600°C had a composition corresponding to that of a mid-ocean ridge basalt as illustrated in Table 1 (see also Cooper and Kohlstedt, 1984a), made up 7.1 wt% of the two-phase material. This melt content corresponds to 8.6 vol% at standard temperature and pressure, if the density of San Carlos olivine (Fo_{92}) is taken as 3310 kg · m^{-3} and that of mid-ocean ridge basalt as 2740 kg · m^{-3}.

Olivine and olivine–basalt powders were cold-pressed at 200 MPa into 10-mm-diameter cylindrical pellets with a length-to-diameter ratio of 1 : 1 or less. These pellets were subsequently isostatically hot-pressed in 25-mm lengths with 0.2 ml of deionized water in iron-jacketed assemblies (Chopra and Paterson, 1981; Paterson *et al.,* 1982) at a confining pressure of 300 MPa and a temperature of 1200°C for periods between 15 and 60 min. Three samples were prepared in this manner: olivine without basalt, olivine with 0.9 vol% basalt, and olivine with 8.6 vol% basalt. In the second sample, the melt fraction decreased from 8.6 to 0.9 vol% during hot-pressing due to melt loss through a hole in the upper end of the jacket.

Cylindrical specimens 7 mm in diameter and 16 mm in length for use in deformation experiments were fashioned from the hot-pressed cylinders by diamond core drilling, combined with grinding and lapping of the ends to ensure parallelism and flatness. These specimens were then rejacketed in iron with an additional 0.08 ml of deionized water and deformed in constant displacement-rate and load-relaxation tests using a gas-medium high-pressure apparatus (Paterson 1970, 1977) at temperatures of 1300 and/or 1200°C under a confining pressure of 300 MPa. At the completion of a deformation experiment, the sample was cooled quickly to 1000°C to preserve deformation-induced microstructural features.

The deformation testing procedures used were similar to those reported by Chopra (1986). However, because of the low flow strength exhibited by the olivine plus basalt sample at 1300°C, the strength of the iron jacket had to be taken into account in calculating the differential stress on the sample from the measured applied load. The fraction of the applied load that was borne by the jacket was calculated using the flow law and data given by Frost and Ashby (1982). Their Eq. (2.19) was used to solve for the flow stress of iron at each instantaneous strain rate in our deformation tests. At 1200°C, the iron accounted for 20% of the load applied during creep of the sample composed of olivine plus 8.6 vol% basalt; this result is similar to that of Karato *et al.* (1986) who used an empirical approach to estimate the strength of the iron jacket. At 1300°C, the iron supported 55% of the applied load during deformation of this two-phase sample. The values of stress reported by Chopra (1986), which had not been corrected for the strength of the jacket, were recalculated for inclusion in Table 2.

In the deformation experiments on water-bearing samples jacketed in iron, the oxygen fugacity was buffered at a value determined by the Fe–FeO solid state buffer. At this oxygen fugacity, olivine near the jacket was reduced, forming precipitates of an iron–nickel alloy (see Fig. 13 of Karato *et al.* (1986)) and pyroxene (Mackwell *et al.*, 1985). In the single-phase samples, the pyroxene fixed the oxide activity at the olivine–enstatite phase boundary. In the partially molten samples, the basaltic melt fixed the oxide activity at about the same value.

After the deformation experiments were completed, specimens were encapsulated in epoxy resin and sliced longitudinally. From these slices, ultrathin sections of $<10\ \mu m$ in thickness, polished on both sides, were prepared for optical microscopy. Sections of such thinness were necessary to minimize the number of grains of different crystallographic orientation superimposed in the light path. Polished planar surfaces were also prepared from these slices for SEM observations, and thin foils were produced by ion-beam thinning for transmission electron microsocpy (TEM) analyses. Samples for the SEM were etched in H_3PO_4/HCl; this etchant dissolves olivine at a rate that depends upon the orientation of the grain and hence produces relief at grain boundaries. This

Table 2

Summary of Experimental Conditions and Results

Material	Run	T (°C)	σ_f (MPa)	$\dot{\epsilon}_f$ (s^{-1})	ϵ_f (%)	d_f (μm)
Olivine	Hot-press (1 h)	1200	—	—	—	7
Olivine[a]	Deformation	1200	121r[b]	1.88×10^{-4}	9.8	—
		1300	49r	1.54×10^{-4}	18.9	12
Olivine + basalt	Hot-press (15 min)	>1200[c]	—	—	—	—
Olivine + 0.9 vol% basalt	Deformation	1200	112	1.40×10^{-4}	13.7	15
Olivine + 8.6 vol% basalt	Hot-press (1 h)	1200	—	—	—	7
Olivine + 8.6 vol% basalt	Deformation	1200	63r	1.40×10^{-4}	8.7	—
		1300	16r	1.42×10^{-4}	15.5	12

[a] From Chopra (1986).
[b] The "r" indicates that a load-relaxation run followed the constant displacement-rate test
[c] Temperature probably exceeded 1200°C, because melt escaping from the sample moved the thermocouple.

treatment greatly improves both the ease and the quality of observations of grain boundaries and at the same time leaves the intergranular basalt relatively unaffected and higher than the olivine grains. The etched material was coated with a gold–palladium alloy to minimize charge buildup under the electron beam.

To examine the water content in the deformed samples, Fourier transformed infrared (FTIR) spectroscopy was carried out on sections that were ~500 μm in thickness and polished on both sides. Room-temperature absorption spectra were collected in the wavenumber range 220–450 mm^{-1}. To remove the contribution resulting from scattering of the infrared beam at grain boundaries, a background correction was made by least-squares fitting a third-order polynomial to the FTIR data outside the range 280–380 mm^{-1}. The concentration of OH was estimated from the height of the OH-stretching bands in the range 300–370 mm^{-1} using the relation proposed by Paterson (1982).

Deformation Results

One sample each of olivine without basalt, olivine plus 0.9 vol% basalt, and olivine plus 8.6 vol% basalt was deformed as part of the present investigation. The conditions and results of the isostatic hot-pressing runs, constant displacement-rate experiments, and load-relaxation tests are summarized in Table 2. In addition to the temperature, T, at which each run was performed, the stress, σ_f, strain rate, $\dot{\epsilon}_f$, and total strain, ϵ_f, at the end of the constant displacement-rate tests are given. Also included is the mean grain size, d_f, at the end of the hot-pressing run and the deformation experiments. The deformation behavior of the single-phase olivine sample has been reported elsewhere (Chopra, 1986).

The differential stress versus strain results from the constant displacement-rate experiments at 1200 and 1300°C are plotted in Figs. 2 and 3, respectively, for all three samples. The specimen with 8.6 vol% basalt is substantially weaker than the basalt-free specimen of the same grain size at both temperatures. At the lower temperature, the strength of the olivine plus 8.6 vol% basalt specimen is approximately one-half that of the single-phase olivine material; at the higher temperature,

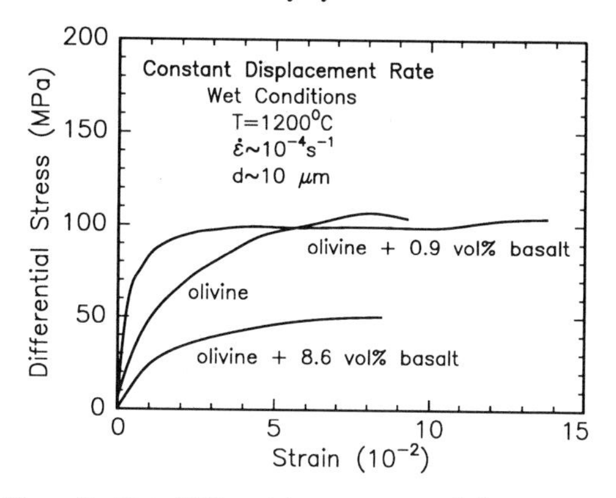

Figure 2 Plot of differential stress versus strain for samples of olivine, olivine plus 8.6 vol% basalt, and olivine plus 0.9% basalt deformed at 1200°C.

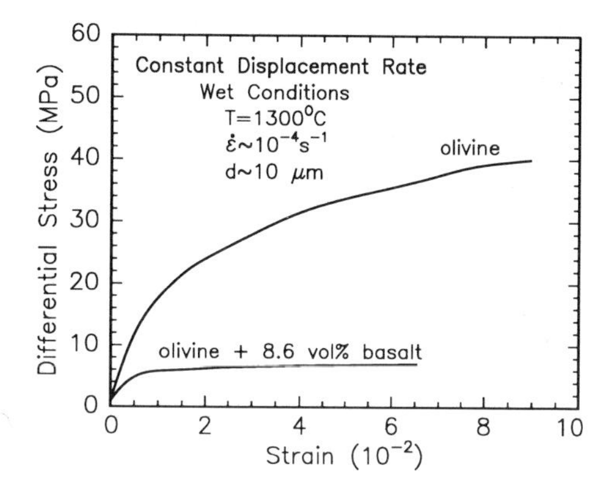

Figure 3 Plot of differential stress versus strain for samples of olivine and olivine plus 8.6 vol% basalt deformed at 1300°C.

the strength of the olivine–basalt aggregate is about one-fifth that of the olivine specimen. The sample containing 0.9 vol% basalt exhibited a strength approximately equal to that of the olivine specimen; however, as described under Discussion, the strengths of these two samples should not be directly compared, because the grain size of the former is ~25% larger than that of the latter.

The results of the load-relaxation tests on the olivine and the olivine plus 8.6 vol% basalt specimens are plotted in Figs. 4 and 5 as the logarithm of the decaying differential stress versus the logarithm of the strain rate. The method for converting the raw load versus time data into differential stress versus strain rate results is described in detail in Kohlstedt and Chopra (1987). Data from the later parts of these experiments (i.e., for strain

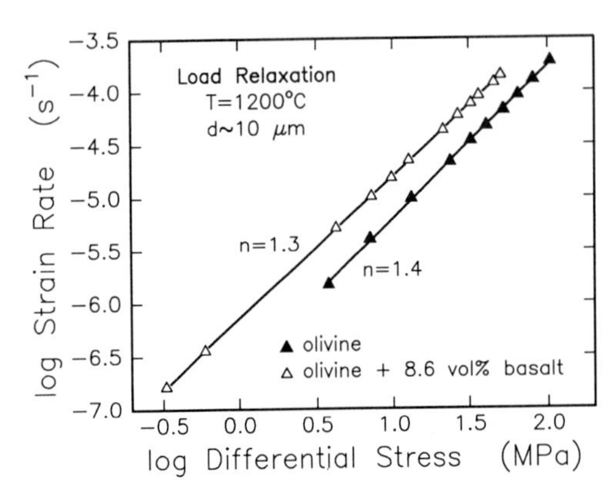

Figure 4 Plot of log strain rate versus log differential stress for load-relaxation experiments carried out at 1200°C.

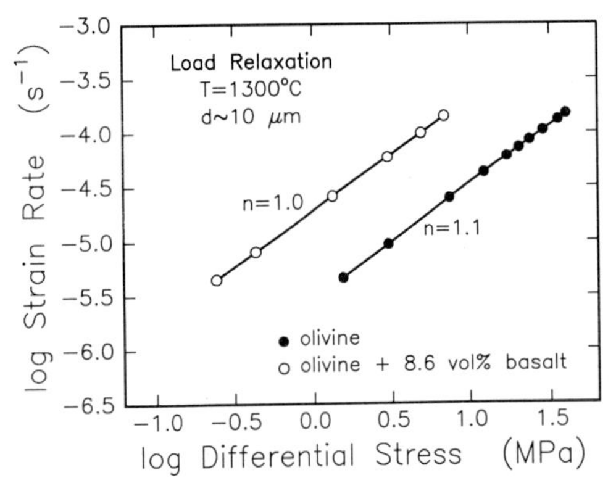

Figure 5 Plot of log strain rate versus log differential stress for load-relaxation experiments carried out at 1300°C.

rates below $\sim 10^{-5.5}\,s^{-1}$) typically were not plotted because of the scatter in the low-stress, low-strain-rate data. This noise is the result of two factors: First, the magnitude of background fluctuations in stress increases relative to the overall differential stress level as the latter decreases. These fluctuations arise from the expansion and contraction of the loading train and specimen resulting from a pressure cycling of about 1.5 MPa and a temperature cycling of about 1°C in the high-pressure deformation apparatus. Second, the magnitude of the elastic distortion of the apparatus, which is a necessary input to the calculation of the instantaneous cross-sectional area of the specimen and hence of the stress, becomes increasingly uncertain with decreasing load because of its variable nonlinearity from one sample assembly to the next at very low loads (Kohlstedt and Chopra, 1987).

The high-temperature deformation data for both the olivine and olivine plus basalt specimens can be described by a power law relation of the form

$$\dot{\epsilon} = A \frac{\sigma^n}{d^m} \exp\left(-\frac{Q}{RT}\right),$$

where $\dot{\epsilon}$ is the strain rate, A a materials parameter, σ the applied differential stress, d the grain size, Q the activation energy for creep at 300 MPa confining pressure, R the gas constant, and T the absolute temperature (in kelvins); in Eq. (1), n and m are the stress exponent and grain size exponent, respectively. The data from each of the load-relaxation tests shown in Figs. 4 and 5 define straight lines, the slopes of which correspond to n. At 1200°C, the value of n is 1.3–1.4 for both the single-phase olivine sample and the two-phase olivine plus 8.6 vol% basalt sample. At 1300°C, n is 1.0–1.1 for both samples. Thus there is no systematic difference in n between the two types of specimen.

With the limited amount of deformation data available here and the observed change in stress exponent with temperature, it is not possible to determine accurately an activation energy. Analysis of the results from the constant displacement-rate and load-relaxation tests in terms of Eq. (1) yields activation energies in the range 300–575 kJ · mol^{-1} (assuming that the grain sizes in the experiments at 1200 and 1300°C are the same). An examination of the results in Table 2 and Figs. 4 and 5 suggests that the activation energy for creep of the olivine plus 8.6 vol% basalt sample may be somewhat larger than that for the single-phase olivine sample, but additional experiments are required to verify this observation.

Microstructural Observations

Hot-Pressed Material

As illustrated by the SEM micrographs in Fig. 6, the basalt in hot-pressed specimens is concentrated in polygonal-shaped pockets at three- and four-grain contacts. The olivine grains are equiaxed with both gently curving and flat grain boundaries. At this scale, olivine–basalt interfaces are frequently straight-edged. Small olivine grains, such as the one in the center of the micrograph in Fig. 6a, which appear in cross section to be largely surrounded by basalt are not uncommon in these isostatically hot-pressed specimens. Voids were not observed.

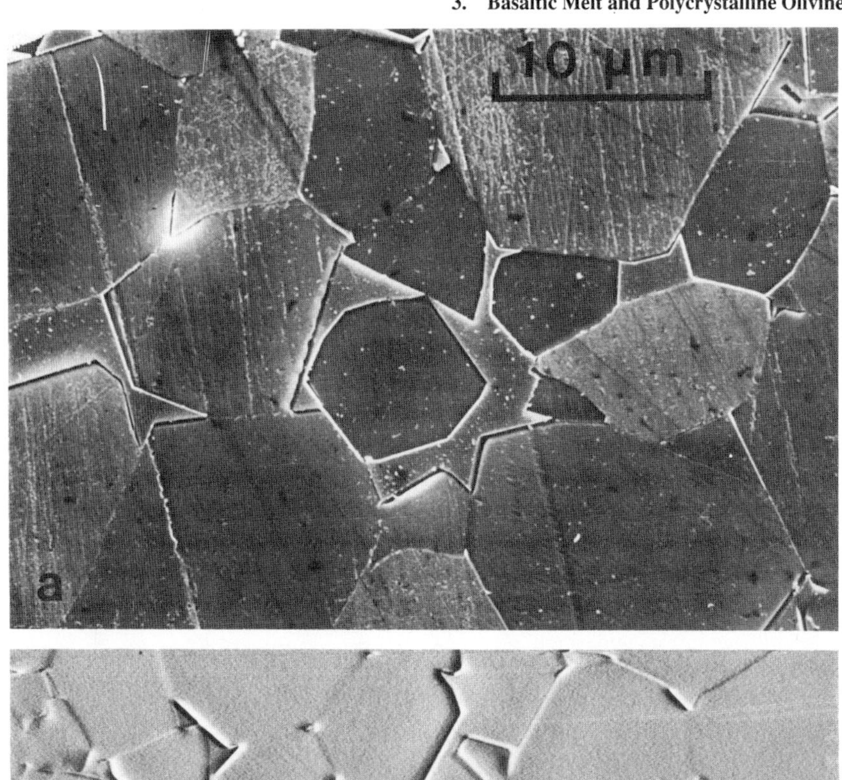

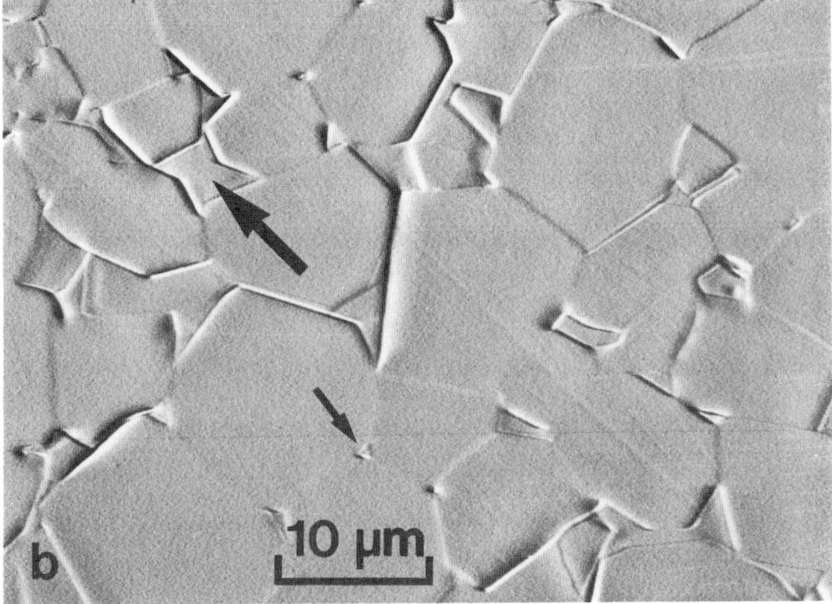

Figure 6 Scanning electron micrographs of olivine plus basalt sample isostatically hot-pressed at 300 MPa and 1200°C. The sample has been etched such that the melt tends to stand higher than the olivine grains. (a) Secondary electron image. At the center of the micrograph, a small grain of olivine is nearly surrounded by melt. (b) Composition (i.e., mixed secondary electron and backscattered electron) mode. One melt-filled triple junction and one large, faceted pocket of melt are marked by the small and large arrows, respectively.

Deformed Material

Optical Microscopy

Longitudinal thin sections of each of the deformed samples were observed with an optical microscope operating at magnifications up to 1000 times. Each thin section was approximately 1.2 × 0.8 mm in area and 10 μm in thickness.

The grains in the deformed single-phase olivine specimen are approximately equant and even-sized with an average size of approximately 12 μm. A few pores are located along the grain boundaries.

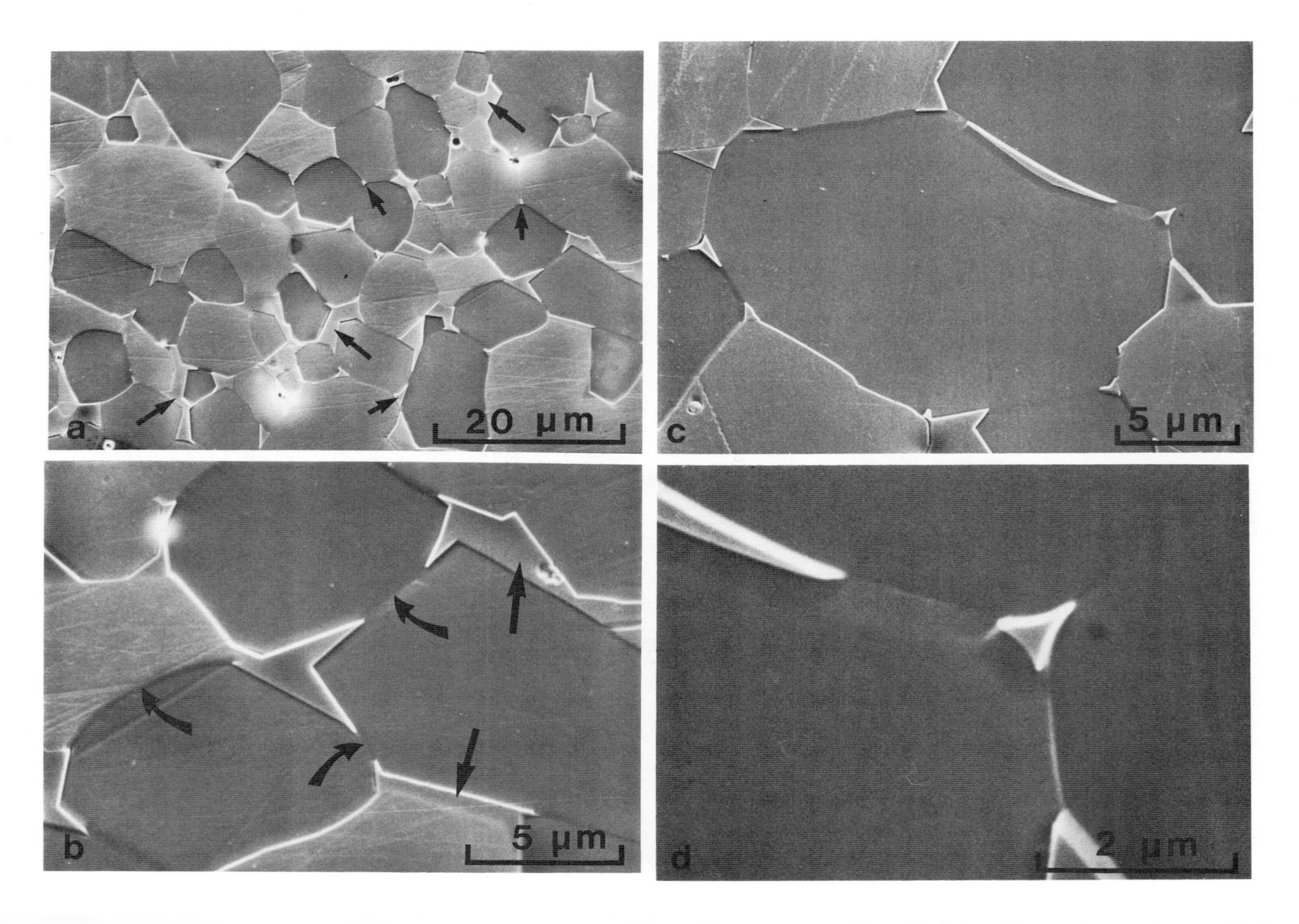
a
20 μm
b
5 μm
c
5 μm
d
2 μm

Over the area of the thin section approximately 100 nearly equant grains with apparent dimensions between 30 and 75 μm are interspersed randomly among this fine-grained material. Many of these larger grains contain within them subspherical shells of pores. In section, these shells appear as rings, similar to those described in Figs. 3 and 5 of Cooper and Kohlstedt (1984a). Some of these rings are nearly equant, while others define elongated or angular shapes. The pores making up these rings were originally located at the boundaries of growing olivine grains but became trapped in their hosts when the mobility of the spreading grain boundaries exceeded their own mobilities. These rings of pores thus define upper limits to the sizes of the original grains from which their hosts grew. The size of the precursors defined in this way is always approximately 20 μm in at least one direction.

Measurements of the size of the fine-grained olivine which makes up the vast majority of the specimen were made using the linear intercept method of Exner (1972) with a sectioning correction factor of 1.5. One thousand measurements made parallel to the direction of the maximum principal stress, σ_1, yielded a grain size of 11.5 μm, while a corresponding number of measurements made perpendicular to σ_1 yielded an estimate of 12.6 μm. This anisotropy in grain shape is oriented in such a way as to be consistent with an origin by flattening during the deformation. However, the extent of this anisotropy, ~9%, is less than one-half the total macroscopic strain of the specimen, ~19%. The microstructure of this deformed single-phase specimen lacks dislocation-associated features (e.g., undulatory extinction, deformation bands, and subgrain walls) although a firm conclusion in this regard is made difficult by the small grain size and the problem, even in this ultrathin section, of superimposed grains.

The deformed olivine plus 8.6 vol% basalt specimen shares a number of microstructural features with the single-phase olivine specimen. The grain size of the olivine is again bimodal. In this case, 1000 intercepts measured parallel to the direction of maximum principal stress yielded a mean grain size of 12.0 μm, while similar measurements perpendicular to σ_1 gave 12.3 μm. The anisotropy in grain shape for the two-phase sample amounts to ~2.5% flattening. The microstructure of this specimen is also characterized by a general lack of dislocation-associated features. The only evidence of dislocations observed was a single tilt wall with a ~4° misorientation in one of the large grains.

There are again approximately 100 large and generally equant grains of largest dimension between 30 and 75 μm. These grains generally contain some captured pores, although the concentrations of these pores are invariably much lower than those in the single-phase olivine sample. In the five grains observed in which the distribution of pores does define a ring, this ring is ~20 μm in its largest dimension. At 1000 times magnification in reflected light, small melt bodies can be seen along many grain boundaries. However, the clarity of the observations is restricted by the limits of resolution of the optical microscope.

The microstructure of the olivine plus 0.9 vol% basalt sample is similar to those of the two samples described above. The primary difference, in this case, is that the grain size (15 μm) is somewhat larger for this sample. Grain growth likely occurred during hot-pressing, since the temperature probably increased substantially above 1200°C when the melt that escaped from the sample moved the thermocouple several millimeters away from the sample.

Scanning Electron Microscopy

As illustrated in the secondary electron images in Fig. 7, melt is located in all of the three- and four-grain junctions; however, some larger melt pockets do exist. Except for these large melt pockets, the melt does not form a continuous film along the

Figure 7 Secondary electron images of olivine plus 8.6 vol% basalt sample deformed at 300 MPa at both 1200 and 1300°C. The sample has been etched so that the melt tends to stand higher than the olivine grains. A region in (a) is viewed at higher magnification in (b), and a region in (c) is viewed at higher magnification in (d). In (a), three melt-filled triple junctions and three large, faceted pockets of melt are marked by short and long arrows, respectively. In (b), three melt-free grain boundaries are marked by curved arrows and two melt pockets separating neighboring grains are marked by straight arrows. In (c), note that the grain boundaries between melt-filled triple junctions are all free of melt; this point is made particularly clear with the high-magnification view in (d).

grain boundaries (i.e., two-grain junctions), within the resolution of the SEM (<10 nm).

The mean dihedral angle θ, determined from tangents drawn to olivine grains in triple junctions, is $\sim 30°$. Hence, the melt is interconnected along triple junctions (e.g., Bulau *et al.*, 1979), a point verified by the presence of melt in all of the triple junctions (see also Daines and Richter, 1988).

Careful examination of a large number of micrographs revealed no apparent influence of differential stress or finite strain on the melt distribution. The melt distribution in the deformed samples appeared identical to that of the undeformed samples. In addition, no anisotropy in melt distribution relative to the direction of σ_1 was observed.

Transmission Electron Microscopy

The dark-field (DF) and bright-field (BF) TEM images of the microstructure of the sample of olivine plus 8.6 vol% basalt reinforce the optical and scanning electron microscopy observations made above. First, the melt is present in all the triple junctions. While some of the melt–solid interfaces are smoothly curved as expected for a two-phase system with isotropic interfacial energies, others are flat or faceted (Fig. 8a; see also Fig. 7a). The flat interfaces involve low-index crystallo-

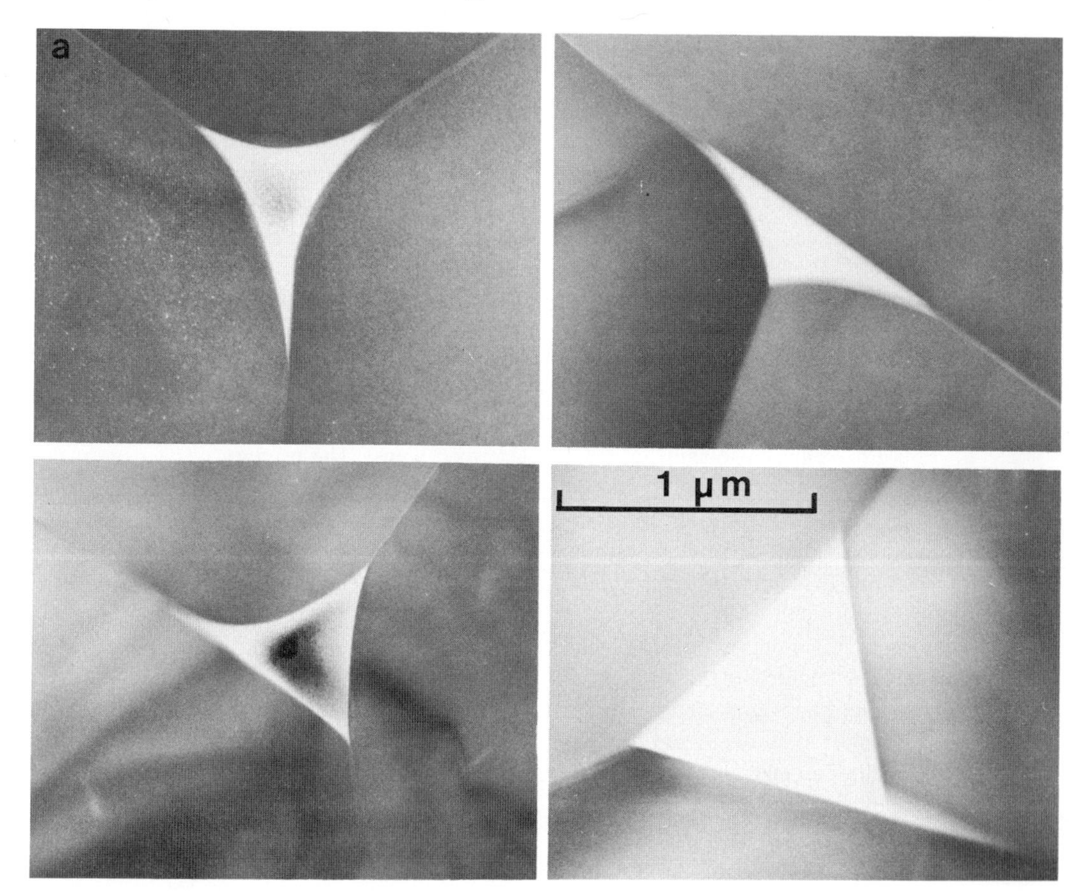

Figure 8 (a) Dark-field TEM images of melt-filled triple junctions. The images were formed using electrons diffusely scattered from the melt, so that the melt phase is lighter than the olivine grains. The dark region near the middle of one of the melt-filled triple junctions (southwest quadrant) is a hole produced by ion thinning. Both straight and smoothly curved basalt–olivine interfaces are present. (b) Bright-field TEM image of grain boundary containing very small pockets of trapped melt, several of which are marked by short arrows. The grain boundary, which runs horizontally in this micrograph, is bounded at either end by a melt-filled triple junction, marked by long arrows. The fringes running parallel to the boundary indicate that the boundary is inclined at an angle of $\sim 30°$ to the incident electron beam. (c) Dark-field TEM image showing the transition from a triangular-shaped melt-filled triple junction to a melt-free grain boundary. The transition in the upper left of the micrograph is marked by a pair of filled arrows, and the triple junction is marked by an open arrow.

continues

graphic planes of olivine, such as (010) and {110}, indicative of marked anisotropy in the crystal structure of olivine (Waff and Faul, 1992).

Second, in rare cases small (<0.1-μm) pockets of melt were trapped in a grain boundary (Fig. 8b). However, high-resolution TEM and high-resolution analytical electron microscopy observations demonstrate that basaltic melt does not generally form a continuous film along the boundaries between olivine grains (Vaughan *et al.*, 1982; Kohlstedt, 1990). Although the transition from the triple junction to the grain boundary is generally

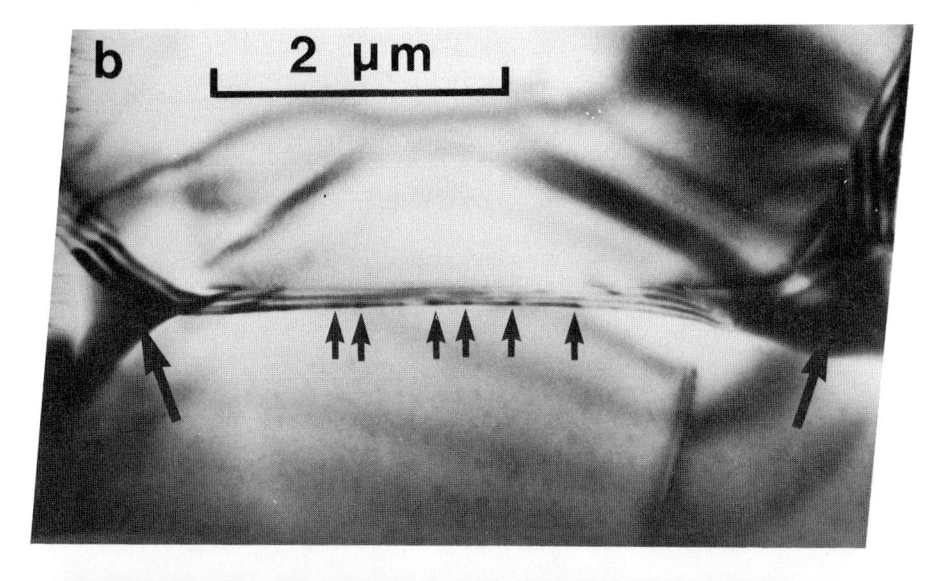

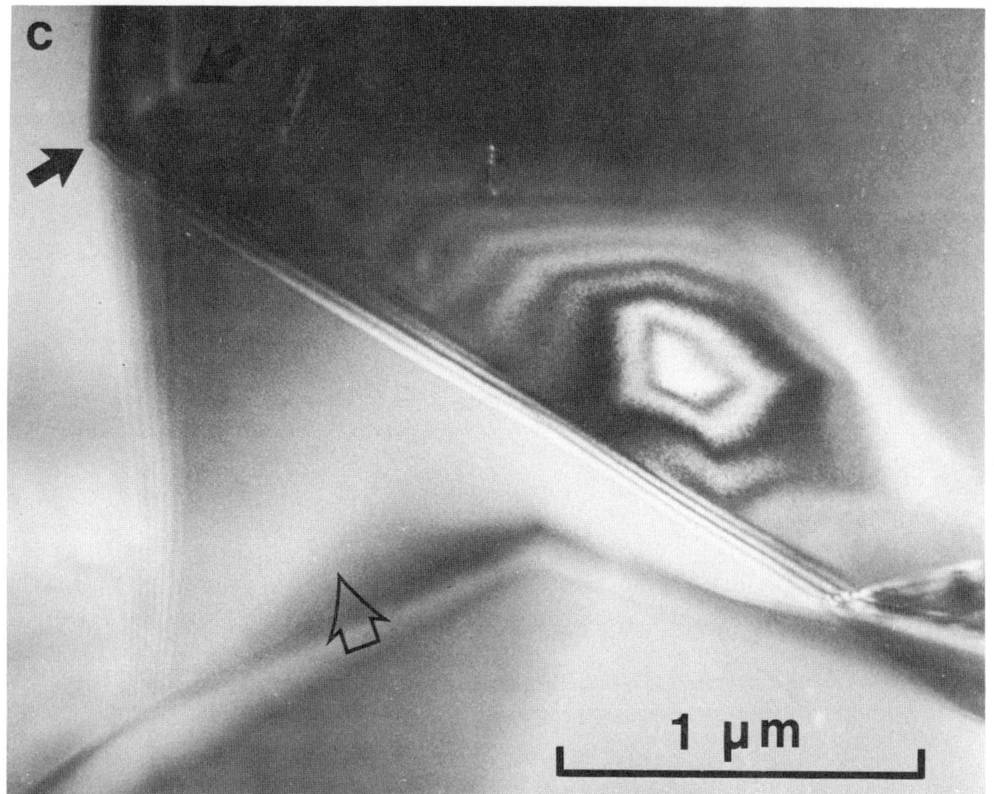

Figure 8 Continued.

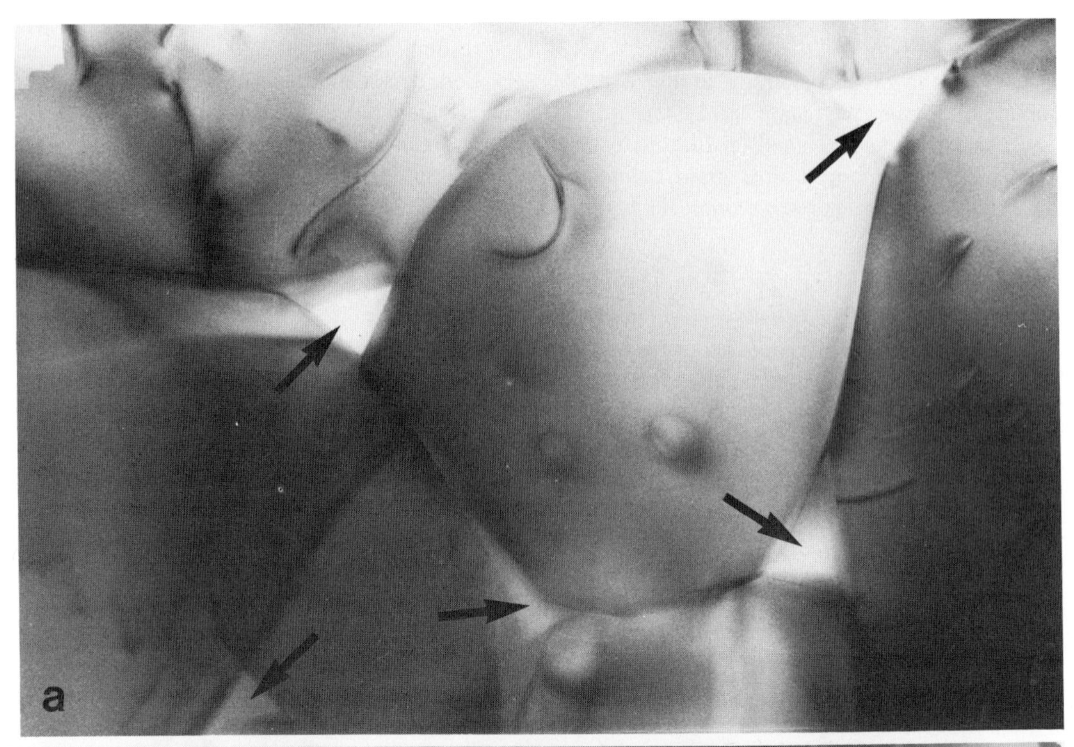

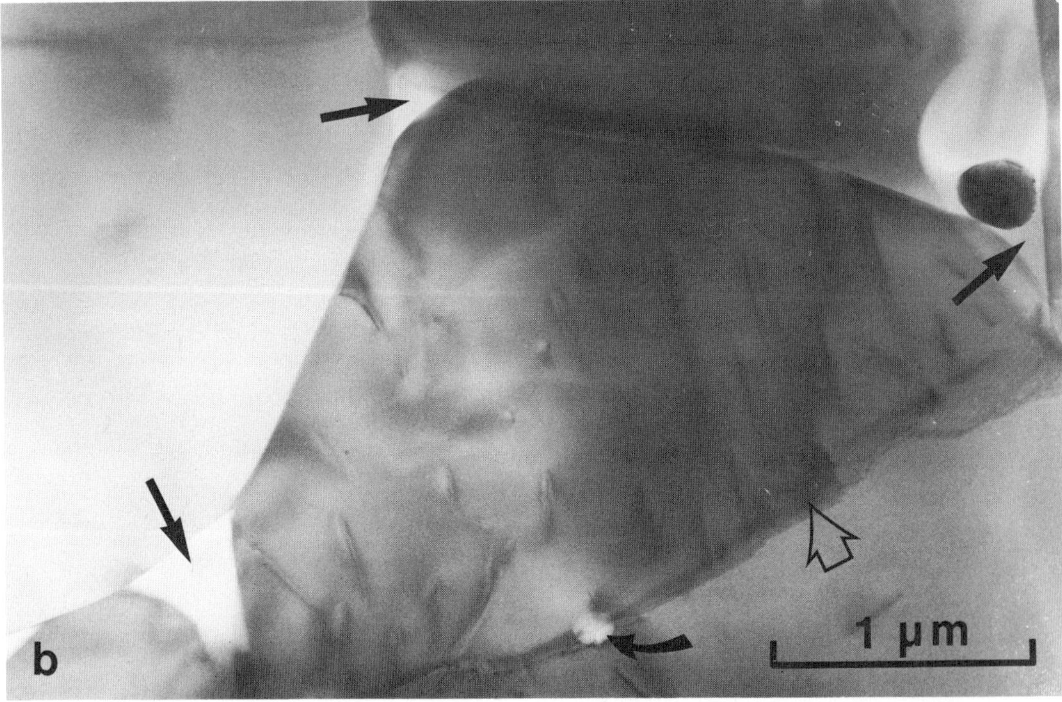

Figure 9 (a) Dark-field, diffuse-scattering TEM images illustrating typical microstructure of the deformed sample of olivine plus 8.6 vol% basalt. Melt-filled triple junctions appear as bright, triangular-shaped regions and are marked by solid arrows. The dislocations are straight or gently curved lines within the olivine grains. (b) In the lower micrograph, a low-angle boundary, marked by an open arrow, divides the grain along a line running NE–SW. A trapped pore, marked by a curved arrow, can be seen along this boundary.

difficult to identify precisely (e.g., Fig. 8a), it is occasionally clearly delineated, as illustrated in Fig. 8c.

Third, dislocations were observed in most, but not all, of the olivine grains. In grains which contained dislocations, the dislocation density was in the range 1×10^{12} to 5×10^{13} m^{-2} (Fig. 9). The dislocation density in natural San Carlos crystals is $\sim 10^{10}$ m^{-2}.

Infrared Analyses

As demonstrated by the FTIR spectra in Fig. 10, all three samples contain significant concentrations of OH-related species. In each case, the total concentration is >2500 H/10^6 Si. The OH-related species are present within the olivine grains, along grain boundaries, in the quenched melt, and in bubbles (Karato *et al.*, 1986). For the olivine without basalt and the olivine with 0.9 vol% basalt samples, the peaks in the wavenumber range 300–370 mm^{-1} are primarily associated with OH-related species in the olivine grains and possibly along grain boundaries. Experiments on single crystals of olivine demonstrate that the maximum concentration in the olivine grains is only

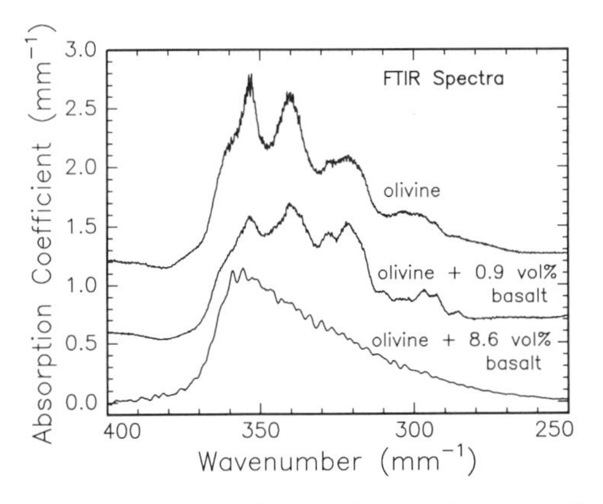

Figure 10 FTIR spectra from the three samples after the deformation experiments. Clear peaks are present near wavenumbers of 340 and 350 mm^{-1} in the spectra from the single-phase olivine and the two-phase olivine plus 0.9 vol% basalt samples, indicating that OH-related species are present in the olivine grains. These peaks are absent from the spectrum from the sample of olivine plus 8.6 vol% basalt, suggesting that most of the water partitioned into the melt. (Note that the sinusoidal oscillations in the spectrum from the sample with 8.6 vol% basalt are interference fringes that occur because this sample is quite thin, $\sim 300 \mu m$.)

~ 250 H/10^6 Si, for the thermodynamic conditions used in the present study (Bai and Kohlstedt, 1992). Hence, the olivine is apparently saturated with water-derived species, and most of the water in these samples is distributed in the melt and trapped in the pores. For the olivine plus 8.6 vol% basalt sample, the absence of these peaks in the 300–370 mm^{-1} range indicates that the concentration of OH-related species within the olivine grains is relatively small. For this sample, most of the OH-related defects have presumably gone into the melt, since the olivine/melt partition coefficient may be as small as 0.01 (Michael, 1988; Dixon *et al.*, 1988).

Discussion

Microstructural Characteristics

The similarity in final grain sizes of our specimens without basalt and with 8.6 vol% basalt from both the hot-pressing experiments and the deformation experiments (Table 2) suggests that under hydrous conditions the presence of basaltic melt in the two-phase specimens does not have a large effect on the kinetics of olivine grain growth. A similar conclusion can be drawn from the observation that the precursors to the large olivine grains, as defined by the size of rings of pores, are similar irrespective of whether the matrix material is single-phase olivine or two-phase olivine plus basalt. However, it must be borne in mind that the melt-free sample was saturated with water, while the melt-rich sample was undersaturated. By comparison, grain growth during hot-pressing under anhydrous conditions is much more rapid in samples of olivine plus basalt than in samples of olivine only (Cooper and Kohlstedt, 1986).

The microstructure of the olivine plus 8.6 vol% basalt sample reflects a marked anisotropy in the solid–melt interfacial energy. A large fraction of the olivine–basalt interfaces are flat (see also Waff and Faul, 1992). For an isotropic solid, the solid–melt interfaces should be curved.

Finally, it should be emphasized that, while melt does not form a continuous film along most of the grain boundaries, it is present in all of the triple junctions. This observation, which is consistent with the measured dihedral angle of $\sim 30°$, demonstrates that the melt forms an interconnected network along the three-grain junctions and through the four-grain junctions. It should also be emphasized that some grains are separated on at least one side from a neighboring grain by large, faceted pockets of melt (Figs. 6 and 7).

Rheological Behavior

The load-relaxation results obtained with both the single-phase olivine specimen and the two-phase olivine plus 8.6 vol% basalt specimen can be adequately described by a power-law relation between strain rate and differential stress, Eq. (1). The stress exponent, n, determined for both the olivine and the olivine plus basalt specimens is 1.0–1.1 at 1300°C (Fig. 5). This Newtonian (i.e., $n = 1$) behavior indicates that diffusion dominates the creep process. However, at 1200°C some contribution to the deformation process due to dislocation creep, for which $n \approx 3.5$ in olivine (Chopra and Paterson, 1984; Karato *et al.*, 1986; Bai *et al.*, 1991), is suggested by the fact that the stress exponents are greater than unity (n = 1.3–1.4, Fig. 4) for both types of samples. This point is reinforced by the flattening of the olivine grains.

For the fine-grained, single-phase olivine sample deformed under hydrous conditions as part of the present study, the flow stress measured at 1300°C and $\sim 10^{-4}$ s^{-1} agrees within 10% with the value reported by Karato *et al.* (1986) for samples crept under similar conditions. Likewise, the stress exponents ($1.0 < n < 1.4$) obtained here are in good agreement with that ($n \approx 1.4$) determined by Karato *et al.* (1986). On the basis of a stress exponent near unity and a grain size exponent in Eq. (1) of $m \approx 3$, Karato *et al.* (1986) concluded that under hydrous conditions their fine-grained samples deformed by grain boundary diffusion creep (Coble, 1963). Hence, we argue that the dominant deformation mechanism in our high-temperature experiments must also be grain boundary diffusion (i.e., Coble creep).

The presence of the 8.6 vol% basaltic melt in the composite specimen results in a pronounced lowering of the flow strength relative to the specimen without basalt and the specimen with only a small amount (0.9 vol%) of basalt. To make a quantitative comparison of the data for the three samples, however, the results in Table 2 must be normalized to a common stress and grain size. On the basis of the discussion in the previous paragraph, values of $n = 1$ and $m = 3$ were used in Eq. (1) for this purpose. At 1300°C, the sample with 8.6 vol% basalt is a factor of ~5 weaker than the melt-free sample. A similar decrease in flow strength (a factor of 2–5) was determined from uniaxial hot-pressing experiments on fine-grain olivine–basalt aggregates under anhydrous conditions (Cooper and Kohlstedt, 1984b). Note that a similar comparison cannot be readily made for the creep results obtained at 1200°C because the grain size for two of the samples was measured only after the experiment at 1300°C.

To describe the creep behavior of partially molten aggregates with $0 < \theta \leq 60°$ by solution–precipitation enhanced diffusion creep, Cooper *et al.* (1989) analyzed the effects of introducing melt into the triple junctions. They concluded that, although the rate of deformation is enhanced by short-circuit diffusion through the melt-filled triple junctions, the deformation process is rate limited by diffusion through the melt-free grain boundaries. Their model is valid as long as the length of grain boundary perpendicular to the triple junction that is lost to the melt phase, Δd, is small compared to the total length of the grain boundary (i.e., as long as θ is not too small and the melt fraction ϕ is not too large). In this case, the strain rate of the two-phase aggregate relative to that of the single-phase aggregate is given by

$$\frac{\dot{\epsilon}_2}{\dot{\epsilon}_1} = \left[\frac{d}{(d - \Delta d)}\right]^4, \tag{2}$$

where Δd is a function of both θ and ϕ, and where $\dot{\epsilon}_1$ is the strain rate of the single-phase aggregate and $\dot{\epsilon}_2$ is the strain rate of the two-phase aggregate. Within the context of this model for solution–precipitation enhanced diffusion creep, the results reported here for creep at 1300°C of aggregates of olivine plus 8.6 vol% basalt and of melt-free olivine are compared in Fig. 11. The reasonably good agreement between the experimental results and the model suggests that the strain-rate enhancement observed in the partially molten olivine plus basalt aggregate under hydrous conditions might indeed be a result of short-circuit diffusion through the melt-filled triple junctions, as described by Cooper *et al.* (1989).

A comparison of the rheological behaviors of the melt-free and partially molten samples may be more complicated, however. The FTIR spectra in

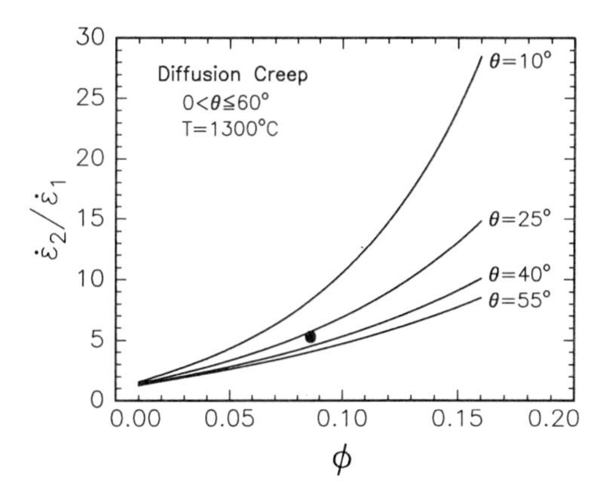

Figure 11 Normalized strain rate of a partially molten, two-phase aggregate ($\dot{\epsilon}_2$) relative to that of the melt-free aggregate ($\dot{\epsilon}_1$) versus melt fraction (ϕ). The solid curves are from the model of Cooper *et al.* (1986) for a variety of dihedral angles, θ. The solid circle was calculated using the creep results obtained in the present study for the olivine plus 8.6 vol% basalt sample and the single-phase olivine sample. Data were corrected to a common stress and a common grain size using $n = 1$ and $m = 3$, respectively, in Eq. (1). The dihedral angle for the sample of olivine plus basalt is ~30°.

Fig. 10 suggest that in the single-phase sample the olivine grains and grain boundaries are probably saturated with water-derived species, whereas in the two-phase sample most of the water partitioned into the melt, leaving the olivine undersaturated. Hence a direct comparison of the two samples may not be appropriate. In fact, for deformation experiments carried out under anhydrous conditions in the diffusion creep regime, partially molten samples containing ~8 vol% of either a natural MORB or a synthetic MORB crept about 16 or 50 times faster, respectively, than their single-phase counterparts of the same grain size (Hirth and Kohlstedt, 1992a). Consequently, the difference between the creep rate of the single-phase sample and that of the two-phase sample might be significantly larger than that measured in our experiments, if the latter were fully saturated with water (i.e., if the effects of melt and water on creep strength were combined). This point was anticipated by Karato (1986). Such a deviation from the model of Cooper *et al.* (1989) might well be expected once the melt fraction exceeds the equilibrium value, which is ~2.3 vol% of the olivine–basalt system (Riley, 1990). At larger melt fractions, a significant portion of the melt is present in large, faceted pockets that separate some grains from at least one of their neighboring grains (Figs. 6 and 7), rather than in triple junctions as assumed in the model.

The suggestion that the single-phase sample deformed under fully hydrous conditions while the two-phase (8.4 vol% basalt) sample deformed under more nearly anhydrous conditions is supported by the observation that the activation energy for creep of the former is smaller than that for the latter. Such a relationship between water content and activation energy for creep has been noted previously for dunite (Chopra and Paterson, 1984; Karato *et al.*, 1986). It should also be noted that many ceramic materials contain a melt phase that forms a thin (1- to 2-μm-thick) film that entirely coats the grain boundaries at high temperatures. For such fine-grained ceramics, the presence of the melt phase results in a very large (>10^2) increase in strain rate, even if the melt fraction is less than 1 vol% (e.g., Hwang and Chen, 1990; Kohlstedt, 1992). In this case, however, the large enhancement in strain rate occurs because diffusion through the melt-filled grain boundaries is orders of magnitude faster than diffusion through melt-free grain boundaries.

At present, it is premature to extrapolate the creep results on partially molten olivine aggregates from laboratory to geological conditions. Substantial uncertainties still exist in the dependence of viscosity on temperature, melt fraction, and water fugacity. Nonetheless, the present study and those of Hirth and Kohlstedt (1992a, 1992b) and Beeman and Kohlstedt (1993) demonstrate that the rheology of partially molten aggregates both in the diffusion creep regime and in the dislocation creep regime will depend sensitively on melt fraction and, hence, on the permeability of the deforming rock. If melt segregation is rapid enough that the amount of melt in a partially molten upwelling mantle rock remains less than about 1%, as suggested by geochemical measurements (Salters and Hart, 1989; Johnson *et al.*, 1990), then the effect of melt on viscosity will be small. However, if melt is prevented from migrating out of the rock mass such that several percent melt accumulates, then the viscosity may decrease substantially more rapidly with increasing melt fraction than predicted by models for deformation of two-phase aggregates (for diffusion creep, see for example Cooper *et al.* (1989); for dislocation creep, see for example Chen and Argon (1979)).

Conclusions

Under hydrous conditions, the melt in aggregates of olivine plus basalt is largely restricted to three-grain and four-grain junctions, which form an interconnected network. Melt does not wet most of the grain boundaries as a thin continuous film, but large pockets of melt do separate some grains from at least one of their neighboring grains.

In the diffusion creep regime at 1300°C, the presence of 8.6 vol% melt reduced the viscosity of an olivine–basalt rock by a factor of ~5 relative to that of a single-phase olivine aggregate of the same grain size. However, while the olivine aggregate was saturated with water, the olivine–basalt sample was undersaturated because the water partitioned strongly into the melt phase. The difference in viscosity would likely be much larger if both samples were deformed at the same water fugacity.

Acknowledgments

Support from the National Science Foundation through Grants EAR-8916438, OCE-9200471, and EAR-9220039 are gratefully acknowledged. The authors are indebted to Mervyn Paterson for the use of his high-pressure rock deformation laboratory and to Martha Daines, Greg Hirth, Shun Karato, Steve Mackwell, and Mark Zimmerman for stimulating discussions and insightful comments. The thoughtful suggestions of two anonymous reviewers helped improve this manuscript.

References

Bai, Q., and Kohlstedt, D. L. (1992). Substantial hydrogen solubility in olivine and implications for water storage in the mantle, *Nature* **357,** 672–674.

Bai, Q., Mackwell, S. J., and Kohlstedt, D. L. (1991). High-temperature creep of olivine single crystals. 1. Mechanical results for buffered samples, *J. Geophys. Res.* **96,** 2441–2463.

Beeman, M. L., and Kohlstedt, D. L. (1993). Deformation of fine-grained aggregates of olivine plus melt at high temperature and pressures, *J. Geophys. Res.* **98,** 6443–6452.

Bulau, J. R., Waff, H. S., and Tyburczy, J. A. (1979). Mechanical and thermodynamic constraints on fluid distribution in partial melts, *J. Geophys. Res.* **84,** 6102–6108.

Chen, I. W., and Argon, A. S. (1979). Steady state power-law creep in heterogeneous alloys with coarse microstructures, *Acta Metall.* **27,** 785–791.

Chopra, P. N. (1986). The plasticity of fine-grained olivine at high pressure and temperature, *in* "Mineral and Rock Deformation: Laboratory Studies. The Paterson Volume" (B. E. Hobbs and H. C. Heard, eds.), pp. 25–31, American Geophysical Union, Washington, DC.

Chopra, P. N., and Paterson, M. S. (1981). The experimental deformation of dunite, *Tectonophys.* **78,** 453–473.

Chopra, P. N., and Paterson, M. S. (1984). The role of water in the deformation of dunite, *J. Geophys. Res.* **89,** 7861–7876.

Coble, R. L. (1963). A model for boundary-diffusion controlled creep in polycrystalline materials, *J. Appl. Phys.* **34,** 1679–1682.

Cooper, R. F., and Kohlstedt, D. L. (1984a). Sintering of olivine and olivine–basalt aggregates, *Phys. Chem. Minerals* **11,** 5–16.

Cooper, R. F., and Kohlstedt, D. L. (1984b). Solution-precipitation enhanced creep of partially molten olivine–basalt aggregates during hot-pressing, *Tectonophys.* **107,** 207–233.

Cooper, R. F., and Kohlstedt, D. L. (1986). Rheology and structure of olivine–basalt partial melts, *J. Geophys. Res.* **91,** 9315–9323.

Cooper, R. F., Kohlstedt, D. L., and Chyung, K. (1989). Solution-precipitation enhanced creep in solid–liquid aggregates which display a non-zero dihedral angle, *Acta Metall.* **37,** 1759–1771.

Daines, M. J., and Richter, F. M. (1988). An experimental method for directly determining the interconnectivity of melt in a partially molten system, *Geophys. Res. Lett.* **15,** 1459–1462.

Dixon, J. E., Stolper, E., and Delaney, J. R. (1988). Infrared spectroscopic measurements of CO_2 and H_2O in Juan de Fuca Ridge basaltic glasses, *Earth Planet. Sci. Lett.* **90,** 87–104.

Exner, H. E. (1972). Analysis of grain and particle-size distributions in metallic materials, *Int. Metall. Rev.* **17,** 25–42.

Frost, H. J., and Ashby, M. F. (1982). "Deformation Mechanism Maps, The Plasticity and Creep of Metals and Ceramics," Pergamon Press, Oxford.

Hirth, J. G., and Kohlstedt, D. L. (1992a). The transition from dislocation creep to diffusion creep in partially molten olivine aggregates, *EOS Trans. Am. Geophys. Union* **73,** 529.

Hirth, J. G., and Kohlstedt, D. L. (1992b). The effect of melt on the strength of olivine aggregates deformed in the diffusion creep regime, *EOS Trans. Am. Geophys. Union* **73,** 517.

Hwang, C., and Chen, I. W. (1990). Effect of a liquid phase on superplasticity of 2-mol% Y_2O_3-stabilized tetragonal zirconia polycrystals, *J. Am. Ceram. Soc.* **73,** 1626–1632.

Johnson, K. T. M., Dick, H. J. B., and Shimizu, N. (1990). Melting in the oceanic upper mantle: An ion microprobe study of diopsides in abyssal peridotites, *J. Geophys. Res.* **95,** 2661–2678.

Karato, S. I. (1986). Does partial melting reduce the creep strength of the upper mantle? *Nature* **319,** 309–310.

Karato, S. I., Paterson, M. S., and FitzGerald, J. D. (1986). Rheology of synthetic olivine aggregates: Influence of grain size and water, *J. Geophys. Res.* **91,** 8151–8176.

Kohlstedt, D. L. (1990). Chemical analysis of grain boundaries in an olivine–basalt aggregate using high-resolution, analytical electron microscopy, *in* "The Brittle-Ductile Transition in Rocks: The Heard Volume" (A. G. Duba, W. B.

Durham, J. W. Handin, and W. F. Wang, eds.), pp. 211–218, American Geophysical Union, Washington, DC.

Kohlstedt, D. L. (1992). Structure, rheology and permeability of partially molten rocks at low melt fractions, *in* "RIDGE Volume on Melt Production and Migration" Geophys. Monogr. 71 (J. Phipps-Morgan, D. K. Blackman, and J. M. Sinton eds.), pp. 103–121, American Geophysical Union, Washington, DC.

Kohlstedt, D. L., and Chopra, P. N. (1987). High temperature apparatuses for rock deformation, *in* "Methods in Experimental Physics—Geophysics" (G. G. Sammis and T. L. Henyey, eds.), pp. 57–87, Academic Press, San Diego.

Mackwell, S. J., Kohlstedt, D. L., and Paterson, M. S. (1985). The role of water in the deformation of olivine single crystals, *J. Geophys. Res.* **90,** 11319–11333.

Michael, P. J. (1988). The concentration, behavior and storage of H_2O in the suboceanic upper mantle: Implications for mantle metasomatism, *Geochim. Cosmochim. Acta* **52,** 555–566.

Paterson, M. S. (1970). A high pressure, high temperature apparatus for rock deformation, *Int. J. Rock Mech. Min. Sci.* **7,** 517–526.

Paterson, M. S. (1977). Experience with an internally heated gas-medium apparatus to 500 MPa, *in* "Proceedings of the 2nd International Conference on High Pressure Engineering, Brighton, 1975," pp. 209–213, Institution of Mechanical Engineers, London.

Paterson, M. S. (1982). The determination of hydroxyl by infrared absorption in quartz, silicate glasses and similar minerals, *Bull. Mineral.* **105,** 20–25.

Paterson, M. S., Chopra, P. N., and Horwood, G. R. (1982). The jacketing of specimens in high-temperature, high-pressure rock-deformation experiments, *High Temp. High Press.* **14,** 315–318.

Riley, G. N., Jr. (1990). "Liquid Distribution and Transport in Silicate Liquid–Olivine Materials," Ph.D. Dissertation, Cornell Univ., Ithaca, NY.

Riley, G. N., Jr., and Kohlstedt, D. L. (1990). An experimental study of melt migration in an olivine–melt system, *in* "Magma Transport and Storage" (M. P. Ryan, ed.), pp. 77–86, Wiley, Chichester/Sussex, England.

Riley, G. N., Jr., and Kohlstedt, D. L. (1991). Kinetics of melt migration in upper mantle-type rocks, *Earth Planet. Sci. Lett.* **105,** 500–521.

Riley, G. N., Jr., Kohlstedt, D. L., and Richter, F. M. (1990). Melt migration in a silicate liquid–olivine system: An experimental test of compaction theory, *Geophys. Res. Lett.* **17,** 2101–2104.

Salters, V. J. M., and Hart, S. R. (1989). The hafnium paradox and the role of garnet in the source of mid-ocean-ridge basalts, *Nature* **342,** 420–422.

Vaughan, P. J., Kohlstedt, D. L., and Waff, H. S. (1982). Distribution of the glass phase in hot-pressed, olivine–basalt aggregates: An electron microscopy study, *Contrib. Mineral. Petrol.* **81,** 253–261.

Von Bargen, N., and Waff, H. S. (1988). Wetting of enstatite by basaltic melt at 1350°C and 1.0- to 2.5-GPa pressure, *J. Geophys. Res.* **93,** 1153–1158.

Waff, H. S., and Bulau, J. R. (1979). Equilibrium fluid distribution in an ultramafic partial melt under hydrostatic stress conditions, *J. Geophys. Res.* **84,** 6109–6114.

Waff, H. S., and Faul, U. H. (1992). Effects of crystalline anisotropy on fluid distribution in ultramafic partial melts, *J. Geophys. Res.* **97,** 9003–9014.

Chapter 4 The Generation and Migration of Partial Melt beneath Oceanic Spreading Centers

David W. Sparks and E. M. Parmentier

Overview

The narrowness of the zone of crustal emplacement at oceanic spreading centers indicates that melt formed in the mantle beneath a spreading center must be focused toward the spreading axis. Focusing can be accomplished by lateral flow in sloping decompacting boundary layers at the top of the melting region. These layers form by decompaction or dilation of the solid matrix, in response to an excess melt pressure that develops beneath an impermeable boundary. The relatively high melt fraction and permeability in the boundary layer allow melt to flow upslope to the spreading axis. The extraction efficiency of decompacting layers is determined by the temperature and upwelling distribution in the mantle.

Mantle flow and melting beneath an offset spreading center has been investigated using three-dimensional numerical experiments. The flow consists of the superposition of passive flow driven by plate-spreading and buoyant flow driven by thermal expansion and compositional density gradients due to melt extraction. Buoyant flow affects the distribution of melt production by increasing the along-axis variation in upwelling beneath the spreading center, and generating melting in the upwelling limbs of off-axis thermally driven convective rolls. Approximate paths for the extraction of melt along decompacting layers at the top of the melting region are calculated. These paths predict the division of the melting region into separate melt extraction regions for different spreading segments. Predicted along-axis distributions of crustal thickness, topography, and gravity have greater variation at slow spreading rates, in agreement with observations.

Notation

		Units
C_p	specific heat	$J \cdot kg^{-1} \cdot C^{-1}$
G	gravitational constant	$m^3 \cdot kg^{-1} \cdot s^{-2}$
L	latent heat of fusion	$J \cdot kg^{-1}$
M	magnitude of mass anomaly	$kg \cdot m^{-2}$
V	along-layer melt velocity	$m \cdot s^{-1}$
W	mantle upwelling velocity	$m \cdot s^{-1}$
X	degree of melting	dimensionless
X_0	maximum degree of melting in upwelling column	dimensionless
T	temperature	$°C$
g	gravitational acceleration	$m \cdot s^{-2}$
h	seafloor topography	m
h_c	oceanic crustal thickness	m
k	matrix permeability	m^2
k_0	constant in permeability relation, Eq. (4)	m^2
n	exponent in permeability relation, Eq. (4)	dimensionless
p	pressure	$kg \cdot m^{-1} \cdot s^{-2}$
t	time	s
$\mathbf{u}$	mantle velocity	$m \cdot s^{-1}$
$\mathbf{u}_f$	melt velocity	$m \cdot s^{-1}$
$\mathbf{u}_s$	solid matrix velocity	$m \cdot s^{-1}$
z	depth	m
Γ	melt production rate	s^{-1}
α	coefficient of thermal expansion	$°C^{-1}$
β	compositional density parameter, Eq. (17)	dimensionless
γ	constant in melting relationship, Eq. (8)	$°C^{-1}$
δ	decompaction length scale	m
δ_{fr}	freezing length scale	m
δ_c	compaction length scale	m
ζ	bulk viscosity of solid–melt aggregate	$Pa \cdot s$
θ	slope of solidus	$°C \cdot m^{-1}$
κ	thermal diffusivity	$m^2 \cdot s^{-1}$
μ_f	melt viscosity	$Pa \cdot s$
μ_s	mantle shear viscosity	$Pa \cdot s$

		Units
ξ	degree of melt depletion	dimensionless
ρ	mantle density	$kg \cdot m^{-3}$
ρ_c	oceanic crustal density	$kg \cdot m^{-3}$
ρ_f	melt density	$kg \cdot m^{-3}$
ρ_w	seawater density	$kg \cdot m^{-3}$
ρ_0	reference mantle density	$kg \cdot m^{-3}$
σ	wavenumber	m^{-1}
ϕ	melt fraction	dimensionless
ϕ_0	melt fraction in decompacting layer	dimensionless
ω	slope of decompacting layer	dimensionless

Introduction

Spreading centers are the largest regimes of melt generation in the Earth. Spreading is relatively continuous in time, and the chemistry of the source material as well as the mantle flow is relatively uniform compared with melting regimes in hotspot plumes and in continental systems. For these reasons, spreading centers are perhaps the simplest systems in which to model the generation and transport of magma. However, most of the available observational information about the nature of melt production and migration at spreading centers is indirect and is based on seismic and gravity surveys, and sonar imaging of the seafloor.

One of the fundamental observations of spreading centers is that over 90% of the oceanic crust is emplaced within a 1- to 2-km-wide neovolcanic zone (Macdonald, 1984). Either the upwelling which produces melt is extremely focused beneath a spreading center or the melt is generated in a tens to hundreds of kilometers wide region and then focused into the neovolcanic zone.

Several studies have investigated mantle flow beneath spreading centers driven by passive rifting and local buoyancy forces. Passive plate-driven flow consists of a broad upwelling region and 100- to 300-km-wide melt production region centered beneath the spreading axis (Reid and Jackson, 1981). The width of the melt production region and the amount of melt produced increases with increasing spreading rate. Passive upwelling and melting are relatively uniform along most of the length of spreading segments and are reduced beneath offsets (Forsyth and Wilson, 1984; Phipps Morgan and Forsyth, 1988; Shen and Forsyth, 1992).

The major sources of buoyancy in the mantle beneath spreading centers are thermal expansion, the presence of low-density melt in pore spaces, and compositional variations caused by the extraction of melt. During the partial melting of peridotite, Fe is preferentially partitioned into the melt phase. When the melt is extracted, the remaining residual mantle is depleted in iron and is compositionally less dense. The density change associated with 25% melt extraction is equivalent to the density change associated with 200°C thermal expansion (Oxburgh and Parmentier, 1977). Mantle that upwells beneath the axis experiences more melting than mantle that upwells off-axis. As this mantle is advected through the melting region, a gravitationally stable density stratification is developed.

The most complete studies of buoyant flow and melting have been two-dimensional. These studies have shown that thermal and compositional buoyancy enhance upwelling rates beneath a spreading center, but do not qualitatively change the form of flow beneath the axis from that of passive upwelling (Sotin and Parmentier, 1989; Scott and Stevenson, 1989; Cordery and Phipps Morgan, 1992). The buoyancy due to the presence of a significant fraction of melt in the mantle can cause the upwelling to become much more focused, due to positive feedback from increased rates of melt production. If the permeability of the mantle is low enough that a melt fraction of a few percent is retained in the mantle, then the melting region narrows to a region that is only 20–40 km wide (Rabinowicz *et al.*, 1984; Scott and Stevenson, 1989; Cordery and Phipps Morgan, 1992). If even higher melt fractions are retained ($\geq$15%) a large reduction in viscosity that can focus upwelling into a very narrow region, on the order of the neovolcanic zone (Buck and Su, 1989), may result. If melt generation were so strongly focused, no lateral melt migration would be required since all melting would occur essentially beneath the neovolcanic zone. However, the interconnectivity of melt at very small melt fractions (<1%, Daines and Richter, 1989), and trace element (Johnson *et al.*, 1990) and ophiolite studies (Ceuleneer, 1992), indicate that the amount of interstitial melt in the mantle is small.

Three-dimensional numerical studies of combined mantle flow and melting have only recently become computationally practical. Buoyant flow increases the amount of along-axis variation. The compositional buoyancy due to melt extraction can cause upwelling that is segmented along-axis into broad zones of enhanced upwelling and reduced zones of decreased upwelling, in which there is little or no melt generated (Parmentier and Phipps Morgan, 1990). Thermal buoyancy increases these along-axis variations, further enhancing and localizing upwelling, but does not significantly narrow the melting region (Sparks and Parmentier, 1993). The major effect of thermal buoyancy on melting is to produce localized regions of melt production far off-axis in the upwelling limbs of thermally driven convective rolls. Initial three-dimensional studies show that several percent of retained melt will produce narrow upwellings, which are segmented in the along-axis direction (Jha *et al.,* 1992). However, as in two-dimensional studies, the upwelling is still at least an order of magnitude wider than the neovolcanic zone. Some mechanism of focusing melt during migration in the mantle is thus required to produce the observed narrow region of crustal formation.

The two proposed mechanisms of melt migration are porous flow through small channels along grain boundaries (cf. Waff and Bulau, 1979; Watson, 1982; Cooper and Kohlstedt, 1986) and flow through a fracture network (cf. Shaw, 1980; Spence and Turcotte, 1985; Sleep, 1988; Ryan, 1988; Stevenson, 1989). Both of these mechanisms must operate in some regions of the mantle. Melt is generated at grain boundaries, so it must initially be transported through the pore spaces around grains. However, near the surface, the mantle is subsolidus. Since melt moving in pore spaces is in thermal equilibrium with the solid matrix, melt cannot exist in pore spaces where the temperature is below the solidus. Therefore, the melt must move through the lithosphere in fractures.

A proposed mechanism for lateral melt transport in the asthenosphere beneath spreading centers arose out of a study of the behavior of a deformable porous medium near an impermeable boundary (Sparks and Parmentier, 1991). Excess melt pressures near the boundary cause melt to collect in a thin layer beneath the impermeable mantle. We refer to the dilation of the matrix of grains to accept a higher melt fraction as "decompaction." Since this boundary slopes away from the axis near a spreading center, it provides a channel within which melt can migrate laterally toward the surface. The existence, position, and efficiency of this boundary at extracting melt will be determined by the temperature and flow field in the mantle.

In this paper we first review the decompacting boundary layer model of melt extraction. We then discuss three-dimensional numerical experiments on thermal and compositional convection and melt production. Thermal and compositional buoyancy enhances the along-axis variations in upwelling rate beneath the axis, and thermal buoyancy drives off-axis convective rolls that align with the spreading direction. Convective rolls can have a large effect on the shape of the melting region, with melting in the upwelling limb extending for well over 100 km from the axis. Melt extraction in decompacting layers will be least efficient in the upwelling limbs, where there is little melting and the melting region is relatively flat. If extracted, this melt will further contribute to along-axis variations in melt production. We then discuss the implied along-axis variations in crustal thickness, and the resulting effects on topography and gravity.

Melt Migration by Porous Flow

The Forces That Drive Porous Flow

The conceptual framework of magma migration through a porous mantle was developed in several early studies (Frank, 1968; Sleep, 1974; Turcotte and Ahern, 1978). The equations that include the role of stresses due to a deformable mantle matrix have been derived and discussed by several authors (McKenzie, 1984; Richter and McKenzie, 1984; Ribe, 1985; Scott and Stevenson, 1986; Fowler, 1990a, 1990b; Spiegelman, 1993a, 1993b):

$$\frac{\partial \phi}{\partial t} + \nabla[\phi \mathbf{u}_f] = \Gamma \quad (1)$$

$$\frac{\partial \phi}{\partial t} - \nabla[(1 - \phi)\mathbf{u}_s] = \Gamma \quad (2)$$

$$(1 - \phi)(\rho - \rho_f)\mathbf{g} = \frac{\mu_f \phi}{k}(\mathbf{u}_f - \mathbf{u}_s) - \mu_s \nabla \times \nabla \times \mathbf{u}_s - \left(\zeta + \frac{4\mu_s}{3}\right)\nabla(\nabla \cdot \mathbf{u}_s) \tag{3}$$

$$k = k_0 \phi^n, \tag{4}$$

where ϕ is melt fraction, ζ is the bulk viscosity of the solid–melt aggregate, Γ is the rate of melt production, $\mathbf{g}$ is the acceleration of gravity, k is the permeability of the solid matrix, $\mathbf{u}$ is a velocity field, ρ is density, μ is shear viscosity, and the subscripts s and f refer to the solid matrix and melt phase, respectively. In this formulation, we make the Boussinesq approximation, neglecting density differences except in the buoyancy term in Eq. (3). The mass balances for the melt and solid phases are given by Eqs. (1) and (2). The force balance, Eq. (3), is expressed in terms of the solid and melt velocities: vertical melt flow is driven by the buoyancy of the melt (left-hand side) and opposed by the viscous interaction of the two phases (first term on right-hand side). These two terms make up Darcy's law for flow in a rigid porous medium. The other terms on the right-hand side of Eq. (3) arise from the viscous deformation of the solid. Equation (4) relates permeability to melt fraction, with n usually taken to be between 2 and 3 (Cheadle, 1989).

The second term on the right-hand side of Eq. (3) describes the effect of stresses arising from the incompressible shear deformation of the matrix. The divergent mantle flow beneath a spreading center creates pressure gradients that tend to direct melt toward the spreading axis (Spiegelman and McKenzie, 1987). However strain rates beneath a spreading center are small enough that large mantle viscosities ($\geq 10^{21}$ Pa · s) are required to create stresses sufficient to cause significant melt focusing. Unless some other mechanism allows lateral melt flow, such as fractures controlled by the local stress distribution (Sleep, 1984) or an anisotropic permeability created by accumulated strain (Phipps Morgan, 1987), porous flow in the melting region is essentially vertical.

The last term in Eq. (3) describes the stresses that arise from volume changes in the solid phase. If melt is extracted from the mantle, then the solid matrix must compact. However, the magnitudes of the stresses that arise from compaction are dependent on the rate at which melt is generated. Beneath a spreading center, about 20% of the mantle is melted during about 60 km of upwelling (cf. Phipps Morgan, 1987). The stresses associated with this slow compaction are small compared to the buoyancy of the melt (Ahern and Turcotte, 1979) so the resistance to the compaction of the matrix is negligible. Compaction stresses are important over a length scale defined by

$$\delta_c = \sqrt{\frac{(\zeta + 4\mu_s/3)k}{\mu_f}} \tag{5}$$

(McKenzie, 1984). In the mantle, this length is typically on the order of tens of meters to 1 km. Ribe (1985) showed that the stresses which arise from compaction are important only if a large amount of melting occurs over a length scale comparable to δ_c. Therefore, porous flow in the melting region beneath spreading centers is closely approximated by Darcy's law, in which melt migrates vertically under the force of gravity. Vertical porous flow is rapid enough so that only small melt fractions can be maintained (Ahern and Turcotte, 1979).

Compaction stresses also allow the propagation of solitary waves in a porous medium (Scott and Stevenson, 1984; Richter and McKenzie, 1984; Barcilon and Richter, 1986). A local maximum in the melt fraction created by a perturbation in melting will propagate through the mantle as a wave in the melt fraction distribution and shed smaller waves in its wake. Solitary waves moving through a slowly melting matrix, such as upwelling mantle, tend to decrease in amplitude (Scott and Stevenson, 1986; Fowler, 1990b), so that small perturbations in the melting rate may not produce a significant time dependence in the amount of magma reaching the top of the melting region. However, solitary waves are amplified by their passage through a region where melt freezes (Spiegelman, 1993c). In the next section, we show that the stresses due to the deformation of the matrix are important near a freezing boundary in the mantle.

Decompacting Boundary Layers

As the temperature of the upwelling mantle drops below its solidus, melt can no longer exist in pore spaces. Solidification, or freezing, of melt reduces the permeability. The effect of this permeability reduction was explored by Sparks and Parmentier (1991). A narrow layer of increased melt fraction forms, bounded by impermeable mantle above and a region of small melt fractions and vertical percolation below.

This result is most easily understood by considering a one-dimensional problem in which freezing is collapsed onto an interface (Fig. 1). A freezing interface would exist if the melting of peridotite were a univariant phase change. All melt present in the pore spaces will freeze as it is carried across the interface, so the permeability vanishes and the melt and solid velocities are equal at the interface. Some distance below the interface, melt is moving faster than the solid, so as the interface is approached, the melt velocity must decrease and the solid velocity increase. These velocity gradients result in a dilation or "decompaction" of the solid matrix. The melt fraction increases from the small value that is supported by Darcy flow in the melting region to the maximum degree of melting at the interface (Fig. 1).

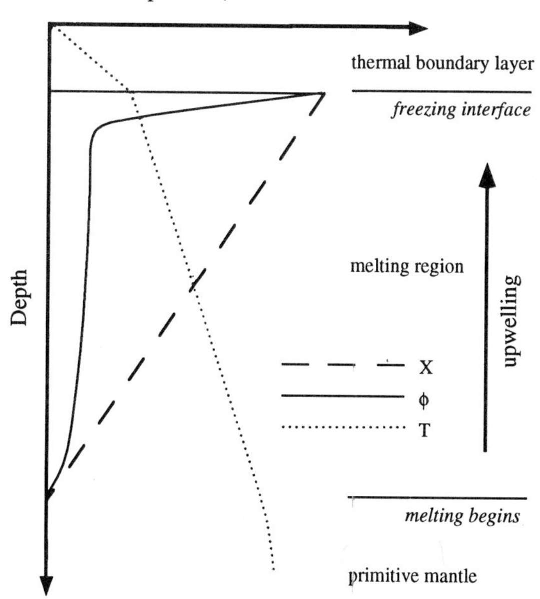

Figure 1 Sketch of the temperature (T), degree of melting (X), and melt fraction (ϕ) in a one-dimensional column of mantle upwelling at constant velocity. The melting process is approximated as univariant so that the temperature remains on the solidus until conduction becomes dominant near the surface. The rate of melting is constant, so X decreases linearly with depth.

Equation (3) can be derived in terms of separate pressures in the solid and melt phases, with the differential pressure proportional to the compaction rate and the bulk viscosity of the system (Scott and Stevenson, 1986). This differential pressure drives compaction and decompaction. Within the melting region, where compaction and extraction occur, the pressure in the melt is less than the pressure in the solid. Near the freezing boundary, the pressure in the melt becomes greater than the pressure in the solid, leading to decompaction of the matrix. The existence of overpressured boundary layers at the top of partially molten regions was suggested by Fowler (1985) as a mechanism for the initiation of lithosphere fracture and subsequent magma eruption. However, if excess fluid pressures always lead to the fracture of the overlying impermeable medium, then the neovolcanic zone at spreading centers would tend to be as wide as the melting region.

The distance over which decompaction occurs is controlled by a balance of the buoyancy forces driving the melt upward, and the viscous forces resisting the deformation of the solid matrix. If the matrix is very strong, it opposes rapid deformation, and the increase in melt fraction is spread over a larger region. The thickness of the layer of increased melt fraction is described by the decompaction length scale, δ,

$$\delta = \sqrt{\frac{(\zeta + 4\mu_f/3)W X_0}{(\rho - \rho_f)g}}, \tag{6}$$

where W is the upwelling velocity of the melt–matrix system, and X_0 is the maximum degree of melting beneath the layer. This length scale is determined by a balance between buoyancy and the compaction stresses and is identical in form to the "reduced compaction length" derived for instantaneous melting at an interface (Ribe, 1985).

Since the melting temperature of peridotite is a function of the degree of melting, due to the changing Fe/Mg ratio (cf. Hess, 1989), freezing in the mantle does not occur on an interface, but is spread over a distance related to the balance among the advection of heat with the solid mantle, conduction to the surface, and the release of latent heat during freezing. We can estimate the characteristic length scale of freezing from the energy balance equation for a supersolidus region. For one-dimensional upwelling at constant velocity, W,

$$W\frac{dT}{dz} = \kappa\frac{d^2T}{dz^2} - \frac{L}{C_p}\gamma\left(\frac{dT}{dz} - \theta\right), \quad (7)$$

where θ is the slope of the mantle solidus, and γ relates the stable melt fraction, X, to the temperature above the solidus:

$$X = \gamma(T - T_{\text{solidus}}). \quad (8)$$

The solution to (7) has the form

$$T(z) = C_1 \exp\left[\frac{W}{\kappa}\left(1 + \frac{L\gamma}{C_p}\right)z\right] + \frac{W}{\kappa}\frac{L\gamma}{C_p}\theta z + C_2. \quad (9)$$

The integration constants, C_1 and C_2, must be determined by matching conditions on the heat flux at the top and bottom of the upwelling column (Sparks and Parmentier, 1991); however, the characteristic length scale of freezing can be determined by inspection. In the melting region the second term dominates, and the temperature increases linearly with depth, with a slope slightly modified from that of the undepleted solidus. In the freezing region, the first term dominates, and the temperature decays exponentially toward the surface. The freezing length scale, δ_{fr}, is

$$\delta_{\text{fr}} = \frac{\kappa}{W}\left(\frac{C_p}{C_p + L\gamma}\right). \quad (10)$$

In the univariant melting problem described above, the freezing length is zero. If δ_{fr} is large compared with the compaction length scale, δ, then melt is frozen as it approaches the boundary layer, and matrix deformation stresses are not important. This is analogous to the behavior in the melting region, where compaction occurs, but the stresses generated are small because melt generation is spread over many kilometers. When the two length scales are equal, then the viscous effects of the impermeable boundary are felt before much melt has frozen, and decompaction occurs. For an upwelling rate of 3 cm/yr, δ_{fr} is about 1 km. Therefore, under most conditions of decompression melting in the mantle, some degree of pooling of melt in the decompacting layers is likely to occur.

Decompacting layers are important for melt extraction at spreading centers because they form along the top of the melting region. Since conductive cooling penetrates progressively further with distance from the axis of a spreading center, the top of the melting region slopes away from the axis. This is true for all models of upwelling, whether passive or buoyant. Since the melt fraction in the decompacting layers is relatively high, the permeability is also high. The decompaction stresses act perpendicular to the layer boundary, and partially counteract the vertical buoyancy force. The component of the buoyancy force that acts in the direction of the layer drives melt laterally (Fig. 2), and melt flows upslope along the layer towards the spreading axis.

The flux of melt in the layer is determined by a balance among the lateral porous flow, proportional to the local slope, the melt supply to the layer from the melting region, and freezing within the layer. Approximate, two-dimensional steady-state solutions for the behavior of the layer show that a significant fraction of melt formed over a wide melting region beneath a spreading center can be delivered to the spreading axis by porous flow within the layer (Sparks and Parmentier, 1991). Lateral melt flow in the layer and subsequent extraction at the spreading axis decreases the melt fraction in the layer to a few percent. The fraction of melt extracted decreases with increasing distance from the spreading axis, as the upwelling rate and total degree of melting decrease.

The formation of decompacting boundary layers has also been observed in highly resolved two-dimensional porous flow calculations (Spiegelman, 1993c). An imposed sloping freezing boundary causes a layer of relatively high melt fraction to form and direct melt flow along the boundary. Less prominent layers of increased melt fraction propagate back from the freezing boundary. One-dimensional analysis also indicates that the melt fraction may increase toward the upper boundary of a melting region in an oscillatory fashion (Fowler, 1990b, Appendix A).

In the one-dimensional upwelling problem described above, matrix shear viscosity and the bulk viscosity of the system appear only in combination. This combination of viscosities was varied over a range of values, but was held to be constant throughout the column. Viscosity is a complex function of temperature, pressure, composition, and melt fraction. Probably the most important of

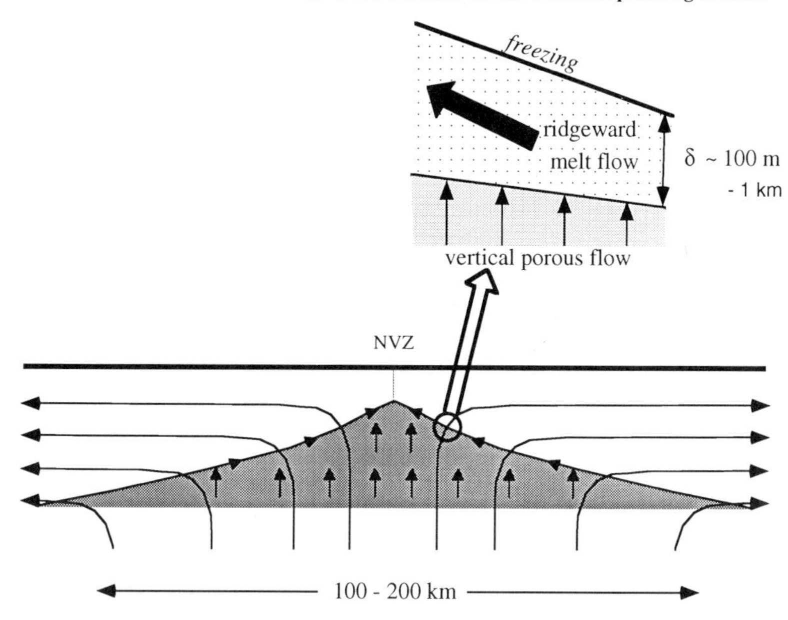

Figure 2 Schematic cross section of a spreading center with a broad region of melt production (stippled). The solid lines are mantle streamlines. The arrows show the direction of melt flow. The inset is a blow up of a melt-rich layer at the top of the melting region, in which melt flows laterally toward the ridge.

these parameters to the behavior near a freezing boundary is melt fraction, since the others vary over relatively long length scales with respect to the compaction length. A reduction in the matrix viscosity within the layer will result in a reduction in the layer thickness (Eq. 6), decreasing the compaction length and the along-layer flux. However, recent measurements of the effect of melt fraction on viscosity suggest that even for several percent melt, the reduction of viscosity is only about an order of magnitude (Kohlstedt and Hirth, 1992). Nonlinear effects associated with variable viscosity could lead to interesting and unpredicted behavior, but these effects still need to be investigated.

The stability of decompacting boundary layers in the presence of a realistic distribution of melting and freezing in two and three dimensions is a fundamental problem that has not been addressed to date. If large local melt fractions develop in parts of the layer, melt could segregate into veins. These veins need to form a connected network in order to focus melt to the axis, since isolated veins will either be carried into the subsolidus mantle, where they will freeze, or propagate upward through the lithosphere and erupt off-axis. Vein formation would have an important effect on basalt composition, since melt segregated into veins is more likely to be out of chemical equilibrium than the melt distributed in pore spaces (Spiegelman and Kenyon, 1992). More work is needed on the properties and physics of partially molten mantle to assess vein formation. In the next section we deal with the long-term results of melting and melt extraction, so we assume the process of melt extraction is steady state, hoping to capture at least the time-averaged behavior.

Effect of Buoyant Mantle Upwelling on Melt Generation and Migration

Three-Dimensional Buoyant Flow beneath an Offset Spreading Center

Numerical Formulation

Three-dimensional numerical experiments were conducted to examine the patterns of buoyant upwelling in the mantle beneath an offset spreading center, and its effect on melt migration. The geometry of the spreading center was taken to be periodic in the along-axis direction, so the region examined consisted of two spreading segments, separated by a transform fault (Fig. 3).

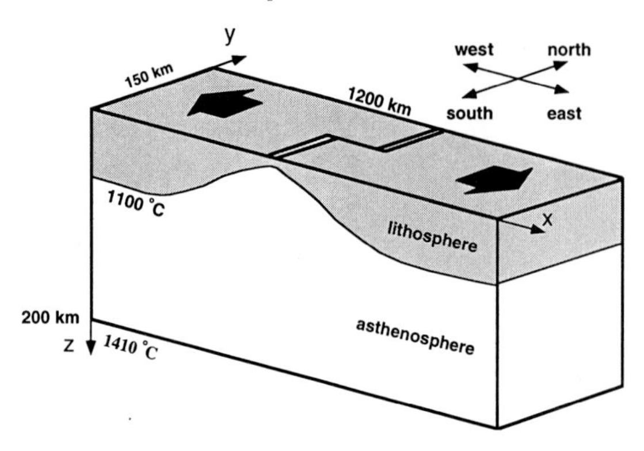

Figure 3 Schematic representation of the region of mantle in the three-dimensional numerical convection experiments. The shaded area is the rigid lithosphere, defined by the 1100°C isotherm. Convection is confined to the asthenosphere (unshaded region). The position of the plate boundary is shown in each plane by double lines for ridge segments and a single line for the transform. For convenience in referring to the different parts of the region in later figures, a directional system is defined by the arrows at the upper right.

This geometry is symmetric about the middle of the spreading segments, so that the calculation need only consider one-half of each segment.

The flow consists of a passive component and a buoyant component. The passive component is driven by the two plates spreading at an imposed velocity along the top of the region. This flow drives upwelling of deep mantle into the bottom of the region, and flow out the ends of the region.

Buoyant flow is driven by density gradients due to thermal expansion and the compositional changes due to melt extraction. Melt is assumed to be extracted rapidly from the mantle, so that the melt fraction in the melting region is very small. Melt-rich, decompacting layers on the order of 1 km in thickness are too small to have an effect on the overall density. Therefore, the melt fraction is neglected in the calculation of buoyant flow. The extraction of melt from the mantle will generate a dilatational component of flow, but this component is small when the melt production region is broad, and it is thus neglected.

The governing equations for temperature, T, the degree of melt depletion, ξ, the rate of melt production, Γ, and the mantle buoyant flow velocities, $\mathbf{u}$, are

$$\frac{\partial T}{\partial t} + \mathbf{u} \cdot \nabla T = \kappa \nabla^2 T - \left(\frac{L}{C_p}\right)\Gamma \quad (11)$$

$$\frac{\partial \xi}{\partial t} + \mathbf{u} \cdot \nabla \xi = \Gamma \quad (12)$$

$$\Gamma = \frac{D}{Dt}\left[\frac{(T - T_{\text{solidus}})}{600°\text{C}}\right] \quad (13)$$

$$T_{\text{solidus}} = 1100\ (°\text{C}) + 4.0\ z\ (\text{km}) \quad (14)$$

$$\nabla \cdot \mathbf{u} = 0 \quad (15)$$

$$\mu_s \nabla^2 \mathbf{u} - \nabla p - \rho \mathbf{g} = 0 \quad (16)$$

$$\rho(T, \xi) = \rho_0(1 - \alpha T - \beta\xi). \quad (17)$$

Here, t is time, z is depth in kilometers, p is pressure, κ is thermal diffusivity, L is the latent heat of melting, C_p is specific heat, ρ is mantle density, η is mantle viscosity, α is the coefficient of thermal expansion, and β describes the decrease in compositional density due to the extraction of iron-rich melt. The values of the parameters used in these experiments are given in Table 1.

The viscosity structure of the mantle is an important factor in determining the pattern of buoyant flow. We approximate the temperature and pressure dependence of viscosity by confining buoyant flow to a uniform viscosity asthenosphere bounded by no-slip boundaries on the top and bottom (see Fig. 3). The top boundary is prescribed by an isotherm (chosen to be 1100°C) representing the bottom of the lithosphere, and the bottom of the asthenosphere is prescribed at a depth of 200 km. The vertical boundaries are symmetry planes for the buoyant flow. Buoyant flow is forced to be two-dimensional at the vertical outflow boundaries by prescribing the temperature field to be independent of distance from the axis near the boundaries. These boundaries are placed far from the spreading center to minimize their effects on flow near the melt production region.

The region considered in the experiments is 200 km deep, 150 km in the along-axis direction, and 1200 km in the spreading direction, with the plate boundary centered on the top of the region. The spreading segments range in length from 75 to 225 km, and the transform offset is 75 km. The temperature is fixed at 0°C at the surface and 1410°C at the bottom of the asthenosphere, the latter value chosen to produce approximately 6 km of crust. Finite difference approximations were used to solve (11)–(17). The buoyant flow equations (15)–(16) were cast in a streamfunction–vorticity formulation and solved using a multigrid

Table 1
Physical Parameters and Values Applicable to the Mantle

Physical parameters	Typical mantle values	Description
L	$6 \times 10^5\ \mathrm{J \cdot kg^{-1}}$	Latent heat of fusion
C_P	$1000\ \mathrm{J \cdot kg^{-1} \cdot {}^\circ C^{-1}}$	Specific heat
κ	$10^{-6}\ \mathrm{m^2 \cdot s^{-1}}$	Thermal diffusivity
k_0	10^{-7}–$10^{-9}\ \mathrm{m^2}$	Constant in permeability relation, Eq. (4)
n	2–3	Exponent in permeability relation, Eq. (4)
ρ	$3300\ \mathrm{kg \cdot m^{-3}}$	Mantle density
ρ_f	$2800\ \mathrm{kg \cdot m^{-3}}$	Melt density
α	$3 \times 10^{-5}\ \mathrm{{}^\circ C^{-1}}$	Coefficient of thermal expansion
β	0.024	Compositional density parameter in Eq. (17)
ζ	10^{19}–$10^{21}\ \mathrm{Pa \cdot s}$	Bulk viscosity of solid–melt aggregate
μ_s	10^{19}–$10^{21}\ \mathrm{Pa \cdot s}$	Mantle shear viscosity
μ_f	0.1–$10\ \mathrm{Pa \cdot s}$	Melt viscosity

iterative Poisson solver. The numerical methods are discussed in more detail in Sparks and Parmentier (1993) and Sparks *et al.* (1993). The multigrid buoyant flow solver is described in Sotin *et al.* (1992).

Structure of Three-Dimensional Flow

Buoyant flow has two effects on the overall flow pattern: the enhancement and localization of upwelling beneath the centers of spreading segments, and the creation of thermal boundary layer instabilities that grow into axis-perpendicular convective rolls (Figs. 4–6), similar to the flow structures that develop beneath a non-offset spreading center (Sparks and Parmentier, 1993). The localized upwelling is driven by temperature gradients across the transform. At fast spreading rates, upwelling beneath the axis is relatively uniform along spreading segments, except near transforms, where the upwelling is weaker and conductive cooling penetrates deeper. At slow spreading rates, buoyant flow is relatively stronger than the passive flow, and causes larger along-axis variations in upwelling rate beneath the entire length of spreading segments.

The axis-perpendicular rolls form in response to along-axis temperature gradients across the fracture zone. In most cases, the upwelling limb of the roll is centered beneath the nearer spreading segment, so that rolls of opposing vorticity are present on either side of the spreading center (Fig. 6). These rolls are thermally driven and advect compositionally less dense, depleted mantle down into the asthenosphere. There is also thermally driven axis-parallel circulation (Fig. 5). This circulation is not continuous along-axis, but is most prominent in the downwelling limbs of axis-perpendicular rolls.

The distance from the spreading axis at which axis-perpendicular rolls (Fig. 6) are well developed is a function of spreading rate and mantle viscosity (Sparks and Parmentier, 1993). Thermal rolls form when the thermal boundary layer has penetrated through the layer of compositionally less dense depleted mantle. Therefore, for a given viscosity, thermal rolls form closer to the spreading axis at slow spreading rates than at fast spreading rates. At lower mantle viscosities, the strength of buoyant flow is increased relative to passive flow, and the rolls are more vigorous and form closer to the axis.

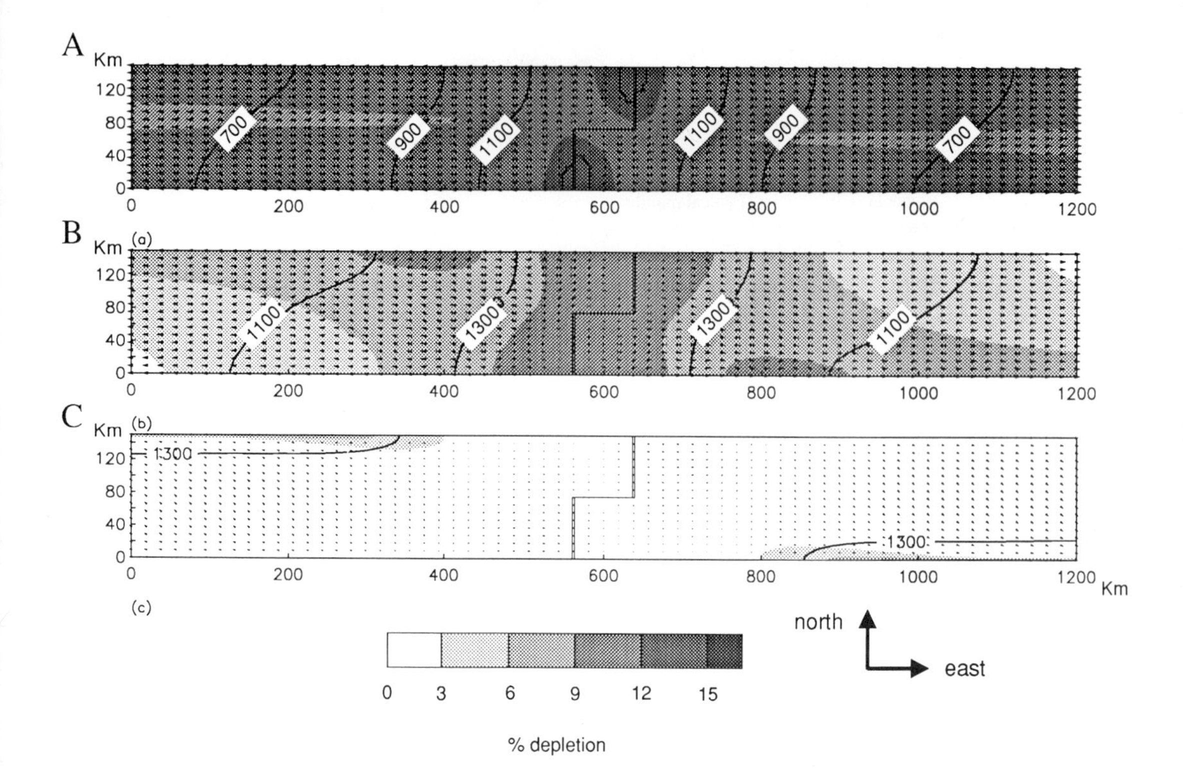

Figure 4 Horizontal cross sections of a numerical convection experiment at depths of (A) 25 km, (B) 43 km, and (C) 100 km. The shading represents the degree of depletion of the mantle, in 3% increments, with darker shades more depleted. Temperature isotherms are plotted at 200°C intervals. The position of the plate boundary is shown in each plane by double lines for ridge segments and a single line for the transform. The spreading half-rate is 2.7 cm/yr, and the mantle viscosity is 5×10^{19} Pa · s. In the two shallower cross sections (A–B), the isotherms reflect cooling with increasing distance from the nearest spreading segment. In the deepest section (C), linear downwelling regions carry cooler, more depleted material to depth.

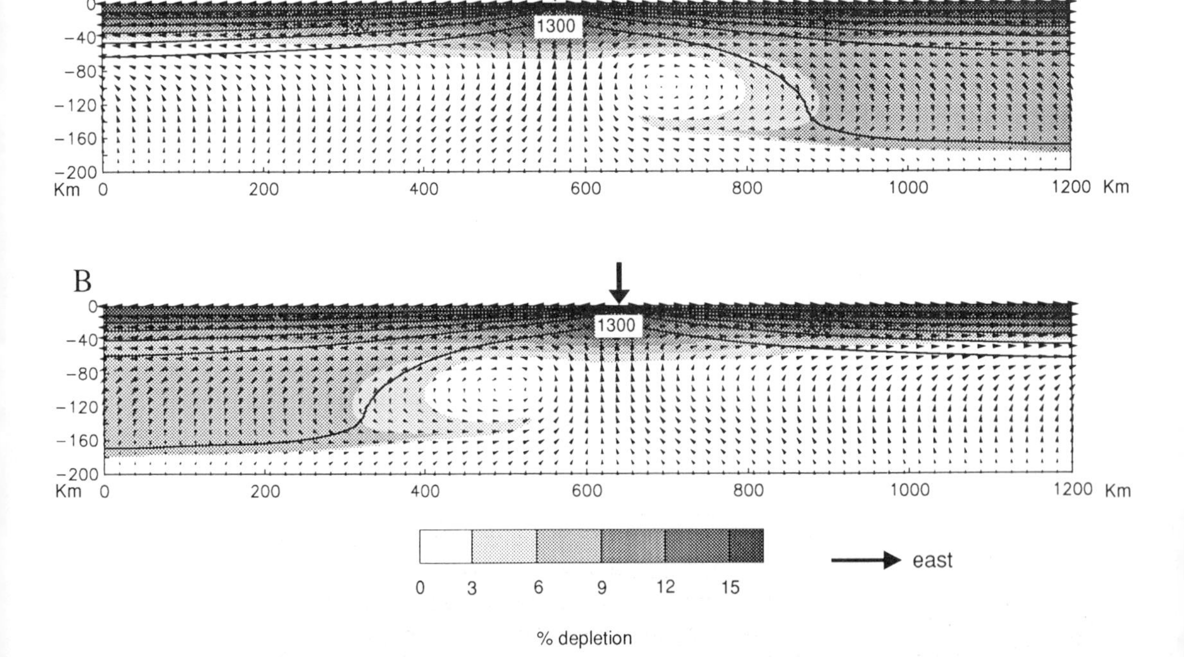

Figure 5 Axis-perpendicular vertical cross sections of the numerical experiment shown in Fig. 4, along the (A) southern and (B) northern edges of the region. The shading and contour interval is as in Fig. 4. The position of the spreading axis is shown by an arrow along the top of each figure. Each section cuts through two convective rolls, with an upwelling on one side of the axis and a downwelling on the other side. The downwellings are evident from the downwarping of the isotherms and depletion contours.

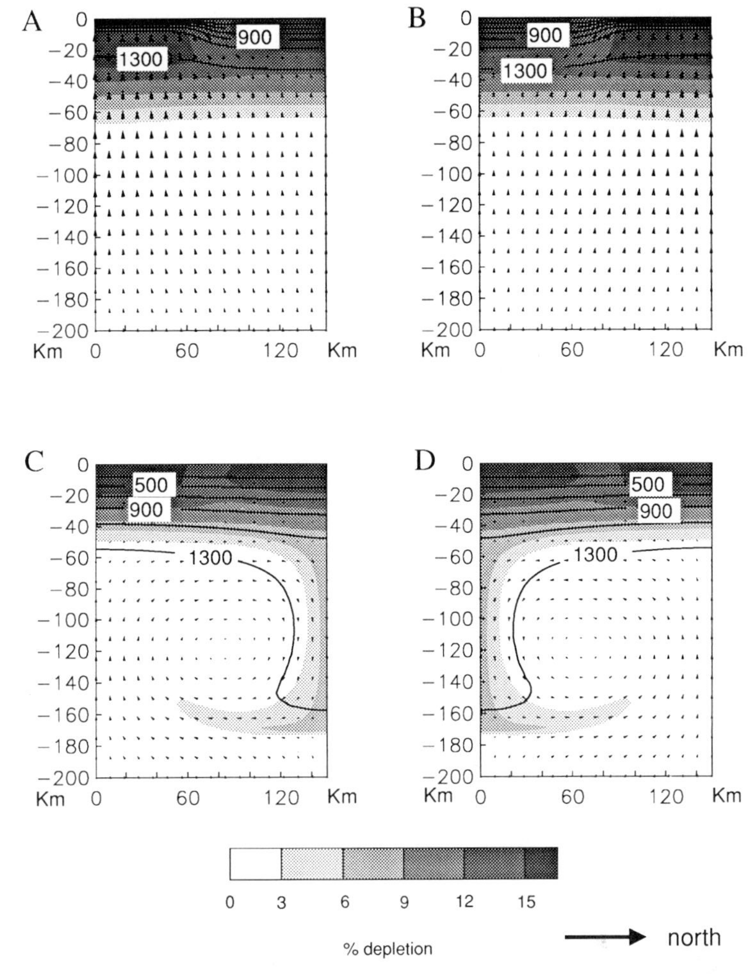

Figure 6 Axis-parallel vertical cross sections of the experiment shown in Figs. 4 and 5: (A) beneath the southern spreading segment, (B) beneath northern spreading segment, (C) 250 km to the west of (A), and (D) 250 km to the east of (B). The shading and contour intervals are as in Fig. 4. The cross sections beneath spreading axes (A–B) show no sign of the convective rolls which form off-axis (C–D).

Melt Migration beneath Spreading Centers

Since decompacting layers form where freezing occurs, the position and slope of the layers are controlled by the temperature distribution. The upwelling and temperature distributions in the numerical experiments described above determine the distribution of melting. The freezing boundary layer, where decompaction occurs, lies along the top of the melting region, so the position of the layer and the direction of melt migration can be estimated from the shape of the melting region.

A description of the shape of the melting region and an approximate model for porous flow within the layer can be combined to take advantage of the very different length scales of melt generation and decompacting layer thickness (Sparks and Parmentier, 1991). This approach was used to estimate the efficiency of melt extraction to the spreading axis in two dimensions, but has not yet been extended to three dimensions. However, we can make simple predictions about the direction and strength of lateral melt migration and the resulting distribution and composition of basaltic crust that may be formed at the spreading axis.

The melting region in these experiments is essentially triangular in cross section; i.e., melting begins at a roughly uniform depth and extends to shallower levels beneath the axis than it does off-axis (Fig. 7). The progressive penetration of con-

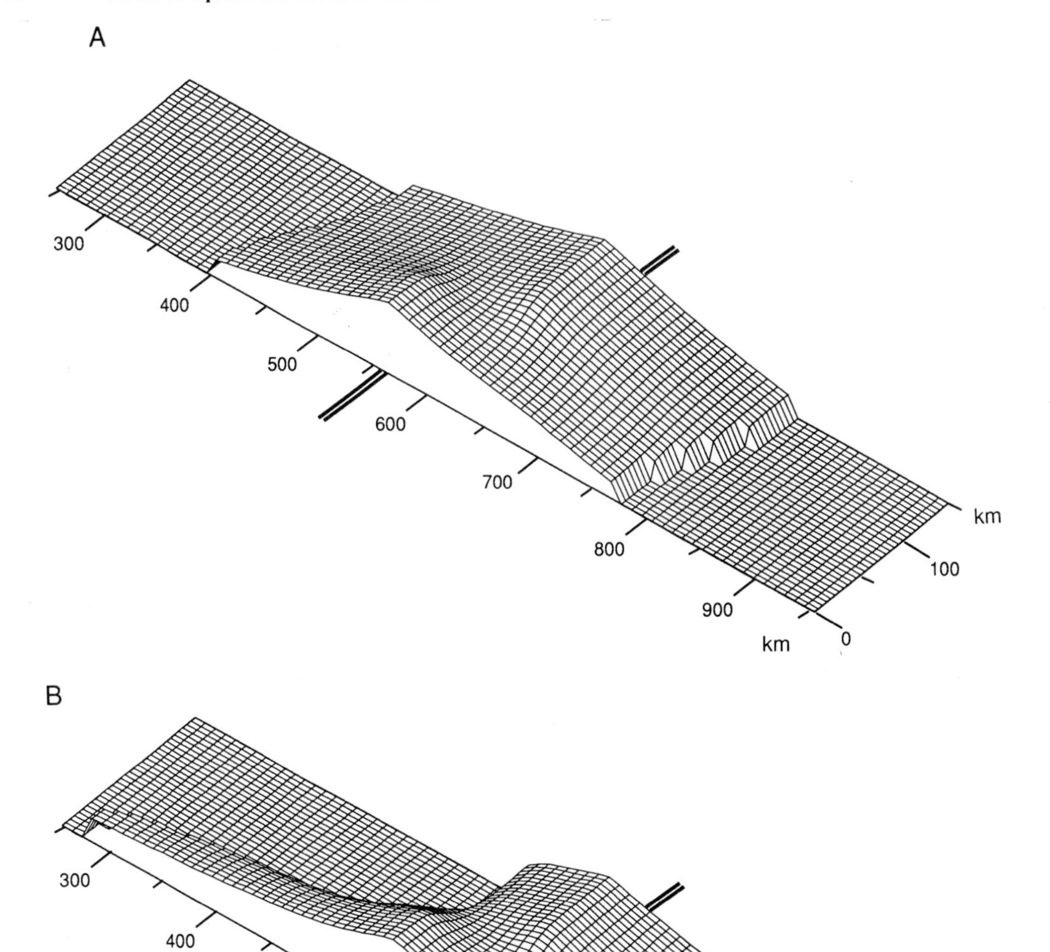

Figure 7 The three-dimensional shape of the top of the melting region in two typical numerical experiments. The orientation of the region is as in Fig. 3. Only part of the 1200-km-long region is shown here, to focus on the region beneath the spreading axis. Where there is no melting the surface drops to a plane at a depth of 120 km. The half-spreading rate is 2.7 cm/yr, and mantle viscosity is (A) 2×10^{20} Pa · s and (B) 5×10^{19} Pa · s. The formation of axis-perpendicular convective rolls in (B) causes significant along-axis variation in the width of the melting region. Melting occurs to a great distance off-axis within the linear upwelling limbs of the convective rolls, while off-axis melting vanishes where downwellings occur.

ductive cooling with increasing distance from the axis causes melting to stop at deeper levels off-axis. Thermal and compositional buoyancies may reduce the across-axis width of the melt region to some extent, but do not change its fundamental shape (cf. Sotin and Parmentier, 1989). If a significant melt fraction (>5%) is retained in the matrix, then the upwelling is focused, and the melting region is not very wide, even at its base (Scott and Stevenson, 1989; Buck and Su, 1989; Jha *et al.*, 1992).

In the case of purely passive spreading (Fig. 7A), at an intermediate spreading rate of 2.7 cm/yr, the melting region extends to nearly 200 km from the axis. This distance decreases slightly near a transform, but the cross section of the melting region remains relatively constant along-axis. When there is strong buoyant flow (Fig. 7B), so that convective rolls form near the melting region, there are large along-axis variations in the width of the melting region. Near the downwelling limb of a roll, the melting region abruptly pinches out relatively close to the axis. In the upwelling limb, the melting region is extended in the spreading direction, with a narrow region of deep melting extending to over 300 km from the axis.

To determine the strength and direction of lateral melt migration in a decompacting layer at the top of the melting region, we examine more closely the expected shape of the layer. Plots of the depth to the top of the melting region show that the presence of a transform offset can have a large effect on the shape of the layer (Fig. 8). The arrows on Fig. 8A show the direction and magnitude of the local slope of a decompacting layer in a purely passive upwelling. The predicted direction is roughly toward the nearest spreading axis, but there is a saddle point in the layer beneath the center of the transform. A line that passes through that saddle point divides the melting region into two parts which provide melt to each spreading segment. If migration in the melting region is vertical, and lateral migration occurs only along the top of the melting region, then melt formed on one side of this melting divide cannot be extracted to the segment on the other side. When strong convective rolls form, melting occurs along the upwelling

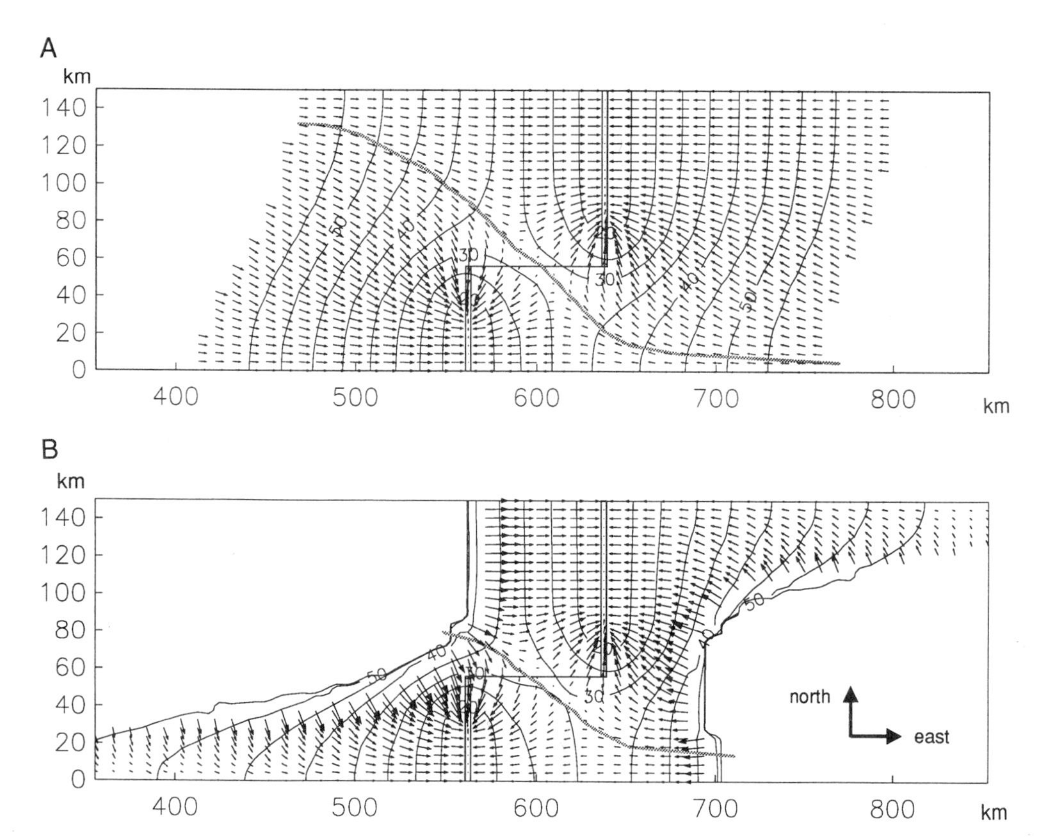

Figure 8 The depth to the top of the melting region, contoured in 5-km intervals. Arrows show the local slope of the surface. The thick gray line divides the area into two melt extraction regions. Melt formed in one region can be extracted only at the spreading axis in the same region. The two cases are for the same experiments shown in Fig. 7, with mantle viscosities of (A) 2×10^{20} Pa · s and (B) 5×10^{19} Pa · s.

limb of the roll (Fig. 8B), but the position of the melting divide is relatively unchanged.

Along-axis variations in melt production and extraction may produce observable variations in the distribution and composition of basaltic crust. To make predictions about the crustal distribution from these numerical experiments, simple assumptions about melt migration are made. Melt must migrate through the subsolidus lithosphere by fracture. Since the neovolcanic zone is very narrow, we assume that fractures are vertical and confined to the plane of the spreading axis. Therefore melt migrates vertically within the melting region into the decompacting layer. Within the layer, melt migrates laterally toward the plane of a spreading axis, where it is extracted from the mantle to form crust. The position of a melting column relative to the melting divide determines at which segment the melt will be extracted. The along-axis distribution of melt production is calculated by summing the melt generation in axis-perpendicular planes along each segment.

Implications for Crustal Thickness and Gravity

Thickness and Composition of the Oceanic Crust

The effect of a transform offset on melt production is shown in Fig. 9, where melt production is expressed as thickness of basaltic crust (melt production rate divided by the spreading rate). The overlaps in the curves are the result of extracting melt as it reaches the plane of a spreading center. The position of the melting divide is chosen by inspection of plots like Fig. 8. The resulting small discretization inaccuracy in the region of overlap, where the amount of melt generated is small, does not affect the results.

At high mantle viscosities, when the flow is

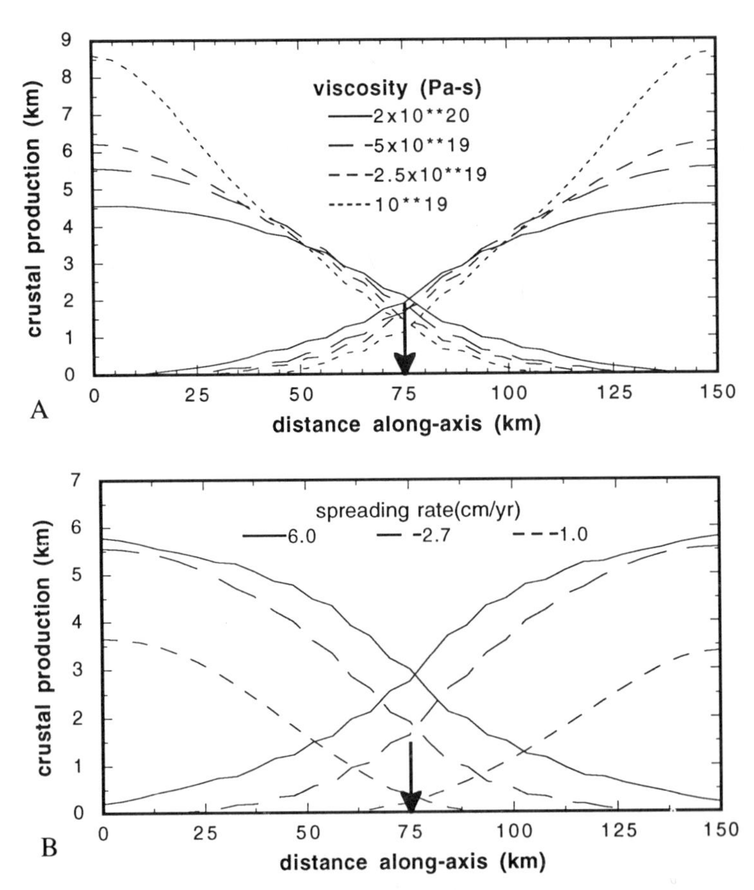

Figure 9 (A) Crustal production as a function of distance *along axis* for a range of mantle viscosities and a spreading rate of 2.7 cm/yr. The bold arrow shows the position of the transform. (B) Crustal production for a range of spreading rates and a viscosity of 5×10^{19} Pa · s.

primarily passive, melt production is relatively constant along the center of the spreading segments and decreases by about a factor of 2 near the transform (Fig. 9A). As viscosity is decreased, and buoyant flow becomes important, the along-axis variation becomes larger in amplitude and is distributed across the entire segment. Decreases in crustal thickness of up to 50% toward transform faults have been measured seismically (Cormier *et al.*, 1984; Purdy and Detrick, 1986). Gravity measurements near several fracture zones (Kuo and Forsyth, 1988; Lin *et al.*, 1990; Blackman and Forsyth, 1991; Morris and Detrick, 1991) are also consistent with about 50% crustal thinning.

For a given viscosity, the along-axis variation is more subdued at higher spreading rates (Fig. 9B). This is in agreement with the observation that the amplitudes of along-axis variations in topography and the mantle Bouguer anomaly increase with decreasing spreading rate (Lin and Phipps Morgan, 1992). Crustal thickness is also much more variable at slow spreading rates (Reid and Jackson, 1981; Chen, 1992; White *et al.*, 1992).

The shape of the melting region also affects the composition of the basalts that form the crust. The partition coefficients of elements between peridotite and basaltic melt can be sensitive to the pressure of melting and the composition of the solid. The composition of an increment of partial melt depends on the instantaneous depth of melting and the integrated melting history of the parcel of mantle that is melting. Therefore, a unique melt composition is generated at each point in the melting region. This melt composition could be determined with exact knowledge of the initial composition of the primitive mantle, and the distribution coefficients for each element, as functions of depth and degree of melting. Experimental petrology provides information on the behavior of some elements, which can be used to predict the composition of melt formed at a given depth and degree of melting (cf. Hanson and Langmuir, 1978; Niu and Batiza, 1991; Kinzler and Grove, 1992a).

Having calculated the distribution of oceanic crust, we can also calculate the average degree of melting and average depth of melting at which the basalt was formed. For each numerical experiment, the degree of melting and depth at each grid node is weighted by the melt production rate at that node. The melt is pooled in planes perpendicular to the axis, using the same division of the melt extraction region as in the previous section, to yield the average depth of melting, $\bar{D}$, and average degree of melting, $\bar{F}$. At high viscosity, $\bar{D}$ and $\bar{F}$ are relatively constant along a spreading segment, with greater depths of melting and smaller degrees of melting near a transform (see Fig. 10). As the viscosity is decreased, the gradient in $\bar{D}$ and $\bar{F}$ near the transform increases. Also, high-$\bar{D}$, low-$\bar{F}$ features develop at the centers of the segments, similar to the features that form near the transform, only broader. This feature is due to the small extents of deep melting in the upwelling limbs of off-axis rolls.

A positive correlation between Na_2O and FeO contents of basalts has been recognized in data collected at slow-spreading centers (Brodholt and Batiza, 1989). This correlation has been suggested to result from imperfect mixing of melts formed at different depths, with deeper melts containing higher Na_2O and FeO contents (Klein and Langmuir, 1989). Our numerical experiments indicate that melts formed beneath a transform fault should have a signature of deeper, smaller extents of melting, particularly at slow spreading centers. While this signature has been found near some transform faults (Langmuir and Bender, 1984; Batiza *et al.*, 1988), smooth along-axis variations in melt composition are not common. If such a variation in basalt chemistry does not exist, it is probably due to some process during crustal emplacement, such as the episodic emplacement of dikes that overlap in space, but have slightly different compositions. A greater number of spreading centers need to be systematically sampled with good spatial resolution to resolve this question.

Melt Extraction Efficiency and Melting in Off-Axis Convective Rolls

A melt-rich decompacting layer will be least effective at extracting melt formed in the upwelling limbs of the convective rolls because of the small degrees of melting and the relatively small slope of the layer in this region. This melt may not be extracted at the axis, but may refreeze into the mantle or pool and then erupt off-axis. In some experiments, melting in the upwelling limbs occurs far enough off-axis that it is separated from the melting region beneath the axis by a subsolidus impermeable mantle. Melt that is not extracted at the axis may account for the production

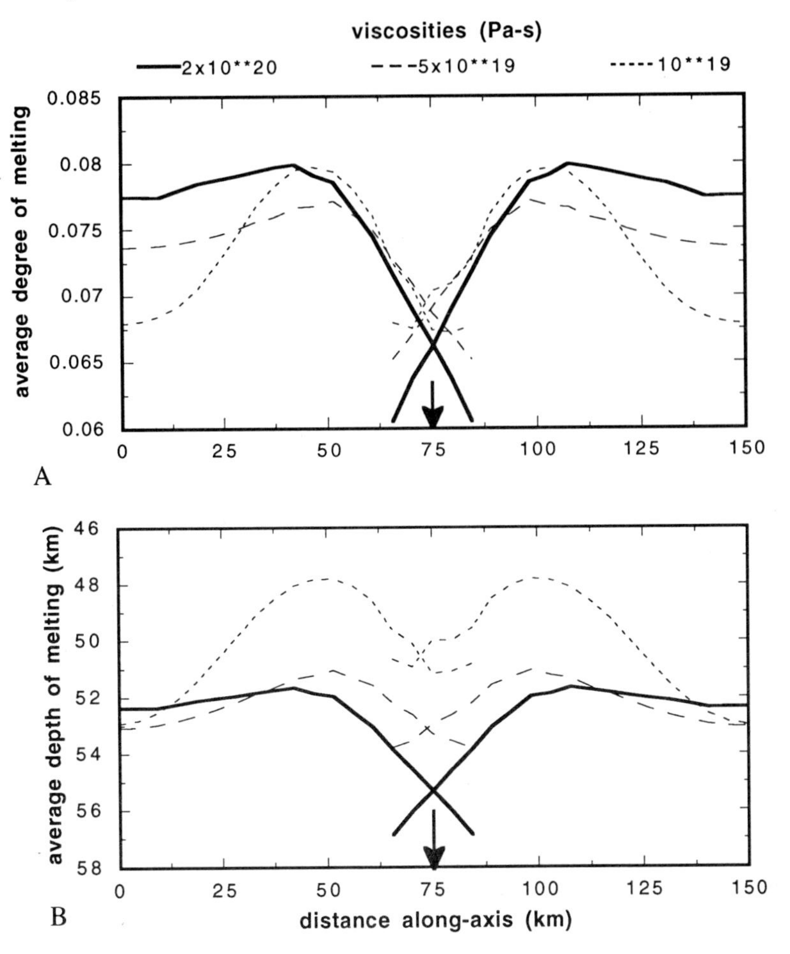

Figure 10 (A) Average degree of melting and (B) average depth of melting for a range of viscosities and a spreading rate of 2.7 cm/yr, as a function of distance *along axis*. The position of the transform is marked by the bold, vertical arrow. The curves are truncated just beyond the transform for clarity.

of off-axis seamount chains (Sparks and Parmentier, 1993).

The formation and extraction efficiency of decompacting layers will depend on a balance among the supply of melt to the layer, the flow along the layer, and the freezing within the layer. Therefore, the magnitude of the melt velocity is not determined solely by the magnitude of the layer slope shown in Fig. 8. We can estimate the spatial variations in the magnitude of the flow within a layer that result from the variations in melt supply and layer slope.

If forces due to gradients in the rate of compaction are ignored, then melt flow in the layer can be treated as one-dimensional Darcy flow. The melt velocity in the upslope direction, V, is

$$V = \frac{k(\rho - \rho_f)g}{\mu_f \phi_0} \sin \omega, \tag{18}$$

where ω is the dip angle of the layer, ϕ_0 is the melt fraction in the layer, and k is given by Eq. (4). The melt fraction at any point in the layer is bounded by the maximum degree of melting within the upwelling mantle directly below. If we substitute the maximum degree of melting into Eq. (18), the resulting melt velocities reflect the maximum spatial variation in the local slope and melt supply (Fig. 11).

The estimated melt velocities in the layer decrease toward the edges of the melting region, and within the convective rolls are one-third or less than the maximum velocities in the layer. The

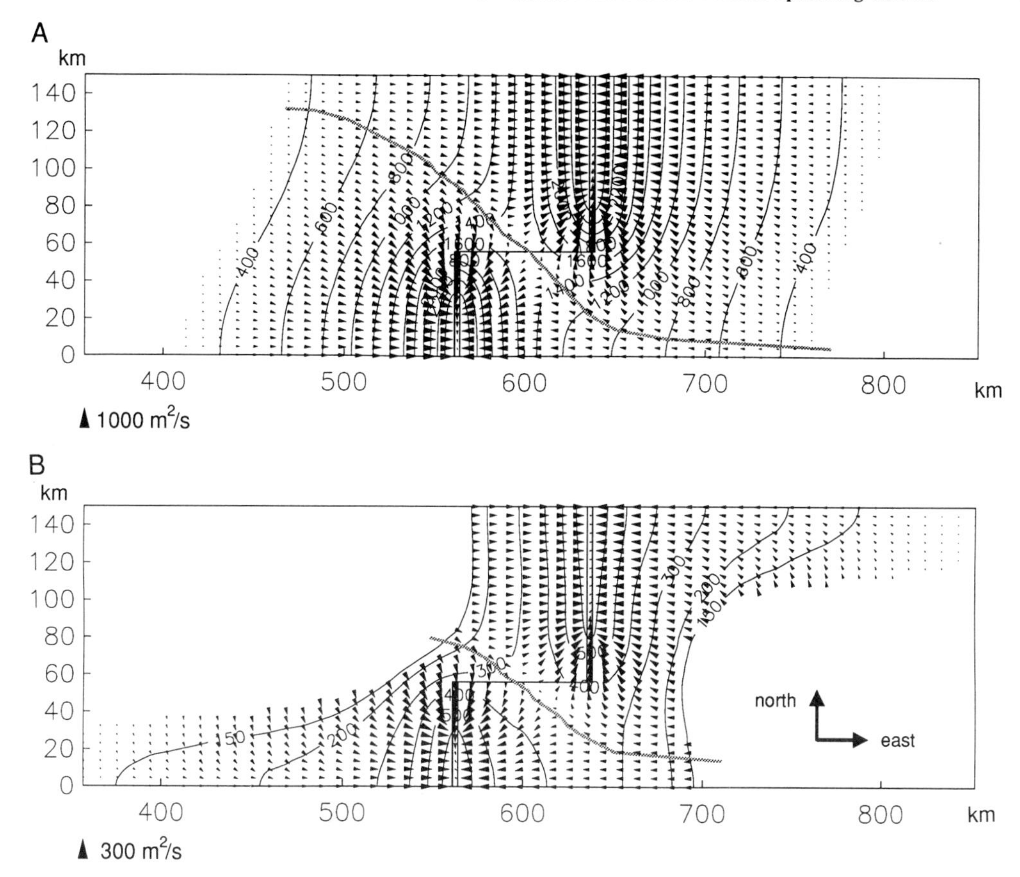

Figure 11 Contours of the decompaction length scale, as given by Eq. (6) for the two experiments shown in Figs. 7 and 8. The contours denote the thickness of the decompacting layer in meters. The arrowheads represent the approximate flux of melt within the layer. The arrows point in the direction of the steepest local slope, as in Fig. 8, but the magnitudes are given by the product of the melt velocity along the layer (Eq. (18)), the porosity in the layer, and the layer thickness. The thick gray line divides the two extraction regions, as in Fig. 8.

thickness of the layer, as estimated from Eq. (6), also decreases toward the edges of the melting region (shown by the contours in Fig. 11) since it is a function of both upwelling velocity and the amount of melt production. Therefore, the melt formed in convective rolls may not contribute to crustal production at the axis. We can attempt to account for this by fixing the width of the region from which melt is extracted.

Figure 12 shows the amount of crustal production in a single numerical experiment for a varying width of melt extraction. Melting occurs more than 50 km from the axis along most of the spreading center, as shown by the difference between the curves for 50 and 100 km. Beyond 100 km from the spreading axis, melting is limited to the upwelling limb of axis-perpendicular rolls.

Predicted Topography and Gravity Signal

When the effects of the two major density interfaces, crust–water and mantle–crust, are subtracted from the free-air gravity anomaly, the resulting mantle Bouguer gravity anomaly (MBA) indicates either mantle density variations or deviations from an assumed constant crustal thickness (see Kuo and Forsyth, 1988). Negative ellipsoidal "bulls-eye"-shaped anomalies indicate that the mantle is less dense and/or the crust is thicker beneath the center of ridge segments. Observed magnitudes typically range from 20 mgal (Kane Fracture Zone: Morris and Detrick, 1991) to 50 mgal (Atlantis Fracture Zone: Lin *et al.*, 1990).

Mantle density variations are insufficient to account for the observed magnitudes of the

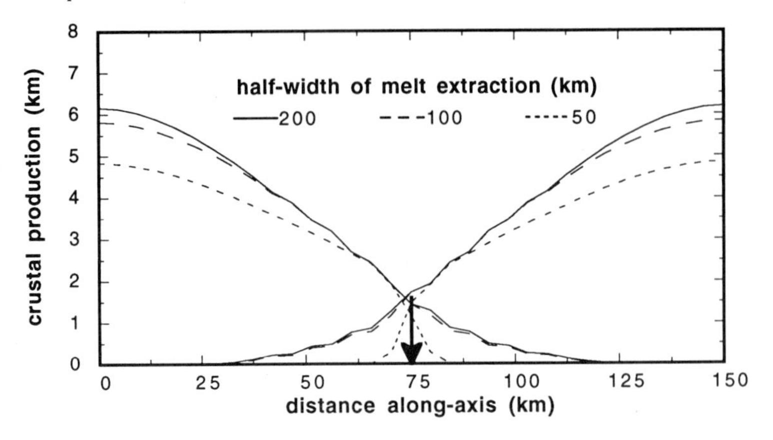

Figure 12 Crustal production in an experiment with a spreading rate of 2.7 cm/yr and viscosity of 2.5×10^{19} Pa · s, for different widths of the region of melt extraction. Melt generated beyond the half-width of melt extraction does not contribute to crust emplaced at the axis.

MBA, producing maximum along-axis variations of about 10 mgal (Sparks *et al.,* 1993). Even the small variations in crustal thickness produced in the high-viscosity experiments double these magnitudes, and at low viscosities, the magnitudes increase tenfold. The magnitude of the along-axis temperature variations is small, since temperature within the melting region is buffered by the absorption of latent heat. These small, deep density variations produce only a small effect on the surface gravity field. However, variations in crustal production may be large, even though the temperature field is relatively uniform, due to along-axis variations in upwelling velocity. Therefore crustal thickness variations dominate the MBA signal.

When buoyant flow is not important, there is not much along-axis variation in topography and MBA, except near the transform. As the viscosity is decreased, patterns similar to the observed "bulls-eyes" emerge (Fig. 13). At intermediate viscosities the mantle Bouguer anomaly increases by 40–50 mgal from the center of the spreading segment to the transform, and the depth of the axis increases by over 500 m. At low viscosities, convective rolls form near the axis causing variations of over 100 mgal and 1.5 km.

Isostatic topography for the numerical experiments is calculated column by column, assuming compensation at the bottom of the asthenosphere. The topography is calculated with reference to a column consisting of 6 km of crust and 200 km of mantle at density $\tilde{\rho}$ (given by Eq. (17) for a temperature of 1410°C and no melt depletion). For a column with crustal thickness, h_c, and mantle density (given by Eq. (17)) averaged over the mantle column, $\bar{\rho}$, the topography, h, is given by

$$h = \frac{200\ \text{km}\ (\tilde{\rho} - \bar{\rho})}{(\bar{\rho} - \rho_w)} + \frac{(h_c - 6\ \text{km})(\bar{\rho} - \rho_c)}{(\bar{\rho} - \rho_w)}, \tag{19}$$

where ρ_c is the density of oceanic crust and ρ_w the density of seawater, taken to be 2800 and 1000 kg · m^{-3}, respectively.

We calculate the MBA by treating horizontal density variations at a each depth as a sheet of anomalous mass. A two-dimensional fast Fourier transform (FFT) transforms the mass variations into wavenumber space. The magnitude of the surface gravity anomaly, Δg, that is created by a sheet of anomalous mass is given by

$$\Delta g = 2\pi G \exp(-\sigma z) M, \tag{20}$$

where M is the magnitude, σ is the wavenumber, and z is the depth of the mass sheet. G is the gravitational constant (6.67×10^{-11} m^3 · kg^{-1} · s^{-2}). The contributions from each wavenumber of the mass sheets at each depth in the grid are summed, and an inverse FFT yields the total gravity anomaly due to mantle density variations. The crust–mantle interface is treated as a mass sheet at the

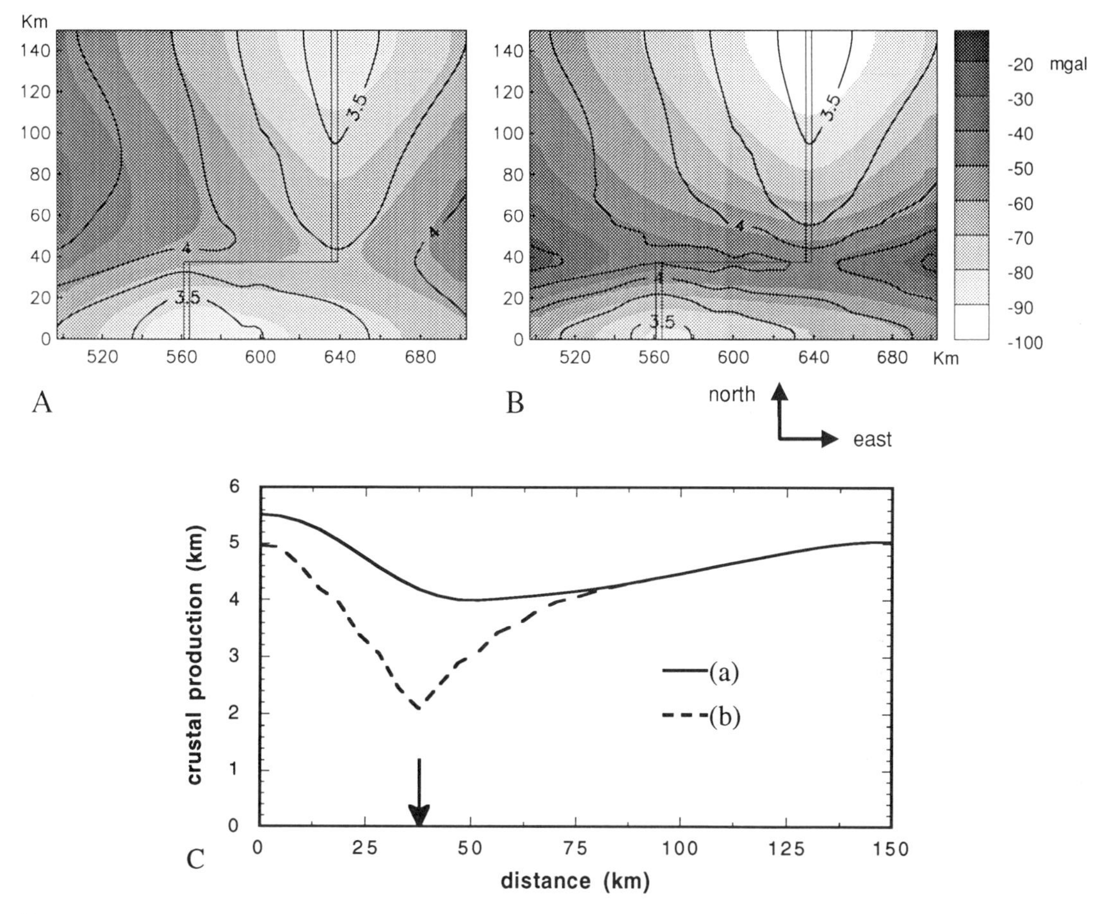

Figure 13 Mantle Bouguer anomaly (mgal, shaded) and bathymetry (km, contoured) for a spreading rate of 2.7 cm/yr, mantle viscosity of 5×10^{19} Pa · s, and ridge length ratio of 3 to 1. (A) Crustal thickness assuming all melt is extracted in axis-perpendicular planes. (B) Crustal thickness assuming two melt extraction regions and no extraction of melt formed beyond the end of the transform. (C) Crustal thickness as a function of distance *along axis,* used in calculating the gravity and topography in (A) and (B). In each case "bulls-eye"-shaped gravity anomalies form over each segment, but in (B) the magnitude is greater.

average depth of the interface from the sea surface, assuming the water depth to be 3 km. Crustal thickness is a function of distance along-axis and is taken to be the total of the crust generated in planes perpendicular to the axis.

To demonstrate the sensitivity of the gravity field to crustal thickness, two different methods of estimating crustal thickness were used for a single numerical experiment (Fig. 13C). First, the crustal production was simply integrated in planes perpendicular to the axis (Fig. 13A); in the second case, the region was divided into two extraction regions to produce crustal thickness curves like those in Fig. 9, and then melt formed beyond the transform fault was neglected. Because the crust beneath the transform is considerably thinner, the variations in gravity and bathymetry are also more pronounced.

Figure 13 also shows the effect of variations in the length of spreading segments. When the melting region is divided, the longer segment is seen to extract more than its proportional share of melt. The resulting thicker crust produces a larger gravity anomaly beneath the longer segment, as is observed south of the Atlantis Fracture Zone (Lin *et al.,* 1990). A greater number of detailed along-axis seismic refraction studies is needed to confirm the relation between gravity and crustal thickness. These studies coupled with modeling of the mantle flow and melt generation can provide important clues to the style and paths of melt migration.

Discussion

Further work is needed to assess the implications of the three-dimensional numerical experiments for observable characteristics of spreading centers. The relationship between plate boundary segmentation and buoyant flow is still unclear. Time dependence develops in some numerical experiments at low viscosities ($\leq 10^{19}$ Pa · s, Sparks *et al.*, 1993). The effect of this time dependence in the mantle flow on observable temporal variability at the ridge crest also needs to be explored. The amount of along-axis variability and the depths of decompacting layers are two unexplored aspects of these experiments that may have important implications for spreading center morphology and episodicity.

Even in numerical experiments at moderate mantle viscosities, the along-axis variations in crustal production are as large as or larger than any observed variations in crustal thickness. Buoyant flow due to retained melt will further increase along-axis variations (Jha *et al.*, 1992). In some experiments crustal production vanishes along parts of the spreading center. This indicates that there must be along-axis melt migration toward transforms that limit the variation of crustal thickness. The predicted migration paths in a decompacting layer (Fig. 8) indicate along-axis flow away from transforms. Therefore, to limit crustal thickness variations, significant migration toward transforms must occur during transport through the lithosphere or emplacement in the crust.

Figure 13 illustrates the degree to which observable quantities are controlled by the distribution of crust. The dynamics of crustal formation are just starting to be addressed in a quantitative way (cf. Phipps Morgan, 1991). A three-dimensional model for the emplacement of the various components of the oceanic crust is the next step necessary to more closely link models of mantle dynamics and melt formation with observations.

The minimum depth to the decompacting layer varies from about 10 km at fast spreading rates to about 25 km at very slow spreading rates. The slightly shallower depths that result from temperature-dependent viscosity (Shen and Forsyth, 1992) do not qualitatively change this result. Therefore, except at East Pacific Rise-type spreading rates, melt must migrate through a significant thickness of subsolidus lithosphere by fracture. Although flow in the decompacting layers may be unsteady, it is likely to be a less time-dependent process than transport through fractures. The greater distance melt must travel in fractures may contribute to the more episodic character of slow spreading centers, e.g., the absence of steady-state crustal magma chambers (Phipps Morgan, 1991; Solomon and Toomey, 1992) and the greater amplitude and spacing of abyssal hill topography (Malinverno and Pockalny, 1990).

Decompacting layers may also have an observable effect on basalt chemistry. The "local trends" in basalt composition have also been interpreted as the result of fractional crystallization at a variety of depths during the ascent of the melt (Kinzler and Grove, 1992b). Since at least a small amount of freezing is occurring within decompacting layers, if the melt in the layers does not fully reequilibrate during its ascent, it should retain a chemical signature of deep fractionation.

The analysis and calculations presented here are a first attempt at describing the three-dimensional character of spreading centers by including processes that occur on a variety of scales: passive mantle flow on the scale of oceanic plates, convection on the scale of individual spreading segments, and porous flow on the scale of decompacting layers. A better understanding of the interaction of processes at all these scales is crucial to describing the nature of mantle flow, melting, and melt migration in the mantle beneath spreading centers.

Acknowledgments

This work was supported by NSF Grants OCE 92-02599 and OCE 93-96097, and by a grant from the University Collaborative Research Program of the Institute for Geophysics and Planetary Physics at Los Alamos National Laboratory. Sparks was supported during part of the preparation of the manuscript by a Postdoctoral Fellowship at Lamont–Doherty Earth Observatory of Columbia University. We thank Neil Ribe, Michael Ryan, and Norm Sleep for thoughtful reviews of the manuscript, and Jason Phipps Morgan and Marc Spiegelman for useful discussions on mantle flow and melt migration. We also thank Jason for providing a program for calculating passive flow and Greg Neumann for a program for generating 3D surface plots.

References

Ahern, J. L., and Turcotte, D. L. (1979). Magma migration beneath an ocean ridge, *Earth Planet. Sci. Lett.* **45,** 115–122.

Barcilon, V., and Richter, F. M. (1986). Non-linear waves in compacting media, *J. Fluid Mech.* **164,** 429–448.

Batiza, R., Melson, W. G., and O'Hearn, T. (1988). Simple magma supply geometry inferred beneath a segment of the Mid-Atlantic Ridge, *Nature* **335,** 428–431.

Blackman, D. K., and Forsyth, D. W. (1991). Isostatic compensation of tectonic features of the Mid-Atlantic Ridge: 25–27°30′S, *J. Geophys. Res.* **96,** 11741–11758.

Brodholt, J. P., and Batiza, R. (1989). Global systematics of unaveraged mid-ocean ridge basalt compositions: Comment on "Global correlations of ocean ridge basalt chemistry with axial depth and crustal thickness" by E. M Klein and C. H. Langmuir, *J. Geophys. Res.* **94,** 4231–4239.

Buck, W. R., and Su, W. (1989). Focussed mantle upwelling below mid-ocean ridges due to feedback between viscosity and melting. *Geophys. Res. Lett.* **16,** 641–644.

Ceuleneer, G. (1992). Distribution of melt-migration structures in the mantle peridotites of Oman: Implications for magma supply processes at mid-ocean ridges, *EOS Trans. Am. Geophys. Union* **73,** 537.

Cheadle, M. (1989). "Properties of Texturally Equilibrated Two-Phase Aggregates," Ph.D. Thesis, Cambridge Univ.

Chen, Y. (1992). Oceanic crustal thickness versus spreading rate, *Geophys. Res. Lett.* **19,** 753–756.

Cooper, R. F., and Kohlstedt, D. L. (1986). Rheology and structure of olivine-basalt partial melts, *J. Geophys. Res.* **91,** 9315–9323.

Cordery, M. J., and Phipps Morgan, J. (1992). Melting and mantle flow beneath a mid-ocean spreading center, *Earth Planet. Sci. Lett.* **111,** 493–516.

Cormier, M. H., Detrick, R. S., and Purdy, G. M. (1984). Anomalously thin crust in oceanic fracture zones: New seismic constraints from the Kane fracture zone, *J. Geophys. Res.* **89,** 10,249–10,266.

Daines, M. J., and Richter, F. M. (1989). An experimental method for directly determining the interconnectivity of melt in a partially molten system, *Geophys. Res. Lett.* **15,** 1459–1462.

Forsyth, D. W., and Wilson, B. (1984). Three-dimensional temperature structure of a ridge-transform-ridge system, *Earth Planet. Sci. Lett.* **70,** 355–362.

Fowler, A. C. (1985). A mathematical model of magma transport in the asthenosphere, *Geophys. Astrophys. Fluid Dyn.* **33,** 63–69.

Fowler, A. C. (1990a). A compaction model for melt transport in the Earth's asthenosphere. Part I. The basic model, *in* "Magma Transport and Storage" (M. P. Ryan, ed.), pp. 3–14, Wiley, Chichester/Sussex, England.

Fowler, A. C. (1990b). A compaction model for melt transport in the Earth's asthenosphere. Part II: Applications, *in* "Magma Transport and Storage," (M. P. Ryan, ed.), pp. 15– 32, Wiley, Chichester/Sussex, England.

Frank, F. C. (1968). Two-component flow model for convection in the Earth's upper mantle, *Nature* **220,** 350–352.

Hanson, G. N., and Langmuir, C. H. (1978). Modelling of major elements in mantle–melt systems using trace element approaches, *Geochim. Cosmochim. Acta* **42,** 725–742.

Hess, P. C. (1989). "Origins of Igneous Rocks," Harvard Univ. Press, Cambridge.

Jha, K., Parmentier, E. M., and Phipps Morgan, J. (1992). Mantle flow beneath spreading centers due to mantle depletion and melt retention buoyancy, *EOS Trans. Am. Geophys. Union* **73,** 291.

Johnson, K. T., Dick, H. J. B., and Shimizu, N. (1990). Melting in the oceanic upper mantle: An ion microprobe study of diopsides in abyssal peridotites, *J. Geophys. Res.* **95,** 2661–2678.

Kinzler, R. J., and Grove, T. L. (1992a). Primary magmas of mid-ocean ridge basalts. 1. Experiments and methods, *J. Geophys. Res.* **97,** 6885–6906.

Kinzler, R. J., and Grove, T. L. (1992b). Primary magmas of mid-ocean ridge basalts. 2. Applications, *J. Geophys. Res.* **97,** 6907–6926.

Klein, E. M., and Langmuir, C. H. (1989). Local versus global variations in ocean ridge basalt compostion: A reply, *J. Geophys. Res.* **94,** 4241–4252.

Kohlstedt, D. L., and Hirth, G. (1992). The effect of melt fraction on the strength of olivine aggregates deformed in the diffusion creep regime, *EOS Trans. Am. Geophys. Union* **73,** 517.

Kuo, B.-Y., and Forsyth, D. W. (1988). Gravity anomalies of the ridge–transform system in the South Atlantic between 31 and 34.5°S: Upwelling centers and variations in crustal thickness, *Marine Geophys. Res.* **10,** 205–232.

Langmuir, C. H., and Bender, J. F. (1984). Petrological and tectonic segmentation of the East Pacific Rise, 5°30′–14°30′, *Earth Planet. Sci. Lett.* **69,** 107–127.

Lin, J., Purdy, G. M., Schouten, H., Sempere, J.-C., and Zervas, C. (1990). Evidence from gravity data for focussed magmatic accretion along the Mid-Atlantic Ridge, *Nature* **344,** 627–632.

Lin, J., and Phipps Morgan, J. (1992). The spreading rate dependence of three-dimensional mid-ocean ridge gravity structure, *Geophys. Res. Lett.* **19,** 13–16.

Macdonald, K. C. (1984). Mid-ocean ridges: Fine scale tectonic, volcanic and hydrothermal processes within the plate boundary zone, *Annu. Rev. Earth Planet. Phys.* **10,** 155–190.

Malinverno, A., and Pockalny, R. A. (1990). Abyssal hill topography as an indicator of episodicity in crustal accretion, *Earth Planet. Sci. Lett.* **99,** 154–169.

McKenzie, D. (1984). The generation and compaction of partially molten rock, *J. Petrol.* **25,** 713–765.

Morris, E., and Detrick, R. S. (1991). Three-dimensional analysis of gravity anomalies in the MARK area, Mid-Atlantic Ridge 23°N, *J. Geophys. Res.* **96,** 4355–4366.

Neumann, G. A., and Forsyth, D. W. (1993). The paradox of the axial profile: Isostatic compensation along the axis of the Mid-Atlantic Ridge? *J. Geophys. Res.* **98,** 17891–17911.

Niu, Y., and Batiza, R. (1991). An empirical method for calculating melt compositions produced beneath mid-ocean ridges: Application for axis and off-axis (seamounts) melting, *J. Geophys. Res.* **96,** 21753–21777.

O'Hara, M. J. (1975). Is there an Icelandic mantle plume?, *Nature Phys. Sci.* **253,** 708–710.

Oxburgh, E. R., and Parmentier, E. M. (1977). Compositional and density stratification in oceanic lithosphere—causes and consequences, *J. Geol. Soc. London* **133,** 343–355.

Parmentier, E. M., and Phipps Morgan, J. (1990). The spreading rate dependence of three-dimensional structure in oceanic spreading centers, *Nature* **348,** 325–328.

Phipps Morgan, J. (1987). Melt migration beneath mid-ocean spreading centers, *Geophys. Res. Lett.* **14,** 1238–1241.

Phipps Morgan, J. (1991). Mid-ocean ridge dynamics: Observations and theory, *Rev. Geophys. Suppl.* **14,** 807–822.

Phipps Morgan, J., and Forsyth, D. W. (1988). Three-dimensional flow and temperature perturbations due to a transform offset: Effects on oceanic crustal and upper mantle structure, *J. Geophys. Res.* **93,** 2955–2966.

Phipps Morgan, J., and Chen, Y. J. (1992). Magma injection, hydrothermal circulation, crustal flow and the genesis of oceanic crust, *EOS Trans. Am. Geophys. Union* **73,** 290.

Purdy, G. M., and Detrick, R. S. (1986). Crustal structure of the Mid-Atlantic Ridge at 23°N from seismic refraction studies, *J. Geophys. Res.* **91,** 3739–3762.

Rabinowicz, M., Nicolas, A., and Vigneresse, J. L. (1984). A rolling mill effect in asthenosphere beneath oceanic spreading centers, *Earth Planet. Sci. Lett.* **67,** 97–108.

Raitt, R. W. (1963). The crustal rocks, *in* "The Sea" (M. N. Hill, ed.), pp. 85–102, Wiley–Interscience, New York.

Reid, I., and Jackson, H. R. (1981). Oceanic spreading rate and crustal thickness, *Marine Geophys. Res.* **5,** 165–172.

Ribe, N. M. (1985). The deformation and compaction of partially molten zones, *Geophys. J. R. Astron. Soc.* **83,** 487–501.

Richter, F. M., and McKenzie, D. (1984). Dynamical models for melt segregation from a deformable matrix, *J. Geol.* **92,** 729–740.

Ryan, M. P. (1988). The mechanics and three-dimensional internal structure of active magmatic systems: Kilauea Volcano, Hawaii, *J. Geophys. Res.* **93,** 3213–4248.

Scott, D. R., and Stevenson, D. J. (1984). Magma solitons, *Geophys. Res. Lett.* **11,** 1161–1164.

Scott, D. R., and Stevenson, D. J. (1986). Magma ascent by porous flow, *J. Geophys. Res.* **91,** 9283–9296.

Scott, D. R., and Stevenson, D. J. (1989). A self-consistent model of melting, magma migration, and buoyancy-driven circulation beneath a mid-ocean ridge, *J. Geophys. Res.* **94,** 2973–2988.

Sempere, J.-C., Rabinowicz, M., Rouzo, S., and Rosemberg, C. (1992). Three-dimensional mantle flow beneath spreading centers: Implications of the models, *EOS Trans. Am. Geophys. Union* **73,** 494.

Shaw, H. R. (1980). The fracture mechanism of magma transport from the mantle to the surface, *in* "Physics of Magmatic Processes" (R. B. Hargraves, ed.), pp. 201–244, Princeton Univ. Press, Princeton, NJ.

Shen, Y., and Forsyth, D. W. (1992). The effects of temperature- and pressure-dependent viscosity on three-dimensional passive flow of the mantle beneath a ridge–transform system, *J. Geophys. Res.* **97,** 19717–19728.

Sleep, N. H. (1974). Segregation of magma from a mostly crystalline mush, *Geol. Soc. Am. Bull.* **85,** 1225–1232.

Sleep, N. H. (1984). Tapping of magmas from ubiquitous mantle heterogeneities: An alternative to mantle plumes?, *J. Geophys. Res.* **89,** 10029–10041.

Sleep, N. H. (1988). Tapping of melt by veins and dikes, *J. Geophys. Res.* **93,** 10255–10272.

Solomon, S. C., and Toomey, D. R. (1992). The structure of mid-ocean ridges, *Annu. Rev. Earth Planet. Sci.* **20,** 329–364.

Sotin, C., and Parmentier, E. M. (1989). Dynamical consequences of compositional and thermal density stratification beneath spreading centers, *Geophys. Res. Lett.* **16,** 835–838.

Sotin, C., Parmentier, E. M., Carey-Gailhardis, E., and Stoclet, P. (1992). A 3D multigrid Poisson solver on the connection machine: Application to convection within planetary interiors, submitted, *J. Comput. Phys.*

Sparks, D. W., and Parmentier, E. M. (1991). Melt extraction from the mantle beneath spreading centers, *Earth Planet. Sci. Lett.* **105,** 368–377.

Sparks, D. W., and Parmentier, E. M. (1993). The structure of three-dimensional convection beneath oceanic spreading centres, *Geophys. J. Int.* **112,** 81–91.

Sparks, D. W., Parmentier, E. M., and Phipps Morgan, J. (1993). Three-dimensional convection beneath a segmented spreading center: Implications for along-axis variations in crustal thickness and gravity, *J. Geophys. Res.* **98,** 21,977–21,995.

Spence, D. A., and Turcotte, D. L. (1985). Magma-driven propagation of cracks, *J. Geophys. Res.* **90,** 575–580.

Spiegelman, M., and McKenzie, D. (1987). Simple 2-D models for melt extraction at mid-ocean ridges and island arcs, *Earth Planet. Sci. Lett.* **83,** 137–152.

Spiegelman, M., and Kenyon, P. (1992). The requirements for chemical disequilibrium during magma migration, *Earth Planet. Sci. Lett.* **109,** 611–620.

Spiegelman, M. (1993a). Flow in deformable porous media. I. Simple analysis, *J. Fluid Mech.* **247,** 17–38.

Spiegelman, M. (1993b). Flow in deformable porous media. II. Numerical analysis—The relationship between shock waves and solitary waves, *J. Fluid Mech.* **247,** 39–63.

Spiegelman, M. (1993c). Physics of melt extraction: Theory, implications and applications, *Philos. Trans. R. Soc. London A* **342,** 23–41.

Stevenson, D. J. (1989). Spontaneous small-scale melt segregation in partial melts undergoing deformation, *Geophys. Res. Lett.* **9,** 1064–1070.

Turcotte, D. L., and Ahern, J. L. (1978). A porous flow model for magma migration in the asthenosphere, *J. Geophys. Res.* **83,** 767–772.

Waff, H. S., and Bulau, J. R. (1979). Equilibrium fluid distribution in an ultramafic partial melt under hydrostatic stress conditions, *J. Geophys. Res.* **84,** 6109–6114.

Watson, E. B. (1982). Melt infiltration and magma evolution, *Geology* **10,** 236–240.

White, R. S., MacKenzie, D., and O'Nions, R. K. (1992). Oceanic crustal thickness from seismic measurements and rare earth element inversions, *J. Geophys. Res.* **97,** 19683–19716.

Chapter 5 | Dike Patterns in Diapirs beneath Oceanic Ridges: The Oman Ophiolite

A. Nicolas, F. Boudier, and B. Ildefonse

Overview

Dike orientation was systematically measured in the vicinity of mantle diapirs exposed in the Oman ophiolite. Collectively, diking reflects a continuous history of melt injection, from the early gabbro and pyroxenite dikes emplaced in a concurrently melting peridotite, to the dominantly gabbro dikes emplaced in peridotites at subsolidus temperatures, and, finally, to microgabbro and diabase dikes emplaced in peridotites that have cooled below 500°C. Preferred orientations of dikes are generally poor, but become progressively better with decreasing wall-rock temperatures in the host peridotite. An orientation that corresponds to steep dikes parallel to the inferred ridge trend (itself determined by the trend of the diabase sheeted dike complex) is first recognized and is ascribed to control of the tensional lithospheric stress field. A second diffuse intrusion orientation corresponds to sills or moderately dipping dikes. It is mainly detected among the early, high-temperature injections. Sill formation is ascribed to the stress field created by the mantle diapir's divergent flow in the sub-ridge asthenosphere just below the newly accreted crust. Advecting melt can circulate either within "lithospheric dikes" or within "asthenospheric sills." Overall, these relations can be explained by a sudden change from a lithospheric to an asthenospheric stress field. The change could result from the relaxation of lithospheric stresses in response to tension fracturing of the lithospheric lid, triggered by a sudden melt surge.

Introduction

Direct knowledge of the diking system in the uppermost mantle below the currently active oceanic ridges is not, unfortunately, possible. This fundamental question, however, is intimately related to the overall problems of melt extraction from the mantle and of the subsequent melt moving toward the active ridge axis and center of crustal accretion. Evidence from ophiolites and from oceanic observations (Auzende *et al.,* 1989) or from ocean crust drilling (Anderson *et al.,* 1982) indicates that below the blanket of basaltic extrusives is a sheeted dike complex composed of vertical diabase dikes that are parallel to the oceanic ridge trend. In addition, the existence of melt lenses at Moho level, 5–20 km away from the East Pacific Rise, has been proposed by Garmany (1989) and Barth *et al.* (1991) on the basis of seismic experiments. If these results are confirmed, they show that both sills and extensional dikes feed the accreting oceanic crust. Vertical fracturing related to extension and underplating by sills are also the two principal modes of basaltic injection envisaged for the deep continental crust, as illustrated by structural studies of diabase dike injection at the mantle–crust interface exposed on Zabargad Island in the Red Sea (Nicolas *et al.,* 1987; Boudier *et al.,* 1988). There, a mantle diapir penetrated the continental crust during an early stage of Red Sea opening. The diapiric peridotites and the surrounding deep crustal gneisses were cut by vertical dikes oriented parallel to the Red Sea trend and thought to represent extensional fractures; emplacement of thick diabase sills followed, with intrusion taking place selectively at the interface between peridotites and overlying rocks.

In this chapter, we explore the problem of magma transport structures and diking at oceanic ridges by reporting a structural study of dike orientations in the peridotite section of the Oman ophiolite. This study is similar to previous work on Zabargad Island. Oman ophiolite is particularly well suited to such a study because, during the course of a long-term project on structural mapping conducted since 1980, mantle diapirs

Magmatic Systems
Edited by M. P. Ryan

through which melt was channeled to the overlying accreting ridge were discovered (Nicolas *et al.*, 1988; Ceuleneer *et al.*, 1988; Ceuleneer, 1991; Nicolas and Boudier, submitted 1993). Diapiric structures in the mantle peridotite are associated with a dense network of gabbro and pyroxenite dikes, melt impregnation veins and clots, mainly within the several-hundred-meters-thick transition zone located just below the Moho and above the relatively homogeneous harzburgitic mantle. It has been suggested that massive melt impregnation within this zone could have reduced the mantle viscosity and thus facilitated the sharp rotation of flow lines within the diapir from a vertical to a horizontal orientation (Rabinowicz *et al.*, 1987). Another favorable circumstance is that these diapirs must have been detached, as part of a future ophiolite, from the ridge of origin while they were still active or in a waning stage of activity; otherwise the high-temperature structures, characteristic of their uprise, would not have been preserved. Thus, a ridge structure and a geologic "moment in time" were frozen and the resulting measured dike system relates directly to the active diapiric structures of the sub-ridge mantle. It should be noted that in most other studies of dike orientations within peridotites, the continuing plastic flow in peridotites after dike emplacement partially blurs the intrusion picture (Fig. 4c, Nicolas and Jackson, 1982).

Finally, although data on dikes from the entire ophiolite belt of Oman are available, we present in this chapter only the results obtained in two massifs from the southeastern part of the belt (Fig. 1) where a detailed structural study was conducted (Nicolas and Boudier, submitted 1993). A similar study, with similar results, is now being conducted in a third massif (Wadi Tayin) of Oman (Ildefonse *et al.*, 1993). These massifs were selected because of the flat-lying and regular attitude of the Moho, which (1) tends to limit the errors related to the rotation necessary to restore the structures in the paleo-ridge reference frame (Moho horizontal) and (2) suggests that these massifs have experienced only moderate deformation during and after their obduction onto the Arabian shield margin. One main conclusion of the study cited above is that the two massifs under consideration (Nakhl-Rustaq and Maqsad) have escaped any visible deformation subsequent to ridge accretion and, thus, can be structurally considered relatively

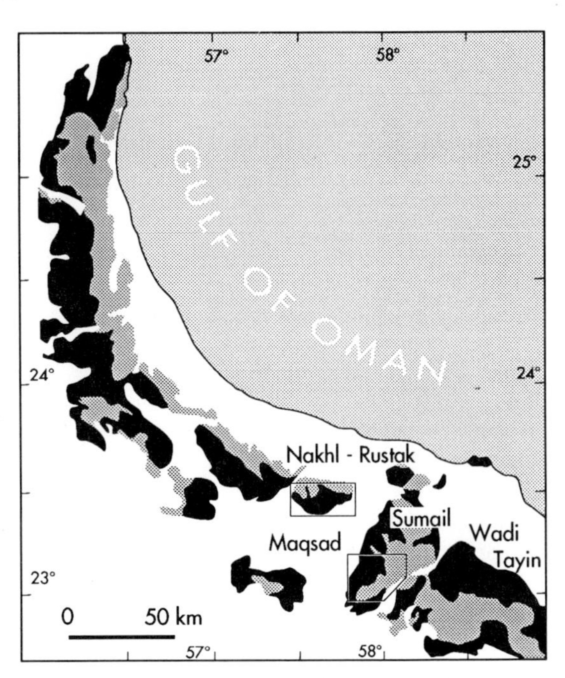

Figure 1 Map of the Oman ophiolite with the locations of the massifs considered in this study. Black pattern, mantle sections; gray pattern, crustal sections.

pristine oceanic spreading centers subsequently dissected by erosion. The near absence of any late deformation in these massifs is also better understood by considering the results of a new gravimetry study, suggesting a keel of relatively strong peridotite that is about 6 km in thickness (Ravaut, 1992).

Within this well-identified ridge frame and in contrast with the other clearly organized structures, the orientation of dikes has been found to be irregular on every scale, as shown by the diffuse patterns of preferred orientation presented in this chapter (see also Nicolas and Boudier, submitted 1993). Because of this dispersion in dike orientations, it is necessary to further analyze these data by individual rotations in the reference frames of both the ridge and the local mantle plastic flow. Consideration of the structure of individual dikes and mutual relationships is also important.

Regions of Diapirism and Sub-ridge Flow

Paleo-ridge structures were reconstructed in the Maqsad (Fig. 2) and Nakhl-Rustaq (Fig. 3) massifs from detailed structural mappings of the high-temperature foliations and lineations induced by solid-state flow in the mantle section and from magmatic foliations and lineations in the layered gabbro unit. The ridge trend itself is defined by the azimuth of the diabase sheeted dike complex. The peridotite foliations and lineations—normally flat-lying and parallel to the Moho—plunge locally and thus define the contours of mantle diapirs. The vertical lineations within the diapir tend to diverge outside this area as seen in Fig. 3. These diapirs, delineated in a map view by isodip lines of the foliations and lineations (30° in Fig. 2A, 45° in Fig. 3A), permit one to locate paleo-spreading centers. As discussed further by Nicolas and Boudier (submitted 1993), this location is independently confirmed by a kinematic analysis (shear sense) of the local mantle flow (thick arrows in Figs. 2A and 3A show the shear flow away from the diapir). Additional information on the spreading axis location comes from the convergence of upper gabbro foliations toward the inferred ridge axis, thus pointing to the along-strike closure of the magma chamber. Finally, the clustering of diabase dikes in the lower gabbros and in peridotites suggests close proximity to the spreading axis. These intrusions have been interpreted as products of the activity of the ridge in its dying phase, with late-stage injections of basaltic melt in a generally cooling spreading center.

In the Maqsad massif, the diapiric area is largely exposed, disappearing only in the southeast below the lower crustal lithologic units (Fig. 2). It is thus

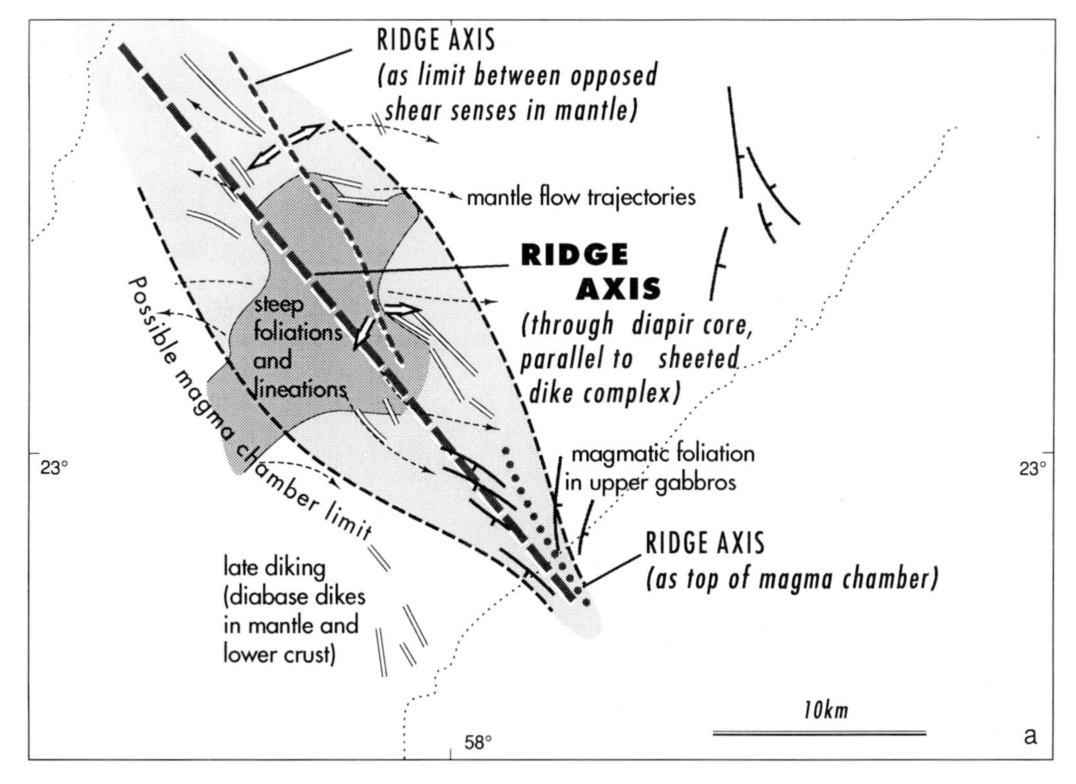

Figure 2 Maqsad area. (a) Paleo-ridge model (location in Fig. 1; detailed map in Fig. 2b). Darker shaded area, mapped contour of the mantle diapir (see text); lighter shaded area, presumed limit of the magma chamber. The ridge axis is independently located as follows: (1) crossing through the center of the diapir, parallel to the sheeted dike trend; (2) separating opposed shear senses of plastic flow in the mantle just below the Moho (thinner dashed line); (3) corresponding to the convergence of magmatic foliations in upper gabbro units (dotted line); and (4) being preferentially intruded by the last melts (diabase dikes) issued from the dying and cooling ridge (modified after Nicolas and Boudier, submitted 1993). (b) Geological map and dike trajectories. (c) Map of individual dike measurements. Note the convergence of dike trajectories toward the ridge axis NW of the diapir center (see Fig. 2a). The data on the sheeted dike attitude come mostly from the NE corner of the map. Continuous black lines represent faults and, dominantly, hydrothermal fracture zones parallel to the sheeted dike complex (Nicolas and Boudier, submitted 1993).

continues

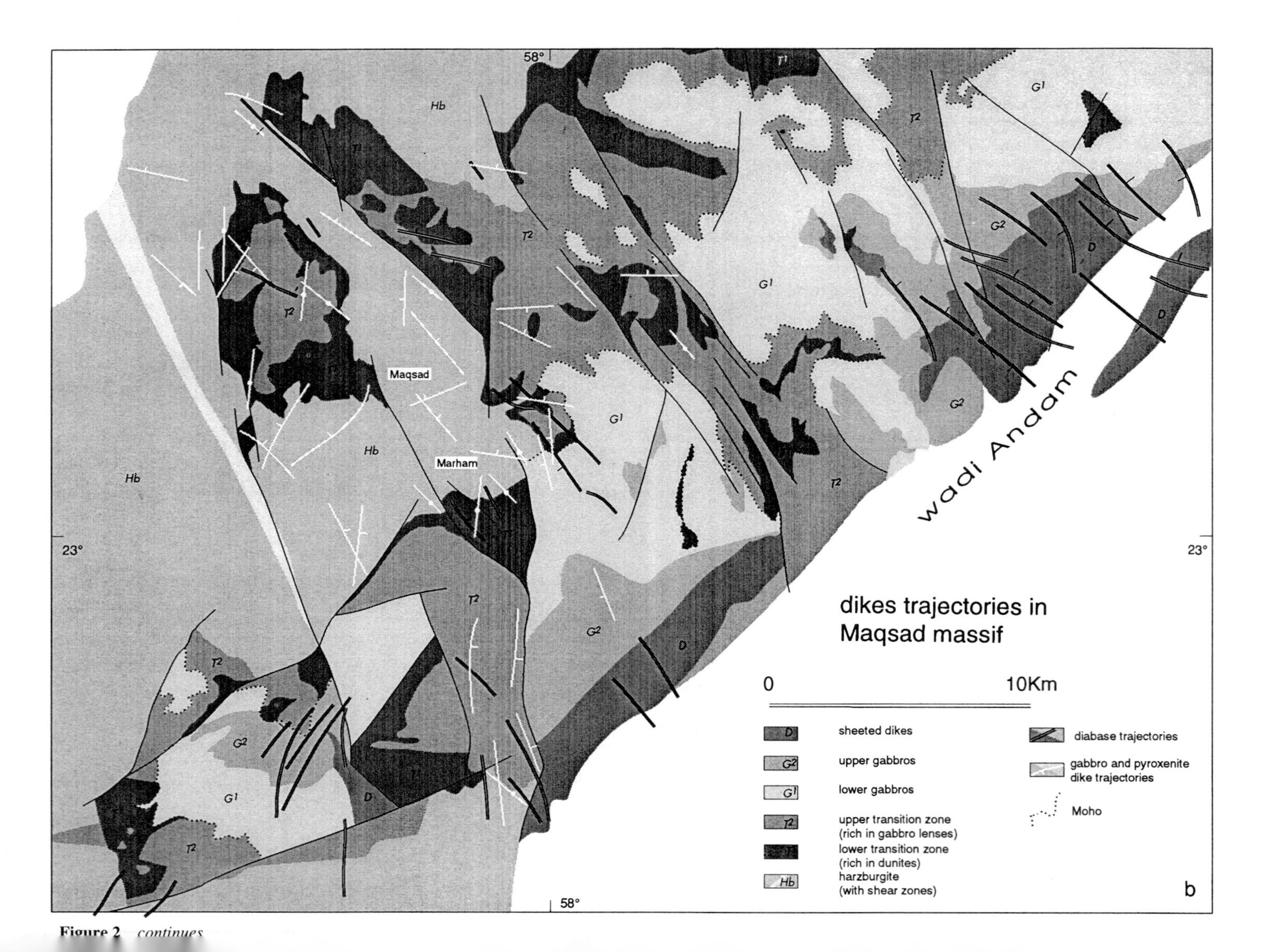

Figure 2 continues

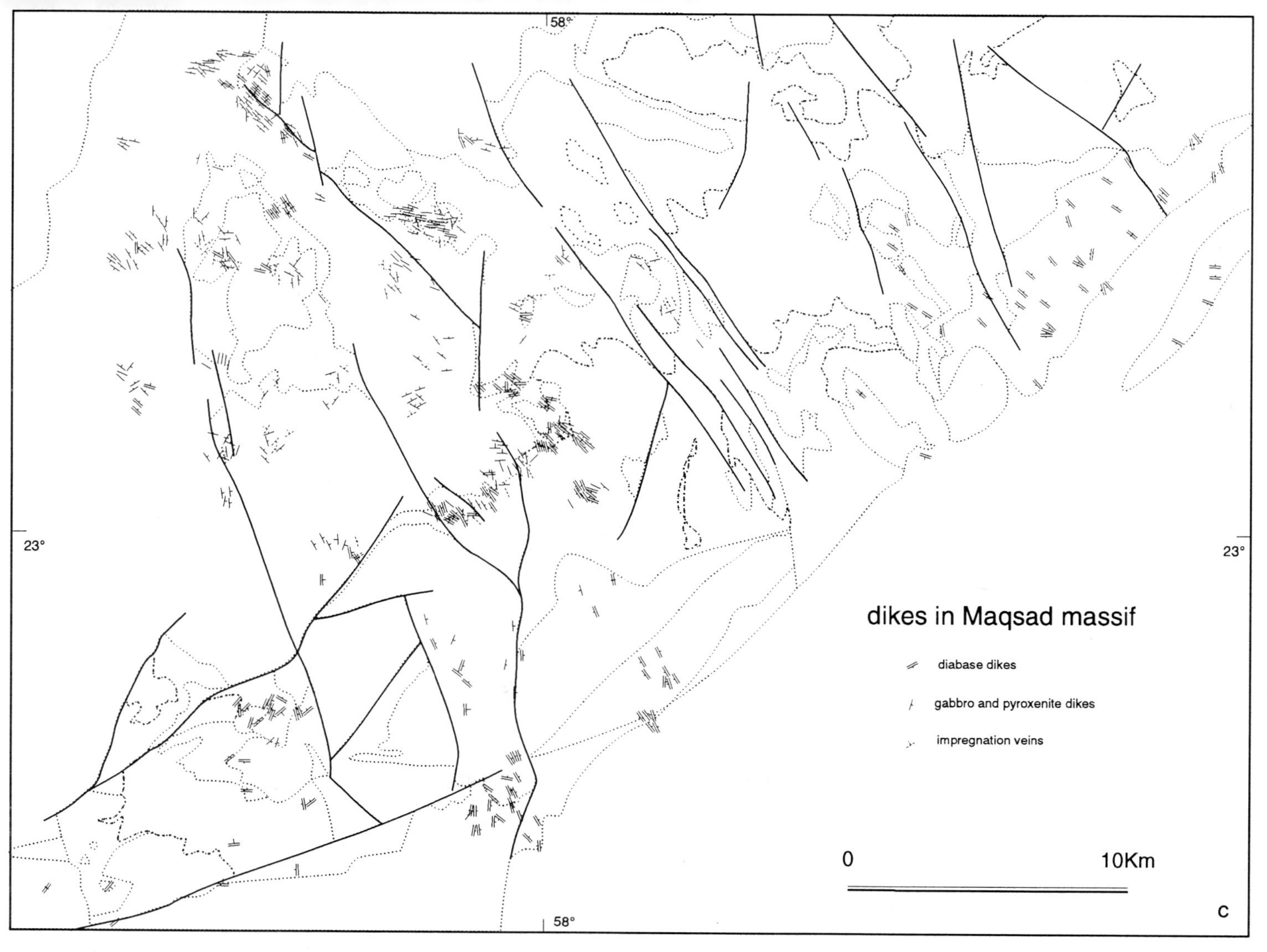

Figure 2 Continued.

possible to compare the dike orientations on the NE and SW sides of the ridge axis. The center of the diapir may also be exposed in the mantle section, making it possible to compare dike orientations along the strike of a ridge segment. Unfortunately, mantle outcrops are limited in the SE part of the segment compared to the northwestern portion, restricting possible comparisons.

In the Nakhl-Rustaq massif (Fig. 3), two diapiric areas have been identified and are separated by a high-temperature shear zone. In the western portion, mantle lithologies are exposed only around the southern termination of a diapir whose detailed relationship with the overall ridge segmentation is unclear. In the eastern area, mantle lithologies are exposed along the southwestern side of the ridge axis, making it possible to compare dike attitudes along the (25-km) half-length of a ridge segment.

Dike Typology in Relation to Diapiric Areas

Several generations of basaltic dikes may be distinguished and mapped separately in peridotite massifs (Nicolas and Jackson, 1982; Nicolas, 1989, pp. 21 and 65–67). Mapping the vast Oman ophiolite and delineating diapiric areas in mantle sections made it possible to clarify the relation between diapirs and successive generations of dikes. Dike density in the mantle sections of this ophiolite is very irregular, but it is always high in diapiric areas. As noted previously, this feature is explained by the channeling of melting asthenosphere and of melt itself by diapirs. In the frozen diapirs considered here, it is still possible to observe this original dike organization. In contrast, in mantle areas far from diapirs such as at the tip of a presumed propagating ridge (Fig. 3), the vigorous mantle flow radiating from the diapir has

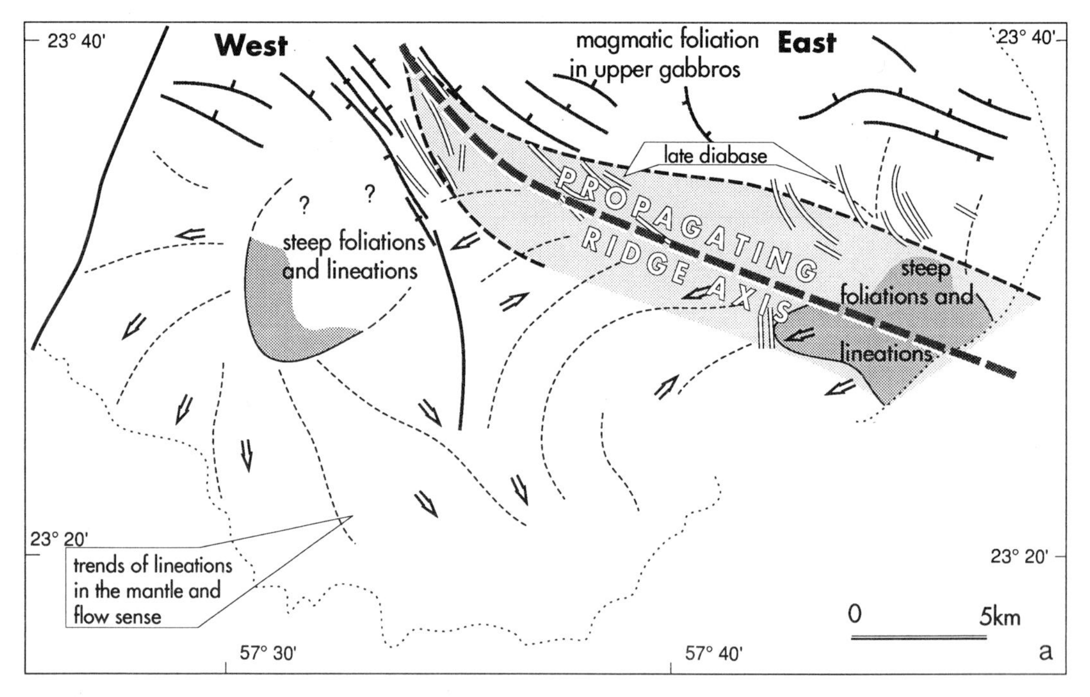

Figure 3 Nakhl-Rustaq massif. (a) Paleo-ridge model (location in Fig. 1; detailed map in Fig. 3b). Two diapiric areas have been identified (dark shading) and treated separately. The dikes in the western diapir (Fig. 9) were measured in the mantle rocks around the southern tip of the diapir; those in the eastern diapir (Fig. 10) along the SW flank of the ridge axis. Same symbols as in Fig. 2a (modified from Nicolas and Boudier, submitted 1993). (b) Geological map and dike trajectories. (c) Map of individual dike measurements. The northerly tilt of the massif exposes the mantle section to the south and the sheeted dike complex to the north. Local northward curvature in the sheeted dike trend is visible in the eastern part of the map and has been related to tectonic activity at the ridge (Nicolas and Boudier, submitted 1993).

continues

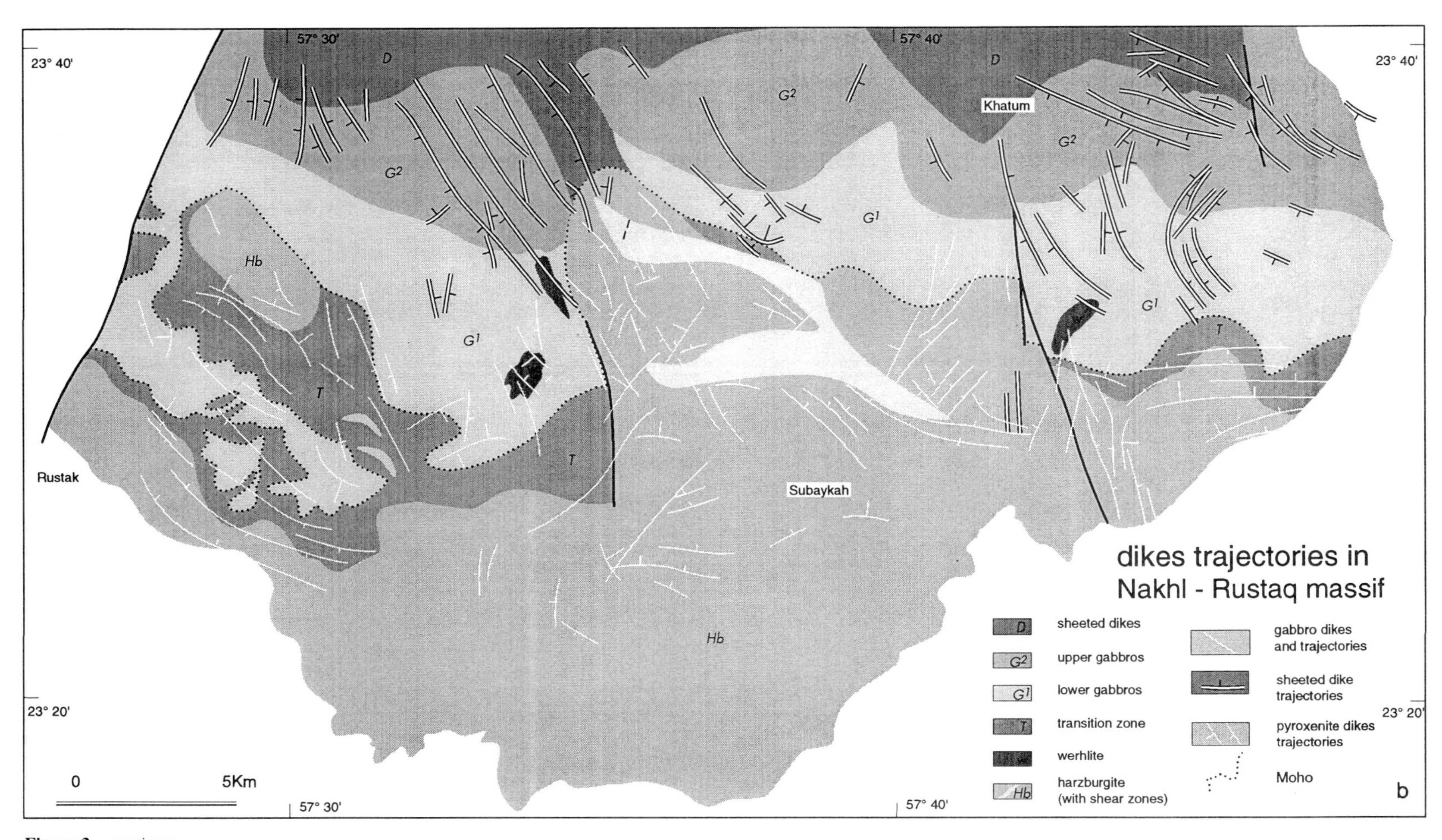

Figure 3 *continues*

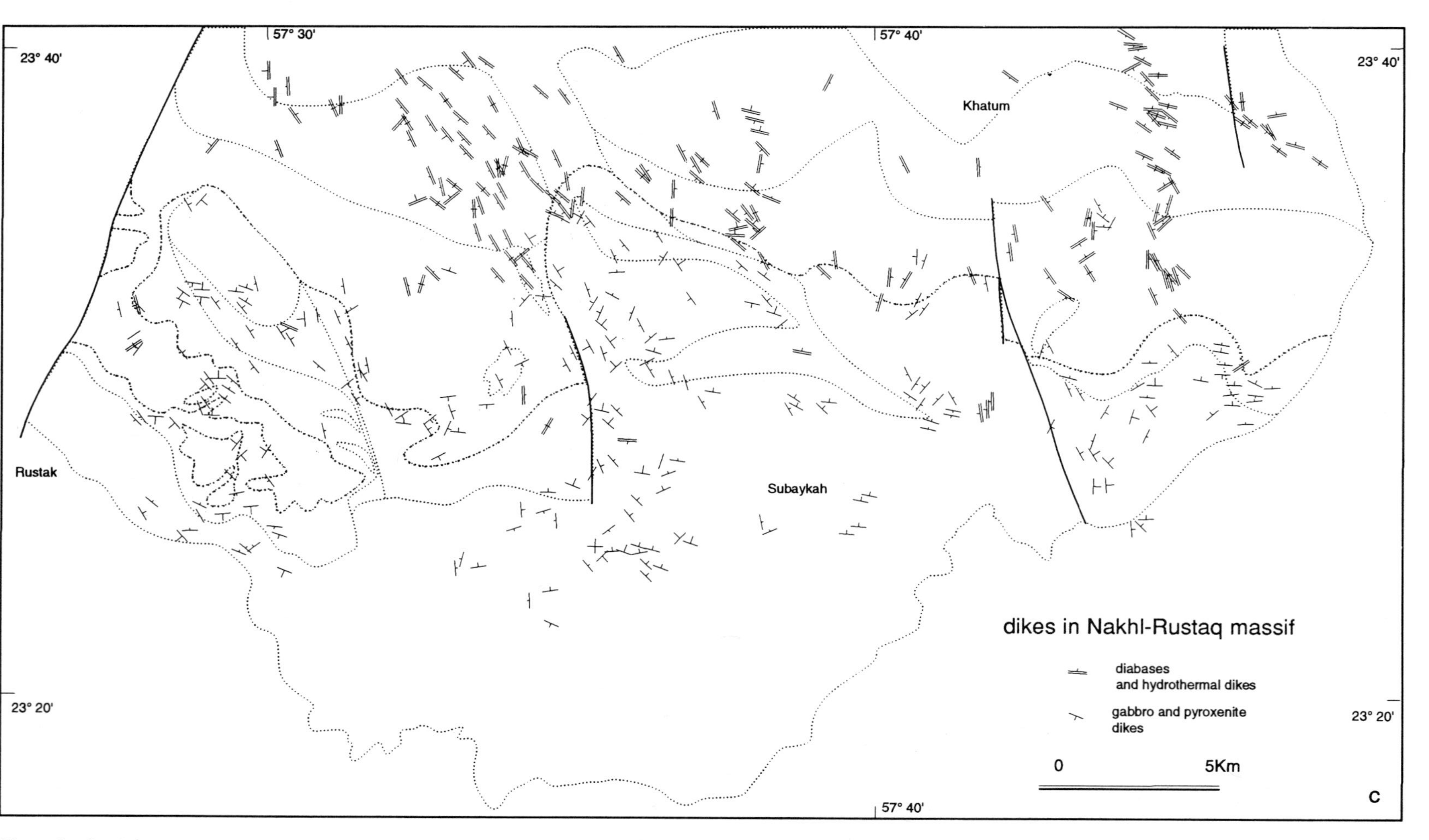

Figure 3 Continued.

deformed and transposed parallel to the flat-lying foliation, all dikes and melt segregations formed inside the diapir; only late dikes are discordant with respect to this foliation.

The first generation of dikes comprise early "indigenous" gabbro and pyroxenite dikes (Figs. 4a, 4c, and 4d), crossing melt-impregnated peridotites and having equilibrated at temperatures above the solidus (1200°C). The pyroxenite dikes are dominantly olivine-bearing websterites or clinopyroxenites and are less commonly orthopyroxenites. Both gabbro and websterite dikes may grade into impregnation clots (Fig. 4c) or into chromite schlierens and pods. Dunite veins are related to these dikes and are formed by wall-rock reaction of the circulating melt in the dike (Figs. 4a, 4b, 4c; see recent reviews by Kelemen, 1990, and Kelemen *et al.*, 1990). Within diapirs, this *first generation* of dikes and sills is often poorly defined in orientation. In contrast, away from diapirs in areas where post-intrusion plastic flow has been intense, these early dikes are deformed and transposed parallel to the foliation and have thus been incorporated into the ubiquitous peridotite banding or layering (Fig. 4d). The next generational type of gabbro and pyroxenite dikes has been called "intrusive" because such dikes have sharp and straight walls without reactions with surrounding peridotites and because they display a comb structure that has been induced by melt crystallizing from the cooler walls toward the dike center (Fig. 5a). Thus, in contrast to the indigenous dikes, these dike types were emplaced in peridotites below their solidus. They commonly constitute swarms of parallel dikes and range in thickness from a few centimeters to a few tens of centimeters (Fig. 5b). They are generally undeformed and discordant with respect to the foliation both within the vicinity of diapirs and far from them. Through microgabbro dikes, showing evidence of brittle fracturing, they grade into diabase dikes, first without chilled margins and finally with chilled margins. Diabase dikes correspond to temperatures in the host rocks lower than 450°C (Nehlig, 1989). Their orientation in the field is generally regular and their thickness is about 1 m.

This classification of dikes as a function of decreasing host rock temperature and, generally, the timing of intrusion should not be viewed too strictly, since there is continuity between the indigenous and the intrusive dikes, suggesting that magma-induced fracture is a continuous process and one that spans the evolving structure of a dynamic diapir and its surroundings. In addition, gabbros or pyroxenites can intergrade into each other and are encountered equally as often within the different categories of high-temperature dikes.

Dike Patterns in the Southeastern Oman Diapirs

In their study of the diapirs of southern Oman, Nicolas and Boudier (submitted 1993) concluded that in the peridotite section, two orientation groups of dikes and veins are defined. This conclusion is based on consideration of all the dikes and veins collectively, regardless of their nature and emplacement timing. These two groups are apparent in maps of the dike trajectories (Figs. 2b and 3b). One well-defined group of dikes that are steep and parallel to the sheeted dike complex is primarily composed of low-temperature dikes: diabase and fine-grained gabbros. It also includes some high-temperature dikes that are indigenous as well as intrusive. Indigenous dikes, however, as well as the dunite veins, have more dispersed orientations than intrusive dikes and, geographically, they tend to be restricted to the diapiric areas. Diabase dikes are not common in the peridotite or in the lower gabbro sections. As mentioned above, these dikes cluster near the inferred ridge axis. The orientation of the second group of dikes is more diffuse than that of the first group. They usually form moderate angles with respect to foliation in the enclosing peridotite, and they grade into sills, thus plotting close to the center of ste-

Figure 4 Photographs of indigenous dikes. (a) Irregular indigenous clinopyroxenite dike (hammer) surrounded by a dark dunite wall in contact with the lighter harzburgite. A coeval dunite sill is present as an off-shoot (parallel to the horizontal foliation and banding of the harzburgite). Another thin and more regular indigenous clinopyroxenite dikelet (left) is cut by an intrusive gabbro dikelet. (b) Dunite vein network in darker harzburgite; the sill and dike orientations are visible to the right. (c) Irregular indigenous gabbro dike grading into melt impregnations (to the left). (d) Melt impregnations, marked by lighter plagioclase, which have been tectonically transposed parallel to the peridotite foliation.

Figure 4 See legend on p. 85.

Figure 5 Photographs of intrusive gabbro dikes in harzburgites. (a) Comb structure perpendicular to the sharp-margin dike wall. (b) Network of parallel dikes emplaced into a harzburgite with a flat-lying foliation.

Figure 6 Relations between sills and dikes. (a) Apparent feeding of gabbro sills in dunites by a vertical dike (to the right); other dikelets are visible near the lower left corner (Buri, Sumail Massif). (b) Network of gabbro sills and dikes with melt that circulated into one another. This photograph is from northern Oman (Wadi Fayd); the dikes are EW and the sills are NNW–SSE, parallel to the foliation and to a locally steeply dipping Moho. Note an olivine segregation in the center of intrusions.

reograms (Figs. 8, 9, 10). As a consequence, in map views, they tend to wrap around the diapir. This second group of dikes is composed of dunite veins, indigenous dikes, and intrusive coarse-grained gabbros. Diabase dikes and fine-grained gabbros intruded into cooling peridotites are present but not common in this group.

An observation, mentioned several times in the field, is that the same melt can be injected in both a dike and a sill configuration. In a few places, dikes oriented parallel to the inferred ridge trend have been observed feeding sills, which are themselves parallel to the Moho within the transition zone (Fig. 6a). More commonly, in areas of heavy diking, dikes and sills show no cross-cutting relationships, and flow structures are continuous and locally wrap around the corner of the intersections (Fig. 6).

A closer analysis of these data is now presented to examine the relationships between dike orientation in the uppermost mantle and either the plastic flow framework or the paleo-ridge framework.

Dikes in the Plastic Flow Reference Frame

These relationships have been studied in the Maqsad area. All dikes, whatever their mineral content, grain size, or structure, have been individually rotated into the structural reference frame of their enclosing peridotites: the high-temperature foliation (oriented perpendicular to the projection plane of the stereograms in Fig. 7) and the mineral lineation (oriented EW in these stereograms). If the orientation of the dikes were controlled by high-temperature plastic flow in the country rocks, one would expect the dikes to display a geometrical relationship with the structural elements. We show in Fig. 7 the results for the (a) intrusive and (b) indigenous dikes. The preferred orientation of intrusive dikes (Fig. 7a) is weak and does not exhibit a pattern consistent with control related to high-temperature foliation. On the other hand, indigenous dikes (Fig. 7b) plot as a weak girdle, with a maximum normal to the foliation that corresponds to fractures parallel to the foliation and may suggest some structural control.

Dikes in the Ridge Reference Frame

To study the dike orientations in the paleo-ridge reference frame, we rotated the Moho in the two massifs considered into a horizontal position. The rotation is 15° down to the west about a north–south axis (symbolized as 0°W15°) in Maqsad and from 110°SW32° to 105°SW30° in Nakhl-Rustaq from west to east. These rotations—deduced from structural maps—have been indirectly checked by the fact that they result in nearly vertical preferred orientations of the diabase dikes from the sheeted dike complex (continuous line in the stereograms of Figs. 8–10). All dikes were rotated following these axes and refer to these NW–SE paleo-ridge directions.

Figure 8 compares the dike orientations from

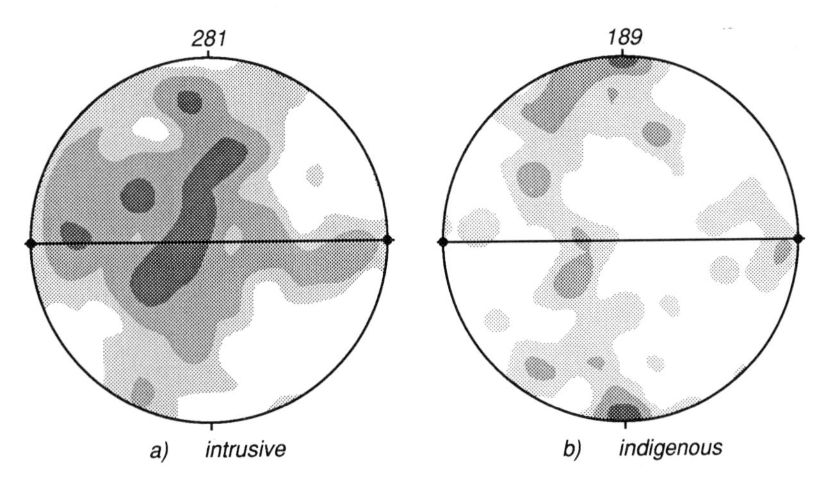

Figure 7 Stereograms (lower hemisphere) showing the pole distribution of (a) intrusive and (b) indigenous dikes from the Maqsad massif, plotted in the plastic flow reference frame: foliation vertical EW (straight line), lineation EW (dots at ends of the line). Contours 1, 2, 4% net area between 100 and 200 measurements and 0.5, 1, 2% net area above 200 measurements. Numbers indicate the hemisphere subtotal of individual measurements.

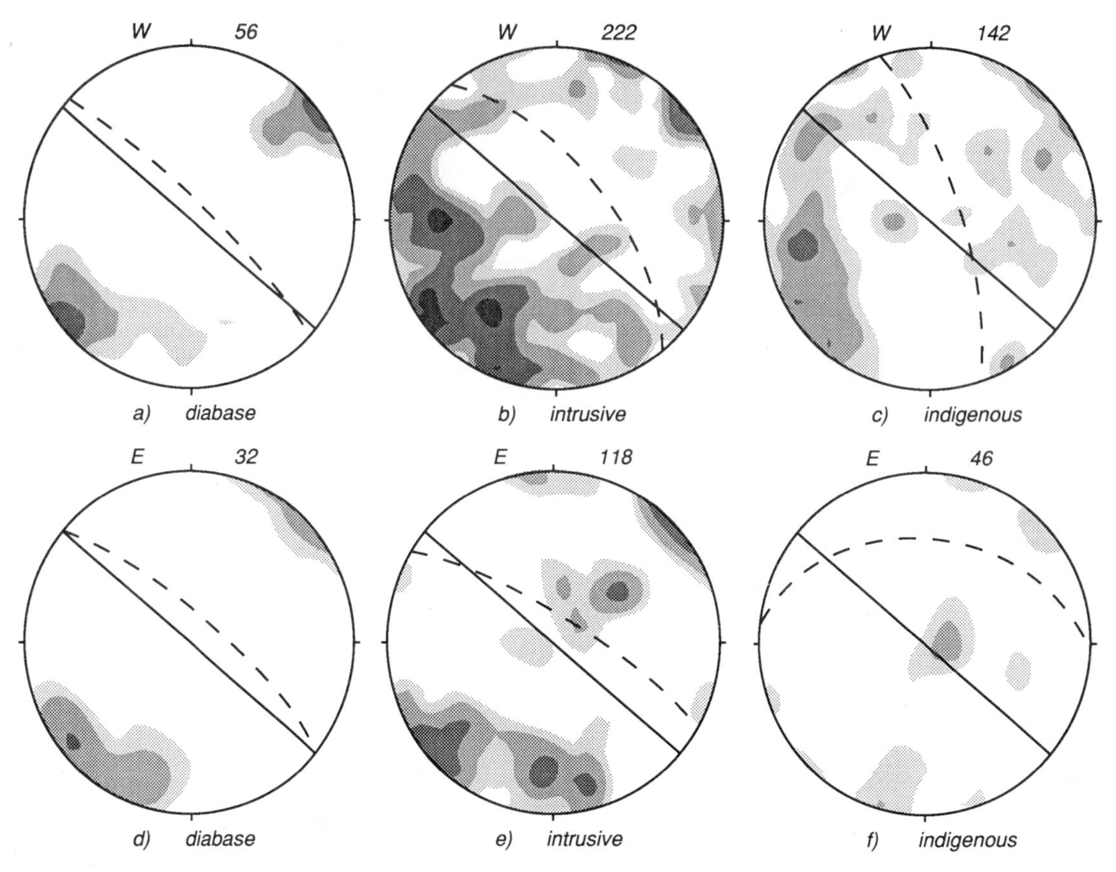

Figure 8 Stereograms (lower hemisphere) of the various dikes in Maqsad massif, plotted into the paleo-ridge reference frame: Moho horizontal, ridge trend defined by that of the sheeted dike complex (continuous line). The computed best plane is projected as dashed lines. Contours: 2, 4, 8% between 30 and 100 measurements; 1, 2, 4, 8% between 100 and 200 measurements; and 0.5, 1, 2, 4% above 200 measurements. Numbers indicate the hemisphere subtotal of individual measurements. W and E symbols refer to data from southwest (W) and northeast (E) of the inferred Maqsad ridge axis.

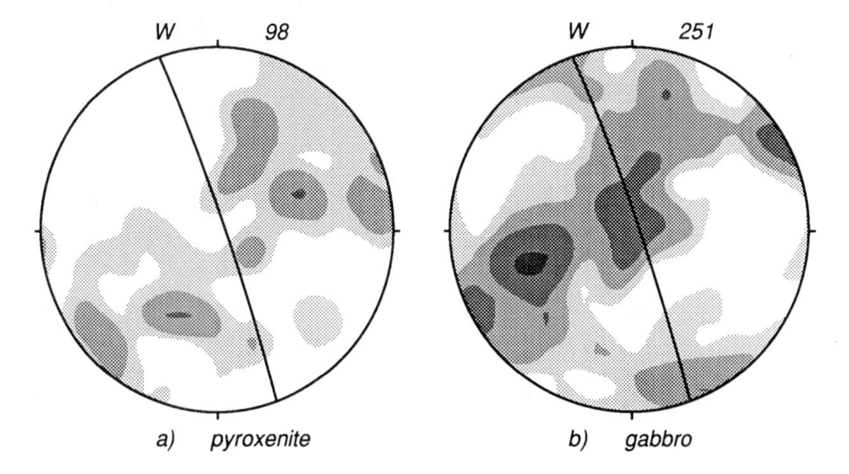

Figure 9 Stereograms (lower hemisphere) of (a) pyroxenite and (b) gabbro dikes around the western diapir of Nakhl-Rustaq massif, plotted into the paleo-ridge reference frame. Same conventions as in Fig. 8.

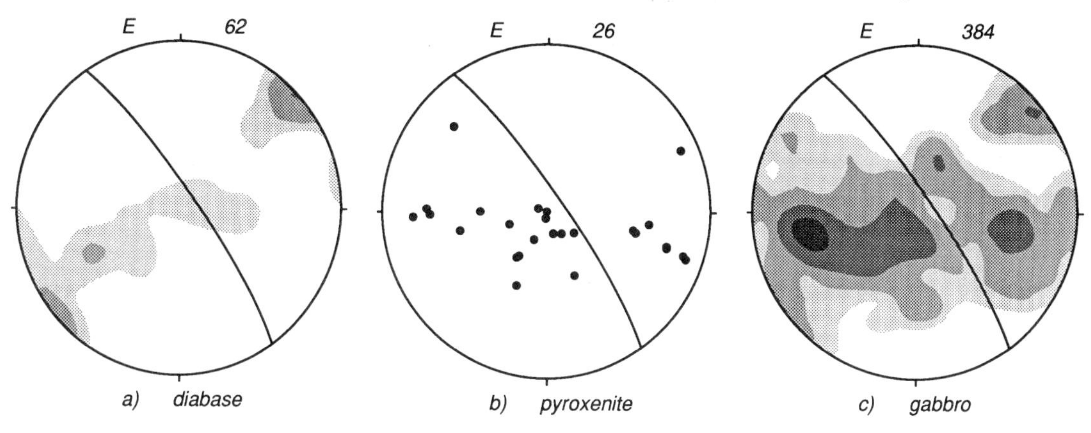

Figure 10 Stereograms (lower hemisphere) of the various dikes along the SW flank of eastern diapir of Nakhl-Rustaq massif, plotted in the paleo-ridge reference frame. Same conventions as in Fig. 8.

the SW side of the Maqsad ridge segment (Figs. 8a–8c) with the orientations from the NE side (Figs. 8d–8f). The diabase dikes emplaced in the peridotites and lower gabbro sequence (Figs. 8a and 8d) plot nearly parallel to the similar diabase dikes of the sheeted dike complex, thus clearly being related to regional lithospheric tensional fracturing. Incidentally, the sheeted dikes used as reference are exposed 25 km from the diapiric area, as seen in Fig. 2, illustrating our previous conclusion that the massif was not significantly deformed after the original ridge accretion events. A first inspection of the intrusive dikes (Figs. 8b and 8e) leads to the conclusion that they are also preferentially emplaced according to a lithospheric tensional orientation. However, their overall orientation is more scattered, with the occurrence of sills plotting near the center of the nets. The best computed plane (dashed lines) of these two stereograms shows that the dike azimuths tend to convergence toward the ridge axis, meaning that dikes from the SW flank plot with a preferred NNW trend, and, conversely, dikes from the NE flank, with a WNW trend; this feature was already visible on the trajectory map (Fig. 2b). Finally, the indigenous dikes (Figs. 8c and 8f) produce diffuse stereograms with, possibly, a similar tendency to split into a group of steep NW–SE dikes and another group of sills. The group of indigenous dikes of NW–SE lithospheric orientation suggests the same convergence toward the ridge axis.

Figure 9 compares the orientations of clinopyroxenite and gabbro dikes in the mantle around the southern termination of the western diapir of Nakhl-Rustaq massif. The intrusive dikes were not distinguished here from indigenous ones. Pyroxenite and gabbro dikes have diffuse but comparable orientations in the ophiolite. This agreement is not surprising since they grade into one another, with a tendency for the pyroxenite dikes to postdate the gabbro dikes. Both categories of dike plot as diluted girdles composed of dikes parallel to the sheeted dike complex and others approaching the attitude of sills parallel to the paleo-Moho, with, however, a large number of intermediate orientations.

In the mantle section SW of the eastern segment, the diabase dikes exposed in the mantle and lower crust close to the paleo-ridge axis (Fig. 10a) plot in a dominant orientation that is parallel to the sheeted dikes, with, however, a significant number of admixed sills. Measurements of pyroxenite dike orientations are simply too scarce to be representative (Fig. 10b). The plot of gabbro dikes (Fig. 10c) produces a weak girdle, again reflecting the dichotomy of dikes oriented approximately parallel to the sheeted dikes and of sills oriented horizontally or with a moderate dip. The girdle obliquity with respect to the sheeted dike orientation shows the same sense as that recorded in the Maqsad massif, consistent with the inferred structural setting on the SW flank of the ridge axis.

Discussion

In considering dike orientations, one must be aware that, except for diabase dikes, there is a spectrum of orientations, even at the scale of a single outcrop. Dikes as remarkably parallel as those in Fig. 5b are uncommon. This spectrum explains why the most representative stereograms, which include a few hundred measurements made over areas of several tens of square kilometers, show diffuse preferred orientations. Therefore, their interpretation, in an effort to understand regional and local contributions to their mechanics, is not straightforward. It is also necessary to observe a large number of individual situations before reaching conclusions. On the basis of the present study and measurements and observations in other areas of the Oman ophiolite (Ildefonse *et al.,* 1993), the following points seem well established.

• Whatever the point of view, there is a continuum between the different dike intrusions. Petrologically, pyroxenites and gabbros grade into one another, coarse gabbros grade into microgabbros, and these, in turn, grade into diabases. In terms of the time sequence and thermal structure of wall-rocks, indigenous dikes, intruded into concurrently melting peridotites, subsequently grade into intrusive dikes with evidence of progressively cooler wall-rocks, the coolest corresponding to diabase dikes with chilled margins. This is also true of their preferred orientations.

• When plotted in the inferred paleo-ridge reference frame (Moho rotated to be horizontal, ridge trend defined by the diabase sheeted dike trend), the intrusions cluster in two main orientations, being either vertical and parallel to the sheeted dike trend or flat-lying, comprising sills that are approximately parallel to the Moho. These two preferred orientations grade into one another and thus define girdles in their resulting stereograms. They are observed in all dike types, although sills are more strongly represented in the indigenous group of intrusions (and are rare in the diabase group). In contrast, vertical dikes dominate in the diabase group, where the preferred orientations are also the strongest.

• Individual observations confirm that a given parcel of melt can be injected either as a sill or as a dike and can flow from one orientation to the other (Fig. 6).

• Dikes tend to deviate slightly from the paleo-ridge trend as defined by the sheeted dikes. In the massifs considered, where the presence of mantle diapirs permitted the location of the ridge axis in the field (Figs. 2a and 3a), it is observed that dikes from the SW flank trend more northerly than the ridge axis and dikes from the NE flank trend more southerly (Fig. 8).

When projected in the structural reference frame of the enclosing peridotites (high-temperature foliation and lineation, Fig. 7), dikes do not exhibit clear preferred orientations. This fact indicates that dike orientation is not controlled by local flow structure or by a mechanical weakness plane, which would relate to these structures. There is, however, a weak tendency for indigenous dikes to plot parallel to the foliation. This tendency may reflect the tectonic transposition of dikes due to high-temperature flow, as described in previous studies (Nicolas and Jackson, 1982). Many dikes were observed to be wholely plastically deformed and close in orientation to the foliation plane. This trait was also found to be true for a large number of chromite deposits (concordant deposits; Cassard *et al.,* 1981). In both situations, these concordant features are explained by the tectonic rotation of the dikes (tectonic transposition), which is induced by a large plastic flow occurring subsequent to the initial dike or chromite pod intrusion. Because our study was conducted mainly in diapiric areas where this plastic flow (related to spreading away from the ridge) is reduced, this problem is partly avoided. However, the deepest indigenous dikes, dunite veins, and chromite pods emplaced in the harzburgite below the transition zone have been tectonically transposed parallel to the peridotite foliation (Ceuleneer and Nicolas, 1985). They contribute to the weak preferred orientation in Fig. 7b, which is parallel to the plastic foliation plane.

Following earlier conclusions (Nicolas and Jackson, 1982), dike orientations in peridotites are believed to be controlled by the ambient stress field, being parallel to the (σ_1, σ_2) principal stress directions and, thus, normal to the σ_3 least principal stress. This conclusion is based on the results of detailed kinematic analyses and on the fact that peridotites are comparatively massive rocks in which the foliation cannot be considered a mechanical weakness plane. Our study of dike orientations in relation to the structural reference frame of high-temperature enclosing peridotites tends to support this conclusion by failing to show any clear relationship between intrusion-preferred orientations and this reference frame.

This conclusion is readily applied to all vertical dikes parallel to the diabase sheeted dike complex, which, being parallel to the presumed paleoridge plane, would be in the orientation of primary lithospheric fracturing. The occurrence of sills and moderately inclined dikes suggests the existence of another stress field, such that σ_3, instead of being horizontal and normal to the ridge, is vertical or nearly vertical. Our two-dimensional modeling of asthenospheric mantle diapirism below ridges (Rabinowicz *et al.*, 1984, 1987) shows that at the outskirts of diapirs, the forced mantle flow generates (just below the Moho) a stress field such that σ_3 is steeply plunging and the (σ_1, σ_2) plane is flat-lying or moderately inclined toward the axis. It is tempting to conclude that the sills are controlled by the asthenospheric stress field related to mantle diapirism. The predominance of sills in the general environment of high-temperature dikes (as contrasted with "low-temperature" intrusions) suggests this tendency, as the former are intruded into asthenospheric mantle and the latter into lithospheric mantle.

A difficulty with the model of dikes and sills that are controlled by two independent stress fields—lithospheric and asthenospheric, respectively—is that the same melt injection can form a sill *and* a dike, melt moving from one into the other. The necessarily rapid change in the orientation of the stress field can be attributed to spatial or temporal changes. Local stress field deviations produced by shear motions are well documented in earth science and have been invoked to explain the synchronous injection of a fluid phase in both shear and tension orientations. This situation is locally observed in our dikes on the scale of a few meters, but we do not believe that it applies generally because within the dikes studied, oblique foliations symptomatic of shear motion are uncommon (Figs. 4–6). These intrusions are also oriented at 90° angles to each other (Fig. 6) and not at the moderate angle that one should expect in a shearing environment.

The temporal change in stress field is an interesting alternative. In previous publications (Nicolas, 1986, 1990), it was proposed that basaltic melt is brought to the spreading center by hydrofractures produced within and issuing from the concurrently rising and melting mantle diapir. Hydrofracturing events are episodic and relatively violent. They bring to the accreting crust a quantity of melt large enough to create 1 m of crust, the width of individual diabase dikes in the sheeted dike complex. Independent pieces of evidence (melt budget, cooling time of dikes) show that a single event lasts only a few weeks. Taking this fact into account, the following scenario is proposed. In the oceanic lithosphere at the ridge, which is already under tensile stress, the sudden release of melt triggers the opening of a fracture, which widens to 1 m during a time lapse of a few weeks. This opening relaxes completely the lithospheric stress. It is speculated that at the onset of the magmatic event, even at the Moho depth beneath the ridge axis, the regional lithospheric stress field predominates over the more local stress field that is related to the relatively small-scale diapir activity. Thus, the first intrusion of melt propagates along regional fractures parallel to the ridge plane (lithospheric fracturing). Once the primary magma batch is injected and lithospheric tensional stresses are thus relaxed, the magma tends to be subsequently channeled into sills controlled by the more local asthenospheric stress field that is associated with the divergence of diapiric flow. This scenario explains that, during the same melt surge, "lithospheric" and "asthenospheric" controls can be successively exerted on the magma-induced fracture orientations.

If the orientation of the sill group is controlled by the stress field related to diapiric activity, one could expect to observe different orientations depending on the location with respect to the diapiric center. Our efforts in Maqsad have been unsuccessful in providing a definitive answer here (see Fig. 8), probably because there is too much scatter in the sill orientations. However, around the Batin diapiric center (Wadi Tayin massif), which is still under study, (Ildefonse *et al.*, 1993) we have detected a tendency for the sills to wrap around the diapir center with a moderate dip away from it. The only consistent figure emerging from the present study in terms of local orientations with respect to diapir location is that the azimuth of vertical dikes tends to converge toward the ridge axis, which means that to the SW of our ridge axes, the dikes have a more northerly azimuth than the ridge itself (Figs. 8a–8c and 10) and, conversely, for those dikes located NE of the ridge (Figs. 8e and 8f), the trend is southerly. This tendency may be ascribed to local deviations of the "lithospheric" stress, but their detailed origin is not well understood.

Conclusions

The diapiric areas of Oman represent exceptionally favorable sites to study the plumbing of a ridge at the Moho level and immediately below. The numerous dikes observed and measured are poorly constrained in their orientations as shown by the stereograms in this chapter. This tendency is mainly true for the earliest or more central dikes emplaced into the melting peridotites of diapirs. In contrast, the diabase dikes—emplaced when the ridge was cooling—have excellent preferred orientations that are mainly parallel to the sheeted dike trend. Whatever the timing and location of the intrusions, a group of dikes is oriented parallel to the sheeted dike complex and ascribed to a relatively regionally organized *lithospheric stress field*. Another group is composed of flat-lying sills and moderately inclined dikes, thought to be controlled by the locally organized *asthenospheric stress field* created by the diapir flow divergence just below the ridge. Melt is injected in either the sills or the dikes. This feature suggests a relatively rapid shift in the orientation of the local stress field, which is attributed to the effects of a large-volume and relatively violent melt surge from the rising diapir into the newly accreting crust.

Because orientations in the high-temperature dikes are, at present, so poorly clustered, more work is necessary before solid conclusions on this interesting class of intrusions may be drawn.

Acknowledgments

This chapter has benefited from reviews and constructive remarks by John S. Pallister, Michael P. Ryan, and an anonymous reviewer. A number of colleagues and students have contributed to field measurements: M. Misseri, S. Crambert, J. L. Bouchez, G. Ceuleneer, I. Reuber, C. Sotin, V. Thomas, C. MacLeod, K. Benn, F. Quatrevaux, and S. Tait. Data were computed by A. Saintenoy and A. Replumaz. This is contribution No. 535 from the Dynamique et Bilans de la Terre program of the Institut National des Sciences de l'Univers-Centre National de la Recherche Scientifique.

References

Anderson, R. N., *et al.* (1982). DSDP Hole 504B, the first reference section over 1 km through layer 3 of the oceanic crust, *Nature* **300,** 589–594.

Auzende, J. M., Bideau, D., Bonatti, E., Cannat, M., Honnorez, J., Lagabrielle, Y., Mallavieille, J., Mamaloukas-Frangoulis, V., and Mével, C. (1989). Direct observation of a section through slow-spreading oceanic crust, *Nature* **337,** 726–729.

Barth, G.A., Mutter, J.C., and Madsen, J.A. (1991). Upper-mantle seismic reflections beneath the East Pacific Rise, *Geology* **19,** 994–996.

Boudier, F., Ceuleneer, G., and Nicolas, A. (1988). Shear zones, thrusts and related magmatism in the Oman ophiolite: Initiation of thrusting on an oceanic ridge, *Tectonophysics* **151,** 275–296.

Boudier, F., and Nicolas, A. (1972). Fusion partielle gabbroïque dans la lherzolite de Lanzo (Alpes piémontaises), *Bull. Suisse Minéral. Pétrol.* **52,** 39–56.

Boudier, F., and Nicolas, A. (1977). Structural controls on partial melting in the Lanzo peridotites, *in* "Magma Genesis" (H. J. B. Dick, ed.), pp. 63–77, State Oregon Dpt. Geol. Min. Ind.

Cassard, D., Moutte, J., Nicolas, A., Leblanc, M., Rabinowicz, M., Prinzhofer, A., and Routher, P. (1981). Structural classification of chromite pods from New Caledonia. *Econ. Geol.* **76,** 805–831.

Ceuleneer, G. (1991). Evidences for a paleo-spreading center in the Oman ophiolite: Mantle structures in the Maqsad area, *in* "Ophiolite Genesis and Evolution of Oceanic Lithosphere" (T. Peters, A. Nicolas, and R. G. Coleman, eds.), pp. 147–173, Kluwer, Dordrecht.

Ceuleneer, G., and Nicolas, A. (1985). Structures in podiform chromite from the Maqsad district (Sumail ophiolite, Oman), *Mineral. Deposita* **20,** 177–185.

Ceuleneer, G., Nicolas, A., and Boudier, F. (1988). Mantle flow patterns at an oceanic spreading centre: The Oman peridotite record, *Tectonophysics* **151,** 1–26.

Garmany, J. (1989). Accumulations of melt at the base of young oceanic crust, *Nature* **340,** 628–632.

Ildefonse, B., Nicolas, A., and Boudier, F. (1993). Evidence from the Oman ophiolite for sudden stress changes during melt injection, *Nature* **366,** 673–675.

Kelemen, P.B. (1990). Reaction between ultramafic rock and fractionating basaltic magma. I. Phase relations, the origin of calc-alkaline magma series, and the formation of discordant dunite, *J. Petrol.* **31,** 51–98.

Kelemen, P.B., Joyce, D.B., Webster, J.D., and Holloway, J.R. (1990). Reaction between ultramafic rock and fractionating basaltic magma. II. Experimental investigation of reaction between olivine tholeiite and harzburgite at 1150–1050°C and 5 kb. *J. Petrol.* **31,** 99–134.

Nehlig, P. (1989). "Etude d'un système hydrothermal océanique fossile: l'Ophiolite de Semail (Oman)," Thèse de Doctorat, Université Brest.

Nicolas, A. (1986). A melt extraction model based on structural studies in mantle peridotites, *J. Petrol.* **27,** 999-1022.

Nicolas, A. (1989). "Structures of Ophiolites and Dynamics of Oceanic Lithosphere," Kluwer, Dordrecht.

Nicolas, A. (1990). Melt extraction from mantle peridotites: Hydrofracturing or porous flow consequence on oceanic ridge activity, *in* "Magma Transport and Storage" (M. P. Ryan, ed.), pp. 160–174, Wiley, Chichester/Sussex, England.

Nicolas, A., and Boudier, F. (1993). Mapping mantle diapirs and oceanic crust segments in Oman ophiolites, submitted for publication.

Nicolas, A., Boudier, F., and Montigny, R. (1987). Structure of Zabargad Island: An early rifting of the Red Sea, *J. Geophys. Res.* **92,** 461–474.

Nicolas, A., Ceuleneer, G., Boudier, F., and Misseri, M. (1988). Structural mapping in the Oman Ophiolites: Mantle diapirism along an oceanic ridge, *Tectonophysics* **151,** 27–56.

Nicolas, A., and Jackson, M. (1982). High-temperature dikes in peridotites: Origin by hydraulic fracturing, *J. Petrol.* **24,** 188–206.

Rabinowicz, M., Ceuleneer, M., and Nicolas, A. (1987). Melt segregation and flow in mantle diapirs below spreading centers: Evidence from the Oman ophiolites, *J. Geophys. Res.* **92,** 3475–3486.

Rabinowicz, M., Nicolas, A., and Vigneresse, J. L. (1984). A rolling mill effect in asthenospheric beneath oceanic spreading centers, *Earth Planet. Sci. Lett.* **92,** 3474–3486.

Ravaut, P. (1992). "Le levé gravimétrique de la chaîne omanaise: Contribution à l'étude des mécanismes de compensation," Diplôme d' Etudes Approfondies, Université Montpellier 2.

Chapter 6 | Neutral-Buoyancy Controlled Magma Transport and Storage in Mid-ocean Ridge Magma Reservoirs and Their Sheeted-Dike Complex: A Summary of Basic Relationships

Michael P. Ryan

Overview

This chapter reviews the neutral-buoyancy phenomenon as it applies to the injection and high-level storage of magma in the mid-ocean ridge environment. Petrophysical data, seismic surveys, and analytic and numerical inversions of geodetic data indicate that the primary mode of along-axis magma transport and storage within volcanic rift systems is regulated by the neutral buoyancy of picritic and tholeiitic melt, and melt and crystal mixtures. The depths of the horizon of neutral buoyancy (HNB) that controls the dynamics of magma storage and lateral injection along the East Pacific Rise, the Valu Fa Ridge–Lau Basin, Kilauea and Mauna Loa volcanoes, Hawaii, and the Krafla Central volcano, Iceland, are remarkably similar, although depths vary somewhat from system to system.

The horizon of neutral buoyancy for MORB may be subdivided into a picritic horizon of neutral buoyancy, HNB_P, and a tholeiitic counterpart, HNB_T. For the East Pacific Rise at 9°N, elastic crack stability relations suggest that $\approx 600 \leq HNB_T \leq \approx 1400$ m, whereas mixtures of picritic melt and olivine crystals are in gravitational equilibrium over $\approx 1400 \text{ m} \leq HNB_P \leq \approx 3000$ m depth beneath the ridge axis. Differentiation processes therefore transform melt in equilibrium at HNB_P into a new equilibrium setting at HNB_T. Thus elastic crack stability relations are consistent with: (i) the paucity of picritic eruptive products on the sea floor; and (ii) the dominantly tholeiitic nature of the sheeted-dike complex and ridge-crest eruptives. For the aggregate density range $2.6 \leq \rho_b \leq 2.82 \text{ g} \cdot \text{cm}^{-3}$, and the compositional range picritic melt + olivine crystals through tholeiitic basalts to ferrobasalts, the transition region between negative buoyancy and neutral buoyancy has an *average* depth of ≈ 1000 m beneath the East Pacific Rise surface.

The *generalized* buoyancy zonation of the East Pacific Rise magma reservoir and its surroundings may be considered in light of the complete compositional and magma density range: $2.6 \leq \rho_b \leq 2.82 \text{ g} \cdot \text{cm}^{-3}$, where ρ_b is the bulk (melt ± crystal) density. The generalized zonation is:

(I) *Negative-buoyancy region.* This region extends from the volcanic surface to ≈ 1 km depth. The density of nonvesiculated magma is greater than that of the country rock such that magma *may descend* under negative-buoyancy forces. This low-density region is produced by high amounts of fracture and grain-scale porosity that also locally lower the elastic moduli and *in situ* elastic wave velocities. It is a gravitational "no-man's-land" for magma, and to erupt, nonvesiculated magma must traverse the region aided by deeper replenishment. Theoretical dike shapes for *descending* magma batches have bulbous (lower) noses and slender (upper) tails. Lithologically, this region corresponds to pillow basalts ± local sediments.

(II) *Neutral-buoyancy region.* This region extends from ≈ 1 to ≈ 3 km depth. The overall density of magma is equal to that of the country rocks, and magma is in local mechanical (gravitational) equilibrium with its surroundings. The differentiation of picritic melt, and the separation of tholeiitic melt from suspensions of melt + crystal mixtures will render magma parcels within this region *positively* buoyant. These tholeiitic parcels will ascend to form a tholeiitic layer at the top of the magma chamber. Rupture of the chamber roof will nucleate a dike-forming event within the dike complex. Thus parcels of magma within the region of neutral buoyancy may flicker between states of neutral and positive buoyancy, depending on their extent of differentiation and their load of suspended crystals. Within the upper portions of this region, the dynamics of lateral magma injections that produce the mid-ocean ridge sheeted-dike complexes mimic their counterparts in active Icelandic and Hawaiian rift zones. Theoretical dike shapes along the HNB have slender tails and enlarged midsections (across-axis profile), whereas the along-axis cross sections show advancing dikes with parabolic noses. Local retardations of along-axis intrusions force dike keels to descend while the dike tops rise toward the axial valley floor, driven by flow rates from the ruptured chamber on the order of $\approx 100-500\ m^3 \cdot s^{-1}$. High flow rates and continued crack-front arrest promote the ascent of the dike top and subsequent eruption. Breakthroughs along the path of lateral advance propel the intrusion forward, bringing its top and bottom back to $\approx 1-\approx 3$ km depth and gravitational equilibrium positions. The overall lateral intrusion process is, therefore, the integrated contribution from each incremental crack advance. An entire sheeted-dike complex is thus produced by successive neutrally buoyant lateral intrusions. The lithology of the HNB is sheeted basaltic dikes and the upper levels of the isotropic gabbro complex.

(III) *Positive-buoyancy region.* This region extends from ≈ 3 to ≈ 75 km depth where melt densities are always less than the country rock. Melt migration begins along a hierarchy of grain-scale microporous networks within the disaggregating parent lherzolite, supplying vein swarms and deep dikes. Theoretical dike shapes for ascending melt batches have bulbous (upper) noses and slender (lower) tails. Lithologically, this region corresponds to garnet and spinel lherzolites, grading further upward into dunite-, websterite-, and gabbro-impregnated harzburgite.

Notation

		Units
D	Melt + matrix material property coefficient	dimensionless
HNB	Horizon of neutral buoyancy	m
HNB_P	Picritic horizon of neutral buoyancy	m
HNB_T	Tholeiitic horizon of neutral buoyancy	m
K_{IC}	Critical (mode I) stress intensity factor	$MPa\sqrt{m}$
$\bar{M}$	Gram formula weight of melt	g
M_c	Summed gram formula weights of components removed from melt	g
M_j	Gram formula weight of the *j*th component	g
N	Number of components	dimensionless
P	Fluid pressure	MPa
P_0	Initial fluid pressure	MPa
Q	Flow rate of injected magma	m^3s^{-1}
V	Volume of injected magma	m^3
$\bar{V}$	Partial molar volume of melt	cm^3
$\bar{V}_c$	Partial molar volume of components removed from melt	cm^3
V_j	Molar volume of the *j*th component	cm^3
V_p	Compressional (primary) elastic wave velocity	$km \cdot s^{-1}$
Vs	Shear (secondary) elastic wave velocity	$km \cdot s^{-1}$
w	Dimensionless fluid-filled fracture half-width	dimensionless
X	Molar fraction of minerals being removed from the melt	dimensionless
X	Cartesian coordinate	m
Y	Cartesian coordinate	m
Z	Cartesian coordinate	m
C	Constant; index for flow regime	dimensionless
C_1	Constant in the melt + material property coefficient D	dimensionless
C_2	Constant in the dike height expression	dimensionless
c	Constant in magma-fracture height evolution	dimensionless
d	Constant in magma-fracture height evolution	dimensionless
e	Constant in magma-fracture height evolution	dimensionless
g	Gravitational acceleration constant	$m \cdot s^{-2}$
h	Height of a fluid-filled fracture	m
m	Elastic modulus	MPa, GPa
n	Exponent	dimensionless
r_j	Mole fraction ratio	dimensionless
s	Adjustable Cartesian coordinate	m
t	Time elapsed since fluid injection	s
t_r	Time of residence of olivine crystals in a MORB and olivine suspension	s
t_0	Time, initial	s
V_c	Summed molar volume of components removed from melt	cm^3
V_j	Molar volume of the *j*th component	cm^3
W	Fluid-filled fracture half-width	m
X_j	Mole fraction of the *j*th component	dimensionless
$\varnothing$	porosity	dimensionless
$\varnothing_0$	Initial reference porosity	dimensionless
Σ	The summation convention	
θ	Density contrast parameter for the horizon of neutral buoyancy	dimensionless
$\bar{\theta}$	Conjugate density contrast parameter	dimensionless
α	Time exponent	dimensionless
ς	Dimensionless fluid-filled fracture height	dimensionless
η	Melt shear viscosity	$Pa \cdot s$
μ	Shear modulus	MPa, GPa

		Units
ν	Poisson's ratio	dimensionless
ρ	Density	$g \cdot cm^{-3}$
ρ_c	Density of the (summed) components removed from melt	$g \cdot cm^{-3}$
ρ_f	Density of the final evolved melt	$g \cdot cm^{-3}$
ρ_g	Density of the matrix grains	$g \cdot cm^{-3}$
ρ_i	Density, initial, of the melt	$g \cdot cm^{-3}$
ρ_l	Density of the region below the HNB	$g \cdot cm^{-3}$
ρ_u	Density of the region above the HNB	$g \cdot cm^{-3}$
ρ_w	Density of sea water	$g \cdot cm^{-3}$
ρ_{is}	*In situ* country rock density	$g \cdot cm^{-3}$
ρ_m	Melt density	$g \cdot cm^{-3}$
$\Delta\rho$	Density contrast	$g \cdot cm^{-3}$
σ	Stress	MPa
σ_3	Minimum principal compressive stress	MPa
∇	Del, nabla, the gradient operator	m^{-1}
∇P_P^-	Vertical gradient in picritic magma pressure (negative-buoyancy region)	$MPa \cdot km^{-1}$
∇P_P^N	Vertical gradient in picritic magma pressure (neutral-buoyancy region)	$MPa \cdot km^{-1}$
∇P_P^+	Vertical gradient in picritic magma pressure (positive-buoyancy region)	$MPa \cdot km^{-1}$
∇P_T^-	Vertical gradient in tholeiitic magma pressure (negative-buoyancy region)	$MPa \cdot km^{-1}$
∇P_T^N	Vertical gradient in tholeiitic magma pressure (neutral-buoyancy region)	$MPa \cdot km^{-1}$
∇P_T^+	Vertical gradient in tholeiitic magma pressure (positive-buoyancy region)	$MPa \cdot km^{-1}$
$\nabla \sigma_H$	Vertical gradient in horizontal component of confining pressure	$MPa \cdot km^{-1}$

Introduction

Why do mid-ocean ridge magma chambers exist? Why do relatively flat-topped sheeted-dike complexes exist? What is it about newly created oceanic crust that enables it to capture and retain magma at shallow depths—and thus makes possible the long-term storage and differentiation of basaltic melt? In this chapter, petrophysical properties of basaltic rock and melt, results from seismic surveys over mid-ocean ridges, and the lithologic variations and compressional wave velocity relationships for ophiolite complexes are combined to provide a physical basis for the formation and sustained existence of mid-ocean ridge magma chambers.

This chapter reviews the buoyancy-zonation structure for mid-ocean ridge magma reservoirs and considers the role of fractional crystallization in altering melt buoyancy, followed by an overview on magma dynamics in the sheeted-dike complex, with comparisons of the analogous dynamics of Icelandic and Hawaiian rift zones.

In Situ Density–Depth Relationships

In situ density–depth relations for the East Pacific Rise were estimated from an evaluation of seismic profiles (Orcutt *et al.* 1976) and consideration of the tabulated velocity–density relations for basaltic, diabasic, and gabbroic rocks (Christensen, 1982). Figure 1 presents *in situ* density vs depth relations based on the velocity–depth profiles of Orcutt *et al.* (1976) for refraction surveys on 2.9×10^6 and 5.0×10^6 yr crust. Density values were assigned on the basis of the density spreads shown by the acoustic results for H_2O-saturated rock samples to 200 MPa confining pressure (Christensen, 1982)—appropriate conditions for the crustal reservoir depth. Both profiles show approximately the same type of density–depth behavior: a nonlinear increase in density over the 0- to 8-km-depth range. Ranges of inferred *in situ* density at specific depth intervals correspond to the variable scatter encountered in V_p–ρ plots of the tabulated data of Christensen (1982): the density range reflecting the total data range at a specific velocity value corresponding to each seismic profile. Superposed in the figure is the band of density values for tholeiitic melt experimentally determined by Fujii and Kushiro (1977) with extension to the total range inferred for picritic melt based on the work of Stolper and Walker (1980), Sparks *et al.* (1980), and Sparks and Huppert (1984). The density bandwidth from 2.6 to 2.8 $g \cdot cm^{-3}$ thus spans the composition range tholeiite to picrite. The *in situ* country rock trend and the melt band show a cross-cutting relationship with melt densities *less than* rock densities beneath ≈3 km depth, melt densities *greater than* rock densities

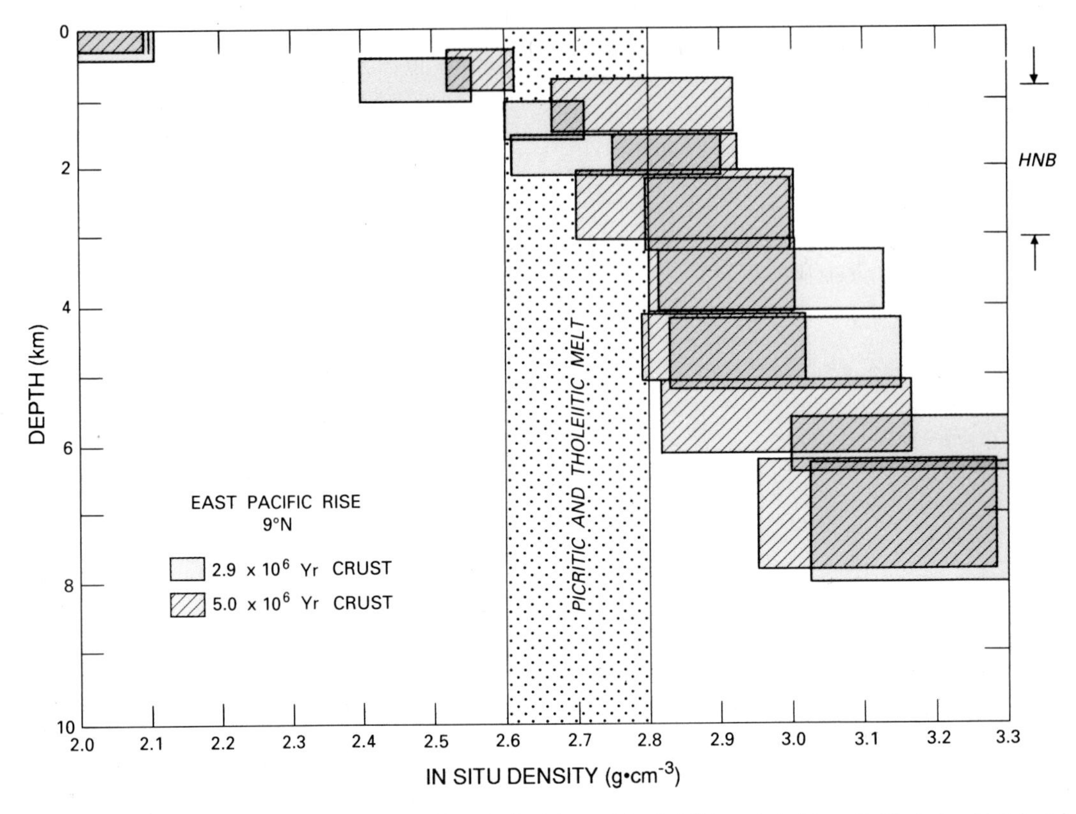

Figure 1 *In situ* density–depth relations for the East Pacific Rise and for tholeiitic and picritic melt. Shaded and hatchured boxes correspond to density ranges based on the seismic profiles of Orcutt *et al.* (1976) and density–velocity distributions for mafic and ultramafic rock types. Between ≈1 and ≈3 km depth, a crossover occurs between crustal and melt densities. This crossover defines the HNB. The density spreads correspond to the ranges in density for laboratory-derived acoustic measurements as compiled by Christensen (1982). The depth ranges for each box correspond to the depth steps for specific inferred seismic velocities in the refraction profiling of Orcutt *et al.* (1976).

above ≈1 km depth, and a rough *equity* in melt and country rock densities in the depth interval ≈1 to ≈3 km. Ryan (1993) has discussed the probable roles of H_2O and suspended olivine phenocrysts in altering magma densities.

In Fig. 1, the horizon of neutral buoyancy (HNB)—by definition—coincides with the ≈1- to ≈3-km-depth interval where *in situ* melt densities are just balanced by the country rock density. Under conditions of *local* density balances between the host rock and magma, the HNB thus represents a mechanical equilibrium position: the net (resultant) integrated forces for inducing magma ascent and descent are in balance, and long-term stability is attained.

Figure 2 compares the *in situ* melt-rock density crossover regions in the Hawaiian, mid-ocean ridge, and Icelandic sections. Density ranges for Hawaii and Iceland are based on combinations of seismic and gravity surveys, and the source references are given in the figure caption and are discussed in Ryan (1987b). All country rock densities were compared with the pressure-corrected tholeiitic melt and picritic melt bands as above. All three sections show: (i) a characteristic nonlinear increase in *in situ* host rock density with depth and (ii) a similar buoyancy zonation with depth.

Definitions

Magma reservoir. For the sub-ridge oceanic crust, a magma reservoir as used in this chapter refers to that domain of fluid-filled matrix that is capable of storing and transmitting magma. This includes

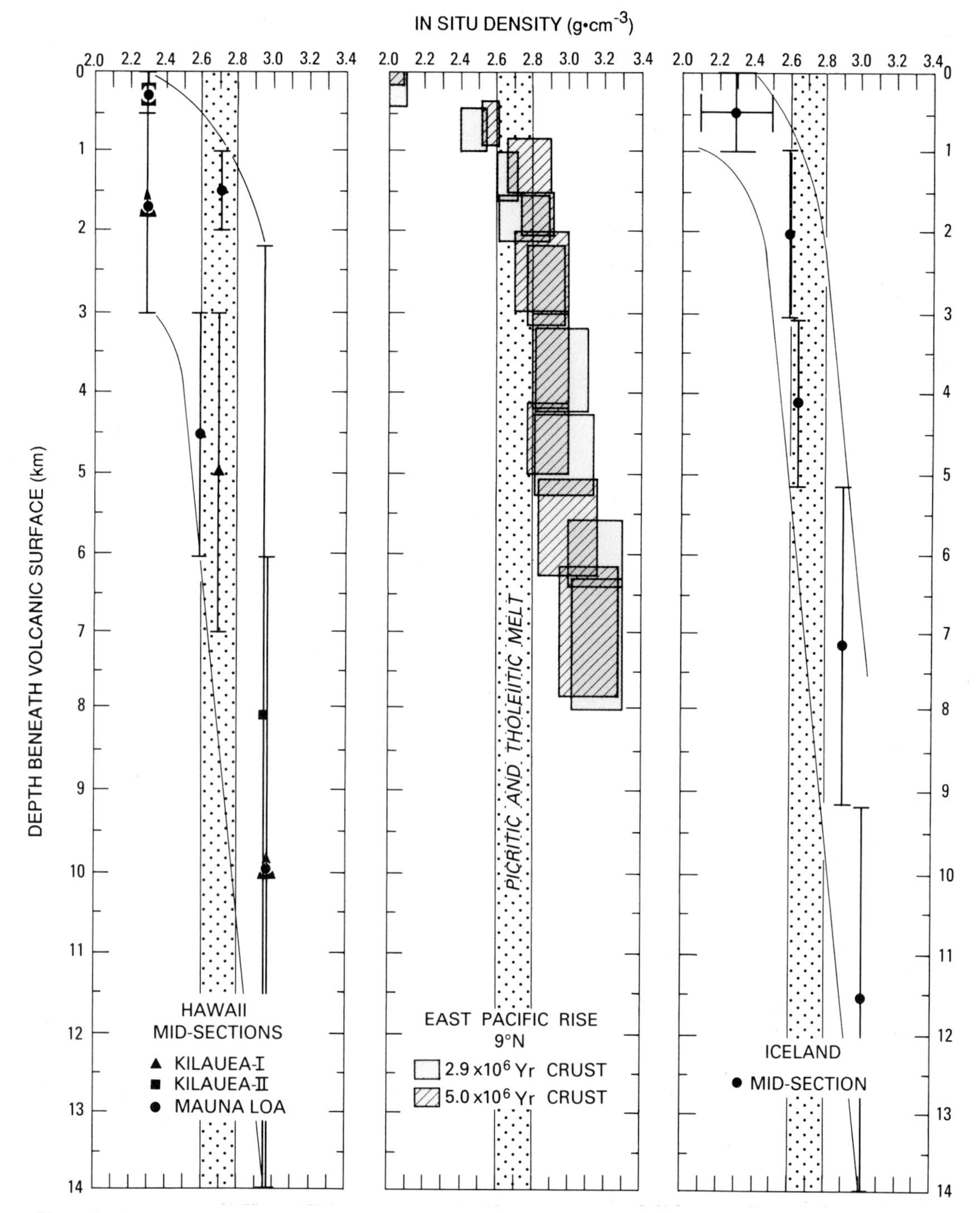

Figure 2 *In situ* density–depth profiles for the East Pacific Rise, in relation to analogous profiles for active Hawaiian and Icelandic rift zones. Collectively, the crossover between the country rock density and that of picritic and olivine tholeiitic melt occurs within the ≈1- to ≈7-km-depth interval. Within an individual system, the crossover defines the horizon of neutral buoyancy and shows some variance between these three systems. *In situ* density values for Hawaii are from the gravity inversions and seismic surveys of Zucca *et al.* (1982) consistent with the seismic refraction surveys of Hill (1969). *In situ* density values for Iceland are from Pálmason (1971) and Pálmason and Saemundsson (1974).

single- and multiple-connected microscopic and macroscopic magma-filled cavities. The reservoir may be logically subdivided into a shallow magma chamber and a deeper more extensive crystal–liquid mush region (Sinton and Detrick, 1992). See Fig. 8.

Magma chamber. The magma chamber is the top of the magma reservoir. It is relatively fluid rich and has a sill-like overall geometry with a ceiling at the base of the sheeted-dike complex. The floor is relatively diffuse and becomes more crystal rich with depth, grading downward into the top of the crystal–melt mush that makes up the volumetrically more significant portion of the reservoir. In the chapter by Phipps Morgan *et al.* (this volume) it is termed the "magma lens." See Fig. 8.

Neutral buoyancy. $\rho_m = \rho_{is}$, where ρ_m is the melt density and ρ_{is} is the *in situ* country rock density. Local contributions to driving forces that tend to induce ascending or descending motion in the melt have been removed, and a state of local mechanical equilibrium exists between the molten region and the subsolidus surroundings.

Horizon of neutral buoyancy (=level of neutral buoyancy (LNB)). The HNB is a layer of narrow vertical extent and wide lateral extent within which melt has achieved mechanical equilibrium. The HNB corresponds to the depth extent of shallow subcaldera magma stabilization and long-term shallow magma accumulation, and the level of lateral dike emplacement in active volcanic systems.

Negative buoyancy. $\rho_m > \rho_{is}$, and gravitational potential energy is released through the *descent* of magma. The region of negative buoyancy lies above the horizon of neutral buoyancy in the Earth's crust, and melt parcels may therefore descend to the HNB throughout this region. The Earth's free surface is the top of the region of negative buoyancy.

Positive buoyancy. $\rho_m < \rho_{is}$, and the gravitational potential energy of the system is reduced by the *ascent* of magma. The region of positive buoyancy is bounded below by the site of magma generation and bounded above by the horizon of neutral buoyancy.

Contractancy. The progressive reduction in macroscopic and microscopic pore space produced by a progressive increase in confining pressure with depth in the Earth's crust. Concomitant increases in density, seismic velocities (V_p, V_s), the bulk modulus (K), and decreases in fluid permeability (K_f) and compressibility (β) accompany the reduction in porosity with increases in depth and confining pressure. States of anelastic volume *decrease* under applied loading. This process may be contrasted with dilatancy—an anelastic volume increase under differential stress. In regions of active basaltic magmatism, rock contractancy and matrix compaction combine with zeolite and greenschist facies hydrothermal mineralization to produce the crossover between melt and country rock density.

The Magma Reservoir Environment and the Region of Neutral Buoyancy

What is the correspondence between regions of low elastic wave velocity in mid-ocean ridges and the neutral buoyancy region for tholeiitic melt? Seismic surveys conducted across and along the ridges provide a means of answering this question. The locations of the seismically defined magma reservoir and the region of neutral buoyancy can be compared on the basis of the region of the crossover in the *in situ* density–depth profiles for rock of the oceanic crust and that of picritic-to-tholeiitic melt. Summary results from several representative surveys are discussed as follows in light of their relationships to the principal buoyancy zonation of the ridge.

In "Webster's Third New International Dictionary of the English Language Unabridged" (Gove, 1964), the term "region" is defined as "one of the major subdivisions into which the body or one of its parts may logically be divided." In this spirit, this paper defines three regions of buoyancy for magma: the region of negative buoyancy (sea floor to $\approx$1 km depth); the region of neutral buoyancy ($\approx$1 to $\approx$3 km depth); and the region of positive buoyancy ($\approx$3 km to the depth of melt separation in the parent lherzolite). As used here, the term "region" is finite or reasonably constricted in its vertical or lateral extent. Thus, the *region of neutral buoyancy* coincides with the upper levels of the magma storage reservoir.

The term "horizon" is defined as (Gove, 1964) "a stratigraphic level or position in the geologic column; a natural . . . layer. Any of the reasonably distinct layers . . . in a vertical section or profile . . . and gradually developed as a result of natural . . . processes." In this chapter, the term "horizon of neutral buoyancy" thus refers to an interval whose vertical extent is finite but very small compared to its great lateral extent. The HNB's great lateral extent guides the relatively rapid lateral magma intrusions that are individual dike-forming events. Thus the *horizon* of neutral buoyancy is important in the dynamics of the lateral intrusion process and contains within it the active region of neutral buoyancy that coincides with sub-ridge magma storage. This usage is consistent with terminology used previously by Ryan (1985, 1987a, 1987b).

The correspondence between the region (horizon) of neutral buoyancy and the preferred depth of magma residence and lateral injection is shown in Figs. 3A and 3B. Depth ranges for sub-ridge magma storage from the East Pacific Rise, and the Valu Fa Ridge, Lau Basin, have been deliberately pooled with ranges for Kilauea and Mauna Loa volcanoes, Hawaii, and the Krafla Central volcano, Iceland, to provide one perspective on the overall similarities of these depth ranges. When considered in light of the density crossover relationships above, the data illustrate that the *preferred depth for magma storage and lateral intrusion is in a virtual 1 : 1 correspondence with the horizon of neutral buoyancy.* Figure 3B subdivides the ranges on a regional basis, illustrating the differences in magma depth between and within these centers.

Seismic refraction profiles have been conducted along the East Pacific Rise crest near lat. 9°N by Orcutt *et al.* (1976) (Fig. 4). Along the crest, compressional wave velocities increase from low values ($\approx$2.5–3.5 km · s^{-1}) near the sea floor, to about 6.7 km · s^{-1} at the top of a low-velocity region at 2 km depth beneath the ridge surface (Fig. 5). Beneath this compressional wave low-velocity zone, V_p increases again to values of about 7.5 km s^{-1} at 7 km depth. Profiles that have resulted from surveys parallel to the rise crest but on 2.9 $\times$ 10^6 and 5.0 $\times$ 10^6 year old crust do not show evidence of a low-velocity zone, but instead show a continuous and progressive increase in V_p through the base of the oceanic crust (Fig. 5). In Fig. 5, the regions of negative and positive buoyancy and the region of neutral buoyancy that correspond to the interval inferred in Fig. 1 have been overlayed.

Multichannel seismic reflection profiles across the East Pacific Rise north of the Siqueiros Fracture Zone have revealed a compressional wave low-velocity region consistent with magma storage at a depth of $\approx$2 km beneath the sea floor (Herron *et al.,* 1978). Compressional wave velocities rise to values in excess of 6.5 km · s^{-1} just above the reflector and then abruptly drop to $\approx$4.5 km · s^{-1} within the 2 to 3-km-depth interval inferred to be magma rich.

Hale *et al.* (1982) have reexamined Lamont–Doherty Earth Observatory multichannel reflection data taken across the East Pacific Rise at lat. 9°N in light of elastic wave velocity values from ophiolite samples. Laboratory-based compressional wave velocity data from samples of the Sumail ophiolite (Christensen and Smewing, 1981) have been combined with the temperature distribution expected for a double spreading rate of 12.2 cm · yr^{-1} (Sleep, 1975) to help constrain the velocity profile and the cross-sectional velocity structure of the rise axis. Figure 6 provides the V_p–depth profile of Hale *et al.* (1982) and compares it with the profile determined by Orcutt *et al.* (1976). Both profiles show a pronounced low-velocity region over the 1- to 3-km—depth core of the region of neutral buoyancy. The upper portion of the high-velocity-gradient region of the magma reservoir roof (0–2 km depth) corresponds in part to heavily fractured and porous pillow basalts and brecciated dikes, where the *in situ* density of these porous rocks is substantially less than—and subsequently approaches with increasing depth—that of nonvesiculated melt. Accordingly, magma within this upper veneer (0–$\approx$1 km) may descend under negative buoyancy forces. Below $\approx$3 km depth, $\rho_m < \rho_{is}$ and melt parcels rise within the magma reservoir driven by positive buoyancy forces. Figure 7 illustrates the across-axis velocity model of Hale *et al.* (1982) and its relation to the principal buoyancy zonations for tholeiitic melt.

Seismic refraction profiling on the crest of the East Pacific Rise at lat. 21°N by Reid *et al.* (1977) also has revealed a region of high shear wave attenuation at a depth of 2.5 km beneath the sea floor. This region is in essential coincidence with the low compressional wave velocity region deter-

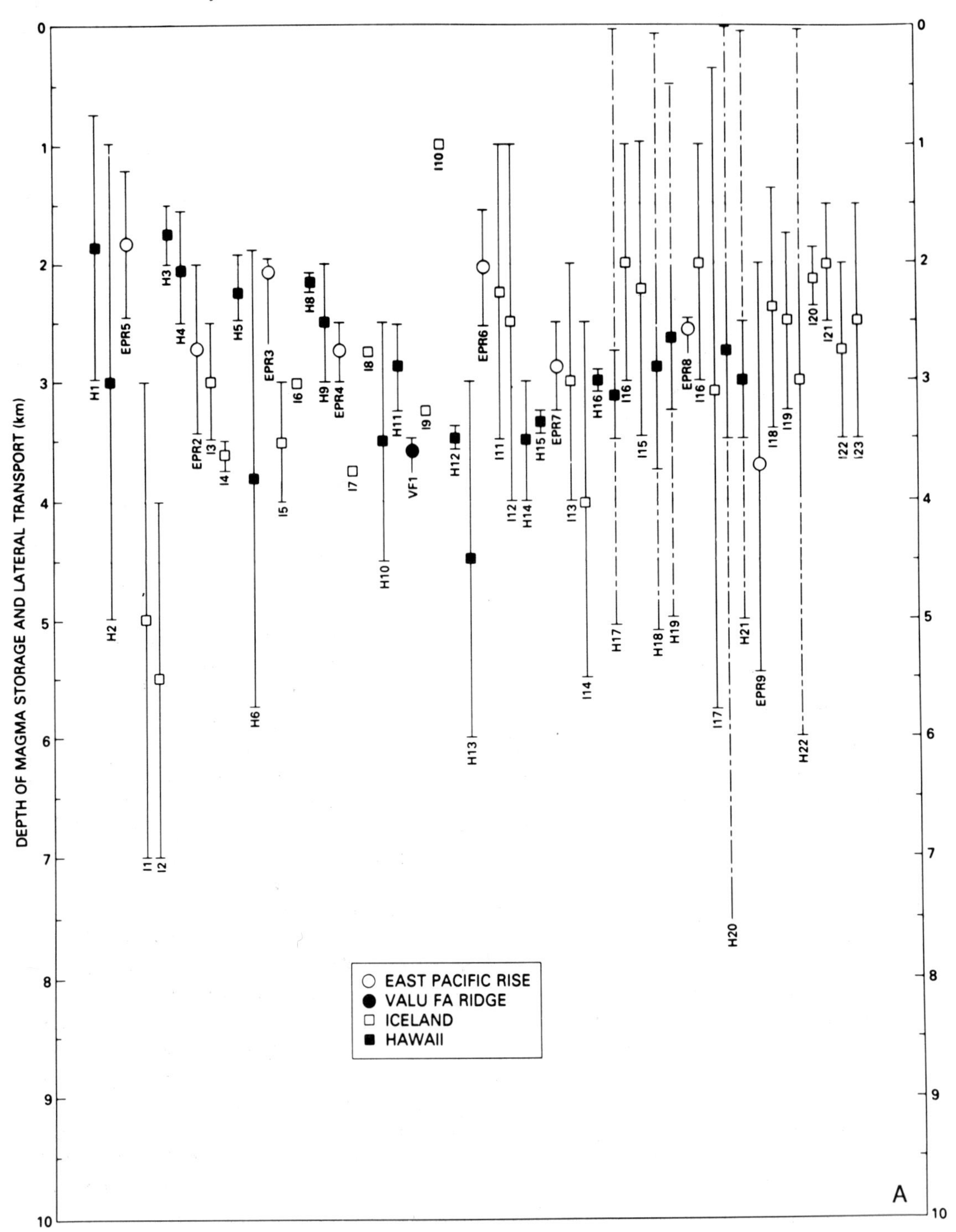

Figure 3A Pooled mean depth ranges for magma storage and transport beneath the rift-zone surface for the East Pacific Rise, the Valu Fa Ridge, the Krafla central volcano, Iceland, and Kilauea volcano, Hawaii. The pooling process illustrates the substantial similarities in the nature of the depth distributions. The collective core of the region of neutral buoyancy, as expressed by the mean depths, lies predominantly within the 1- to 4-km-depth interval. High-level, mid-ocean ridge magma storage and intrusion within the sheeted dike complex is in a 1 : 1 correspondence with the region of neutral buoyancy. (Data sources are provided in the appendix.)

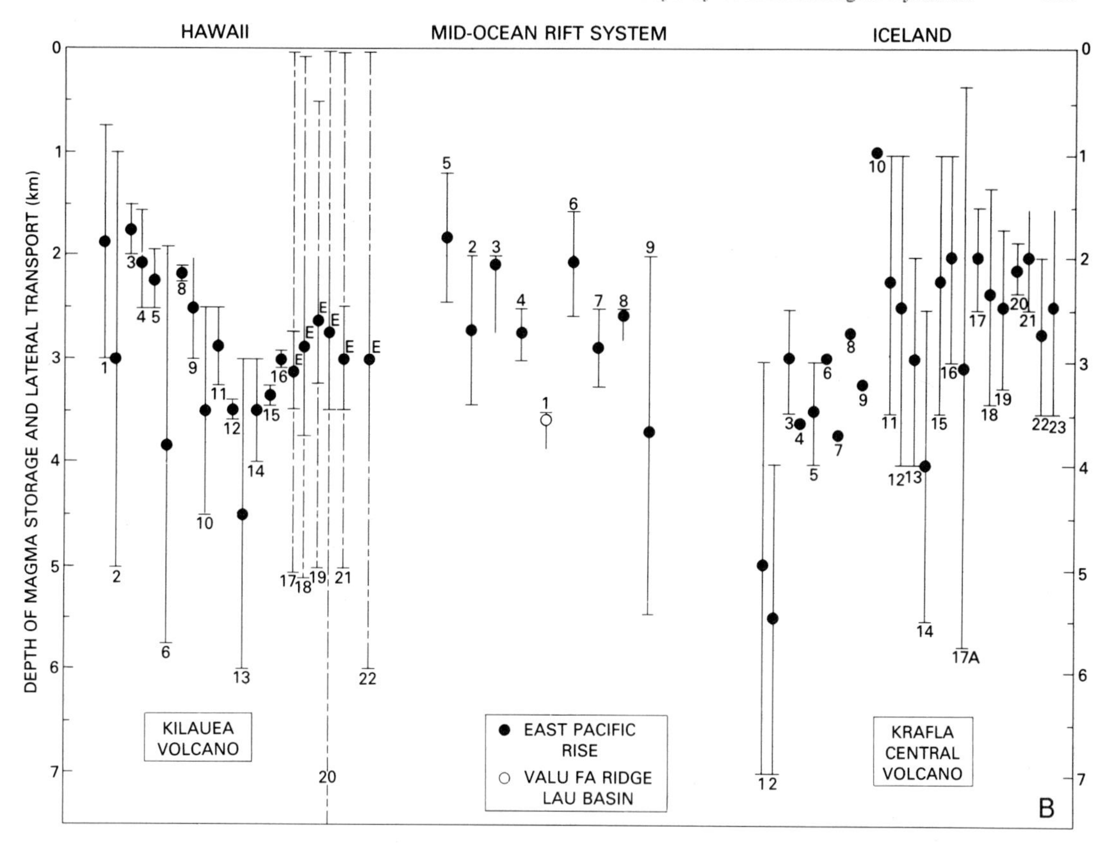

Figure 3B Regional patterns in depths of magma reservoirs and lateral dike-forming magma injections. Note that the total range in depths of mid-ocean ridge reservoirs is less than the vertical variation for Kilauea volcano alone. Symbols are the same as in Figure 3A. "E" signifies the initial and final gravitational equilibrium position of magma after a series of wave-like oscillations about the horizon of neutral buoyancy, induced by the periodic halting of the advancing magma fracture front. The vertical dashed lines above the "E" position are thus transient nonequilibrium excursions during lateral intrusion and are damped down upon the completion of intrusion. Note that the final resting place is always at the HNB (equilibrium) position after an intrusion transient. For Kilauea, the HNB is centered at 3 km depth.

mined by Orcutt *et al.* (1976) at 9° N and also corresponds to the region of neutral buoyancy in Fig. 1.

Multichannel seismic reflection surveys across the Valu Fa Ridge (Lau Basin) by Morton and Sleep (1985) have revealed a reflector at a depth of 3.5 km beneath the ridge. This reflector has been interpreted as the relatively flat-lying roof of a magma chamber. The roof width is about 2 to 3 km. The Valu Fa Ridge is a back-arc spreading center with a spreading rate of about 70 mm · yr^{-1} (Weissel, 1977). Like the East Pacific Rise, the Valu Fa Ridge magma chamber coincides with the (slightly deeper) position of neutral buoyancy for picritic-to-tholeiitic melt (Fig. 3A, VF1 at left-center of plot).

Structure of the Ridge Magma Reservoir in Relation to the Region of Neutral Buoyancy

Detrick *et al.* (1987) have conducted multichannel seismic surveys along the East Pacific Rise between lat. 8°50′N and lat. 13°30′N. The study was designed to resolve questions related to along-axis magma continuity as well as to constrain the depth–width variations of the axial magma reservoir. The reservoir width (the chamber portion) was resolved at its top as a relatively high-amplitude reflection, with a maximum of 2 to 3 km. The reflector is relatively flat-lying, further suggesting a roof structure that dips gently outward from the ridge axis. Reflections from the Moho, however, extend to 2 to 3 km on either side

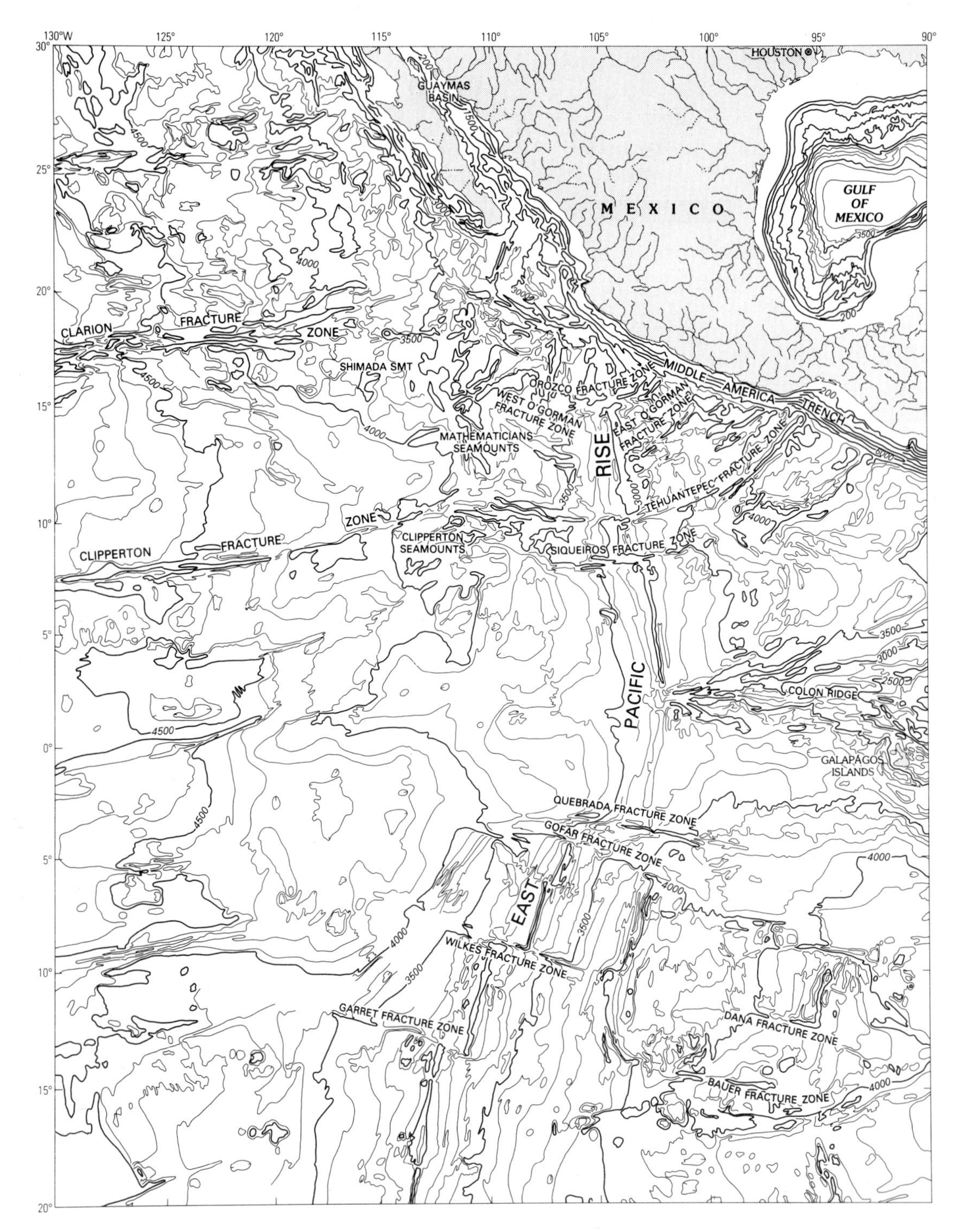

Figure 4 Marine bathymetry of the East Pacific Rise. The horizon of neutral buoyancy for the EPR lies at the ≈1- to ≈3-km-depth interval beneath the ridge crest; it controls the dynamics of lateral dike-forming injections in the sheeted-dike complex and modulates magma storage in the uppermost portion of the reservoir. The EPR surveys discussed in this chapter were not plotted to avoid crowding the map and obscuring bathymetric contours. They are: (i) 9° N (Orcutt *et al.*, 1976; Hale *et al.*, 1982); (ii) 9°–10° N (Toomey *et al.*, 1990); (iii) south of the Clipperton Fracture Zone (Vera *et al.*, 1990); (iv) north of the Siqueiros Fracture Zone (Herron *et al.*, 1978); (v) 8°50′N–13°30′N (Detrick *et al.*, 1987); and (vi) 21° N (Reid *et al.*, 1977). Based on the compilation of Mammerickx and Smith (1980).

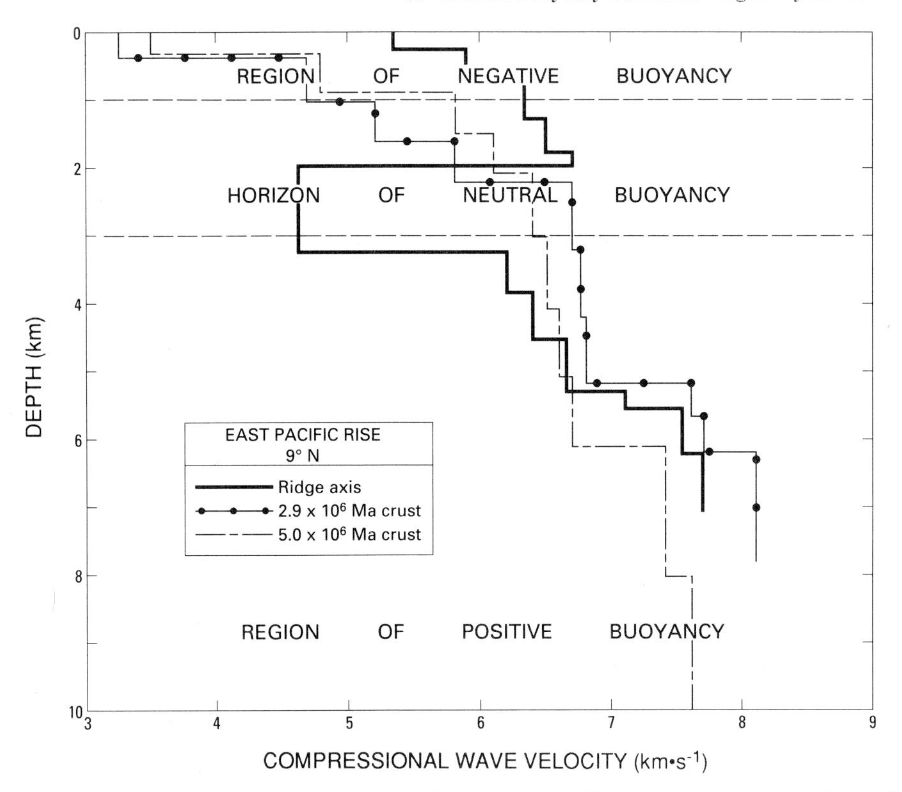

Figure 5 Compressional wave velocity profiles for the East Pacific Rise at lat. 9°N in relation to the region of neutral buoyancy for tholeiitic melt. For the zero-age rise axis profile, most of the low compressional wave velocity region is contained within the region of neutral buoyancy. The dashed lines at 1 and 3 km depth denote the approximate limits of the region of neutral buoyancy. Modified after Orcutt *et al.* (1976).

of the Rise axis, suggesting a magma reservoir base within a range 4 to 6 km wide. The depth to the top of the magma chamber is constrained by the two-way travel time between the ridge surface and the chamber roof and is generally 1.2–2.4 km beneath the sea floor (see also Figs. 5 and 6). Minimum estimates of roof thickness tend to correlate with the shallowest positions of the ridge axis beneath sea level. These positions tend to be near the segment center and attest to the relatively high magma budgets (Macdonald *et al.,* 1984) that have constructed these topographic highs.

The lat. 8°50′N to lat. 13°30′N study area contains prominent overlapping spreading centers as well as several along-axis deviations from axial linearity (*devals*). Apparent along-axis continuity of the magma chamber was observed for an aggregate length of ≈350 km (or ≈61%) of the 500 km of ridge crest surveyed, with laterally continuous reflections on 40- to 50-km-long traverses. Most devals appear to be superficial features with respect to the reservoir, and the reservoir roof passes continuously beneath some 70% of them.

The East Pacific Rise just south of the Clipperton Fracture Zone (Fig. 4) contains a bright upper crustal reflector suggestive of the roof of a volume with high magma-to-rock ratios (Fig. 8; Vera *et al.,* 1990) and lies beneath a well-developed axial graben (not illustrated). Three-dimensional seismic tomography by Toomey *et al.* (1990) has related the low compressional wave velocity core of the along-axis magmatic region to the surficial structures and to the surrounding distribution of relatively high- and low-velocity material within the newly created oceanic crust. Inversion of compressional wave travel time residuals has produced a succession of sections through the compressional wave velocity structure in horizontal depth slices from the sea floor to 3 km depth, as well as in cross sections across and along the rise

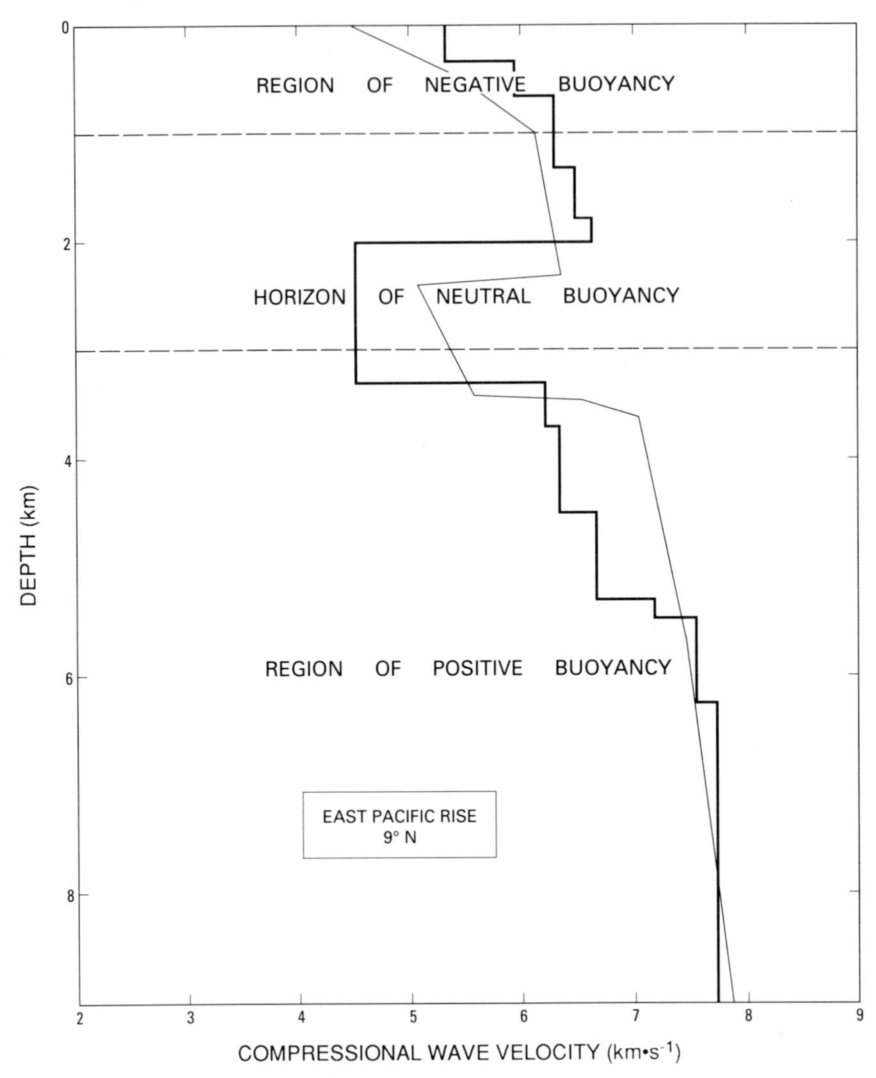

Figure 6 Compressional wave velocity profiles for the East Pacific Rise at lat. 9°N in relation to the region of neutral buoyancy for tholeiitic melt. For both profiles, the low compressional wave velocity region approximates a 1:1 correspondence with the region of neutral buoyancy. The square-stepped (heavy line) profile has been derived by Orcutt *et al.* (1976) whereas the piecewise-linear (fine line) profile is the product of the multichannel reflection survey reexamined by Hale *et al.* (1982).

Figure 8 Compressional wave velocity contours for the East Pacific Rise at lat. 9°N in cross section, and their relation to the region of neutral buoyancy, and the regions of negative and positive buoyancy. The lowest velocity core is totally contained within the region of neutral buoyancy. The magma chamber corresponds to the lowest velocity upper portion of the region, where the fluid/rock ratios are highest and the configuration is sill-like (Sinton and Detrick, 1992). The magma reservoir includes the chamber, but extends downward into the high-velocity ($V_p = 5.5–6.0$ km · s^{-1}) cumulates that may be sufficiently porous and permeable to permit the accumulation (and transmission) of melts. *Magma in the lower half of the reservoir is always positively buoyant, whereas magma in the upper half may be either neutrally buoyant or positively buoyant depending on the suspended crystal load and the degree of differentiation.* Diverging asthenospheric flow patterns (diverging arrows beneath 7 km) correspond to mineral lineations in the harzbugite of the sub-Moho sections of ophiolite complexes. Vertical long and short rulings denote the sheeted-dike complex. Based in part on contour distributions provided in Vera *et al.* (1990), and modified after Ryan (1993).

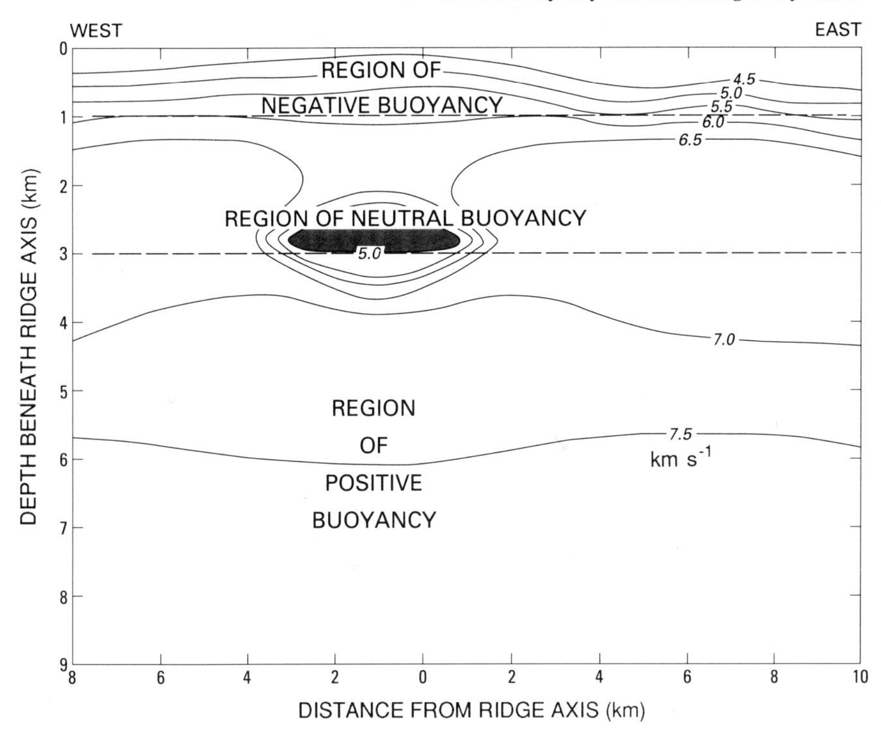

Figure 7 Cross-sectional compressional wave velocity structure of the East Pacific Rise at lat. 9°N based on the multichannel seismic survey of Hale *et al.* (1982) and employing the thermal model of Sleep (1975) with laboratory-determined V_p measurements and the temperature derivatives of compressional wave velocities.

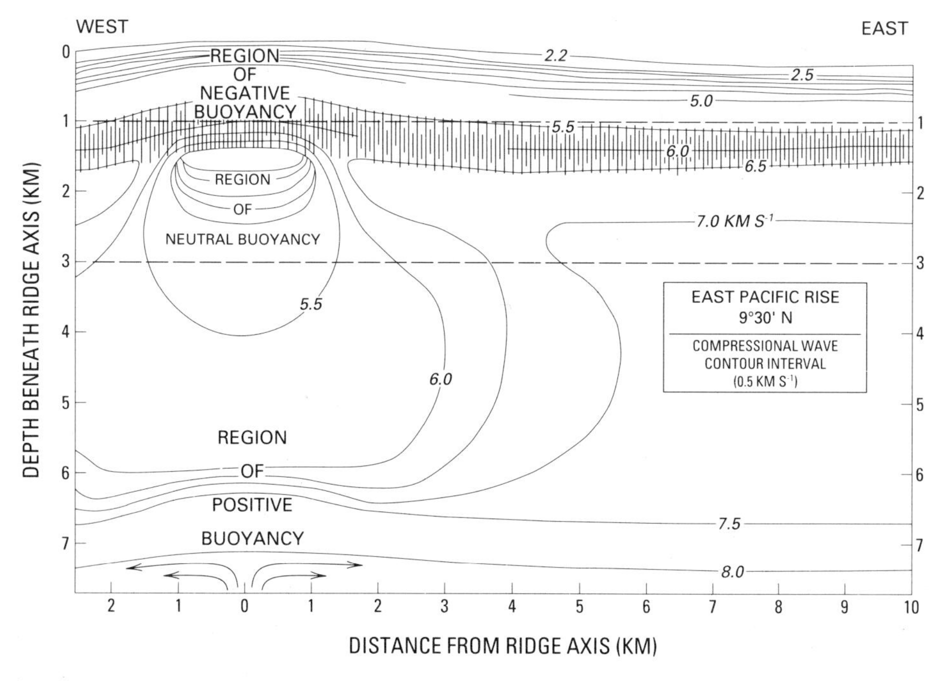

Figure 8

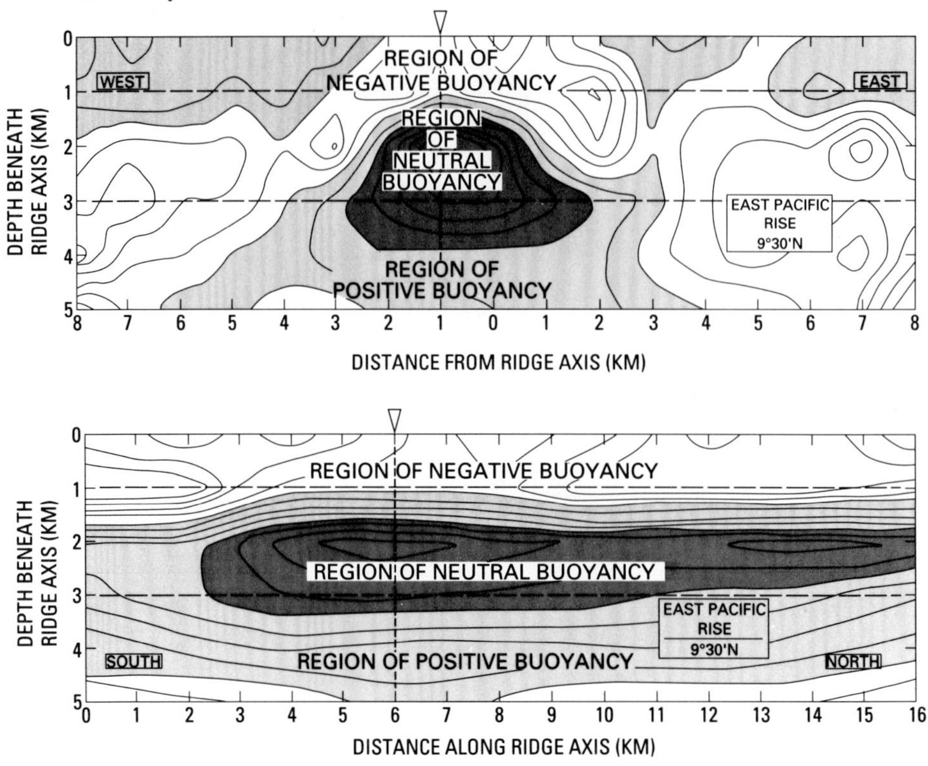

Figure 9 East Pacific Rise compressional wave velocity structure in relation to the region of neutral buoyancy from lat. 9°N to 10°N (darkest shading). V_p contour intervals are in units of 0.2 km · s^{-1}, and contours begin at the 3 km · s^{-1} minimum centered in the region of neutral buoyancy. (A, Top) Cross-sectional structure showing the core region (darkest shading) of lowest velocity and presumably melt enrichment centered along the axis; regions of reduced velocity beneath and bilaterally flanking the reservoir above, and the relatively high-velocity newly created oceanic crust. The vertical dashed line indicates the along-axis, cross-section of Figure 9B. (B, Bottom) Longitudinal along-axis cross section. For both the across-axis and along-axis sections, the low-velocity core region is in approximate correspondence with the region of neutral buoyancy. Vertical dashed line indicates the across-axis slice location of Figure 9A. Figures modified after Toomey *et al.* (1990). (From Ryan, 1993; reproduced with permission of the American Geophysical Union.)

axis. Figures 9A and 9B relate the low-velocity core region of the ridge to the regions of neutral, negative, and positive buoyancy.

Lithologic Associations in Ophiolite Complexes

The regions (horizons) of negative buoyancy and neutral buoyancy and the region of positive buoyancy each have a distinctive igneous association. Studies of ophiolite complexes, combined with the density zonations observed for active mid-ocean ridges, reveal the relationships between rock type and buoyancy zonation. These relationships are illustrated in Figs. 10 and 11 for the Blow-Me-Down Massif, Bay of Islands Complex, Newfoundland, as well as Troodos, Cyprus.

Region of Negative Buoyancy

Pillow basalts and sheeted basalt flows compose the dominant lithology. Low *in situ* densities are produced by pervasive macroscopic fractures and microscopic porosity. The resulting high fluid permeabilities (Nelig and Juteau, 1988) promote the hydrothermal penetration of this layer, with fluids sporadically reaching into the horizon of neutral buoyancy, below. Locally, brecciated dikes make up a portion of the lower part of the region of negative buoyancy and attest to sporadic hydrothermal explosions. The contact with the top of

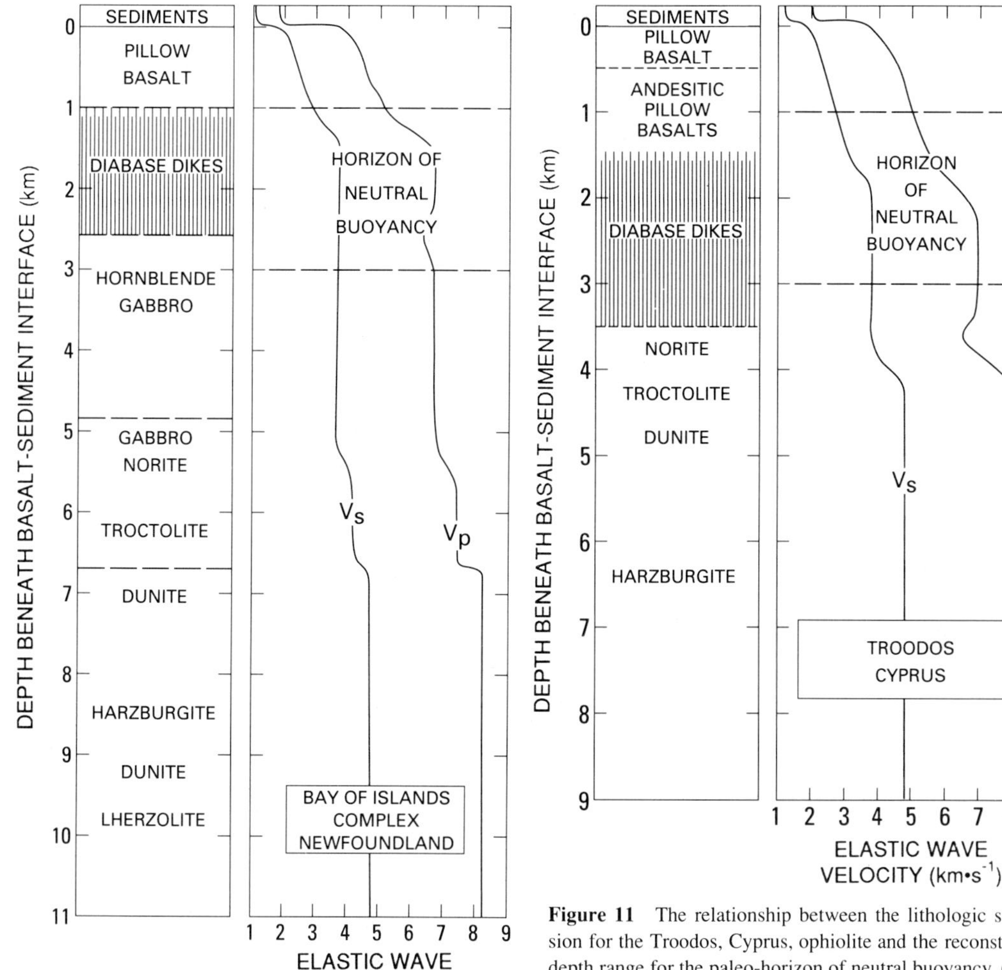

Figure 10 The relationship between the lithologic succession for the Bay of Islands ophiolite complex and the reconstructed depth ranges for the paleo-horizon of neutral buoyancy. The paleo-horizon of negative buoyancy lies above the sheeted-dike complex, within the pillow basalts. (Profiles are based on laboratory-derived acoustic velocities and have been summarized in Christensen, 1978.)

Figure 11 The relationship between the lithologic succession for the Troodos, Cyprus, ophiolite and the reconstructed depth range for the paleo-horizon of neutral buoyancy. (Compressional and shear wave velocity profiles are based on laboratory-derived acoustic measurements as summarized in Christensen, 1978.)

the sheeted dike swarm may be relatively sharp. Beneath the "geotimes" volcanics of the Oman ophiolite (Lippard *et al.,* 1986; Nicolas, 1989), for example, the transition region from the basalts to the top of the sheeted dikes is tens of meters to ≈100 m thick. The upper portions of both the vertically and laterally propagating dikes thus would encounter the influences of the region of negative buoyancy—inhibiting upward magma movement—within a fairly restricted transition zone. This would correspond to the transition interval between the region of negative buoyancy (porous, fractured basalts) and the region of neutral buoyancy (sheeted diabase dikes).

Region of Neutral Buoyancy

The sheeted diabase dike swarms and the uppermost portions of the gabbro complex make up the region of neutral buoyancy. It is at this level that rising melt begins to spread laterally, and the isotropic gabbros thus must have experienced significant along-axis melt migration in response to replenishment from below. Periodic reservoir rupture produces the sheeted-dike complex, and, by analogy with well-documented centers such as Krafla, Iceland, the dikes were formed dominantly during lateral along-axis emplacement. The roots of ≈1-m-thick fine-grained dolerite dikes have been mapped in the Maydan syncline area of Oman. They may be traced downward and show crosscutting relations with high-level gabbros, where their chilled margins flare outward, attaining widths of 15–20 m. Progressing downward, they grade into doleritic-textured microgabbro and then more coarsely crystalline massive gabbro (Rothery, 1983). They thus appear to represent the magma chamber connection where subvertical flow has drained the overpressured chamber compartment beneath. In other areas, members of the base of the dike swarm do not root in the plutonic complex, but crosscut the subsolidus gabbros beneath and show true dike keels. This relationship should be a common one and would be a natural consequence of neutrally buoyant (and negatively buoyant) lateral injection episodes from reservoirs that are somewhat offset from each other in plan view. Intermittent and low-volume reservoirs maintained at low spreading rates would be expected to rupture at their up-rift and down-rift margins (Ryan, Fig. 17, 1987b). At high spreading rate ridges such as the East Pacific Rise, the sheeted-dike complex is underlain nearly continuously by stored magma, and the rupture is expected to occur in the chamber roof itself and directly over the centrally located mantle replenishment source (Whitehead *et al.*, 1984; Ryan, 1987b). In such cases, the dike attempts to grow upward into the region of negative buoyancy, but negative buoyancy forces require it to "roll over," and—in cockscomb fashion—to propagate laterally along the magma chamber roof axis. Examples of the transition from vertical injection through the chamber roof with subsequent roll over to a dominantly lateral injection mode have been well documented at Krafla, Iceland, during the September 8, 1977, multiple intrusion-eruption episode (Brandsdottir and Einarsson, 1979; Ryan, 1987b). Table 1 lists a number of magma transport and physical property correlations within a neutral buoyancy context.

Table 1

Neutral Buoyancy Magma Transport Correlations: Intrusion Dynamics, the *In Situ* Environment, and the Development of Igneous Structure

- World-wide correlations of the depth of magma storage and intrusion in basaltic rift zone environments—independent of local tectonic settings
- Physical existence of basaltic magma chambers
- Long rift zone development
- Long (bladed) dike generation
- Upward evolution of active magma reservoirs
- Starting and stopping position (depths) of dike-forming magma intrusion
- Noneruptive nature of most shallow intrusions
- Depth location of the sheeted dike swarm
- Episodic reequilibration to the HNB dike emplacement depth during incremental intrusion
- Preferred horizon of magma mixing (intra- and inter-reservoir)
- Location of subcaldera magma reservoirs
- Location of rift zone dike-induced microseismicity
- Geodetic signatures of dike intrusion and subcaldera magma storage
- First-order fracture porosity and fracture networks in rift zones
- Eruption-drainback phenomenology: dynamics and volumes
- *In situ* density–depth profiles
- *In situ* V_p, V_s—depth profiles
- Experimental crack closure data at high pressures
- Experimental V_p V_s data as a function of confining pressure
- Depth location of fluid pressure center (maximum magma driving pressure) in geodetic inversions

Differentiation and Melt Density

Mid-ocean ridge tholeiites are the differentiation products of picritic liquids, as suggested by studies of crystal–melt phase equilibria (O'Hara, 1968a, 1968b), and of the crystallization products of ophiolite complexes, especially the ultramafic rocks at and beneath the crust–mantle interface (e.g., Nicolas, 1989). How does the crystallization of a picritic melt influence the density of the evolving liquid, and how would changes in melt density potentially interact with the ambient density structure of an active magma chamber?

Figure 12 illustrates the changes in melt density produced in picritic and tholeiitic melts as a function of olivine, plagioclase, clinopyroxene, and ilmenite crystallization. Picritic liquids (triangles) are based on both rock sample analyses (Clarke, 1970; Elthon, 1979) and modeled com-

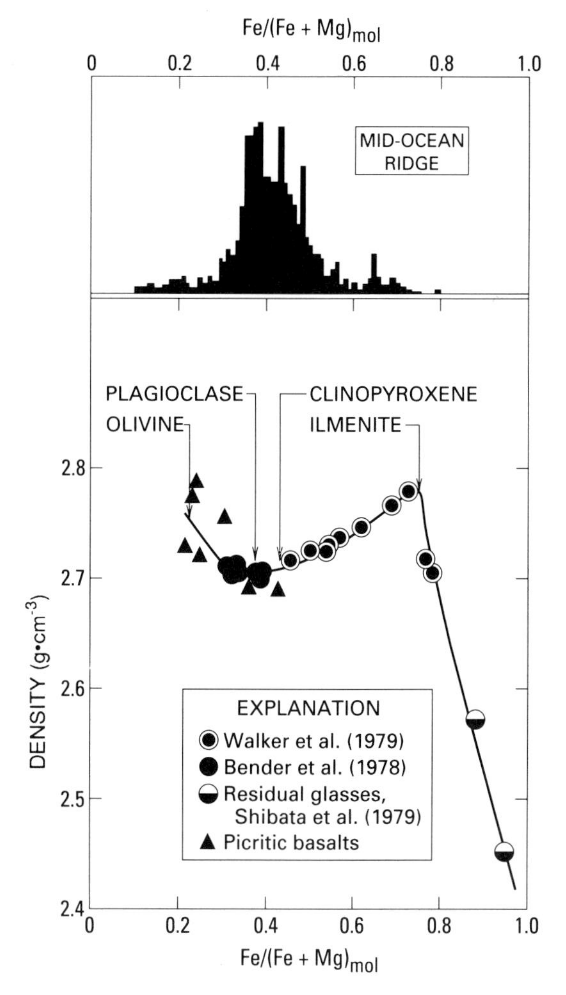

Figure 12 (Lower) Melt-density changes produced by fractional crystallization of picritic and tholeiitic liquids, as illustrated by changes in Fe–Mg content. Points of appearance of liquidus mineral phases are indicated by arrows. (Upper) The preponderance of mid-ocean ridge basalts is associated with the density minimum produced by olivine fractionation. Based on a presentation of Stolper and Walker (1980).

positions (Irvine, 1977). Liquidus temperatures were used to infer density values following the approach of Roeder (1974) and the empirical partial molar volume–density relations of Bottinga and Weill (1970). For the fractional crystallization–density relations plotted in Fig. 12, the changes in $(\mathrm{Fe}/(\mathrm{Fe} + \mathrm{Mg}))_{\mathrm{mol}}$ was used as an index of differentiation. That portion of the liquid line of descent spanning the interval Fe/[Fe + Mg] (molar) from 0.3 to 0.8 is from the 1-atm melting experiments of Bender *et al.* (1978) and Walker *et al.* (1979). At the most fractionated portion of the curve, analyses of glasses in mid-ocean ridge tholeiite samples were combined with an inferred liquidus temperature of 1050° C to generate the "residual glasses" track (semisolid circles in Fig. 12).

The *factionation density* is defined (Sparks and Huppert, 1984) as the ratio of the gram-formula weight to molar volume of the liquid-phase chemical components selectively removed from the melt by fractional crystallization. Thus, for a melt consisting of N components with mole fractions X_j ($j = 1, 2, 3, \ldots, N$), the initial melt density is:

$$\rho_{\mathrm{i}} = \frac{\sum_{j=1}^{N} X_j M_j}{\sum_{j=1}^{N} X_j V_j} \equiv \frac{\overline{M}}{\overline{V}} \tag{1}$$

where X_j is the mole fraction, M_j is the gram formula weight, and V_j is the molar volume of the jth component, and $\overline{M}$ and $\overline{V}$ are the gram formula weight and the molar volume of the melt, respectively. Fractional crystallization processes may remove minerals composed of components in a mole fraction ratio r_j, where $\Sigma\, r_j = 1$. Sparks and Huppert (1984) defined the density of the components in the fluid that are selectively removed through the fractional crystallization process as

$$\rho_{\mathrm{c}} = \frac{\sum r_j M_j}{\sum r_j V_j} = \frac{M_{\mathrm{c}}}{V_{\mathrm{c}}}, \tag{2}$$

where M_{c} and V_{c} are the summed gram formula weights and molar volumes, respectively.

For a molar fraction $\mathbf{X}$ of minerals that are crystallizing and being removed from the original melt, the density of the final evolved melt, ρ_{f}, is

$$\rho_{\mathrm{f}} = \frac{\sum (X_j - r_j\mathbf{X})M_j}{\sum (X_j - r_j\mathbf{X})V_j}, \tag{3}$$

which may be reformatted as

$$\rho_{\mathrm{f}} = \frac{\rho_{\mathrm{i}}[1 - ((\rho_{\mathrm{c}}\overline{V}_{\mathrm{c}})/(\rho_{\mathrm{i}}\overline{V}_{\mathrm{i}}))\mathbf{X}]}{[1 - (\overline{V}_{\mathrm{c}}/\overline{V}_{\mathrm{i}})\mathbf{X}]}. \tag{4}$$

Figure 13A illustrates the variations in individual fractionation densities for the components forming the olivine, clinopyroxene, orthopyroxene, and plagioclase solid–solution series. Note that ρ_{c} is a *fictive* parameter and relates to the chemical components of an individual solid–solution series and not to the actual minerals. For

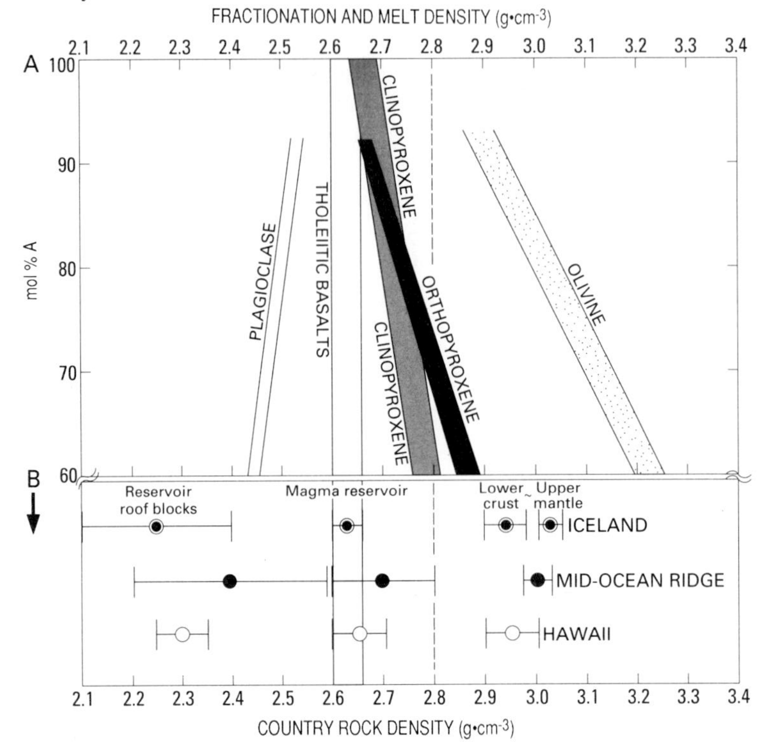

Figure 13 (A) Fractionation densities as a function of depletion of mol% *A*, where *A* represents the high-temperature component in each mineral solid–solution series. Also shown is the reference range for tholeiitic melt density. The vertical dashed line represents the density limit for picritic melts. (B) *In situ* density ranges for magma reservoir roof blocks, lateral magma reservoir country rocks, and the subreservoir oceanic crust and upper mantle for Iceland, the mid-ocean ridges, and Hawaii. As in (A), the density range that includes picritic melts extends to the dashed line. Modified after Sparks and Huppert (1984).

those components forming olivine, clinopyroxene, and orthopyroxene, fractional removal *reduces* the remaining melt density, whereas the removal of plagioclase components increases melt density. These results are in accord with Fig. 12 and the results of Stolper and Walker (1980) and Sparks *et al.* (1980). (See also related work in Koyaguchi, 1990, and in Huppert and Sparks, 1980.

Forced and free convection within the upper fluid-rich chamber will entrain olivine phenocrysts within a flowing suspension. Thus the bulk (melt + crystal) density, ρ_b, of a picritic melt and olivine mixture has a natural place in neutral buoyancy assessments. Incremental additions of olivine (Fo_{90}) to a picritic melt may be expected to raise the crystal-free density (ρ_m=2.723 g · cm^{-3}) through the range ρ_b = 2.746, 2.770, 2.795, and 2.820 g · cm^{-3}, for additions of 5, 10, 15, and 20% olivine (Huppert and Sparks, 1980; Ryan, 1993). Thus bulk densities at least to $\rho \simeq 2.82$ g · cm^{-3} are expectations for the flowing suspension.

It is important to observe that the 2.723 g · cm^{-3} density for the melt phase, ρ_m, reference point should be regarded as a *minimum* value, which has the effect of making the 2.82 g · cm^{-3} value for the melt and 20% olivine crystal suspension a conservative density estimate for the mixture. The picritic melt densities of Fig. 12 approach 2.8 g · cm^{-3}, suggesting that the bulk density for a new estimate of the melt and olivine suspension at 10 vol% would approach $\rho_b \cong 2.847$ g · cm^{-3}, whereas at 20% crystals, the revised estimate is $\rho_b \approx 2.897$ g · cm^{-3}.

The long-term maintenance of a suspension of picritic melt and olivine crystals depends on the strength of the buoyant lift provided by rising fluid in natural and/or forced convection. This lift occurs through a balance between the positive buoyancy of the melt and the Stokes settling of olivine (Stommel, 1949; Marsh and Maxey, 1985).

The residence times (t_r) expected for olivine suspensions in a MORB melt matrix cover roughly the range $\approx 10^{-2} \leq t_r \leq \approx 10$ yr (Martin and Nokes, 1988), and increases in crystal growth and diminished temperature contrasts within the chamber (that drive convection) will thus enhance the likelihood of olivine sedimentation (e.g., Sparks *et al.*, 1993). Within the chamber, the states of magma buoyancy are expected to alternate between neutral buoyancy and positive buoyancy as the mixture(s) dynamically evolve. Thus rounds of sedimentation and differentiation may yield progressive changes from neutral to positive buoyancy, and those magma parcels affected are expected to rise progressively toward the chamber roof in response. Continued differentiation to tholeiitic "minimum-density" melts may be expected to correlate with magma chamber roof fracture events and injection in the sheeted-dike complex.

Density reductions due to H_2O additions must be understood within the perspective of the relatively low H_2O contents found for MORB. The infrared spectroscopy of Dixon *et al.* (1988) suggest dissolved H_2O contents over the range 0.07–0.48 wt% (Endeavor segment of the Juan de Fuca Ridge), whereas the Cobb offset lavas yielded 0.15–0.36 wt%. Density reductions induced by adding 1.08 wt% H_2O to anhydrous basalt (I. Kushiro, personal communication, 1986) produce a $\Delta\rho \simeq 0.05$ g · cm^{-3}. Thus the shaded melt density bands of Figs. 1 and 2 contain all conceivable perturbations due to water in mid-ocean ridge basalts.

Hooft and Detrick (1993) estimated the rock density–depth profile for the 0- to 1400-m-depth interval of the EPR on the basis of three approaches. Their initial approach made use of published porosity–depth profiles (Berge *et al.*, 1992) based on Christeson *et al.* (1992) and Vera *et al.* (1990) seismic velocity profiles. This initial approach used the relation $\rho(Z) = \rho_g - \varnothing(Z)(\rho_g - \rho_w)$, with an assumed grain density of $\rho_g = 2.95$ g · cm^{-3}. The resulting high-gradient profiles attain inferred densities of 2.75 g · cm^{-3} over 200–500 m depth. Next, employing an exponential relationship between depth and porosity, $\varnothing(Z) = \varnothing_0 e^{-\lambda z}$, yielded the estimate $\rho = 2.75$ g · cm^{-3} by about 500 m depth. Finally, application of the empirical velocity–density relation of Christensen and Shaw (1970), $\rho = 1.85 + 0.165 V_p$, to the profile of Vera *et al.* (1990) yields $\rho \cong 2.80$ g · cm^{-3} by 1000 m depth, and $\rho \cong 2.85$ g · cm^{-3} by 1200 m depth. Hooft and Detrick (1993) next assumed that the contents of the EPR magma reservoir may be characterized by a density of 2.70 ± 0.02 g · cm^{-3}, and in a comparison with their profiles outlined previously, then concluded that the neutral buoyancy level lies at a depth 100–400 m below the rise axis. This unfortunate assumption has implicitly confined attention to only *minimum density tholeiitic* melts. The complete compositional range for MORB includes ferrobasalts as well as picrites, however. In addition, it is important to consider potential contribtions from suspended crystal phases (e.g., olivine), which have been shown (Huppert and Sparks, 1980; Ryan, 1993) to impact significantly the bulk or aggregate density of the flowing melt and crystal mixture. Considering now the complete compositional and density range appropriate for EPR magmas, we can return to the curves of Hooft and Detrick (1993) and see that their density estimate of the Vera *et al.* (1990) profile (their Fig. 4a, p. 425) is compatible with that of a picritic melt and olivine crystal mixture through the 1000- to 1200-m-depth interval. Indeed, it is also compatible with a *tholeiitic* melt and olivine crystal mixture at depths in excess of 1000 m.

The inferred interactions between evolving melt densities with fractional crystallization and *in situ* rock density must be consistent with the observed lithologic succession in ophiolite complexes as well as the seismic velocity structure of the active mid-ocean ridges. Figure 14 schematically illustrates the crosscutting relationship between the density band for picritic-to-tholeiitic melts and the country rock density profile(s). Three rock profiles were drawn and range from high to low values of $d\rho/dz$. Note that while the range in melt densities is constrained to be between 2.6 and 2.8 g · cm^{-3}, the surroundings of the magma reservoir may have a high or low gradient in *in situ* density depending on factors such as the maturity of the crust (thickness of the low-density basalt units), spreading rate(s), and associated thermal structures. For "high-gradient" crust (i.e., $d\rho/dz$ large), the density profile cuts the melt band at relatively shallow depth (Z_1) and has a correspondingly thin horizon of neutral buoyancy HNB_1 (Fig. 14). Thus the *equilibrium* magma chamber would be expected to be relatively shallow and of small floor-to-ceiling height. Conversely, for "low-gradient" crust (i.e., $d\rho/dz$ small), the pro-

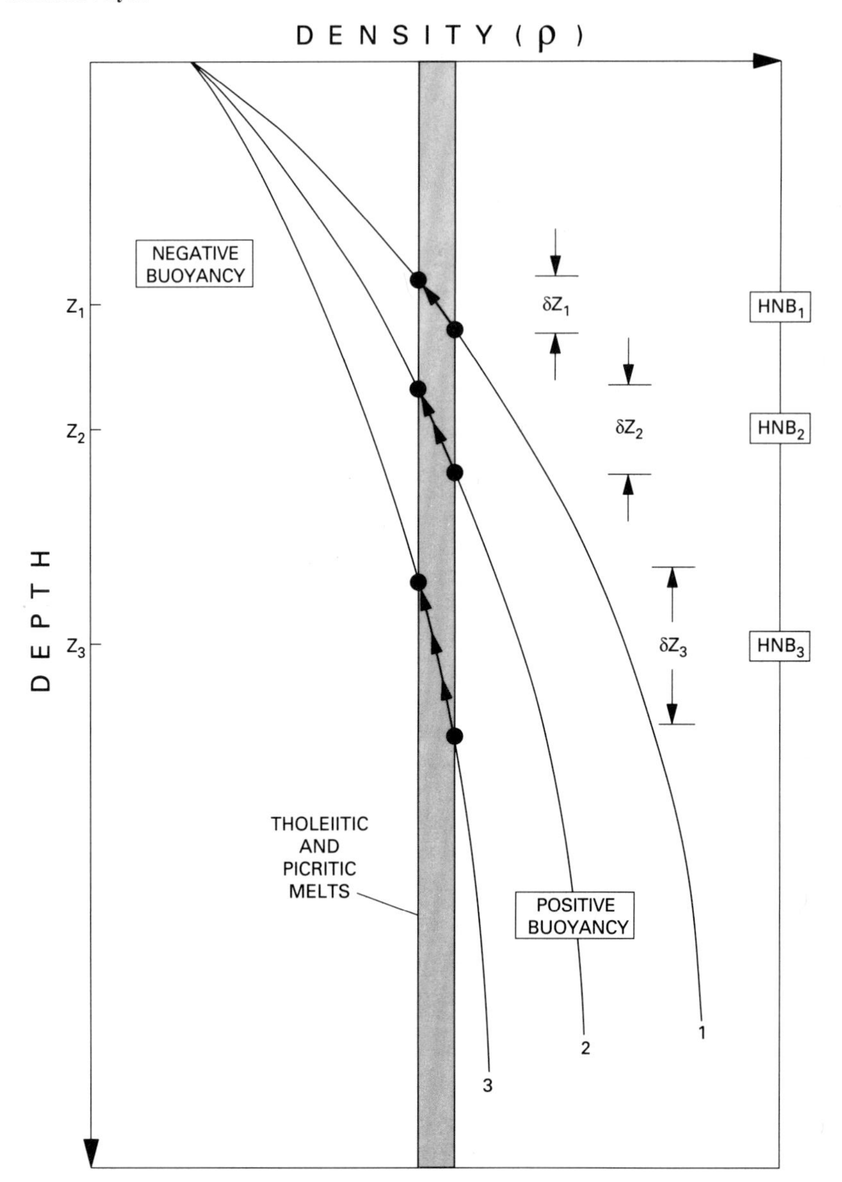

Figure 14 Density–depth relations between mafic and ultramafic melts (shaded) and country rock (solid curves) for three schematic host rock profiles. For each profile, cross-cutting relationships between melt (ρ-Z) and host rock (ρ-Z) define an HNB whose thickness and depth depend sensitively on the gradient of *in situ* density with depth. For example, δZ_1, and HNB_1, correspond to the thickness and the position of the horizon of neutral buoyancy for profile 1. Arrows along a profile connect the picritic values (right dot) with the tholeiitic values (left dot) in schematic fractional crystallization paths. While these relations are general, applications to the oceanic crust suggest that the magma density band lies over the 2.6–2.8 g · cm^{-3} interval.

file cuts the melt band at greater depth (Z_3) and thus has a thicker horizon of neutral buoyancy HNB_3 (Fig. 14), tending to promote a more vertically extensive magma chamber.

As discussed earlier, the fractional crystallization of a picritic melt continuously changes the density of the remaining liquid; thus interactions should be expected between melt and the subsolidus surroundings. Figure 15 is a synoptic density-difference map for the depth evolution of equilibrium melt parcels during fractional crystallization. The equilibrium position of the picritic melt is

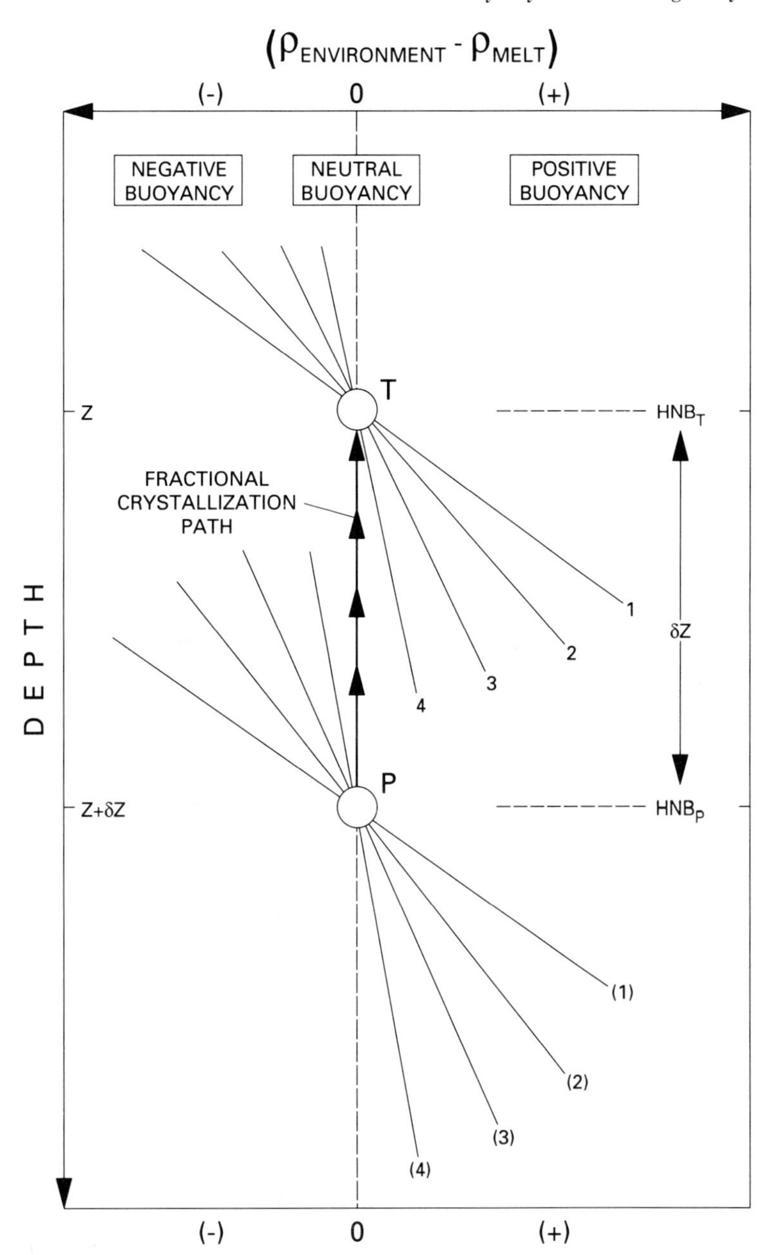

Figure 15 Density difference map showing schematic density contrasts between the environment and newly fractionated melt batches during fractional crystallization from picritic (P) to tholeiitic (T) compositions. Radiating lines of potential density contrast results from a spectrum of potential *in situ* host rock density contrasts above and below the picritic and tholeiitic horizons of neutral buoyancy. The map centerline is defined by $(\rho_{\mathrm{ENVIRONMENT}} - \rho_{\mathrm{MELT}}) = 0$ and is the line of neutral buoyancy. Melts on the right half are positively buoyant, whereas those on the left are negatively buoyant.

point P and rests at the picritic horizon of neutral buoyancy. A family of potential density contrast lines passes through point P and describes the relative magnitudes of positive or negative buoyancy for picritc melt parcels if suddenly displaced upward or downward from the equilibirum position at P. These lines are, in turn, generated by the potentially high (line 1) or low (line 4) values of $d\rho/dz$, of the country rock as discussed above. As fractional crystallization proceeds, the equilibrium pathway from picrite to tholeiite lies directly along the line of neutral buoyancy, and a final equilibrium position (at depth Z) is obtained at the tholeiitic horizon of neutral buoyancy, HNB_T. As before, there is a family of *potential* density contrast lines (1–4) that passes through the tholeiitic position, and mechanical equilibrium (and sudden departures from it) may be related to any one line (but only one). Eruptions are considered nonequilibrium events, and volumetric displacements (influxes of deeper melt) are generally expected to be required to lift tholeiitic liquids through the region of negative buoyancy for extrusion at the surface. [A second condition for "summit" eruption (directly above the replenishment site) is high values of crack-tip K_{IC} in the lateral sheeted dike complexes and high values of σ_3 normal to the lateral intrusion pathway.] Thus the density minima (tholeiitic) melts of Fig. 12 are the recurrent and preferred products.

Fractional crystallization progresses by the sequential appearance of liquidus phases, with each appearance inducing a change in melt density. Figure 16 is a highly schematic illustration of one mode in which fractionation can displace a melt batch from its former gravitational equilibrium position. Removal of olivine, for example, from a picritic batch reduces the derivative melt density (as shown in Figs. 12 (lower) and 13), producing an increment of density reduction ($\delta\rho$). Such reduction renders the melt parcel unstable, and a new mechanical equilibrium must be sought at a higher level (δZ). Continued episodes of crystallization and ascent are expected to progressively elevate the parcel; however, the removal of relatively low-density components—through plagioclase fractionation, for example—may tend to increase the density of the remaining melt and lead to negative buoyancy.

The overall horizon of neutral buoyancy may be subdivided into its ultrabasic and basic components. Gravitational equilibrium in the lower portion requires matching the local gradient in confining pressure, $\nabla\sigma_H$, with the local gradient in picritic magmastatic pressure, ∇P_p. For the East Pacific Rise, the nonlinear $\rho_{is}(Z)$ gradient (Fig. 1) may be approximated by three piecewise-linear segments: a near-surface 22.75 MPa · km^{-1} segment, a central 27 MPa · km^{-1} portion, and a lower 29 MPa · km^{-1} gradient, respectively. Comparison with $\nabla P_p \cong 28$ MPa · km^{-1} provides a best match within the interval ≈1400 to ≈3000 m beneath the volcanic surface. This is the approximate picritic horizon of neutral buoyancy (HNB_P) for melt and crystal mixtures. Thus the summary relationships are

Horizon of negative buoyancy:	$\nabla P_P^- > \nabla\sigma_H$	magma-filled cracks descend	(5)
Horizon of neutral buoyancy:	$\nabla P_P^N \cong \nabla\sigma_H$	magma-filled cracks are gravitationally stable	(6)
Region of positive buoyancy:	$\nabla P_P^+ < \nabla\sigma_H$	magma-filled cracks ascend.	(7)

In relations (5) thru (7), the superscripts (−, N, +) indicate a magma pressure gradient within the negative-, neutral-, and positive-buoyancy regimes, respectively. Elastic fracture stability conditions suggest that above the HNB_P, $\nabla P_P^- > \nabla\sigma_H$, and negative-buoyancy forces promote the descent of picritic melt and crystal mixtures. This conclusion is in accord with the relative rarity of ocean floor picritic eruptives.

For tholeiitic minimum-density melts (see Fig. 12), $\nabla P_T \cong 27$ MPa · km^{-1}, and gravitational equilibrium is approximated in the interval ≈600 to ≈1400 m depth below the volcanic surface. This position corresponds to the sheeted-dike complex, in broad terms, and accords with the generally tholeiitic nature of these dikes. In the figures of this chapter, the depth ≈1000 m represents the *mean* transition, therefore, between the negative-buoyancy region and the HNB for the combined MORB density range 2.6 to 2.8 g · cm^{-3}. As in relations (5) thru (7), an analogous set of relations describe the stability of tholeiitic minimum-density melts: $\nabla P_T^- > \nabla\sigma_H$, $\nabla P_T^N \cong \nabla\sigma_H$, and $\nabla P_T^+ < \nabla\sigma_H$, for the regions of negative, neutral, and positive buoyancy, respectively.

Icelandic and Hawaiian Analogs

Episodes of rock fracture that accompany lateral magma intrusions in active Icelandic rift systems produce microearthquakes that are routinely re-

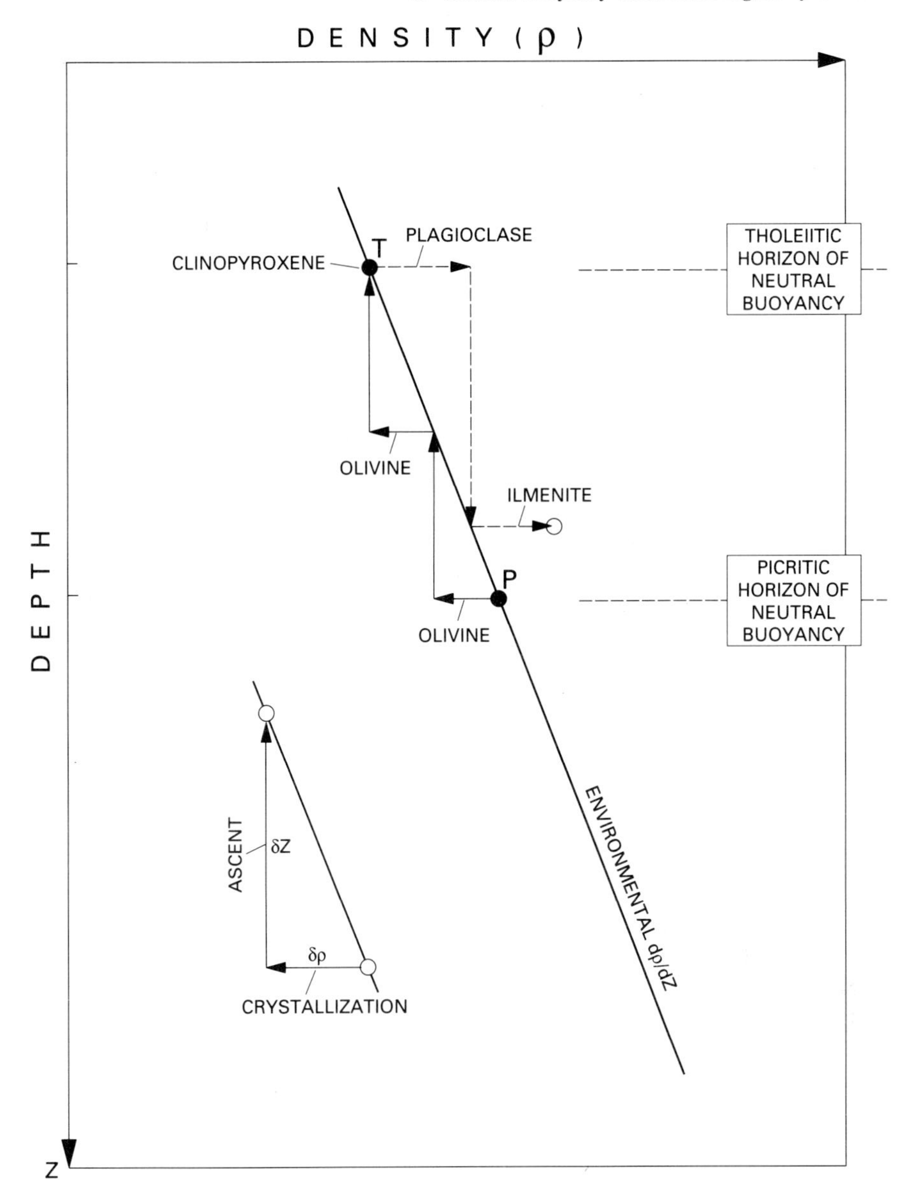

Figure 16 Density–depth relations for the depth evolution of the equilibrium position of a melt parcel during fractional crystallization. "P" denotes picritic depth whereas "T" denotes tholeiitic depth. Increments of $\delta\rho$ correspond to liquidus phases whose fractional removal induces increments of δZ (ascent). Picritic and tholeiitic horizons of neutral buoyancy are HNB_P and HNB_T, respectively.

corded on portions of the Icelandic seismic network. If the seismicity associated with the lateral advance of the magma fracture front is systematically tracked during a dike-formation episode, the geometric relationships of dike evolution about the horizon of neutral buoyancy then can be evaluated. Such a plot is shown in Fig. 17 and is based on the microseismicity produced during and following the rupture of the Krafla magma reservoir on September 8, 1977 (Brandsdottir and Einarsson, 1979) as shown in Ryan (1987b)). The sequence began with the vertical ascent of magma from the subcaldera inflation center that fed a small fissure eruption in the Gjástykki section of Krafla's fissure swarm. Within 19 h, the intrusion had "rolled over" the magma reservoir roof, and, driven by the continued overpressure in the ruptured reservoir and modulated by negative-

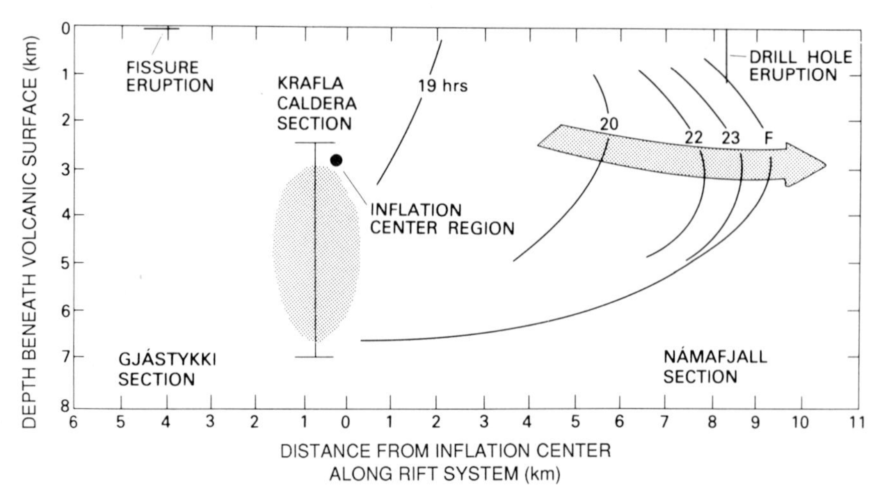

Figure 17 Depth–time–distance evolution of the magmatic fracture front at the Krafla central volcano, Iceland. The parabolic fracture front continually adjusts its position to conform to the equilibrium position at the *local* horizon of neutral buoyancy. Original data from the September 8, 1977, episode as reported in Brandsdottir and Einarsson (1979). (From Ryan, 1987b.) Numerals on the progressive locations of the front positions are times (in hours) after the initial magma reservoir rupture. The shaded arrow is the parabola peak pathway as the equilibrium density–depth position is approached. "F" denotes the final position of the newly formed dike snout. (After Ryan (1987b).)

buoyancy forces in the melt in the upper 2 km of the rift, began a southward-directed lateral intrusion into the Námafjall section. By $t_0 + 20$ h, the intrusion had developed a characteristic parabolic magma fracture front and by $t_0 + 22$ h, it had reached a position centered at the horizon of neutral buoyancy. The final resting position ("F" in Fig. 17) was at the equilibrium neutral-buoyancy level. It is coincident with the *in situ* melt country rock density crossover position in the Icelandic crust. Therefore, flowing melt tends to attempt to achieve a *local* mechanical equilibrium with its surroundings, which in effect makes the lateral intrusion process a natural moving *in situ* "density meter."

Analogous behavior is commonly shown during lateral magma intrusion episodes in Kilauea's rift zones. Figure 18 is from Ryan (1987b) and illustrates depth–time portraits for dike-formation episodes. They are based on the depth–time swarms of microseismicity as depicted by the shaded envelopes. These earthquakes were generated by rock fracture events induced by the hydraulic pressure transients during intrusion. Two broad classes of intrusion are illustrated: (A) and (B) are single-increment intrusions and (C)–(F) are multiple-increment events.

Single-increment intrusions, as illustrated in frame (A), typically begin at the HNB and advance until the fracture front is gradually arrested and magma begins to be impounded behind the crack tip. The arrested front forces melt above and below the HNB in ways that induce the dike keel to descend while the dike top grows toward the volcanic surface. Eventual lateral breakthrough along the horizontal fracture path induces a simultaneous lowering and rising of the magmatic column above and below the HNB, respectively. At the conclusion of the intrusion episode, the parabolic fracture front comes to rest at the equilibrium position. Frame (B), on the other hand, shows that an intrusion has experienced an *early* fracture front arrest and thus has a keel and top that begin their motion with a rapid descent and ascent, respectively. Like intrusion (A), however, intrusion (B) also begins and ends at the horizon of neutral buoyancy and displays a crude top and bottom symmetry about the HNB.

Multiple-increment intrusion cycles show a more complex depth–time portrait that, however, has several features in common with the single-increment mode. For example, in (F), the initial intrusion event beneath the Puú Oó eruption site contained three discrete path blockage events that correspondingly induced three depth–time pulsations in the height of the within-dike magma

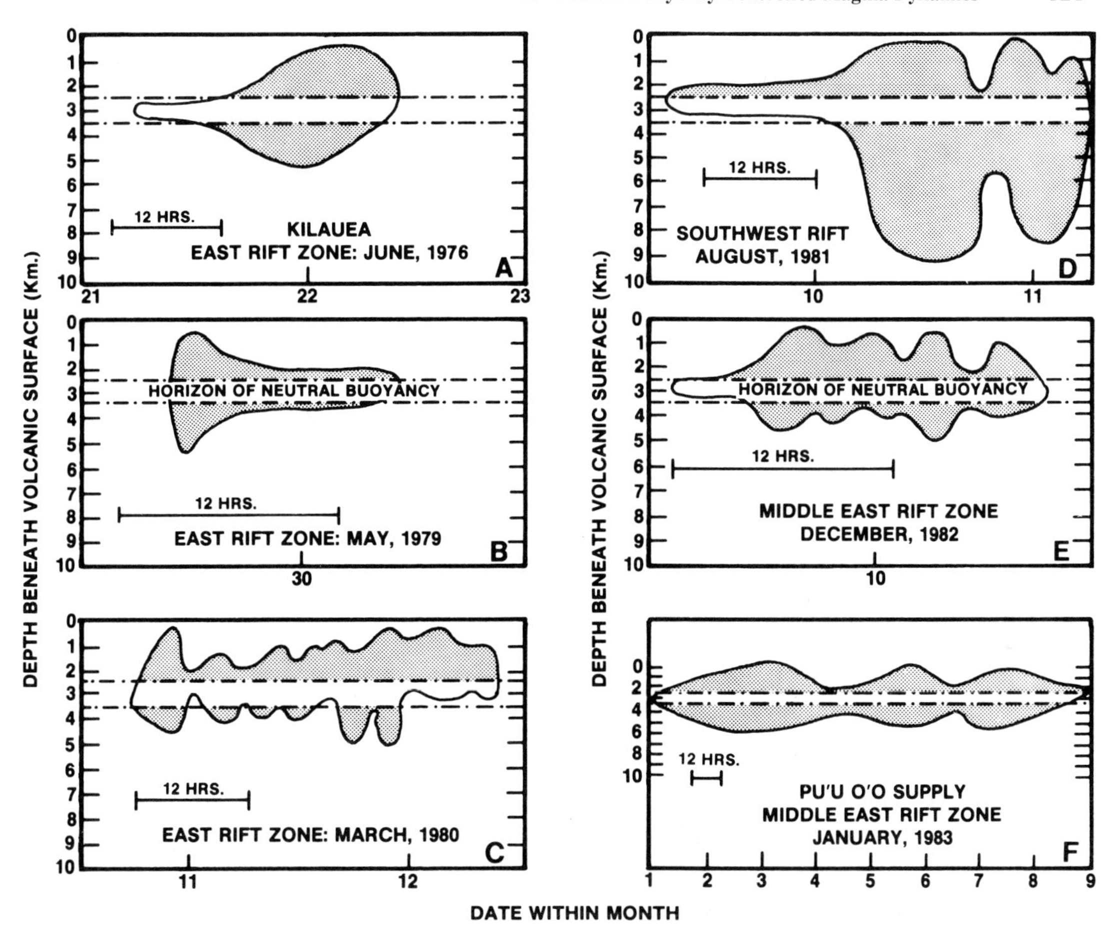

Figure 18 Depth–time portraits of neutrally buoyant dike-formation episodes at Kilauea volcano, Hawaii. In all cases, the intrusion stops and starts at the horizon of neutral buoyancy and has a mean height that is regulated by the HNB equilibrium position. All horizontal time bars correspond to a 12-h duration. Frames (A) and (B) illustrate single-increment intrusions, whereas frames (C)–(F) illustrate multiple-increment intrusion cycles. The timing of the three wave-like oscillations in frame (F) corresponds to changes in the summit subsidence rate as recorded by the Uwekahuna tiltmeter (J. Dvorak, written communication). Envelopes are approximate, and are influenced by the standard errors in hypocenter locations. Only frame (F) was associated with surface outbreaks of eruptive activity January 1983, corresponding to the maximum vertical surges of magma, as illustrated. In general, multiple wave-like fluctuations in the dike height are responses to multiple incidents of crack-front arrest during intrusion, each arrest producing oscillations above and below the equilibrium HNB, as magma is impounded behind the crack tip. (From Ryan 1987b.)

column (i.e., the dike height). Upon breakthrough at each blockage point, the vertical extent of the dike contracted symmetrically about the horizon of neutral buoyancy. Similarly, the multiple-increment cycles depicted in frames (C)–(E) show the *pressure buildup—dike inflation*—patterns behind the arrest site (increases in dike height) and the local *pressure reduction—dike deflation*—pattern that accompanies breakthrough events and subsequent dike-height reductions along the intrusion pathway. Decreases in dike height are also associated with abrupt increases in dike length, due to conservation of mass. When time-integrated through an entire set of intrusion cycles, a wave-like rise and fall of the vertical magma column in the growing dike is a characteristic feature. *These oscillations may have substantial variations in the amplitude (intrusion height) and frequency (time variations in dike height); however, they always exhibit approximate symmetry about the horizon of neutral buoyancy.* Thus wave-like patterns (in depth–time sections) as illustrated in Fig. 18 are

Figure 19 Mid-ocean ridge environments at 9°–14°N (1) and at 21°N (2) along the East Pacific Rise where the highest levels of magma storage and lateral injection in the sheeted dike complex is controlled by neutral buoyancy. Additional regions with

the result of magma encountering several obstructions along the rift zone path, with each obstruction forcing an increase in dike height as impounded magma rose above and descended below the horizon of neutral buoyancy. Figure 19 summarizes the well-documented locations where magma storage and intrusion dynamics are controlled by neutral buoyancy.

Neutral-Buoyancy Control in Lateral Intrusion Dynamics: The Sheeted-Dike Complex

Figure 20 illustrates the structural and kinematic context within which the lateral high-level intrusion of magma takes place. Integrated through time, this process builds the sheeted-dike com-

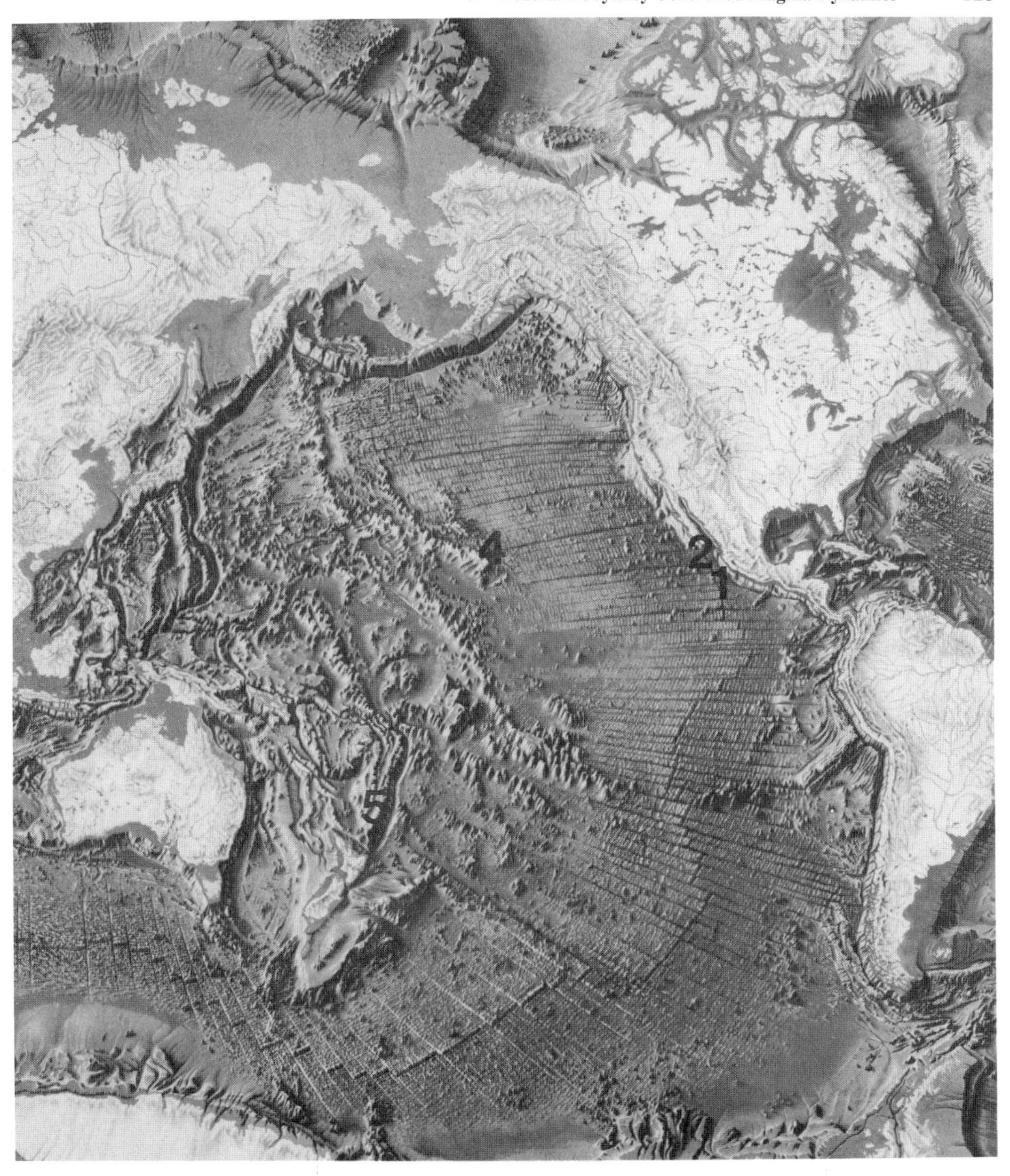

demonstrated neutral buoyancy control are Iceland (3) and Hawaii (4). The back-arc spreading center of the Valu Fa Ridge (5) in the Lau Basin is also inferred to have similar control. (Modified after the Mercator projection of Heezen and Tharp, 1977.)

plex. The environment has two important attributes, both essential for lateral intrusion: nonlinear increases in the *in situ* density structure with depth and available magma driving pressures that reach maximum values at the HNB and can thus counteract the (horizontal) σ_3 stress component normal to the rift zone axis. Near the surface and beneath the axial valley floor, volumetrically high amounts of fracture porosity produce a region of negative buoyancy, such that for nonvesiculated tholeiitic melt, $\rho_m > \rho_{is}$. This negative buoyancy condition will be, therefore, accentuated if the melt is picritic. Increases in confining pressure with depth progressively eliminate this fracture porosity, such that a crossover is produced between the density of the melt and that of

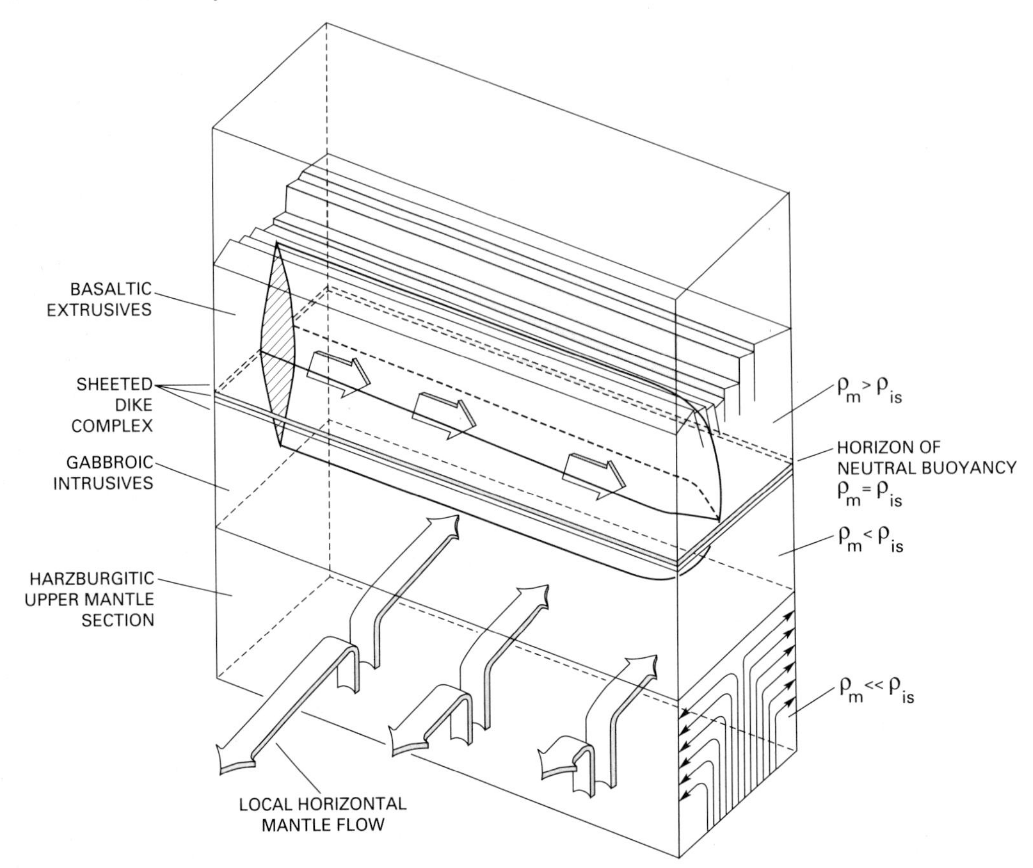

Figure 20 Schematic isometric relationships between the evolving lateral magma injection pathway (short arrows) and the structure of the axial valley and ridge axis. The orientation of the injection pathway is orthogonal to the sub-Moho flow of harzburgite. Sea level corresponds to the top surface of this reference volume.

the surrounding rock, generating a horizon of neutral buoyancy. As discussed earlier, it is along this horizon that magma is injected laterally. Beneath the HNB, continued increases in confining pressure, combined with progressive changes to ultramafic lithologies, produce the upper portions of the region of positive buoyancy. Thus in Fig. 20, the basaltic extrusives blanket the sheeted-dike complexes, whereas the layered gabbros and the harzburgites of the uppermost mantle form an environment that produces positive melt buoyancy.

Coordinates for lateral intrusion are shown in Fig. 21 and illustrate the range of melt dynamics that produce the characteristic parabolic profile of the magma fracture front as well as the important melt-height perturbations induced by crack-front arrest. Symmetric intrusion modes are expected, with the HNB serving as the plane of symmetry. Thus negatively buoyant fluid elements above the HNB counterbalance positively buoyant contributions to dike-height establishment from fluid beneath the HNB. When combined with the positive pressure differential from the ruptured reservoir that drives the fracture front forward, these negative and positive contributions to melt buoyancy continuously modulate the elevation of the dike centerline during lateral growth. Profiles 1 and 2 (Fig. 21) thus correspond to an unobstructed—and unarrested—mode of forward growth. By profile 3, the intrusion front has been arrested by locally high values of σ_3, or high K_{IC} values along the crack-advance path. Magma then impounds behind the crack tip. Continued melt outflow from the ruptured reservoir now begins to inflate the dike, and the walls move outward, swelling the dike and increasing its width. Conservation of mass now also forces magma up above the previous dike top, while the keel begins a simultane-

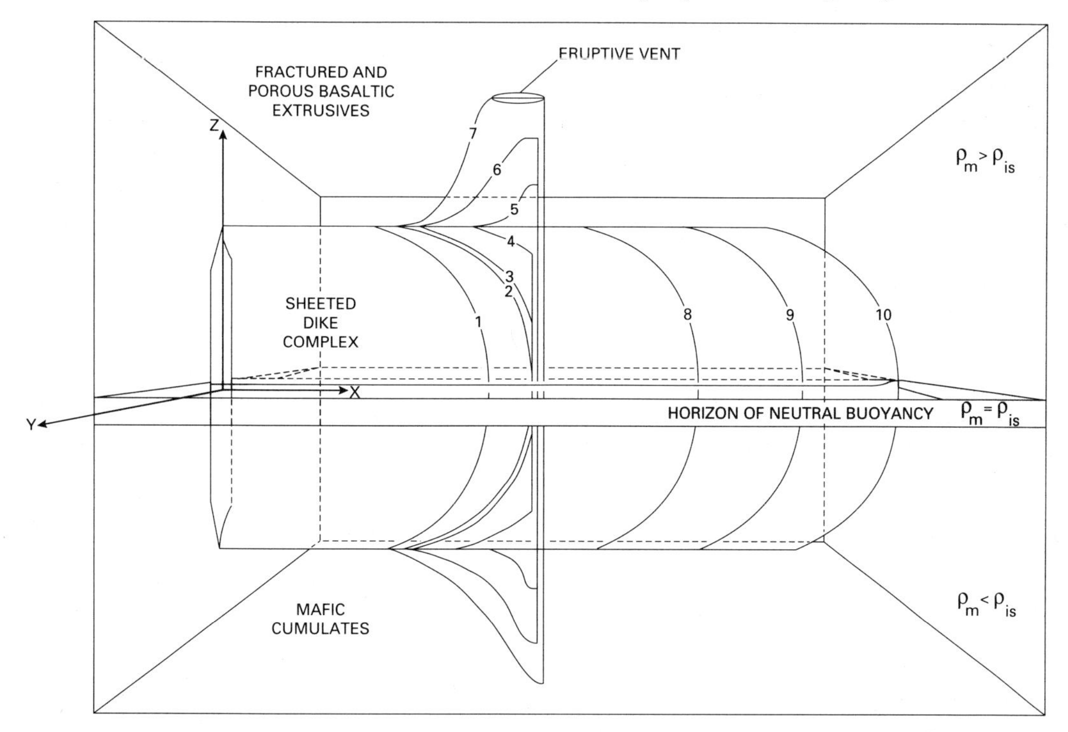

Figure 21 Symmetry relationships and temporal evolution of the magma fracture front during lateral dike emplacement along the horizon of neutral buoyancy. Numbers along profiles correspond to sequential positions of the front. At position 3, high resistance to continued lateral advance has arrested the crack and impounded magma behind the crack tip. Continued out-flow from the ruptured reservoir inflates (widens) the dike and forces magma up above and down below the equilibrium depth level (curves 4, 5, and 6). Curve 7 breaks the surface, producing an eruption. Eventual rupture of the path-advance barrier tends to return parcels of relatively shallow and deep magma to the neutrally buoyant equilibrium position, and curves 8, 9, and 10 therefore track subsequent magma crack tip positions; the eruption stops with intrusion being the only mode of magma movement. Repeated episodes of the lateral injection process build the sheeted-dike complex. (The horizon of neutral buoyancy was schematically compressed into a plane of symmetry for this illustration. In nature, it has a finite vertical extent much larger than that implied here.)

ous descent, thus providing profiles 4, 5, and 6. Should the fracture front continue to be arrested, melt is forced further up along profile 7, which may subsequently intersect the Earth's surface, producing an eruption. Breakthrough at the arrest region once again advances the dike, draws down melt from the strongly negatively buoyant region high above the HNB—while simultaneously shutting off the eruption—and simultaneously draws up relatively positively buoyant melt from the dike keel. This process reestablishes the parabolic fracture front, and profiles 8, 9, and 10 then track the continued down-rift progression of the intrusion. Deviatoric stress states induce important modulations in the crack-advance pathway. These modulations may include inflections in the crack plane, fracture-front breakups, bifurcations and dike-splitting events, and the transition from dike to sill formation. Each accommodates the rapidly varying spatial changes in the resolved least compressive stress orientation (σ_3) and/or heterogeneously distributed elastic moduli.

Consider a fluid-filled crack in an isotropic and linear elastic solid where the flow regime is laminar and undergoes Poiseulle flow between plane parallel walls. The flow is driven by the primary pressure differential ∇P within the fluid, and the temporal evolution of the fracture width is given by (Lister, 1990)

$$\frac{dw}{dt} = \frac{1}{3\eta} \nabla \cdot (w^3 \nabla P), \tag{8}$$

where η is the viscosity of the magma and w is the fluid-filled crack half-width. The relations of Lister (1990) and Lister and Kerr (1990, 1991) are summarized in the following outline.

The elastic pressure in the solid is given by

$$p = -m\,\mathcal{H}\left(\frac{dw}{ds}\right), \tag{9}$$

where m $= \mu/(1 - \nu)$, μ is the shear modulus of the country rock, ν is Poisson's ratio, and $\mathcal{H}$ is the Hilbert transform (Muskhelishvili, 1963; Erdelyi *et al.*, 1954). The total pressure is the buoyancy pressure resulting from density contrasts between the melt and the country rock, as transmitted from the magma reservoir, in addition to the elastic pressure of the crack walls. Taking the vertical coordinate as Z, the pressure is

$$p = -\Delta\rho g Z - m\,\mathcal{H}\left(\frac{dw}{dz}\right). \tag{10}$$

For the laterally directed intrusion along the HNB, under density *step* conditions, the hydrostatic pressure differences between the fluid and the solid are

$$p = p_0(x) - \bar{\theta}\,(\rho_l - \rho_u)\,gZ \tag{11}$$

(above the HNB)

and

$$p = p_0(x) + \theta\,(\rho_l - \rho_u)\,gZ \tag{12}$$

(below the HNB),

where θ is the density contrast parameter (given in (10)) and the conjugate-density contrast parameter $\bar{\theta}$ is

$$\bar{\theta} = 1 - \theta = \frac{\rho_m - \rho_u}{\rho_l - \rho_u}. \tag{13}$$

The notations ρ_l and ρ_u refer to country rock densities below (lower) and above (upper) the horizon of neutral buoyancy, respectively; they are thus a further subdivision of the (ρ_{is}) *in situ*

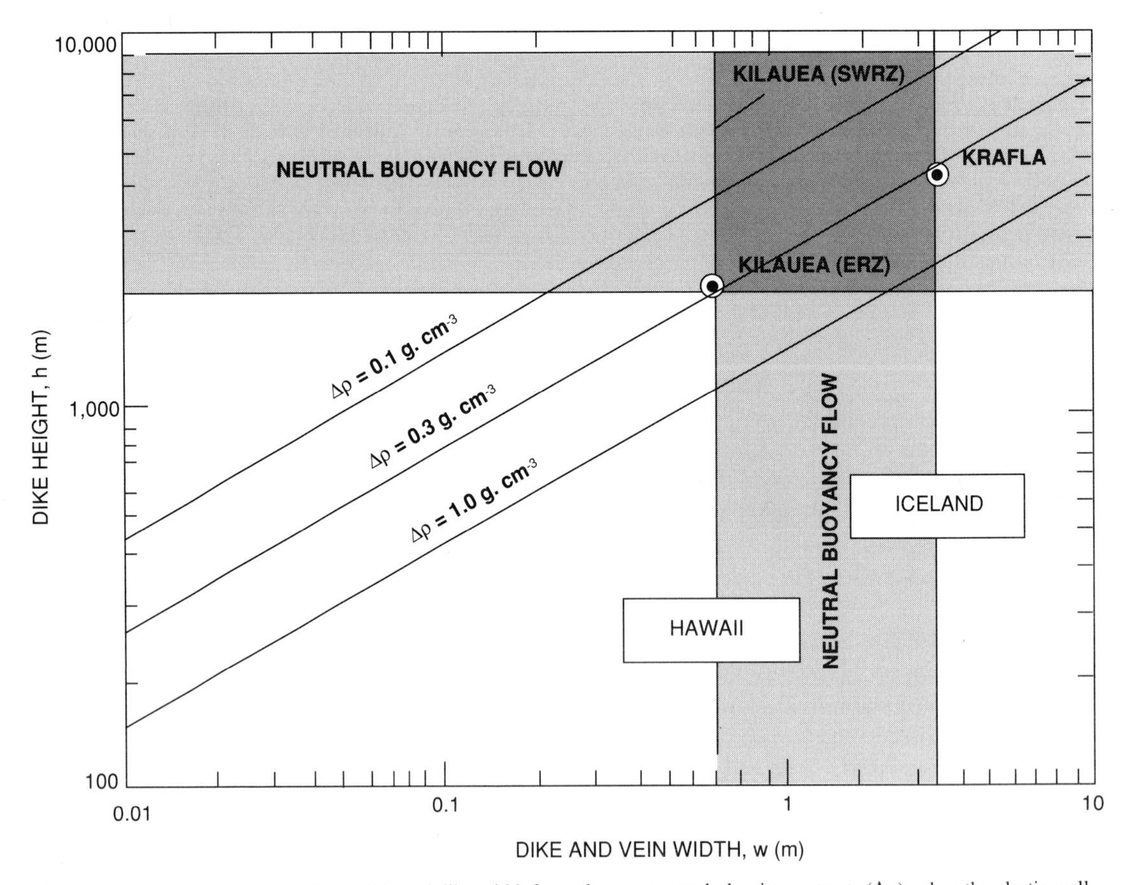

Figure 22 Relations between dike height and dike width for melt–country rock density contrasts ($\Delta\rho$), when the elastic wall rock stress is just balanced by fluid pressure in the flowing magma. Average dike/vein widths for Hawaii and Iceland as well as the maximum range for dike heights are given by the shaded bands. At Kilauea, the two major rift zones are the East rift zone (ERZ) and the Southwest rift zone (SWRZ).

density. *Above* the HNB, the melt is more dense than the country rock, and the condition $\rho_m - \rho_u > 0$ helps to drive the flow laterally along the HNB. Similarly, *below* the HNB, the condition $\rho_l - \rho_m < 0$ drives fluid upward, and then outward along the horizon of neutral buoyancy. Both conditions in (11) and (12) combine to modulate the flow and to provide an overall ∇P to drive lateral intrusion of magma. In addition to the primary fluid pressure accumulation in the magma reservoir, long rift zone intrusions produced by high-batch volume injections will also include a pressure contribution produced by the topographically derived hydraulic head.

Lister and Kerr (1990, 1991) and Lister (1990) have provided similarity solutions for the first-order fracture parameters under conditions that include neutral-buoyancy controlled crack advance. This section makes use of those solutions, with application to intrusion in mid-ocean ridge, Icelandic, Hawaiian, and ophiolite rift systems. In particular, Figs. 22–26 were adapted from Lister and Kerr (1991), modified for the applications of this chapter.

Relationships between dike height and dike and vein widths for conditions approaching neutral-buoyancy flow are illustrated in Fig. 22. The relationship is given for density-contrast values ($\Delta\rho$) spanning the range 1.0–0.1 g · cm^{-3}. These curves move progressively from positively buoyant conditions ($\Delta\rho$ = 1.0–0.3 g · cm^{-3}), appropriate for ascent through the basal region of a mid-ocean ridge magma reservoir, to the virtually neutrally buoyant conditions ($\Delta\rho \cong 0.1$ g · cm^{-3}), appropriate for the uppermost isotropic gabbro and the sheeted-dike regimes of the reservoir and rift systems. Superposed are the fields of neutrally buoyant dike widths and dike heights for Hawaii and Iceland, since such dike dimensions for active mid-ocean ridges are not available.

The relationship between the dimensionless dike height (ς) and dimensionless dike width (w) is illustrated in Fig. 23, for end-member classes of the buoyancy parameter θ. Collectively, they re-

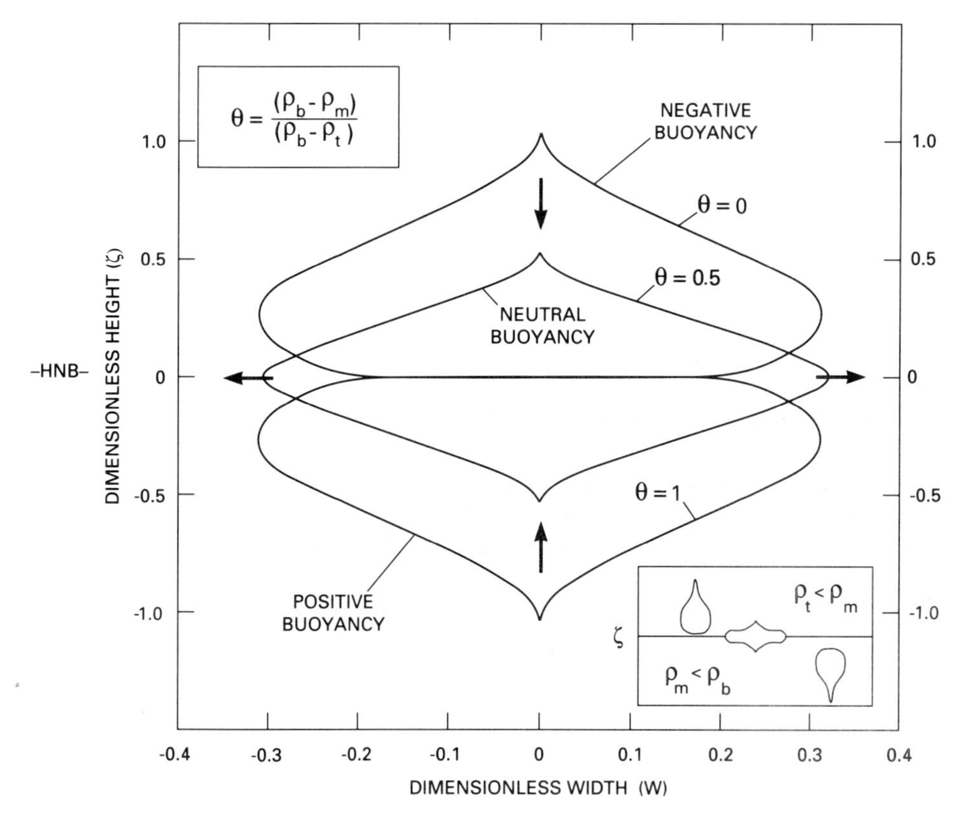

Figure 23 Cross-sectional aspect ratios of dikes as a function of dimensionless height (ζ) and dimensionless width (w), for select values of the density contrast parameter θ.

spectively portray the cross-sectional aspect ratios of inflated dikes in the strongly positively buoyant, the fully developed negatively buoyant, and the neutrally buoyant intrusion modes. θ reflects the density contrast between rocks that lie immediately below and above the horizon of neutral buoyancy, as well as the contrast between the melt density and the surrounding country rocks during initial ascent. It is defined (Lister, 1990) as

$$\theta = \frac{(\rho_l - \rho_m)}{(\rho_l - \rho_u)}. \tag{14}$$

θ values that approach 1 denote strongly positively buoyant ascent modes associated with dikes that tend to develop bulbous tops as they approach the HNB and have tails that are progressively squeezed shut by local components of the confining pressure. Similarly, as θ approaches zero, dikes above the HNB may descend with bulbous (lower) noses and tails squeezed closed by the local values of the effective stress. One could further refine the mechanical equilibrium–non-equilibrium arguments by defining a picritic θ and a tholeiitic θ.

Along the horizon of neutral buoyancy, $\theta \approx 0.5$ and is associated with dikes that have a maximum dilation (width) *centered on the HNB*, and both a keel and a top that are progressively (and symmetrically) squeezed shut by the horizontal component of the effective stress. Each of the three major intrusion modes with respect to θ is illustrated in Fig. 23 and schematically sketched in the figure inset box. The dimensionless nature of the plot, combined with the scaling chosen, results in cross-sectional aspect ratios that, while preserving an accurate and faithful set of *ratios* (ζ, w), also exaggerate the inflated widths so that one may illustrate the relevant trends.

From a general similarity theory for buoyancy-driven flows (Lister, 1990), the height evolution of a vertically oriented fluid-filled fracture is given by

$$\frac{dh^e}{dt} = D\frac{d}{dx}\left[h^{d+e-c}\left(\frac{dh}{dt}\right)^c\right] \tag{15}$$

with a relation between the flow rate and the fracture height integrated along the length and given by

$$\int_0^{x_n} h^e\, dx = Qt^\alpha, \tag{16}$$

where c, d, and e are constants, and D is a coefficient combining material-property parameters for the fluid and matrix:

$$D = \frac{c_1[(\rho_l - \rho_u)g]^3}{\eta[\mu/(1 - \nu)]^2}. \tag{17}$$

For conditions of laminar flow, $c = 1$, while the problem posed for a density step condition at the HNB requires $e = 3$, in (15), and $d = 5$.

For a lateral injection along the horizon of neutral buoyancy, Lister and Kerr (1991) have related the crack height to fluid and rock properties as a specialization of expression (15),

$$h(x, t) = \mathscr{E}_n^{2/5}\left[\frac{\eta[\mu/(1 - \nu)]^4\, Q^2 t^{2\alpha-1}}{c^1 c_2^2[(\rho_l - \rho_u)g]^5}\right]^{1/11} H(\mathscr{E}), \tag{18}$$

where

$$\mathscr{E} = \left[\frac{x}{\mathscr{E}n}\right]\left[\frac{c_2^5\eta^3[\mu/(1 - \nu)]}{c_1^3[(\rho_l - \rho_u)g]^4 Q^5 t^{5\alpha+3}}\right]^{1/11} \tag{19}$$

and

$$\mathscr{E}n = \left[\int_0^1 H^3\, d\mathscr{E}\right]^{-5/11}. \tag{20}$$

In (20), H satisfies the ordinary differential equation

$$\alpha H^3 - \frac{5\alpha + 3}{11}[\mathscr{E}H^3]' = (H^7 H)'' \tag{21}$$

and boundary condition

$$H(1) = 0. \tag{22}$$

Physical-property values recommended for the modeling were derived from a combination of laboratory measurements and geophysical surveys. Values for the rigidity (μ = 20 GPa) reflect the *in situ* high temperatures at depth under conditions of partial fracture closure and were determined by torsional resonance spectroscopy (see Ryan, 1987a). The Poisson's ratio was taken as 0.25. The shear viscosity of tholeiitic melt ($\eta = 10^2$ Pa · s) was determined by Couette viscometry (Ryan and Blevins, 1987, pp. 455–457, 466). The *in situ* density ($\rho_u = 2.3$ g · cm^{-3}) of the near-surface rocks (0.0–2 km depth) was inferred from the combined seismic refraction and gravity studies of David Hill and co-workers (Zucca *et al.*, 1982) and was summarized in (Ryan, 1987b, Fig. 7b). It is an average value for the horizon of negative buoyancy. Similarly, the density (ρ_l) for the deeper region of positive buoyancy (*ibid.*) is 2.9 g · cm^{-3}. The density of the magma was inferred from the high-pressure falling-sphere melt density measurements of Fujii and Kushiro (1977) on olivine tholeiite and is 2.6 g · cm^{-3}. These values have also been adopted by Lister (1990) and by Lister and Kerr (1990, 1991). Importantly, note that melt densities for the reservoir *and* sheeted-dike complex will be in the range 2.6–2.8 g · cm^{-3}, reflecting a compositional range that includes picrites as well as tholeiites.

Neutrally buoyant dike heights as a function of flow duration are compared in Fig. 24 with heights inferred during observatory-based real-time intrusion monitoring in Iceland and Hawaii. Analogous to the dike-width relationships, the constant flow rate curves show a gradually increasing height—but at a progressively diminished rate—as the flow progresses. Similarly, the constant batch volume conditions suggest a progressive decrease in height that is consistent with the conservation of mass requirement: the continued lateral spread along the HNB (dike lengthing) will correspondingly diminish the dike height. The measured flow durations at Krafla and Kilauea are representative, but by no means exhaustive. Thus, both the constant flow rate and the constant batch predictions

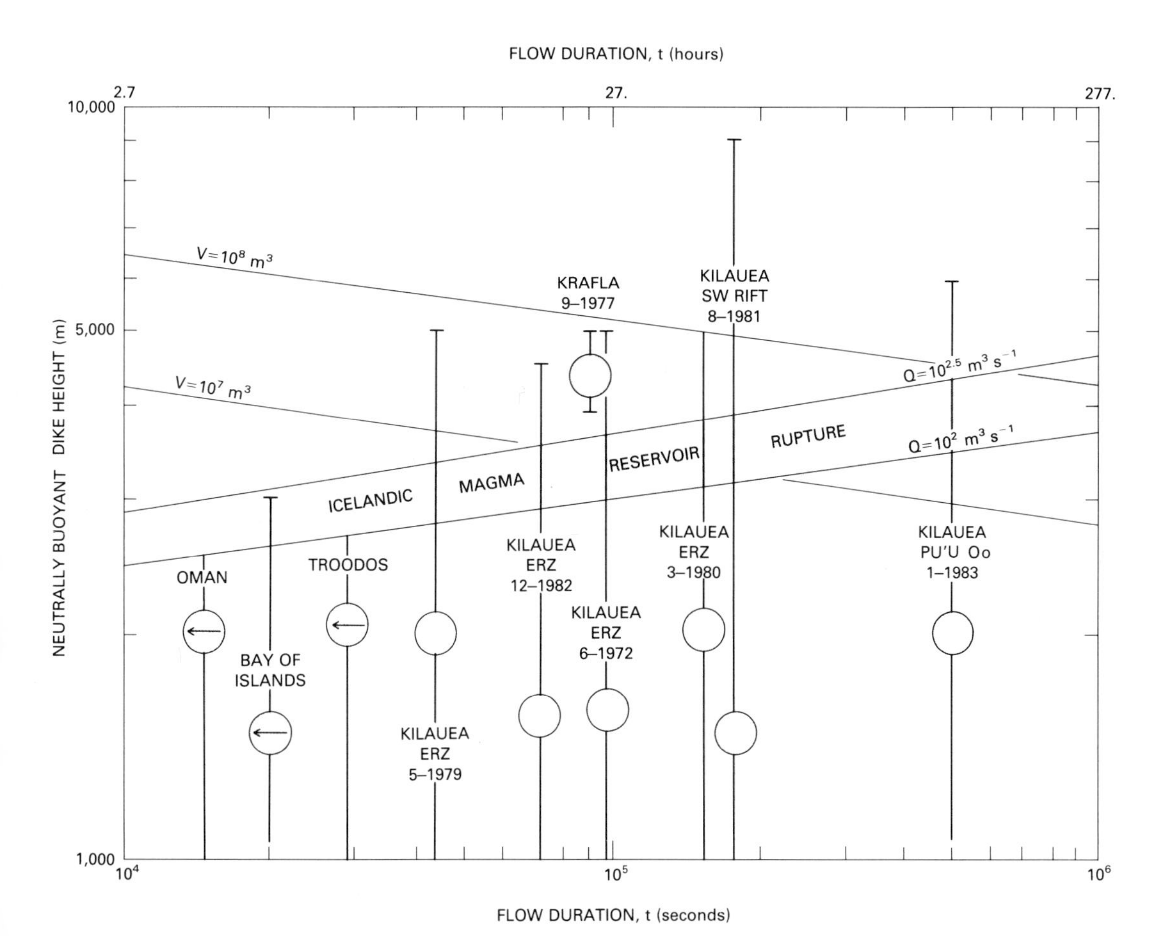

Figure 24 Neutrally buoyant dike-height variations as a function of flow duration. Constant flow rates (10^2 m^3 · s^{-1} $\leq Q \leq$ $10^{2.5}$ m^3 · s^{-1}) appropriate for Icelandic magma reservoir rupture as well as the rupture of mid-ocean ridge reservoirs illustrate the declining rates of dike height evolution. Constant magma batches (10^7 m^3 $\leq V \leq$ 10^8 m^3) produce initially high dikes that progressively shrink in height as melt more closely approaches the horizon of neutral buoyancy and becomes increasingly spread out laterally along this equilibrium depth level. The Kilauea and Krafla data have a dike height bar that corresponds to the maximum depth range (heights) observed, while the shaded circles lie at the average height. The Oman, Bay of Islands, and Troodos dikes are constrained in height only; however, their dimensions are compatible with the short-to-moderate flow durations shown here. These ophiolite points have thus been tentatively plotted to be compatible with the predicted dike heights.

correspond with at least some part of virtually all observed ranges of dike height. For Kilauea, yet lower values of both the flow rates and the magma batch volumes appear required to pick up the entire range of observed heights. Ophiolite dike dimensions and geometric flow parameters for Oman (Juteau *et al.*, 1988; Christensen and Smewing, 1981), the Bay of Islands complex, Newfoundland (Salisbury and Christensen, 1978), and the Troodos, Cyprus, complex (Christensen and Salisbury, 1975) are constrained in height only, and not, of course, in flow duration. Their modest heights, however, are roughly compatible with both the Icelandic and Hawaiian flow rates during early flow periods, immediately after magma reservoir rupture. They have, therefore, been tentatively plotted as illustrated. Ryan (1988, 1990) has discussed the kinematics of rift zone intrusion along the HNB in Hawaii and in Iceland, respectively.

Neutrally buoyant dike widths as a function of flow duration are compared with field data in Fig. 25. The widths were estimated by assuming two types of "end member" flow regimes: cases of constant flow rate ($Q = 10^2$ to $10^{2.5}\ m^3 \cdot s^{-1}$) and constant magma batch volume ($10^7 \leq V \leq 10^8\ m^3$). Experience in monitoring Kilauea and Krafla has demonstrated that episodes of reservoir rupture have outflow rates that are of order 10^2 to $5 \times 10^2\ m^3 \cdot s^{-1}$; thus the plotted flow rates have a scaling appropriate for dike-forming reservoir rupture in Hawaii, Iceland, and, by analogy, the mid-ocean ridge system. Plots of water-tube tiltmeter and electronic tiltmeter signals during the

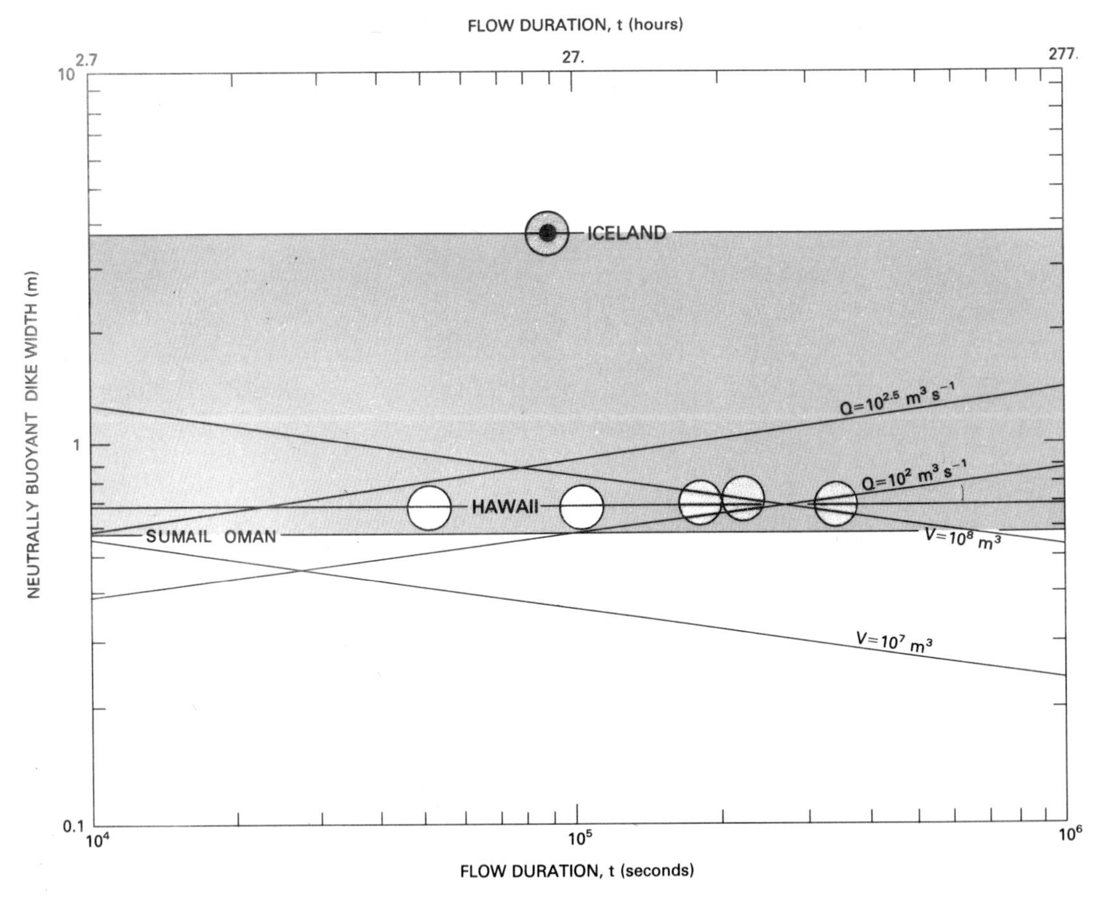

Figure 25 Neutrally buoyant dike widths as a function of flow duration. End-member cases of constant flow rate ($10^2\ m^3 \cdot s^{-1} \leq Q \leq 10^{2.5}\ m^3 \cdot s^{-1}$) and constant volume ($10^7\ m^3 \leq V \leq 10^8\ m^3$) span ranges appropriate for the mid-ocean ridges, Hawaii and Iceland. The gray band brackets average Icelandic dike widths (circle + dot) and average Sumail ophiolite, Oman dike widths. Inferred widths for Hawaiian dikes are plotted (circles) for their measured flow durations at Kilauea and are compatible with the predictions based on both constant flow rate and constant magma batch (volume) conditions.

rupture event, however, clearly demonstrate that magma reservoir-exiting flow rates are exponentially decreasing functions of time (e.g., Dvorak and Okamura, 1987), and thus the constant rates illustrated in Fig. 25 must be understood to be highly idealized. Constant batch volumes have a physical basis that is solidly grounded in observation: the abrupt rupture of the fluid-swollen reservoir releases, in certain cases, a discrete volume of magmatic fluid into the rift system. After release, the reservoir deflation process may reseal fluid passageways, and thus a *finite* volume of magma is introduced into the sheeted-dike complex. Superposed on Fig. 25 are the flow durations for closely monitored Hawaiian and Icelandic intrusions, respectively representing Kilauea and Krafla. Average dike widths for Icelandic and Hawaiian rift systems as well as the Sumail ophiolite, Oman, define a broad field that is believed to include the active mid-ocean ridge sheeted-dike swarms. A consideration of that portion of the plot that is well-populated by data points suggests that there is a broad correspondence between the ophiolite and Hawaiian widths and the constant volume/flow rate predictions. The larger Icelandic dikes require yet higher flow rates and volumes, however. The constant flow-rate curves reflect an increase in width with flow duration, but at a greatly reduced rate of growth after the initial 27-h period. Conversely, the release of constant batch volumes into a growing dike produces a diminution in widths with flow duration, reflecting the conservation of mass consequences of increased lateral spreading (dike lengthening) along the HNB.

Neutrally buoyant dike lengths as a function of flow duration are compared with real-time flow durations and dike length data for Iceland and Hawaii in Fig. 26. Again, the observed values are believed to be representative of each location, but are not exhaustive. As expected, the constant flow-rate conditions predict ever-increasing dike lengths through time. Constant batch volumes also suggest increases in dike length, as the finite volume becomes increasingly spread out and attenuated along the finite thickness horizon of neutral buoyancy. For both the constant batch and constant flow-rate conditions, fairly good agreement is found with the overall spread of observed intrusions at Krafla and at Kilauea. For ophiolite and mid-ocean ridge comparisons, a dike length that scales with the segment half-length has been assumed, consistent with a dike-forming intrusion that has a centrally located origin within a spreading center segment magma reservoir (Whitehead and Helfrich, 1990; Ryan, 1987b). Thus points 8 (Oman; Nicolas, 1989), 9 (Juan de Fuca; Nicolas, 1989), and 10 (East Pacific Rise average; Bonatti, 1985) are roughly constrained in length only, while the flow durations are, of course, unknown. Nevertheless, the confidence gained in the Icelandic and Hawaiian comparisons, which *are* well constrained, suggests that the upper right-hand side of the graph is broadly consistent with their lengths. If this agreement is true, then one has the physical basis for making some rough flow duration estimates for observationally inaccessible or ancient magma intrusions.

Notes on Hydrothermal Interactions

The interrelationships between spreading rate, neutral-buoyancy zonation depths, and mid-ocean ridge thermal structure are as yet unknown, but such relationships are now required to complete our understanding of the detailed balances that promote the long-term stability of shallow magma chambers. That the position of the shallow sub-ridge magma chambers and their sheeted-dike complexes approximate a 1 : 1 correspondence with the horizon of neutral buoyancy is evident from this chapter and from Ryan (1985, 1987a, 1987b, 1993). What is not clear, however, is the additional detailed range of nonequilibrium effects associated with hydrothermal heat withdrawal and the details of the solidification process.

Magma positive buoyancy and heat losses are competing influences in modulating the ascent process. Understanding this phenomenon within the context of long-term magma storage requires, among other things, a three-dimensional model of a spreading center that properly incorporates the fracture mechanics of deep dilatant cracking in the magma reservoir roof. Fracture processes are an important function of spreading rate and thermal structure, bearing in mind that much of the density-reducing porosity in the roof is fracture porosity. It is fracture porosity, that, for example, dramatically lowers the *in situ* V_p, V_s values in the upper 2 km of the oceanic crust. Moreover, fracture porosity and fracture permeability dominate

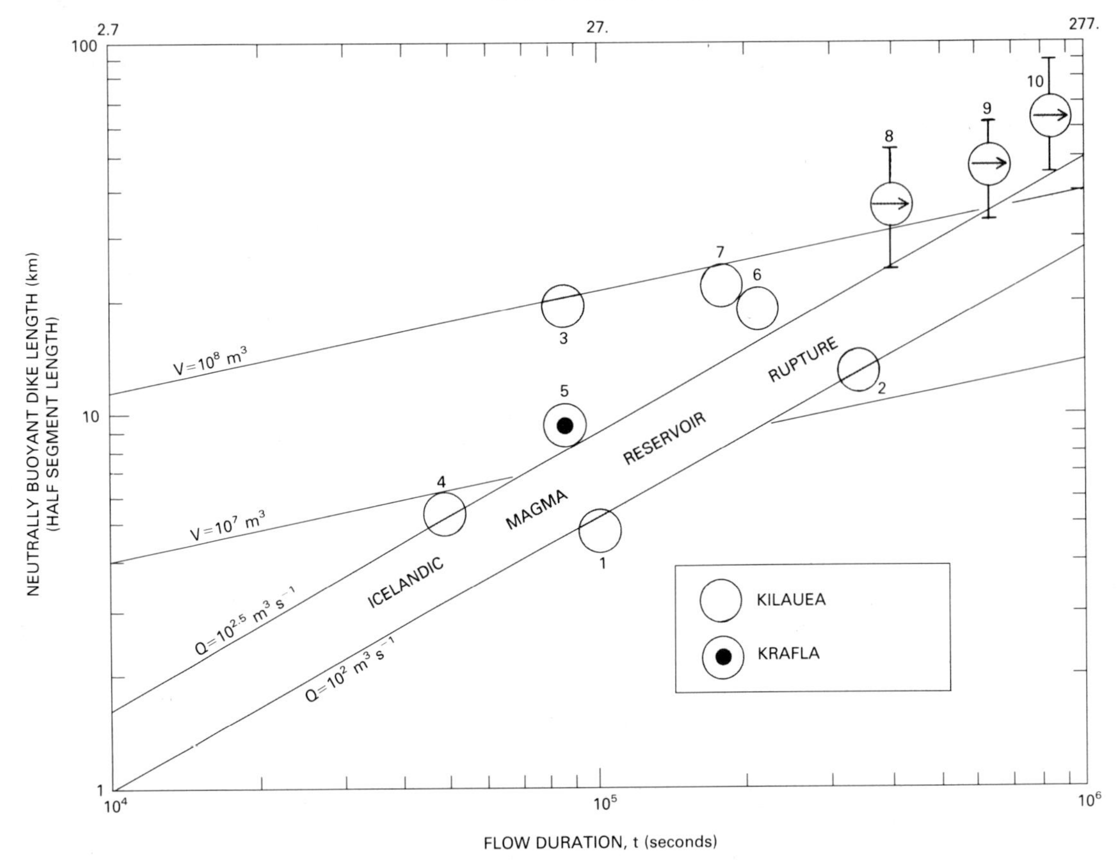

Figure 26 Neutrally buoyant dike length variations as a function of flow duration. For constant flow rates, mid-ocean ridge dike lengths are expected to increase at rates compatible with the documented growth rates at Kilauea, Hawaii, and Krafla, Iceland. Constant magma batch volumes (10^7 m^3 $\leq V \leq 10^8$ m^3) are similarly compatible with both the Hawaiian and the Icelandic rift zone injection environments. Ranges (and averages) at positions 8, 9, and 10 are for Oman, the Juan de Fuca Ridge, and the East Pacific Rise, respectively, and are constrained in segment length only.

the efficacy of the hydrothermal system and its depth extent.

Phipps Morgan *et al.* (1994, this volume) and Phipps Morgan and Chen (1993) have found broad agreement between the apparent depth dependence of the tholeiite solidus (predicted from numerical experiments with Nusselt numbers (Nu) over the range $8 \leq \text{Nu} \leq 12$) and half-spreading rates, assuming hydrothermal penetration temperatures of 600°C (see their Fig. 9 (Phipps Morgan *et al.*, 1994) and Fig. 7 (Phipps Morgan and Chen, 1993), respectively). They have also plotted the depth extents of axial earthquake centroids, on the basis of the teleseismic study of Huang and Solomon (1988) in Fig. 9 of their chapter in this volume. These results indicate deep fracturing (to depths of 8–12 km) at slow spreading rates and a progressive shallowing of the fracture front as the spreading rate increases. Because much of the density reductions that produce a negatively buoyant environment (that overlies and induces the neutral-buoyancy horizon) are associated with fracture porosity, the Huang and Solomon (1988) data suggest that the horizon of negative buoyancy may have a greater depth extent over slow spreading centers—thus tending to deepen somewhat the horizon of neutral buoyancy and thus the sheeted-dike complex and the uppermost magma reservoir. This implication should, in turn, be consistent with the progressive elevation of the brittle–ductile transition as expected as spreading rates increase. Such a relationship is borne out by the results of their numerical experiments. An overlay of the ≈1- to ≈3-km-depth range inferred for the horizon of neutral buoyancy of this chapter, with the modeled magma lens depth of Phipps Morgan and Chen (1993) or Phipps Morgan *et al.* (1994, this volume), shows complete agreement throughout the half-spreading rate range 30–80 mm · yr^{-1}. This agreement included all EPR points plotted. Comparisons with other areas (e.g., Juan de Fuca and the Lau Basin) should utilize seismic–velocity profiles specific to the region for detailed study. I conclude here that the gross density zonation of mid-ocean ridges over the full spreading rate range 20–160 mm · yr^{-1}, and the position of the horizon of neutral buoyancy in particular, provides fundamental control of the location of the uppermost portions of the magma reservoir as well as the location and the intrusion dynamics within the sheeted-dike complex. Within this overall picture, thermal effects that include both solidification and hydrothermal penetration may provide significant local variations on what is dominantly a magma-buoyancy and gravitational equilibrium theme. Hydrothermal interactions that locally retard the achievement of equilibrium neutral-buoyancy positions (depths) are expected to be most pronounced where the spreading rates are low and the deep fracture network promotes deep hydrothermal penetration. The Mid-Atlantic Ridge fits into this category. Ryan (1987b) has demonstrated, however, that the magma chamber of the Krafla central volcano is coincident with the HNB, and this region has a double spreading rate of 20 mm · yr^{-1}. Thus the purely hydrothermal-dominated portion of the spreading rate range must be confined to low rates indeed.

Summary Relationships and Evolution of the Oceanic Crust

The region of negative buoyancy in the newly created lithosphere begins at the ocean floor and extends to ≈1 km depth. It comprises the pillow-basalt flow units (± sediments) and the uppermost brecciated dikes and corresponds to oceanic crustal layers 1 and 2A (Fig. 27). *In situ* densities are lower than nonvesiculated tholeiitic melt and are kept low by deeply penetrating fractures and large amounts of grain-scale porosity. For tholeiitic melt, this region represents a gravitational "no

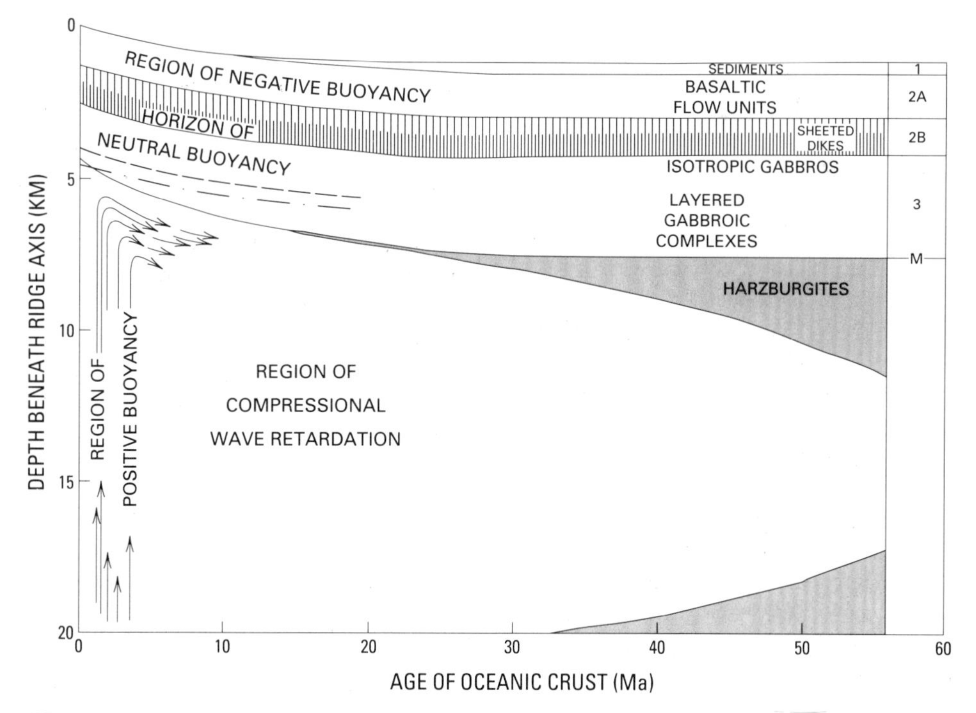

Figure 27 Evolution of the horizons of neutral buoyancy and negative buoyancy and their lithologic products as sea floor spreading progresses astride the ridge. "M" denotes the Mohorovicic discontinuity. Seismic layer numbers (1, 2, and 3) are given in the right-most column (modified after Christensen and Salisbury, 1975).

man's land," and to erupt, magma must traverse the region of negative buoyancy aided by continued volumetric displacements and reservoir replenishment from positively buoyant melt, providing vertically integrated lift from depth.

The *generalized* HNB begins at ≈1 km and extends to ≈3 km depth. Lithologically, it produces the sheeted-dike swarms, as well as the uppermost isotropic gabbros. It corresponds to layer 2B and to the uppermost portion of oceanic layer 3 (Fig. 27). From data derived from seismic reflection and refraction surveys of the East Pacific Rise, for example, the HNB approximates a 1 : 1 correspondence with the low-velocity region lying just beneath the rise axis. In tomographic resolutions of the compressional wave velocity structure, the horizon of neutral buoyancy corresponds to the lowest velocity core region and suggests the highest fluid-to-rock ratios. Late-stage differentiates such as trondjhemite intrusions, are contained within the horizon of neutral buoyancy at the base of the sheeted-dike root region. Proper discussion of buoyancy zonation must incorporate the *complete* expected compositional range: picritic to tholeiitic to ferrobasaltic melts, with due consideration of suspended phenocryst phases.

As the oceanic crust cools and is rafted aside on a flowing asthenosphere, the region of negative buoyancy and the horizon of neutral buoyancy ride with it (Fig. 27). Hydrothermal alteration due to continued zeolite and greenschist facies metamorphism will gradually increase the density of permeable basaltic flow units (Carlson and Raskin, 1984; Carlson and Herrick, 1990; Nelig and Juteau, 1988); however, *in situ* densities that are lower than tholeiitic melt densities should persist to at least 25–30 Myr and become progressively elevated (Carlson and Herrick, 1990, p. 9168). Therefore, upward migrating off-axis melts may continue to be trapped in oceanic crust as old as 25–30 Myr, and *high-level* sills may be expected at the most shallow depths. Thus sea mounts developing in off-ridge locations, for example, may be expected to be underlain at relatively shallow depths by rather small magma storage reservoirs. These relationships are consistent with the downhole sonic logging inferences of Carlson and Herrick (1990), which indicate that *in situ* densities match those of tholeiitic melts at depths of 200–450 m within the 110-Ma Atlantic ocean crust of DSDP hole 418A. [Their Figs. 5 and 6 (pp. 9161–9162) contain an order of magnitude error in the depth axis labeling.] Table 1 summarizes several correlations of the neutral-buoyancy hypothesis.

Table 2

A World *Without* Neutral Buoyancy-Controlled Magma Storage and Along-Axis Transport in the Mid-Ocean Ridge

- Roughly steady-state artesian-like eruption of basaltic melt *directly above* regions of melt production and sub-Moho melt focusing
- Extrusion of basaltic melt only *above* the site of production, and thus an excessive concentration at the segment center
- No *lateral* injections of magma from crustal reservoirs
- No well-developed sheeted-dike complexes in ophiolites and oceanic crust
- No long-lived magma reservoirs beneath mid-ocean ridges with high melt/crystal ratios
- No significant development of isotropic gabbro
- *Marked* increases in the ratio of extrusives to intrusives in ridge and ophiolite environments

Comments at Closure: A World without Neutral Buoyancy

It is interesting to contemplate a world *without* neutral buoyancy (Table 2) and its consequences. In such a world, crustal rocks are everywhere of greater density than the melts that pass through them and, accordingly, only a universal regime of positive buoyancy exists—from the Earth's surface to sites of grain-scale melt mobilization. Table 2 lists several consequences of such a world and, taken collectively, they suggest a very different picture than that observed for both the rock record in ophiolite complexes, and the eruptive, intrusive, and gross geomorphic character of mid-ocean ridges.

Acknowledgments

Discussions with Charles Carrigan (Lawrence Livermore National Laboratory), Robert Coleman (Stanford University), and Robert Crosson (University of Washington) have contributed to this work. John Lister (University of Cambridge) kindly provided a preprint of his JGR paper with Ross Kerr. Helpful review comments by Paul Delaney, John Dvorak, Terrence Edgar, John Lister, Janet Morton, Wayne Shanks, and Robert Tilling improved progressive revisions of the manuscript. Shirley Brown, Lendell Keaton, and Nancy Polend of the U.S. Geological Survey assisted in the preparation of several of the figures. Michelle Urie and Debby Pasquale prepared the typescript. John Sinton and Jason Phipps Morgan are thanked for preprints. Parts of this work were presented at the Princeton–Conoco Symposium in Earth Sciences, Princeton University; the V. M. Goldschmidt Conference, Reston, Virginia; the Symposium on the Generation, Segregation, Ascent and Storage of Magma at the 29th International Geological Congress, Kyoto, Japan; the Geophysical Institute of Kyoto University; the Earthquake Research Institute of the University of Tokyo; and the Department of Earth and Planetary Sciences of Harvard University.

Appendix

Data sources for depth distributions of neutrally buoyant melt beneath the East Pacific Rise, the Valu Fa Ridge, Lau Basin, the Krafla Central volcano, NE Iceland, and Kilauea volcano, Hawaii. Symbols correspond to those in Figs. 3A and 3B.

I. *East Pacific Rise*

EPR 2. (9°N) Seismic refraction profile, Orcutt *et al.* (1976)
EPR 3. (9°N) Multichannel seismic reflection, Herron *et al.* (1978)
EPR 4. (9°N) Multichannel seismic reflection, Hale *et al.* (1982)
EPR 5. (9°–14°N) Multichannel seismic reflection, Detrick *et al.* (1987)
EPR 6. (9°–10°N) Multichannel seismic reflection, Vera *et al.* (1990)
EPR 7. (9°N) Seismic reflection, Sleep *et al.* (1983)
EPR 8. (21°N) Seismic refraction profile, Reid *et al.* (1977)
EPR 9. (9°N) Seismic refraction, Rosendahl *et al.* (1976)

II. *Valu Fa Ridge, Lau Basin*

VF 1. Multichannel seismic reflection, Morton and Sleep (1985)

III. *Krafla central volcano, Iceland*

I1. Einarsson (1978)
I2. Einarsson (1978)
I3. Björnsson *et al.* (1979)
I4. Tryggvason (1981, 1986)
I5. Ryan, unpublished numerical results (1983)
I6. Johnsen *et al.* (1980)
I7. Marquart and Jacoby (1985)
I8. Marquart and Jacoby (1985)
I9. Marquart and Jacoby (1985)
I10. Larsen *et al.* (1979); Björnsson and Sigudsson (1978)
I11. Marquart and Jacoby (1985)
I12. Marquart and Jacoby (1985)
I13. Marquart and Jacoby (1985)
I14. Brandsdottir and Einarsson (1979)
I15. Brandsdottir and Einarsson (1979)
I16. Tryggvason (1980)
I17. Pollard *et al.* (1983)
I18. Ewart *et al.* (1990)
I19. Ewart *et al.* (1990)
I20. Ewart *et al.* (1990)
I21. Ewart *et al.* (1990)
I22. Ewart *et al.* (1990)
I23. Ewart *et al.* (1990)

IV. *Kilauea volcano, Hawaii*

H1. Dieterich and Decker (1975)
H2. Duffield *et al.* (1982)
H3. Walsh and Decker (1971)
H4. Mogi (1958)
H5. Jackson *et al.* (1975)
H6. Fiske and Kinoshita (1967)
H8. Davis *et al.* (1974)
H9. Ryan *et al.* (1981)
H10. Dvorak *et al.* (1983)
H11. Ryan *et al.* (1983)
H12. Eaton (1962)
H13. Koyanagi *et al.* (1976)
H14. Ryan *et al.* (1983)
H15. Ryan *et al.* (1983)
H16. Ryan *et al.* (1983)
H17. Moore and Fiske (1969)
H18. Swanson *et al.* (1976)
H19. Jackson *et al.* (1975)
H20. Wright and Weiblen (1968)
H21. Swanson *et al.* (1976)

References

Bender, J. F., Hodges, F. N., and Bence, A. E. (1978). Petrogenesis of basalts from the project FAMOUS area: Experimental study from 0 to 15 kbars, *Earth Planet. Sci. Lett.* **41,** 277–302.

Berge, P., Fryer, G., and Wilkens, R. (1992). Velocity–porosity relationships in the upper oceanic crust: Theoretical considerations, *J. Geophys. Res.* **97,** 15239–15252.

Björnsson, A., and Sigurdsson, O. (1978). Hraungos úr borholu i Bjarnarflagi, *Nátturu Fraedingurinn,* 19–23.

Björnsson, A., Johnsen, G., Sigurdsson, S., Thorbergsson, G., and Tryggvason, E. (1979). Rifting of the plate boundary in north Iceland: 1975–1978, *J. Geophys. Res.* **84,** 3029–3038.

Bonatti, E. (1985). Punctiform initiation of seafloor spreading in the Red Sea during transition from a continental to an oceanic rift, *Nature* **316,** 33–37.

Bottinga, Y., and Weill, D.F. (1970). Densities of liquid silicate systems calculated from partial molar volumes of oxide components, *Am. J. Sci.* **269,** 169–182.

Brandsdóttir, B., and Einarsson, P. (1979). Seismic activity associated with the September 1977 deflation of the Krafla central volcano in northeastern Iceland, *J. Volcanol. Geotherm. Res.* **6,** 197–212.

Carlson, R. L.. and Raskin, G. S. (1984). Density of the ocean crust. *Nature* **311,** 555–558.

Carlson, R. L., and Herrick, C. N. (1990). Densities and porosities in the oceanic crust and their variations with depth and age, *J. Geophys. Res.* **95,** 9153–9170.

Christensen, N. I. (1978). Ophiolites, seismic velocities and oceanic crustal structure, *Tectonophysics* **47,** 131–157.

Christensen, N. I., and Salisbury, M. H. (1975). Structure and constitution of the lower oceanic crust, *Rev. Geophys. Space Phys.* **13,** 57–86.

Christensen, N. I., and Shaw, G. H. (1970). Elasticity of mafic rocks from the mid-Atlantic ridge, *Geophys. J. Roy. Astron. Soc.* **20,** 271–284.

Christensen, N. I., and Smewing, J. D. (1981). Geology and seismic structure of the northern section of the Oman ophiolite, *J. Geophys. Res.* **86,** 2545–2555.

Christensen, N. I. (1982). Seismic velocities, *in* "Handbook of Physical Properties of Rocks" (R. S. Carmichael, ed.), Vol. II, pp. 1–228, CRC Press, Boca Raton, FL.

Christeson, G. L., Purdy, G. M., and Fryer, G. J. (1992). Structure of young upper crust at the East Pacific Rise near 9°30′N, *Geophys. Res. Lett.* **19,** 1045–1048.

Clarke, D. B. (1970). Tertiary basalts of Baffin Bay: Possible primary magma from the mantle, *Contrib. Mineral. Petrol.* **25,** 203–224.

Davis, P. M., Hastie, L. M., and Stacey, F. D. (1974). Stresses within an active volcano—with particular reference to Kilauea, *Tectonophysics* **22,** 355–362.

Detrick, R. S., Buhl, P., Vera, E., Mutter, J., Orcutt, J., Madsen, J., and Brocher, T. (1987). Multi-channel seismic imaging of a crustal magma chamber along the East Pacific Rise, *Nature* **326,** 35–41.

Dieterich, J. H., and Decker, R. W. (1975). Finite element modeling of surface deformation associated with volcanism, *J. Geophys. Res.* **80,** 4094–4102.

Dixon, J. E., Stolper, E., and Delaney, J. R. (1988). Infrared spectroscopic measurements of CO_2 and H_2O in Juan de Fuca Ridge basaltic glasses, *Earth Planet. Sci. Lett.* **90,** 87–104.

Duffield, W. A., Christensen, R. L., Koyanagi, R. Y., and Peterson, D. W. (1982). Storage, migration and eruption of magma at Kilauea volcano, Hawaii, 1971–1972, *J. Volcanol. Geother. Res.* **13,** 273–307.

Dvorak, J., Okamura, A., and Dieterich, J. H. (1983). Analysis of surface deformation data, Kilauea volcano, Hawaii, October 1966 to September 1970, *J. Geophys. Res.* **88,** 9295–9304.

Dvorak, J., and Okamura, A. T. (1987). "A Hydraulic Model to Explain the Variations in Summit Tilt Rate at Kilauea and Mauna Loa Volcanoes, U.S. Geological Survey Prof. Paper 1350, pp. 1281–1296.

Eaton, J. (1962). Crustal structure and volcanism in Hawaii, *in* "Crust of the Pacific Basin," pp. 13–29. Am. Geophys. Union Monograph.

Einarsson, P. (1978). S-wave shadows in the Krafla caldera in NE-Iceland, evidence for a magma chamber in the crust, *Bull. Volcanol.* **41–3,** 1–9.

Elthon, D. (1979). High magnesia liquids as the parental magma for ocean floor basalts, *Nature* **278,** 514–518.

Erdelyi, A., Mangus, W., Oberhettinger, F., and Tricomi, F. G. (eds). (1954). "Tables of Integral Transforms," McGraw-Hill, New York.

Ewart, J. A., Voight, B., and Björnsson, A. (1990). Dynamics of Krafla caldera, north Iceland: 1975–1985, *in* "Magma Transport and Storage" (M. P. Ryan, ed.), Wiley, Chichester/Sussex, England.

Fiske, R. S., and Kinoshita, W. T. (1967). Inflation of Kilauea volcano prior to its 1967–1968 eruption, *Science* **165,** 341–349.

Fujii, T., and Kushiro, I. (1977). Density, viscosity, and compressibility of basaltic liquid at high pressures, *Carnegie Inst. Washington Yearbook* **76,** 419–424.

Gove, P. B. (1964). "Webster's Third New International Dictionary of the English Language Unabridged," G.&C. Merriam Co., Springfield, MA.

Hale, L. D., Morton, C. J., and Sleep, N. H. (1982). Reinterpretation of seismic reflection data over the East Pacific Rise, *J. Geophys. Res.* **87,** 7707–7717.

Heezen, B. C., and Tharp, M. (1977). (Map of) the world's ocean floor, Mercator Projection, Scale: 1:23,230,300, United States Navy, Office of Naval Research.

Herron, T. J., Ludwig, W. J., Stoffa, P. L., Kan, T. K., and Buhl, P. (1978). Structure of the East Pacific Rise crest from multichannel seismic reflection data, *J. Geophys. Res.* **83,** 798–804.

Hill, D. P. (1969). Crustal structure of the Island of Hawaii from seismic-refraction measurements, *Bull. Seis. Soc. Am.* **59,** 101–130.

Hill, D. P., and Zucca, J. J. (1987). Geophysical constraints on the structure of Kilauea and Mauna Loa volcanoes and some implications for seismomagmatic processes, *in* "Volcanism in Hawaii" (R. W. Decker, T. L. Wright, and P. H. Stauffer, eds.), U.S. Geological Survey Prof. Paper 1350, 917.

Hooft, E. E., and Detrick, R. S. (1993). The role of density in the accumulation of basaltic melts at mid-ocean ridges, *Geophys. Res. Lett.* **20,** 423–426.

Huang, P. Y., and Solomon, S. C. (1988). Centroid depths of mid-ocean ridge earthquake: Dependence on spreading rate, *J. Geophys. Res.* **93,** 13445–13477.

Huppert, H. E., and Sparks, R. S. J. (1980). Restrictions on the compositions of mid-ocean ridge basalts: A fluid dynamical investigation, *Nature* **286,** 46–48.

Irvine, T. N. (1977). Definition of primitive liquid compositions for basic magmas, *Carnegie Inst. Washington Yearbook* **76,** 454–461.

Jackson, D. B., Swanson, D. A., Koyanagi, R. Y., and Wright, T. L. (1975). The August and October 1968 East Rift Eruptions of Kilauea Volcano, Hawaii," U.S. Geological Survey Prof. Paper 890.

Johnsen, G. V., Björnsson, A., and Sigurdson, S. (1980). Gravity and elevation changes caused by magma movement beneath the Krafla caldera, northeast Iceland, *J. Geophys. Res.* **47,** 132–140.

Juteau, T., Beurrier, M., Dahl, R., and Nehlig, P. (1988). Segmentation at a fossil spreading axis: The plutonic sequence of the Wadi Haymiliyah area (Haylayn Block, Sumail Nappe, Oman), *Tectonophysics* **151,** 167–197.

Koyaguchi, T. (1990). Graphical estimations of magma densities and compositional expansion coefficients, *Contrib. Mineral. Petrol.* **105,** 173–176.

Koyanagi, R. Y., Unger, J. D., Endo, E. T., and Okamura, A. T. (1976). Shallow earthquakes associated with inflation episodes at the summit of Kilauea volcano, Hawaii, *Bull. Volcanol.* **39,** 621–631.

Larsen, G., Grönvold, K., and Thorarinsson, S. (1979). Volcanic eruption through a geothermal borehole at Námafjall, Iceland, *Nature* **278,** 707–710.

Lippard, S. J., Shelton, A. W., and Gass, I. G. (1986). The ophiolite of Northern Oman, *Geol. Soc. London Memoir,* **11.**

Lister, J. R. (1990). Buoyancy-driven fluid fracture: Similarity solutions for the horizontal and vertical propagation of fluid-filled cracks, *J. Fluid Mech.* **217,** 213–239.

Lister, J. R., and Kerr, R. (1990). Fluid-mechanical models of dyke propagation and magma transport, *in* "Mafic Dyke and Emplacement Mechanisms" (A. J. Parker, P. C. Rickwood, and D. H. Tucker, eds.), pp. 69–80, A. A. Balkema Publishers, Rotterdam.

Lister, J. R., and Kerr, R. (1991). Fluid-mechanical models of crack propogation and their application to magma transport in dykes, *J. Geophys. Res.* **96,** 10049–10077.

Macdonald, K. C., Sempere, J. C., and Fox, P. J. (1984). East Pacific Rise from Siqueiros to Orozco fracture zones: Along-axis-strike continuity of axial neovolcanic zone and structure and evolution of overlapping spreading centers, *J. Geophys. Res.* **89,** 6049–6069.

Mammerickx, J., and Smith, S. M. (1980). General bathymetric chart of the oceans (GEBCO), (Mercator Projection, Scale: 1 : 10,000,000 at equator), Canadian Hydrographic Service, Ottawa, Canada.

Marquart, G., and Jacoby, W. (1985). On the mechanism of magma injection and plate divergency during the Krafla rifting episode in NE-Iceland, *J. Geophys. Res.* **90,** 10178–10192.

Marsh, B. D., and Maxey, M. R. (1985). On the distribution and separation of crystals in convecting magma, *J. Volcanol. Geother. Res.* **24,** 95–105.

Martin, D., and Nokes, R. (1988). Crystal settling in a vigorously convecting magma chamber, *Nature* **332,** 534–536.

Mogi, K. (1958). Relations between the eruptions of various volcanoes and the deformation of the ground surfaces around them, *Bull. Earthquake Res. Inst. Tokyo Univ.* **36,** 99–134.

Moore, J. G., and Fiske, R. S. (1969). Volcanic substructure inferred from dredge samples and ocean bottom photographs, Hawaii, *Geol. Soc. Am. Bull.* **80,** 1191–1202.

Morton, J. L., and Sleep, N. H. (1985). Seismic reflections from a Lau Basin magma chamber, *in* "Geology and Offshore Resources of Pacific Island Arcs—Tonga Region (D. W. Scholl and T. L. Vallier, eds.), Vol. 2, pp. 441–453, Circum—Pacific Council for Energy and Mineral Resources, Earth Sciences Series, Houston, TX.

Muskhelishvili, N. I. (1963). "Some Basic Problems of the Mathematical Theory of Elasticity," Noordhoff, Leiden, The Netherlands.

Nelig, P., and Juteau, T. (1988). Flow porosities, permeabilities and preliminary data on fluid inclusions and fossil thermal gradients in the crustal sequence of the Sumail ophiolite (Oman), *Tectonophysics* **151,** 199–221.

Nicolas, A. (1989). Structures of ophiolites and dynamics of oceanic lithosphere, Kluwer, New York.

O'Hara, M. J. (1968a). Are any ocean floor basalts primary magma? *Nature* **220,** 683–686.

O'Hara, M. J. (1968b). The bearing of phase equilibria studies in synthetic and natural systems on the origin and evolution of basic and ultrabasic rocks, *Earth Sci. Rev.* **4,** 69–1330.

Orcutt, J. A., Kennett, B. L. N., and Dorman, L. M. (1976). Structure of the East Pacific Rise from an ocean bottom seismometer survey, *Geophys. J. Roy. Astron. Soc.* **45,** 305–320.

Pálmason, G. (1971). "Crustal Structure of Iceland from Explosion Seismology," Soc. Sci. Islandica, **R.I.T. XL.**

Pálmason, G., and Saemundsson, K. (1974). Iceland in relation to the mid-Atlantic ridge, *Annu. Rev. Earth Planet. Sci.* **2.**

Phipps Morgan, J., and Chen, Y. J. (1993). The genesis of oceanic crust: Magma injection, hydrothermal circulation, and crustal flow, *J. Geophys. Res.* **98,** 6283–6297.

Phipps Morgan, J., Harding, A., Orcutt, J., Kent, G., and Chen, Y. J. (1994). An observational and theoretical synthesis of the magma chamber geometry and of crustal genesis along a mid-ocean spreading center, *in* "Magmatic Systems" (M. P. Ryan, ed.), Academic Press, San Diego.

Pollard, D. D., Delaney, P. T., Duffield, W. A., Endo, E. T., and Okamura, A. T. (1983). Surface deformation in volcanic rift zones, *Tectonophysics* **94,** 541–584.

Reid, I., Orcutt, J. A., and Prothero, W. A. (1977). Seismic evidence for a narrow zone of partial melting underlying the East Pacific Rise at 21°N, *Geol. Soc. Am. Bull.* **88,** 678–682.

Roeder, P. L. (1974). Activity of iron and olivine solubility in basaltic liquids, Earth Planet. Sci. Lett. **23,** 397–410.

Rosendahl, B. R., Raitt, R. W., Dorman, L. M., Bibee, L. D., Hussong, D. M., and Sutton, G. H. (1976). Evolution of oceanic crust. Part I. Physical model of the East Pacific Rise crest derived from seismic refraction data, *J. Geophys. Res.* **81,** 5294–5304.

Rothery, D. A. (1983). The base of a sheeted dyke complex, Oman ophiolite: Implications for magma chambers at oceanic spreading centers, *J. Geol. Soc. London* **140,** 287–296.

Ryan, M. P. (1985). The contractancy mechanics of magma reservoir and rift system evolution: *EOS Trans. American Geophysical Union,* vol. 66, no. 46, p. 854.

Ryan, M. P. (1987a). The elasticity and contractancy of Hawaiian olivine tholeiite, and its role in the stability and structural evolution of sub-caldera magma reservoirs and rift systems, *in* "Volcanism in Hawaii" (R. W. Decker, T. L. Wright, and P. H. Stauffer, eds.), Vol. 2, U.S. Geological Survey Prof. Paper 1350.

Ryan, M. P. (1987b). Neutral buoyancy and the mechanical evolution of magmatic systems, *in* "Magmatic Processes: Physicochemical Principles" (B. O. Mysen, ed.), Geochemical Society Special Publication No. 1, pp. 259–287, Univ. Park, PA.

Ryan, M. P. (1988). The mechanics and three-dimensional internal structure of active magmatic systems: Kilauea volcano, Hawaii, *J. Geophys. Res.* **93,** 4213–4248.

Ryan, M. P. (1990). The physical nature of the Icelandic magma transport system, *in* "Magma Transport and Storage" (M. P. Ryan, ed.), Wiley, Chichester/Sussex, England.

Ryan, M. P. (1993). Neutral buoyancy and the structure of mid-ocean ridge magma reservoirs, *J. Geophys. Res.,* **98,** 22, 321–22, 338.

Ryan, M. P., and Blevins, J. Y. K. (1987). "The Viscosity of

Synthetic and Natural Silicate Melts and Glasses at High Temperatures and One Bar (10^5 Pascals) Pressure and at Higher Pressures," U.S. Geological Survey Bulletin 1764.

Ryan, M. P., Blevins, J. Y. K., Okamura, A. T., and Koyanagi, R. Y. (1983). Magma reservoir subsidence mechanics: Theoretical summary and application to Kilauea volcano, Hawaii, *J. Geophys. Res.* **88,** 4147–4181.

Ryan, M. P., Koyanagi, R. Y., and Fiske, R. S. (1981). Modeling the three-dimensional structure of magma transport systems: Application to Kilauea volcano, Hawaii, *J. Geophys. Res.* **86,** 7111–7129.

Salisbury, M. H., and Christensen, N. I. (1978). The seismic velocity structure of a traverse through the Bay of Islands ophiolite complex, Newfoundland, and exposure of oceanic crust and upper mantle, *J. Geophys. Res.* **83,** 805–817.

Shibata, T., DeLong, S. E., and Walker, D. (1979). Abyssal tholeiites from the oceanographer fracture zone. I. Petrology and fractionation, *Contrib. Mineral. Petrol.* **70,** 89–102.

Sinton, J. M., and Detrick, R. S. (1992). Mid-Ocean ridge magma chambers, *J. Geophys. Res.* **97,** 197–216.

Sleep, N. H., Morton, J. L., Burns, L. E., and Wolery, T. J. (1983). Geological constraints on the volume of hydrothermal flow at ridge axes, *in* "Hydrothermal Processes at Sea Floor Spreading Centers" (P. Rona, K. Bostrom, L. Lambier, and K. Smith, eds.), NATO Conference Series II, Marine Sciences 12, Plenum Press, New York.

Sleep, N. (1975). Formation of oceanic crust: Some thermal constraints, *J. Geophys. Res.* **80,** 4037–4042.

Sparks, R. S. J., and Huppert, H. E. (1984). Density changes during the fractional crystallization of basaltic magmas: Fluid dynamic implications, *Contrib. Mineral. Petrol.* **85,** 300–309.

Sparks, R. S. J., Huppert, H. E., Koyaguchi, T., and Hallworth, M. A. (1993). Origin of modal and rhythmic igneous layering by sedimentation in a convecting magma chamber, *Nature*, **361,** 246–249.

Sparks, R. S. J., Meyer, P., and Sigurdsson, H. (1980). Density variation amongst mid-ocean ridge basalts: Implications for magma mixing and the scarcity of primitive lavas, *Earth Planet. Sci. Lett.* **46,** 419–430.

Stolper, E., and Walker, D. (1980). Melt density and the average composition of basalt, *Contrib. Mineral. Petrol.* **74,** 7–12.

Stommel, H. (1949). Trajectories of small bodies sinking slowly through convection cells, *J. Marine Res.* **8,** 24–29.

Swanson, D. A., Jackson, D. B., Koyanagi, R. Y., and Wright, T. L. (1976). "The February 1969 East Rift Eruption of Kilauea Volcano, Hawaii," U.S. Geological Survey Prof. Paper 891.

Toomey, D. R., Purdy, G. M., Solomon, S., and Wilcox, W. (1990). The three-dimensional seismic velocity structure of the East Pacific Rise near latitude 9°30′N *Nature* **347,** 639–644.

Tryggvason, E. (1980). "Observed Ground Deformation during the Krafla Eruption of March 16, 1980," Nordic Volcanol. Inst. Rept. 80–05, Univ. of Iceland.

Tryggvason, E. (1981). "Pressure Variations and Volume of the Krafla Magma Reservoir," Nordic Volcanol. Inst. Rept. 81–05, University of Iceland.

Tryggvason, E. (1986). Multiple magma reservoirs in a rift zone volcano: Ground deformation and magma transport during the September 1984 eruption of Krafla, Iceland, *J. Volcanol. Geotherm. Res.* **28,** 1–44.

Vera, E. E., Mutter, J. C., Buhl, P., Orcutt, J. A., Harding, A. J., Kappus, M. E., Detrick, R. S., and Brocher, T. M. (1990). The structure of 0- to 0.2-m.y.-old oceanic crust at 9°N on the East Pacific Rise from expanded spread profiles, *J. Geophys. Res.* **95,** 15529–15556.

Walker, D., Shibata, T., and DeLong, S. E. (1979). Abyssal tholeiites from the oceanographer fracture zone. II. Phase equilibria and mixing, *Contrib. Mineral. Petrol.* **70,** 111–125.

Walsh, J. B., and Decker, R. W. (1971). Surface deformation associated with volcanism, *J. Geophys. Res.* **76,** 3291–3302.

Weissel, J. K. (1977). Evolution of the Lau Basin by the growth of small plates, *in* "Island Arcs, Deep Sea Trenches and Back-Arc Basins" (M. Talwani and W. C. Pitman, III, eds.), American Geophysical Union, Maurice Ewing Series 1, pp. 429–436.

Whitehead, J. A., Dick, H. J. B., and Schouten, H. (1984). A mechanism for magmatic accretion under spreading centers, *Nature* **312,** 146–147.

Whitehead, J. A., and Helfrich, K. R. (1990). Magma waves and diapiric dynamics, *in* "Magma Transport and Storage" (M. P. Ryan, ed.), pp. 53–76, Wiley, Chichester/Sussex, England.

Wright, T. L., and Weiblen, P. W. (1968). "Mineral Composition and Paragenesis in Thoeliitic Basalt from Makaopuhi Lava Lake, Hawaii [abs]," Geol. Soc. Amer. Special Paper 115, pp. 242–243.

Zucca, J. J., Hill, D. P., and Kovach, R. L. (1982). Crustal structure of Mauna Loa Volcano, Hawaii, from seismic refraction and gravity data, *Bull. Seismol. Soc. Am.* **72,** 1535–1550.

Chapter 7 An Observational and Theoretical Synthesis of Magma Chamber Geometry and Crustal Genesis along a Mid-ocean Ridge Spreading Center

J. Phipps Morgan, A. Harding, J. Orcutt, G. Kent, and Y. J. Chen

Overview

In this chapter we review seismological evidence and other geophysical evidence that the axial magma chamber beneath a fast spreading ridge is a narrow (~1-km-wide), thin (~50- to 200-m-thick), magma lens that lies at the sheeted dike/gabbro cumulate transition region roughly 1.2–1.5 km beneath the seafloor and overlies a broader region of "hot rock" with at most ~3–5% partial melt fraction. This axial magma chamber appears to contradict earlier ophiolite-based studies that used the dip and dip relations within the "cumulate" gabbro layer to argue for a broad, gabbro-layer-thickness magma body that deposited cumulates along its base and sides. However, it is compatible with an emerging theoretical paradigm that crustal accretion occurs by magma emplacement and solidification within this magma lens, with cumulates subsiding and flowing to form the lower crust as initially proposed by N. H. Sleep (*J. Geophys. Res.* **80,** 4037–4042, 1975). Here we present a theoretical thermal and mechanical model for crustal genesis that incorporates this paradigm and appears to explain successfully the observed depth dependence of the axial magma lens with spreading rate (and the fact that no axial magma lens has been seen in a ridge with a median valley morphology), as well as observed relationships among axial morphology, spreading rate, and magma supply. We suggest that a fairly delicate balance between magmatic heat input and hydrothermal heat removal determines the thickness (i.e., yield strength) of the axial lithosphere, which in turn controls the axial morphology associated with plate boundary extension. Thus the depth (and existence) of an axial magma lens and the axial morphology along a spreading center share a common thermal origin that is a function of spreading rate and magma supply. "Local" trend geochemical systematics may also share this common thermal origin.

Notation

		Units
D_t	Lagrangian derivative	s^{-1}
K	permeability	m^2
L	latent heat	$J \cdot kg^{-1}$
L_{ik}	velocity gradient tensor	s^{-1}
Nu	Nusselt number	dimensionless
Q	attenuation	dimensionless
T	temperature	K, °C
T_m	mantle temperature	K, °C
T_{cutoff}	temperature limit for hydrothermal circulation	K, °C
T_{seawater}	seawater temperature	K
b	crack spacing	m
p	pressure	Pa
q_{hydro}	axial hydrothermal heat flux	$W \cdot m^{-2}$
t	time	s
$\mathbf{u}$, u_i	velocity vector	ms^{-1}
$\mathbf{x}$, x_i	position vector	m
$\mathbf{y}$, y_i	particle displacement vector	m
z_{cutoff}	depth limit for hydrothermal circulation	m
z_{hydro}	axial depth of hydrothermal penetration	m
α	thermal expansivity	K^{-1}, $°C^{-1}$
ϕ	porosity	dimensionless
κ	thermal diffusivity	$m^2 \cdot s^{-1}$
λ	bulk viscosity	$Pa \cdot s$
μ	shear viscosity	$Pa \cdot s$
τ_{ij}	deviatoric stress tensor	Pa
$\dot{\psi}$	magma emplacement rate	$m^3 \cdot s^{-1}$
$\mathfrak{F}$, $\mathfrak{F}_{ij}$	finite strain tensor	dimensionless

Magmatic Systems
Edited by M. P. Ryan

Notation for Oceanic Crust

Seismic layer	Seismic velocity (km · s^{-1})	Analogous ophiolite crustal layer	Ophiolite structure	Typical depth range (m)
2a	2.5–5	٢a	Pillow basalt	0–500
2b	5–6.5	٢b	Sheeted dike	500–1500
3	6.5–7	٣	Cumulate gabbro	1500–6000

Note. We use a separate notation for seismic and ophiolite crustal layering to keep these very distinct observations conceptually distinct. Western numerals are used for seismic layering (the standard convention) and Arabic numerals are used for ophiolite layering.

Acronym Glossary

	Expanded Abbreviation	Definition
AMC	*A*xial *m*agma *c*hamber	The melt lens + underlying mush that makes up an EPR magma chamber.
CDP	*C*ommon *d*epth *p*oint gather	Standard sorting technique where shot-receiver pairs with a shared "midpoint" are gathered.
DEVAL	*Dev*iation from *a*xial *l*inearity	A kink in the ridge axis.
EPR	*E*ast *P*acific *R*ise	Type-example of a fast-spreading ridge.
ESP	*E*xpanding *s*pread *P*rofile	Seismic reflection technique where source and receiver pairs are sited equal distances from a central point to image the structure beneath the central point at multiple "look" angles. This is essentially a CDP gather extending to tens of kilometers, but involves two ships steaming tens of kilometers away from this common point.
MAR	*M*id-*A*tlantic *R*idge	Type-example of a slow-spreading ridge.
MCS	*M*ulti*c*hannel *s*eismic reflection technique	Seismic technique that uses a multiple-source/multiple-receiver geometry to enhance observation of seismic reflectors.
Mg#	Magnesium number	[MgO/(MgO + FeO) moles]
MORB	*M*id-*o*cean *r*idge *b*asalt	Most common basalt extruded at a mid-ocean ridge, less Si-rich than ocean island or island arc basalts.
^{8}Na	"Sodium 8"	A geochemical observable is corrected for the effects of shallow crustal fractionation along a "proper" liquid line of chemical evolution to its appropriate value when the MgO content of the liquid was 8%. [cf. Klein and Langmuir, 1987]
OBS	*O*cean *b*ottom *s*eismograph	Self-contained seafloor seismograph with instrument and recording package.
OSC	*O*verlapping *s*preading *c*enter	A (small) offset in the ridge axis where two ridge segments overlap instead of being linked by a transform fault offset.
P-wave	Pressure wave	Compressional mode of seismic energy propagation.
PmP	P-moho-P	P-wave arrival that has reflected from the Moho.
S-wave	Shear wave	Shear mode of seismic energy propagation.

Note. Marine geologists, seismologists, and geochemists share a fondness for creating acronyms that may confuse a newcomer. We hope this glossary makes amends for our introduction of a separate notation for seismic and ophiolite oceanic crustal layering by describing the acronyms that are liberally used in this chapter.

Introduction

The mid-ocean ridge system is the most productive volcanic system on Earth, annually adding an average of 20 km^3 of new material to the oceanic crust. Almost all the material added to the oceanic crust is thought to experience some degree of low-pressure fractionation within a crustal magma chamber. However, the nature of the magma chamber and of crustal accretion at fast- and slow-spreading ridges is fundamentally different. Early thermal models of ridge crest structure (Sleep, 1975; Kusznir and Bott, 1976) predicted that the heat supplied by magma rising from the mantle would lead to a large (~5-km-thick, ~15-km-wide at its base) magma chamber beneath the axis of a fast-spreading ridge. Sleep (1975, 1978) and Dewey and Kidd (1977) proposed much smaller molten magma bodies, but these ideas were apparently contradicted by early interpretations of ophiolite observations (see Fig. 1A), which were used to support the crustal-scale magma chamber hypothesis, with magma deposition at the base of this large molten magma chamber leading to the formation of cumulate layered gabbro structures (Cann, 1974; Smewing, 1981; Pallister and Hopson, 1981). The "fly in the ointment" came from early seismic measurements that showed no large molten zone present beneath the axis of present day ridges. Subsequent seismic studies have

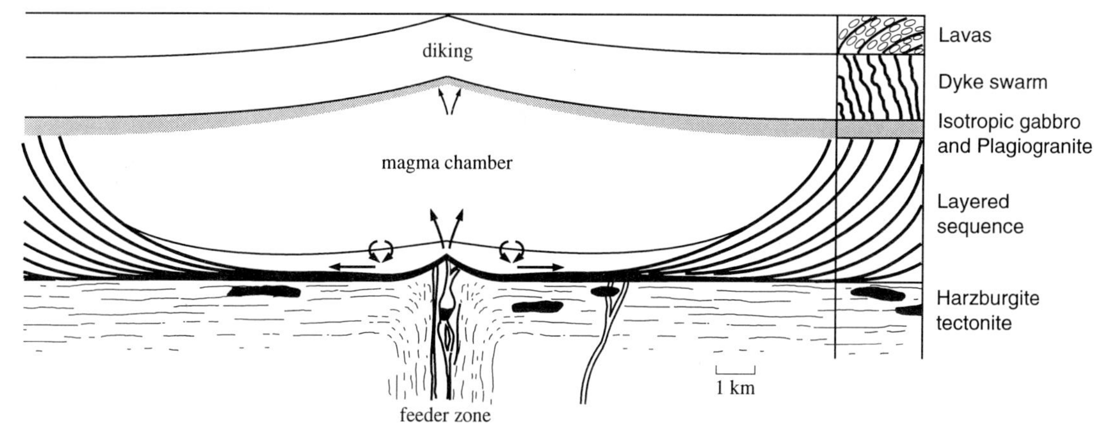

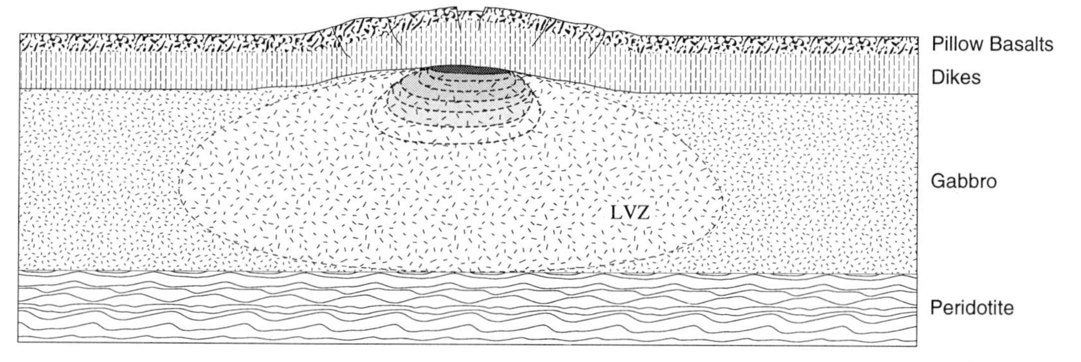

Figure 1 (A) Schematic model of a mid-ocean ridge spreading center derived from Oman ophiolite studies (after Smewing, 1981). The stratigraphic sequence below the pillow and sheeted dike complexes is shown on the right side of the figure. Isotropic gabbros just beneath the sheeted dike complex grade into gabbros with a (weakly developed) near vertical dip that becomes both more developed (Nicolas *et al.*, 1988; Nicolas, 1989) and more shallowly dipping as you move deeper into the gabbro section. (The layering is best developed and parallel to the Moho directly above the gabbro-peridotite "petrologic Moho" contact.) This dip structure was used by Smewing (1981) to infer that gabbro layering reflects cumulate deposition on the floor of the large magma chamber sketched here. (B) Model of magma chamber structure for the East Pacific Rise derived from seismic studies. Molten magma is concentrated in a lens approximately 1 km wide that resides at the base of the sheeted dike complex. Beneath the magma lens and extending to mid-crustal depths is a broader region of rock at elevated temperatures that contains a few percent partial melt. Details of the structure immediately above the Moho are unclear but it is known that the seismic Moho forms within a few kilometers of the axis. The surficial layer of pillow basalts and sheet flows increases in thickness away from the axis as a result of successive volcanic eruptions.

continues

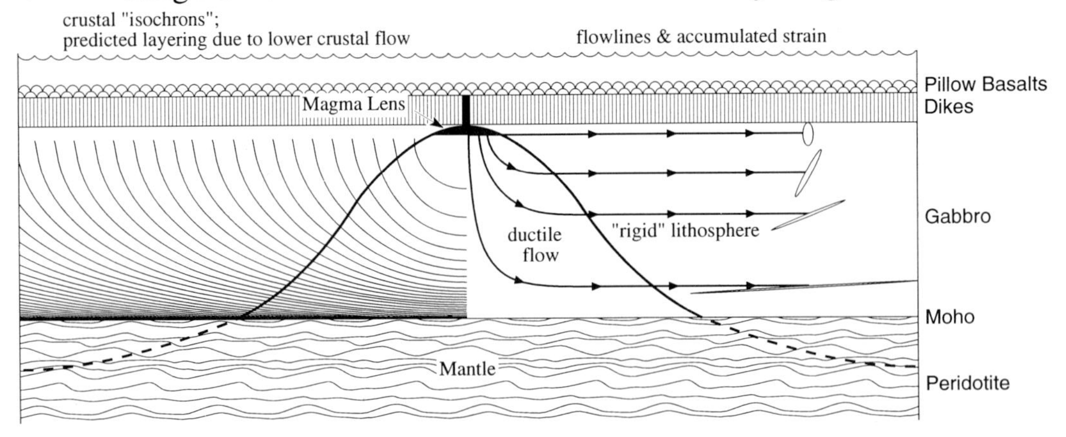

Figure 1 continued. (C) Schematic theoretical model of crustal extension and flow showing flowlines, accumulated strain, and the layering predicted by lower crustal flow away from a shallow injection site at the sheeted dike/gabbro interface. The left side of the figure shows predicted layering "isochrons" generated from crustal flow away from a magma lens at the base of the sheeted dike complex. The right-hand side shows several typical crustal flowlines and the accumulated strain along each flowline. Strain is most intense in the lowermost part of the gabbro section. Note the similarity in form between the crustal-flow-generated layering and the Oman ophiolite layering shown in (A).

found that a very small (50 to hundreds of meters thick, 1-km-wide) magma chamber structure (see Fig. 1B) is ubiquitously present beneath the axis of a fast-spreading ridge (Detrick *et al.*, 1987; Harding *et al.*, 1989; Vera *et al.*, 1990), but a magma body of even this size has yet to be detected along the axis of a slow-spreading ridge (Detrick *et al.*, 1990). Sleep's (1975) original "magma chamber" and crustal flow scenario has been revived and its consequences have been further explored (cf. Fig. 1C). Magma injection into a small magma lens at the base of the sheeted dike complex appears to have the potential to explain both the flow patterns seen in the ophiolite record and the small magma lens that seismic methods image at active fast-spreading ridges (Phipps Morgan and Chen, 1993; Quick and Denlinger, 1993; Henstock *et al.*, 1993). Thermal predictions and geophysical observations are reconciled by including the effects of hydrothermal circulation in the heat budget. Early thermal models (Sleep, 1975; Kusznir and Bott, 1976) cooled the crust purely by a combination of conductive cooling through the seafloor and advection of heat away from the ridge crest by the diverging plates. Hydrothermal circulation may enhance the cooling through the roof of a magma chamber by a factor of about 8–12 compared to conduction (Morton and Sleep, 1985; Phipps Morgan and Chen, 1993; Henstock *et al.*, 1993) and thus significantly reduce the size of any magma chamber or cause it to freeze (Lister, 1983). Ophiolite layering can also be reconciled with a small shallow magma lens if most lower crustal material is initially injected into this shallow lens—the subsequent flow to make the lower crust will impart a strain-induced fabric that is similar in form to the fabric seen in the ophiolite record (Phipps Morgan and Chen, 1993; Quick and Denlinger, 1993; Henstock *et al.*, 1993). In this chapter we first review the current body of geophysical constraints on ocean ridge magma chamber structure. Then we present the theoretical foundations and implications of this emerging physical paradigm for oceanic crustal genesis.

The Seismic Record: Seismic Constraints on Magma Emplacement and Crustal Accretion at a Fast-Spreading Ridge Axis

Geophysical experiments, particularly seismic experiments, conducted in the last decade support a basic division between a temporally variable accretionary pattern at slow-spreading ridges with cycles of magmatic injection and amagmatic extension, and an essentially steady-state magma chamber sustaining crustal accretion at fast-spreading ridges. So far, experiments at slow-spreading ridges have failed to detect anything that could be considered to be the seismic signature of a molten magma body (Detrick *et al.*, 1990), although seismic velocity anomalies that can be considered the thermal afterglow of the most recent site of magmatic injection have been found in isolated locations beneath the median valley of the Mid-Atlantic Ridge (Purdy and Detrick, 1986; Kong, 1990; Kong *et al.*, 1992). In contrast, seismic experiments conducted along the fast-spreading East Pacific Rise have found a largely continuous shallow reflector (Detrick *et al.*, 1987, 1993) that is interpreted as the top of a similarly continuous magma body. The magma body is underlain by a broader region of reduced seismic velocity that is even more continuous along-axis and is probably disrupted only by larger ridge discontinuities such as transforms. In the following sections we review the observations that constrain the current seismic picture of magma chamber structure at fast-spreading ridges.

Northern East Pacific Rise

The geometric constraints on magma chamber structure at fast-spreading ridges are in large part the result of a diverse series of seismic experiments conducted along the northern East Pacific Rise during the 1980s. These experiments include the MAGMA *o*cean *b*ottom *s*eismograph (OBS) tomography experiment at 12°N (McClain *et al.*, 1985), a 1985 two-ship multichannel seismic experiment that included a regional reflection survey between 9° and 13°N, plus detailed refraction and reflection experiments at two sites at 9° and 13°N (Harding *et al.*, 1989; Vera *et al.*, 1990), and an ocean bottom hydrophone tomography experiment at 9°N (Toomey *et al.*, 1990). During the previous decade, refraction experiments conducted as part of project RISE at 21°N (Reid *et al.*, 1977) and project ROSE at 12°N (Lewis and Garmany, 1982; Ewing and Meyer, 1982; Bratt and Solomon, 1984) as well as at 9°N (Orcutt *et al.*, 1975; Rosendahl *et al.*, 1976) had all found evidence for an axial magma chamber, as had a pair of reflection lines shot across the rise axis at 9°N (Herron *et al.*, 1978; Hale *et al.*, 1982). However, no generally accepted model for the structure of the axial magma chamber emerged from these experiments. Although these experiments demonstrated that a magma chamber could exist beneath the East Pacific Rise they simultaneously demonstrated that an *extensive* body of magma *could not exist,* thereby confounding expectations based on thermal and ophiolite models.

Individually the seismic results from the northern East Pacific Rise are not unique. Magma chamber reflections have been recorded within the Lau Basin (Morton and Sleep, 1985; Collier and Sinha, 1990) and at the Juan de Fuca Ridge (Morton *et al.*, 1987; Rohr *et al.*, 1988). Zones of reduced seismic velocity, a sign of at least elevated temperatures, have been reported beneath the Juan de Fuca Ridge (White and Clowes, 1990) and also the median valley of the Mid-Atlantic Ridge. Rather, the significance of the northern East Pacific Rise results lies in the fact that they can be united into a coherent and consistent picture, albeit still somewhat blurred, of the axial magma chamber structure. In particular, reflection, refraction, and tomography experiments were collocated within the 9°N segment so that results of one experiment could be compared directly to another.

The 9°N segment of the East Pacific Rise is bounded at its northern end by the Clipperton transform at 10°10′N, where the axis is offset by 85 km, and at its southern end by the 9°03′N *o*verlapping *s*preading *c*enter (OSC), where the rift zones are separated by 8 km. A finer-scale segmentation into 10 or more segments has been proposed on the basis of ridge crest morphology and detailed near-bottom mapping of the axial summit graben or caldera (Haymon *et al.*, 1991a). In addition to the geophysical experiments, the segment has been the site of systematic dredging, an extensive submersible dive program, and a recent attempt at bare rock drilling by the Ocean Drilling Program. The segment was also the site of a volcanic eruption during the early part of 1991 (Haymon *et al.*, 1991b).

Correlation between Seismic and Ophiolite Observations of Crustal Structure

A detailed interpretation of seismic velocity results in terms of the structure and petrology of the oceanic crust is the subject of a constantly evolving debate. The basis for interpretation is the association of the seismically determined layering

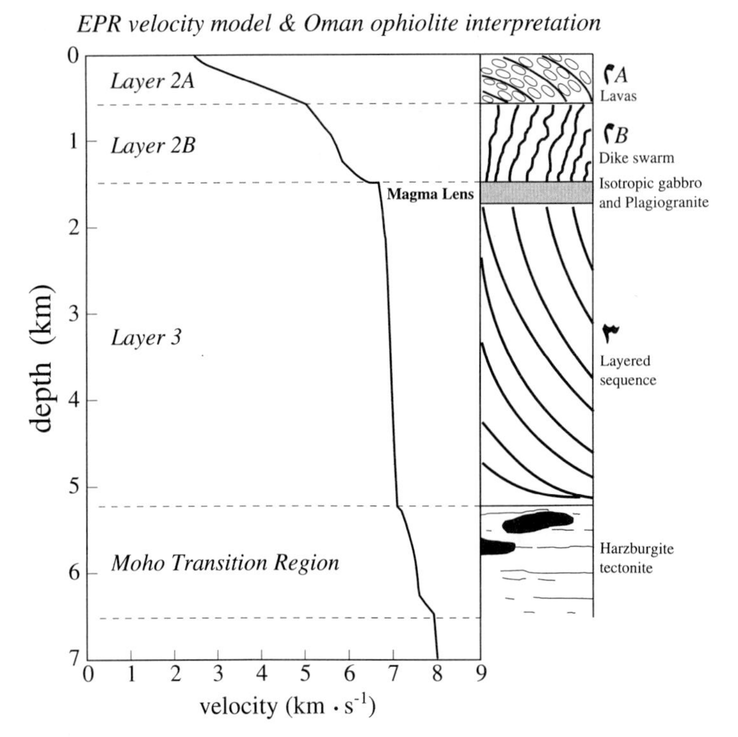

Figure 2 A common means of inferring oceanic crustal structure from seismic measurements is to relate them to ophiolite observations. This figure shows a possible correlation between an East Pacific Rise velocity model and a stratigraphic column from the Oman ophiolite (cf. Figure 1A). The correlation of layer 2 with the upper volcanics, and layer 3 with the plutonic section is generally accepted. However, the relationship of the subdivisions of seismic layer 2 to structure is less certain and is almost certainly not uniform across all spreading rates.

of the oceanic crust with the structural sequence found in ophiolites as shown in Fig. 2. For a fast-spreading ridge, the following "simple" layering pattern appears to be appropriate. Seismic layer 2 is associated with the volcanic basalt section (layer ٢) of the ophiolite crust; 2A is associated with pillow lavas, and layer 2B with the sheeted dike complex of the upper part of the ophiolite sequence. (Seismic layer 1 is associated with deep marine sediments that cap the pillow basalts in the ophiolite section. However, sediment cover is at most only a few meters thick in local sediment ponds at the extremely young seafloor near a fast-spreading ridge axis.) Seismic layer 3 is associated with the plutonic gabbro section (layer ٣) of ophiolite crust (e.g., Kempner and Gettrust, 1982). Several reasonably successful attempts were made to correlate seismic velocities of ophiolite rocks and structural associations with the marine crustal velocity structure leading to ophiolite-derived velocity structures much like that shown in Fig. 2 (Christensen and Smewing, 1981; Kempner and Gettrust, 1982; Karson *et al.*, 1984). The association is complicated by, among other factors, the strong dependence of seismic velocity on porosity (Spudich and Orcutt, 1980), the relatively weak dependence of seismic velocity on mineralogy, and uncertainty as to how representative ophiolites are of oceanic crust (e.g., Moores, 1982). Support for the ophiolite model of oceanic crust comes from drilling at DSDP/ODP hole 504B, which has penetrated the pillow lavas and sheeted dikes that form the upper part of the ophiolite sequence (Becker *et al.*, 1989), and from recent submersible observations at Hess Deep (Francheteau *et al.*, 1992). The observations at Hess Deep, which is a rifted portion of normal East Pacific crust, give structural thicknesses for the upper crust that are comparable to seismic layer thicknesses derived from rise-axis experiments. These recent comparisons make it reasonably certain that the base of layer 2 does indeed lie at the base of the sheeted dike complex and that this is the level of the top of the seismically imaged axial magma chamber at a fast-spreading ridge.

Current Seismic Picture of an Oceanic Magma Chamber at a Fast-Spreading Ridge

In this section, the current geometrical model of the structure of a fast-spreading magma chamber is presented as fact without many of the necessary qualifications and caveats that are inevitable for a model based on indirect measurements. Subsequent sections more thoroughly discuss the geophysical evidence for the model.

In cross section, the magma chamber at a fast-spreading ridge is a narrow (~1-km-wide), thin (~50 to hundreds of meters thick) melt lens that overlies a broader (~6-km-wide), thicker (2- to 4-km) region of hot rock that may include a small (<3–5%) melt fraction (Fig. 1B). The lens accumulates at the base of the sheeted dike complex through the trapping of buoyant magma that is injected into the crust from the mantle and is rereleased from the (initially cumulate) crystal mush as the mush subsides and shears to form the lower part of the oceanic crust. Vigorous cooling of the lens from above by hydrothermal circulation causes freezing or plating of gabbro from the lens onto the base of the sheeted dike complex and cumulates freezing and settling to a subsiding floor of the lens. Only the top of the magma chamber, where a lens or sill of predominantly molten magma accumulates, is well defined. While the top edges of the magma lens are quite distinct, both the sides and the base of the magma chamber are indistinct with melt percentages within a crystal mush zone falling toward the edges of the chamber and grading into plutonic rocks with residual elevated temperatures. The maximum width of the magma chamber with a mean melt fraction less than ~3% is typically on the order of 6–8 km, whereas the base of the chamber lies at mid-crustal depths, 3–4 km below the seafloor, within the gabbroic section.

It is evident that this idea of a magma "chamber," in the conventional sense of a well-defined cavity within the host rock, is something of a misnomer at fast-spreading ridges. Petrologists have been loath to restrict the label magma chamber to the magma lens, even though this may be the closest approximation to the one in the conventional physical sense, as the interaction between the magma lens and the underlying mush zone with the exchange of melt and crystals may be important in the chemical differentiation of the oceanic crust (Langmuir, 1989; Sinton and Detrick, 1992). From a seismological viewpoint it is convenient to divide the magma chamber into two distinct components, a magma lens and an underlying low velocity region. The low velocity zone contains hot rock with a possible core of a crystal mush zone that surrounds the magma lens—both elevated temperatures and partial melt cause a reduction in seismic velocities. The seismological division reflects the fact that different elements of the magma chamber structure are detected by different seismological methods. Evidence for the magma lens comes principally from reflection data while evidence for the low velocity zone comes from refraction and tomography data. The next sections review the primary seismic constraints on the structure of this type of magma chamber.

There Is an Axial Magma Lens at the Base of the Sheeted Dike Complex that Contains a Large (>25%) Fraction of Melt

The best evidence for a magma lens or sill beneath the axis of the East Pacific Rise comes from multichannel seismic reflection data. When properly migrated, the mid-crustal reflection interpreted as the magma lens typically appears on cross-axis reflection profiles as a bright, narrow event centrally located about 0.6 s beneath the rise-axis (Fig. 3). The same event, termed the axial magma chamber reflection or AMC reflection, also appears at the same location on intersecting along-axis reflection profiles and is continuous for tens of kilometers along strike (Fig. 3). When correlated with velocity studies, it can be shown that this reflection marks the top of a crustal low velocity zone and that the reflection is caused by a negative acoustic impedance contrast.

Further information on conditions within the magma lens can be deduced by detailed examination of the amplitude and phase characteristics of the AMC reflection, which depends on the contrast of material properties across the top of the magma lens. Data on the variation of the seismic

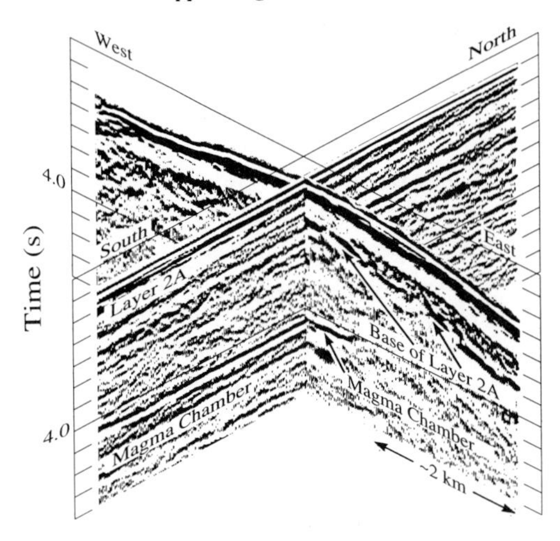

Figure 3 An intersecting pair of migrated seismic sections from 9°40′N on the East Pacific Rise. On the cross-axis profile, CDP 29, the magma chamber reflection appears as a narrow, <1-km-wide, event at about 4.0 s two-way travel time directly beneath the axial summit caldera and the thinnest portion of layer 2A. The base of layer 2A is interpreted as marking the transition between the extrusive unit and the sheeted dike complex. On the intersecting along-axis profile, CDP 41, the magma chamber reflection is continuous along-axis for tens of kilometers although reflection strength is variable due in part to the wandering of the profile with respect to the narrow reflection.

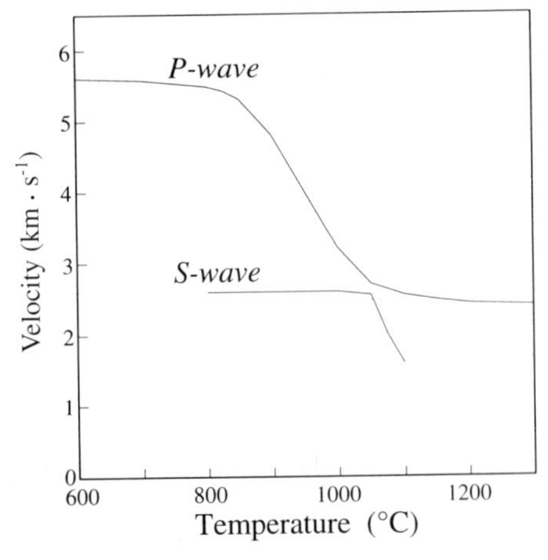

Figure 4 The variation of seismic velocity with temperature of a basalt (modified from Murase and McBirney, 1973). Although the general behavior is correct, the detailed variation of velocity between the solidus and liquidus as a function of melt fraction is not well resolved.

velocities of basalts with temperature and degree of partial melting is limited, particularly for pressures approximate to mid-ocean ridges. Those studies that do exist (Murase and McBirney, 1973; Christensen, 1979; Khiratov *et al.*, 1983; Manghnani *et al.*, 1986) plus appropriately scaled (Henstock *et al.*, 1993) analogous data on the high-pressure melting of peridotite (Sato *et al.*, 1989) can be used to infer the temperature and melt fraction distribution within and around the magma lens.

Over the temperature range from 0°C to the solidus at around 1000°C, P-wave velocities are reduced on the order of 0.5 km · s^{-1} (Fig. 4). The velocity variation is probably approximately linear at subsolidus and immediately supersolidus temperatures and there is no discontinuous change in velocity at the solidus. At greater temperatures there is a knee in the velocity–temperature curve with P-wave velocities decreasing rapidly above the knee, and at the liquidus P-wave velocities have been approximately halved to less than 3 km · s^{-1}. The variation of crystal fraction with temperature shows a similar knee to the velocity function (Marsh, 1981; Sinton and Detrick, 1992) and there may be a more nearly linear relationship between velocity and melt fraction over a larger supersolidus temperature range (Sato *et al.*, 1989; Sato and Sacks, 1990). Murase and McBirney's results show shear wave velocities decrease only slightly as P-wave velocities are reduced and then fall discontinuously to zero, suggesting that it is not possible to have large changes in S-wave velocity without also having large changes in P-wave velocity. The combined results indicate that a velocity anomaly on the order of 0.5 km · s^{-1}, a ~10% reduction, requires only a small percentage of melt and that a fully molten magma chamber should produce a large amplitude reflection. However, it is not possible, at present, to quantify precisely such phrases as "nearly completely molten" or "partially solidified" in terms of melt fraction and it may be possible for a magma to have a crystal fraction of 25% or greater, not be able to convect (Marsh, 1989), and still produce a large-amplitude reflection indicative of a nearly completely molten magma chamber.

If P-wave velocities are halved at an interface, then the normal incidence reflection is phase-reversed and has an amplitude of 0.33 relative to the incident wave. Estimates of normal incidence reflection coefficients from the northern EPR data range up to 0.3–0.4 (Barth *et al.*, 1987), while the AMC reflection often appears to be phase-reversed relative to the seafloor reflection (e.g., Fig. 3). Both observations are thus consistent with the existence of a large negative velocity contrast at an AMC. Estimates of normal incidence reflection coefficients tend to be highly variable due to, among other factors, the sensitivity of recorded amplitudes to small bathymetric features that can focus or defocus seismic energy. A more reliable estimate of material properties at the top of the magma chamber may be made from the amplitude versus offset behavior of the AMC reflection. Normal incidence reflection coefficients are sensitive primarily to P-wave velocities. To a lesser extent reflection coefficients are sensitive to density (the density decrease associated with melting is on the order 5–10%). However, the amplitude versus offset behavior of the reflection is strongly dependent on the difference in shear modulus across an interface (Harding *et al.*, 1989). A large decrease in shear modulus, as would be the case at the top of the magma chamber, causes a rapid decrease in the reflection amplitude with offset and the reflection can switch signs becoming a low-amplitude normal polarity arrival at larger offsets. On the other hand, a small decrease in shear modulus, even if accompanied by a small drop in P-wave velocities, results in a reflection that has reversed polarity at all offsets and can increase in amplitude at larger offsets. Amplitude versus offset behavior of the AMC reflection in rise-axis ESPs and CDP data at 9°N falls into the former pattern (Vera *et al.*, 1990) and has a very rapid decrease with increasing offset while similar data from 13°N follows the latter pattern, leading to the inference that the top of the magma chamber at 9°N is nearly completely molten while the magma chamber at 13°N is partially solidified.

The Magma Lens Is Usually a Continuous Feature along the Axis of a Fast-Spreading Ridge

The 1985 regional reflection survey also provided the first convincing geophysical evidence that an axial magma chamber was a reasonably continuous and steady-state feature beneath a fast-spreading ridge. A magma chamber reflection was found along more than 60% of the rise-axis between 9° and 13°N with the major gaps occurring near major ridge discontinuities such as the Clipperton transform. Even at these "gaps," it is possible that the ship wandered off the narrow magma lens rather than the lens itself being absent. Given such an observation, it is reasonable to think in terms of a typical magma chamber structure for a fast-spreading ridge and to interpret seismic results within such a framework rather than viewing each seismic measurement as potentially unique. Similar continuity of the magma lens is seen in recent seismic experiments along the southern East Pacific Rise. To a first approximation, crustal accretion along the northern East Pacific Rise appears remarkably uniform both spatially and temporally, particularly when compared to slow-spreading environments. The majority of well-constrained crustal thickness determinations for the East Pacific Rise cluster around 6 km with a spread of ± 1 km (Chen, 1992), and along-strike variations in mantle Bouguer anomalies between 9° and 13°N, an indicator of crustal variations, are small, on the order of 10 mGal (Madsen *et al.*, 1990).

The Magma Lens is Thin: ~50–200 m Thick

At present no reflection has been identified in the CDP data that can unambiguously be associated with the bottom of a magma chamber, nor for that matter has a side-wall reflection. The absence of a bottom reflection can be explained by a gradual increase in the crystal content with depth within the magma chamber and a concomitant gradual increase in seismic velocity, which would not produce a distinct reflection. Alternatively, if the magma chamber is a thin lens, the top and bottom reflection will coalesce into a single arrival effectively masking the bottom of the chamber. Figure 5 plots the primary reflection response for a thin layer with equal but opposite polarity reflections at either boundary as the layer thickness varies using an appropriate source wavelet. For small thicknesses, the reflection response has reduced amplitude and appears to have reversed polarity. At larger separations, there is a constructive interference or tuning of the two reflections, which increases the peak amplitude by about 50%, and even a simple classification between either normal or reversed polarity is difficult. Finally the top and bottom reflections separate into two distinct arrivals. Morton and Sleep (1985) reported evidence for closely spaced negative and

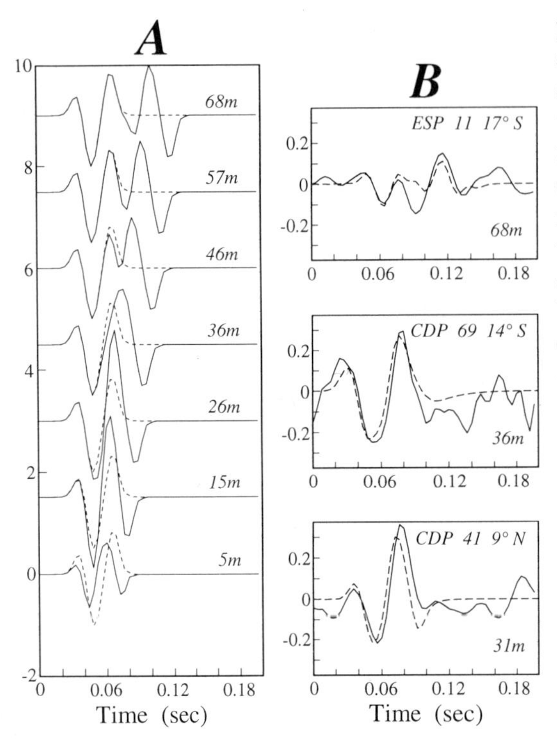

Figure 5 (A) Response of a thin magma layer with equal amplitude but opposite polarity top and bottom reflections as a function of separation between the top and bottom of layer (solid line) for an airgun source wavelet. Dashed line is the response of the top of the layer alone. Separations are based on a nominal velocity of 2.6 km · s^{-1}. At separations of about 15 m the amplitude of the response is enhanced relative to a single interface by tuning of the layer, and the secondary positive pulse exceeds the primary negative pulse in amplitude. At around 30 m positive and negative pulses have approximately equal amplitudes and at greater separations the wavelets separate in distinct arrivals. (B) Comparison of magma reflections at different localities (solid line), with the predicted response of a thin-layer model (dashed line). Source wavelets in each case were estimated from seafloor reflection response. Response based upon a 1D model that includes attenuation within the upper crust.

positive polarity reflections in data from the Lau Basin. Kent *et al.* (1990) demonstrate that a magma lens with a thickness between 10 and 50 m will mask the bottom reflection and produce a calculated response that is comparable to the data at 9°N (cf. Fig. 5).

The Magma Lens Is (Usually) Narrow: ~1 km Wide

When carefully stacked, cross-axis reflection profiles characteristically show diffraction tails emanating from the AMC reflection (Fig. 6). The presence of diffractions is consistent with the abrupt cross-axis termination of a magma lens or chamber. Time migration of the stacked section with the appropriate velocities derived from refraction data results in a collapsed reflection with a width of around 0.7 km. Migration using stacking velocities results in only a partial collapse of the diffraction tails and a larger apparent magma chamber. Stacking velocities over the rise-axis are biased low by the axial bathymetry and the thickening of layer 2A while stacking velocities increase along the diffraction tails in a predictable manner because of the offset reflection of the reflection midpoint from the edge of the magma chamber (Kent *et al.*, 1993a). When the AMC reflection is narrow and centrally located beneath the rise-axis, widths estimated from migrated sections are comparable to more careful estimates made by forward modeling the diffraction hyperbolas (Kent *et al.*, 1993a). It should be noted that the width of the top of the low velocity zone as determined by refraction and tomography results exceeds that of the AMC, which may be explained as truncation of the base of the magma lens against the dipping roof of the chamber or as a rapid lateral gradient in melt fraction. Subsidence of the sheeted dike complex combined with the freezing or plating of gabbro on the roof of the chamber should ensure a roof with dips on the order of tens of meters, rather than hundreds of meters, over widths typical of the AMC reflection.

Sometimes the Magma Lens Is Offset from the Ridge Axis—Then It Is Wider (2–4 km Wide) and Slopes up toward the Ridge

A strikingly different AMC reflection is present in Fig. 7, which shows a stacked cross-axis profile, CDP 33, shot about 35 km to the south of the previously discussed line at 9°19′N. The image of layer 2A is not as clear or continuous as in the previous example and the AMC reflection is considerably more complex and is displaced to the west of the axis. Layer 2A deepens asymmetrically away from the rise-axis, being thicker to the west of the axis than to the east. With such a flat rise-axis profile the location of the layer 2A minimum thickness is a more reliable indicator than bathymetry of the location of the rise-axis. The large variation in the thickness of layer 2A above the AMC reflection is responsible for its complex appearance in the stacked section. The

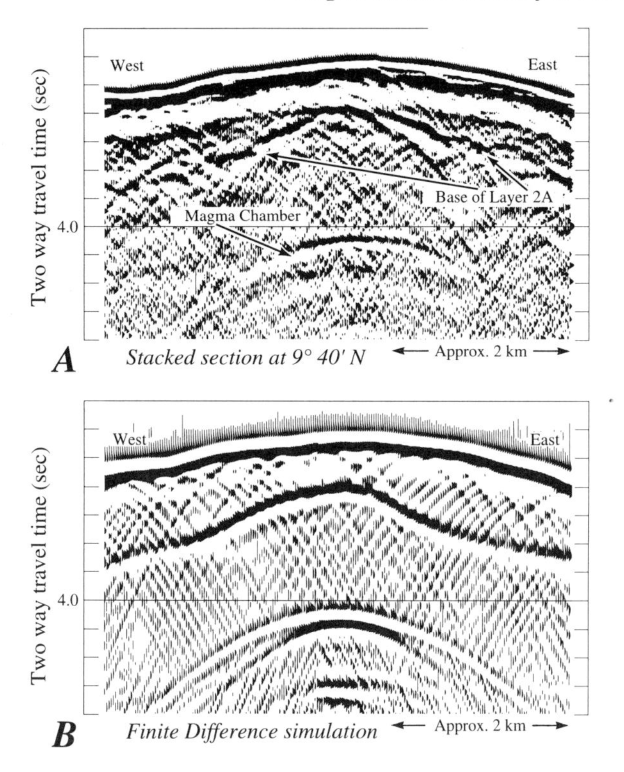

Figure 6 (A) Unmigrated version of CDP 29 profile, cf. Figure 3. The magma chamber reflection has extended diffraction tails, suggesting an abrupt cross-axis termination to the magma chamber reflector. (B) A finite difference simulation of the profile including a narrow elliptical magma lens, 750 m wide and 50 m thick, reproduces the general character of the reflection response including the extended diffraction tails. The increased thickness of the magma lens in the simulation compared to that estimated for the profile extends the reflection response in the simulation.

low velocities within layer 2A cause the AMC reflection to be pulled up beneath the rise-axis and pushed down west of the axis (Harding *et al.*, 1993; Kent *et al.*, 1993b). In this case, estimates of AMC width and depth have been derived through the use of image rays after time migration (Hubral, 1977) and checked with forward modeling (Kent *et al.*, 1993b). The resultant AMC reflector lies at a nearly constant depth below the base of layer 2A and has a total width of 4.2 km. The average dip on the reflector is estimated as 10° with one edge located directly beneath the rise-axis. The asymmetry and dip of the reflector may be possibly explained as a melt migration path; buoyant melt collects off-axis at the base of the sheeted dike complex and then flows upward and across-axis to erupt along the axial rift zone. This interpretation of the AMC reflector suggests a displacement of the underlying magma source region relative to the rise-axis. The AMC reflection in CDP 33 is rare and is thought to be a consequence of the adjustment of magma supply across the 9°03′N OSC. Away from the OSC the magma supply from the mantle is concentrated directly beneath the rise-axis. This configuration of mantle supply and rift results in the prototypical cross-axis reflection profile with an approximately 1-km-wide AMC reflection centered beneath the rise-axis and Moho reflections that can be traced to within a few kilometers of the edge of the AMC reflection.

Synthetic Experiments Can Reproduce These Basic Magma Lens Features Quite Well

The justification for describing the molten part of the axial magma chamber as typically a narrow lens rests in part on its satisfying the constraints imposed by the reflection data and in part on such a geometry being physically simple to achieve, requiring only that the magma is buoyantly, and usually stably, trapped against an impermeable lid. The consistency of the magma lens model

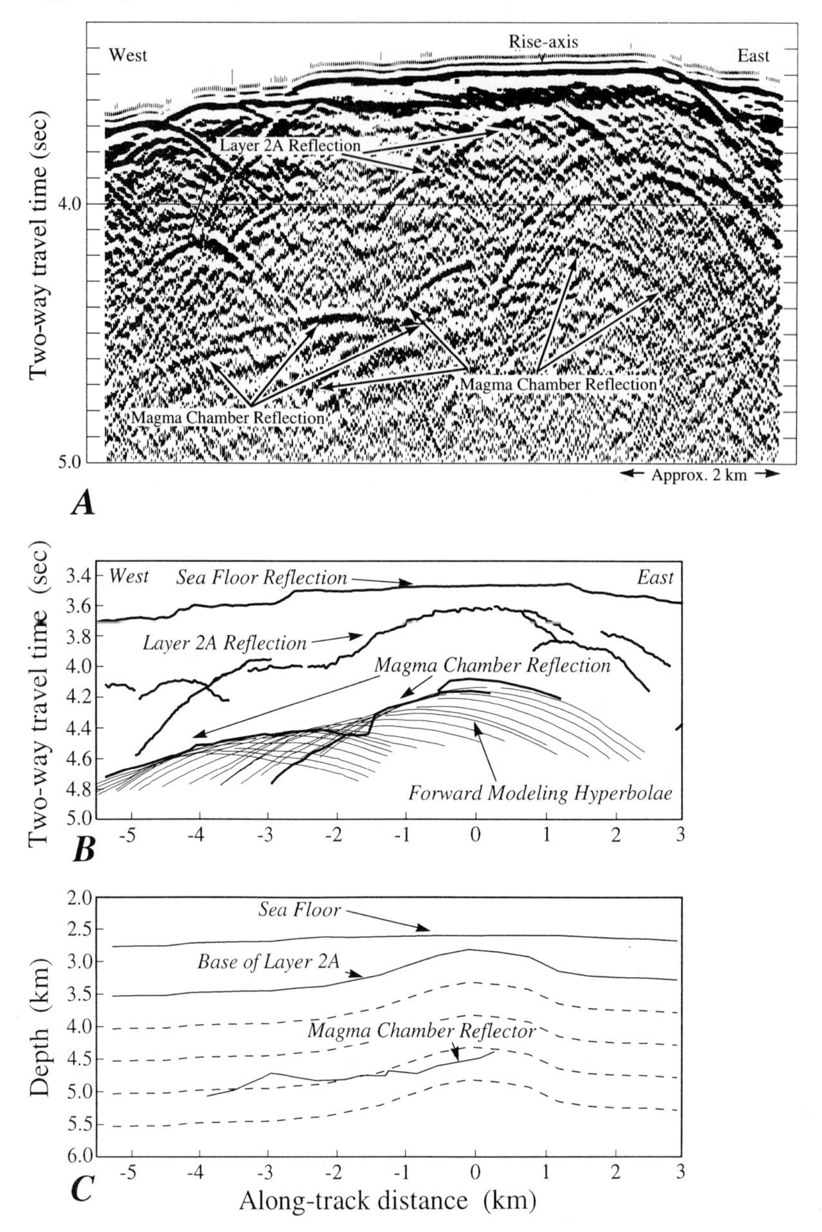

Figure 7 (A) Cross-axis profile, CDP 33, at 9°19′N. Profile is unmigrated to aid identification of reflection arrivals. The magma chamber reflection is displaced to the west of the axis and is considerably wider than the CDP 29 profile. The time separation of the magma chamber reflection from the base of layer 2A is approximately constant, indicating that the complexity of the reflection is a consequence of propagation through a thickening extrusive layer. (B) Line drawing of the reflections in (A) plus the travel time hyperbola resulting from forward modeling the magma chamber reflector as a series of point diffractors. (C) Magma chamber reflector determined from ray theoretical depth migration. The reflector is 4 km wide with the eastern edge located beneath the rise-axis. The reflector is located approximately 1.5 km beneath the rise-axis and has a dip of about 10°.

with the data can be demonstrated through a finite difference simulation of the cross-axis profile, CDP 29, Fig. 6B. The magma chamber in the simulation was an elliptical lens with a width of 0.75 km and a maximum thickness of 50 m. The response of the lens in the simulated section is comparable to that of the AMC in the stacked data, with diffractions originating from the edge of the lens. There is no obvious bottom response from the lens although the finite thickness has extended the reflection signature.

Other Seismic Constraints on Crustal Accretion at a Fast-Spreading Ridge

Crustal Accretion and Moho Formation Occur in a Narrow Zone about the Ridge Axis

Several lines of evidence support the hypothesis that the magma lens is a fairly stable feature and that crustal accretion is occurring within a narrow region indeed. A distinct Moho reflection can be traced to within ~2 km of the ridge axis (Detrick *et al.*, 1987); i.e., a distinct seismic reflection Moho is present immediately after one moves away from the near-axis ranges where a noticeable shallow AMC diffraction tail obscures any deeper, weaker reflections. As well as being evidence for a region of (crustal) magma supply directly beneath the axis, the proximity of Moho reflections to the axis suggests that the basic structure of the lower crust is formed close to the rise-axis. Shallow crustal extrusion also appears to occur within a narrow zone directly above the magma lens. Seismic studies along the northern East Pacific Rise have found evidence of a systematic variation of upper crustal structure within 1–2 km of the axis. The tomography results from 9°N found an axial positive velocity anomaly within the upper 1 km of crust. By correlating ESP-derived velocity models for the upper crust with CDP reflection images, it can be demonstrated (Harding *et al.*, 1993) that the magnitude of the velocity anomaly can be accounted for by the deepening of the surficial low velocity layer, layer 2A, seen in the CDP profiles (Fig. 6). It has been proposed that the base of layer 2A marks the top of the sheeted dike complex and thus that the thickening of layer 2A is a consequence of thickening of the extrusive section to a full thickness of ~400–600 m within 1–2 km of the axis (Harding *et al.*, 1993; Christensen *et al.*, 1992). The layer 2A reflection reaches its minimum depth at the rise-axis directly above the AMC reflection and directly below the axial summit graben or axial summit caldera. The axial summit caldera marks the loci of the most recent volcanic eruptions and is believed to a syn- or post-eruptive feature resulting from the collapse of the volcanic carapace above the magma chamber (Haymon *et al.*, 1991a). Thus, the layer 2A reflection is interpreted as marking the top of the sheeted dike complex; i.e., the dikes that feed eruptive activity lie within 100–200 m of the seafloor at the rise-axis. The symmetric deepening of the layer 2A reflection away from the rise-axis is interpreted as the result of the progressive burial of the sheeted dike complex by successive lava flows. Near the rise-axis, seismic velocities within the upper part of layer 2A lie between 2.5 and 3.0 km · s^{-1}, velocities that can reasonably be explained only by large porosities within the upper crust. The transition to velocities in excess of 5.5 km · s^{-1} below the base of layer 2A, values more nearly comparable to those of unfractured basalt, can be accounted for solely by a downward decrease in porosity, and it is conceivable that the base of layer 2A is simply a porosity horizon that deepens as the result of tectonic fracturing and is not structurally controlled. However, recent submersible dives at Hess Deep, on tectonically exposed sections through normal East Pacific crust, found a total extrusive section on the order of 400–600 m comparable to the off-axis thickness of layer 2A. In addition, the off-axis thickening of layer 2A observed in sections such as CDP 33 (Fig. 7) does not occur at the expense of the rest of layer 2 above the axial magma chamber, as would be the case for a deepening of a porosity horizon within a constant thickness crustal layer, but rather indicates an overall thickening of layer 2. The 2- to 4-km width of the neovolcanic zone inferred from layer 2A thickening is comparable in magnitude to earlier estimates based on the transition width of seafloor magnetic lineations (Sempere *et al.*, 1988).

The Magma Lens Is Underlain by a Broader Low Velocity Zone of Hot Rock and Possible Small Amounts of Melt (<3–5%)

A largely solidified cumulate mush has only a small impedance contrast relative to the host rocks and is essentially invisible to reflection methods. However, such a chamber can be detected by refraction and tomography methods that are sensitive to seismic velocity variations. The divisions between refraction and tomography are somewhat arbitrary but in the context of studies of the East Pacific Rise most seismic refraction lines have been shot parallel to the rise-axis to a single receiver and have been used to determine a one-dimensional depth-dependent profile. In this context, a low-velocity zone is a decrease in seismic velocity with depth, which creates a shadow zone in the arrivals. Seismic tomography experiments use the travel times from multiple shots to multiple receivers to create two- and three-dimensional velocity models of the rise-axis. Tomographic results are usually expressed in terms of velocity anomalies and a low velocity zone is a negative velocity anomaly relative to a 1D reference model. Vertical velocity gradients within layer 3 of the oceanic crust are sufficiently small, on the order of 0.1 $(km \cdot s^{-1}) \cdot km^{-1}$, that the refraction-derived and tomographic axial low-velocity zones roughly coincide.

The axial low velocity zone determined by tomographic results from 9°N (Toomey *et al.*, 1990) and 12°N (McClain *et al.*, 1985; Burnett *et al.*, 1989; Caress *et al.*, 1992) as well as the *e*xpanding *s*pread *p*rofile (ESP) refraction results from 9° and 13°N (Harding *et al.*, 1989; Vera *et al.*, 1990) is 4–5 km wide at its top, broadening to 8–10 km at its base. Maximum velocity perturbations are on the order of 0.5–1 $km \cdot s^{-1}$, corresponding to absolute velocities of around 6 $km \cdot s^{-1}$, and are confined principally to the upper part of the crust above a depth of 4 km. In the ESP data, wide-angle AMC reflections are asymptotic to upper crustal refractions at the ranges just before the start of the shadow zone (Fig. 8), showing clearly that the AMC reflection lies at the top of the axial low velocity zone and corroborating the interpretation of the AMC reflection as being from the top of a molten magma body. The ESP results also demonstrate that the relatively flat-lying top of the axial low velocity zone is wider than the AMC reflection at 9°N and also that a low velocity zone can exist where there is no distinct AMC reflection visible in the CDP data (Harding *et al.*, 1989).

Although the crude dimensions are known, the detailed structure of the low velocity zone is poorly constrained by the seismic data. The horizontal and vertical resolution of the tomography data is at its best on the order of 1 km for resolving shallow crustal structure and much poorer than this in the lower crust (Caress *et al.*, 1992). Sharply delineated structures will be smeared out in the tomographic images. For example, a magma lens is not resolved by the tomographic images although the largest low velocity anomaly is centered at the location of the magma lens. The ESP data have good vertical resolution down to the top of the low velocity zone lid and can resolve the depth of the lid to within a few hundred meters. However, the velocities and depths of structures within and beneath the low velocity zone are poorly constrained. No rays pass through the lowest velocity region in tomographic experiments, which is essentially why the lowest velocities are not imaged. (Were it a high velocity region like a salt dome, then the tomographic image of the high velocity structure would be much better because seismic ray paths preferentially sample a high velocity anomaly.) Bounds may be placed on the vertically integrated velocity anomaly within the low velocity zone using the arrival times and position of wide-angle arrivals from the Moho, PmP, in the ESP data (Harding *et al.*, 1989). These bounds are consistent with the tomographic results, indicating that a large 0.5- to 1-$km \cdot s^{-1}$ ve-

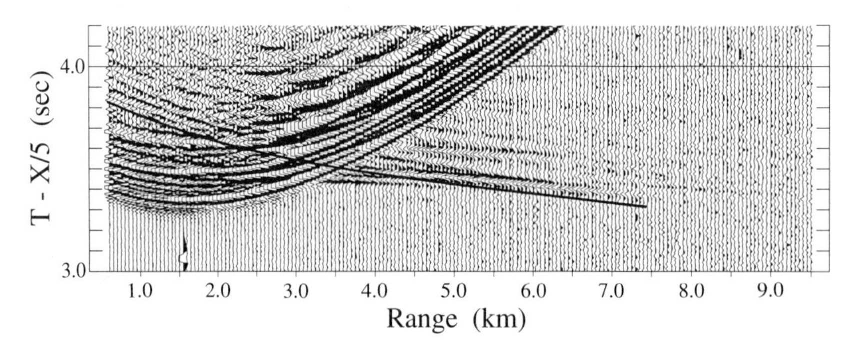

Figure 8 An expanding spread profile at 17°S on the East Pacific Rise. The solid line marks the reflection from the axial magma lens, which can be traced from a vertically incident reflection out to a distance of 7.5 km from the axis where it asymptotically approaches the upper crustal refraction arrival that emerges from the seafloor reflection at 3.5 km. There is a shadow zone at ranges greater than 7.5 km that indicates that a low velocity zone underlies the axial magma lens.

locity anomaly can extend only to mid-crustal depths and that the vertically integrated strength of the anomaly decreases with distance from the axis. An additional difficulty with proposing a detailed structure for the low velocity zone is that, as discussed earlier, the relationship between seismic velocity and partial melt fraction is relatively poorly known even for small hand samples at the appropriate pressures. Extrapolation of hand sample measurements to crustal scales suggests that on average the low velocity zone is largely solidified with only a small percentage of partial melt although at elevated temperatures. However, isolated conduits, dikes, and sills that are of petrologic importance could be contained within the average structure but not be resolvable from the seismic data. A 1-km $\cdot$ s^{-1} velocity anomaly at mid-crustal depths implies that melt, if uniformly distributed over this depth range, must fill less than 3% of the rock volume at these depths (Caress *et al.*, 1992).

The tomography data set at 9°30′N has also been used to obtain models of crustal P-wave attenuation (Wilcock *et al.*, 1992). Spectral estimates of the attenuation of P-waveforms are inverted for the reciprocal of the quality factor Q (a measure of anelasticity or elastic wave attenuation), assuming the velocity structure and ray paths obtained by delay time tomography (Toomey *et al.*, 1990). Q is not imaged within the magma lens, since no rays pass through this region. Below the magma lens, Q values are 20–50 compared with values of ~500 at similar depths well off-axis. The low on-axis values are very similar to Q_P values inferred from torsional oscillation experiments on gabbros at solidus temperatures and seismic frequencies (Kampfmann and Berckhemer, 1985). Thus, the attenuation models also suggest that the melt fraction at mid- and lower crustal depths beneath the melt lens is no more than a few percent.

A "Hot Zone" beneath the Magma Lens Is Needed to Support the Axial Topography at Fast-Spreading Ridges

If one accepts that the base of layer 2A marks the top of the sheeted dike complex, then the behavior of this horizon may be used to infer structural information about the nature of the axial ridge along the East Pacific Rise. At 13°N on the EPR, the axial high is dominated by a narrow, triangular ridge that is 1–2 km wide and 100–200 m high (Fig. 10b). One explanation for such ridges at intermediate spreading rates is that they are constructional features that are built during periods of increased magmatic activity (Kappel and Ryan, 1986). However, a constructional origin for the ridge at 13°N is precluded by the uniform thinness of layer 2A along-axis and its increased thickness off-axis (Kappus, 1991). Instead, the seismic results support the conjecture that the axial high is buoyantly supported and the relatively modest thickness increase of layer 2A, by 60%, is consistent with the notion that the triangular cross section of the 13°N segment is indicative of relative magmatic starvation (Macdonald *et al.*, 1984; Macdonald and Fox, 1988).

Assuming relatively low densities for layer 2A of 2.35 Mg $\cdot$ m^{-3} (Stevenson, 1992), and "eruptable" densities of ~2.6 Mg $\cdot$ m^{-3} for the magma, the triangular neovolcanic zone can be Airy compensated near mid-crustal depths by a narrowly confined magma chamber. As an order of magnitude estimate, this value is consistent with the tomographic images that show the main low velocity anomaly extending to mid-crustal depths and indicates that the axial topography is supported by a thicker buoyant and low viscosity region than just the thin magma lens. Isostatic support is consistent with the axial high that exists within 1–2 km of the axis (Kappus *et al.*, 1992). Crustal density variations may also be a much larger source of compensation for the broader region of anomalous elevation within ~15 km of the ridge axis than inferred by recent studies by Wilson (1992) and Wang and Cochran (1993); these results are more consistent with the large component of crustal compensation of axial relief found in Madsen *et al.* (1990). A simple estimate of these crustal effects considers the isostatic effects of two crustal columns, one on-axis and the other located ~15 km off-axis. The isostatic effect of crustal cooling due to hydrothermal circulation can be estimated from the thermal model for a fast-spreading ridge shown in Fig. 16A. This thermal structure would predict the axial crust to be 4.5 km of ~1200°C crust capped by ~1.5 km of ~300°C. Fifteen kilometers off-axis, the lithosphere is a ~7-km-high hydrothermally cooled column of crust and mantle with an average temperature of ~300°C. This results, assuming a thermal expansivity $\alpha = 3 \times 10^{-5}\ K^{-1}$, in a thermal subsidence of ~150 m for this 5.5-km-thick lithosphere column of crust and mantle that cools by 900°C within 10 km of the ridge axis, a number that is ~120 m larger than that used in the Wilson (1992) and Wang and Cochran (1993) studies. Similarly, a 6-km-high axial crust with an average of 3% retained melt will produce an isostatic uplift of ~50 m, and a 100-m-thick magma lens will lead to an axial isostatic uplift of ~25 m. These isostatic effects account for roughly 225 of the ~350–400 m of relief associated with the near-axis bathymetric high, leaving at most ~175 m of relief to be produced by density variation (due to the presence of melt?) beneath the Moho, i.e., leaving only about one-half the anomalous topographic high assumed in Wang and Cochran (1993) and Wilson (1992) to infer that a deep low-density root is needed beneath a fast-spreading ridge. This important question clearly merits further investigation.

Delivery of Melt from the Mantle to Supply the Magma Lens

Geophysical information on the nature of the mantle delivery of magma to the crust is at present extremely limited. There is some evidence from time delay and Q tomography measurements of anomalous structure at the base of the crust beneath the rise-axis (Solomon and Toomey, 1992; Wilcock *et al.*, 1992). Madsen *et al.* (1984) used compensation depths of gravity anomalies to argue for a magma chamber at Moho depths, while Garmany (1989) has suggested that certain converted shear phases are due to isolated pockets of melt residing at Moho depths. However, none of these measurements can shed light on the geometry of magma delivery to the crust.

One means of inferring information about magma delivery is from the segmentation of the crustal magma chamber along-axis. At 9°N, information on segmentation comes from the tomography results (Toomey *et al.*, 1990) and the comparatively densely spaced set of MCS profiles (Kent *et al.*, 1993a). The low velocity anomaly imaged by the tomography experiment narrows and thins toward a pair of ridge axis kinks or DEVALs at 9°28′N and 9°35′N where the local ridge trend changes abruptly. Magma lens width estimates show distinct jumps in widths at the 9°35′N and 9°17′N DEVALs. North of 9°35′N, widths are less than 0.7 km, from 9°35′N to 9°17′N widths vary between 1.0 and 1.2 km, whereas south of 9°17′N to the 9°03′N OSC widths exceed 3 km. South of the 9°03′N OSC the width of the reflector is reduced once again to between 0.7 and 1.3 km. The increased widths coincide with increasing displacement of the AMC reflection to the west and, as discussed previously, are believed to be due to accommodation of the mantle supply to the 9°03′N OSC. South of the 9°35′N DEVAL, the westward displacement of the low velocity zone is comparable to that of AMC reflection. Thus at 9°35′N there is evidence for a segment boundary in the low velocity zone, in the AMC reflection, and in the axial morphology. However, the 9°N section will, in detail, be more segmented than the MCS results indicate as the MCS data do not, for example, resolve a segment boundary at 9°28′N.

The seismic results do not support the idea of a single focus of magma supply for the 9°N segment. Instead they provide evidence for segmentation of the axial magma chamber on length scales of 10–30 km, and, by inference, of the underlying magma supply. Additionally, there is no evidence for a decrease in magma supply toward the ends of the segment as would be predicted by a single injection model; to the contrary, gravity measurements, seismic estimates of crustal and extrusive thicknesses, and the AMC width estimates indicate a relative increase in magma supply near the 9°03′N OSC. The extent and importance of along-axis magma migration at fast-spreading ridges in smoothing out fluctuations in crustal structure are still unresolved. However, the petrologic boundaries at certain DEVALs as well as the discrete jumps in AMC width indicate that along-axis mixing of magma is incomplete.

Spreading Rate Dependence of the Magma Lens: Seismic and Gravity Constraints

A similar set of seismic experiments have been performed along the southern East Pacific Rise where spreading rates are close to the upper end of the spectrum. Preliminary results from these experiments (Detrick *et al.*, 1993; Kent *et al.*, 1993c) support the magma chamber model developed from the northern East Pacific Rise experiment and demonstrate that the size of the magma chamber is not strongly dependent on spreading rate. Once the transition to an essentially two-dimensional accretion pattern has been made at spreading rates between 25 and 35 mm · yr^{-1} (half-rate), the basic magma chamber structure appears to be set.

There is a small spreading rate dependence to the depth of the magma lens where it has so far been observed as shown in Fig. 9. The magma lens is slightly shallower at the southern East Pacific Rise than the northern East Pacific Rise,

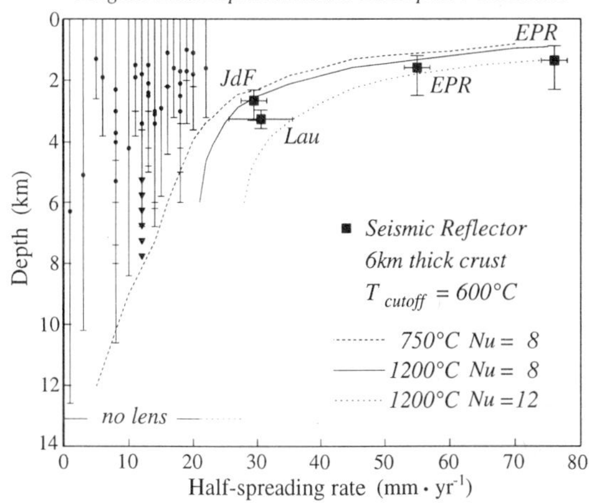

Figure 9 Depth to the top of the magma lens as a function of spreading rate, all other parameters being held constant. Solid squares (and associated uncertainties from Purdy *et al.*, 1992) show multichannel seismic observations of the depth of the magma lens along intermediate and fast-spreading ridges. Curves show results from a suite of numerical experiments with Nu = 8 and Nu = 12 and T_{cutoff} = 600°C. For these hydrothermal cooling parameters a steady-state magma lens (existing when a 1200°C isotherm lies within the crust) can exist only within the crust at half-spreading rates greater than about 20 (or 30) mm · yr^{-1}. A well-developed shallow magma lens exists only for spreading rates greater than 30 mm · yr^{-1}. There is good agreement between model predictions of the depth dependence of the magma lens with spreading rate and multichannel seismic observations. Depths to the 750°C isotherm are also shown for a suite of numerical experiments with Nu = 8 and T_{cutoff} = 600°C. The model 750°C isotherm correlates well with the spreading rate dependence inferred from axial earthquake centroid depths (shown by small dots and associated bars representing inferred total rupture depth) of Huang and Solomon's (1988) teleseismic study. Triangles are microearthquake focal depths beneath the axis of the northern Mid-Atlantic Ridge (Toomey *et al.*, 1988; Kong *et al.*, 1992). There is good agreement between model isotherms and the spreading rate dependence of the depth of the seismically observed brittle–ductile transition.

which is spreading at a slower rate. The intermediate-rate Juan de Fuca and Lau Basin ridges are spreading at lower rates and have deeper magma lenses than the northern East Pacific Rise. A magma lens has yet to be observed on any slow-spreading ridge. There is a similar spreading rate dependence of the thickness of layer 2 supporting the observed on-axis spreading rate dependence (Purdy *et al.*, 1992). However, at slow spreading rates, the thickness of layer 2 appears to plateau at a maximum thickness of roughly 3 km or ~50% of the crustal thickness (Purdy *et al.*, 1992). The thickness of the axial seismogenic zone, or brittle lithosphere, is also a sensitive function of spreading rate as shown in Fig. 9.

As noted above, seismic experiments to date have found no direct evidence for an axial magma chamber beneath a slow-spreading ridge. This nonobservation suggests that magmatic emplacement occurs in a much more episodic fashion at slow-spreading ridges, in agreement with petrologic observations that argue for small, intermittent magma bodies at these spreading centers (e.g., Natland, 1980; Sinton and Detrick, 1992). Slow-spreading ridges typically differ significantly from fast-spreading ridges in their axial morphology—they have a 15- to 30-km-wide, 1- to 2-km-deep median valley as their axial expression instead of a 1- to 2-km-broad, several hundred meter high axial high as shown by characteristic axial bathymetry in Fig. 10. In addition, the axial relief along the axis of a slow-spreading ridge is much more variable than at a fast-spreading ridge. Figure 11 shows profiles of axial bathymetry and axial Bouguer gravity as a function of spreading rate. The axial relief along a fast-spreading ridge is extremely uniform in contrast to the ~1-km variation in median valley depths along a single segment of a slow-spreading ridge. The axial mantle Bouguer anomaly can be used to infer along-axis variations in crustal thickness. The along-axis mantle Bouguer gravity profiles in Fig. 10 suggest ~3 km of along-axis variation in crustal thickness at a typical segment of a slow-spreading ridge in contrast to at most $\frac{1}{2}$ km of along-axis crustal thickness variation where an axial magma chamber is present (e.g., Lin and Phipps Morgan, 1992). This dramatic change in along-axis crustal variation could reflect a fundamental change in upwelling between fast- and slow-spreading ridges, with fast-spreading ridges

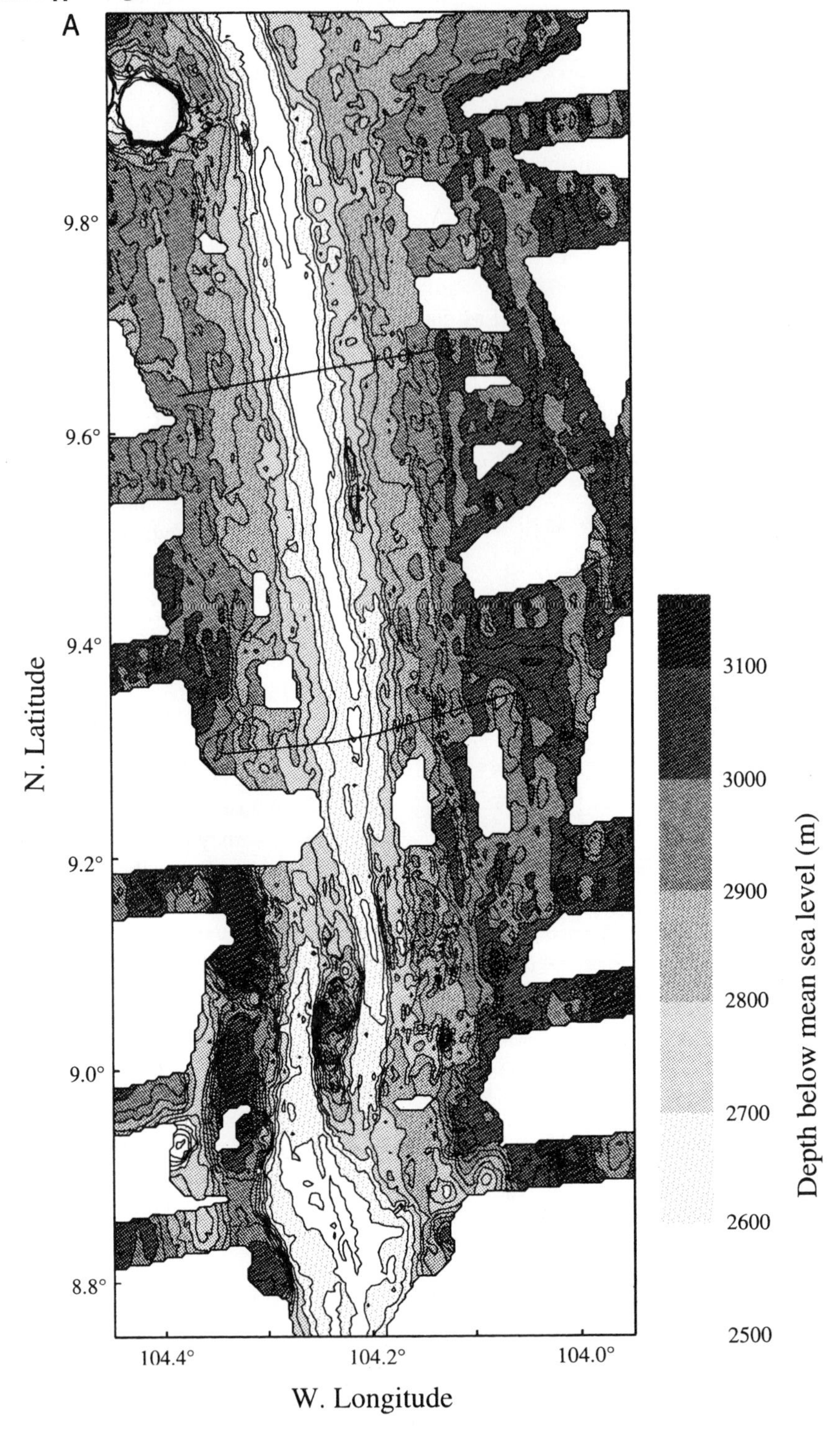

Figure 10 (A) Regional axial high bathymetry of the relatively fast spreading East Pacific Rise. This section is near 9°N. Selected profiles are shown by solid lines. Note the changes in axial segmentation between this typical fast- (EPR) and slow- (MAR, Figure 10C) spreading ridge.

continues

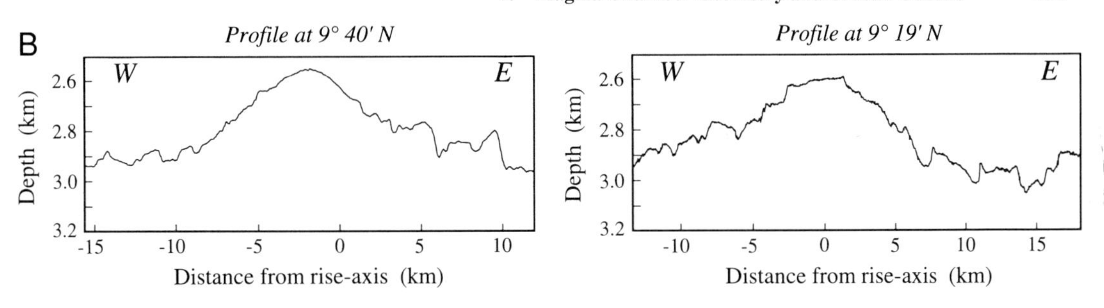

Figure 10 *continues* (B) Axial bathymetry across the EPR profiles noted in Figure 10A. Note the changes in axial relief between these typical fast- (EPR) and slow- (MAR, Figure 10C) spreading ridges.

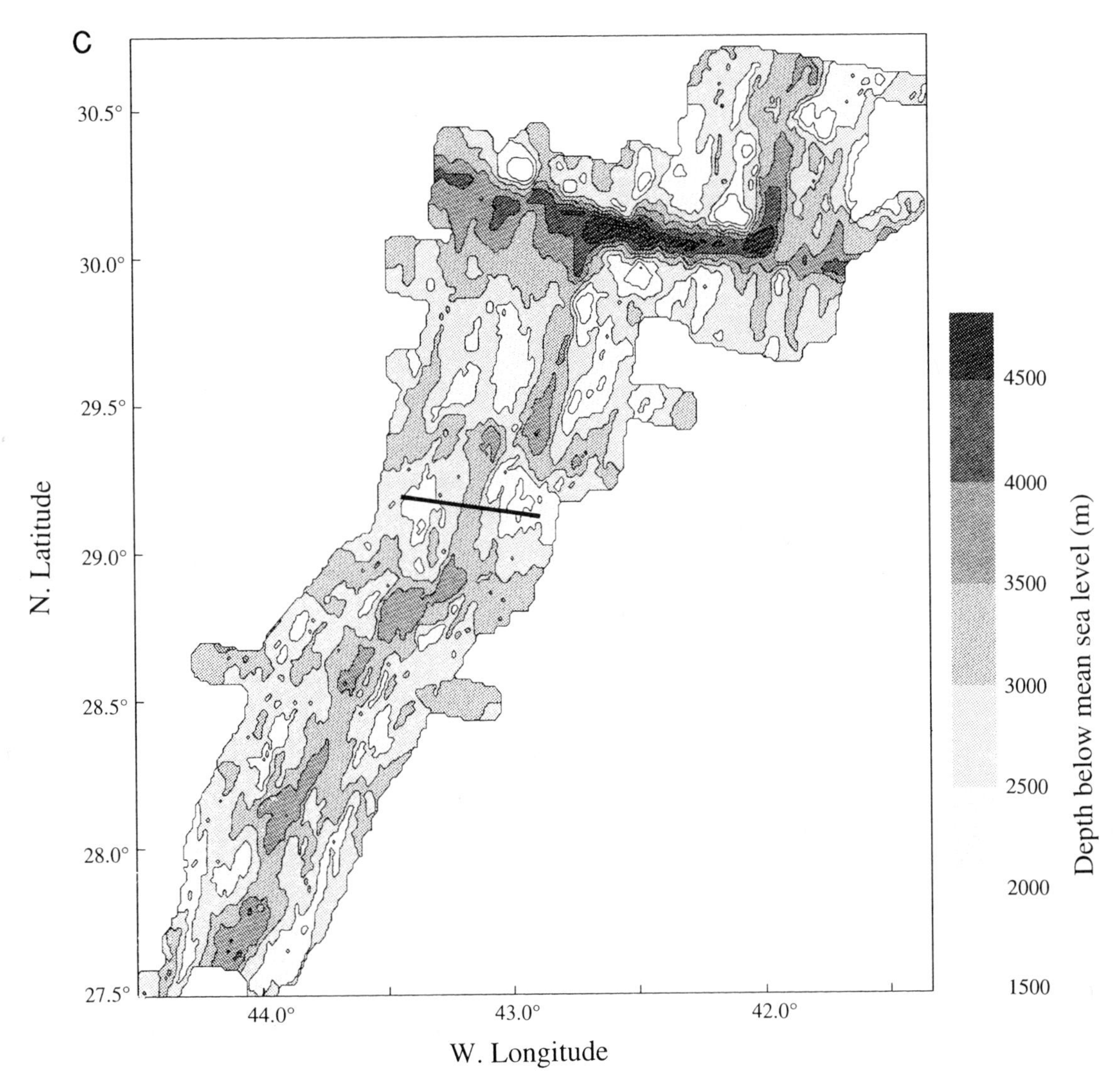

Figure 10 *continues* (C) Regional axial bathymetry and across-axis bathymetric profiles for the relatively slow-spreading Mid-Atlantic Ridge, at the 29°N segments. The median valley bathymetry is illustrated. Selected profile is shown by a solid line.

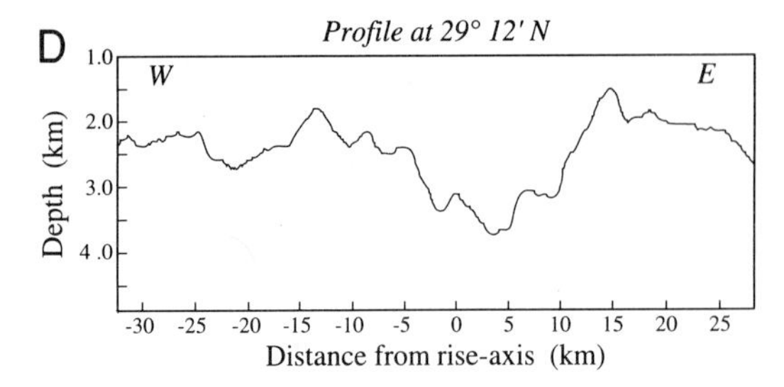

Figure 10 continued. (D) Axial bathymetry across the profile noted in Figure 10C.

having a 2D form of upwelling and melting while slow-spreading ridges are underlain by diapiric 3D upwelling points along-axis (Parmentier and Phipps Morgan, 1990; Lin and Phipps Morgan, 1992). However, these observations are also compatible with three dimensional upwelling and melting at both slow- and fast-spreading ridges with a magma lens acting to smooth along-axis crustal thickness variations where it is present (Phipps Morgan, 1991; Lin and Phipps Morgan, 1992).

One consequence of the type of fast-spreading axial magma chamber inferred from seismic studies is that the low-viscosity melt lens and hot crustal rocks beneath the lens will be an extremely efficient zone for along-axis crustal flow. (This will certainly be true if the magma lens is truly "molten." Marsh (1989) suggests that a ~55% crystal fraction is the rheological equivalent of a "viscous solid." This type of strong Marsh mush may still be a good seismic reflector. It is still almost certain to be the lowest viscosity region in the axial crustal section.) Thus plate extension due to plate spreading is likely to force significant along-axis flow if crustal emplacement from the mantle is focused at several sites along a spreading segment. This flow will preferentially occur in the magma lens if a molten zone is a persistent feature. If not, this along-axis flow is likely to be concentrated in the "hot mush" zone beneath the magma lens. In contrast, where no magma lens and associated basal mush zone is present, crustal rocks will be much colder and stronger so that they can support large lateral variations in axial crustal structure. This inference suggests that the lack of an observed axial magma chamber at slow-spreading ridges is not mere bad luck—slow-spreading ridges may be fundamentally colder accretion sites than fast-spreading ridges.

Closure on Observational Constraints on the Structure of an Axial Magma Chamber

Although many of the details of magma chamber structure have not been elucidated by current seismic experiments, it is important to recognize that these models are fundamentally different from earlier models that proposed a large, well-mixed, molten magma chamber that was 4–5 km deep and extended throughout the lower crust. Even a narrow molten magma chamber with a width on the order of the AMC reflection is not supported by the seismic data. Similarly, the low amplitude of the mantle Bouguer anomaly, ~10 mGal, at the axis of the EPR is incompatible with a large molten magma body, although a narrow magma body would be permitted by the gravity data (Madsen *et al.*, 1990; Stevenson, 1992). The basic 3D interpretation of the previous colocated seismic studies is shown in Fig. 12 (after Kent *et al.*, 1993b). The seismic experiments in the 1980s have proposed this picture as a challenge to the theoretical and ophiolite-based models, which presumed a large crustal magma body. In the next sections we present an emerging theoretical paradigm that may be able to explain successfully this apparent discrepancy and lead to a picture of crustal accretion in striking concordance with seismic constraints.

A Theoretical Model for Crustal Genesis

Introduction

In this section we construct a thermal and mechanical model for the genesis of oceanic crust that we feel is a good candidate to integrate the above observations into a coherent synthesis. Its most important conceptual ingredient is that it is the interplay between magmatic crustal injection and hydrothermal cooling that is responsible for the presence of a quasi-steady-state magma lens beneath a fast-spreading ridge and the dramatic change in axial morphology between a fast- and slow-spreading ridge axis (because axial morphology directly reflects the lithospheric strength or thickness across a ridge axis that is quite sen-

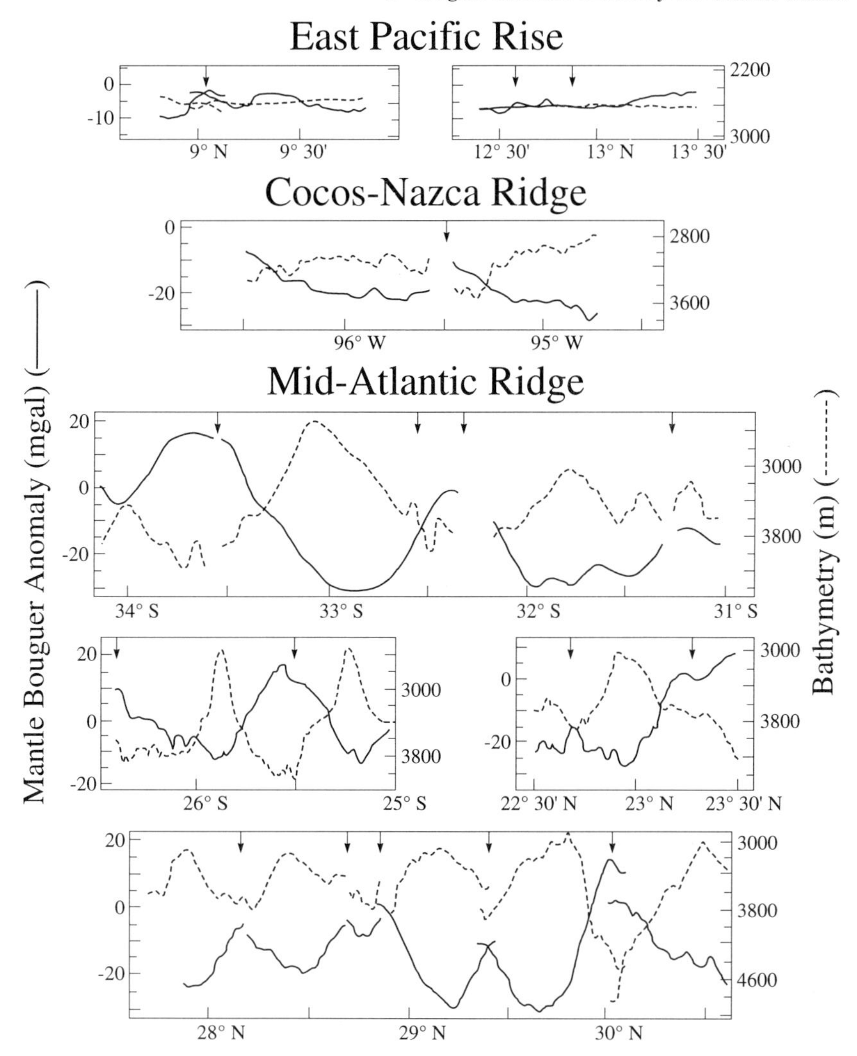

Figure 11 Along-axis profiles of seafloor depth (dotted lines) and mantle Bouguer anomalies (solid lines) for ridge segments spreading at fast (EPR), intermediate (Cocos-Nazca), and slow (MAR) rates. Arrows at the top show locations of transform, nontransform, and overlapping spreading center ridge axis offsets. Zero level for mantle Bouguer anomalies is arbitrary. Mantle Bouguer anomalies were generated by subtracting from the free-air anomaly the attraction of seafloor topography and the attraction of relief on the crust–mantle interface. The crust was assumed to be 6 km thick. Densities of 1.03, 2.7, and 3.3 Mg · m^{-3} were assumed for sea water, crust, and mantle, respectively. The resulting mantle Bouguer anomaly predominantly reflects along-axis variations in crustal thickness and density but also reflects along-axis variations in mantle density. All calculations use gravity reduction techniques developed at Brown University (cf. Kuo and Forsyth, 1988).

sitive to this thermal balance). The basic model geometry is shown in Figs. 1C and 13. The resulting magma injection and crustal flow structure that we envision has been the subject of at least three recent efforts, all apparently inspired by field trips and discussion at the 1989 Oman Ophiolite Conference. We follow in this chapter the treatment of Phipps Morgan and Chen (1993). Henstock *et al.* (1993) explore the effects on seismic velocity structure and potential episodicity, and Quick and Denlinger (1993) explore the petrologic implications of this form of crustal genesis. This emerging paradigm is extremely similar to the qualitative crustal accretion scenario sketched by Sleep (1975, 1978) and also similar to the subsiding magma chamber floor model of Dewey and Kidd (1977) and Browning (1984). The common thread in these studies is the realization that crustal

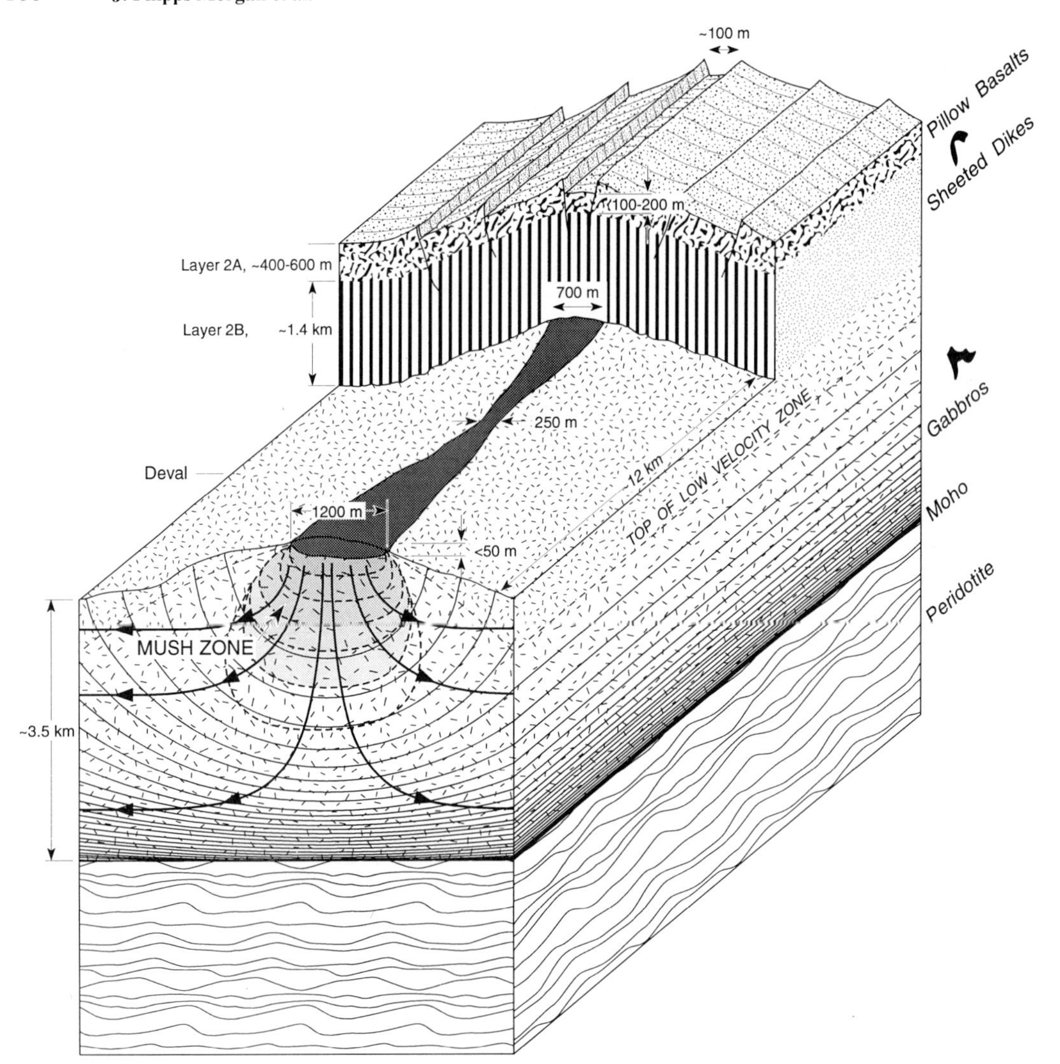

Figure 12 A schematic cross section of the proposed melt-lens model for crustal accretion at a fast-spreading ridge axis. The upper crust is partially stripped away to show segmentation of the magma lens near a small DEVAL "kink" in the spreading axis. Off-axis thickening of the extrusive section occurs within the first few kilometers of the ridge axis. The low velocity zone, corresponding to a region of elevated temperatures with at most a few percent partial melt, originates at the base of the magma lens and extends through the upper half of the gabbro section. Crustal strain of cumulate gabbro formed in the magma lens leads to the development of gabbro layering and fabric, which becomes progressively more intense near the Moho. Schematic crustal layering is shown by "isochron" lines of similar crustal age, and the crustal flow pattern is shown by the flow lines with embodied arrowheads.

subsolidus flow away from a small, shallow level magma lens is potentially able to generate the gabbro "layering" seen in the Oman ophiolite—cumulate freezing at the sloping sides of a crustal-thickness-scale magma body is not necessary to produce this general layering pattern. (This flow pattern also provides an appealing explanation for the observation that cryptic geochemical variation within the gabbro section has a maximal 200-m cyclicity without an apparent depth dependence (Browning, 1984). There should be no depth dependence if all the gabbro section is emplaced through the same magma lens (Browning, 1984).) Phipps Morgan and Chen (1993) propose that a fairly delicate balance between magma input and hydrothermal cooling at the ridge axis leads to the observed spreading rate dependence of the depth of the magma lens. To consider the time-averaged

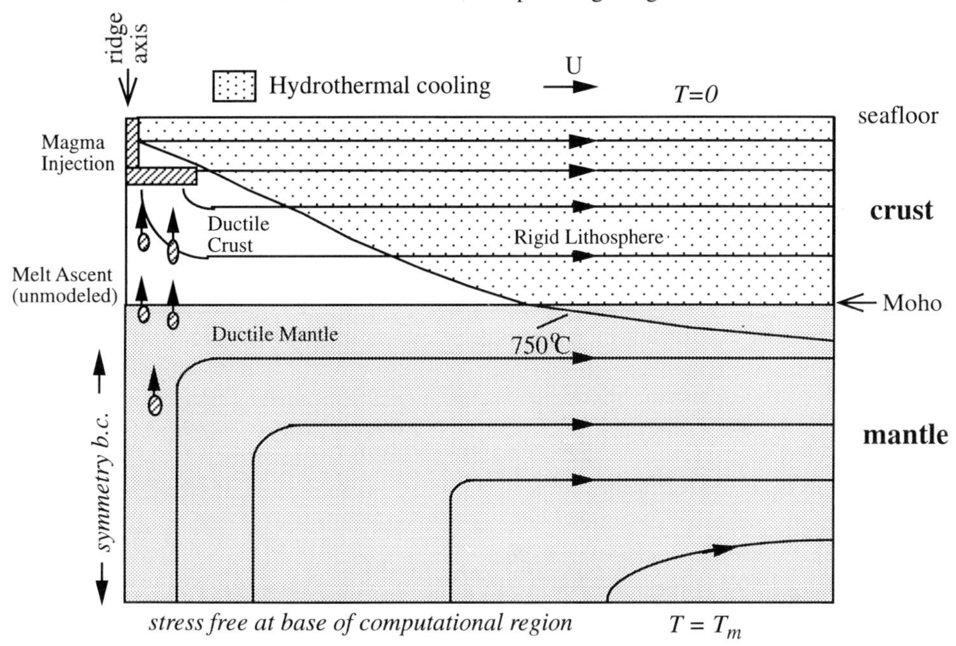

Figure 13 Geometry of the theoretical model.

thermal structure at a ridge it is useful to parameterize these processes in a reasonably simple manner, which we next describe.

Magma Ascent and Emplacement

Magma ascent within the oceanic crust is probably controlled by two complementary processes: (*a*) Magma will ascend only if buoyancy forces (or, if large enough, viscous pressure gradients) cause it to; thus magmas may rest in their ascent for a while at zones of neutral buoyancy until they have fractionated and reacted to the point where the magma's density is less than that of surrounding host rock (this idea has been explored in particular by Ryan (1987, and this volume). (*b*) Magma can also be halted in its ascent when it reaches a freezing horizon, where the dilational volume change associated with magma freezing leads to viscous stresses that favor magma ponding within roughly one viscous "compaction length" of the freezing horizon (term coined by McKenzie (1984)). This idea has been recently proposed in a particularly simple form by Sparks and Parmentier (1991, and this volume) to explain melt focusing to a narrow neovolcanic zone at a spreading center.

Henstock *et al.* (1993) suggest that neutral buoyancy is responsible for the fact that the observed magma lens is at roughly a constant depth where magmas fractionate until they are light enough to reach the surface. However, a recent assessment of this hypothesis by Hooft and Detrick (1993) suggests that the neutral buoyancy level is at most 500 m deep at the East Pacific Rise, in contradiction with the observed ~1500-m depth of the axial magma lens. In contrast, Phipps Morgan and Chen (1993) were stuck by the apparent spreading rate dependence of the depth of the magma lens, which suggested to them that magma freezing may be the process that more strongly limits the depth of magma ascent at a fast-spreading ridge. Here we initially consider only the effects of magma freezing in limiting melt ascent and ignore density effects on lower crustal melt transport, reaction, and segregation. Later we lightly touch on potential observational tests of whether magma freezing or neutral buoyancy play the more important role in limiting magma ascent. Since magma freezing must occur (by definition) within the sheeted dike and extrusive sections of oceanic crust, these accretion processes, unlike a magma lens, must be fundamentally transient in nature. To model the steady-state thermal impact of crustal accretion via dike injection and pillow flows we treat all crustal accretion in rocks below a magma "solidus" to occur in a narrow 250-m-wide dike-like region centered about the ridge axis. Within this region magma emplacement rates are taken to be equal to the spreading rate divided by the width of the diking region—i.e., all shallow crust is modeled to be emplaced within this region. Although correctly treating heat injection within the sheeted dike section, this approximation does somewhat overemphasize the importance of magma heat injection within the pillow section. Since pillow flows are extruded on the seafloor, burying previous flows, they rapidly cool to the ambient seawater temperature and essentially advect this cold boundary layer downward with subsequent pillow burial and subsidence as discussed previously. In addition, seismic observations discussed previously suggest that the thickening of the pillow layer occurs within roughly ~1–2 km of the axis, instead of within the central 250 m. The neglect of these effects implies that this model will tend to underestimate the depth of the magma lens isotherm, by an amount that is no more than the thickness of the pillow section above the magma lens, i.e., ~200 m (e.g., Fig. 6).

The rest of the oceanic crust is modeled to be emplaced as a steady-state magma lens directly beneath the solidus freezing horizon. We take the ~1-km-wide, ~250-m-thick prismatic shape inferred from recent EPR seismic studies discussed previously to be a kinematic constraint on the shape of this magma lens. Once we have this shape, the steady-state emplacement rate is determined by the constraint that this lens supply all crust not emplaced through diking/extrusion above the magma lens. The depth of this lens is controlled by the depth of the magma solidus (here taken to be 1200°C) determined from a self-consistent thermal structure for a spreading center. Thus the lens will cease to exist in this model if the steady-state thermal structure places this solidus isotherm beneath the crust. In addition, the injection rate within the lens will diminish as the lens moves deeper into the crust, since magma injection into the lens supplies magma only for crustal sections below the sheeted dike complex. Note that this solidus temperature is more properly viewed as the effective temperature at which the magma is sufficiently crystallized to behave mechanically as a strong, viscous fluid—this extreme physical simplification must be improved if we wish to model the chemical evolution of the crust associated with magma fractionation and cumulate segregation. For the same reason we do not address convection and magma fractionation processes within the magma lens. See Quick and Denlinger (1993) for an analysis of some of the petrologic consequences of crustal genesis by predominant melt injection within a small magma lens. The thermal structure at a spreading center is predominantly influenced by two factors in this model: (*a*) the depth and injection rate within a potential steady-state magma lens; (*b*) the efficiency of hydrothermal circulation in removing heat through rocks that are cool enough to permit cracking and hydrothermal heat transport.

Hydrothermal Cooling

The primary remaining uncertainty in these numerical experiments is how to parameterize appropriately the form and magnitude of the effects of hydrothermal circulation on shaping heat transport within the crust. We choose the formulation developed in Phipps Morgan *et al.* (1987) which uses the results of Combarnous and Bories (1975) and Combarnous (1978) to treat hydrothermal heat transport as an enhanced thermal conductivity within the temperature and depth range where hydrothermal activity occurs. The enhanced conductivity is parameterized by the Nusselt number Nu, which is defined as the ratio of hydrothermal heat transport within a permeable layer to heat transport by heat conduction alone. Rock that is either at temperatures greater than 600°C or at a depth greater than 6 km is assumed to be impermeable. Since water as hot as 400°C discharges from vents on the seafloor, this is a minimum value of the maximum temperature through which water must circulate. Phipps Morgan *et al.* (1987) and Sleep (1991) use a value of 400–450°C for this cutoff temperature, while Morton and Sleep (1985) and Wilson *et al.* (1988) prefer a hotter 600°C cutoff for hydrothermal circulation. Morton and Sleep (1985) suggest that the limit temperature should be at least that obtained by extrapolating the surface hydrothermal venting temperature down to 45 MPa pressures at a shallow magma lens, which yields a 465°C cutoff (Bischoff and Rosenbauer, 1984). In addition, they suggest that hydrothermal cooling will rapidly cool the region close to a fluid-filled crack so that the average rock temperature can be ~600°C with fluid present in locally cooler cracks. For the purposes of this study we can sidestep this question to some degree because the hydrothermal heat loss will be governed by an approximate product of the hydrothermal heat transport enhancement factor Nu times the hydrothermal cutoff temperature T_{cutoff}, e.g.,

$$q_{\text{hydro}} \approx \text{Nu}(T_{\text{cutoff}} - T_{\text{seawater}})/z_{\text{hydro}}, \quad (1)$$

where z_{hydro} is the maximum depth of axial hydrothermal penetration. Thus for a lower T_{cutoff} we find that we will need a higher Nu value to produce a magma lens at a given depth for a fast-spreading numerical experiment. Figure 14 shows the strong tradeoff between T_{cutoff} and Nu in cooling the axial upper crust, illustrating graphically that once we can determine T_{cutoff} from rock and water chemistry observations, we can use this approach to determine the effective additional heat transport by hydrothermal circulation and so estimate the effective permeability within the hydrothermal system. Gregory and Taylor (1981) report that subsolidus oxygen isotope exchange occurred mainly within the upper 5–6 km of the Oman ophiolite, thus giving an estimate of the maximum

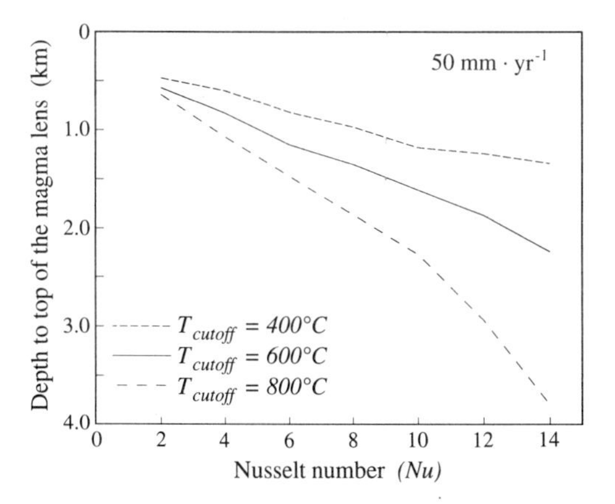

Figure 14 Plot of the depth to the top of an axial magma lens described in the text for a half-spreading rate of 50 mm · yr^{-1} and various hydrothermal heat transport enhancement factors (Nu) and cutoff isotherms (T_{cutoff}) above which hydrothermal flow ceases. In this model, the effectiveness of hydrothermal cooling determines how deep a steady-state magma lens will reside. Thus if we can determine T_{cutoff} by geochemical means, then the depth of a steady-state magma lens will directly constrain Nu.

depth of water penetration. Nehlig and Juteau (1988) find a similar depth limit to near-ridge hydrothermal circulation and also find that the highest recorded hydrothermal fluid inclusion trapping temperatures are ~400–530°C. In these studies we generally choose to model hydrothermal heat transport as T_{cutoff} = 600°C and Nu = 8. We choose these values because they lead to a solution where a steady-state magma lens can exist at 1.2- to 1.5-km depths beneath a fast-spreading ridge and not exist beneath a slowly spreading ridge (cf. Fig. 9). Note, however, that the results in Fig. 14 show that this parameterization of heat transport leads to the same hydrothermal heat flow as T_{cutoff} = 530°C and Nu = 9, or any pair of Nu and T_{cutoff} with $\text{Nu}(T_{\text{cutoff}} - T_{\text{seawater}}) \approx 4800$°C. Studies of the relationship between Rayleigh number and Nusselt number of hydrothermal convection (Combarnous and Bories, 1975; Combarnous, 1978) imply that a Nusselt number of 10 corresponds to a Rayleigh number of ~400, about 10 times the critical Rayleigh number for the onset of convection. The rock permeability needed for this hydrothermal flow through an axial hydrothermal layer that is 1200 m thick is $K = 8 \times 10^{-15}$, roughly ~800× larger than the measured value of 10^{-17} in DSDP hole 504B at depths greater than 600 m in 6.5-Myr-old crust (Becker *et al.*, 1989). The low measured value at DSDP hole 504B would clearly not account for convective hydrothermal heat transfer with a Nusselt number as large as 10. However, near the ridge axis where active faulting and cracking associated with rapid cooling occur (Lister, 1974), it is not unreasonable to expect significantly larger permeabilities, which are rapidly reduced by cracks and inactive faults being filled by hydrothermal mineral deposition. Simple geometric models (e.g., Turcotte and Schubert, 1982) for the rock permeability K suggest that

$$K \cong b^2\phi^2/72\pi, \quad (2)$$

where b is the crack spacing and ϕ is the crack porosity of a cubic matrix of circular tubes (other regular geometrical forms, e.g., a network of flat cracks, will differ by geometrical factors of order 1 (Turcotte and Schubert, 1982). For a crack spacing of 1 m (a typical sheeted dike thickness), this implies an active crack porosity of 10^{-4}%. This porosity is equivalent to the volume change due to a secular cooling of 0.3°C, or the porosity opened by 1 day of amagmatic extension over an axial width of 250 m for a typical East Pacific Rise spreading rate of 100 mm · yr^{-1}.

As noted above, deep hydrothermal alteration within the Oman ophiolite is concentrated near faults. This observation suggests that it may be a reasonable hypothesis that crustal extension through faulting opens channels for deep hydrothermal flow—channels which lead to a higher effective hydrothermal cooling enhancement in slow-spreading environments where large median valley bounding normal faults and 6- to 10-km-deep seismically active faults are present (Fig. 9). In this study we only briefly explore a spreading rate dependence on the efficiency of hydrothermal heat transport. If this dependence does exist (which we feel is likely), it is likely to be an enhancement from Nu = 8–10 at fast-spreading ridges to Nu = 12–15 at median valley ridges. We choose not to include this effect because it would only enhance the already strong trends that are seen in the following suite of numerical experiments.

Model Formulation

To assess this conceptual model, we have implemented it as a finite-element code that models the flow and heat transport associated with incompressible flow for a crust and mantle that have a small magma lens and dike region of magma injection. The system of equations governing the resulting steady-state heat and mass transport is

Conservation of Energy

$$L\dot{\psi} + \nabla \cdot (\mathbf{u}T) = \kappa \nabla^2 T, \qquad (3)$$

where $\dot{\psi}$ is the magma injection rate, L is the magma's latent heat of cooling (334 kJ · kg^{-1}) converted into a "superheat" of 320°C, the thermal diffusivity $\kappa = 10^{-6} m^2 \cdot s^{-1}$, $\mathbf{u}$ is the velocity, and T is the temperature.

Conservation of Mass

$$\nabla \cdot \mathbf{u} = \dot{\psi}. \qquad (4)$$

Conservation of Momentum

$$\tau_{ij} = \mu \left[\frac{\partial u_i}{\partial x_j} + \frac{\partial u_j}{\partial x_i} \right] + \lambda \delta_{ij} \nabla \cdot \mathbf{u} \qquad (5)$$

$$p_{,i} + \tau_{ij,j} = 0,$$

where p is the pressure, and μ and λ are the shear and bulk viscosities, respectively (cf. Tritton, 1977). In the preceding equations we use a comma to represent differentiation by the subsequent index (i = 1,2 correspond to x- and z-directions), and the summation convention applies to an implicit sum of repeated indices. While the shear viscosity of mantle and crustal rocks is poorly known, the bulk viscosity is even less well known, so for this suite of numerical experiments we choose $\lambda = \mu$. In addition, in this suite of numerical experiments we do not treat buoyant contributions to mantle upwelling that are discussed in Scott (1992) and Turcotte and Phipps Morgan (1992). Here we try to limit our model complexity to the oceanic crust. The viscous crust and mantle rheology that we use is a simplification of that presented in Chen and Morgan (1990). We assume a Newtonian lithosphere, mantle, and lower crustal viscosity structure with a lithosphere viscosity that is 10^4 greater than the asthenosphere (mantle) viscosity, which in turn is 10^3 times more viscous than crust that is hotter than 750°C (both crust and mantle that are colder than 750°C have a lithosphere rheology). This results in a total viscosity contrast that can be as large as 10^7, the maximum strength range that can be treated by the penalty finite-element solution algorithm that we employ to solve the flow problem (Hughes, 1987). The dike injection region has a strength 10 times less than that of the lithosphere. These strength contrasts approximate the rheology in Chen and Morgan (1990) and were chosen to eliminate an additional nonlinear solution iteration needed to solve the non-Newtonian flow problem studied in Chen and Morgan (1990). The finite-element code that we use has a flow-solver that is an extension of the standard penalty formulation for incompressible flow that uses a penalized Lagrange multiplier term to satisfy the incompressibility constraint. Reddy (1984) and Hughes (1987) provide good summaries of this technique. The difference between the code in this study and standard penalty formulations is that this code prescribes the dilation to have a nonzero value in regions where magma injection is occurring. We solve for heat transport using a standard streamline-upwind Petrov–Galerkin technique that is well summarized in Brooks and Hughes (1982). The advantages of this formulation are that it has no cross-wind artificial diffusion that would introduce a "spurious" numerical diffusion in the direction perpendicular to fluid flow and that it also provides a numerically consistent weighting for the energy source $L\dot{\psi}$ in the numerical formulation of the equation describing energy conservation. Additional details of our finite-element implementation are discussed in the appendix to Phipps Morgan and Chen (1993).

Although we are primarily interested here in solving for crustal flow and thermal structure, we also need to solve for the mantle flow to properly treat the thermal and mechanical effects of a growing lithosphere. Thus we solve for heat and mass transport within a 90 × 140-km region on one side of a symmetric ridge as shown in Fig. 15. All subsequent figures of the temperature and flow structure near a spreading axis are extracted from this larger computational region. The temperature boundary conditions for this problem are that the mantle is flowing into this region at a constant temperature T_m = 1350°C, that heat is free to move out of the sides of the box, and that the top (seafloor) temperature is 0°C. For flow, the vertical velocity and horizontal shear stress are zero at the seafloor; on the sides of the box we assume that passive plate spreading occurs beneath the rigid part of the lithosphere, and the bottom of the box is a shear- and normal-stress free surface. See Chen and Morgan (1990) for further discussion of these boundary conditions. The boundary condition that is unique to the magma lens problem is that for the influx of magma in the zone of crustal accretion. We treat this magma, for the geometry described above, as inflowing at a steady-state rate, which exactly balances the rate that crust leaves the box due to plate spreading. The effective temperature of this inflowing material in the axial dike region is set to be higher than the magma solidus by an amount equal to the energy released as latent heat of cooling. See Phipps Morgan *et al.* (1987) for further discussion of this type of boundary condition. For magma lens accretion, all of the latent heat of cooling for the crust that lies below the depth of the magma lens is released at the top of the magma lens. The depth of the lens is determined by the depth of the axial 1200°C isotherm, which is determined after each solution iteration. This procedure leads to a stable solution after O(20) iterations. A typical numerical experiment takes ~1 h on a Sun Sparcstation.

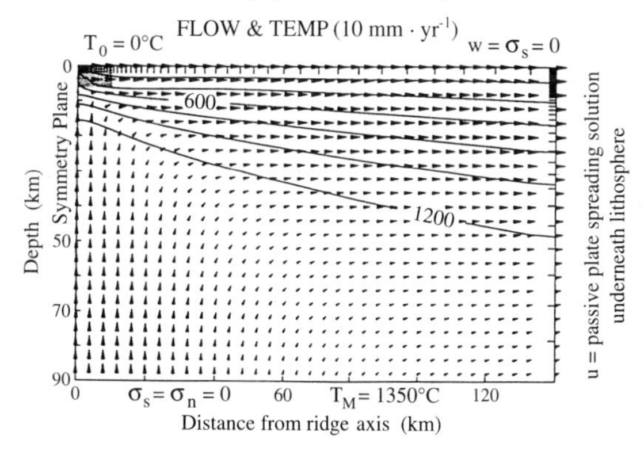

Figure 15 Problem geometry and boundary conditions for the numerical experiments performed in this study. The problem region that we consider is a 140-km-wide by 90-km-deep region on one side of a symmetric ridge axis. Mantle flow is driven solely by plate spreading, and crust is emplaced at the ridge axis according to thermal and geometrical criteria developed in the text. We solve the problem on a 63 (x-dir) × 80 (y-dir) variable spacing tensor-product grid with an x- and y-nodal spacing shown by tick marks along the top and right-hand sides of the region. The problem boundary conditions are shown on each side of the box. The lightly shaded box is the subregion from which the solution is extracted to make the detail plots shown in Figure 16. This subregion contains 43 vertical by 30 horizontal grid points. The sample temperature and flow field shown here is for a half-spreading rate of 10 mm · yr^{-1}, where no steady-state magma lens exists within the crust (the run from which Figure 16 was extracted). Solution isotherms are contoured at 200°C intervals.

Numerical Determination of Accumulated Crustal Strain

Once we have solved for a steady-state flow field, the accumulated crustal strain associated with crustal flow is found using the formulation and techniques summarized in McKenzie (1979). If we imagine the vector $\mathbf{y}'(t)$, which joins two nearby particles in a fluid element at a time t, then this vector is related to a previous vector $\mathbf{y}$ joining these two particles at $t = 0$ by

$$\mathbf{y}'(t) = \Im(t)\mathbf{y}. \tag{6}$$

The matrix $\Im$ is initially the unit matrix. At time t,

$$\mathbf{y}'(t) = \mathbf{x}_2'(t) - \mathbf{x}_1'(t) \tag{7}$$

and at a later time $t + \delta t$,

$$\mathbf{y}'(t + dt) = \mathbf{x}_2'(t + dt) - \mathbf{x}_1'(t + dt). \tag{8}$$

Subtracting and going to the limit as $\delta t \to 0$ gives

$$D_t\mathbf{y}'(t) = \mathbf{v}'(\mathbf{x}_2', t) - \mathrm{v}'(\mathbf{x}_1', t), \tag{9}$$

where $\mathbf{v}'$ is the velocity of the fluid at position $\mathbf{x}'$ and time t, and D_t is the Lagrangian derivative. If we approximate $\mathbf{v}'(\mathbf{x}_2',t)$ with a first-order Taylor expansion about $\mathbf{x}_1'$, then we arrive at

$$D_t y_i'(t) = L_{ik} y_k'(t) = \left[\frac{\partial v_i'}{\partial x_k'}\right] y_k'(t), \quad (10)$$

where summation over repeated subscripts is implied. Substitution of Eq. (6) into Eq. (10) gives

$$D_t \Im_{ij} = L_{ik} \Im_{kj}. \quad (11)$$

We use this relation to determine $\Im(t)$ by tracing the flowline and accumulated strain from seedpoints just outside the region of magma injection. Equation (6) is numerically integrated with a variable time step that is chosen to place ~3 evaluation points within each element.

Model Results

At a fast-spreading ridge with Nu = 8, T_{cutoff} = 600°C, magma freezing at 1200°C, and a latent heat of solidification of 334 kJ · kg^{-1}, a steady-state magma chamber exists 1.35 km below the seafloor 0°C bounding isotherm (see Fig. 16A). All magma that forms the lower crust rises to this level, solidifies, and then flows to deeper crustal levels.

Crustal Strain Can Lead to the Development of Gabbro Layering

Figure 16A shows that the accumulated strain during this flow process is most intense at deeper crustal levels for a ridge with a shallow axial region of crustal emplacement. Strain becomes progressively more intense and more flat-lying as the Moho is approached—the strain within the lowermost kilometer of the crust is too strong to effectively show with the "stretched ellipse" convention shown in Fig. 16. A comparison of Fig. 16A (or Fig. 1C) with Fig. 1A clearly shows that crustal strain can produce layering with the dip and layer development seen in the Oman ophiolite, thus providing an appealing explanation for the orientation and strength of lower crustal layering. In contrast, for a slow-spreading ridge (Fig. 16B), no steady-state magma lens exists within the crust for the same hydrothermal cooling enhancement. We can use Fig. 16 to assess the implications of a deeper magma injection lens beneath a fast-spreading ridge. If magma injection is relatively uniform with depth (like in Fig. 16B) then there will be no major differences in accumulated strain with depth in a crustal or ophiolite section. In the accumulated strain hypothesis for the development of layering, it is where flow streamlines turn sharply that the straining is most intense; thus if injection occurred at the bottom of the crust with crustal flow to shallower levels, then we would expect a sense of layer development opposite that seen in the Oman ophiolite. It is only a shallow-level intrusive center that leads to isotropic gabbros at shallow stratigraphic levels underlain by progressively more deformed gabbro sections. The layering development, intensity, and gabbro dips in the Oman ophiolite support a scenario where crustal injection to form this crust occurred by predominant magma freezing within a narrow sill near the sheeted dike-isotropic gabbro contact—in agreement with previous assertions that this crust was created at an analog to a fast-spreading ridge like the present East Pacific Rise spreading center (Nicolas, 1989). (Note that Nicolas (1989) favors an interpretation of the Oman ophiolite record where the ophiolite section lies on the side of the spreading axis opposite that determined by Pallister and Hopson (1981) and Smewing (1981). In this case, mantle must flow faster than the lower crust for the accumulated strain to have the observed pattern. If future work can resolve on which side of the ridge the ophiolite was created, then the crustal and mantle strain seen in the Oman ophiolite will give us a strong constraint on the relative importance of buoyant vs plate-spreading induced mantle upwelling beneath a ridge.)

Theoretical Spreading Rate Dependence of the Existence and Depth of a Magma Lens

Figure 9 shows that theoretical depth to a magma injection lens plotted vs spreading rate where Nu and the solidus temperature are fixed, and only the spreading rate is varied. In this case there is a fairly abrupt transition with spreading rate from a shallow steady-state magma lens at a 30 mm · yr^{-1} half-spreading rate to no steady-state magma lens within the crust at a 20 mm · yr^{-1} half-spreading rate. There is good agreement with seismic observations of the depth to a magma lens as a function of spreading rate, which are also plotted in Fig. 9. (Note, however, that a higher Nu = 12 is more consistent with the 30–35 mm ·

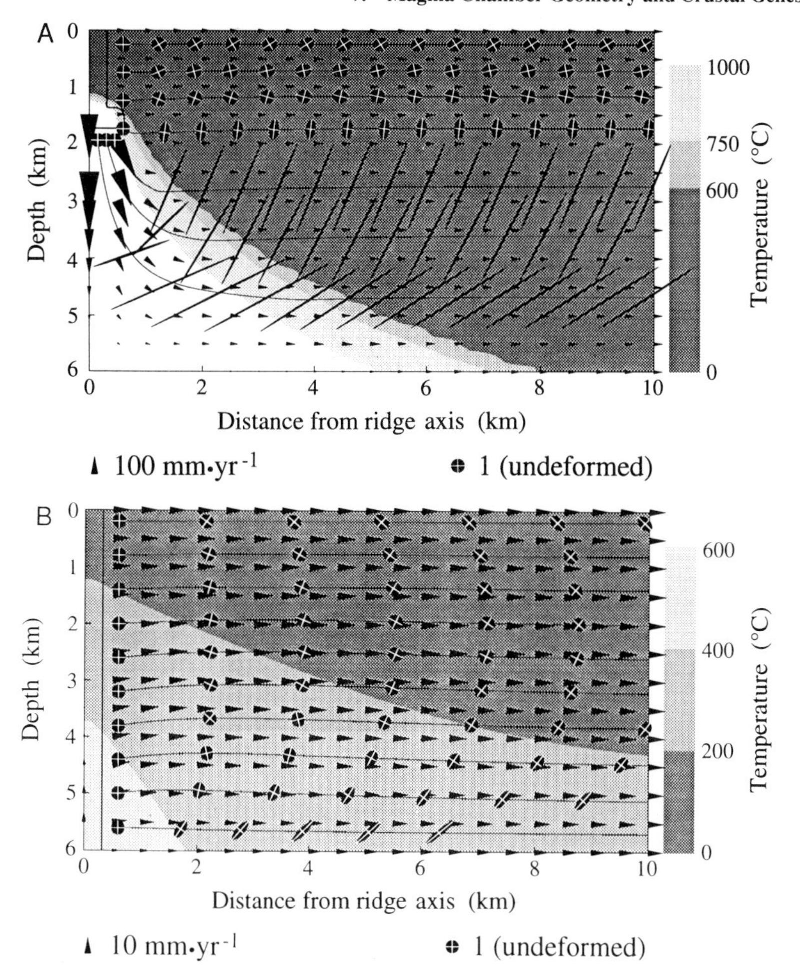

Figure 16 (A) Temperature, flow, and crustal strain for a model spreading center with an opening half-rate of 50 mm · yr^{-1}, Nu = 8, and T_{cutoff} = 600°C. Thermal structure is shown by levels of gray shading; crustal flow by arrows. The extent of the magma intrusion zone is shown by the heavy contour—in this case intrusion is limited to the "dike" section and a shallow magma lens directly beneath the diking zone. Crustal flow lines away from the intrusion zone are shown by light lines, and accumulated strain is shown at 15,000-year time steps along the crustal flow lines. We suggest that the gabbro fabric will reflect the accumulated strain pattern resulting in a fabric development like that seen in Figure 1A that is a consequence of a crustal flow away from a quasi-steady-state shallow-level intrusion zone. This figure was extracted from a calculation done in the larger computational domain shown in Figure 15. (B) Temperature, flow, and crustal strain for a model spreading center with an opening half-rate of 10 mm · yr^{-1}, Nu = 8, and T_{cutoff} = 600°C. Same notation conventions as in (A). In this case intrusion extends completely through the crust since none of the crust is hot enough to sustain a quasi-steady-state magma lens. Crustal flow lines away from the intrusion zone are shown by light lines, and accumulated strain is shown at 160,000-year time steps along the crustal flow lines. There is not nearly as strong a gradient in accumulated crustal strain in (B) as that for the fast-spreading case in (A) because the intrusion rate is constant with depth. A small gradient exists because passive mantle flow near the Moho beneath the ridge axis moves more slowly than the plate opening velocity. This flow pattern occurs because the weak lower crust effectively acts as a near-ridge stress free boundary condition on mantle upwelling in response to plate spreading. This figure was extracted from a calculation done in the larger computational domain shown in Figure 15.

yr^{-1} spreading half-rate, which is roughly observed (Macdonald, 1986; Small and Sandwell, 1989) as the transitional spreading rate between median valley and axial high relief. However, it is less consistent with the depth of the magma lens at fast-spreading ridges—results which support a slight hydrothermal enhancement associated with median valley extension.) We performed a suite of numerical experiments in which the magma lens was assumed to be a 2-km-wide, 500-m-thick body, i.e., twice the width and more than four times the volume of the lens in the preceding numerical experiments. We found that the depth of the lens is most strongly controlled by the balance between the rate of magma injection within the lens and hydrothermal cooling—to first order a bigger lens does not influence the net rate of magma injection and hence does not affect the depth of the lens. To second order, a wider lens is more efficiently cooled at the axis, resulting in a slightly (200–300 m) deeper 2-km-wide lens than a 1-km-wide lens for the same Nu, hydrothermal cut-off temperature, and spreading rate.

Implications for Axial Morphology

Tapponier and Francheteau (1978) proposed that the extension of a strong ridge axis lithosphere layer may be the origin of median valley topography. This hypothesis has been extended in more recent work by Phipps Morgan *et al.* (1987), Lin and Parmentier (1990), and Chen and Morgan (1990). Phipps Morgan *et al.* (1987) showed that moments due to lithospheric stresses within a brittle plate that is 8 km thick at the ridge axis and thickens by only a few kilometers within the 30-km half-width of the axial valley can produce the typical axial topography of a slow-spreading ridge. Lin and Parmentier (1990) developed an elastic/plastic idealization of plate extension that allows them to explore the transient development of extensional rift valley topography. They found that the form of the rift valley depends on the thickness and thickness variations in the stretching lithosphere and that the lithosphere stress-supported topography remains after extension stops, successfully explaining the persistence of failed rift topography. This emerging paradigm for axial accretion incorporates the idea that it is the balance between magmatic heat input and hydrothermal heat removal that determines the thickness ($\approx$ yield strength) of the axial lithosphere, which in turn controls the axial morphology associated with plate boundary extension. The integrated axial strength of the lithosphere is a qualitative measure of the magnitude of the horizontal extensional stress that can be supported during ridge axis extension. Figure 17 shows the axial yield strength as a function of spreading rate for two crustal and mantle rheologies: (i) where the crust and mantle rheologies are both described by an olivine brittle–ductile rheology, and (ii) where the crust is described by a weaker diabase rheology while the mantle is described by an olivine rheology. (See Chen and Morgan (1990) for more discussion of an appropriate ridge axis rheological structure.) Again in all numerical experiments the only physical parameter that varies is the spreading rate. Independent of detailed crustal rheology, there is a strong increase in axial yield strength once the half-spreading rate drops below ≈ 20 mm $\cdot$ yr^{-1}. The large variation in integrated axial yield strength with spreading rate shown in Fig. 17 is a likely reason for the typical presence of a strong lithosphere extension-generated median valley at a slow-spreading ridge and its absence at a fast-spreading ridge where a shallow melt lens is commonly seen. Figure 9 shows that the depth to the 750°C isotherm found in these experiments correlates well with the spreading-rate-dependent maximum earthquake slip depths inferred from teleseismic and microseismic studies.

Effects of Variations in Crustal Thickness at a Given Spreading Rate

These model results suggest that a fairly delicate balance exists between magmatic heat injection during crustal accretion and hydrothermal heat removal, which leads to a strongly different crustal thermal structure at fast- and slow-spreading ridge axes—a difference in thermal regime that is directly responsible for the observed differences in axial morphology. This suggests a strong observational test for this model—can it explain known axial variability at a given spreading rate? Hydrothermal cooling is a function of temperature difference and (fault-) permeability structure, which are not *intrinsically* a function of spreading rate. (The fault-extension induced component of the permeability structure is quite likely a function of axial thermal structure, which we predict to be strongly dependent on spreading rate. As we discuss later, this effect will enhance the basic predictions of this model.) In contrast, the rate of crustal injection is potentially variable at a given spreading rate. Along-axis gravity variations discussed earlier suggest that crustal accretion typically varies by $\sim$3 km along the axis of a given slow-spreading ridge segment (Kuo and Forsyth, 1988; Lin *et al.*, 1990; Blackman and Forsyth, 1991; Lin and Phipps Morgan, 1992).

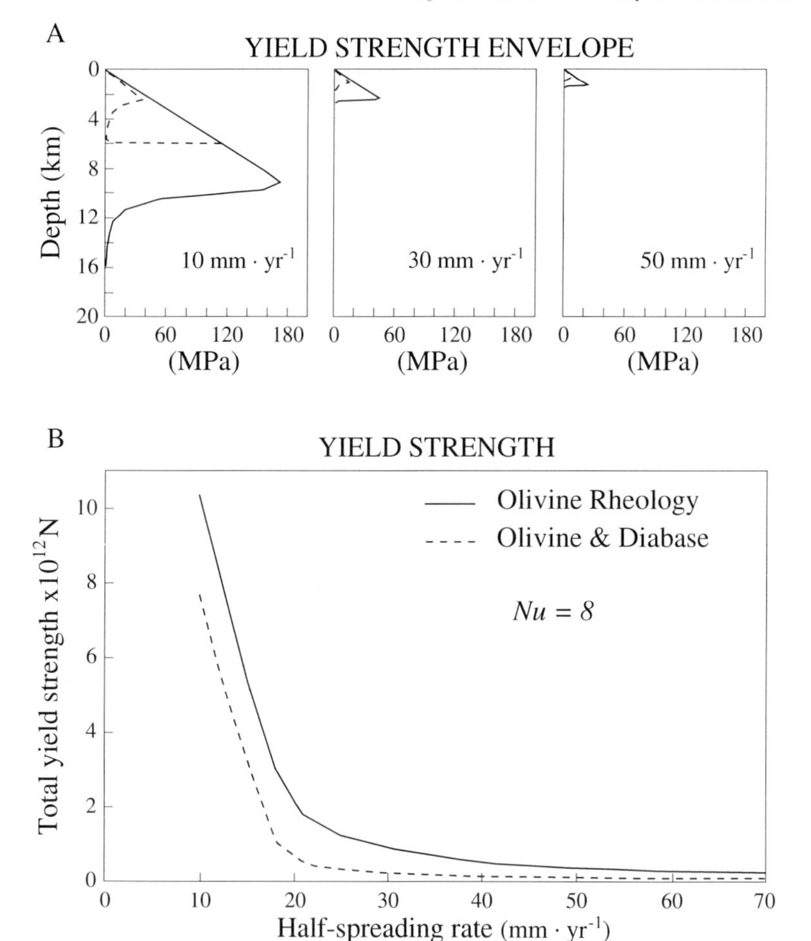

Figure 17 Integrated axial lithosphere strength for a suite of numerical experiments plotted as a function of spreading rate. (A) Mantle yield strength envelopes used in the calculations for 10, 30, and 50 mm · yr^{-1}. Solid lines show the yield strength envelope for a crust and mantle with an assumed brittle–ductile olivine rheology. Dashed lines show the yield strength envelope where the crust has a diabase rheology and the mantle has an olivine rheology (see Chen and Morgan, 1990, for more details of this rheological structure). (B) Depth-integrated total axial yield strength as a function of spreading rate. There is a strong increase in total axial yield strength at spreading half-rates below 20 mm · yr^{-1}, which suggests that extension of strong axial lithosphere is a likely mechanism for the presence of a median valley at slow-spreading ridges and the absence of a median valley relief along fast-spreading ridges.

To explore the effect of crustal thickness variation at a given spreading rate we have performed a suite of experiments at half-spreading rates of 10, 30, and 50 mm · yr^{-1} for a range of potential crust thicknesses. Figure 18 shows these results for T_{cutoff} = 600°C, Nu = 8 and 12, and two assumptions about the maximal depth of penetration of hydrothermal circulation: (a) Hydrothermal penetration is limited by the depth or pressure at which cracks close, i.e., is limited to ~6 km independent of crustal thickness. (b) Hydrothermal penetration is limited by the chemical contrast between crust and mantle; i.e., it always has the potential to reach (but just reach) the Moho where cracks become rapidly filled with precipitates.

Figures 18A and 18B show the depth of the 1200°C isotherm as a function of crustal thickness at half-spreading rates of 10, 30, and 50 mm · yr^{-1}. Figures 18C and 18D show the resulting integrated axial yield strength as a function of crustal thickness at these spreading rates. The axial "strength" of the lithosphere is the depth-integral of the axial yield-strength envelope. It is a measure of the magnitude of the horizontal extensional stress that can be supported during ridge axis extension. According to the lithosphere

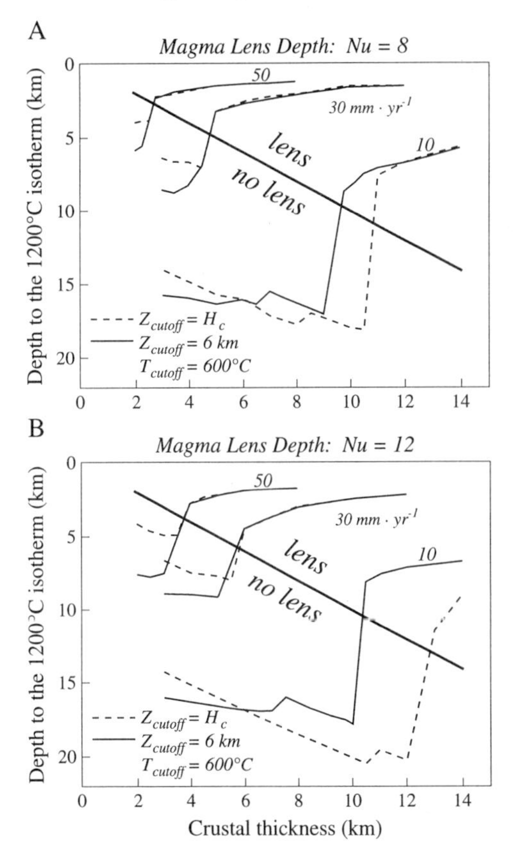

Figures 18A and 18B Magma lens depth as a function of crustal thickness at spreading rates of 10, 30, and 50 mm · yr^{-1}. Figures show the depth of the 1200°C isotherm plotted as a function of crustal thickness for Nu = 8 and 12, and for two different hydrothermal penetration limits Z_{cutoff} of 6 km and the crustal thickness H_c. The region above the lens/no-lens transition is where the 1200°C isotherm will lie within the crust. Note the sharp change in the depth of this isotherm for small variations in crustal thickness around a "critical" crustal thickness.

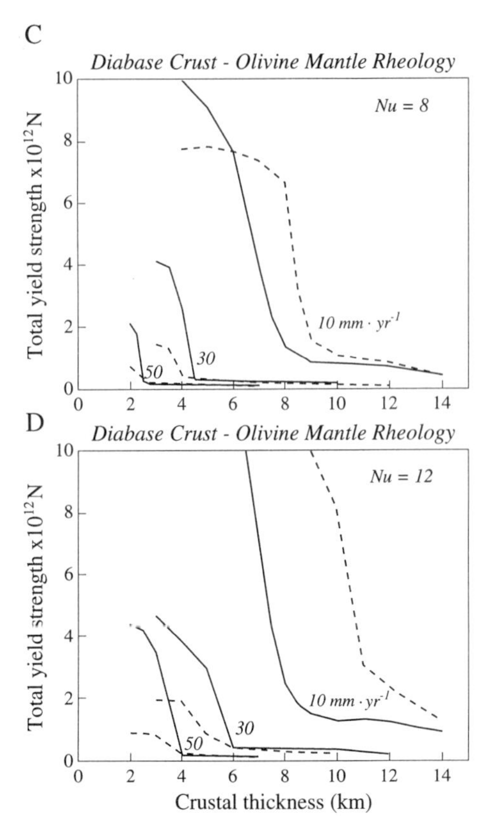

Figures 18C and 18D Axial yield strength plotted as a function of crustal thickness at spreading rates of 10, 30, and 50 mm · yr^{-1}. Figures show the corresponding integrated axial yield strength plotted as a function of crustal thickness. The axial yield strength also changes dramatically near this "critical" crustal thickness.

stretching hypothesis for the origin of axial topography (Phipps Morgan *et al.*, 1987; Lin and Parmentier, 1990; Chen and Morgan, 1990), the net horizontal axial force is proportional to the amplitude of the resulting median valley relief. At a half-spreading rate of 30 mm · yr^{-1} we find that a magma lens is present for a crustal thickness of 6 km but disappears when the crust is slightly thinner (4.5 km, Fig. 18A). This result suggests that the intermediate spreading rate Australian–Antarctic Discordance, which has a median valley morphology, can be produced by crust that is 1–1.5 km thinner than the crust of ridge segments to the east and west that have an axial high morphology. Similar results for a 20 mm · yr^{-1} half-rate typical for the southern Mid-Atlantic Ridge suggest that the along-axis crustal thickness variations of 4.5–7.5 km inferred from axial seismic studies (Tolstoy *et al.*, 1992) and axial gravity studies (cf. Lin and Phipps Morgan, 1992; Neumann and Forsyth, 1992) are sufficiently large to change the ridge from an axial high to a median valley morphology. In general, for transitional spreading rates of 20–30 mm · yr^{-1} a dramatic change in axial strength will be associated with these along-axis changes in crustal thickness and thermal structure as shown in Figs. 18C and 18D.

At a half-spreading rate of 10 mm · yr^{-1} a very large crustal thickness change from a "normal" 6-km-thick crust is needed before a quasi-steady-state magma lens appears. At this crustal thickness of 9–12 km (the exact value depends on which assumption is used for Nu or the intrinsic depth

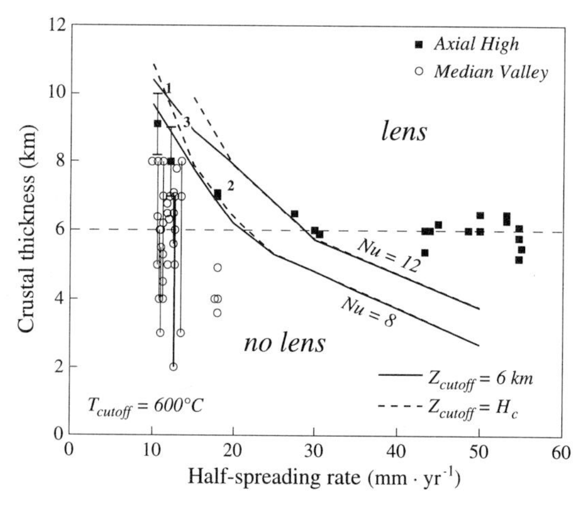

Figure 18E The minimal crustal thickness where a steady-state magma lens will exist within the crust plotted against spreading rate. Small variations about the "normal" crustal thickness of 6 km (Chen, 1992) are most significant for the lens/no-lens transition at intermediate half-spreading rates of 20–30 mm · yr^{-1}.

limit for hydrothermal penetration) the transition is abrupt. This result roughly agrees with a seismic refraction determination of 8- to 10-km-thick crust along a 9-Ma isochron at the slow-spreading axial high Reykjanes Ridge. (Bunch and Kennett, 1980). Finally, at half-spreading rates of 50 mm · yr^{-1} or greater these results suggest that it is difficult to ever generate median valley relief—the crust must be less than ~3 km thick before the magma lens structure disappears. This crustal thickness is lower than any oceanic crustal thickness measurement to date (Chen, 1992). One striking feature of these results is the sharp change in axial thermal structure with small changes in crustal thickness near the "threshold" crustal thickness for the existence of a steady-state crustal magma lens. For the 10 mm · yr^{-1} experiments, a change in crustal thickness from 9 to 9.5 km (at Nu = 8) leads to a shallowing of the 1200°C isotherm from 16 to 8 km! This effect can be understood from the nature of the latent heat release in a magma lens—it releases the latent heat of the magma–cumulate phase change for *all* of the "cumulate gabbro" crust that resides below the magma lens. Thus an upward displacement of the level of the magma lens because of the increase of the total heat input for thicker crust, or the increase of spreading rate, would result in more latent heat release (positive feedback), an effect that would favor a lens at an even shallower depth. This positive-feedback effect is why these model thermal structures are more sensitive to spreading rate or crustal thickness variations than the thermal structures produced in previous thermal models which treat crustal accretion always as a vertical crustal-thickness dike (Sleep, 1975; Phipps Morgan *et al.*, 1987; Lin and Parmentier, 1989; Chen and Morgan, 1990).

The results in Figs. 18A–D show the existence of a "threshold" crustal thickness at a given spreading rate about which small changes in crustal thickness can produce a dramatic change in axial thermal structure. Figure 18E shows a plot of this threshold crustal thickness as a function of spreading rate for the several hydrothermal cooling assumptions discussed above. (This threshold crustal thickness is defined as the point where the 1200° isotherm reaches the base of the crust.) This figure supports the result found in Figs. 18A–18D—variations in axial thermal structure are most sensitive at intermediate spreading rates of 20–30 mm · yr^{-1} half-rate to small fluctuations in crustal input about the normal crustal thickness of 6 km. Greater hydrothermal heat removal increases the spreading rate of this transition and, at slow spreading rates, greater depths of hydrothermal penetration require a larger crustal thickness (magmatic heat input) for a quasi-steady-state crustal magma lens to form. The observed Reykjanes Ridge axial high for a crustal thickness of ~8–10 km (Bunch and Kennett, 1980) implies that T_{cutoff} = 600°C and Nu = 8 is a good description of hydrothermal heat transport at a slow-spreading range. The observed variation in axial relief at intermediate spreading rates (Macdonald, 1986) is also better fit in Figs. 18C and 18D by models where the depth of hydrothermal circulation is limited by the depth at which cracks close (pressure) instead of the depth of the crust–mantle transition. These results lead to a (weakly) preferred parameterization of hydrothermal circulation by a cutoff depth Z_{cutoff} of ~6 km, thermal limit T_{cutoff} = 600°C, and heat transport enhancement Nu = 8.

Axial Variability along a Single Segment

Currently, there is a debate about whether magma emplacement along a fast-spreading ridge is fairly continuous along-axis or confined to a few discrete volcanic centers that are foci of axial accretion processes (e.g., Macdonald *et al.*, 1990;

Phipps Morgan, 1991). Lin and Phipps Morgan (1992) note that while gravity and topography data show that slow-spreading ridges have a clear along-axis variation in crustal thickness (i.e., integrated magma supply varies along-axis), the much smaller along-axis gravity and topography variation at a fast-spreading ridge can be explained either by a more 2D pattern of upwelling and melting beneath a fast-spreading ridge or by a well-connected, temporally persistent magma lens (or low viscosity zone) that smooths the along-axis crustal structure at a fast-spreading ridge. These results suggest that magmatic heat input associated with a mean 6-km crustal thickness is sufficient for a quasi-steady-state magma lens to form at fast-spreading ridges. At slower-spreading rates with 6-km-thick crust, a continuous magma lens cannot be a quasi-steady-state feature along an entire segment. Again, along-axis crustal flow will preferentially occur in the hottest, weakest regions since plate extension tends to confine flow of stronger regions to the plate-spreading direction. However, the crustal thickness at slow-spreading ridges is (usually) not strongly smoothed by along-axis flow, resulting in an observed ~3 km along-axis crustal thickness variation (e.g., Blackman and Forsyth, 1991; Lin and Phipps Morgan, 1992; Neumann and Forsyth, 1993). The resulting along-axis variations in across-axis thermal structure lead to dramatic variations in axial yield strength (Figs. 18C and 18D) and hence in the along-axis relief of median valley topography. This model predicts that along-axis variations in axial yield strength will be strongly correlated with crustal thickness—in agreement with observations of strongly correlated axial mantle Bouguer anomaly (a poor man's proxy for crustal thickness) and axial depth profiles at slow- and transitional-spreading ridges as seen in Fig. 11 (Phipps Morgan, 1991; Lin and Phipps Morgan, 1992; Neumann and Forsyth, 1993).

Neutral Buoyancy versus Magma Freezing as a Fundamental Limit to Magma Ascent

We favor the hypothesis that magma ascent at a fast-spreading ridge is primarily limited by axial thermal structure—magma tends to pond beneath a freezing horizon until sufficient melt buoyancy builds up for the magma to fracture to sheeted dikes and the surface. This mechanism naturally leads to a spreading rate dependence of the depth of the magma lens, which is comparable to that observed. At its simplest expression, neutral buoyancy arguments would suggest that magma intrusion depths are independent of spreading rate. Hooft and Detrick (1993) point out that the observed magma lens depths are invariably deeper than predicted neutral buoyancy depths of ~100–400 m below the seafloor. However, Ryan (this volume) suggests that Hooft and Detrick (1993) use a minimum magma density when inferring a too-shallow neutral buoyancy depth, whereas they should use an average magma density (i.e., including ferrobasaltic magmas and entrained crystals) when inferring a magma's neutral buoyancy level, which would predict neutral buoyancy depths that are consistent with observed EPR magma lens depths. The neutral buoyancy hypothesis is also hard-pressed to explain the 4-km-wide, *ridgeward sloping* lens observed near the 9°03′N overlapping spreading center along the East Pacific Rise (cf. Fig. 7), a feature that would be a natural consequence of magma freezing-limited melt ascent if the primary magma source is offset from the ridge axis. Joann Stock (personal communication) has pointed out that if the depth of hydrothermal circulation is controlled by a spreading-rate-dependent balance between the heat of magma injection and cooling by hydrothermal processes, then hydrothermal alteration may change the shallow axial density structure so that magmas intrude to a deeper level at a slow-spreading ridge. Ryan (this volume) explores this effect in more detail. (This mechanism for crustal density reduction would imply pervasive crustal serpentinization as opposed to fault-concentrated hydrothermal alteration.)

These hypotheses may have different predictions for the relationship between basalt composition and the depth of the magma lens. If the depth is buoyancy controlled, then basalts erupted from a ridge with a deeper magma lens should, in general, show more fractionation than basalts erupted from shallower magma lenses. In contrast, the magma freezing hypothesis would predict that basalts freeze a (slightly) thicker cumulate section as the lens depth shallows, which would imply an inverse correlation between the average concentration of incompatible elements in erupted basalts and the depth to the magma lens. Petrologists are just starting to explore whether there are critical geochemical tests for these two mechanisms of limiting melt ascent.

The effects of neutral buoyancy and axial thermal structure may, in fact, be strongly linked in real mid-ocean ridge magmatic systems. If the floor of the subsiding magma lens consists predominantly of dense cumulate phases that flow away to form the lower crust, while the roof is compositionally lighter, then the magma lens itself may be a strong mechanism to shape the density structure of the crust, which in turn will limit magma ascent. A key ingredient to resolving some of these questions will be to understand better the evolution of the crystallate fraction of ascending magma, as it plays a key role in shaping the magma's density. In addition, if we can compare basalts that are erupted from the flank with those from the ridge crest of the sloping magma lens at the 9°03′N OSC, then these rocks may provide a means to test if either neutral buoyancy or axial thermal structure is the primary limit to magma ascent.

Geochemical Variability between Ridge Segments

Petrologists have long noted a difference in chemical systematics between fast- and slow-spreading ridges—fast-spreading ridge MORBs (mid-ocean ridge basalts) have both a lower bulk Mg#, suggesting more cumulate fractionation and magma evolution within the crust, and less magma diversity. Both observations support the existence of a persistent magma reservoir at fast, but not slow-, spreading ridges (cf. Sinton and Detrick, 1992). MORB systematics appear to show three distinct effects: (i) shallow crystal fractionation, which workers usually "correct" for by referring to an element concentration at a reference magma MgO content, which is found by extrapolating a shallow fractionation trend through a suite of surface samples (e.g., ^{8}Na in Klein and Langmuir, 1987; (ii) local chemical variability, which cannot be accounted for by low-pressure crystallization, which Klein and Langmuir (1989) refer to as a *local trend;* and (iii) global chemical variability of average composition at a given MgO content, which Klein and Langmuir (1987) refer to as a *global trend.* Niu (1992) and Niu and Batiza (1993) have determined how well chemical systematics along 32 different individual ridge segments fit either a global or a local trend. Figure 19 shows these data. In general, MORB variability along fast-spreading ridge segments follows the "global" trend, whereas chemical variability along slow-spreading ridge segments follows the "local" trend (Niu and Batiza, 1993). We suggest that the local trend may be due to magma crystal fractionation (Grove *et al.,* 1992) and/or reaction within the uppermost, cooler mantle beneath a slow-spreading ridge. In this case the local trend should be strongly correlated with axial topography—both the local trend and a median valley will exist only where the uppermost mantle beneath a ridge is "cold." The correlation of median valley vs axial high morphology with global vs local trend is even better than the correlation with spreading rate. At fast-spreading ridges with an axial high the global trend dominates the chemical systematics along a spreading segment, at slow-spreading ridges with a median valley the local trend dominates, and at intermediate spreading rates the global or local trend correlates extremely well with both axial morphology and the presence of a magma lens seismic reflector as shown in Fig. 19. In addition, the three propagating ridge segments in the Niu and Batiza (1993) data set follow the local trend, in accord with previous work that suggests that the growth of a propagating ridge segment is associated with a transient "colder" accretion environment than that of a similarly spreading stable ridge segment. (This idea was suggested by Sinton and Christie in the early 1980s to explain the petrologic gradients seen approaching a propagating ridge tip and the diverse suite of basalts seen near the propagator tip proper (Christie and Sinton, 1981; Sinton *et al.,* 1983). We feel that these results support the hypothesis that the presence of a local trend and an axial median valley reflects the same systematic variations in the axial thermal structure with spreading rate and with variations in crustal thickness at a given spreading rate.

Closure

This chapter has attempted to summarize an emerging theoretical and observational paradigm for magma chamber structure at a fast- and slow-spreading ridge. This paradigm relates magma chamber structure and depth dependence to the

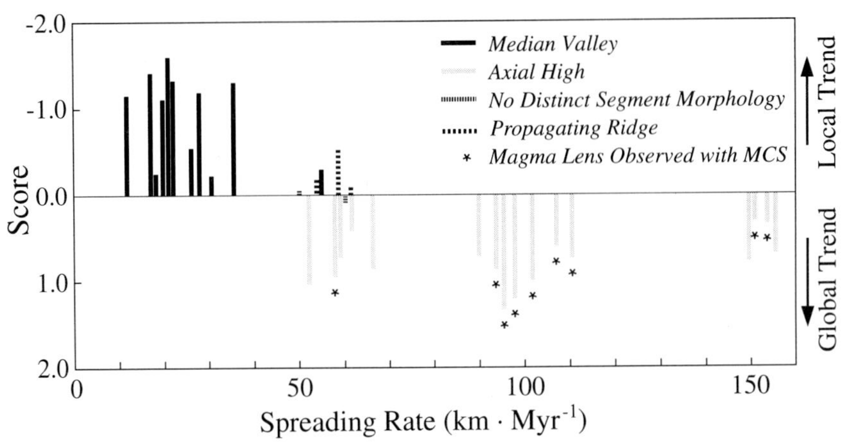

Figure 19 Geochemical systematics along 32 ridge segments (Niu and Batiza, 1993) are plotted versus spreading rate. The length of each line segment is an index to how well the variability along a segment parallels the Klein and Langmuir (1989) "global" or "local" trends. Dark lines note where the axial segment morphology is a median valley, light lines where the segment morphology is an axial high. Short-dashed lines denote ridge segments with no distinct segment morphology. Long-dashed lines are segments with current/recent ridge propagation. Stars show axial segments that have an axial magma lens reflector that has been observed in an axial seismic reflection study along at least part of the ridge segment (Detrick *et al.*, 1987, 1993). This figure clearly shows that chemical systematics along fast-spreading ridges follow Klein and Langmuir's (1987) global trend while slow-spreading ridges follow their local trend (Niu and Batiza, 1993). All axial high segments mapped so far with reflection seismic methods have had a crustal magma lens as predicted by our model; the axial high segments without a star in this plot are segments that have not yet been mapped with seismic reflection methods. Segment scale chemical systematics correlate even better with axial morphology than with spreading rate—axial high segments follow the global trend while median valley segments follow the local trend. This correlation suggests a common dependence of median valley relief and local trend chemical systematics on a cold shallow mantle directly beneath the ridge axis. Propagating ridge segments also tend to follow a local trend, in agreement with previous suggestions that propagating ridge tips have an anomalously cold initial accretion environment.

thermal structure of a mid-ocean ridge, which is controlled by a delicate balance between heat input by crustal injection and removal by hydrothermal circulation so that the presence or absence of axial topography directly reflects the axial thermal structure along a ridge. This paradigm is consistent with the melt lenses recently imaged along fast-spreading ridges but that have not been found where an axial valley morphology is present. It suggests that much of the variability in axial relief at a given spreading rate is due to local variations in crustal magmatic injection. It also suggests that many of the differences in basalt composition and chemical systematics between "slow-" and "fast"-spreading ridges are due to differences in shallow mantle and crustal thermal structure beneath these ridges—there is a fast-spreading "hot" environment that has little magma freezing and reacting with the mantle as it ascends beneath the ridge and a temporally continuous magma lens within the crust, and a slow-spreading "cold" environment with significant magma freezing/reaction within the shallow mantle and no temporally persistent crustal magma reservoir. Finally it appears to raise several further predictions on the relationships among basalt chemistry, crustal thickness, and spreading rate, questions that will let us actively explore and test the viability of this emerging paradigm.

Acknowledgments

We thank Mike Ryan and Bob Detrick for helpful reviews of this chapter and for giving us preprints of their recent work on the role of neutral buoyancy in magma ascent at a fast-spreading ridge. We also thank Mike Ryan for his editorial suggestions. Breck Betts created and drafted many of the figures. Maya Tolstoy and Bob Detrick generously allow us to quote their prepublication seismic and gravity observations of crustal thickness variations shown in Fig. 18E, Yaoling Niu permitted us to use his and Batiza's prepublication measurements of axial geochemistry used in Fig. 19, and Jian Lin and J.-C. Sempere provided the median valley bathymetric data shown in Fig. 10. Their help is gratefully acknowledged, as are the years of collaboration with Don Forsyth, Jason Morgan, and Marc Parmentier during the formulation of the theoretical synthesis that is presented here. The development of these observations and ideas was supported by the National Science Foundation.

References

Barth, G. A., Mutter, J. C., and Detrick, R. W. (1987). Along-axis variability in East Pacific Rise seismic structure, *EOS Trans. Am. Geophys. Union* **44,** 1491–1492.

Becker, K., Sakai, H., Adamson, A. C., Alexandrovich, J., Alt, J. C., Anderson, R. N., Bideau, D., Gable, R., Herzig, P. M., Houghton, S., Ishizuka, H., Kawahata, H., Kinoshita, H., Langseth, M. G., Lovell, M. A., Malpas, J., Masuda, H., Merrill, R. B., Morin, R. H., Mottl, M. J., Pariso, J. E., Pezard, P., Phillips, J., Sparks, J., and Uhlig, S. (1989). Drilling deep into young oceanic crust, Hole 504B, Costa Rica rift, *Rev. Geophys.* **27,** 79–102.

Bischoff, J. L., and Rosenbauer, R. J. (1984). The critical point and two-phase boundary of seawater, 200–500°C, *Earth Planet. Sci. Lett.* **68,** 172–180.

Blackman, D. K., and Forsyth, D. W. (1991). Isostatic compensation of tectonic features of the Mid-Atlantic Ridge: 25–27°S, *J. Geophys. Res.* **96,** 11741–11758.

Bratt, S. R., and Solomon, S. C. (1984). Compressional and shear wave structure of the East Pacific Rise at 11°20′N: Constraints from three-component ocean-bottom seismometer data, *J. Geophys. Res.* **89,** 6095–6110.

Browning, P. (1984). Cryptic variation within the cumulate sequence of the Oman ophiolite: Magma chamber depth and petrologic implications, *in* "Ophiolites and Oceanic Lithosphere." (I. G. Gass, S. J. Lippard, and A. W. Shelton, eds.), Geol. Soc. London, Spec. Publ., 71–82.

Bunch, A. W. H., and Kennett, B. L. N. (1980). The crustal structure of the Reykjanes Ridge from seismic refraction, *Geophys. J. R. Astron Soc.* **61,** 141–166.

Burnett, M. S., Caress, D. W., and Orcutt, J. A. (1989). Tomographic image of the magma chamber at 12°50′N on the East Pacific Rise, *Nature* **239,** 206–208.

Cann, J. R. (1974). A model for oceanic crustal structure developed, *Geophys. J. R. Astron. Soc.* **39,** 169–187.

Caress, D. W., Burnett, M. S., and Orcutt, J. A. (1992). Tomographic image of the axial low velocity zone at 12°50′N on the East Pacific Rise, *J. Geophys. Res.* **97,** 9243–9264.

Chen, Y., and Morgan, W. J. (1990). A nonlinear-rheology model for mid-ocean ridge axis topography, *J. Geophys. Res.* **95,** 17583–17604.

Chen, Y. J. (1992). Oceanic crustal thickness versus spreading rate, *Geophys. Res. Lett.* **19,** 753–756.

Christensen, N. I. (1979). Compressional wave velocities in rocks at high temperatures and pressures, critical thermal gradients, and crustal low velocity zones, *J. Geophys. Res.* **84,** 6489–6857.

Christensen, N. I., and Smewing, J. D. (1981). Geology and seismic structure of the northern section of the Oman ophiolite, *J. Geophys. Res.* **86,** 2655–2555.

Christenson, G. L., Purdy, G. M., and Fryer, G. J. (1992). Structure of young oceanic crust at the East Pacific Rise near 9°30′N, *Geophys. Res. Lett.* **19,** 1045–1048.

Collier, J., and Sinha, M. (1990). Seismic images of a magma chamber beneath the Lau Basin back-arc spreading centre, *Nature* **348,** 646–648.

Combarnous, M., and Bories, S. A. (1975). Hydrothermal convection in saturated porous media, *in* "Advances in Hydroscience." (V. T. Chow, ed.), pp. 231–307. Academic Press, New York.

Combarnous, M. (1978). Natural convection in porous media and geothermal systems, *in* "Proceedings, 6th Int. Heat Transfer Conf.," pp. 45–59.

Detrick, R. S., Buhl, P., Vera, E., Mutter, J., Orcutt, J., Madsen, J., and Brocher, T. (1987). Multi-channel seismic imaging of a crustal magma chamber along the East Pacific Rise, *Nature* **326,** 35–41.

Detrick, R. S., Mutter, J. C., Buhl, P., and Kim, I. I. (1990). No evidence from multichannel reflection data for a crustal magma chamber in the MARK area on the Mid-Atlantic Ridge, *Nature* **347,** 61–64.

Detrick, R. S., Harding, A. J., Kent, G. M., Orcutt, J. A., Mutter, J. C., and Buhl, P. (1993). Seismic structure of the southern East Pacific Rise, *Science* **259,** 499–503.

Dewey, J. F., and Kidd, W. S. F. (1977). Geometry of plate accretion, *Geol. Soc. Am. Bull.* **88,** 960–968.

Ewing, J. I., and Meyer, R. P. (1982). Rivera ocean seismic experiment (ROSE) overview, *J. Geophys. Res.* **87,** 8345–8358.

Francheteau, J., Armijo, R., Cheminee, J. L., Hekinian, R., Lonsdale, P., and Blum, N. (1992). Dyke complex of the East Pacific Rise exposed in the walls of Hess Deep and the structure of the upper oceanic crust, *Earth Planet. Sci. Lett.* **111,** 109–121.

Garmany, J. (1989). Accumulation of melt at the base of young oceanic crust, *Nature* **340,** 628–632.

Gregory, R., and Taylor, H. P. (1981). An oxygen isotope profile in a section of Cretaceous oceanic crust, Samail ophiolite, Oman: Evidence for d^{18}O buffering of the oceans by deep (>5 km) seawater-hydrothermal circulation at mid-ocean ridges, *J. Geophys. Res.* **86,** 2737–2755.

Grove, T. L., Kinzler, R. L., and Bryan, W. B. (1992). Fractionation of mid-ocean ridge basalt (MORB): Evidence for fractionation in the uppermost oceanic mantle, *EOS Trans. Am. Geophys. Union* **73,** 615.

Hale, L. D., Morton, C. J., and Sleep, N. H. (1982). Reinterpretation of seismic reflection data over the East Pacific Rise, *J. Geophys. Res.* **87,** 7707–7717.

Harding, A. J., Orcutt, J. A., Kappus, M. E., Vera, E. E., Mutter, J. C., Buhl, P., Detrick, R. S., and Brocher, T. M. (1989). The structure of the young oceanic crust at 13°N on the East Pacific Rise from expanding spread profiles, *J. Geophys. Res.* **94,** 12163–12196.

Harding, A. J., Kent, G. M., and Orcutt, J. A. (1993). A multichannel seismic investigation of upper crustal structure at 9°N on the East Pacific Rise: Implications for crustal accretion, *J. Geophys. Res.* **98,** 13925–13944.

Haymon, R. M., Fornari, D. J., Edwards, M. H., Carbotte, S., Wright, D., and Macdonald, K. C. (1991a). Hydrothermal vent distribution along the East Pacific Rise crest 9°.09′N–9°54′N and its relationship to magmatic and tectonic processes on fast-spreading mid-ocean ridges, *Earth Planet. Sci. Lett.* **104,** 512–534.

Haymon, R., Fornari, D., Von Damm, K., Edmond, J., Lilley,

M., Perfit, M., Shanks, W. C., III, Grebmeier, J., Lutz, R., Carbotte, S., Wright, D., and Smith, M., McLaughlin, E., Beedle, N., Seewald, J., Reudelhuber, D., Olson, E., Johnson, F. (1991b). Eruption of the EPR crest at 9°45′–54′N since late 1989 and its effects on the hydrothermal venting: Results of the ADVENTURE program, an ODP site survey with Alvin, *EOS Trans. Am. Geophys. Union* **72,** 480.

Henstock, T. J., Woods, A. W., and White, R. S. (1993). The accretion of oceanic crust by episodic sill intrusion, *J. Geophys. Res.* **98,** 4143–4162.

Herron, T. J., Ludwig, W. J., Stoffa, P. L., Kan, T. K., and Buhl, P. (1978). Structure of the East Pacific rise crest from multichannel seismic data, *J. Geophys. Res.* **83,** 798–804.

Hooft, E. E., and Detrick, R. S. (1993). The role of density in the accumulation of basaltic melts at mid-ocean ridges, *Geophys. Res. Lett.* **20,** 423–426.

Huang, P. Y., and Solomon, S. C. (1988). Centroid depths of mid-ocean ridge earthquakes: Dependence on spreading rate, *J. Geophys. Res.* **93,** 13445–13477.

Hubral, P. (1977). Time migration—Some ray theoretical aspects, *Geophys. Prosp.* **25,** 738–745.

Kampfmann, W., and Berckhemer, H. (1985). High temperature experiments on the elastic and anelastic behaviour of magmatic rocks, *Phys. Earth Planet. Int.* **40,** 223–247.

Kappel, E. S., and Ryan, W. B. F. (1986). Volcanic episodicity and a non-steady state rift valley along northeast Pacific spreading centers: Evidence from SeaMARC I, *J. Geophys. Res.* **91,** 13925–13940.

Kappus, M. E. (1991). "A Baseline for Upper Crustal Velocity Variations along the East Pacific Rise," Ph.D. Thesis, Univ. of California, San Diego.

Karson, J. A., Collins, J. A., and Casey, J. F. (1984). Geologic and seismic velocity structure of the crust/mantle transition in the Bay of Islands ophiolite complex, *J. Geophys. Res.* **89,** 6126–6138.

Kempner, W. C., and Gettrust, J. F. (1982). Ophiolites, synthetic seismograms, and ocean crustal structure. 1. Comparison of ocean bottom seismometer data and synthetic seismograms for the Bay of Islands ophiolite, *J. Geophys. Res.* **87,** 8447–8462.

Kent, G. M., Harding, A. J., and Orcutt, J. A. (1990). Evidence for a smaller magma chamber beneath the East Pacific Rise at 9°30′N, *Nature* **344,** 650–653.

Kent, G. M., Harding, A. J., and Orcutt, J. A. (1993a). Distribution of magma beneath the East Pacific Rise between the Clipperton transform and the 9°17′N Deval from forward modeling of CDP data, *J. Geophys. Res.* **98,** 13945–13970.

Kent, G. M., Harding, A. J., and Orcutt, J. A. (1993b). Distribution of magma beneath the East Pacific Rise near the 9°03′N overlapping spreading center from forward modeling of CDP data, *J. Geophys. Res.* **98,** 13971–13998.

Kent, G. M., Harding, A. J., and Orcutt, J. A. (1993c). The uniform accretion of oceanic crust south of the Garrett transform at 14°15′S on the East Pacific Rise, submitted for publication, *J. Geophys. Res.*

Khiratov, N. I., Lebedev, E. B., Dorfman, A. M., and Bagdassrov, N. S. (1983). Study of process of melting of the Kirgurich basalt by the wave method, *Geochimica* **9,** 1239–1246.

Klein, E. M., and Langmuir, C. H. (1987). Global correlations of ocean ridge basalt chemistry with axial depth and crustal thickness, *J. Geophys. Res.* **92,** 8089–8115.

Klein, E. M., and Langmuir, C. H. (1989). Local versus global variation in ocean ridge basaltic composition: A reply, *J. Geophys. Res.* **94,** 4241–4252.

Kong, L. S. L. (1990). "Variations in Structure and Tectonics along the Mid-Atlantic Ridge, 23°N and 26°N," Ph.D. Thesis, Mass. Inst. of Technol./Woods Hole Oceanogr. Inst., Woods Hole, MA.

Kong, L. S. L., Solomon, S. C., and Purdy, G. M. (1992). Microearthquake characteristics of a mid-ocean ridge along-axis high, *J. Geophys. Res.* **97,** 1659–1685.

Kuo, B.-Y., and Forsyth, D. W. (1988). Gravity anomalies of the ridge-transform system in the South Atlantic between 31 and 34.5°S: Upwelling centers and variations in crustal thickness, *Mar. Geophys. Res.* **10,** 205–232.

Kusznir, F. D., and Bott, M. H. P. (1976). A thermal study of oceanic crust, *Geophys. J. R. Astron. Soc.* **47,** 83–95.

Langmuir, C. H. (1989). Geochemical consequence of *in situ* crystallization, *Nature* **340,** 199–205.

Lewis, B. T. R., and Garmany, J. D. (1982). Constraints on the structure of the East Pacific Rise from seismic refraction data. *J. Geophys. Res.* **87,** 8417–8425.

Lin, J., and Parmentier, E. M. (1990). A finite amplitude necking model of rifting in brittle lithosphere, *J. Geophys. Res.* **95,** 4909–4923.

Lin, J., and Phipps Morgan, J. (1992). The spreading rate dependence of three-dimensional mid-ocean ridge gravity structure, *Geophys. Res. Lett.* **19,** 13–16.

Lin, J. Purdy, G. M., Schouten, H., Sempere, J.-C., and Zervas, C. (1990). Evidence from gravity data for focussed magmatic accretion along the Mid-Atlantic Ridge, *Nature* **344,** 627–632.

Lister, C. R. B. (1983). On the intermittency and crystallization mechanisms of sub-seafloor magma chambers, *Geophys J. R. Astron. Soc.* **73,** 351–365.

Macdonald, K. C., Sempere, J.-C., and Fox, P. J. (1984). East Pacific Rise from Siqueiros to Orozco Fracture Zones: Along-strike continuity of axial neovolcanic zone and structure and evolution of overlapping spreading centers, *J. Geophys. Res.* **89,** 6049–6069.

Macdonald, K. C. (1986). The crest of the Mid-Atlantic Ridge: Models for crustal generation processes and tectonics, *in* "The Geology of North America: The Western North Atlantic Region, v. M." (P. Vogt. and B. Tucholke, eds.), pp. 51–68, Geol. Soc. Am., Boulder, CO.

Macdonald, K. C., and Fox, P. J. (1988). The axial summit graben and cross-sectional shape of the East Pacific Rise as indicators of axial magma chambers and recent volcanic eruptions, *Earth Planet. Sci. Lett.* **88,** 119–131.

Macdonald, K. C., Fox, P. J., Perram, L. J., Eisen, M. F., Haymon, R. M., Miller, S. P., Carbotte, S. M., Cormier, M.-H., and Shor, A. N. (1990). A new view of the mid-ocean ridge from the behaviour of ridge-axis discontinuities, *Nature* **335,** 217–225.

Madsen, J. A., Forsyth, D. W., and Detrick, R. S. (1984). A new isostatic model for the East Pacific rise crest, *J. Geophys. Res.* **89,** 9997–100015.

Madsen, J. A., Detrick, R. S., Mutter, J. C., and Orcutt, J. A.

(1990). A two- and three-dimensional analysis of gravity anomalies associated with the East Pacific Rise at 9°N and 13°N, *J. Geophys. Res.* **95,** 4967–4987.

Manghnani, M. H., Sato, H., and Chandra, S. R. (1986). Ultrasonic velocity and attenuation measurements on basalt melts to 1500°C: Role of composition and structure in the viscoelastic properties, *J. Geophys. Res.* **91,** 9333–9342.

Marsh, B. D. (1981). On the crystallinity, probability of occurrence and rheology of lava and magma, *Contrib. Mineral. Petrol.* **78,** 85–98.

Marsh, B. D. (1989). Magma chambers, *Annu. Rev. Earth Planet. Sci.* **17,** 439–474.

McClain, J. S., Orcutt, J. A., and Burnett, M. (1985). The East Pacific Rise in cross section: A seismic model, *J. Geophys. Res.* **90,** 8627–8639.

McKenzie, D. P. (1979). Finite deformation during fluid flow, *Geophys. J. R. Astron. Soc.* **58,** 689–715.

McKenzie, D. (1984). The generation and compaction of partially molten rock, *J. Petrol.* **25,** 713–765.

Moores, E. M. (1982). Origin and emplacement of ophiolites, *Rev. Geophys. Space Phys.* **20,** 735–760.

Morton, J. L., and Sleep, N. H. (1985a). A mid-ocean ridge thermal model: Constraints on the volume of axial hydrothermal flux, *J. Geophys. Res.* **90,** 11345–11353.

Morton, J. L., and Sleep, N. H. (1985b). Seismic reflections from the Lau Basin magma chamber, *in* "Geology and Offshore Resources of Pacific Island Arcs—Tonga Region." (D. W. Scholl and T. L. Vallier, eds.), pp. 441–453, Circum Pacific Council for Energy and Mineral Resources Earth Science Series.

Morton, J. L., Sleep, N. H., Normark, W. R., and Tompkins, D. H. (1987). Structure of the Southern Juan de Fuca Ridge from reflection records, *J. Geophys. Res.* **92,** 11315–11326.

Murase, T., and McBirney, A. R. (1973). Properties of some common igneous rocks and their melts at high temperatures, *Geol. Soc. Am. Bull.* **84,** 3563–3592.

Mutter, J. C., Barth, G. A., Buhl, P., Detrick, R. S., Orcutt, J., and Harding, A. (1988). Magma distribution across ridge-axis discontinuities on the East Pacific Rise from multichannel seismic images, *Nature* **336,** 156–158.

Natland, J. H. (1980). Effects of axial magma chambers beneath spreading centers on the compositions of basaltic rocks, *in* "Initial Report of the Deep Sea Drilling Project," (B. R. Rosendahl and R. Hekinian, eds.), v. 54, pp. 833–850.

Nehlig, P., and Juteau, T. (1988). Flow porosities, permeabilities, and preliminary data on fluid inclusions and fossil thermal gradients in the crustal sequence of the Sumail ophiolite (Oman), *Tectonophysics* **151,** 199–221.

Neumann, G. A., and Forsyth, D. W. (1993). The paradox of the axial profile: Isostatic compensation along the axis of the Mid-Atlantic Ridge? *J. Geophys. Res.* **98,** 17891–17904.

Nicolas, A., Reuber, I., and Benn, K. (1988). A new magma chamber model based on structural studies in the Oman ophiolite, *Tectonophysics* **151,** 87–105.

Nicolas, A. (1989). "Structures of Ophiolites and Dynamics of Oceanic Lithosphere," Kluwer, New York.

Niu, Y. (1992). "Mid-Ocean Ridge Magmatism: Style of Mantle Upwelling, Partial Melting, Crustal Level Processes, and Spreading Rate Dependence—A Petrologic Approach," Ph.D. Thesis, Univ. of Hawaii.

Niu, Y., and Batiza, R. (1993). Chemical variation trends at fast- and slow-spreading mid-ocean ridges, *J. Geophys. Res.* **98,** 7887–7902.

Orcutt, J. A., Kennett, B. L. N., Dorman, L. M., and Prothero, W. A. (1975). Evidence for a low velocity zone underlying a fast spreading rise crest, *Nature* **256,** 475–476.

Pallister, J. S., and Hopson, C. A. (1981). Samail ophiolite plutonic suite: Field variations, phase variations, cryptic variation and layering, and a model of a spreading ridge magma chamber, *J. Geophys. Res.* **86,** 2593–2644.

Parmentier, E. M., and Phipps Morgan, J. (1990). The spreading rate dependence of three-dimensional structure in oceanic spreading centres, *Nature* **348,** 325–328.

Phipps Morgan, J. (1991). Mid-ocean ridge dynamics: Observations and theory, *Rev. Geophys., U.S. National Report to IUGG, Suppl.,* 807–822.

Phipps Morgan, J., Parmentier, E. M., and Lin, J. (1987). Mechanisms for the origin of mid-ocean ridge axial topography: Implications for the thermal and mechanical structure at accreting plate boundaries, *J. Geophys. Res.* **92,** 12823–12836.

Phipps Morgan, J., and Chen, Y. J. (1993). The genesis of oceanic crust: Magma injection, hydrothermal circulation, and crustal flow, *J. Geophys. Res.* **98,** 6283–6298.

Purdy, G. M., and Detrick, R. S. (1986). Crustal structure of the Mid-Atlantic Ridge at 23°N from seismic refraction studies, *J. Geophys. Res.* **91,** 3739–3762.

Purdy, G. M., Kong, L. S. L., Christeson, G. L., and Solomon, S. (1992). Relationship between spreading rate and the seismic structure of mid-ocean ridges, *Nature* **355,** 815–817.

Quick, J. E., and Denlinger, R. P. (1993). Ductile deformation and the origin of layered gabbro in ophiolites, *J. Geophys. Res.* **98,** 14015–14028.

Reid, I. D., Orcutt, J. A., and Prothero, W. A. (1977). Seismic evidence for a narrow zone of partial melting underlying the East Pacific Rise at 21°N, *Geol. Soc. Am. Bull.* **88,** 678–682.

Rohr, K. M., Milkereit, B., and Yorath, C. J. (1988). Asymmetric deep crustal structure across the Juan de Fuca ridge, *Geology* **16,** 533–537.

Rosendahl, B. R., Raitt, R. W., Dorman, L. M., Bibee, L. D., Hussong, D. M., and Sutton, G. H. (1976). Evolution of oceanic crust. 1. A physical model of the East Pacific Rise crest derived from seismic refraction data, *J. Geophys. Res.* **81,** 5294–5304.

Ryan, M. P. (1987). Neutral buoyancy and the mechanical evolution of magmatic systems, *in* "Magmatic Processes: Physiochemical principles," The Geochemical Society, Special Publication 1, pp. 259–287.

Ryan, M. P. (1994). Neutral-buoyancy controlled magma transport and storage in mid-ocean ridge magma reservoirs and their sheeted dike complex: A summary of basic relationships, *in* "Magmatic Systems" (M. P. Ryan, ed.), Academic Press, San Diego.

Sato, H., Sacks, I. S., and Murase, T. (1989). The use of laboratory velocity data for estimating temperature and partial melt fraction in the low-velocity zone: Comparison with heat flow and electrical conductivity studies, *J. Geophys. Res.* **94,** 5689–5704.

Sato, H., and Sacks, I. S. (1990). Magma generation in the upper mantle inferred from seismic measurements in peridotite at high pressure and temperature, *in* "Magma Transport and Storage" (M. P. Ryan, ed.), pp. 277–292. Wiley, Chichester, England.

Scott, D. S. (1992). Small-scale convection and mantle melting beneath mid-ocean ridges, *in* "Mantle Flow and Melt Generation at Mid-ocean Ridges," (J. Phipps Morgan, D. K. Blackman, and J. Sinton, eds.), AGU Monograph **71,** 327–352.

Sempere, J.-C., Macdonald, K. C., Miller, S. P., and Shure, L. (1988). Detailed study of the Brunhes/Matuyama boundary on the East Pacific Rise at 19°30′S: Implications for crustal emplacement processes at an ultra fast spreading center, *Mar. Geophys. Res.* **9,** 1–25.

Sinton, J. M., and Detrick, R. S. (1992). Mid-ocean ridge magma chambers, *J. Geophys. Res.* **97,** 197–216.

Sleep, N. H. (1975). Formation of oceanic crust: Some thermal constraints, *J. Geophys. Res.* **80,** 4037–4042.

Sleep, N. H. (1978). Thermal structure and kinematics of the mid-ocean ridge axis, some implications to basaltic volcanism, *Geophys. Res. Lett.* **5,** 426–428.

Sleep, N. H. (1991). Hydrothermal circulation, anhydrite precipitation, and thermal structure at ridge axes, *J. Geophys. Res.* **96,** 2375–2387.

Small, C., and Sandwell, D. T. (1989). An abrupt change in ridge axis gravity with spreading rate, *J. Geophys. Res.* **94,** 17383–17392.

Smewing, J. D. (1981). Mixing characteristics and compositional differences in mantle-derived melts beneath spreading axes: Evidence from cyclically layered rocks in the ophiolite of North Oman, *J. Geophys. Res.* **86,** 2645–2659.

Solomon, S. C., and Toomey, D. R. (1992). The structure of mid-ocean ridges, *Annu. Rev. Earth Planet. Sci.* **20,** 329–364.

Sparks, D. W., and Parmentier, E. M. (1991). Melt extraction from the mantle beneath spreading centers, *Earth Planet. Sci. Lett.* **105,** 368–377.

Sparks, D. W., and Parmentier, E. M. (1994). The generation and migration of partial melt beneath spreading centers, *in* "Magmatic Systems," (M. P. Ryan, ed.), Academic Press, San Diego.

Spudich, P., and Orcutt, J. A. (1980). Petrology and porosity of an oceanic crustal site: Results from wave form modeling of seismic refraction data, *J. Geophys. Res.* **85,** 1409–1433.

Stevenson, J. M. (1992). "Applications of Marine Gravimetry to Mid-oceanic Spreading Centers and Volcanos," Ph.D. Thesis, Univ. of California, San Diego.

Tapponier, R., and Francheteau, J. (1978). Necking of the lithosphere and the mechanics of slowly accreting plate boundaries, *J. Geophys. Res.* **83,** 3955–3970.

Tolstoy, M., Harding, A. J., and Orcutt, J. A. (1992). An explanation for "bull's eye" mantle bouguer anomalies on the southern Mid-Atlantic Ridge, *EOS Trans. Am. Geophys. Union (Suppl.)* **73,** 495.

Toomey, D. R., Solomon, S. C., and Purdy, G. M. (1988). Microearthquakes beneath the median valley of the Mid-Atlantic Ridge near 23°N: Tomography and tectonics, *J. Geophys. Res.* **93,** 9093–9112.

Toomey, D. R., Purdy, G. M., Solomon, S. C., and Wilcock, W. S. D. (1990). The three-dimensional seismic velocity structure of the East Pacific Rise near latitude 9°30′N, *Nature* **347,** 639–645.

Turcotte, D. L., and Schubert, G. (1982). "Geodynamics," Wiley, New York.

Turcotte, D. L., and Phipps Morgan, J. (1992). The physics of magma migration and mantle flow beneath a mid-ocean ridge, *in* "Mantle Flow and Melt Generation at Mid-ocean Ridges" (J. Phipps Morgan, D. K. Blackman, and J. Sinton, eds.), AGU Monograph **71,** pp. 155–182.

Vera, E. E., Mutter, J. C., Buhl, P., Orcutt, J. A., Harding, A. J., Kappus, M. E., Detrick, R. S., and Brocher, T. (1990). The structure of 0- to 0.2-m.y.-old oceanic crust at 9°N on the East Pacific Rise from expanded spread profiles, *J. Geophys. Res.* **95,** 15529–15556.

Wang, X., and Cochran, J. R. (1993). Gravity anomalies, isostasy, and mantle flow at the East Pacific Rise crest, submitted for publication, *J. Geophys. Res.*

White, D. J., and Clowes, R. M. (1990). Shallow crustal structure beneath the Juan de Fuca ridge from 2-D seismic refraction tomography, *Geophys. J. Int.* **100,** 349–376.

Wilcock, W. S. D., Solomon, S. C., Purdy, G. M., and Toomey, D. R. (1992). The seismic attenuation structure of a fast spreading mid-ocean ridge, *Science* **258,** 1470–1474.

Wilson, D. S., Clague, D. A., Sleep, N. H., and Morton, J. L. (1988). Implications of magma convection for the size and temperature of magma chambers at fast spreading ridges, *J. Geophys. Res.* **93,** 11974–11984.

Wilson, D. S. (1992). Focussed upwelling beneath mid-ocean ridges: Evidence from seamount formation and isostatic compensation of topography, *Earth Planet. Sci. Lett.* **113,** 41–55.

Chapter 8 | Deep Structure of Island Arc Magmatic Regions as Inferred from Seismic Observations

Akira Hasegawa and Dapeng Zhao

Overview

Recent seismic observations provide possible evidence for deep-seated magmatic activity in some of the subduction zones of the world. Tomographic inversions for seismic wave velocity structures delineate low-velocity zones in the crust and in the mantle wedge beneath active volcanoes. Seismic attenuation tomography also delineates similar zones of low-Q value in the crust and mantle wedge beneath active volcanoes, although they are less clearly imaged due to lower spatial resolution. The most typical example is that beneath the northeastern Japan arc, where inclined P-wave low-velocity zones have been clearly imaged using data acquired through microearthquake observations with dense networks. The low-velocity zones with 2–6% velocity lows are continuously distributed from the upper crust right under active volcanoes to a depth of 100–150 km in the mantle wedge, their thicknesses being about 50 km. They are approximately parallel to the dip (~30°) of the underlying subducted Pacific plate and their lower edges are 30–60 km apart from the top of the subducted plate. These low-velocity zones probably reflect the pathway of magma ascent from a deeper part of the mantle to the Earth's surface, along which additional evidence for deep magmatic activity has been found.

Notation

		Units
C_{ij}	coefficient of power series expressing configuration of ith velocity discontinuity	
$\mathbf{G}$	matrix consisting of partial derivatives of travel time	
H_i	depth to ith velocity discontinuity	km
Q	quality factor that is a dimensionless measure of anelasticity	dimensionless
Q_p	P-wave quality factor	dimensionless
Q_s	S-wave quality factor	dimensionless
T_{ij}^{obs}	observed arrival time for ith earthquake at jth station	s
T_{ij}^{cal}	calculated arrival time based on a velocity model	s
V	seismic wave velocity	$km \cdot s^{-1}$
V_n	seismic wave velocity at nth grid point	$km \cdot s^{-1}$
$\mathbf{d}$	column vector consisting of travel time residuals	
$\mathbf{e}$	error vector	
e_{ij}	higher-order term of perturbation and observation errors	
h	depth from the Earth's surface	km
t_{ij}	travel time residual for ith earthquake at jth station	
Δh_i	correction term for focal depth of ith earthquake	
$\Delta\mathbf{m}$	column vector consisting of correction terms for source and medium parameters	
ΔT_{0i}	correction term for origin time of ith earthquake	
ΔV_n	correction term for seismic wave velocity at nth grid point	
$\Delta\phi_i$	correction term for latitude of ith earthquake	
$\Delta\lambda_i$	correction term for longitude of ith earthquake	
ϕ	latitude	°, degrees
λ	longitude	°, degrees

Introduction

New oceanic plates are formed at mid-ocean ridges by the upwelling of hot mantle material, producing mid-ocean ridge magmatism. Conservation of mass suggests that an equal amount of material returns downward, back into the Earth's mantle. This consumption of the oceanic plates takes place mainly in subduction zones, where heavier oceanic plates subduct into the mantle beneath the lower-density continental plates. The subduction of cold, and hence relatively heavy, oceanic plates causes high seismic and volcanic activity along the subduction zones. The fundamental paradox of subduction zone magmatism is that thus both the high heat-flow values and the abundance of melt generation occur just above the relatively cold subducting plate, which is considered to be an enormous heat sink (e.g., Uyeda, 1982; Davies and Stevenson, 1992).

Seismic waves, passing through the Earth's interior and finally arriving at seismic stations on the surface, provide information on the physical properties of the materials along their ray paths within the Earth. Many studies using seismic waves have been conducted for estimating both the seismic wave velocity structure and the seismic attenuation (anelasticity) structure of the Earth's interior. Seismic properties of rocks and minerals, which compose the crust and the mantle, at high pressures and temperatures have also been studied in the laboratory. Laboratory measurements show that seismic wave velocity in the mantle rocks decreases with increasing temperature, and in particular, it drops abruptly when the degree of partial melting exceeds some threshold amount (e.g., Murase and Kushiro, 1979; Murase and Fukuyama, 1980; Sato and Sacks, 1990). Seismic anelasticity in peridotite also depends on temperature, and Q values decrease rapidly with increasing temperature even in the subsolidus temperature range (Sato and Sacks, 1990). Of course, in some cases, decreases in seismic wave velocity or in the Q value are caused not by high temperatures but by other factors such as chemical composition changes. Nevertheless, estimations of the detailed seismic structure are critical to develop an understanding of the thermal and mechanical state within the Earth, and they thus yield important information on the deep structure of magmatic regions.

Recent seismological studies, and new seismic imaging techniques such as seismic tomography, have revealed a more precise image of the seismic structure of the subducting plate and the mantle wedge above it (e.g., Hasegawa *et al.*, 1991; Zhao *et al.*, 1992). Possible evidence for deep-seated magmatic activity in subduction zones has been obtained from these studies. For example, seismic low-velocity zones extending from the upper crust just beneath active volcanoes to the mantle wedge are clearly delineated from tomographic studies. Taking into account the velocity–temperature relation of the mantle rocks obtained from the laboratory measurements, this fact suggests the existence of an ascending flow of hot mantle wedge material beneath active volcanoes in subduction zones. In the present chapter, we briefly describe some of the results of these recent seismological studies on the deep structure of island arc magmatic regions.

Three-Dimensional Seismic Velocity Structure

A number of studies of the three-dimensional seismic velocity structure on both local and regional scales have been conducted in various regions of the world since the pioneering works of seismic tomography by Aki and Lee (1976) and Aki *et al.* (1977). The most thorough studies of the three-dimensional seismic velocity structure of subduction zones conducted so far are those of the Japanese subduction zone, partly because dense seismic networks have been deployed by several institutions in this region. On a regional scale, Hirahara (1977) and Hirahara and Mikumo (1980) investigated the three-dimensional P-wave velocity structure of the crust and upper mantle beneath the Japanese Islands by using teleseismic and local earthquake data, and pointed out the existence of a P-wave high-velocity zone corresponding to the subducting Pacific plate.

The inclined high-hvelocity zone corresponding to the subducting plate has also been detected by the work of Kamiya *et al.* (1989) and Zhou and Clayton (1990), who estimated the three-dimensional P-wave velocity structure to a depth of 1200 km beneath the Japanese Islands by using teleseismic data. On the basis of the estimated tomographic images, they discussed the prospects for further penetration of the subducting Pacific plate into the lower mantle, although the conclusions of the two studies opposed each other. Van der Hilst *et al.* (1991) have imaged the P-wave velocity structure of the mantle to a depth of 1200 km beneath northwest Pacific island arcs by

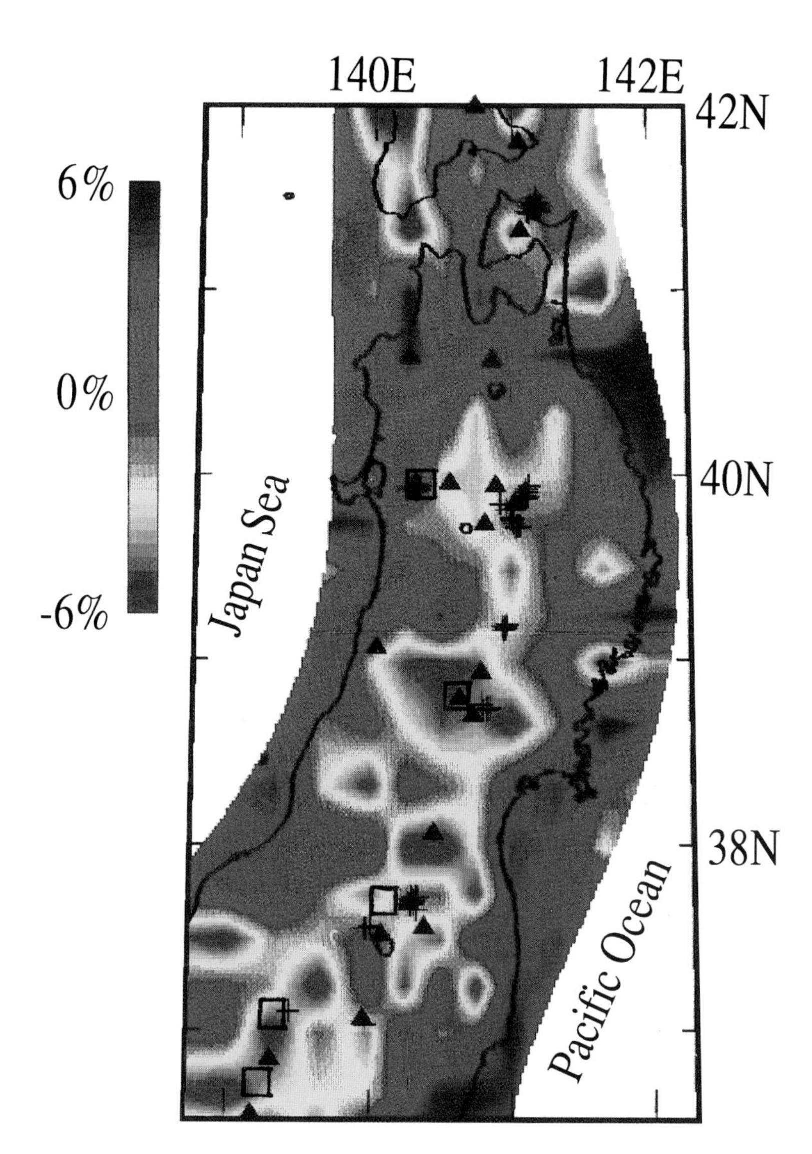

Figure 1 Horizontal section of the fractional P-wave velocity perturbations at the 40-km-depth level beneath northeastern Japan (Hasegawa *et al.*, 1991; Zhao *et al.*, 1992). Velocity perturbations are shown as percentages by color scale on the left. Red and blue indicate low and high P-wave velocities, respectively. Solid triangles, crosses, and open squares are the locations of active volcanoes, deep low-frequency microearthquakes, and midcrustal S-wave reflectors, respectively.

Figure 2 Vertical cross sections of fractional P-wave velocity perturbations along the lines (a) AA′, (b) BB′, and (c) CC′ in the insert map (Hasegawa *et al.*, 1991; Zhao *et al.*, 1992). Velocity perturbations are shown as percentages according to the same color scale as in Fig. 1. Locations of the trench axis, active volcanoes, and land area are shown at the top by inverted triangles, red triangles, and thick horizontal lines, respectively. The Conrad and Moho discontinuities and the top and bottom of the subducting Pacific plate are denoted by thick lines. Open and red circles denote microearthquakes and deep low-frequency microearthquakes within a 60-km width along each line.

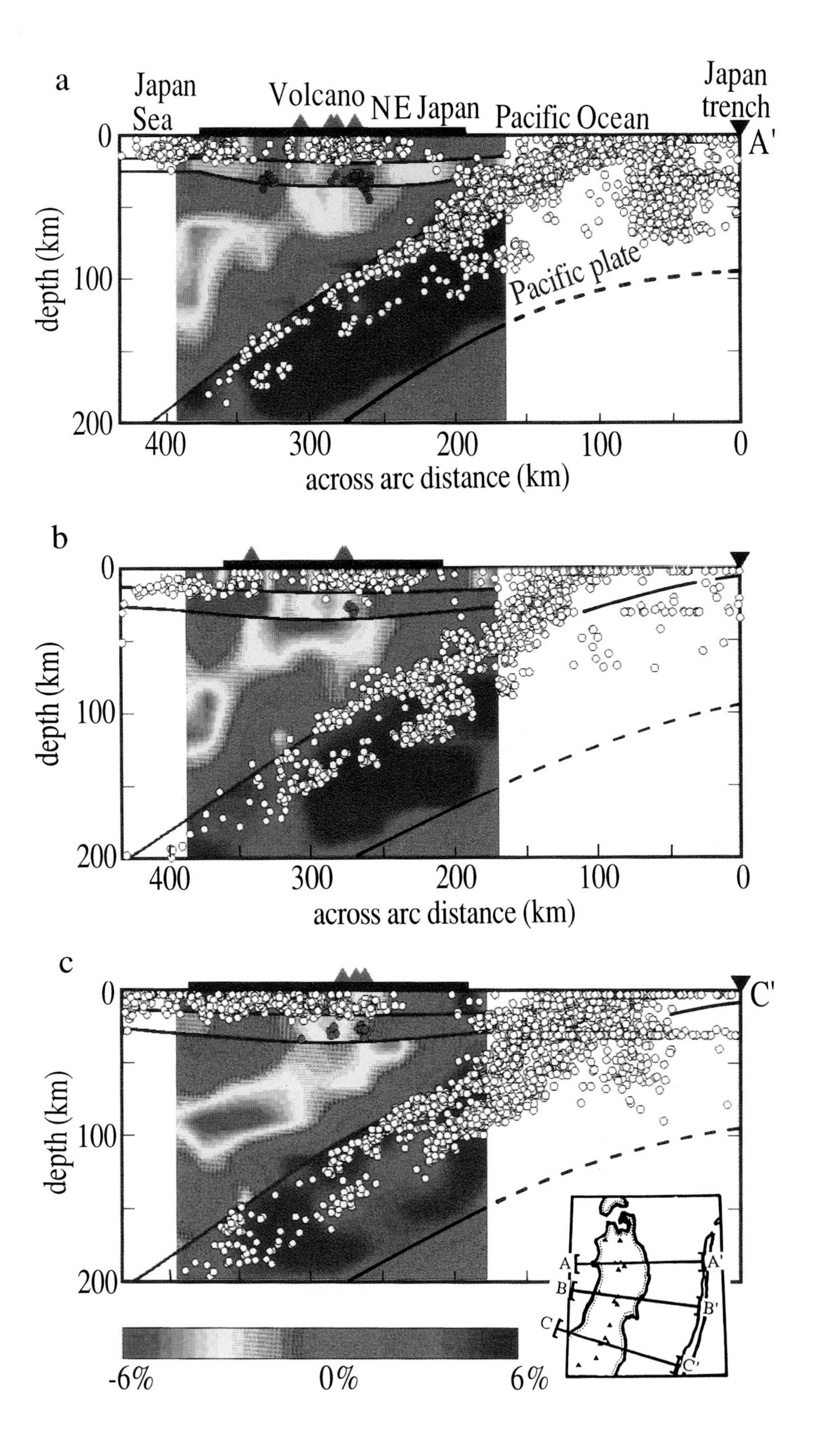
a
Japan Sea
Volcano
NE Japan
Pacific Ocean
Japan trench
A'
Pacific plate
depth (km)
0
100
200
400
300
200
100
0
across arc distance (km)
b
depth (km)
across arc distance (km)
c
C'
depth (km)
-6%
0%
6%
A
A'
B
B'
C
C'

using a more realistic background earth model and surface-reflected seismic phase data as well as direct phase data. Their result has yielded higher-quality images of the subducted Pacific plate and has shown that the subducted plate beneath the northeastern Japan arc and the Izu–Bonin arc does not sink into the lower mantle but is deflected at the boundary between the upper and lower mantle. Their images of slab deflection at the 670-km discontinuity are in good agreement with the result of Fukao *et al.* (1992), who have also yielded higher-quality images by solving simultaneously for updates of the one-dimensional background earth model and for aspherical variations in P-wave velocity.

Many tomographic studies on a local scale have been carried out in several regions beneath the Japanese Islands by using arrival time data from local earthquakes (e.g., Miyamachi and Moriya, 1984; Nakanishi, 1985; Hasemi *et al.*, 1984; Obara *et al.*, 1986; Horie and Aki, 1982; Ishida and Hasemi, 1988; Hirahara *et al.*, 1989). These studies have also indicated the existence of the inclined high-velocity zones corresponding to the subducting Pacific and Philippine Sea plates beneath the Japanese Islands. Hirahara *et al.* (1989) investigated the P-wave velocity structure to a depth of about 200 km beneath central Japan and detected low-velocity bodies in the mantle wedge beneath the active volcanoes. One of these low-velocity bodies coincides with an S-wave anisotropic body as estimated from shear wave splitting analyses, suggesting the presence of magma-filled crack alignment in the body (Ando *et al.*, 1983). Tomographic studies by Hasemi *et al.* (1984) and by Obara *et al.* (1986) give a detailed three-dimensional P-wave velocity structure to a depth of about 200 km beneath northeastern Japan and have revealed low-velocity zones continuously distributed in the crust and in the mantle wedge just beneath the active volcanoes.

Zhao *et al.* (1992) have obtained more distinct P-wave tomographic images of the crust and the upper mantle beneath northeastern Japan by developing a new tomographic method (Figs. 1 and 2). This study has updated the works by Hasemi *et al.* (1984) and Obara *et al.* (1986) by improving the resolution. It has clearly delineated P-wave low-velocity zones, which are inclined to the west and are continuously distributed from the uppermost mantle to the upper crust beneath the active volcanoes in northeastern Japan. The results of this study are described in some detail in the next section because they provide the most physically obvious and most typical images of the crust and upper mantle velocity structure beneath subduction zones.

Compressional wave low-velocity zones in the crust and in the mantle wedge beneath active volcanoes similar to those in the Japanese Islands have also been found in the Cascadia subduction zone and in the Alaska–Aleutian subduction zone, where extensive studies of three-dimensional seismic velocity structures on regional or local scales have been initiated. Inversion of teleseismic P-wave arrival time data has led to the successful detection of an inclined high-velocity zone that corresponds to the subducting Juan de Fuca plate beneath Washington and Oregon (Michaelson and Weaver, 1986; Rasmussen and Humphreys, 1988; Harris *et al.*, 1991). Harris *et al.* (1991) have also detected a low-velocity zone in the crust and partly in the uppermost mantle beneath the Crater Lake volcano region. By inverting teleseismic P-wave travel time data, Benz *et al.* (1992) have imaged a steeply dipping portion of the subducting Gorda plate as a +5% velocity high to a depth near 270 km and a low-velocity zone in the crust and in the mantle wedge to a depth of about 100 km beneath the active volcanic field that includes Mt. Lassen, of northern California. In related Cascadia work, Iyer *et al.* (1990) have imaged the three-dimensional distribution of low-velocity regions beneath the Newberry volcano in Oregon to a depth of 26 km by the tomographic inversion of explosion-generated compressional wave travel time residuals.

Zhao and Christensen (1992) have applied the tomographic approach of Zhao *et al.* (1992) to local earthquake data in central and southern Alaska and have estimated the three-dimensional seismic velocity structure of the crust and upper mantle to a depth of 200 km. They have clearly delineated the subducting Pacific plate having a 4–6% P-wave velocity high and a thickness of about 50 km. Low-velocity zones are also imaged in the crust as well as the mantle wedge beneath active volcanoes. Although crustal low-velocity zones and mantle wedge low-velocity zones beneath volcanic fields have not been clearly imaged in other subduction zones, future studies of seismic tomography on regional or local scales with high spatial resolution will open the possibility for detection, just as that described in this report.

P-Wave Velocity Structure beneath the Northeastern Japan Arc

In this section, a P-wave tomographic study in the northeastern Japan arc (Hasegawa *et al.*, 1991; Zhao *et al.*, 1992) is described briefly as an example showing a typical image of the crust and upper mantle in subduction zones. Details of the tomographic inversion are shown in Zhao *et al.* (1992).

Most of the conventional tomographic studies conducted on local scales have neglected the effect of the complex shape of seismic velocity discontinuities such as the Moho, the Conrad discontinuity, and the upper boundary of the subducting plate itself. This neglect distorts the estimated tomographic images, especially near the discontinuities. Calculated ray paths of seismic waves based on a simple one-dimensional velocity model usually adopted in the conventional tomographic studies deviate considerably from the real paths at large hypocentral distances and in highly inhomogeneous regions such as subduction zones. This deviation also seriously distorts the estimated tomographic images.

A new method of seismic tomography has been developed by Zhao *et al.* (1992) to solve these problems. This method copes with a general velocity structure with complex velocity discontinuities in the modeling space and with three-dimensional velocity variations in each layer bounded by the velocity discontinuities. Depth distributions of the velocity discontinuities are expressed in two ways. One is to define the velocity discontinuity by using power series of latitude and longitude. The depth to the ith velocity discontinuity H_i is expressed as

$$H_i(\phi, \lambda) = C_{i1} + C_{i2}\phi + C_{i3}\lambda + C_{i4}\phi^2 + C_{i5}\lambda^2 + C_{i6}\phi\lambda + \cdots, \qquad (1)$$

where ϕ and λ are latitude and longitude, respectively, and C_{ij} are coefficients of power series for the ith discontinuity. The other approach is to define the discontinuity by using a two-dimensional grid. Once a depth distribution of grids is given for a discontinuity, the depth to the discontinuity at any location can be calculated by linearly interpolating the depths at four surrounding grid nodal points.

Three-dimensional grids are arranged individually in every layer bounded by the discontinuities, and the velocity at each grid point is taken to be an unknown parameter. A velocity $V(\phi, \lambda, h)$ at any point (ϕ, λ, h) in the modeling space can be calculated by linearly interpolating the velocities $V(\phi_i, \lambda_j, h_k)$ at eight surrounding grid points (ϕ_i, λ_j, h_k) as

$$V(\phi, \lambda, h) = \sum_{i=1}^{2}\sum_{j=1}^{2}\sum_{k=1}^{2} V(\phi_i, \lambda_j, h_k) \cdot \left[\left(1 - \left|\frac{\phi - \phi_i}{\phi_2 - \phi_1}\right|\right) \cdot \left(1 - \left|\frac{\lambda - \lambda_j}{\lambda_2 - \lambda_1}\right|\right) \cdot \left(1 - \left|\frac{h - h_k}{h_2 - h_1}\right|\right)\right], \qquad (2)$$

where h is the depth from the Earth's surface. An efficient three-dimensional ray tracing algorithm that iteratively uses a pseudo-bending technique (Um and Thurber, 1987) and Snell's law has been developed. This algorithm calculates the ray paths and the travel times of seismic waves rapidly and accurately in the general velocity structure mentioned above.

Starting with the initial hypocenter locations, the origin times for a set of earthquakes, and an initial velocity model, the correction terms for source and medium parameters are estimated iteratively in a way that the observed arrival time data are best explained in the least-squares sense. The observation equation is written as

$$T_{ij}^{\rm obs} = T_{ij}^{\rm cal} + \left(\frac{\partial T}{\partial \phi}\right)_{ij}\Delta\phi_i + \left(\frac{\partial T}{\partial \lambda}\right)_{ij}\Delta\lambda_i + \left(\frac{\partial T}{\partial h}\right)_{ij}\Delta h_i + \Delta T_{0i} + \sum_n \frac{\partial T}{\partial V_n}\Delta V_n + e_{ij}, \qquad (3)$$

where $T_{ij}^{\rm obs}$ is the observed arrival time for ith earthquake at jth station; $T_{ij}^{\rm cal}$ is the calculated arrival time based on the velocity model; ϕ_i, λ_i, h_i, T_{0i} are the latitude, longitude, focal depth, and origin time of ith earthquake; $\Delta\phi_i$, $\Delta\lambda_i$, Δh_i, ΔT_{0i}

are the correction terms for the above parameters; $(\partial T/\partial \phi)_{ij}$, $(\partial T/\partial \lambda)_{ij}$, $(\partial T/\partial h)_{ij}$ are the partial derivatives of travel time with respect to latitude, longitude, and focal depth; V_n is the velocity at nth grid point; ΔV_n is the correction term for velocity at nth grid point; $(\partial T/\partial v_n)$ is the partial derivative of travel time with respect to velocity parameters; and e_{ij} is the higher-order term of perturbation and observation errors. Travel times and ray paths are calculated by the three-dimensional ray tracing algorithm. Partial derivatives with respect to hypocenters can be calculated analytically. Velocity parameter derivatives are calculated by using a linear interpolation function (2) expressing the velocity field.

The travel time residual t_{ij} is written as

$$t_{ij} = T_{ij}^{\text{obs}} - T_{ij}^{\text{cal}}. \tag{4}$$

The travel time residuals form a whole set of data with a column vector $\mathbf{d}$ of dimension N, where N is the number of arrival time data. Correction terms for source and medium parameters, defined by a column vector $\Delta\mathbf{m}$, are expressed as

$$\begin{aligned}\Delta\mathbf{m}^T = (&\Delta\phi_1, \Delta\lambda_1, \Delta h_1, \Delta T_{01}, \cdots, \\ &\Delta\phi_M, \Delta\lambda_M, \Delta h_M, \Delta T_{0M}, \\ &\Delta V_1, \Delta V_2, \cdots, \Delta V_K),\end{aligned} \tag{5}$$

where M and K are the numbers of earthquakes and grid points, respectively. The data vector $\mathbf{d}$ and the unknown parameter vector $\Delta\mathbf{m}$ are related as

$$\mathbf{d} = \mathbf{G}\Delta\mathbf{m} + \mathbf{e}, \tag{6}$$

where $\mathbf{e}$ is an error vector, and $\mathbf{G}$ is a matrix with dimension $N \times (4M + K)$ whose elements consist of the partial derivatives. There are several approaches to solving Eq. (6). Here, a conjugate gradient solver, the LSQR algorithm of Paige and Saunders (1982), is used to solve the extremely large and sparse system of observation equations arising from the inversion problems. The algorithm has been used by several researchers (e.g., Nolet, 1985; Spakman and Nolet, 1988; van der Hilst *et al.*, 1991) and is confirmed to be an efficient algorithm to solve large inversion problems.

Zhao *et al.* (1992) have obtained P-wave tomographic images of the crust and upper mantle to a depth of 200 km beneath northeastern Japan by applying the method described earlier to the arrival time data of seismic waves observed by the seismic networks of Tohoku University and those of several other national universities in Japan (Hokkaido, Hirosaki, Tokyo, Nagoya, Kyoto, Kochi, and Kyushu University). The arrival time data used are 42,494 first P-wave data, 8141 first S-wave data, and 284 P-to-S and S-to-P converted wave data at the velocity discontinuities from 1200 shallow, intermediate-depth, and deep earthquakes. The medium under study is divided into four layers by the Conrad discontinuity, the Moho, and the upper boundary of the subducted Pacific plate. The four layers correspond to the upper crust, the lower crust, the mantle wedge, and the mantle below the upper plate boundary. The depth distributions of the Conrad, the Moho, and the top of the plate are fixed and expressed by continuous functions of spatial locations so as to coincide with the result of previous studies (Hasegawa *et al.*, 1983; Matsuzawa *et al.*, 1986, 1990; Zhao *et al.*, 1990). One, one, sixteen, and sixteen layers of grid nets are arranged in the upper crust, the lower crust, the mantle wedge, and the mantle below the upper plate boundary, respectively. The separation between grid points is 25–33 km in both vertical and horizontal directions.

Two kinds of resolution tests are made to evaluate the resolution of the tomographic images obtained. One is checkerboard resolution tests with various wave lengths of velocity change, where the basic idea of the testing is given in Humphreys and Clayton (1988). In the tests, positive and negative velocity perturbations are assigned at regular intervals to the three-dimensional blocks (here, grid points) of a homogeneous velocity model. Inversion of synthetic arrival time data calculated from this checkerboard velocity model by using the inversion algorithm provides images from which it is easy to understand where the resolution is good or poor. Figure 3 shows the result of the checkerboard resolution test. P-wave velocity perturbations obtained by the resolution test are plotted on the horizontal sections at depths of 10, 24, 40, 65, 90, and 115 km, respectively. The originally assigned patterns of the checkerboard are well reconstructed for the study area. Reconstructed amplitudes of velocity anomalies are more than 80% of the original amplitudes for most grid points.

The other resolution test is to take the tomographic images obtained from the actual data set as the synthetic velocity model. Comparing the inverted images from the synthetic arrival time data

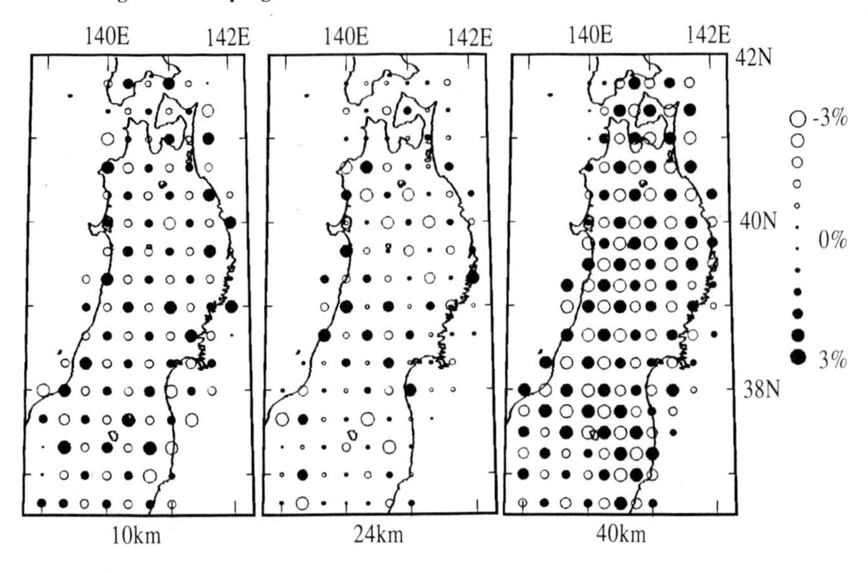

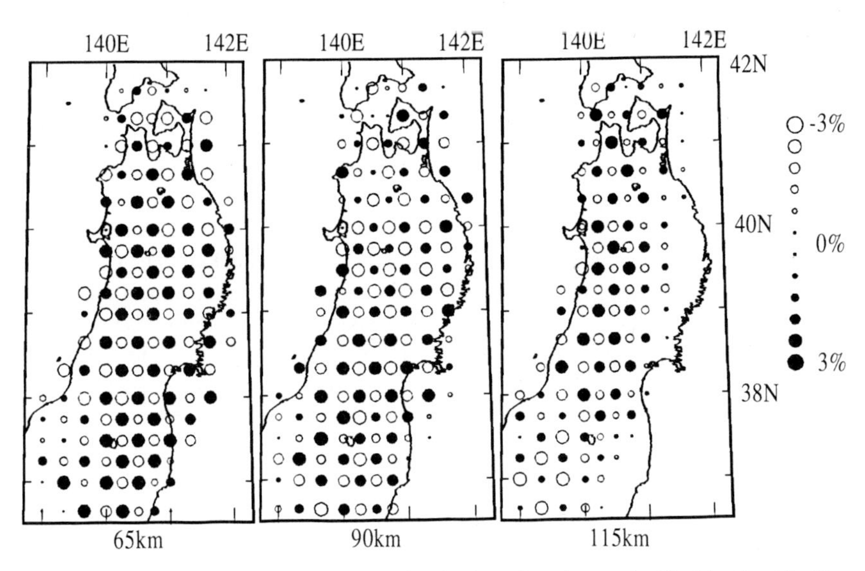

Figure 3 Result of the checkerboard resolution test plotted on horizontal sections at the 10-, 24-, 40-, 65-, 90-, and 115-km-depth levels (Zhao *et al.*, 1992). In this test alternating positive and negative velocity perturbations are assigned to the three-dimensional grid points of a homogeneous velocity model. Inverted P-wave velocity perturbation for this synthetic checkerboard velocity model is shown at each grid point by the scale on the right.

with the original ones, we can see how the original velocity structure is reconstructed. The result, which is not shown here, indicates that the original tomographic images are well reconstructed for most parts of the study area, although the images are distorted to some degree at the edge of the area as expected. These resolution tests indicate that a meaningful solution is accurately obtained in the present tomographic study.

The estimated three-dimensional P-wave velocity structure of the crust and upper mantle in northeastern Japan is shown in Figs. 1 and 2. Figure 1 shows a horizontal section of fractional P-wave velocity perturbations at a 40-km depth. Figures 2A–2C are three vertical cross sections of fractional P-wave velocity perturbations along the profiles of AA′, BB′, and CC′ in the inset map, which are nearly perpendicular to the trench axis. The fractional velocity perturbation is from the mean value of estimated velocities at each given depth level and is shown by the color scale. Red and blue colors correspond to low and high velocities, respectively, the perturbation scale being from −6 to 6%. Microearthquakes located within a 60-km width along each profile are plotted as circles in Figs. 2A–2C. Also shown by the thick curves in the figure are the locations of the Conrad discontinuity, the Moho, and the top of the subducting plate, which are fixed in the tomographic inversion procedure.

An inclined high-velocity zone corresponding to the subducting Pacific plate is clearly delineated in all the three vertical sections shown in Fig. 2. The bottom of the high-velocity Pacific plate can be clearly recognized, the estimated location of the bottom also being drawn by a solid (and broken) curve in the figure. The thickness of the subducting Pacific plate can be estimated to be 80–90 km. A reliable estimate of the plate thickness has not been made from the previous tomographic studies (e.g., Hirahara, 1977; Hirahara and Mikumo, 1980; Kamiya *et al.,* 1989) because of the inherent lower spatial resolution. Umino *et al.* (1990) have detected a reflected and S-to-P-converted wave at the bottom of the subducting Pacific plate in seismograms of both intermediate-depth and deep earthquakes. They have estimated the thickness of the plate to be 80–90 km by using arrival times of this reflected wave, which agrees well with the estimation from the tomographic inversion. The bottom of the subducting Pacific plate as imaged by the inversion (Fig. 2) has a sharp velocity contrast, which again agrees with the existence of the reflected SP wave at the bottom. Figure 2 also shows that the earthquakes forming the double seismic zone (Hasegawa *et al.,* 1978) occur in the upper half of the subducting Pacific plate and that the lower half of the plate is evidently incapable of generating earthquakes.

In northeastern Japan, many active volcanoes are distributed on the island arc's surface, occurring primarily along the volcanic front that passes through the middle of the land area and runs parallel to the trench axis. As seen in Fig. 1, active volcanoes plotted at the top as red triangles are mostly located above the low-velocity zones in the uppermost mantle wedge. Horizontal sections of velocity perturbations in the crust, although not shown here, indicate that the low-velocity zones are also distributed in the crust beneath active volcanoes. This distribution is partly seen on the three vertical sections of velocity perturbations (Figs. 2A–2C). The vertical sections further show that the low-velocity zones in the crust and the uppermost mantle wedge beneath active volcanoes dip to the west and extend to a depth of 100–150 km. Although the work by Suyehiro and Sacks (1983) and that by van der Hilst *et al.* (1991) suggest the possibility that the low-velocity zones continue to depths larger than 100–150 km, the present result shows this is not the case. The low-velocity zones are nearly parallel to the dip of the subducting Pacific plate. This feature can be seen in all the vertical sections along the three profiles perpendicular to the trench axis. Davies (this volume) has computed the thermal structure and the flow field induced in the slab and upper mantle wedge to 140 km depth in a general treatment of subduction zone magma generation.

Three-Dimensional Seismic Attenuation Structure

The seismic attenuation (anelasticity) structure of the crust and the upper mantle is estimated from observed data of seismic wave attenuation, which provides additional information on the physical properties of the subduction zone magmatic system. Highly attenuated seismic waves have been found beneath several volcanic fields in the world, suggesting the presence of magmatic or hydrothermal bodies beneath active volcanoes. The detection and location of strongly attenuating bodies can be achieved by using a method similar to seismic velocity tomography. Attenuating bodies reduce the amplitude or change the spectral content of seismic waves that pass through them, just as low-velocity bodies increase the travel time of seismic waves. Tomographic methods used in travel time tomography can also be applied to amplitude data or to amplitude spectrum data to determine the attenuation structure of the medium.

Highly attenuating bodies have been tomographically imaged mainly in the upper crust beneath active volcanoes or beneath geothermal areas in western North America (e.g., Young and Ward, 1980; Ho-Liu *et al.,* 1988; Evans and Zucca, 1988; Clawson *et al.,* 1989; Iyer *et al.,* 1990). For attenuation tomography, Ho-Liu *et al.* (1988) used observed S- to P-wave amplitude ratios determined from records of 16 earthquakes in the Coso–

Indian Wells region, southern California. The attenuation inversion with a small block size of 2 × 2 × 0.2 km delineates a highly attenuating body (S-wave quality factor $Q_s \cong 30$) at depths of 3–5 km beneath the Coso–Indian Wells region. The location of this body coincides with a slow P-wave velocity anomaly mapped by Walck and Clayton (1987), suggesting the existence of a magmatic or hydrothermal body beneath this region.

Evans and Zucca (1988) determined the P-wave velocity and attenuation structure of the upper crust beneath Medicine Lake volcano, northern California, by using a high-resolution active source P-wave travel time and attenuation tomography method. The P-wave attenuation structure was estimated by inverting observed amplitude spectral ratios between stations. A P-wave low-velocity low-Q body not larger than a few tens of cubic kilometers in volume is tomographically imaged at depths 1–3 km beneath the eastern caldera, possibly corresponding to the magma chamber feeding several of the youngest summit silicic eruptions (Evans and Zucca, 1988). By using a method similar to that of Evans and Zucca (1988), Clawson *et al.* (1989) imaged a relatively high P-wave attenuation body (P-wave quality factor $Q_p \cong 40$) in the upper crust beneath the northeastern Yellowstone caldera, which also corresponds to a slow P-wave velocity anomaly.

Attenuation tomographic studies of the mantle wedge also have been performed, although their spatial resolution is much lower than those of the upper crust described previously. By applying a three-dimensional block inversion method to a set of observed S- to P-wave spectral ratios, Umino and Hasegawa (1984) estimated the attenuation structure of the crust and upper mantle beneath northeastern Japan. They obtained low-Q_s values in the crust beneath the active volcanoes, a low-Q_s value ($Q_s \cong 150$) in the mantle wedge on the back-arc side of the volcanic front, an intermediate-Q_s value ($Q_s \cong 500$–800) in the mantle wedge on the fore-arc side of the volcanic front, and a high-Q_s value ($Q_s \cong 1500$) in the subducting Pacific plate itself.

A similar attenuation structure in the crust and in the upper mantle of the subduction zone has also been obtained beneath the Kanto–Tokai area, in central Japan, by Sekiguchi (1991). The method he used is similar to that of Umino and Hasegawa (1984). Figure 4 shows an east–west vertical cross section of the estimated Q_p^{-1} structure of the central Kanto–Tokai area. The upper boundaries of the subducting Pacific plate and the Philippine Sea plate are indicated by large solid circles in this figure. Low-Q_p values (shaded area) are clearly seen in the mantle wedge on the back-arc side of the volcanic front, which is shown by a solid triangle at the top. Finally, Sato and Sacks (1990) have estimated the thermal structure of the mantle wedge beneath northeastern Japan on the basis of Q_s and Q_p values determined by Umino and Hasegawa (1984). For $Q_s \approx 96 \pm 36$ and $Q_p \approx 210 \pm 80$, an approximate temperature of $980 \pm 30°$C has been inferred at a depth of 31–56 km.

Similar attenuation structures of subduction zones have been estimated even from observed data of the seismic intensity, that is, the degree of ground shaking caused by earthquakes (Hashida and Shimazaki, 1987; Hashida, 1989; Satake and Hashida, 1989). Hashida and Shimazaki (1987) and Hashida (1989) applied a three-dimensional block inversion method to observed seismic intensity data and estimated the attenuation structures of the crust and the upper mantle beneath northeastern Japan and beneath whole parts of the Japanese Islands, respectively. Their results show the presence of low-Q zones in the mantle wedge (to a depth of ~90 km) beneath active volcanoes and high-Q zones corresponding to the subducting Pacific plate. Low-Q zones in the mantle wedge beneath active volcanoes are also detected from inversion of seismic intensity data to a depth of ~100 km beneath North Island, New Zealand (Satake and Hashida, 1989). Figure 5 shows a NW–SE vertical cross section of the New Zealand attenuation structure that they estimated along a profile nearly perpendicular to the trench axis. A low-Q zone imaged beneath the volcanic front (triangle) dips to the northwest and attains a depth of about 100 km.

Comparison of the estimated attenuation structures with the seismic velocity structures shows that low- (high-) Q regions generally correspond to low- (high-) velocity regions, although the images obtained of low-Q zones in the mantle wedge beneath the active volcanoes and the high-Q zones corresponding to the subducted slab are less clear than those resolved by travel time tomography. This difference is due to the lower spatial resolution of the attenuation tomography. Considering the velocity–temperature relation and the Q-

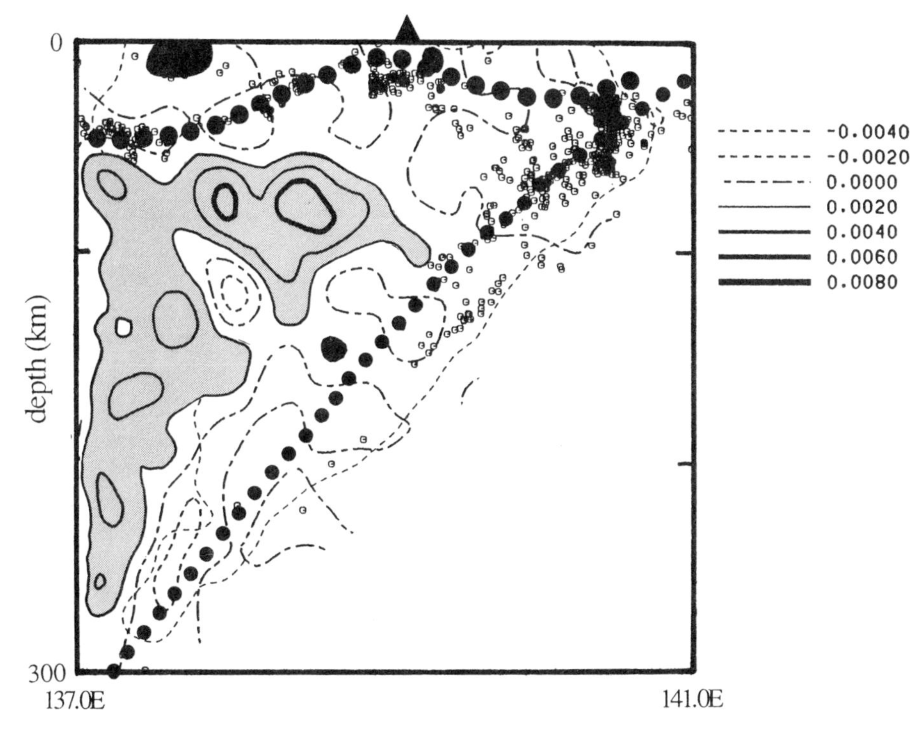

Figure 4 E–W vertical cross section of the Q_p^{-1} structure in the Kanto–Tokai area of central Japan. Contours indicate estimated values of Q_p^{-1} (at an interval of 0.002). Shaded areas indicate low-Q_p regions ($Q_p^{-1} > 0.002$). Large solid circles denote the tops of the subducting Pacific and Philippine Sea plates. The triangle at the top is the location of the volcanic front. Microearthquakes located within a 33-km width along the profile are also denoted by open circles. (After Sekiguchi, 1991.)

temperature relation of mantle rocks, these seismic observations suggest the existence of high temperatures and partially molten materials in the low-velocity low-Q zones of the mantle wedge beneath active volcanoes.

Midcrustal Magma Bodies Detected by Reflected Seismic Waves

A typical example of magma bodies seismically detected so far in the midcrust are those beneath

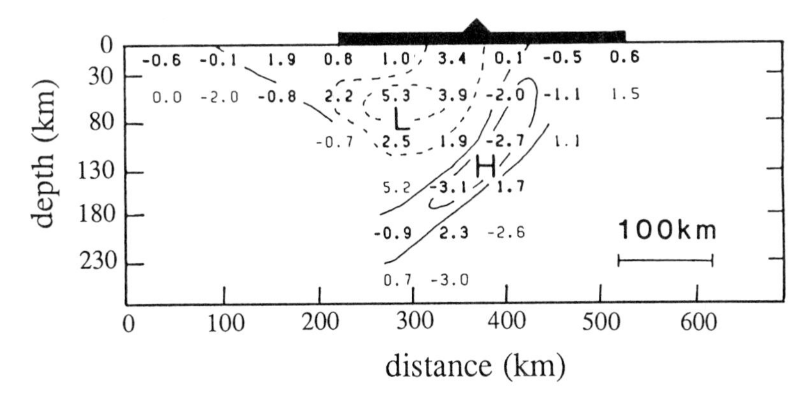

Figure 5 NW–SE cross section of attenuation structure in the northeastern part of North Island, New Zealand. Numerals are the estimated deviations from the initial values of the attenuation coefficient in units of $10^{-2}\ s^{-1}$, and the contours are drawn at an interval of $2 \times 10^{-2}\ s^{-1}$. Locations of the volcanic front and land area are denoted by a triangle and a thick horizontal line at the top. (After Satake and Hashida, 1989.)

the Long Valley Caldera (Sanders, 1984; Hill *et al.*, 1985; Dawson *et al.*, 1990; Iyer *et al.*, 1990). Sanders (1984) estimated the locations and configurations of magma bodies beneath the caldera from S-wave attenuation analyses. The estimated magma bodies, i.e., anomalously high S-wave attenuation bodies, are located at depths of 5–15 km. The magma bodies inferred from teleseismic travel time inversions have a volume of 500–1000 km^3 (Iyer *et al.*, 1990). Reflected waves are also found both from the top and from the bottom of the estimated magma bodies (Hill *et al.*, 1985; Luetgert and Mooney, 1985; Zucca *et al.*, 1987).

Unlike the mid-ocean ridge or the intraplate volcanoes, almost no clear evidence for the existence of magma bodies in the midcrust has been reported for subduction zone volcanoes from seismic observations. An exception is the distinct S-wave reflectors in the midcrust detected at several locations near active volcanoes in the Japanese Islands. Figure 6 is an example of three-component short-period seismograms of a shallow microearthquake that occurred beneath Nikko–Shirane volcano in central Japan. One can see a sharp impulsive phase denoted by SxS following the direct S-wave at the two stations located just above the earthquake focus. This anomalous phase has very large amplitudes and is most clearly defined on the horizontal component seismograms.

Phase identification by arrival time analyses shows that this phase is a reflected S-wave (SxS phase) from a strong velocity discontinuity existing in the midcrust. A ray path of this SxS phase is schematically illustrated on a vertical section by a bold broken line in Fig. 7. The phase identification is confirmed by detection of a reflected and S-to-P-converted wave (SxP phase) from the same velocity discontinuity, a signature that can be clearly seen on the vertical component at station GZD in Fig. 6. A ray path for the SxP phase is also shown by a bold broken and solid line in Fig. 7. Arrival time analyses of the SxS phase observed by a dense seismic network show that the reflector is distributed over an area of 15 × 10 km^2 at depths of 8–15 km and becomes shallow toward the north at an angle of about 30°, in the direction of the Nikko–Shirane volcano (Matsumoto and Hasegawa, 1991).

A distinct S-wave reflector in the midcrust,

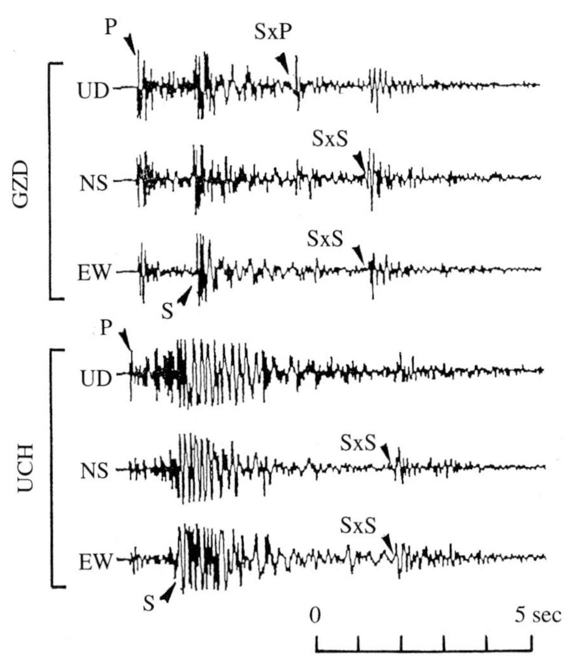

Figure 6 An example of three-component seismograms of a shallow microearthquake that occurred near Nikko–Shirane volcano, Japan. These seismograms were recorded at two stations, GZD and UCH, located just above the earthquake focus. Arrival times of direct P- and S-waves are denoted by P and S. Later arrivals, reflected waves from a midcrustal thin magma body, are clearly seen and are indicated by SxS and SxP.

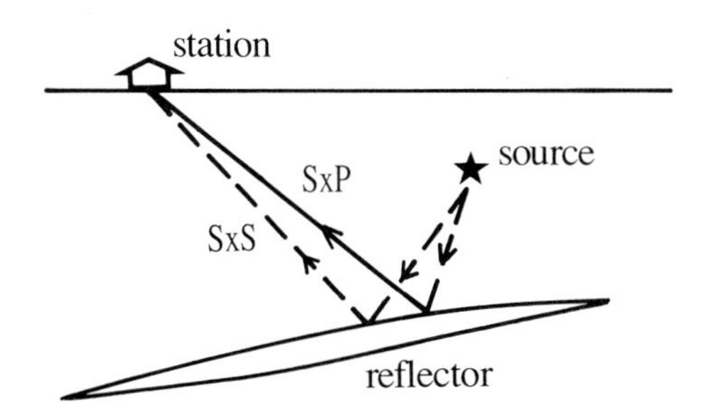

Figure 7 Schematic illustration of ray paths for a reflected S-wave (SxS) and a reflected and S-to-P converted wave (SxP) from a thin midcrustal magma body. Solid and broken lines denote P- and S-waves, respectively.

identical with that described previously was first detected and located beneath the central Rio Grande Rift near Socorro, New Mexico (Sanford *et al.*, 1973). The large amplitude of the SxS phase relative to the direct S phase and the ratio of SxP to SxS amplitudes can be explained by a large velocity contrast across the discontinuity underlain by very-low-rigidity material such as a magma body. Ake and Sanford (1988) estimated the thickness of the magma body by modeling the observed spectra of the reflected phase. The results show a thin (~70-m) layer of nonrigid, low-velocity material underlain by a second, thin (~60-m) layer of slightly higher-velocity material, the total thickness of the low-velocity region being less than 150 m.

Spectral ratios of SxS to direct S phases observed in the Nikko–Shirane volcanic area have three peaks in the frequency range 3–20 Hz, which can be explained by a very thin (~100-m) magma body with low-rigidity material (Matsumoto and Hasegawa, 1991). A very thin magma body model is also supported by the observations that direct S-waves, passing vertically through the body and observed at stations just above it, do not have spectra different from those at other stations. These observations indicate that passing vertically through the body does not attenuate S-waves significantly, suggesting a very thin thickness of the body (Hasegawa *et al.*, 1991).

Distinct S-wave reflectors in the midcrust, similar to that described above, have been detected so far at nine locations beneath the Japanese Islands (Mizoue, 1980; Mizoue *et al.*, 1982; Horiuchi *et al.*, 1988; Iwase *et al.*, 1989; Nishiwaki *et al.*, 1989; Hori and Hasegawa, 1991; Inamori *et al.*, 1992; Hasegawa *et al.*, 1991). Five of these reflectors are found in northeastern Japan, and their locations are plotted as large squares on the horizontal section of P-wave velocity perturbations (Fig. 1). As is obvious from this figure, all the reflectors are located near active volcanoes (red triangles) and/or in or around the P-wave low-velocity zones (red or yellow areas). The features of SxS and SxP phases described previously indicate that these are the reflected phases from thin magma bodies existing in the midcrust beneath active volcanoes, or in close proximity to the P-wave low-velocity zones.

Additional seismic evidence for deep-seated magmatic activity in northeastern Japan is the deep low-frequency microearthquakes that occur beneath the active volcanoes (Hasegawa *et al.*, 1991; Hasegawa and Yamamoto, 1994). Most shallow earthquakes beneath the surface in northeastern Japan are known to occur in the upper 15 km of the crust, forming a brittle seismogenic zone. Exceptionally deep microearthquakes, well below the base of the brittle seismogenic zone, also occur at depths of 22–40 km beneath the land area, although their occurrence is rather rare. All of the 151 deep events found thus far at 10 locations in northeastern Japan have extremely low dominant frequencies (1.5–3.5 Hz) for both compressional and shear waves. Figure 8a shows an example of three-component short-period seismograms selected from the 151 events. Another example of seismograms for an event with nearly the same epicenter location but with a normal focal depth is shown in Fig. 8b. It is obvious by comparing the two that the deep event has anomalously low dominant frequencies for both the P- and the S-waves. The epicenters of all 151 deep low-frequency events are plotted as crosses on the horizontal section of P-wave velocity perturbations (Fig. 1). The deep events within a 60-km width along the profiles AA′, BB′, and CC′ are plotted as red circles on the vertical sections of velocity perturbations (Figs. 2A–2C). The deep events are clearly isolated from the main activity of normal focal-depth events and are located approximately under the active volcanoes (red triangles) or around the low-velocity zones (red or yellow areas).

Focal mechanisms of these low-frequency events have not been determined since the magnitudes of these events are small. A preliminary estimation by a moment-tensor inversion using

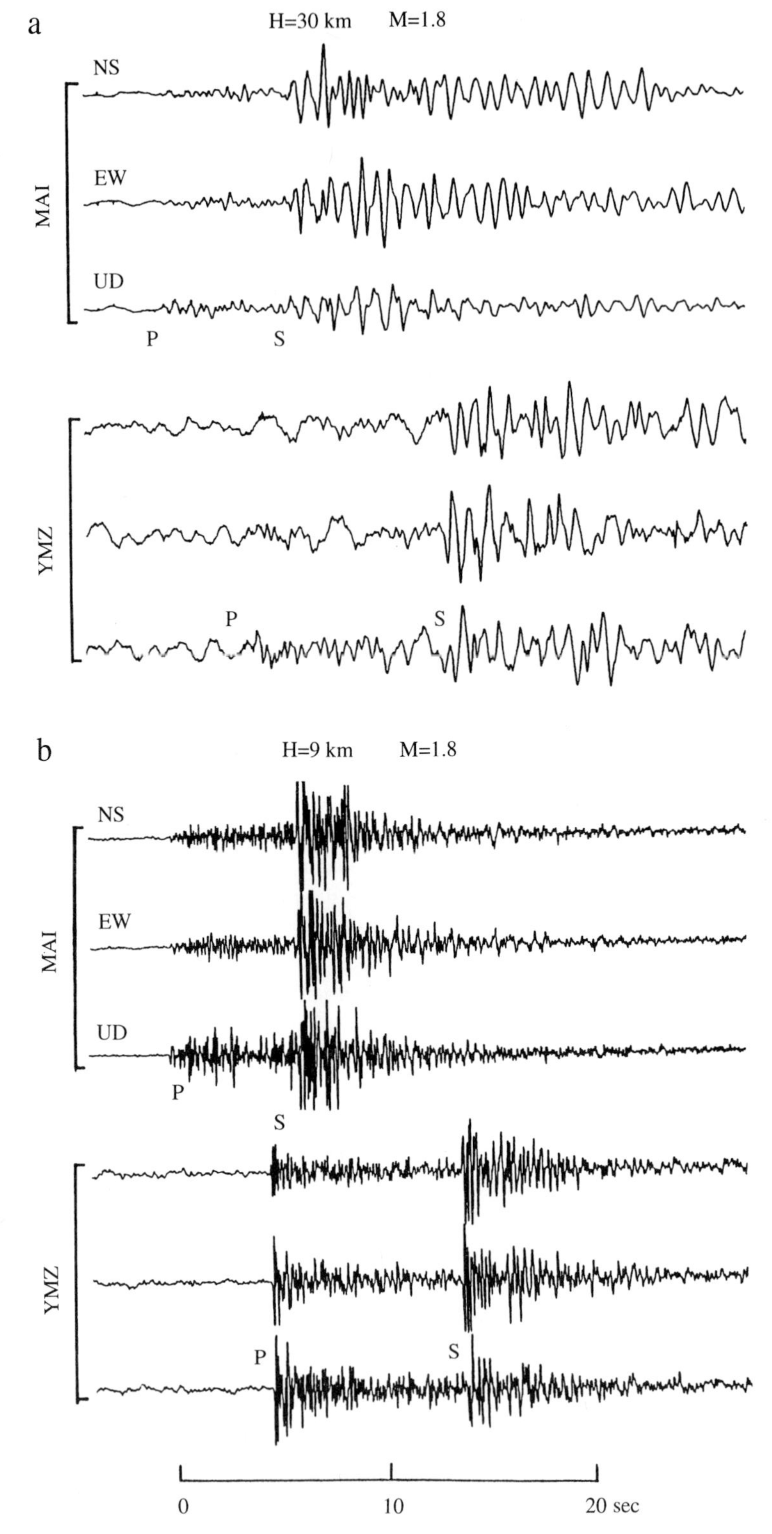

Figure 8 (a) An example of three-component seismograms of a deep (focal depth H = 30 km), low-frequency microearthquake that occurred beneath Hiuchidake volcano and (b) that of a normal focal depth (H = 9 km) microearthquake with nearly the same epicenter location. These seismograms are recorded by short-period (1 s) seismographs at two stations, MAI and YMZ, about 45 and 75 km east of the two earthquakes, respectively. Arrival times of P- and S-waves are indicated by P and S, respectively.

P- and S-waveforms prefers a non-double-couple mechanism to a normal double-couple mechanism, which is expected for ordinary earthquakes caused by fault slip (Kosuga and Hasegawa, 1992). This fact and the anomalous features of these events described previously suggest that these deep low-frequency events are generated by deep-seated magmatic activity, such as the rapid movement of magma accompanied by a fracture of crustal or mantle rocks, either in the lower crust or in the uppermost mantle beneath the active volcanoes. The distinct S-wave reflectors described previously are located at shallower depths in the midcrust above these low-frequency events.

Studies detecting deep low-frequency microearthquakes at other locations in the Japanese Islands have been reported in rapid succession. These microearthquakes occur beneath the Tokachi volcano in Hokkaido (Suzuki *et al.*, 1992), beneath the Izu–Oshima volcano in central Japan (Ukawa and Ohtake, 1987), beneath the Hida mountainous range in central Japan (Yamauchi *et al.*, 1992), and beneath the Sakurajima volcano in southern Kyushu (Goto *et al.*, 1992). All these events are located beneath active volcanoes and have features similar to those detected in northeastern Japan. Similar deep low-frequency microearthquakes have not been detected to date in other subduction zones of the world, possibly due to the lower overall resolution of seismic networks in those regions. The extensive distribution of the deep low-frequency events throughout Japan suggests that the occurrence of deep low-frequency events beneath active volcanoes is a phenomenon common to all the subduction zones, and a fundamental indicator of relatively deep magmatism.

Deep Structure of Arc Volcanoes

The cross-arc vertical cross-sectional model of the crust and upper mantle beneath northeastern Japan, inferred from the recent seismic observations described in the previous sections, is shown schematically in Fig. 9. It illustrates the generalized

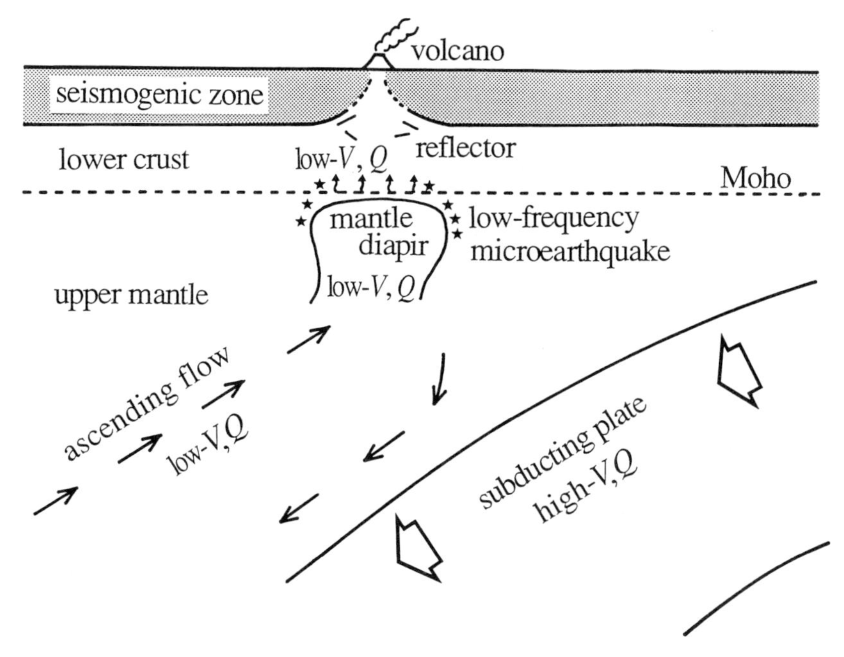

Figure 9 Schematic illustration of across-arc vertical cross section of the crust and upper mantle beneath the northeastern Japan arc. The subduction of the oceanic plate with a high seismic velocity (high V) and a low seismic attenuation (high Q) generates a secondary mechanically induced convection in the overlying mantle wedge. Decompression melting within the ascending mantle wedge flow produces low V and low Q. Magma within the ascending flow finally reaches the top of the mantle and is then segregated from the stagnated mantle diapir, which also has low V and low Q. Magma further migrates upward into the crust due to its positive buoyancy, again producing low V and low Q around it. The temporary storage of magma at midcrustal levels appears as distinct S-wave reflectors, which are considered to be thin magma bodies. Repeated discharges of magma, by its further upward migration to the Earth's surface, form the arc volcanoes.

deep structure of island arc magmatic regions. The subduction of the oceanic plate generates a secondary mechanically induced convection in the overlying mantle wedge (e.g., McKenzie, 1969; Sleep and Toksöz, 1973; Toksöz and Bird, 1977). We infer that the low-velocity zones that are continuously distributed in the mantle wedge and parallel to the dip of the subducting plate (clearly delineated in Fig. 2) are images of the ascending flow of hot mantle material from depth and are a portion of the secondary subduction-induced convection. Decompression melting within the ascending mantle wedge flow produces low seismic velocities and high attenuations (low Q). Magma within the ascending mantle flow finally reaches the top of the mantle and is then segregated from the stagnated diapir. The low-velocity zones just beneath the Moho and imaged by seismic velocity tomography are the manifestation of mantle diapirs, and their magmatic activity may generate deep low-frequency microearthquakes around them, which are inferred to be related to the rapid movement of magma.

Magma bodies may further migrate upward into the crust because of their positive buoyancy, at which time they suffer fractional crystallization and chemical reaction with the surrounding crustal rocks. This portion of the upward migration of magma in the crust again produces a seismic low-velocity region and the resulting high attenuation: effects that are clearly imaged in vertical cross sections of the P-wave velocity perturbations (Fig. 2) and in that of the resulting attenuation structure (Fig. 5). The temporary storage of magma at midcrustal levels appears as distinct S-wave reflectors. This upward migration of magma raises the temperature of crustal materials around it and consequently causes the local elevation of the bottom of the brittle seismogenic zone. Magma finally reaches the Earth's surface by its further upward migration, possibly along preexisting conduits in the upper crust, and repeated discharges to the surface form the arc volcanoes.

Figure 2 shows that the low-velocity zones in the mantle wedge penetrate into the crust and finally reach the root regions of the active volcanoes at the surface, thus providing a pathway of magma ascent as described previously. Beneath the northeastern Japan arc, in addition to the main volcanic chain along the volcanic front itself, some of the active volcanoes are distributed on the back-arc side of the volcanic front and approximately parallel to it. The two separate chains of volcanoes similar to the northeastern Japan arc are present in most subduction zones, and systematic differences in magma compositions between the two chains have been pointed out (e.g., Sakuyama and Nesbitt, 1986; Marsh, 1979). A volcano belonging to the back-arc side volcanic chain can be seen along the profile BB′ in Fig. 2; Chokai volcano is located near the Japan Sea coast. We can see from Fig. 2B that a side path of the low-velocity zone originating in the mantle wedge ultimately reaches the root region of this volcano.

Many of the petrological or geochemical studies extensively conducted to date attribute the magma generation in subduction zones to the progressive dehydration of the oceanic plate during its descent (e.g., McBirney, 1969; Anderson *et al.*, 1976; Kushiro, 1983; Tatsumi *et al.*, 1983; Tatsumi, 1989), implying direct vertical ascent of magma to the volcanoes on the surface rather than the inclined ascending flow. The presence of the two volcanic chains with systematic differences in their chemical compositions has been explained by this vertical ascent of magma from the subducted plate. Strong S-wave reflections at 8 Hz from the top of the subducted plate detected beneath the southern Kanto district indicate the existence of a narrow zone containing many liquid bodies (Obara and Sato, 1988). This work offers seismic evidence for the presence of free water or melts on the upper surface of the subducted plate. If the subduction-induced mantle wedge flow exists as suggested by the seismic tomographic images of the mantle wedge, H_2O from the subducted plate may not ascend vertically. Davies and Stevenson (1992) and Davies (this volume) propose that the combination of vertical ascent of water as a free phase and the transport of hydrous phases by the subduction-induced flow leads to the net transport of H_2O being horizontal, across the mantle wedge from the subducted plate. It is quite probable that this kind of horizontal transport, as well as a vertical mode of H_2O migration, is actually occurring in the source region of subduction zone magmatism.

In any case, the seismic structure imaged by the tomographic studies provides an important constraint for understanding the processes and three-dimensional structure of magma generation in subduction zones.

Acknowledgments

We appreciate the critical reviews of M. P. Ryan, R. van der Hilst, and two anonymous referees who helped to improve this manuscript.

References

Ake, J. P., and Sanford, A. R. (1988). New evidence for the existence and internal structure of a thin layer of magma at mid-crustal depths near Socorro, New Mexico, *Bull. Seism. Soc. Am.* **78,** 1335–1359.

Aki, K., and Lee, W. H. K. (1976). Determination of three-dimensional velocity anomalies under a seismic array using first P arrival times from local earthquakes. 1. A homogeneous initial model, *J. Geophys. Res.* **81,** 4381–4399.

Aki, K., Christoffersson, A., and Husebye, E. S. (1977). Determination of the three-dimensional seismic structure of the lithosphere, *J. Geophys. Res.* **82,** 277–296.

Anderson, R. N., Uyeda, S., and Miyashiro, A. (1976). Geophysical and geochemical constraints at converging plate boundaries. I. Dehydration in the down-going slab, *Geophys. J. R. Astron. Soc.* **44,** 333–357.

Ando, M., Ishikawa, Y., and Yamazaki, F. (1983). Shear wave polarization anisotropy in the upper mantle beneath Honshu, Japan, *J. Geophys. Res.* **88,** 5850–5864.

Benz, H. M., Zandt, G., and Oppenheimer, D. H. (1992). Lithospheric structure of northern California from teleseismic images of the upper mantle, *J. Geophys. Res.* **97,** 4791–4807.

Clawson, S. R., Smith, R. B., and Benz, H. M. (1989). P wave attenuation of the Yellowstone caldera from three-dimensional inversion of spectral decay using explosion source seismic data, *J. Geophys. Res.* **94,** 7205–7222.

Davies, J. H. (1994). Lateral water transport across a dynamic mantle wedge: A model for subduction zone magmatism, *in* "Magmatic Systems" (M. P. Ryan, ed.), Academic Press, San Diego, CA.

Davies, J. H., and Stevenson, D. J. (1992). Physical model of source region of subduction zone volcanics, *J. Geophys. Res.* **97,** 2037–2070.

Dawson, P. B., Evans, J. R., and Iyer, H. M. (1990). Teleseismic tomography of the compressional wave velocity structure beneath the Long Valley region, California, *J. Geophys. Res.* **95,** 11021–11050.

Evans, J. R., and Zucca, J. J. (1988). Active high-resolution seismic tomography of compressional velocity and attenuation structure at Medicine Lake volcano, northern California Cascade Range, *J. Geophys. Res.* **93,** 15016–15036.

Fukao, Y., Obayashi, M., Inoue, H., and Nenbai, M. (1992). Subducting slabs stagnant in the mantle transition zone, *J. Geophys. Res.* **97,** 4809–4822.

Goto, K., Hasegawa, S., and Kanjo, K. (1992). Deep low-frequency earthquakes beneath the volcanoes in Kyushu, *in* "Abstracts, 1992 Japan Earth and Planetary Science Joint Meeting," p. 243.

Harris, R. A., Iyer, H. M., and Dawson, P. B. (1991). Imaging the Juan de Fuca plate beneath southern Oregon using teleseismic P wave residuals, *J. Geophys. Res.* **96,** 19879–19889.

Hasegawa, A., and Yamamoto, A. (1994). Deep, low-frequency microearthquakes in or around seismic low-velocity zones beneath active volcanoes in north-eastern Japan. *Tectonophysics,* in press.

Hasegawa, A., Umino, N., and Takagi, A. (1978). Double-planed deep seismic zone and upper mantle structure in the northeastern Japan arc, *Geophys. J. R. Astron. Soc.* **54,** 281–296.

Hasegawa, A., Umino, N., Takagi, A., Suzuki, S., Motoya, Y., Kameya, S., Tanaka, K., and Sawada, Y. (1983). Spatial distribution of earthquakes beneath Hokkaido and northern Honshu, Japan, *J. Seism. Soc. Japan* **36,** 129–150.

Hasegawa, A., Zhao, D., Hori, S., Yamamoto, A., and Horiuchi, S. (1991). Deep structure of the northeastern Japan arc and its relationship to seismic and volcanic activity, *Nature* **352,** 683–689.

Hasemi, A. H., Ishii, H., and Takagi, A. (1984). Fine structure beneath the Tohoku District, northeastern Japan arc, as derived by an inversion of P-wave arrival times from local earthquakes, *Tectonophysics* **101,** 245–265.

Hashida, T. (1989). Three-dimensional seismic attenuation structure beneath the Japanese Islands and its tectonic and thermal implications. *Tectonophysics* **159,** 163–180.

Hashida, T., and Shimazaki, K. (1987). Determination of seismic attenuation structure and source strength by inversion of seismic intensity data: Tohoku District, northeastern Japan arc, *J. Phys. Earth* **35,** 67–92.

Hill, D. P., Bailey, R. A., and Ryall, A. S. (1985). Active tectonic and magmatic processes beneath Long Valley caldera, eastern California: An overview, *J. Geophys. Res.* **90,** 11111–11120.

Hirahara, K. (1977). A large-scale three-dimensional seismic structure under the Japan Islands and the Sea of Japan, *J. Phys. Earth* **25,** 393–417.

Hirahara, K., and Mikumo, T. (1980). Three-dimensional seismic structure of subducting lithospheric plates under the Japan Islands, *Phys. Earth Planet. Int.* **21,** 109–119.

Hirahara, K., Ikami, A., Ishida, M., and Mikumo, T. (1989). Three-dimensional P-wave velocity structure beneath central Japan: Low-velocity bodies in the wedge portion of the upper mantle above high-velocity subducting plates, *Tectonophysics* **163,** 63–73.

Ho-Liu, P., Kanamori, H., and Clayton, R. W. (1988). Applications of attenuation tomography to Imperial Valley and Coso–Indian Wells region, southern California, *J. Geophys. Res.* **93,** 10501–10520.

Hori, S., and Hasegawa, A. (1991). Location of a mid-crustal magma body beneath Mt. Moriyoshi, northern Akita Prefecture, as estimated from reflected SxS phase, *J. Seism. Soc. Japan* **44,** 39–48.

Horie, A., and Aki, K. (1982). Three-dimensional velocity structure beneath the Kanto District, Japan, *J. Phys. Earth* **30,** 255–281.

Horiuchi, S., Hasegawa, A., Takagi, A., Ito, A., Suzuki, M., and Kameyama, H. (1988). Mapping of a melting zone near Mt. Nikko–Shirane in northern Kanto, Japan, as inferred from SxP and SxS reflections, *Tohoku Geophys. J.* **31,** 43–55.

Humphreys, E., and Clayton, R. W. (1988). Adaptation of back projection tomography to seismic travel time problems, *J. Geophys. Res.* **93,** 1073–1085.

Inamori, T., Horiuchi, S., and Hasegawa, A. (1992). Location of mid-crustal reflectors by a reflection method using aftershock waveform data in the focal area of the 1984 Western Nagano Prefecture earthquake, *J. Phys. Earth* **40,** 379–393.

Ishida, M., and Hasemi, A. H. (1988). Three-dimensional fine velocity structure and hypocentral distribution of earthquakes beneath the Kanto–Tokai District, Japan, *J. Geophys. Res.* **93,** 2076–2094.

Iwase, R., Urabe, S., Katsumata, K., Moriya, M., Nakamura, I., and Mizoue, M. (1989). Mid-crustal magma body in southwestern Fukushima Prefecture detected by reflected waves from microearthquakes, *in* "Programme and Abstracts of Seism. Soc. Japan," No. 1, p. 185.

Iyer, H. M., Evans, J. R., Dawson, P. B., Stauber, D. A., and Achauer, U. (1990). Differences in magma storage in different volcanic environments as revealed by seismic tomography: Silicic volcanic centers and subduction-related volcanoes, *in* "Magma Transport and Storage" (M. P. Ryan, ed.), Wiley, Chichester/Sussex, England.

Kamiya, S., Miyatake, T., and Hirahara, K. (1989). Three-dimensional P-wave velocity structure beneath the Japanese Islands, *Bull. Earthq. Res. Inst. Univ. Tokyo* **64,** 457–485.

Kosuga, M., and Hasegawa, A. (1992). Moment tensor analysis of low-frequency microearthquakes, "Abstracts, 1992 Japan Earth and Planetary Science Joint Meeting," p. 244.

Kushiro, I. (1983). On the lateral variation in chemical composition and volume of quaternary volcanic rocks across Japanese arcs, *J. Volcanol. Geotherm. Res.* **18,** 435–447.

Luetgert, J. H., and Mooney, W. D. (1985). Crustal refraction profile of the Long Valley caldera, California, from the January 1983 Mammoth Lakes earthquake swarm, *Bull. Seism. Soc. Am.* **75,** 211–221.

Marsh, B. D. (1979). Island arc development: Some observations, experiments and speculations, *J. Geol.* **87,** 687–713.

Matsumoto, S., and Hasegawa, A. (1991). Characteristics of mid-crustal S-wave reflector in Nikko–Ashio region, *in* "Programme and Abstracts of Seism. Soc. Japan," p. 204.

Matsuzawa, T., Umino, N., Hasegawa, A., and Takagi, A. (1986). Upper mantle velocity structure estimated from PS-converted wave beneath the northeastern Japan arc, *Geophys. J. R. Astron. Soc.* **86,** 767–787.

Matsuzawa, T., Kono, T., Hasegawa, A., and Takagi, A. (1990). Subducting plate boundary beneath the northeastern Japan arc estimated from SP converted waves, *Tectonophysics* **181,** 123–133.

McBirney, A. R. (1969). Andesitic and rhyolitic volcanism of orogenic belts, *in* "The Earth's Crust and Upper Mantle" (P. J. Hart, ed.), Geophys-Monogr. Ser, Vol. 13, AGU Washington, DC, pp. 501–506.

McKenzie, D. P. (1969). Speculations on the consequences and causes of plate motions, *Geophys. J. R. Astron. Soc.* **18,** 1–32.

Michaelson, C. A., and Weaver, C. S. (1986). Upper mantle structure from teleseismic P arrivals in Washington and northern Oregon, *J. Geophys. Res.* **91,** 2077–2098.

Miyamachi, H., and Moriya, T. (1984). Velocity structure beneath the Hidaka mountains in Hokkaido, Japan, *J. Phys. Earth* **32,** 13–42.

Mizoue, M. (1980). Deep crustal discontinuity underlain by molten material as deduced from reflection phases on microearthquakes seismograms, *Bull. Earthq. Res. Inst. Univ. Tokyo* **55,** 705–735.

Mizoue, M., Nakamura, I., and Yokota, T. (1982). Mapping of an unusual crustal discontinuity by microearthquake reflections in the earthquake swarm area near Ashio, northwestern part of Tochigi Prefecture, central Japan, *Bull. Earthq. Res. Inst. Univ. Tokyo* **57,** 653–686.

Murase, T., and Fukuyama, H. (1980). Shear wave velocity in partially molten peridotite at high pressures, *Carnegie Inst. Washington Year Book* **79,** 307–310.

Murase, T., and Kushiro, I. (1979). Compressional wave velocity in partially molten peridotite at high pressures, *Carnegie Inst. Washington Year Book* **78,** 559–562.

Nakanishi, I. (1985). Three-dimensional structure beneath the Hokkaido–Tohoku region as derived from a tomographic inversion of P-arrival times, *J. Phys. Earth* **33,** 241–256.

Nishiwaki, M., Morita, Y., Nagare, S., Kakishita, T., Osada, Y., and Nagai, N. (1989). Detection of S-wave reflector beneath the Matsushiro array, central Japan, *in* "Programme and Abstracts of Seism. Soc. Japan," No. 1, p. 184.

Nolet, G. (1985). Solving or resolving inadequate and noisy tomography system, *J. Comp. Phys.* **61,** 463–482.

Obara, K., Hasegawa, A., and Takagi, A. (1986). Three-dimensional P and S wave velocity structure beneath the northeastern Japan arc, *J. Seism. Soc. Japan* **39,** 201–215.

Obara, K., and Sato, H. (1988). Existence of an S wave reflector near the upper plane of the double seismic zone beneath the southern Kanto district, Japan, *J. Geophys. Res.* **93,** 15037–15045.

Paige, C. C., and Saunders, M. A. (1982). LSQR: An algorithm for sparse linear equations and sparse least squares, *ACM Trans. Math. Software* **8,** 43–71.

Rasmussen, J., and Humphreys, E. (1988). Tomographic image of the Juan de Fuca plate beneath Washington and western Oregon using teleseismic P-wave travel times, *Geophys. Res. Lett.* **15,** 1417–1420.

Sakuyama, M., and Nesbitt, R. W. (1986). Geochemistry of the quaternary volcanic rocks of the northeastern Japan arc, *J. Volcanol. Geotherm. Res.* **29,** 413–450.

Sanford, A. R., Alptekin, O., and Toppozada, T. R. (1973). Use of reflection phases on microearthquake seismograms to map an unusual discontinuity beneath the Rio Grande Rift, *Bull. Seism. Soc. Am.* **63,** 2021–2034.

Sanders, C. O. (1984). Location and configuration of magma bodies beneath Long Valley, California, determined from anomalous earthquakes signals, *J. Geophys. Res.* **89,** 8287–8302.

Satake, K., and Hashida, T. (1989). Three-dimensional attenuation structure beneath North Island, New Zealand, *Tectonophysics* **159,** 181–194.

Sato, H., and Sacks, I. S. (1990). Magma generation in the upper mantle inferred from seismic measurements in peridotite at high pressure and temperature, *in* "Magma Transport and Storage" (M. P. Ryan, ed.), Wiley, Chichester/Sussex, England.

Sekiguchi, S. (1991). Three-dimensional Q structure beneath the Kanto–Tokai district, Japan, *Tectonophysics* **195,** 83–104.

Sleep, N., and Toksöz, M. N. (1973). Evolution of marginal basins, *Nature* **233,** 548–550.

Spakman, W., and Nolet, G. (1988). Imaging algorithms, accuracy and resolution in delay time tomography, *in* (N. J. Vlaar, G. Nolet, M. J. R. Wortel, and S. A. P. L. Cloetingh, eds.), pp. 157–187, Mathematical Geophysics, Reidel.

Suyehiro, K., and Sacks, I. S. (1983). An anomalous low velocity region above the deep earthquakes in the Japan subduction zone, *J. Geophy. Res.* **88,** 10429–10438.

Suzuki, S., Miyamura, J., and Nishimura, Y. (1992). Low-frequency microearthquakes occurring near the Moho boundary beneath Tokachidake volcano in relation to the 1988–89 eruption, *in* "Abstracts 1992 Japan Earth and Planetary Science Joint Meeting," p. 242.

Tatsumi, Y. (1989). Migration of fluid phases and genesis of basalt magmas in subduction zones, *J. Geophys. Res.* **94,** 4697–4707.

Tatsumi, Y., Sakuyama, M., Fukuyama, H., and Kushiro, I. (1983). Generation of arc basalt magmas and thermal structure of the mantle wedge in subduction zones, *J. Geophys. Res.* **88,** 5815–5825.

Toksöz, M. N., and Bird, P. (1977). Formation and evolution of marginal basins and continental plateaus, *in* "Island Arcs, Deep Sea Trenches and Back-Arc Basins" (M. Talwani and W. C. Pitman, eds.), Amer. Geophys. Union.

Ukawa, M., and Ohtake, M. (1987). A monochromatic earthquake suggesting deep seated magmatic activity beneath Izu–Ooshima Volcano, Japan, *J. Geophys. Res.* **92,** 12649–12663.

Um, J., and Thurber, C. H. (1987). A fast algorithm for two-point seismic ray tracing, *Bull. Seism. Soc. Am.* **77,** 972–986.

Umino, N., and Hasegawa, A. (1984). Three-dimensional Q_s structure in the northeastern Japan arc, *J. Seism. Soc. Japan* **37,** 217–228.

Umino, N., Matsuzawa, T., and Hasegawa, A. (1990). X phases after P arrivals from deep earthquakes beneath the northeastern Japan arc, "Abstracts Ann. Meet. Seism. Soc. Japan," No. 2, p. 200.

Uyeda, S. (1982). Subduction zones: An introduction to comparative subductology, *Tectonophysics* **81,** 133–159.

Van der Hilst, R., Engdahl, R., Spakman, W., and Nolet, G. (1991). Tomographic imaging of subducted lithosphere below northwest Pacific island arcs, *Nature* **353,** 37–43.

Walck, M. C., and Clayton, R. W. (1987). P wave velocity variations in the Coso region, California, derived from local earthquake travel times, *J. Geophys. Res.* **92,** 393–406.

Yamauchi, K., Ando, M., and Wada, H. (1992). Low-frequency earthquakes in the northern Chubu region(2), *in* "Abstracts, 1992 Japan Earth and Planetary Science Joint Meeting," p. 240.

Young, C. Y., and Ward, R. W. (1980). Three-dimensional *Q* model of Coso Hot Springs known geothermal resource area, *J. Geophys. Res.* **85,** 2459–2470.

Zhao, D., and Christensen, D. (1992). Tomographic imaging of the crust and upper mantle beneath central and southern Alaska, *Seism. Res. Lett.* **63,** 64.

Zhao, D., Hasegawa, A., and Horiuchi, S. (1992). Tomographic imaging of P and S wave velocity structure beneath northeastern Japan, *J. Geophys. Res.* **97,** 19909–19928.

Zhao, D., Horiuchi, S., and Hasegawa, A. (1990). 3-D seismic velocity structure of the crust and the uppermost mantle in the northeastern Japan arc, *Tectonophysics* **181,** 135–149.

Zhou, D., and Clayton, R. W. (1990). P and S wave travel time inversions for subducting slab under the island arcs of northwest Pacific, *J. Geophys. Res.* **95,** 6829–6851.

Zucca, J. J., Kasameyer, P. W., and Mills, J. M., Jr. (1987). Observation of a reflection from the base of a magma chamber in Long Valley caldera, California, *Bull. Seism. Soc. Am.* **77,** 1674–1687.

Chapter 9 Lateral Water Transport across a Dynamic Mantle Wedge: A Model for Subduction Zone Magmatism

J. Huw Davies

Overview

A model for subduction zone magmatism wherein the mantle wedge melts by hydrous fluxing from the subducting slab is presented. The subducting slab induces secondary flow in the overriding mantle wedge. This induced flow leads to high temperatures attained near, but not at, the subducting slab. Water is released from amphibole in the oceanic crust at a depth of around 80 km. On entering the mantle, it reacts to form amphibole, which is carried down by the induced flow as part of the solid matrix. At a depth of around 100 km it breaks down and releases its water. The water ultimately rises vertically through the mantle, passing through the amphibole-saturated mantle until it reaches dry mantle. There it reacts to form amphibole, which is once more carried down by the induced flow. This process repeats, with the net effect being the lateral transport of water from the oceanic crust out into the mantle wedge. Finally, the water reaches the high temperatures near the slab where amphibole is unstable. At this point the hydrous melt leads to ever greater degrees of melting as it rises vertically. The process ultimately leads to such substantial degrees of melting that some of the melts can segregate in sufficiently large cracks to emplace magma in the lithosphere. These processes describe the proposed source region for subduction zone magmatism.

This chapter discusses details, variations, and uncertainties surrounding this lateral transport mechanism. This model can potentially explain many of the features of subduction zone magmas, including the location of the volcanic front and the composition of proposed primary magmas.

Notation

		Units
F	Total mass flux of water from slab into wedge per meter along strike	$kg \cdot m^{-1} \cdot s^{-1}$
P	Difference between total pressure, p, and hydrostatic pressure; i.e., $P = p - \rho g z$, where z is the depth in meters	Pa
T	Temperature	°C
$\mathbf{V}$	Mantle velocity	$m \cdot s^{-1}$
V_x	Horizontal component of mantle velocity	$m \cdot s^{-1}$
V_z	Vertical component of mantle velocity	$m \cdot s^{-1}$
a	Grain size	m
b	Constant in permeability versus porosity relationship	dimensionless
d	Thickness of mechanical lithosphere	m
f	Porosity	dimensionless
g	Acceleration due to gravity	$m \cdot s^{-2}$
h	Thickness of amphibolitized layer	m
k	Permeability	m^2
t	Time	s
w	Weight percentage water content of amphibolitized mantle	dimensionless
$\Delta\rho$	Density difference between fluid and matrix	$kg \cdot m^{-3}$
η	Viscosity	$Pa \cdot s$
κ	Thermal diffusivity	$m^2 \cdot s^{-1}$
ρ_m	Density of mantle	$kg \cdot m^{-3}$
ρ_f	Density of fluid (melt or water)	$kg \cdot m^{-3}$

Introduction

Subduction and subduction zone magmatism are fundamental processes in the evolution of the Earth. They play critical roles in the present day differentiation of the Earth, where subduction zones are believed to be the major sites of generation of continental crust. Subduction is also sig-

Magmatic Systems
Edited by M. P. Ryan

nificant in the water and carbon cycles, and because it is one of the most significant geodynamic processes, an improved understanding of its nature should lead to insights into the driving forces of plate tectonics. In this chapter, we present a model for subduction zone magmatism that combines slab and wedge flow kinematics, thermal structure, and a magma source region consistent with geophysical and geochemical constraints.

In contrast to magmatism at mid-ocean ridges, where it is largely agreed that the responsible process is adiabatic upwelling, magmatism at subduction zones has led to a range of proposals for the source region and processes. With the advent of plate tectonics it was proposed that shear and frictional heating led to melting of the oceanic crust (Oxburgh and Turcotte, 1970; Turcotte and Schubert, 1973). Hsui *et al.* (1983) suggested that the oceanic crust melted as a result of high temperatures arising from the hot induced mantle flow impinging on it. Some have proposed that the melting is the result of adiabatic upwelling, of either eclogitic crust (Brophy and Marsh, 1986; Ida, 1987) or mantle wedge (Plank and Langmuir, 1988; Tatsumi *et al.*, 1983), while many other workers have argued that the melting results from fluxing of the peridotitic mantle wedge by water from the oceanic crust (Davies and Stevenson, 1992; Gill, 1981; Tatsumi, 1989).

One of the striking observations of subduction zone volcanism is that the active volcanoes nearest the trench seem to describe nearly straight lines parallel to the trench: the volcanic front (Marsh, 1979b; Sugimura, 1960). In addition, these volcanoes have been found to lie some 120 km above the Benioff zone of the subducting plate (Tatsumi, 1986). This constraint is quite robust and one that any model hoping to explain subduction zone magmatism must be able to fulfill.

We argue that the subducting plate induces a flow in the mantle wedge. This flow leads to high temperatures near the subducting slab and also to a mechanism for the lateral transport of H_2O across the dynamic wedge from the subducting oceanic crust. This transport results in melting of the hot wedge peridotite due to the lowering of the solidus by the influx of water. In the next section we describe the results of thermal modeling and comment on the different proposals mentioned previously. The results help motivate the mechanism for the lateral transport of water.

Thermal Model

Brief Review of Previous Thermal Models

To place our thermal model in context we briefly review previous thermal models. One of the earliest thermal models of subduction was the analytic model of McKenzie (1969). The model addressed the thermal state of the slab and its relationship to Wadati–Benioff seismicity. It did not address the thermal state of the mantle wedge, which in fact was assumed to be at a constant hot temperature. Soon afterward, models tried to account for the subduction zone magmatism; initially they assumed that the magmatism was the result of frictional/viscous heating at the slab–wedge interface (Oxburgh and Turcotte, 1968, 1970; Turcotte and Schubert, 1968, 1973). Clearly this assumption dominated the preceding thermal models and all other models where it was significant (Minear and Toksoz, 1970a, 1970b; Toksoz *et al.*, 1971). The preceding thermal models are not considered realistic since frictional and viscous heating are now believed to be of only secondary importance. This fact is expanded upon in the section Frictional Heating at the Slab–Wedge Interface.

The subducting plate induces a secondary flow in the mantle wedge, which advects heat into the mantle wedge corner (Anderson *et al.*, 1980; Andrews and Sleep, 1974; Bodri and Bodri, 1978; Honda, 1985; Hsui *et al.*, 1983; McKenzie, 1969). Hsui *et al.* (1983) argued that the induced flow could be sufficient to melt the subducting slab significantly. Davies and Stevenson (1992) have shown with a similar model, but one that includes the slab as well as the mantle wedge, that the subducting slab is unlikely to melt extensively. Honda (1985) modeled the Tohoku subduction zone in Japan, and he needed to propose relatively high shear stresses on the interface to satisfy the surface heat flow. Subsequent work though has shown that substantial heat is generated in the continental crust in Japan, and as a result high shear stresses are not required (Furukawa and Uyeda, 1989; Nagao and Uyeda, 1989). Thermal models that include induced secondary mantle flow and ignore other heat sources have also been used to investigate seismic signatures of subduction (Creager and Jordan, 1984; Helffrich *et al.*, 1989).

All the preceding models have prescribed the subduction process kinematically. Dynamical modeling of subduction is only just beginning and it is not as robust as kinematic modeling, especially since the initiation of subduction is so poorly understood. Workers have attempted to combine the kinematic description of subduction with a dynamic model in the mantle wedge (free convection). Many problems with interpreting and constraining these models exist, as discussed in Davies and Stevenson (1992). First, the dynamical effects can be very sensitive to the rheology; for example, a large increase of viscosity at depth could lead to dramatic consequences (Jurdy and Stefanick, 1983). Unfortunately, the details of the actual rheology are not well known. The second problem is more fundamental for these hybrid models: the negative thermal buoyancy in the mantle wedge near the slab is counted twice because it contributes to the motion of the subducting slab (which is prescribed kinematically), as well as the dynamic motions in the mantle wedge. Therefore we suggest that free convection involving thermal buoyancy can be rigorously entertained only in dynamic models of subduction. Otherwise, it is impossible to know how to modify the thermal buoyancy of the mantle wedge to account for the kinematic description of the subducting slab. Jurdy and Stefanick (1983) have demonstrated that the possible effects of free convection can be significant given low viscosities. Since the effects are currently poorly constrained, investigators have favored the better constrained end-member case of forced convection. This case assumes that sources of buoyancy in the mantle wedge will only modulate the overall flow field rather than dominate it (Davies and Stevenson, 1992). Clearly this assumption needs to be checked by a dynamic model of subduction constrained by good estimates of mantle rheology.

Outline of Thermal Model

In this section we present an outline of the thermal model, and in subsequent sections we discuss some of the more important assumptions in more detail. Before presenting the results we also discuss theoretical and numerical details. The fundamental assumptions are that the subducting plate induces a secondary flow in the overriding mantle wedge and that frictional heating is not considered significant. In addition, the thermal lithosphere is assumed to be 100 km thick whereas the mechanical lithosphere is assumed to be only 40 km thick. We discuss in the section Ablation why this difference might be reasonable. No significant heat sources or body forces are assumed to exist. The mantle is assumed to have a Newtonian constant viscosity rheology. Given a kinematically driven flow field and the lack of body forces, temperature-dependent viscosity was found to make little difference. Temperature-dependent viscosity could have significant effects in dynamic models where body forces are included.

The thermal results presented are for steady state—thus the results describe steady-state thermal structures. Hence it must be remembered that before the steady state is reached, temperatures in the mantle wedge will probably be higher than those presented here, since subduction zones at the time of their initiation are expected to be at least as hot as average mantle. Subduction zones probably reach thermal equilibrium relatively quickly, being largely controlled by the time it takes for heat to conduct through the mechanical lithosphere of the overriding plate. For a lithospheric thickness of 40 km, this fact suggests a time of around 5 Myr ($t = d^2/\pi^2\kappa$). The attainment of thermal equilibrium will probably take longer if the ablation of the mechanical lithosphere is achieved partly by the conduction of heat. For instance, if the mechanical lithosphere were 80 km in thickness, then the equilibration time would approach 20 Myr.

Induced Flow in the Mantle Wedge

We first present arguments as to why the subducting slab must induce a secondary flow in the mantle wedge. The mantle wedge must be mobile since a stagnant mantle wedge could not provide the source material from which to generate the magma to produce some of the large volcanic structures (e.g., the Andes) found above subduction zones. We expect the slab to induce flow in the mantle wedge by shear coupling across the wedge–slab interface. Even if a very weak shear zone that limited shear coupling existed between the subducting plate and the overriding mantle, the neighboring mantle wedge would still be cooled by the slab and its net resultant negative buoyancy would drive a downflow. In addition, a stagnant mantle wedge would be cooled progressively by the underthrusting subducting slab and would be-

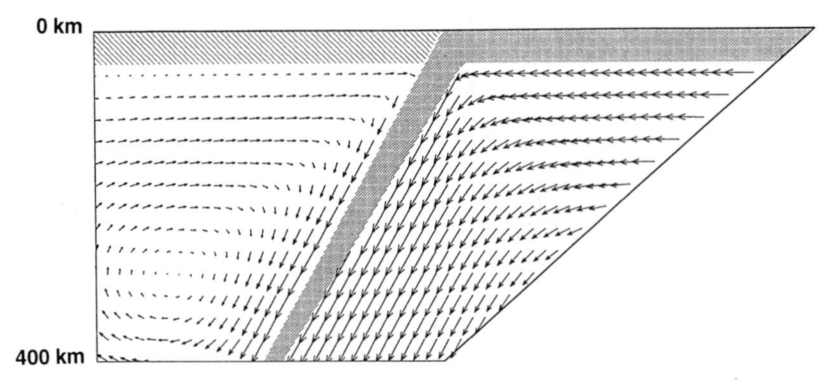

Figure 1 The velocity field induced by a subducting slab. Note no absolute velocity scale is provided, since the scale is arbitrary. The longest vectors correspond to the subduction velocity of the descending slab. The descending slab and overriding mechanical lithosphere are shaded. From Davies and Stevenson (1992).

come ever cooler. Ultimately, without the induced flow advecting in heat, the region beneath the volcanic front would become so cold that no magmatism or heat flow anomaly could exist. Figure 1 illustrates an example of the flow induced in the mantle by a subducting slab, where local thermal buoyancy in the mantle wedge is insignificant, and the influence of rheology on the flow is well approximated by a fluid of constant viscosity.

Frictional Heating at the Slab–Wedge Interface

As a second consideration, we present arguments that support our neglect of frictional and shear heating. Yuen *et al.* (1978) have shown that hot rocks are so weak that little heat is produced; hence, frictional heating cannot play the critical role in generating subduction zone magmatism. In a detailed study, van den Beukel and Wortel (1987) have shown that the effect due to frictional heating is relatively local and can lead to temperatures along the thrust interface in the subduction zone being some 50 to 200°C hotter than those derived in the model of this chapter, depending upon rheology. Their work also shows that ignoring frictional heating has little or no effect on the temperatures evaluated within the mantle wedge. Peacock (1992) also argues for low shear stresses at depths of 15–50 km due to the low temperatures recorded by blueschist-facies metamorphic rocks.

Ablation

A significant aspect of the model is its rheology. In this model we have assumed a rigid lithosphere overriding a viscous asthenosphere. Clearly this model is a major simplification but probably describes the behavior of this system well on a timescale of 10 Myr or greater. The depth of the transition is somewhat arbitrary. For the following model we have assumed that this transition occurs at a depth of 40 km. It is argued that a thin mechanical lithosphere might be expected near the volcanic front due to the effect of ablation. The mantle flow is focused slightly upward in this region if it is like classic corner flow (as observed weakly in Fig. 1; Batchelor, 1967). This focusing will then lead to higher temperatures, which will lead to less viscous rheologies, leading to a further focusing of the hot advective flow (Bodri and Bodri, 1978). This ablative process will be limited when the conduction of heat and the temperature-dependent rheology combine to ultimately produce a steady-state geometry. Another contributing factor to the ablation is the very high stresses implicit in the flow through such a confined geometry; remember that high stresses lower the effective viscosity in materials described by power-law rheologies, such as those undergoing dislocation creep. Figure 2 is a schematic of the possible geometry for the mantle wedge corner, reflecting the largest changes in rheology. In the thermal modeling of subduction zones, the mechanical boundary layer has been assumed to be of constant thickness, rather than thickening away from the wedge corner. We have used a thickness more appropriate for the ablated wedge corner since we are more interested in modeling the temperatures in the source region of the subduction zone magmas. If the lithosphere were thickened away from

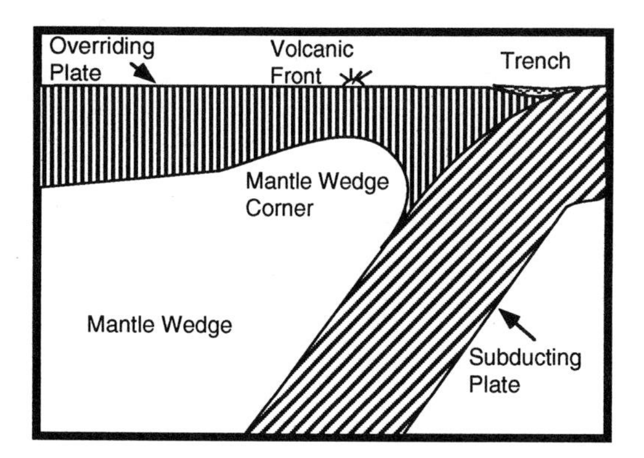

Figure 2 Summary diagram of the possible geometry of wedge corner following ablation.

the trench, then one would expect higher temperatures since the material being advected into the mantle wedge would be coming from a greater depth. In addition, while a sharp wedge corner is common in thermal models, in reality it is likely to be smooth. A smooth corner in the model presented would lead to slightly lower temperatures at the slab–wedge interface at shallow depths, since the induced flow would not be able to impinge as strongly on the shallow slab.

Trench Migration

There is much evidence that subducting slabs generally do not subduct in a direction parallel with their dip, but rather move at an angle between their angle of dip and the downward vertical (Garfunkel *et al.,* 1986; Hamilton, 1988). In all thermal models to date, the subducting slabs have been modeled as "sliding down slots," i.e., moving parallel to their dip. The difference can be largely accounted for by considering the subduction velocity (the velocity at which the subducting plate enters the asthenosphere) rather than the convergence velocity (the velocity at which the subducting plate approaches the rigid part of the overriding plate—the rigid part is usually beyond the arc and any back-arc basin). Clearly how the mantle flow accommodates rollback (trench migration) is potentially quite significant, but if, during thermal assimilation at depth, the subducting slab is incorporated into the mantle on the overriding plate side, then the thermal models are probably only weakly affected. On the other hand, if the mantle accommodates the process by flowing from beneath or around the subducting slab, then there is a greater impact. It is interesting to note that a wide range of convergence velocities exist, but for slabs that subduct deeply (>400 km), subduction velocities seem to be largely constant (Otsuki, 1989; Otsuki *et al.,* 1990). As seen in the following section, relatively high subduction velocities are required for hot mantle wedges, and hence in this respect rollback and back-arc spreading are critically important in raising the values of some low convergence velocities.

Theoretical Background and Numerical Implementation

We have solved the advection–conduction heat equation assuming an incompressible material, where the subduction process is enforced by means of kinematic boundary conditions. The material is assumed to have a Newtonian constant viscosity rheology. Note the cold rigid lithosphere of the overriding plate is enforced by internal boundary conditions where the appropriate nodes are held fixed. Similarly, the mechanical lithosphere of the downgoing plate is modeled kinematically by prescribing the subduction velocity to appropriate nodes. The velocity boundary conditions are illustrated in Fig. 3. The following equations were solved.

The equation for the conservation of mass (continuity equation),

$$\nabla \cdot \mathbf{V} = 0, \tag{1}$$

where $\mathbf{V}$ is the velocity of the mantle, and ∇ is the gradient operator.

The equation for the conservation of momentum (Stokes equation),

$$\nabla P = \nabla \cdot (\eta \nabla \mathbf{V}), \tag{2}$$

where P is the difference between the total pressure and the hydrostatic pressure, and η is the viscosity of the mantle.

The equation for the conservation of energy (heat equation),

$$\partial T/\partial t + \mathbf{V} \cdot \nabla T = \kappa \nabla^2 T, \tag{3}$$

where T is temperature, t is time, and κ is the thermal diffusivity.

We have used the Boussinesq approximation (Tritton, 1977), which implies we ignore all variations

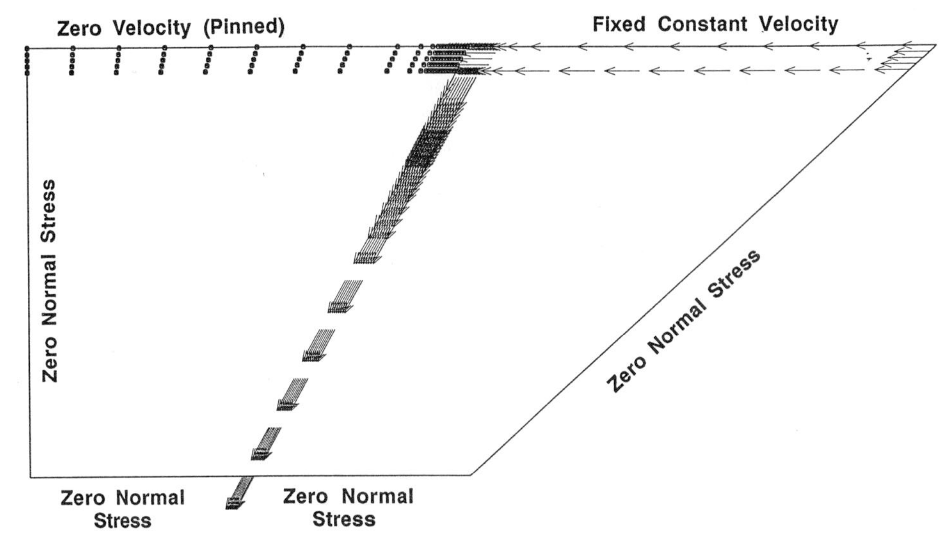

Figure 3 The velocity and stress boundary conditions for the momentum equation, for the 60° dipping slab; similar boundary conditions applied for the 30° dipping slab model. From Davies and Stevenson (1992).

in density other than as a body force in the Stokes equation; this assumption leads to the continuity equation reducing to a requirement for incompressibility. Since we solve for steady-state temperature structures, the heat equation becomes

$$\mathbf{V} \cdot \nabla T = \kappa \nabla^2 T. \tag{4}$$

The equations were nondimensionalized as follows: $x' = x/d, z' = z/d, t' = \kappa t/d^2, P = \eta\kappa P'/d^2$, where d is a characteristic length scale, e.g., the depth of the model (400 km). The primes correspond to nondimensionalized quantities. Dropping the primes we get the nondimensionalized equations

$$\nabla P = \nabla^2 \mathbf{V} \tag{5}$$

and

$$\mathbf{V} \cdot \nabla T = \nabla^2 T. \tag{6}$$

The thermal boundary conditions are illustrated in Fig. 4, together with the finite-element grid used to solve the preceding equations. A compari-

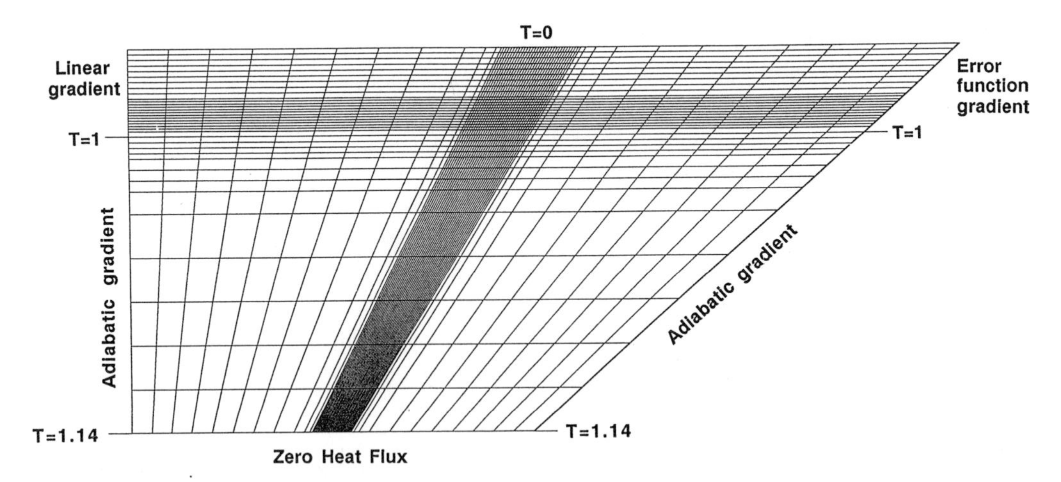

Figure 4 The temperature and heat flow boundary conditions, and the finite-element grid used to solve both the energy and momentum equations. This grid is for the 60° dipping slab; the grid for the 30° dipping slab model had even higher resolution. From Davies and Stevenson (1992).

son of the thermal boundary equations in Fig. 4 and the velocity boundary conditions in Fig. 3 shows the difference in the assumed thicknesses of the thermal and mechanical lithospheres already discussed. The finite-element grid allows us to reduce the influence of the boundary conditions selected by keeping the boundaries far from the region of interest. It also allows good resolution in the region of interest without incurring unnecessary additional computational expense.

The momentum equation was solved using a penalty method to enforce incompressibility (Hughes *et al.*, 1979a), whereas the energy equation is solved implicitly (Hughes *et al.*, 1979b) using a streamline upwind Petrov–Galerkin method (Brooks, 1981). An explicit version of the code has been adapted for vectorizing machines and is described in King *et al.* (1990).

The equations solved are nondimensional, but to present specific results we assumed the following constants. The thermal diffusivity κ was assumed to be $10^{-6}\,m^2 \cdot s^{-1}$ (Fujisawa *et al.*, 1968) and $T = 1$ (the temperature at the base of the thermal lithosphere) was assumed to be 1325° C (McKenzie and Bickle, 1988). The actual value of the matrix viscosity was not required since it does not appear in the nondimensionalized equations, other than as a scaling for the pressure, which is never interpreted. This simplification is a consequence of the fact that the velocity field is kinematically prescribed.

Results of Thermal Modeling

In Fig. 5 we illustrate the results of the thermal modeling (Davies and Stevenson, 1992). What is demonstrated most clearly is how the subduction zone is appreciably cooler than the surrounding mantle. This difference is expected since subduction zones involve the advection of oceanic crust and mantle that has cooled near the Earth's surface and is brought back into the mantle. The obvious question and a seeming paradox is, Why should there be melting at all in subduction zones? We now try to describe the physical processes that control some of the details of the results.

In thermal models where local heat sources, as well as advection of heat by migration of aqueous fluids or melts, have been ignored, the only processes left are advection by flow of solid matrix and conduction. Frictional heating was argued previously to be of only limited importance and the migration of fluids (Peacock, 1987b) has been shown to be unimportant for the large-scale thermal structure of subduction zones. Advection is the transport of heat by the movement of material; and in the subduction zone case, the advection of the ocean floor into the mantle is of the greatest importance. Since heat conduction in rocks is very inefficient, relatively cool temperatures may extend deep into the upper mantle (McKenzie, 1969). Since a secondary flow is induced in the mantle wedge, there is also significant advection of heat

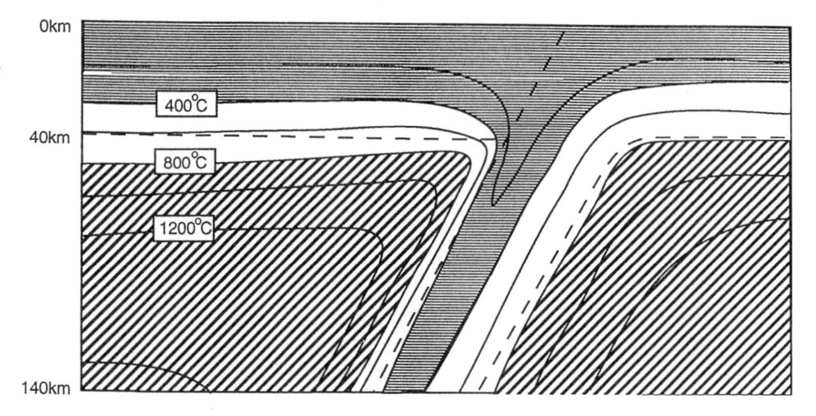

Figure 5 Thermal field of a subduction zone with induced flow of the mantle wedge. The dashed lines outline the mechanical lithospheres. The coolest and hottest temperatures have been shaded for emphasis, with the contours 200° C apart. Note how cool temperatures extend down into the mantle. The contour between the two shaded regions is the 600° C isotherm. Note how the cold region of the slab grows inward away from the interface, and the hot region also moves slowly away from the interface. These regions are termed "boundary layers" in the text. Adapted from Davies and Stevenson (1992).

toward the wedge corner. The induced flow though must ultimately turn and flow down-dip alongside the slab, and the heat flow across the slab–wedge boundary is by conduction. Hence, due to the inefficiency of conduction, the slab–wedge boundary remains relatively cool, while at the same time the mantle wedge can remain hot, even though it is close to the subducting slab. From a close inspection of Fig. 5 one can observe two thermal "boundary layers" developing very slowly from a point just below where the slab enters the wedge; one grows out into the wedge, and the other grows into the slab, as we go deeper. They can be thought of as thermal boundary layers since the flow of heat perpendicular to them is nearly totally by conduction, with little or no contribution by advection. The presence of two boundary layers reduces the cooling effect of the slab on the wedge, and the heating effect of the wedge on the slab; in this way very steep thermal gradients can be maintained.

Dependence on the Slab Velocity

Since advection is all important, the velocity at which slab material is subducted clearly plays a dominant role in controlling the thermal structure through the thicknesses of the thermal boundary layers. The effect of subduction velocities on the thermal structure is illustrated in Fig. 6. Slower velocities lead to less advection of heat, and hence, conduction is relatively more significant, leading to weaker thermal gradients. Hence, the subducting slab is hotter and the mantle wedge is cooler, for slower subduction velocities. Otsuki's (1989) observation, noted previously, that subduction velocities are practically constant and high at 7 cm · yr^{-1} even though convergence velocities vary widely is quite significant in this respect.

Dependence on the Age of Subducting Plate

Irrespective of its age, the surface of the oceanic crust is always near 273 K just before subduction. Therefore, material this cool is always advected into the mantle wedge. Hence, one always expects a relatively cool tongue to penetrate somewhat into the mantle. Clearly, the age of the subducting plate plays little role in the temperatures in the overriding mantle wedge since it is largely a balance between the advection of heat by induced secondary wedge flow and conduction into the cold oceanic crust of the downgoing slab. If the subducting plate is very young, then it is possible for the oceanic crust to be heated by its underlying hot mantle in such a way that the cold tongue disappears by the time the oceanic crust reaches a depth of 100 km. In this case, clearly the mantle wedge is not cooled as dramatically, and both the slab and the wedge are much hotter. Since local heat sources such as frictional heating and dehydration play a role in deciding at what age this effect is significant, a firm answer cannot be given from this model, but we are confident that the thermal structure in the mantle wedge is practically independent of the age of the subducting plate for ages greater than 30 Myr, and probably much younger.

Dependence on the Dip of the Subducting Plate

The dip of subduction was found to play a negligible role, provided it was not very shallow. The dip only begins to affect the thermal structure in the mantle wedge when it is so shallow that the thermal boundary layer under the overriding plate and the thermal boundary layer growing out from the subducting plate into the mantle wedge begin to interact and squeeze out the hot tongue from the corner of the mantle wedge.

Implications of the Results

From the results of the thermal modeling, one can see that the induced secondary flow does not lead to high temperatures at the surface of the subducting oceanic crust as proposed by Hsui *et al.* (1983). This result occurs because the velocity of the induced flow normal to the oceanic crust must be zero, and it approaches zero over an appreciable distance (~10 km). Therefore, the subducting oceanic crust is not likely to be the major source region of subduction zone magmas.

Adiabatic Upwelling

Adiabatic upwelling could play a role in subduction zone magmatism. Rather than address individually the specific models that have been proposed, we shall concentrate on presenting arguments that show the general difficulty encountered by all such processes. Although extension combined with decompression melting in ascending mantle flow is the cause of melting at mid-

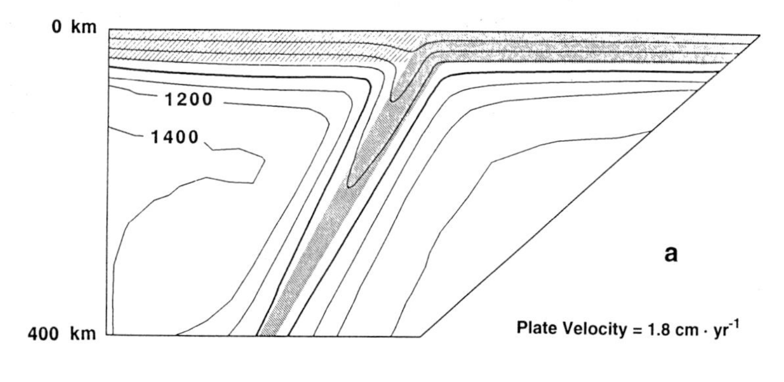

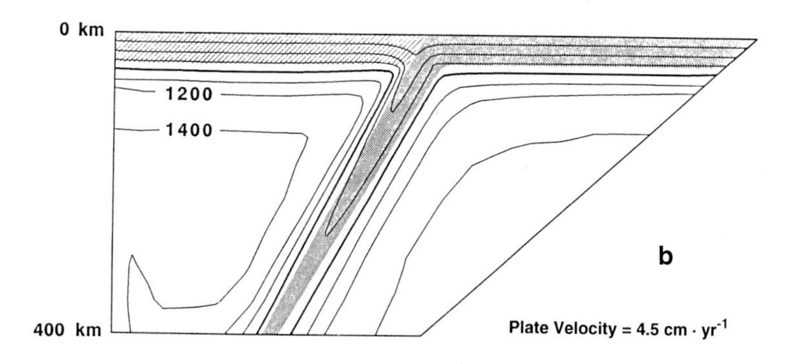

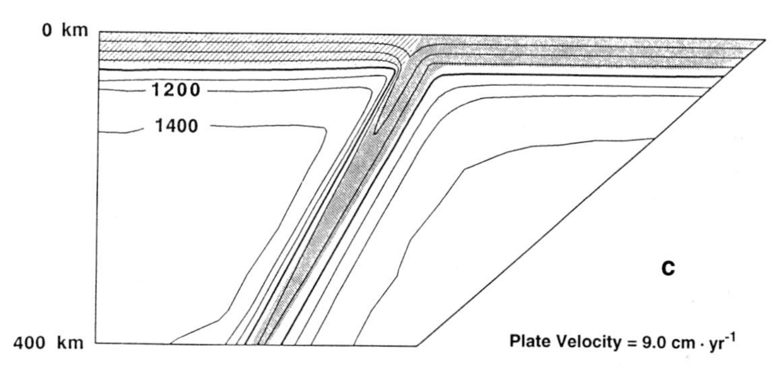

Figure 6 Effect of subduction velocities on thermal structures. (a) 1.8 cm · yr^{-1}, (b) 4.5 cm · yr^{-1}, (c) 9.0 cm · yr^{-1}. The mechanical lithospheres are shaded, and the thermal contours are spaced 200° C apart. The heavy contour corresponds to 800° C. Note the higher subduction velocity leads to the hotter mantle wedge. From Davies and Stevenson (1992).

ocean ridges, in back-arc basins, and in the splitting of some volcanic fronts, it generally is not the *dominant mode* of behavior in subduction zones. Adiabatic upwelling is also the probable explanation for melting at "hotspots," where hotter than average mantle upwells in plumes due to its own intrinsic thermal positive buoyancy. Subduction zones are cooler than surrounding mantle and hence exhibit only negative thermal buoyancy. Melting can potentially produce major sources of buoyancy, the buoyant residue and melt. Clearly though, adiabatic upwelling driven by melting cannot initiate melting. Other potential sources of buoyancy are the presence of water and hydrated mantle; their impact though is much reduced by their presence in cooler material, which is both more rigid and intrinsically negatively buoyant.

Subduction zones not only lack the enabling

factors of adiabatic upwelling of other tectonic environments but also possess characteristics that intrinsically make generating a substantial upflow difficult. This trend is fundamentally because subduction zones are regions of large-scale downflow of mantle material. The slab couples to the overriding mantle wedge leading to an induced downflow in the mantle wedge near the slab. This result will occur to some degree even if they are rheologically decoupled by a weak zone because, as mentioned earlier, the dramatic cooling effects would extend across the decoupling zone and give the surrounding mantle an intrinsic negative buoyancy. Davies and Stevenson (1992) have shown that it would require large amounts of positive buoyancy, a low-viscosity mantle wedge, and a decoupling of the mantle wedge to allow for substantial upwelling. The geometrical constraints also make the development of an effective upflow difficult since a cold rigid subducting slab exists at a depth of around 120 km below the volcanic front, and at around 30–50 km a rigid mechanical lithosphere tends to inhibit further upwelling closer to the surface.

The upflow that complements the downflow in the mantle wedge for corner flow is much more diffuse and more nearly horizontal than vertical (Fig. 1), so that it is not adiabatic. As seen from the results of the thermal modeling, this upflow leads to only a modest rising of isotherms in a restricted region (Figs. 5 and 6). This limited effect is important for the model proposed, but is not sufficient to lead to melting directly by adiabatic upwelling.

Given that magma and a residue of melting are ultimately produced, positively buoyant body forces and hence some net upwelling will clearly exist. We argued previously that its influence will probably be only a modulation. Even if the local buoyancy forces were to become dynamically significant, its influence would wane since the resulting upflow in the mantle wedge would form a small enclosed convection cell that would become progressively less fertile and colder, until finally no melt could be formed (Davies and Stevenson, 1992). Hence, even in the scenario of buoyancy-driven upwelling, adiabatic upwelling could not continue once initiated; rather it would be only the second and dependent one-half of a cycle, initiated by melting in a regime of induced flow in the mantle wedge. Therefore, adiabatic upwelling is of debatable but probably minor importance, whereas melting with induced flow in the mantle wedge is always significant and probably all important. If substantial matrix flow exists parallel to the arc (along-strike), then the preceding statement of an expected closed cell is not rigorous. Such a complicated matrix flow pattern should lead to nearly regularly spaced alternate regions of active (upflowing) and inactive (downflowing) arc. Little evidence for such a complicated model exists.

Phase Equilibria for Hydrous Peridotite

In Fig. 7 we present a simplified phase diagram for both the peridotite and the basalt systems, based on the work of Green and Wyllie (Green, 1973; Wyllie, 1979). The diagram presents the solidus for anhydrous systems, i.e., no water present; the solidus for hydrated systems with no free fluid, i.e., amphiboles present; and the solidus for water-saturated systems, i.e., both amphiboles and free aqueous phase present. The most striking fea-

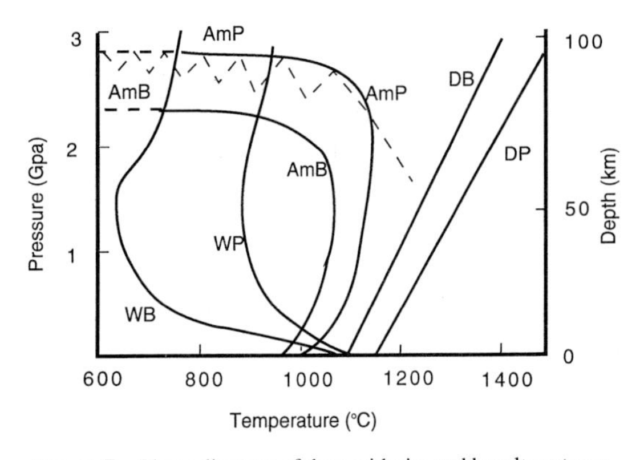

Figure 7 Phase diagram of the peridotite and basalt systems. The second letter identifies the bulk composition: P, peridotite; B, basalt. The first letter describes the amount of water present. D represents dry conditions, i.e., anhydrous. W represents wet conditions; i.e. water-saturated, free aqueous fluid is present in addition to amphiboles. Am represents conditions where there is no free water, but the system is hydrated with the water held in amphiboles. A possible path of a water molecule through the phase diagram is illustrated by the thin dashed line; it progresses from low to high temperatures, where it is ultimately carried away in melts. The up-pressure parts of the path are where the water is carried in the solid matrix fixed in hydrated minerals (probably the amphibole pargasite), while the down-pressure parts of the path are where water moves as a free phase, either by fracture or by porous flow, either as an aqueous fluid or as a water-rich melt. Based on the experimental work of Green (1973) and Wyllie (1979).

ture of this diagram is the substantial reduction in the solidus due to the presence of water. It has been argued for some time that it is the presence of water that is responsible for the generation of melt in subduction zones. What has generally been glossed over is the means by which the water reaches that part of the mantle wedge that is sufficiently hot to produce melt. This omission has been used to cast doubt on the whole process of melting by fluxing; e.g., Marsh (1979a) rightly points out that diffusion alone is too slow to allow sufficient migration of water from the slab into the mantle.

Basic Mechanism of Lateral Water Transport

From the results of the thermal modeling (presented in more detail in Figs. 8a and 8b), it is clear that the subducting oceanic crust is relatively cool

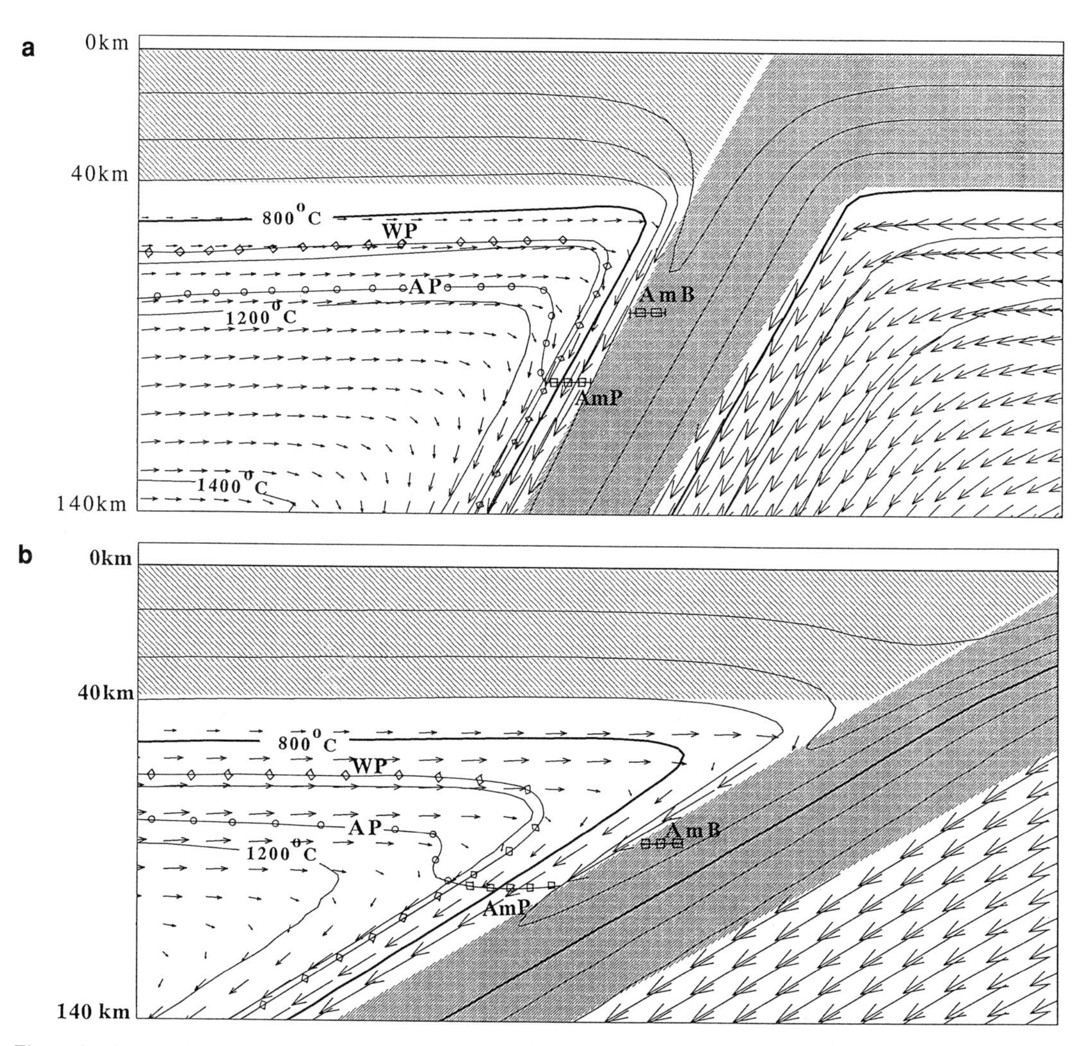

Figure 8 Computed thermal and flow structures (a) for a 60° dipping slab and (b) for a 30° dipping slab. The heavy contour represents 800° C. An interpretation of the phase boundaries is provided and labeled as in Figure 7, but with AP representing AmP above the wet solidus, where we get dehydration melting. The lines with squares are the breakdown limits of amphibole in peridotite (AmP) and in basalt (AmB), leading to subsequent dehydration. The lines with circles are the breakdown limits of amphibole leading to dehydration melting. The line with diamonds represents water-saturated melting (wet peridotite, WP). The isotherms are separated by 200° C. The directed arrows show the induced flow in the mantle wedge as well as the outboard upper mantle. Note that the lateral transport mechanism is effective out to the thermal stability maximum of amphibole; i.e., the secondary induced flow still has a horizontal component away from the wedge corner, at the base of the amphibole ledge. The largest vectors correspond to the subduction velocity 7.2 cm · yr^{-1}. From Davies and Stevenson (1992).

and the mantle wedge is also cooler than the surrounding mantle. Hence no obvious reason exists why there should be magmatism. We have further argued earlier that although adiabatic decompression might contribute, it cannot be a factor until there is significant melting. Hence, the only well-accepted mechanism left for initiating melting is the input of water into the mantle wedge. Tatsumi (1986) was one of the first to address the problem of how water reaches the hot part of the mantle. He points out that amphiboles dehydrate at shallower depths in the basaltic bulk composition than in a peridotitic bulk composition. He proposes that the amphiboles in the oceanic crust dehydrate by a depth of 80 km (see curve AmB in Fig. 7) and that the water migrates into the mantle wedge where it reacts with the anhydrous peridotite to form amphibole again. He then proposes that this hydrated mantle is carried down by the induced flow until it reaches a depth of around 100 km, at which point the water is released and generates melting (see curve AmP in Fig. 7). From the results of the thermal models in Fig. 8 we see that the mantle next to the subducting plate is not hot enough to melt. Tatsumi (1989) envisages that the water migrates vertically, inducing ever increasing melting as it encounters higher temperatures. He then requires that diapirs be spawned off the top, rise, and be heated to yet higher temperatures (Tatsumi *et al.,* 1983). Ida (1987) points out that the diapirs cannot be small enough to be effectively heated, yet still be large enough to overcome the induced flow. We extend Tatsumi's ideas for the migration of water, which avoids the requirements for diapirism.

As shown in Fig. 8, a hot wedge of mantle exists not far from the oceanic crust. Clearly, if water were transported into this region of the mantle wedge, then it would be possible to generate large amounts of melting even given the downward induced flow. As Tatsumi argues, the water is released from the amphiboles at a depth of about 100 km in the mantle wedge (as seen in Fig. 7, and interpreted in Fig. 8). Provided the water moves vertically more quickly than it is carried back down by the induced secondary flow, it will rise through the amphibole-saturated mantle until it reaches higher dry mantle. Here it reacts with the hot country rocks to form amphibole. It then again becomes part of the induced flow and is carried back downward and away from the mantle wedge

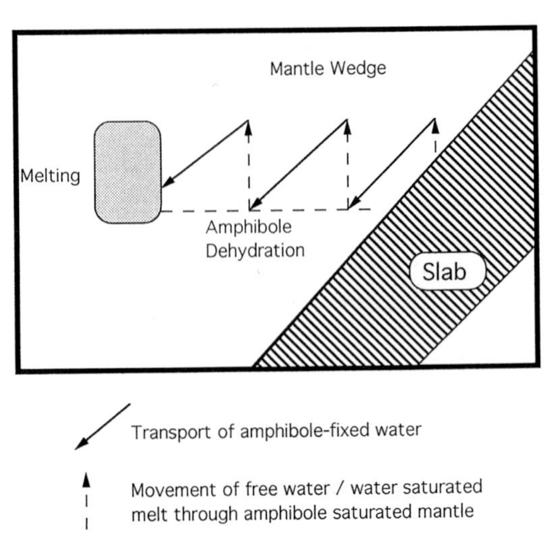

Figure 9 Synopsis of the lateral water transport mechanism. We schematically follow one water molecule. To emphasize the two main components of the mechanism, we have assumed that the water can always migrate rapidly vertically. From Davies and Stevenson (1992).

corner. Once more, it reaches a pressure at which the water is released, and the cycle is repeated. As seen in Fig. 9, a schematic of the fundamental process, the net effect is the lateral transport of water across from the slab and into the hot mantle wedge. Watson *et al.* (1990) have shown that water establishes an interconnected network with olivine under higher pressures and temperatures (e.g., ≈1100° C at 1 GPa, and probably around 1000° C at 3 GPa), allowing the vertical transport of fluids by intergranular porous flow. They note that surface energy anisotropies seem to give partial connectivity for water at temperatures as low as 900° C (at 1 GPa). Connectivity at such low temperatures is also suggested in Watson (1991). At some point, the water is not released as a free phase from the amphiboles on their high-pressure breakdown, but is dissolved in the melts formed by their breakdown in dehydration melting (Beard and Lofgren, 1990). These melts similarly rise through the amphibole-saturated mantle and react above with dry mantle to yield amphibole. Fujii *et al.* (1986) have shown that hydrous basaltic melts readily lead to permeable interconnections along grain boundaries in peridotite, so even if crack propagation is not dominant, the melts in this region are still expected to rise rapidly by porous media flow. At some distance into the wedge though, amphibole is no longer stable, and in such a case one expects appreciable melting. If sufficient melting does occur, then some melt can segregate from the source region and be emplaced in the lithosphere, leading ultimately to volcanism.

Clearly, important requirements of the water transport mechanism are (i) that the water or water-rich melt can at some point migrate rapidly in a near-vertical direction and (ii) that the induced flow has a significant component of velocity away from the mantle wedge corner, at least out as far as the point at which amphibole is no longer stable.

Hence, this mechanism suggests that the location of the source region is well out in the interior of the mantle wedge, initiating at the depth corresponding to the maximum thermal stability of amphibole, and rising more or less vertically with probably the bulk of the melting occurring at the shallowest regions of the source region. It is unclear where the top of the source region will be. Davies and Stevenson (1992) assumed that the top of the source region corresponds to the maximum temperatures encountered by the rising melts.

One question that must be addressed is, Why do the very hydrous melts (which are part of the lateral transport mechanism) rehydrate the mantle when they reach dry mantle, but the melts leaving the source region do not all stop and rehydrate the shallow mantle? A very important aspect of this apparent selectivity must be that for melts to reach the near surface in subduction zones, the magmas must segregate into large cracks. Only then can they propagate sufficiently rapidly so that they do not cool, freeze, and react with their mantle wall rock to form hydrous minerals. On the other hand, for the very hydrous melts not to segregate away from the mantle wedge, they must move sufficiently slowly in small cracks or more slowly still in small veins or pores by porous flow. In this case chemical equilibrium is apparently sufficiently maintained so that the hydrous melts react with their wall rocks to form hydrous minerals (assuming that they are not already saturated), while migrating only a short distance. The potentially different styles of behavior must reflect the appreciably lower porosity away from the source region, which may not allow the development of major dikes and cracks.

Clearly it is possible that even some of the melt formed in the source region is not emplaced in the mechanical lithosphere. Rather it might stop below the mechanical lithosphere and freeze, reacting with its surrounding mantle; then it would be entrained back into the wedge corner where it would have to be recycled quickly. Morris *et al.* (1990) have argued from the variation of $^{10}Be/Be$ versus B/Be that the mantle wedge is uncontaminated with boron; i.e., its residence time in the subarc mantle is comparable to, or less than, the ^{10}Be decay time, ~5 Myr. Since boron would probably be found in any melts that could not escape the mantle wedge, this constraint suggests that either all melts do leave the mantle wedge or they are quickly recycled. Melts that do not reach the crust but are emplaced in the mantle part of the mechanical lithosphere also satisfy this boron constraint.

Discussion of the Lateral Transport Mechanism

In this section we expand on our mechanism, considering the process in more detail and then clouding it somewhat by considering potential complications.

Transition from a Silica-Rich Fluid to a Hydrous Melt

As mentioned previously, at some point the mobile phase changes from a silica-rich hydrous phase near the subducting plate to a hydrous silicate melt near the source region. At pressures of 3 GPa, they are two separate phases and cannot be changed continuously from one to the other (Eggler, 1987). The melt can hold up to 25 wt% water (Green, 1973) and the water can hold more than 30 wt% silica (Walther and Helgeson, 1977).

Due to the potential two-phase nature of the fluid in this transition region, permeability could drop and lead to a concomitant increase in the amount of fluid present. The water-rich phase is expected to have a very low viscosity, and the hydrous melt phase to have a higher viscosity. Three factors compete to control its viscosity: most importantly the high water content that will lower the viscosity and the high silica content that will increase the viscosity, and less importantly the relatively low temperatures that will lead to higher viscosities. Clearly the lower permeability and the possibility of higher viscosity could lead to decreased velocities and hence a buildup in fluid-filled porosity at the transition. This buildup could lead to a substantial body force, but its impact may well be limited by the fact that permeability is such a strong function of porosity (proportional to the second or third power; Cheadle, 1989). Even though we do not know the detailed behavior of the system during this potentially fundamental transition, it is plausible that the permeability is not greatly reduced due to the ability of heat conduction at grain scales to produce only a single phase in any portion of the porous network. In this case, the two phases will not interfere and hence the permeability might not be reduced dramatically. In addition, the high water contents of the melts probably lead to relatively low viscosities, such that the increase in porosity as we cross the transition will be so small that the buildup of porosity and positive melt-enhanced buoyancy is insufficient to control the overall dynamics of the mantle wedge.

Transport of Water-Dominated Fluid

Watson *et al.* (1990) have shown that water exhibits lower dihedral angles with olivine (i.e., is more likely to set up an interconnected fluid network along crystal edges) under higher pressures and temperatures. The water probably does not form an interconnected network upon initial release into the mantle wedge due to relatively low temperatures. It does at some point however form an interconnected network since the water is continually taken to conditions of higher pressure and higher temperature. The texture at equilibrium depends upon the nature of both the matrix and the fluid, through the surface interactions. Since the dissolved contents of the hydrous phase can change the nature of these surface interactions they could be very significant. Fluids in subduction zones have been interpreted to have appreciable amounts of dissolved solutes (Philippot and Selverstone, 1991), which would probably increase the probability of interconnection (Watson *et al.*, 1990). Experimentally Watson *et al.* (1990) have found only a small effect for halides with olivine as opposed to the large effect of halides with quartz. Dissolved silica might be more effective because it would make the fluid more melt-like (melts do interconnect with olivine), and the level of dissolved silica could be very high. Provided that the bulk of the water released at 100 km depth on breakdown of pargasite is not incorporated in hydrous minerals that are stable to much greater depths and temperatures, then it is not critical if the water does not interconnect immediately following dehydration.

The presence of CO_2 reduces the possibility of interconnection (Watson *et al.*, 1990). The amount of CO_2 relative to H_2O in fluids released from the slab is probably small (Giggenbach, 1992). This assumption is also suggested by the fact that calcite is stable under the conditions present shallow in the subduction zone (Huang *et al.*, 1980).

Unfortunately no textural experiments have been undertaken between water and garnet, amphibole, or pyroxenes, so we do not know the depth at which an interconnected network will be formed in basalt/eclogite. Water has been speculated to have a smaller dihedral angle with garnet than with olivine (Davies and Stevenson, 1992). This speculation is based solely on the observation that garnet tends to float in pools of melt in amphibolite melting experiments, which suggests strongly that at least basalt readily interconnects with garnet. It must be noted that the lateral transport mechanism does not require that all the water actually enter the wedge at a depth of 80 km or shallower. We speculate as follows on potential mechanisms and pathways for water mobility in the mantle wedge.

If olivine is a good analog for eclogite minerals, then the cold temperatures of the slab suggest that there is no interconnection at textural equilibrium. The cold temperatures could prevent the attainment of textural equilibrium, but it must be remembered that the slab is undergoing phase changes, and releasing water by dehydration; both processes lead to more rapid kinetics.

If experiments with eclogite minerals show no interconnection, even at higher pressures, then we need to consider other processes that are currently poorly understood. The existence of magmatism itself is the strongest argument for the mobility of water from the cold slab and across the cold mantle wedge because it is easier to accept the hypothesis of water mobility under these conditions than to discard the hypothesis of hydrous fluxing of the mantle wedge as the cause of subduction zone magmatism. Clearly, this aspect of the migration of water from the slab to the wedge is the most poorly understood and is the weakest link to the hydrous fluxing hypothesis. For those who do not see hydrous fluxing as a strong hypothesis and believe the mobility of water under cold conditions to be impossible, we present some further evidence for the mobility of water.

Large serpentine diapirs have been observed in the Marianas fore-arc (Fryer, 1992), veining in the Catalina schist fore-arc melange (Bebout, 1991; Bebout and Barton, 1989), and pervasive hydration and metasomatism in the ultramafic hanging wall of the Trinity thrust (Peacock, 1987a), as well as appreciable flow deep in accretionary prisms (Vrolijk *et al.,* 1988); all point to the mobility of water under these conditions. Mysen *et al.* (1978) have experimentally measured an aqueous infiltration rate in peridotite of 23 $m \cdot yr^{-1}$, at 2 GPa and 850° C, although the identity of the active micromechanisms and their relevance for the mantle remain uncertain.

An additional process is probably some form of microcracking, but first we mention factors that might make the development of a porous network more likely than suggested solely by dihedral angle data. Interconnection could result from aligned, needle-shaped amphiboles with anisotropic surface energies, such that there is interconnection in a special direction (Wolf and Wyllie, 1990). Alternatively, interconnection could be the result of heterogeneous and localized hydration along fractures at mid-ocean ridges, leading naturally to interconnection at dehydration, assuming that the heterogeneous hydration was not homogenized during the subsequent metamorphism as the result of subduction. One interesting possibility in the wedge is that the water sets up an interconnected network by forming pathways around amphibole grains that form during mantle hydration. This process requires sufficient volume fractions of amphibole grains so that the resulting network could cross the percolation threshold. If so, then the water could be generating its own permeability by amphibolitizing the mantle wedge.

Microcracking must be seriously considered since water at these temperatures and pressures is very reactive and hence could initiate subcritical cracking by means of stress corrosion (Anderson and Grew, 1977; Atkinson, 1984). Ultimately the microfractures coalesce to form larger cracks that could propagate efficiently and temporarily leave a porous network in their wake. Since the cracks are always propagating into conditions of higher temperature, they suffer less from blockage by the precipitation of solutes. Watson *et al.* (1990) state the difficulties inherent in the consideration of volatiles transported by cracks. Experiments by Brenan and Watson (1988) suggest that crack propagation is a plausible mechanism for fluid migration at high P and T, at least at the grain scale.

Clearly of significance is the associated deformation, which could lead to aligned textures and also to the syntectonic migration of fluid (Philippot and Selverstone, 1991). For example, the continuous increase of water pore fluid pressures may ultimately produce hydrofracturing. It is possible then that the volume increase produced by the dehydration of some minerals could lead to higher pressures if the fluid cannot readily escape. This situation is unlikely for amphibole dehydration in its high-pressure breakdown region since that involves a volume decrease. The increase in temperature experienced by the fluids as they are dragged down in the wedge flow field and especially as they propagate upward could lead to increases of pressure, which could hydraulically open fractures. Are intermediate-depth earthquakes related to the presence of overpressured fluids (Blanpied *et al.,* 1992)? If so, it may be possible that some fluids are differentially compressed, and perhaps because of the temporary presence of a seal, for example, cannot readily migrate, leading to a pore pressure increase, and facilitating subsequent faulting. The possibilities are numerous. Pressurization could reach a level where one would get hydrofracturing in several plausible scenarios. The direction of least compressive stress (Davies and Stevenson, 1992) is such that a hydrofracture would be expected to be directed outward and upward into the mantle wedge.

From the phase diagram in Fig. 7, we see that the solidus of water-saturated basalt is between about 650 and 750° C. Even though the temperatures at the boundary between the subducting oceanic crust and the mantle wedge in the models presented are generally lower than this, it must be remembered that frictional heating is ignored, which will increase temperatures by up to 50° C (Peacock, 1990) or possibly by as much as 200° C, depending upon rheological conditions (van den Beukel and Wortel, 1987). In addition, temperatures are higher for younger subducting crust and for younger subduction zones, which have not yet established a steady-state thermal structure. The slab–wedge interface gets hotter as we go deeper. Therefore, there are certain situations in which the water leaves the slab as a hydrous melt and not as a water-rich fluid. Since most silicate melts examined experimentally to date form interconnected networks with most silicate minerals, it is probable that this melt would also have no problem migrating from the subducting crust (Cooper and Kohlstedt, 1982; Waff and Bulau, 1979, 1982).

Given that most processes of migration are probably favored by higher temperatures and pressures, two other possible geometries for the lateral water transport mechanism are illustrated in Fig. 10.

Estimate of the Water Flux That Enters the Mantle Wedge

The water starts its journey from the ocean to the mantle wedge during hydrothermal activity at the mid-ocean ridge. Alt *et al.* (1986) have logically argued that the hydrothermal alteration of oceanic crust is strongly controlled by permeability. Following this logic, the greatest alteration should be expected in the extruded basalts, decreasing as we descend through the sheeted dikes, and being dramatically lower in the gabbros. Peacock (1990) assesses the degree of alteration and suggests 2 wt% water in the upper 2.5 km of basalt and 1 wt% water in the lower 5 km of gabbro.

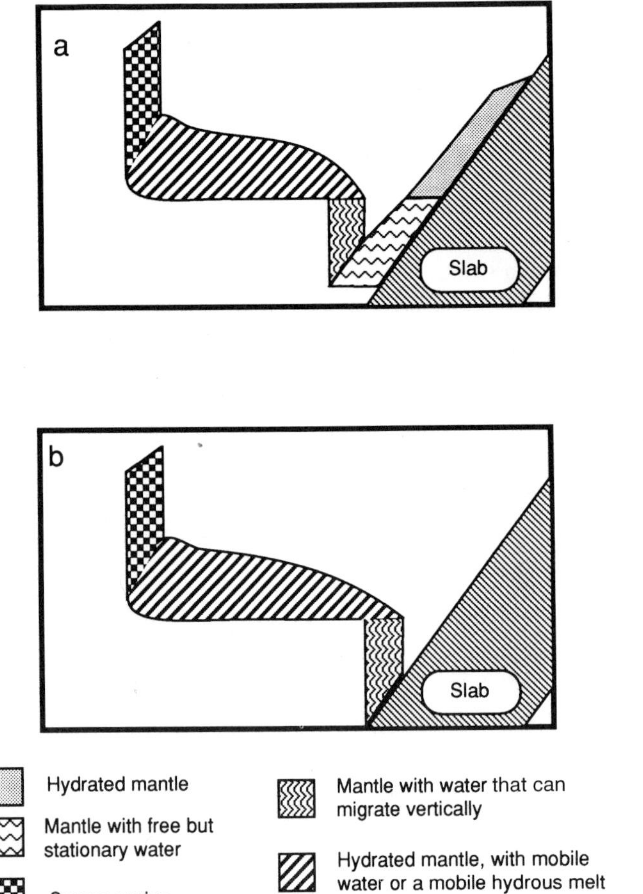

Figure 10 Two possible paths for the lateral transport of water out into the mantle wedge. (a) Water leaves the slab easily, but cannot migrate vertically in the mantle wedge until it is dragged to higher temperatures and pressures. (b) Water cannot leave the slab until it is dragged to higher temperatures and pressures, but can migrate easily upon entering the mantle wedge.

Adjusting the figures to allow for the rapid drop in permeability with depth, we suggest that the upper 1.2 km might be widely altered, containing 1.5 wt% water, as observed in hole 504B (Becker *et al.*, 1989), and the rest of the sheeted dikes and gabbros might be negligibly altered, only suffering alteration locally. Other deep sea drilling holes quoted by Peaçock (332B and 418A; 1990) averaged 1 wt% water in their drilled sections (<600 m of pillow basalt); the only substantial drilling of gabbro (a fracture zone where more hydration might be expected because of greater access) in hole 735B gave 0.8 wt% water, averaged over 500 m. This estimate implies substantially less water than Peacock's estimate, which was already less than many previous estimates (Ito *et al.*, 1983, 2 wt% throughout the oceanic crust; Fyfe, 1978, and Anderson *et al.*, 1976, who estimate 3 wt% through the whole thickness of the oceanic crust).

Gregory and Taylor (1981) found in the Oman ophiolite evidence from oxygen isotopes for extensive water–rock interaction to a depth greater than 5 km. It is possible that the Oman ophiolite corresponded to a back-arc basin rather than an oceanic ridge, and clearly that it was obducted rather than subducted implies that it is not typical. Since ophiolites are affected by later regional metamorphism, doubts exist about whether they are truly representative of alteration at mid-ocean ridges.

As the oceanic crust subducts beneath the overriding plate in the trench, most of the sediments are added to the foot of the accretionary prism or underplated beneath. The free pore water and weakly bound water are released and driven to the surface by compaction processes. Hence, most of the water in the sediments and the oceanic crust does not reach the thrust; in fact, it might be the presence of large amounts of water that prevents a brittle thrust extending all the way to the surface.

We assume that all water released at a shallower level than the bottom of the interplate thrust returns to the near surface up the thrust or enters the fore-arc (Fryer, 1992; Peacock, 1987a). Then we need only consider the water that is released below this depth. The question of whether flow is allowed along a fault is difficult to answer; but the evidence for substantial flow along the decollement beneath accretionary prisms suggests that the source of some of this fluid may be the thrust itself (Vrolijk *et al.*, 1988). Further evidence for the mobility of water in subduction zones and accretionary prisms is provided by abundant evidence of pore fluid expulsion (Kastner *et al.*, 1991) and the alteration of the Catalina schist (Bebout and Barton, 1989), which was in the overhanging wall of a subduction zone.

Clearly, as the oceanic crust proceeds to higher pressures and temperatures, so the metamorphic facies change from blueschist to greenschist through amphibolite to eclogite. Blueschists have been estimated to hold 3 wt% water, so it is possible that some of the water released in the sediments is incorporated into the blueschists below, since the initial degree of hydration is probably less than 3 wt%. Generally the water content of each succeeding higher pressure facies is lower. We suggest that all the water released in the transition from the blueschist facies to greenschist facies, and a proportion of water released during the greenschist to amphibolite facies transition, escapes back up the thrust or into the fore-arc since it occurs at a shallower level than the base of the thrust (around 60 km; Tichelaar and Ruff, 1989). The amount of water released into the mantle wedge is therefore probably much lower than the estimate of water fixed in hydrous minerals at the ridge. Ultimately most of the water is held in amphiboles, probably pargasite.

Importance of Hydrous Minerals Other Than Amphiboles

Clearly it is possible that a hydrous mineral more stable than amphibole could take some water to an appreciable depth and that some of this water might reappear in back-arc basin volcanism. Candidates include the nominally anhydrous minerals (e.g., garnet, clinopyroxene, olivine, orthopyroxene; Bell and Rossman, 1992), as well as K-amphibole and phlogopite (Niida and Green, 1990; Sudo and Tatsumi, 1990). They all could take the water deep into the mantle and balance the output from the mantle at mid-ocean ridges and at hotspots. The amounts of phlogopite, K-amphibole, and K-richterite all depend on the amount of potassium available. Since the amount is likely to be very low in a depleted mantle, phlogopite and the other K-hydrous minerals are unlikely to be sufficiently common to clear the subduction zone wedge of all its water. They might be responsible for the water in back-arc basin basalts though (Stolper and Newman, 1992).

Other candidates proposed to remove water from the subduction zone should be discussed. The dense hydrous magnesian silicates, or "alphabet minerals," can hold large amounts of water and are stable to high pressures. It seems probable though that they are not common in hydrated arc peridotites (Wunder and Shreyer, 1992). Serpentine is stable to high pressures, but not to high temperatures, so it could be very important locally in cold regions of the mantle wedge nearest the slab due to its high water content, but is unlikely to be important in removing water from the subduction zone region. Talc is another important hydrous mineral, with relatively high water content; it is also stable to high pressures (beyond 5 GPa), but its thermal stability is limited to less than 900° C. Chlorite stability is temperature dependent, and at 500° C chlorite is stable to 5 GPa, but at 800° C it is stable to only 3 GPa (Goto and Tatsumi, 1990). Like serpentine, it is improbable that talc or chlorite will carry much water out of the subduction zone but again they could be important in the cooler regions of the mantle wedge and could carry some water beyond the source region of the volcanic front. For further discussion of the distribution of water in the upper mantle, see Thompson (1992) and Tatsumi (1989).

How Much Amphibole Is Formed?

If the mantle is dry next to the oceanic crust, then the water reacts to form amphibole (probably pargasite). The mantle entrained nearest the slab is very shallow, possibly directly beneath the crust. The probable depth from which this mantle is ablated is a strong function of the composition and temperature of the mantle, and also the previous tectonic history in terms of stress release, work hardening, etc. Further from the subducting oceanic crust, the entrained mantle comes from greater depths and is probably more fertile. Hence, it is possible that little amphibole can be formed in the harzburgite close to the slab, but it increases volumetrically as we progress away from the slab into more lherzolitic compositions. Clearly, what hydrous phases are formed depends not only on the composition of the wedge but also on the amount of water, the composition of its carrier phase, and the temperature and pressure (Tatsumi, 1989; Thompson, 1992). From experiments with peridotite, pargasite was found to be the primary hydrous mineral formed (Green, 1973). Green suggests that one could get up to 30 wt% pargasite, i.e., a fixed water content in the wedge (w) of 0.4 wt% H_2O. By equating the flux from the oceanic crust into the mantle wedge (F) with the flux carried horizontally by the induced flow of the amphibole-saturated mantle wedge, we can evaluate the thickness of the amphibolitized layer (for further details see Appendix B, Davies and Stevenson, 1992); i.e.,

$$F = V_x w \rho_m h, \tag{7}$$

where F is the flux of H_2O, V_x the horizontal component of mantle velocity, w the weight percentage of H_2O in the amphibolitized mantle, ρ_m the mantle density, and h the thickness of the amphibolitized layer. We have ignored the water in the free fluid porosity, which is also carried horizontally and is shown as follows to be less than w. Assuming $V_x = 2\ \text{cm} \cdot \text{yr}^{-1}\ (6 \times 10^{-10}\ \text{m} \cdot \text{s}^{-1})$, $w = 0.4\%$, and a flux (F) of water entering the mantle wedge, which is the result of the release of 1 wt% water from a 1-km thickness of oceanic crust subducting at $7\ \text{cm} \cdot \text{yr}^{-1}\ (2 \times 10^{-9}\ \text{m} \cdot \text{s}^{-1})$, then the thickness ($h$) is 8.75 km. Since the horizontal component of the induced mantle wedge velocity decreases with distance from the slab, the thickness of amphibolitized mantle increases with distance from the slab. This trend might be countered by the fact that the mantle nearest the oceanic crust possibly cannot hold as much water in hydrous minerals since it might be more depleted due to its shallower origin (i.e., $w < 0.4$ wt%). If the flux instead had been 2 wt% water released from 7 km of oceanic crust then the thickness of the amphibolitized mantle at a point at which its horizontal velocity is $2\ \text{cm} \cdot \text{yr}^{-1}$ would be around 120 km. Since the mantle wedge is not this thick, the mechanism is not able to transport all the water laterally, and the excess water returns to the near surface. As stated previously, we prefer lower fluid fluxes entering the mantle wedge than previously assumed, so we expect the thickness of amphibolitized mantle to be closer to 10 km than 100 km.

How Much Free Water Is There?

The amount of free phase can be calculated by assuming that the system is steady state and by equating the vertical fluxes (Appendix B, Davies and Stevenson, 1992). The vertically downward flux, carried by the hydrous minerals in the induced wedge flow, must be balanced by the vertical flux in the upward direction resulting from the flow of the free positively buoyant phase. If the ascent and vertical transport are the result of hydrofracture, then they are very rapid and very little free phase is required. To evaluate an upper bound we assume that the upward transport is a result of the slower porous flow. The relevant equation is

$$\rho_m V_z(w + (f\rho_f/\rho_m)) = \rho_f a^2 f^2 \, \Delta\rho g/\eta b, \quad (8)$$

where ρ_m is the mantle density, V_z is the vertical component of mantle velocity, w is the weight percentage of H_2O in the amphibolitized mantle, f is the fluid-filled porosity, ρ_f is the density of the fluid, a is the grain size, $\Delta\rho$ is the density contrast between the fluid and the matrix, g is the gravitational acceleration, η is the viscosity of the fluid, and b is the permeability–porosity constant. Here we have assumed that the permeability (k) is proportional to the square of both the grain size (a) and porosity (f), i.e., $k = a^2f^2/b$ (as predicted in detailed calculations by Cheadle (1989) for low porosity, with $b = 3 \times 10^3$). If the water has a viscosity (η) of 10^{-4} Pa · s then we find that the porosity (f) is 6×10^{-5}. We have assumed a vertical velocity (V_z) of 3 cm · yr^{-1}, a grain size (a) of 1 mm, and a density difference ($\Delta\rho$) between the fluid and the matrix of 2.0×10^3 kg · m^{-3}. Clearly such a low porosity has no effect on the phase diagram; i.e., it is unclear whether even the experimental charges of nominally anhydrous runs could demonstrate that they had less than 0.01 wt% water present, due to surface-adsorbed water. Applying the same calculation to the region where wet basalt is the free phase carrying water vertically upward, we find a value for the porosity (f) of 10^{-2}. Here we have assumed a viscosity of 10 Pa · s, a grain size of 3 mm, a density difference of 0.5 k · g · m^{-3}, a vertical velocity downward for the induced flow of 3 cm · yr^{-1}, and a solubility of 20 wt% for water in the melt. A larger grain size might be expected due to Ostwald ripening at the higher temperatures in the presence of a liquid phase. This porosity is sufficiently high to have a limited influence on the phase diagram. Amphibole in this case is not stable to as high temperatures as those for the case with no excess water, which is the assumption for the amphibole stability curve in the phase diagram of Fig. 7. The peridotite here would have a water content of 0.6 wt% (0.4 wt% (w) + 0.2 wt% (20% of f)); Green (1973, see his Fig. 3) suggests that amphibole should still be stable until around 1100° C. Hence, we expect the source region to be initiated at slightly lower temperatures (compared to around 1150–1175° C), closer to the slab, which also implies that the peak temperature of the source region is proportionally lower. Many uncertainties evident in these calculations, including the permeability versus porosity relationship, the viscosity of the fluid, and the grain size, would affect estimates of the porosity of the free water-carrying phase in the amphibole-saturated regions. Provided the porosity does not become so high as to change the phase diagram, then the basic mechanism will still be effective. The flux of water into the wedge and the water content of amphibolitized mantle are also uncertain and affect the height (thickness) of the region that is amphibolitized.

Chemistry of Amphiboles

We have continuously assumed that pargasite is the dominant amphibole and hydrous mineral at these depths. It is critically important (as mentioned previously) that there is no other hydrous mineral that is sufficiently stable and so common that it can take all the water released by pargasite. The amphibole family is a very broad family with a range of cation sites, enabling it to accommodate a wide range of ions. This flexibility leads to the possibility that slight variations exist in the chemical composition of the amphiboles as we go to higher temperatures and pressures so that their stability is increased somewhat. Can this increase help explain the occasional observation that the potassium contents of lavas tend to increase as we progress away from the trench (Dickinson and Hatherton, 1967)? The assumption would be that more K is present at greater depths since amphiboles richer in potassium may be more stable than Na-rich amphiboles.

Model Predictions

Presence and Location of the Volcanic Front

From thermal modeling we have seen that sufficiently high temperatures can be achieved quite close to the slab at a depth of 60–100 km. The 1200° C isotherm can get to within 20 km of the oceanic crust. The lateral H_2O transport mechanism would then suggest a source region located vertically above the subducting slab at a depth of around 120–140 km. If the melts were to rise vertically then the volcanic front would be 130–150 km above the Benioff zone, which is somewhat deeper than estimates. Davies and Stevenson (1992)—assuming that the least compressive stress controls the direction of crack propagation in the asthenosphere—show that the stress regime induced by a corner flow focuses the cracks toward the mantle wedge corner, leading to estimates for the height of the volcanic front above the Benioff zone that agree well with observations. Another possibility is that higher temperatures are achieved closer to the slab. Such temperatures could occur if the mantle surrounding subduction zones is hotter than the mantle surrounding mid-ocean ridges, or because the flow is more focused than that produced by the uniform viscosity assumed in this model, due to variable rheology, or comes from greater depth due to thickening of the mechanical lithosphere of the overriding plate away from the mantle wedge corner. The thermal stability of pargasite in the mantle is probably slightly lower than assumed in the phase diagram of Fig. 7 due to the limited presence of free water, as discussed in the preceding section. Hence, the presence and location of the volcanic front can be rationalized by this model for the lateral transport of water to the source region. The high heat flow at the volcanic front is partly attributed to heat flow resulting from advection by the magma, but such advection can have no more than a local effect. We ascribe this anomaly in general to the thinness of the mechanical lithosphere in this region due to ablation leading to a much higher conductive geothermal gradient.

Uncertainties in Geometric Predictions

The uncertainties in our model are very difficult to assess because many of the component processes are poorly understood. It is unclear what differences free convection in a dynamic thermal model would make relative to the present forced convection in a kinematic model of subduction; it clearly would lead to a modulation and possible temporal variability (Davies and Stevenson, 1992). The thermal effects of melting were not included; they should not have a large effect on the total thermal structure, but could lead to significant local changes in and around the source region. For example, they might lead to differences in the source width.

Our lack of understanding with regard to the depth of the asthenosphere–lithosphere boundary (thickness of the mechanical boundary layer) affects the accuracy of the shallow thermal structure. This uncertainty probably has a larger effect on our lack of understanding of the location of the top of the source region than does our uncertainty with regard to magma segregation. The top of the source region is at least as shallow as the maximum temperature achieved vertically and no shallower than the solidus. Since these levels are not very far apart (<10 km), the uncertainty in the location of the top of the source region resulting from uncertainty in the control of magma segregation is small.

There is some uncertainty about whether the local stress field controls the direction of crack propagation. The difference between crack propagation controlled by the local stress field and vertically propagating cracks limits the possibilities. The resulting uncertainty is a small but systematic difference in the prediction of the location of the volcanic front. As mentioned earlier, the vertically propagating cracks lead to volcanic fronts about 20 km too high above the Benioff zone for this model.

The uncertainty in the phase diagrams is difficult to judge, given the large number of variables and the differences between some studies and similarities between others (Basaltic Volcanism Study Project, 1981; Olafsson and Eggler, 1983; Wallace and Green, 1988). The maximum dehydration melting temperature of amphibole (pargasite) from 2–3 GPa, which exerts strong control of the location of the source region, seems relatively robust, e.g., 1050° C (Wyllie, 1979), 1150° C (Green, 1973), 1080° C (Wallace and Green, 1988), and 1070° C (Olafsson and Eggler, 1983).

Many of the uncertainties inherent in the lateral

water transport mechanism have been discussed previously. The uncertainty in the velocity of vertical propagation of the fluid depends first on whether it is by porous flow or crack propagation. If it is by porous flow, what is the permeability, and what is the viscosity of the fluid? These questions are very difficult to answer, but clearly the slower the velocity the higher the fluid content. Provided the water content does not reach levels such that the phase diagrams considered are inapplicable, then the remaining predictions, e.g., source width, source region position, are not greatly affected. In conclusion, many of the individual components of this model are very uncertain; even so, the major processes are surprisingly robust. The melting of a water fluxed mantle wedge is the result of a two-component lateral transport mechanism for the water; it combines the near-vertical transport of a free phase with transport in hydrated minerals carried down by the induced secondary flow.

Major Element Composition of Primary Magmas

Davies and Bickle (1991) have undertaken a simple and crude model for melting by fluxing, where it was assumed that the source region was in hydrous equilibrium; i.e., the water content of the melt is the water content of the whole local system. With that model they generated predictions of primary magmas that fall in the middle of current estimates. They found that the volume of magma was sensitive to the water flux and to the thickness of the melting column (i.e., the thickness of mechanical lithosphere). Stolper and Newman (1992) have measured the water contents of back-arc basin magmas and found a strong correlation between water content and composition. They have extended the study to include volcanic front magmas in the Marianas and argue that the magmas are the result of hydrous fluxing, following percolation through the mantle wedge. Fluids fluxing the back-arc basins have traveled a longer path such that they are in equilibrium with the mantle, while fluids fluxing the mantle to produce the volcanic front magmas have not equilibrated completely with the mantle.

Slab Trace Element Signature

Hawkesworth *et al.* (1993) have assumed a percolation model based on the lateral transport mechanism that is the focus of this chapter. They considered only the part of the transport involving a fluid phase, assumed chemical equilibrium, and used the partition coefficients of Brenan and Watson (1991). The study ignored the difference between the spinel and garnet facies of the peridotite. The study demonstrated that expected trace element inputs from the subducting slab followed by percolation through the mantle wedge could produce an arc trace element signature, provided the partition coefficients were two to three orders of magnitude less than those currently suggested by experimental data, or that the water fluxes were around three orders of magnitude higher than currently estimated. Other possibilities include an imperfect chemical equilibrium since the water possibly migrates by high-velocity fracturing processes. In addition, it should be noted that the actual partition coefficients are very sensitive to the exact nature of the fluid and solid matrix.

Other Predictions

The hydrous fluxing model of Davies and Bickle (1991) based on this lateral transport mechanism suggests reasonable segregation temperatures and reasonable water contents. Their predictions of arc growth compared to previous estimates (Reymer and Schubert, 1984) suggest low fluid fluxes from the oceanic crust or that not all the melt reaches the arc crust. The average degree of melting predicted is 2 to 8%, which again may not reflect the melts observed at the surface since probably not all the primary magmas reach the arc crust. The assumption of the input of water from the slab allows one to explain the enhanced Sr isotopes (from the Sr input into the oceanic crust by hydrothermal circulation) and the ^{10}Be and B signature (Morris *et al.*, 1990). It also explains the more hydrous and explosive nature of subduction zone volcanics.

This process is expected to produce a locally heterogeneous seismic velocity and attenuation structure in the source region and the strongly hydrated regions. The excellent tomographic seismic studies of Hasegawa *et al.* (1991) and Hasegawa and Zhao (1994) seem to image the hot mantle tongue predicted by the thermal models. There is also a weak suggestion of a feature at around 100 km depth near the slab surface, which one might speculate could be related to the release of water. We probably need to be conservative in interpreting all tomographic images of deep structures, especially due to the difficulty of correcting for shallow poorly resolved heterogeneous structures. In addition, interpretation of seismic signatures is nonunique and difficult, but the preceding study seems largely to support the mechanism presented. Limited spatial resolution (>25 km) though makes it unlikely that the individual components (source region, hydrated layer, region with water-rich low-porosity melts) of the mechanism can be imaged clearly. The seismic attenuation study of Umino and Hasegawa (1984) also suggests a similar thermal structure, with Sato (1992) and Sato and Sacks (1990) interpreting the intermediate attenuation in the fore-arc as evidence of hydration.

Discussion of Estimated Magma Segregation Temperatures

Not all observations can be explained easily by this model. One observation concerns the pressure and temperature conditions under which estimated primary basalts are in equilibrium with olivine, orthopyroxene, and clinopyroxene (Nye and Reid, 1986; Tatsumi *et al.*, 1983). Tatsumi *et al.* (1983) suggest that the magmas segregate at a temperature of 1320° C and at pressures ranging from 11 to 23 kbar. It is argued that these temperatures are so high that they are due to a diapir, and due to the absorption of the latent heat of melting in the diapir, it must have passed through a region with temperatures higher than 1400° C. A temperature of 1400° C is higher than those envisaged in this model. Interpretation of the results of these equilibration experiments in terms of conditions in the mantle wedge assumes that the primary magmas on segregation were in equilibrium with the three residual minerals. The primary magma is produced over a depth interval and not at one point (McKenzie and Bickle, 1988) and probably does not continue to reequilibrate as it migrates up through the melting column (Spiegelman and Kenyon, 1992). Hence, the assumptions implicit in the interpretation of Tatsumi *et al.* (1983) are questionable and the constraints provided by the results are equally questionable.

Conclusions

The source areas of subduction zone magmatism are regions within the mantle wedge that are fluxed by a fluid component carrying water inward from the subducting oceanic crust. The only exceptions are possibly the subduction of very young, hot oceanic crust or where subduction has just initiated.

The subducting slab induces a corner flow in the mantle wedge that advects heat into the wedge corner, leading to relatively high temperatures close to the subducting slab. Most water released at shallower levels than the bottom of the interplate thrust is assumed to migrate back up the thrust or into the fore-arc, which is part of the mechanical lithosphere. The remaining water released, largely from amphiboles, is argued to enter the mantle wedge where it reacts to form additional amphiboles. The induced secondary flow carries these amphiboles down through their breakdown pressure (depth) range, where the water is subsequently released. It rises vertically through the amphibole-saturated mantle until it reaches dry mantle, with which it again reacts to form additional amphibole. The cycle thus repeats with the net integrated result being the lateral transport of water away from the oceanic crust and inward to the hot regions of the mantle wedge. The process is assumed to continue even when it is a near water-saturated melt that is rising vertically since at low water contents, amphiboles are stable to some extent above the water-saturated solidus.

In this chapter, we hope we have enforced the robustness of the general mechanism while discussing the uncertainties in critically assessing it. The major question is, How does water leave the slab and move across the cold region of the mantle wedge? There is little question in my mind that such movement occurs, and a selection of possible processes have been identified, with microcracking a probable candidate; however, as long as this uncertainty remains, models of hydrous fluxing of mantle wedge will be incomplete. Such models can explain the presence and location of a volcanic front and also give a reasonable estimate for the composition of the primary magma.

Acknowledgments

I acknowledge discussions with Prof. Dave Stevenson and Dr. Mike Bickle, who have contributed to aspects of this work. The reviews of two anonymous referees helped improve this chapter. The NSF (US) and NERC (UK) supported much of the work presented here.

References

Alt, J. C., Honnorez, J., Laverne, C., and Emmermann, R. (1986). Hydrothermal alteration of 1km section through the upper oceanic crust, Deep Sea Drilling Project Hole 504B: Mineralogy, chemistry and evolution of seawater–basalt interactions, *J. Geophys. Res.* **91,** 10309–10336.

Anderson, O. L., and Grew, P. C. (1977). Stress corrosion theory of crack propagation with applications to geophysics, *Rev. Geophys. Space Phys.* **15,** 77–104.

Anderson, R. N., Delong, S. E., and Schwarz, W. M. (1980). Dehydration, asthenospheric convection and seismicity in subduction zones, *J. Geol.* **88,** 445–451.

Anderson, R. N., Uyeda, S., and Miyashiro, A. (1976). Geophysical and geochemical constraints at converging plate boundaries. Part 1. Dehydration in the downgoing slab, *Geophys. J. R. Astron. Soc.*, **44,** 333–357.

Andrews, D. J., and Sleep, N. H. (1974). Numerical modelling of tectonic flow behind island arcs, *Geophys. J. R. Astron. Soc.* **38,** 237–251.

Atkinson, B. K. (1984). Subcritical crack growth in geological materials, *J. Geophys. Res.* **89,** 4077–4114.

Basaltic Volcanism Study Project (1981). "Basaltic Volcanism Study Project, Basaltic Volcanism on the Terrestrial Planets," Pergamon, New York.

Batchelor, G. K. (1967). "An Introduction to Fluid Dynamics," Cambridge Univ. Press, Cambridge.

Beard, J. S., and Lofgren, G. E. (1990). Dehydration melting and water-saturated melting basaltic and andesitic greenstones and amphibolites, *J. Petrol.* **32,** 365–401.

Bebout, G. E. (1991). Field-based evidence for devolatilization in subduction zones: Implications for arc magmatism, *Science* **251,** 413–416.

Bebout, G. E., and Barton, M. D. (1989). Fluid flow and metasomatism in a subduction zone hydrothermal system: Catalina schist terrane, California, *Geology* **17,** 976–980.

Becker, K., *et al.* (1989). Drilling deep into young oceanic crust at hole 504B, Costa Rica Rift, *Rev. Geophys.* **27,** 79–102.

Bell, D. R., and Rossman, G. R. (1992). Water in Earth's mantle: The role of nominally anhydrous minerals, *Science* **255,** 1391–1397.

Blanpied, M. L., Lockner, D. A., and Byerlee, J. D. (1992). An earthquake mechanism based on rapid sealing of faults, *Nature* **358,** 574–576.

Bloxham, J., and Jackson, A. (1991). Fluid flow near the surface of Earth's outer core, *Rev. Geophys.* **29,** 97–120.

Bodri, L., and Bodri, B. (1978). Numerical investigation of tectonic flow in island-arc areas, *Tectonophysics* **50,** 163–175.

Brenan, J. M., and Watson, E. B. (1988). Fluids in the lithosphere. 2. Experimental constraints on CO_2 transport in dunite and quartzite at elevated P-T conditions with implications for mantle and crustal decarbonation processes, *Earth Planet. Sci. Lett.* **91,** 141–158.

Brenan, J. M., and Watson, E. B. (1991). Partitioning of trace elements between olivine and aqueous fluids at high P-T conditions: Implications for the effect of fluid composition on trace element transport. *Earth Planet. Sci. Lett.* **107,** 672–688.

Brooks, A. (1981). "A Petrov–Galerkin Finite-Element Formulation for Convection Dominated Flows," Ph.D. Thesis, California Institute of Technology.

Brophy, J. G., and Marsh, B. D. (1986). On the origin of high-alumina arc basalts and the mechanics of melt extraction, *J. Petrol.* **27,** 763–789.

Cheadle, M. J. (1989). "Properties of Texturally Equilibrated Two-Phase Aggregates," Ph.D. Thesis, Univ. of Cambridge.

Cooper, R. F., and Kohlstedt, D. L. (1982). Interfacial energies in the olivine–basalt system, *in* "High Pressure Research in Geophysics," (S. Akimoto and M. N. Manghnani eds.), Adv. Earth Planet. Sci. 12, pp. 217–228, Center for Academic Publications, Tokyo.

Creager, K. C., and Jordan, T. H. (1984). Slab penetration into the lower mantle, *J. Geophys. Res.* **89,** 3031–3049.

Davies, J. H., and Bickle, M. J. (1991). A physical model for the volume and composition of melt produced by hydrous fluxing above subduction zones, *Philos. Trans. R. Soc. London A* **335,** 355–364.

Davies, J. H., and Stevenson, D. J. (1992). Physical model of source region of subduction zone volcanics, *J. Geophys. Res.* **97,** 2037–2070.

Dickinson, W. R., and Hatherton, T. (1967). Andesitic volcanism and seismicity around the Pacific, *Science* **157,** 801–803.

Eggler, D. H. (1987). Solubility of major and trace elements in mantle metasomatic fluids: Experimental constraints *in* "Mantle Metasomatism" (M. A. Menzies and C. J. Hawkesworth, eds.), pp. 21–41, Academic Press, San Diego.

Fryer, P. (1992). Mud volcanoes of the Marianas, *Sci. Am.* **Feb.,** 26–32.

Fujii, N., Osamura, K., and Takahashi, E. (1986). Effect of water saturation on the distribution of partial melt in the olivine–pyroxene–plagioclase system, *J. Geophys. Res.* **91,** 9253–9260.

Fujisawa, H., Fujii, N., Mizutami, H., Kanamori, H., and Akimoto, S. (1968). Thermal diffusivity of Mg_2SiO_4, Fe_2SiO_4, and NaCl at high pressure and temperature, *J. Geophys. Res.* **73,** 4727–4733.

Furukawa, Y., and Uyeda, S. (1989). Thermal state under the Tohoku Arc with consideration of crustal heat generation, *Tectonophysics* **164,** 175–187.

Fyfe, W. S., Price, N. J., and Thompson, A. B. (1978). "Fluids in the Earth's Crust," Elsevier, Amsterdam.

Garfunkel, Z., Anderson, C. A., and Schubert, G. (1986). Mantle circulation and the lateral migration of subducted slabs, *J. Geophys. Res.* **91,** 7205–7223.

Giggenbach, W. F. (1992). Isotopic shifts in waters from geothermal and volcanic systems along convergent plate boundaries and their origin, *Earth Planet. Sci. Lett.* **113,** 495–510.

Gill, J. (1981). "Orogenic Andesites and Plate Tectonics," Springer-Verlag, New York.

Goto, A., and Tatsumi, Y. (1990). Stability of chlorite in the upper mantle, *Am. Mineral.* **75,** 105–108.

Green, D. H. (1973). Contrasted melting relations in a pyrolite upper mantle under mid-oceanic ridge, stable crust and island arc environments, *Tectonophysics* **17,** 285–297.

Gregory, R. T., and Taylor, H. P. (1981). An oxygen isotopic profile in a section of Cretaceous oceanic crust, Samail ophiolite, Oman: Evidence for ^{18}O buffering of the oceans by deep (>5km) seawater-hydrothermal circulation at mid-ocean ridges, *J. Geophys. Res.* **86,** 2737–2756.

Hamilton, W. B. (1988). Plate tectonics and island arcs, *Geol. Soc. Am. Bull.* **100,** 1503–1527.

Hasegawa, A., and Zhao, D. (1994). Deep structure of island arc magmatic regions as inferred from seismic observations *in* "Magmatic Systems" (M. P. Ryan, ed.), Academic Press, San Diego.

Hasegawa, A., Zhao, D., Hori, S., Yamamoto, A., and Horiuchi, S. (1991). Deep structure of the northeastern Japan arc and its relationship to seismic and volcanic activity, *Nature* **352,** 683–689.

Hawkesworth, C. J., Gallagher, K., Hergt, J. M., and McDermott, F. (1993). Trace element fractionation processes in the generation of island arc basalts, *Phil. Trans. R. Soc. London A.,* **342,** 179–191.

Helffrich, G. R., Stein, S., and Wood, B. J. (1989). Subduction zone thermal structure and mineralogy and their relationship to seismic wave reflections and conversions at the slab/mantle interface, *J. Geophys. Res.* **94,** 753–763.

Honda, S. (1985). Thermal structure beneath Tohoku, Northeast Japan—A case study for understanding the detailed thermal structure of the subduction zone, *Tectonophysics* **112,** 69–102.

Hsui, A. T., Marsh, B. D., and Toksoz, M. N. (1983). On the melting of the subducted oceanic crust: Effects of subduction induced mantle flow, *Tectonophysics* **99,** 207–220.

Huang, W.-L., Wyllie, P. J., and Nehru, C. E. (1980). Subsolidus and liquidus phase relationships in the system CaO–SiO_2–CO_2 to 30kbar with geological applications, *Am. Mineral.* **65,** 285–301.

Hughes, T. J. R., Liu, W. K., and Brook, A. (1979a). Finite element analysis of incompressible viscous flows by the penalty function formulation, *J. Comput. Phys.* **30,** 1–60.

Hughes, T. J. R., Pister, K. S., and Taylor, R. L. (1979b). Implicit–explicit finite elements in nonlinear transient analysis, *Comp. Methods Appl. Mech. Eng.* **17,** 159–182.

Ida, Y. (1987). Structure of the mantle wedge and volcanic activities in the island arcs, *in* "High-Pressure Research in Mineral Physics" (M. H. Manghnani and Y. Syono, eds.), pp. 473–480, Terra Scientific Publishing Co., Tokyo; American Geophysical Union, Washington, DC.

Ito, E. I., Harris, D. M., and Anderson, Jr, A. T. (1983). Alteration of oceanic crust and geologic cycling of chlorine and water, *Geochim. Cosmochim. Acta* **47,** 1613–1624.

Jurdy, D. M., and Stefanick, M. (1983). Flow models for backarc spreading, *Tectonophysics* **99,** 191–206.

Kastner, M., Elderfield, H., and Martin, J. B. (1991). Fluids in convergent margins, What do we know about their composition, origin, role in diagenesis and importance for oceanic chemical fluxes? *Philos. Trans. R. Soc. London A* **335,** 243–259.

King, S. D., Raefsky, A., and Hager, B. H. (1990). ConMan: Vectorizing a finite element code for incompressible two-dimensional convection in the Earth's mantle, *Phys. Earth Planet. Int.* **59,** 195–207.

Marsh, B. D. (1979a). Island arc development: Some observations, experiments and speculations, *J. Geol.* **87,** 687–713.

Marsh, B. D. (1979b). Island-arc volcanism, *Am. Sci.* **67,** 161–172.

McKenzie, D. P. (1969). Speculations on the consequences and causes of plate motions, *Geophys. J. Roy. Astron. Soc.* **18,** 1–32.

McKenzie, D. P., and Bickle, M. J. (1988). The volume and composition of melt generated by extension of the lithosphere, *J. Petrol.* **29,** 625–680.

Minear, J. W., and Toksoz, M. N. (1970a). Thermal regime of a downgoing slab, *Tectonophysics* **10,** 367–390.

Minear, J. W., and Toksoz, M. N. (1970b). Thermal regime of a downgoing slab and new global tectonics, *J. Geophys. Res.* **75,** 1397–1419.

Morris, J. D., Leeman, W. P., and Tera, F. (1990). The subducted component in island arc lavas: Constraints from Be isotopes and B–Be systematics, *Nature* **344,** 31–36.

Mysen, B. O., Kushiro, I., and Fujii, T. (1978). Preliminary experimental data bearing on the mobility of H_2O in crystalline upper mantle, *Carnegie Inst. Washington Year Book,* 793–797.

Nagao, T., and Uyeda, S. (1989). Heat flow measurements in the northern part of Honshu, Northeast Japan, using shallow holes, *Tectonophysics* **164,** 301–314.

Niida, K., and Green, D. H. (1990). Pargasitic dehydration solidus of peridotites hydrated in subduction wedge mantle, *EOS Trans. Am. Geophys. Union* **71,** 949.

Nye, C. J., and Reid, M. R. (1986). Geochemistry of primary and least fractionated lavas from Okmok volcano, central Aleutians: Implications for arc magmagenesis, *J. Geophys. Res.* **91,** 10271–10287.

Olafsson, M., and Eggler, D. H. (1983). Phase relations of amphibole, amphibole–carbonate, and phlogopite–carbonate peridotite: Petrologic constraints on the asthenosphere, *Earth Planet. Sci. Lett.* **64,** 305–315.

Otsuki, K. (1989). Empirical relationships among the convergence rate of plates, rollback rate of trench axis and island-arc tectonics: "laws of convergence rate of plates," *Tectonophysics* **159,** 73–94.

Otsuki, K., Heki, K., and Yamazaki, T. (1990). New data which support the "laws of convergence rates of plates" proposed by Otsuki, *Tectonophysics* **172,** 365–368.

Oxburgh, E. T., and Turcotte, D. L. (1968). Problems of high heat flow and volcanism associated with zones of descending mantle convective flow, *Nature* **216,** 1041–1043.

Oxburgh, E. T., and Turcotte, D. L. (1970). Thermal structure of island arcs, *Geol. Soc. Am. Bull.* **81,** 1665–1688.

Peacock, S. M. (1987a). Serpentinization and infiltration metasomatism of the Trinity peridotite, Klamath province, northern California: Implications for subduction zones, *Contrib. Mineral. Petrol.* **95,** 55–70.

Peacock, S. M. (1987b). Thermal effects of metamorphic fluids in subduction zones, *Geology* **15,** 1057–1060.

Peacock, S. M. (1990). Fluid processes in subduction zones, *Science* **248,** 329–337.

Peacock, S. M. (1992). Blueschist-facies metamorphism, shear heating and P–T–t paths in subduction shear zones, *J. Geophys. Res.* **97,** 17693–17707.

Philippot, P., and Selverstone, J. (1991). Trace-element-rich brines in eclogitic veins: Implications for fluid composition and transport during subduction, *Contrib. Mineral. Petrol.* **106,** 417–430.

Plank, T., and Langmuir, C. H. (1988). An evaluation of the global variations in the major element chemistry of arc basalts, *Earth Planet. Sci. Lett.* **90,** 349–370.

Reymer, A., and Schubert, G. (1984). Phanerozoic addition rates to the continental crust and crustal growth, *Tectonics* **3,** 63–77.

Sato, H. (1992). Thermal structure of the mantle wedge beneath northeastern Japan: Magmatism in an island arc from combined data of seismic anelasticity and velocity and heat flow, *J. Volcanol. Geotherm. Res.* **51,** 237–252.

Sato, H., and Sacks, I. S. (1990). Magma generation in the upper mantle inferred from seismic measurements in peridotite at high pressure and temperature, *in* "Magma Transport and Storage" (M. P. Ryan, ed.), pp. 277–292, Wiley, Chichester/Sussex, England.

Spiegelman, M., and Kenyon, P. (1992). The requirements for chemical disequilibrium during magma migration, *Earth Planet. Sci. Lett.* **109,** 611–620.

Stolper, E., and Newman, S. (1992). The role of water in the petrogenesis of Mariana trough magmas, submitted for publication.

Sudo, A., and Tatsumi, Y. (1990). Phlogopite and K-amphibole in the upper mantle: Implication for magma genesis in subduction zones, *Geophys. Res. Lett.* **17,** 29–32.

Sugimura, A. (1960). Zonal arrangement of some geophysical and petrological features in Japan and its environs, *J. Fac. Sci. Univ. Tokyo Sec. 2* **12**(2), 133–153.

Tatsumi, Y. (1986). Formation of the volcanic front in subduction zones, *Geophys. Res. Lett.* **13,** 717–720.

Tatsumi, Y. (1989). Migration of fluid phases and genesis of basalt magmas in subduction zones, *J. Geophys. Res.* **94,** 4697–4707.

Tatsumi, Y., Sakuyama, M., Fukuyama, H., and Kushiro, I. (1983). Generation of arc basalt magmas and thermal structure of the mantle wedge in subduction zones, *J. Geophys. Res.* **88,** 5815–5825.

Thompson, A. B. (1992). Water in the Earth's upper mantle, *Nature* **358,** 295–302.

Tichelaar, B. W., and Ruff, L. (1989). Variability in depth of seismic coupling along the Chilean subduction zone, *EOS Trans. Am. Geophys. Union* **70,** 398.

Toksoz, N., Minear, J. W., and Julian, B. R. (1971). Temperature field and geophysical effects of a down-going slab, *J. Geophys. Res.* **76,** 1113–1138.

Tritton, D. J. (1977). "Physical Fluid Dynamics," Van Nostrand Reinhold, New York.

Turcotte, D., and Schubert, G. (1968). A fluid theory for the deep structure of dip-slip fault zones, *Phys. Earth Planet. Int.* **1,** 381–386.

Turcotte, D., and Schubert, G. (1973). Frictional heating of the descending lithosphere, *J. Geophys. Res.* **78,** 5876–5886.

Umino, N., and Hasegawa, A. (1984). Three-dimensional Q_s structure in the northeastern Japan arc, *J. Seismol. Soc. Japan (Zisin)* **37,** 217–228.

Van den Beukel, J., and Wortel, R. (1987). Temperatures and shear stresses in the upper part of a subduction zone, *Geophys. Res. Lett.* **14,** 1057–1060.

Vrolijk, P., Myers, G., and Moore, J. C. (1988). Warm fluid migration along tectonic melanges in the kodiak accretionary complex, Alaska, *J. Geophys. Res.* **93,** 10313–10324.

Waff, H. S., and Bulau, J. R. (1979). Equilibrium fluid distribution in an ultramafic partial melt under hydrostatic stress conditions, *J. Geophys. Res.* **84,** 6109–6114.

Waff, H. S., and Bulau, J. R. (1982). Experimental determination of near-equilibrium textures in partially molten silicates at high pressures, *in* "High Pressure Research in Geophysics" (S. Akimoto and M. H. Manghnani, eds.), Adv. Earth Planet. Sci., Vol. 12, pp. 229–236, Center for Academic Publications, Tokyo.

Wallace, M. E., and Green, D. H. (1988). An experimental determination of primary carbonatite composition, *Nature* **335,** 343–345.

Walther, J. V., and Helgeson, H. C. (1977). Calculations of the thermodynamic properties of aqueous silica and the solubility of quartz and its polymorphs at high pressures and temperatures, *Am. J. Sci.* **277,** 1315–1351.

Watson, E. B. (1991). Diffusion in fluid-bearing and slightly-melted rocks: Experimental and numerical approaches illustrated by iron transport in dunite, *Contrib. Mineral. Petrol.* **107,** 417–434.

Watson, E. B., Brenan, J. M., and Baker, D. R. (1990). Distribution of fluids in the continental mantle, *in* "Continental Mantle" (M. A. Menzies, ed.), pp. 111–125, Clarendon Press, Oxford.

Wolf, M. B., and Wyllie, P. J. (1991). Dehydration-melting of solid amphibolite at 10 Kbar: Textural development, liquid interconnectivity and applications to the segregation of magmas, *Mineral. Petrol.* **44,** 151–179.

Wunder, B., and Shreyer, W. (1992). Metastability of the 10A phase in the system $MgO–SiO_2–H_2O$ (MSH): What about hydrous MSH phases in subduction zones? *J. Petrol.* **33,** 877–889.

Wyllie, P. J. (1979). Magmas and volatile components, *Am. Mineral.* **64,** 469–500.

Yuen, D. A., Fleitout, L., Schubert, G., and Froidevaux, C. (1978). Shear deformation zones along major transform faults and subducting slabs, *Geophys. J. R. Astron. Soc.* **54,** 93–119.

Chapter 10 | Buoyancy-Driven Fracture and Magma Transport through the Lithosphere: Models and Experiments

Moritz Heimpel and Peter Olson

Overview

Magma generated in partially molten regions in the upper mantle may ascend to crustal magma chambers and ultimately to the Earth's surface via fracture propagation alone or by fracture-assisted diapirs. We present the results of experiments on buoyancy-driven fluid transport by fracture propagation, using various fluids injected into gelatin. The buoyancy and volume of fluids have been varied over three orders of magnitude, and several concentrations of gelatin, which correspond to a range of shear velocities and fracture toughnesses, are used. The resulting crack propagation velocities vary over four orders of magnitude. Two regimes of propagation are identified: (*a*) slow propagation characterized by cracks with a subcritical stress intensity; and (*b*) fast propagation characterized by a supercritical stress intensity and a fracture profile that narrows smoothly to a thin conduit at the tail. Crack propagation velocities are found to depend on the fluid buoyancy, the yield strength and fracture toughness of the solid medium, and the size of the fluid-filled fracture. Because none of the existing models of fracture propagation predict the observed velocities, we derive a new model, in which fracture at the tip and closure at the tail are coupled by elastic wave propagation down the crack surface. The experimental data bound fracture velocities derived from this model and the predicted velocities of buoyancy-driven magma fracture in the lithosphere are consistent with velocities inferred for magmatic systems. In addition, crack-tip fracture and the consequent production of seismic radiation from propagating dikes are consistent with our model. Also in this chapter, the role of the fracture toughness on magma fracture is briefly reviewed. Fracture toughness estimates obtained from the dimensions of igneous dikes are shown to be two to three orders of magnitude higher than laboratory fracture toughness estimates, indicating that the field-derived and laboratory-derived estimates are probably appropriate for magma fracture under differing conditions.

Notation

		Units
E	Young's modulus	Pa, MPa
K_I	mode I (purely tensile) stress intensity	$Pa \cdot m^{1/2}$
K_{Ic}	mode I critical stress intensity (fracture toughness)	$Pa \cdot m^{1/2}$
K_-	lesser fracture toughness of two crack tips	$Pa \cdot m^{1/2}$
K_+	greater fracture toughness of two crack tips	$Pa \cdot m^{1/2}$
P	internal overpressure	Pa, MPa
P_o	overpressure near crack tip	Pa, MPa
Q_s	source flux	$m^3 \cdot s^{-1}$
U_E	elastic wave velocity	$m \cdot s^{-1}$
U_F	fracture resistance velocity	$m \cdot s^{-1}$
U_P	Poiseuille velocity	$m \cdot s^{-1}$
a	crack or dike half-thickness	m
a_1	crack or dike half-thickness at $s = r$	m
a_{max}	maximum crack or dike half-thickness	m
a_∞	crack or dike conduit half-thickness	m
b	crack half-length	m
c	fracture resistance velocity coefficient	dimensionless
g	gravity	$m \cdot s^{-2}$
g'	buoyancy	$m \cdot s^{-2}$
r	crack-tip radius	m
r	crack or dike head radius	m
$2h$	crack or dike height	m
s, x, y, z	Cartesian coordinates	m
t	lag time for crack closure	s
u	average fluid velocity	$m \cdot s^{-1}$
ΔP_h	hydrostatic pressure drop	Pa, MPa

Magmatic Systems
Edited by M. P. Ryan

		Units
ΔP_v	viscous pressure drop	Pa, MPa
Θ	crack geometric parameter	dimensionless
η	fluid viscosity	Pa · s
μ	shear modulus	Pa, MPa
ν	Poisson's ratio	dimensionless
ρ	solid density	$kg \cdot m^{-3}$
ρ_l	fluid density	$kg \cdot m^{-3}$
σ_b	extensional stress at crack tip	Pa, MPa
σ_y	extensional yield strength of solid	Pa, MPa

Introduction

Fracture propagation is a mechanism by which positively buoyant magmas, generated in mantle upwellings, are transported through the lithosphere. Magma injection (and dike formation) is of primary importance in the building of intraplate oceanic islands, in continental rifting, and in the generation of the oceanic crust at mid-ocean ridges. While mechanisms of magma transport such as diapirs of partial melt or magma (Whitehead and Luther, 1975; Marsh, 1982) and solitary wave propagation within partially molten regions (McKenzie, 1984; Scott and Stevenson, 1984; Olson and Christensen, 1986) are likely to be important in the viscous asthenosphere, a mechanism involving fracture propagation seems necessary to transport magma through the elastic lithosphere.

Most dikes that intersect the surface of the Earth have their intermediate origin in crustal magma chambers a few kilometers in depth. However, the generation of melt takes place in the upper mantle below the base of the lithosphere at depths over the range of approximately 100 km in intraplate provinces to approximately 30 km at mid-ocean ridges. The low yield strength of partially molten rock suggests that magma fractures can readily initiate in partially molten regions (Fowler, 1990; Sleep, 1988). Crosscutting dikes in the mantle peridotites of ophiolites, which are inferred to have formed while the peridotites were partially molten, provide field evidence of magma fracture nucleation in partial melt regions (Nicolas, 1986, 1990).

Although fracture propagation is the generally accepted mechanism for magma transport through the lithosphere, the issue of what controls the shape and the velocity of buoyant propagating fractures remains unresolved. To address this issue, we have performed experiments designed to investigate the mechanics of buoyancy-driven fluid fracture and test existing models of this phenomenon. While some elements of each of these existing models are identified in the experiments, none of the models explains the propagation velocities we observe.

Magma Fracture Models

Buoyancy-driven magma fracture has been investigated from several points of view. First, there is the "quasi-static fracture" model, which utilizes the two-dimensional shape of a static volume of fluid embedded in an elastic solid (Weertman, 1971; Pollard and Muller, 1976). According to this model, when a crack is sufficiently large and the fluid sufficiently buoyant, the stress intensity at the leading edge exceeds the fracture toughness of the country rock, and the fracture propagates upward. In order to maintain the shape predicted by elasticity theory, the fracture must close at the tail. This static model does not predict propagation velocities.

A second model is based on the stress corrosion theory of materials science, as applied to magma fracture (Anderson and Grew, 1977). In this model, cracking is controlled by environmental factors such as temperature, chemical reaction kinetics, and the fluid viscosity near the crack tip, and the stress intensity is assumed to be less than the fracture toughness. The resulting crack propagation velocities are typically less than 0.1 $mm \cdot s^{-1}$. The importance of this process in magma fracture is probably limited to special circumstances, such as dike initiation and the slow cracking of high fracture toughness barriers that separate melt volumes at depth.

Investigators have also considered the role of fluid (magma) viscosity in limiting the propagation velocity (Spence *et al.*, 1987; Lister, 1990; Turcotte, 1990; Lister and Kerr, 1991). In this approach, fracture mechanics determines the crack-tip shape, but the effect of the crack-tip fracture resistance on the crack propagation velocity is assumed to be negligible. Models based on this concept predict a magma fracture with a slightly bulging head that tapers to a narrow conduit. With a constant source flux the propagation velocity in these models is equivalent to the Poiseuille velocity of the conduit fluid.

The flow of magma directly into dikes in partial melt regions is a complex process involving concurrent porous flow, matrix compaction, and the fracture of the rock matrix so that flux into the dike is limited by rock permeability, rock viscosity, and the melt viscosity (Sleep, 1988). In such a process, two source conditions are plausible idealizations: (*a*) the *constant volume source condition,* where the magma source is depleted rapidly compared to the time required for magma transport through the newly formed magma fracture, and (*b*) the *constant flux source condition,* appropriate if flow into the magma fracture proceeds long after fracture nucleation. In this chapter we discuss the shape and velocity of buoyant fractures with constant fluid volume that close at the tail and of fractures with constant fluid flux in which the leading pulse of fluid tapers to a conduit of uniform thickness.

Experiments on Buoyancy-Driven Fracture

Among the previous experimental investigations of buoyancy-driven fracture propagation (Fiske and Jackson, 1972; Maaloe, 1987; Takada, 1990), only Takada presented propagation velocity data. Our experimental results extend the original experimental results of Takada. We also propose a new model for buoyancy-driven crack propagation, in which the velocity depends on the fracture resistance of the solid. From dimensional analysis based upon the experimental results we derive a fracture resistance velocity U_F and show that it collapses the velocity data for our experiments. Also, by applying U_F to rock properties, we show that the fracture resistance is likely to be important for buoyancy-driven magma fracture propagation in the lithosphere.

Experimental Method and Results

Gelatin is a clear, brittle, viscoelastic solid with a low rigidity and a Poisson's ratio of nearly 0.5. It has been useful as a modeling material in geotechnical engineering applications because its low rigidity allows gravity to be significant in laboratory-scale models (Richards and Mark, 1966). Table 1 shows the concentrations and corresponding material properties of the gelatin mixtures used. High-clarity, 250-bloom, pigskin-derived gelatin in granular form was supplied by Kind and Knox Company. Table 2 shows the properties of the fluids injected into the solidified gelatin mixtures.

Table 1

Experimental Gelatin Properties

Concentration (%)	Shear modulus (Pa)	Shear velocity ($m \cdot s^{-1}$)	Yield strength (Pa)	Fracture toughness ($Pa \cdot m^{1/2}$)
1.4	190	0.44	980	15
1.6	276	0.53	1300	19
1.8	355	0.60	1650	23
4.0	2150	1.47	8100	114

The yield strengths of the gelatin gels were obtained directly by measuring the force required pull an object of known surface area out of the gelatin samples. Shear modulus data were obtained by comparing the theoretical and experimental shapes of fluid-filled cracks, and gelatin fracture toughness values were obtained from plots of stress intensity versus crack velocity.

The apparatus shown in Fig. 1 consists of a transparent right circular cylinder, filled with gelatin, inside a transparent fluid-filled jacket. The jacket controls the gelatin temperature and also corrects for optical distortion. The cylinder can be rotated to observe the fractures in any orientation. For each experiment the gelatin was allowed to set for 24 h at 8° C. Known volumes of several types of positively buoyant fluid were injected into the gelatin at the bottom of the cylinder. The injected

Table 2

Experimental Fluid Properties

Working fluid	Buoyancy ($m \cdot s^{-2}$)	Shear viscosity ($Pa \cdot s$)
Air	+9.8	10^{-5}
Hexane	+3.4	10^{-4}
1 Cst Si oil	+1.8	10^{-3}
Mineral oil	+1.5	0.1
Corn syrup solution	−4.1	5.0
Mercury (Hg)	−126.0	10^{-3}

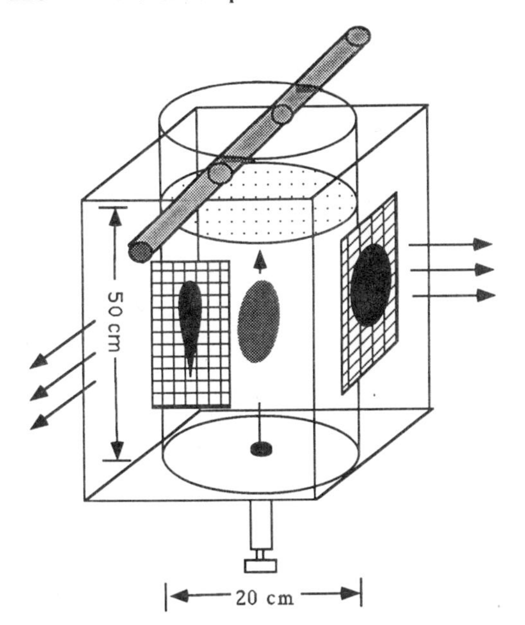

Figure 1 Schematic sketch of the experimental apparatus. Cylinder of height 50 cm and diameter 20 cm contains gelatin and may be rotated within the cooling tank. Constant volumes of positively buoyant fluids are injected at the bottom. Constant flux of negatively buoyant fluid is injected at the top of the cylinder. Cross-sectional (left) and in-plane (right) shadows of a typical constant volume crack are shown projected onto the tank walls by parallel light.

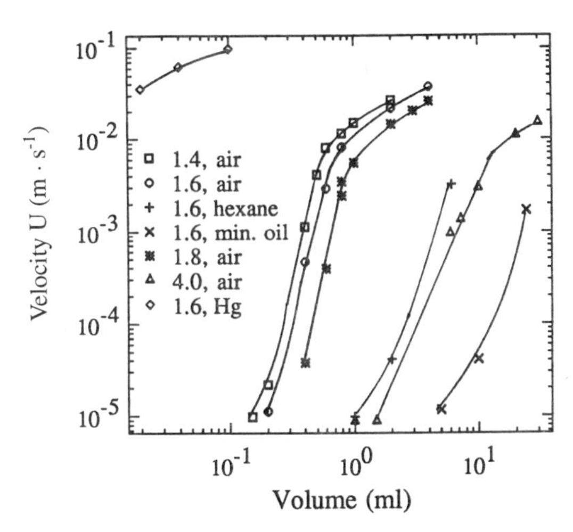

Figure 2 Summary plot of crack propagation velocity versus volume of injected fluid for various gelatin concentrations and fluids.

fluid temperature matched the gelatin temperature. In some experiments, negatively buoyant fluid was injected at constant volume and at constant flux at the cylinder top.

The positively buoyant fluids ascended (negatively buoyant fluids descended) through the solid gelatin by fracture. Most of the fractures propagated with smooth fracture surfaces. Many fractures ascended with step-like motion of the tail, which resulted in roughly evenly spaced notches on an otherwise smooth fracture surface. The mineral oil-filled fractures tended to display rough fracture surfaces, possibly indicating chemical reactivity between the mineral oil and the gelatin gel. Fluid viscosities were sufficiently low so that the volume of fluid lost to the trailing conduit from the propagating crack through the tail was small compared to the injected volume. We obtained images of the ascending cracks by projecting parallel light through the tank onto a gridded screen and photographing the shadowgraph image. Crack dimensions and velocities were measured from the resulting photographic images. The raw data from the constant volume experiments consist of the observed crack propagation velocities versus injected fluid volume. These data are shown in Fig. 2.

In those experiments using low-viscosity fluids, we find that the crack propagation velocity is evidently limited by the fracture resistance of the solid. Accordingly, an appropriate measure of the loading parameter for crack propagation is the purely tensile, or *mode I,* stress intensity factor K_I. A circular or "penny-shaped" crack of radius r is a good approximation for the shape of the constant volume fractures near the crack tip. The crack tail tends to extend so that the crack height $2h$ from tail to tip is somewhat greater than $2r$ (see Figs. 7a and 8). Hence the appropriate formulation for K_I is

$$K_I = \frac{2}{\pi} P_o \sqrt{\pi r} = \frac{4}{\pi} \rho g' h \sqrt{\pi r}, \qquad (1)$$

where $P_o = \rho g' 2h$ is the overpressure near the leading edge of the crack, $2h$ is the crack height, r is the crack head radius, $g' = g(\rho - \rho_l)/\rho$ is the fluid buoyancy, g is the gravitational acceleration, ρ is the solid density, and ρ_l is the fluid density.

Figure 3 shows the observed crack propagation velocity U versus K_I for the constant volume experiments. We identify two regimes of propagation. In *regime* 1, U is small ($<0.7\ \text{cm} \cdot \text{s}^{-1}$) and the crack velocity is highly sensitive to the loading ($U \propto K_I^n$, where $5 < n < 8$). This type of dependence of crack propagation velocity on the

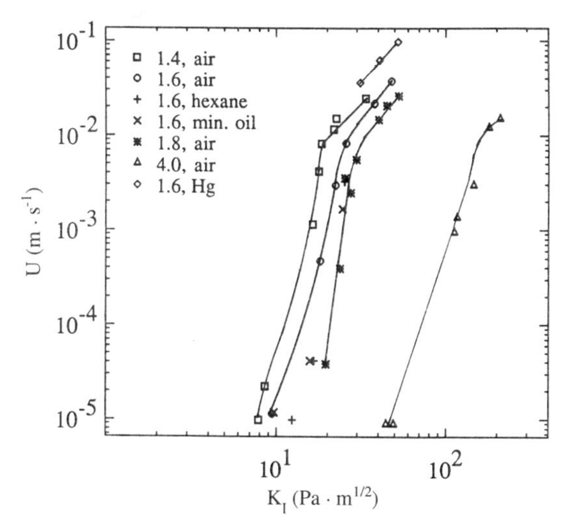

Figure 3 Plot of the stress intensity factor K_I versus the experimentally observed crack propagation velocity U for the various gelatin concentrations and fluids used in the constant volume experiments. Two regimes of crack propagation are distinguishable. In regime 1, cracks propagate slowly and $U \propto K_I^n$, where $5 < n < 8$. In regime 2, fast propagation occurs and $U \propto K_I^2$.

loading is analogous to that exhibited by other brittle solids such as ceramics and rocks during subcritical crack growth (Evans and Langdon, 1976; Atkinson and Meredith, 1987). In *regime* 2, U is large (>0.7 cm · s^{-1}) and $U \propto K_I^2$. This regime of crack propagation resembles region 2 of the three regions of subcritical crack growth described by Atkinson and Meredith (1987; see Fig. 3). In our experiments, however, the transition from $5 < n < 8$ at low propagation velocity (regime 1) to $n \approx 2$ at high propagation velocity (regime 2) is not followed by a third regime analogous to region 3 for subcritical crack growth where n increases again (Aktinson and Meredith, 1987). In fact Fig. 3 shows that regime 2 in our experiments extends to propagation velocities approaching the shear velocity of the medium ($U/U_s \approx 0.2$). We therefore arrive at the hypothesis that the transition from regime 1 to regime 2 marks the onset of dynamic fracture propagation (see the next two sections for dimensional and physical justification of this assertion).

The critical stress intensity factor K_{Ic} associated with the transition from subcritical to dynamic propagation can be obtained graphically from the U versus K_I data. Determination of the K_{Ic} values of the gelatin preparations used in our experiments is more clearly obtained from Fig. 4, which shows the observed crack propagation versus K_I on linear axes. We define K_{Ic} for our gelatin gel experiments as the value K_I (Eq. (1)) where the transition from slow propagation in regime 1 to fast propagation in regime 2 is observed. We have graphically determined K_{Ic} from Fig. 4 at the intersection of the best-fit line of velocity vs stress intensity factor with the horizontal axis.

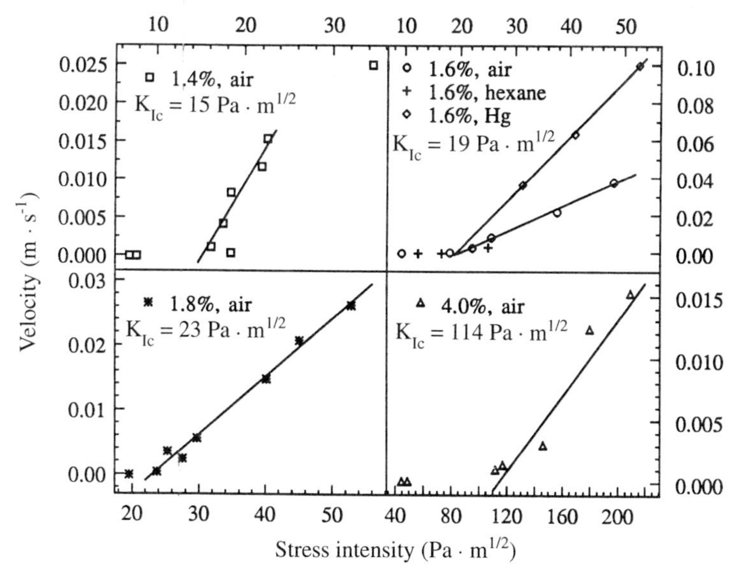

Figure 4 Crack propagation velocity versus stress intensity factor for four gelatin concentrations. For each concentration, the critical stress intensity K_{Ic} is determined by extrapolating the curve to zero crack velocity.

Dimensional Analysis of Results

A major goal of our experiments is to characterize the functional dependence of the buoyancy-driven crack propagation velocity on the physical parameters involved. In the previous section, we argue the crack propagation velocity U_F is a function of the applied loading K_I, which depends on the fluid buoyancy. U_F may also depend on the viscosity η in the crack cavity, the rigidity (characterized by the shear velocity U_E), the local stress σ_y required to fracture the solid medium near the crack tip, and length $2h$ and thickness $2a$ of the propagating fracture. Hence,

$$U = F(K_I, \eta, U_E, \sigma_y, 2h, 2a). \qquad (2)$$

Dimensional analysis then yields the following set of four dimensionless groups,

$$(\Pi_1, \Pi_2, \Pi_3, \Pi_4) = \left(\frac{U}{U_E}, \frac{a}{h}, \frac{K_I}{\sigma_y\sqrt{2h}}, \frac{\eta U_E}{\sigma_y 2h}\right), \qquad (3)$$

such that

$$\Pi_1 = f(\Pi_2, \Pi_3, \Pi_4). \qquad (4)$$

Letting

$$\Pi_1 \propto \frac{\Pi_3 \Pi_2^2}{\Pi_4} \qquad (5)$$

and combining Eqs. (1), (3), and (5) yield the Poiseuille velocity

$$U_P \propto \frac{\rho g' a^2}{\eta}. \qquad (6)$$

This is the appropriate crack propagation velocity when the resistance to crack propagation is controlled by the fluid viscosity. In our experiments, however, we find that Eq. (6) does not fit the data. If the crack propagation velocity is limited by the fracture resistance then a velocity appropriate to our experiments should be a function of the crack loading and the properties of the solid. Figure 3 implies that in the fast propagation regime the crack propagation velocity $U \propto K_I^2$. Hence we choose

$$\Pi_1 \propto \Pi_3^2, \qquad (7)$$

which yields a fracture resistance velocity

$$U_F \propto U_E \frac{K_I^2}{\sigma_y^2 2h} \propto U_E \frac{(\rho g' 2h)^2}{\sigma_y^2}. \qquad (8)$$

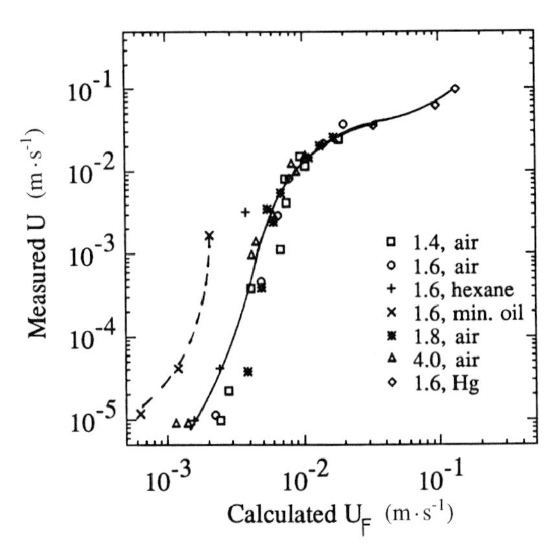

Figure 5 Experimental crack propagation velocity versus calculated fracture resistance velocity given by Eq. (8) for $c = 1$. Most of the data collapse onto a line of varying slope, indicating that $U \propto U_F^n$. The data for constant volumes of mineral oil plot away from the line. The mineral oil fractures tended to have anomalously rough fracture surfaces, possibly indicating chemical reaction between the oil and the gelatin gel.

Figure 5 plots U_F given by Eq. (8) versus the observed crack propagation velocity U and shows that the most of the data collapse onto a line of varying slope indicating that $U = f(U_F)$.

The velocity proportionality given by Eq. (8) as illustrated in Fig. 5 relates the crack propagation velocity to crack parameters for both the slow and fast regimes of propagation in our experiments. However, Eq. (8) does not include the critical stress intensity factor K_{Ic} associated with transition between these regimes. Since we interpret K_{Ic} for our experiments as the fracture toughness of gelatin, the fracture resistance velocity U_F for crack propagation in regime 2 should include K_{Ic} so that U_F approaches zero as K_I approaches K_{Ic} from above. To accomplish this, we introduce a factor into Eq. (8) and obtain a corrected expression for the fracture resistance velocity U_F in the fast propagation regime of our experiments,

$$U_F = c\frac{U_E K_I^2}{2h\,\sigma_y^2}\left|\left(1 - \frac{K_I^2}{K_{Ic}^2}\right)\right|, \qquad (9)$$

where c is a dimensionless constant of proportionality and K_I is given by Eq. (1).

Figure 6 shows a comparison of all of the experimental propagation velocity data and the frac-

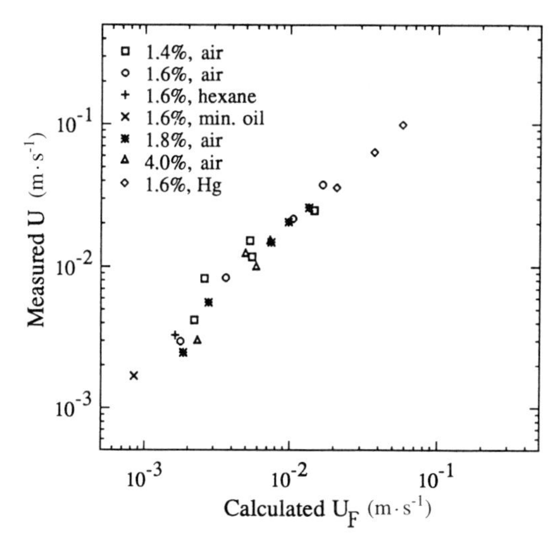

Figure 6 Experimental crack propagation velocity versus the calculated fracture resistance velocity given by Eq. (9) for $c = 1$. Only fractures with supercritical stress intensity are shown since subcritical fractures are undefined in Eq. (9).

ture resistance velocity given by Eq. (9). The value of c is shown to be of order one, indicating good agreement between the observed crack propagation velocity and U_F. Fractures with stress intensity factor $K_I < K_{Ic}$, which we interpret as propagating in the subcritical regime, are not described by Eq. (9) and hence do not appear on the plot.

A New Model for the Crack Propagation Velocity

In the previous sections we obtained a crack velocity that fits the experimental data using dimensional analysis. We can now ask, What physical process results in this crack velocity? The empirically derived crack velocity U_F is a function of the loading term K_I, crack dimensions r and h, and the material properties of the solid U_E, σ_y, and K_{Ic}. Perhaps surprisingly, Eq. (9) implies that the crack loading given by K_I routinely exceeds the critical stress intensity K_{Ic} required for dynamic crack propagation. Furthermore U_F implies $K_{Ia} > K_{Ic}$ for cracks propagating at velocities that can be several orders magnitude less than the elastic wave velocity of the solid medium.

Here we present an argument that *the loading configuration for buoyancy-driven fracture leads to a dynamical constraint on the crack propagation velocity.* The resulting formula for the crack propagation velocity U_F is then identical to Eq. (9).

It is usually assumed in dynamic fracture mechanics that the crack propagation velocity very rapidly approaches the elastic wave velocity of the solid once K_I exceeds K_{Ic}. During dynamic propagation, the dynamic stress intensity factor K_{Id} is a function of the loading and the crack extension velocity $K_{Id} = f(P, r, dr/dt)$ (Freund, 1990). Morita (1993) has pointed out that for the dynamic, viscoelastic fracture problem the stress intensity factor approximated by an elastostatic fracture equation (termed a "pseudo-stress intensity factor") varies by several-fold exceeding the fracture toughness depending upon the loading rate.

It is important to note that the crack propagation velocity in the formulation of the dynamic stress intensity factor is generally equal to the rate of crack growth dr/dt. This means that the loading geometry associated with K_{Id} varies as the crack propagates in the case where the load is applied externally to the solid medium (e.g., in a double torsion experiment tractions are applied to the crack faces at one end of the specimen). In contrast, for constant volume buoyancy-driven crack propagation the load on the crack tip is due to the body force $\rho g'V$, which moves along with the propagating crack so that $dr/dt = 0$ even when the propagation velocity $U \neq 0$. Therefore, for any volume V of fluid in the crack, we expect (and we observe experimentally) that the loading configuration reaches a time-averaged steady state in which the crack-tip propagation velocity is matched by the crack-tail closure velocity.

For the case where the fluid volume in the buoyancy-driven fracture is sufficient for K_I to exceed K_{Ic}, unstable dynamic propagation of the crack tip results, but is constrained by the condition that the loading configuration must maintain a nearly steady state.

This process is best illustrated by considering an incremental crack propagation model. Given a crack initially at rest in a brittle solid that is then filled with fluid so that K_I slightly exceeds K_{Ic}, how will propagation proceed? Since $K_I > K_{Ic}$ at the crack tip, unstable dynamic propagation (accompanied by elastic wave radiation) should occur so that the crack tip propagates upward at a velocity $U \sim U_E$, where U_E is the elastic wave velocity of the solid. However, the entire length of the crack cannot instantaneously respond to motion of the crack tip. As a result, the stress intensity K_I at the crack tip drops below K_{Ic} after some distance d of crack-tip propagation, until the loading configuration readjusts. Since the crack is initially at rest, crack-tail closure can proceed only after information from the crack-tip propagation event, in the form of elastic waves, arrives at the crack tail. This requirement is equivalent to the constraint that the crack maintain a time-averaged constant shape and that the crack tail close upward at a rate equivalent to the crack-tip propagation velocity. Hence the loading configuration for constant fluid volume buoyancy-driven cracks leads to the requirement that propagation events at the crack tip and closure at the tail have an associated lag time $t = 2h/U_E$, which is the time required for information to travel from the crack tip to the crack tail.

For this incremental model the average velocity of crack propagation is then

$$U_F \sim \frac{d}{t} \sim \frac{dU_E}{2h}. \quad (10)$$

Here we made the approximation that the lag time $t >> d/U_E$, where d/U_E is the crack-tip dynamic propagation time for one cycle of incremental crack propagation. The incremental crack-tip propagation length d is the distance ahead of the crack tip where the stress exceeds the local effective yield strength of the solid. This is similar to the fracture mechanics definition of the process zone radius (e.g., Pollard, 1987) or the Barenblatt cohesive zone size (Barenblatt, 1962; Rubin, 1993), $d \propto K_I^2/\sigma_y^2$. However, a definition for the process zone size relevant to this incremental propagation model must include the result that dynamic propagation occurs only when $K_I > K_{Ic}$. Hence the modified definition for the process zone size becomes

$$d \propto \frac{K_I^2}{\sigma_y^2}\left|\left(1 - \frac{K_I^2}{K_{Ic}^2}\right)\right|, \quad (11)$$

where the local effective yield stress of the solid σ_y is the sum of the cohesive strength of the solid and the regional stress acting on the crack-tip process zone (Rubin, 1993). Combining Eqs. (10) and (11) we recover our empirically derived fracture resistance velocity

$$U_F = c\frac{U_E}{2h}\frac{K_I^2}{\sigma_y^2}\left|\left(1 - \frac{K_I^2}{K_{Ic}^2}\right)\right|. \quad (12)$$

Equation (12) implies that for materials with very low shear velocity (e.g., gelatin) or for very long cracks (e.g., magma fractures) buoyancy-driven fractures may propagate at velocities $U_F << U_E$ even when the loading gives a stress intensity $K_I > K_{Ic}$.

Shape of Fractures with Constant Fluid Volume

Our experiments indicate that the shape of the propagating constant fluid volume fractures approximates the fluid-filled crack shape predicted by elastostatic theory (see Fig. 8). This result verifies the assertion that the fluid pressure gradient is nearly hydrostatic and is consistent with the findings of Takada (1990). In terms of the fracture propagation model just presented, it implies that (for $U_F << U_E$) the crack adjusts dynamically to maintain an equilibrium shape that approximates the elastostatic shape.

Figure 7a shows the geometry considered in the

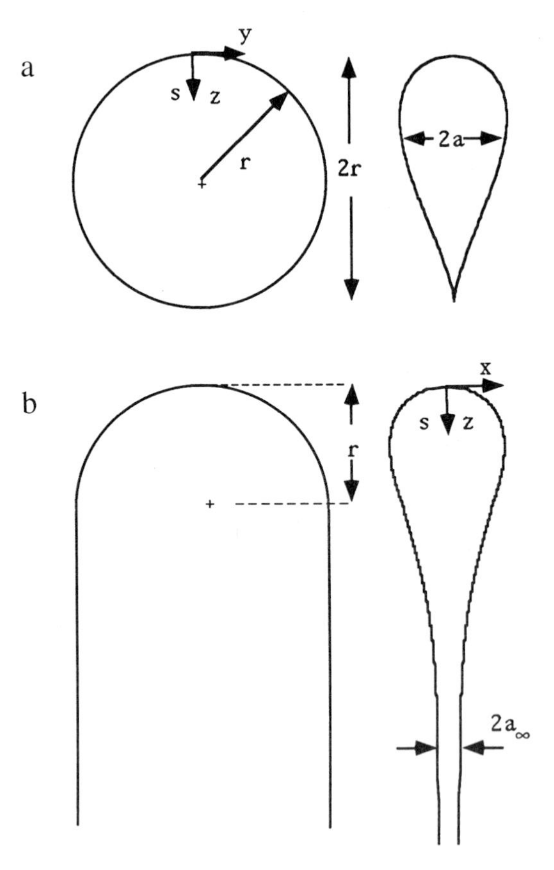

Figure 7 Shape of propagating buoyancy-driven fluid-filled fractures. Idealized in-plane crack contour used (left) in the calculation of the cross-sectional shape (right) resulting from (a) a buoyant fluid with negligible viscosity (Eq. (14)) and (b) a buoyant fluid with significant viscosity (Eqs. (18)–(20c)).

following analysis. According to Sneddon (1946), the thickness $2a$ of a static pressurized circular crack with radius r in an elastic medium is

$$2a(y, z) = \frac{4}{\pi} P\left(\frac{1 - \nu}{\mu}\right) \cdot (r^2 - y^2 - z^2)^{1/2}, \quad (13)$$

where P is the fluid overpressure, ν is Poisson's ratio, μ is the shear modulus of the solid, r is the crack radius, and y, z are Cartesian coordinates. We compare the cross-sectional experimental shape of the propagating cracks to the theoretical shape of an internally pressurized fracture with a linear pressure gradient. Since the fracture closes at the tail as it propagates upward, the condition of zero stress intensity at the tail ($da/dz = 0$ at $z = r$) is thus appropriate. This requirement results in the overpressure being zero at the tail. Hence the thickness along $y = 0$ of a static, circular crack filled with buoyant fluid is

$$2a(z) = \frac{4}{\pi} \rho g' \left(\frac{1 - \nu}{\mu}\right) \cdot (r + z)(r^2 - z^2)^{1/2}, \quad (14)$$

where $g' = g(\rho - \rho_1)/\rho$ is the fluid buoyancy, g is the gravitational acceleration, ρ is the solid density, and ρ_1 is the fluid density. The maximum width occurs at $z = r/2$ and is given by

$$2a_{\mathrm{m}} = \frac{3\sqrt{3}}{\pi} \frac{(1 - \nu)}{\mu} \rho g' r^2. \quad (15)$$

Except for a factor of $2/\pi$, the result given by Eqs. (14) and (15) is identical to the shape for a two-dimensional crack in the presence of a linear pressure gradient and with zero stress intensity at the tail (Weertman, 1971; Pollard and Muller, 1976).

It is interesting to note that contrary to the two-dimensional case, the fluid overpressure near the tail of a static circular crack is never negative. This is a consequence of the finite breadth of the circular crack. The curvature of the crack face indicates the relative pressure. A convex outward shape (positive curvature) indicates a positive fluid overpressure and a concave (negative curvature) crack face indicates a fluid pressure less than the ambient pressure in the solid medium. Even though the curvature in z given by Eq. (14) is negative near the tail, that total curvature in y and z is always positive and goes to zero only at $y = 0$, $z = -r$.

Figure 8 compares the experimental shape of propagating fluid-filled fractures and the theoretical shape given by Eq. (14). The fact that the observed cross-sectional shape closely matches the shape for a static crack shows that dynamic viscous forces do not significantly alter the hydrostatic pressure in the fracture. This is another indication that fluid viscosity plays a negligible role in controlling the overall propagation velocity of these fractures.

Shape of Fractures with Constant Fluid Flux

Constant flux experiments were performed using negatively buoyant corn syrup, Cadmium chloride and water solution (see Table 2 and Fig. 9). The experiments show that, after broadening rapidly near the source, a crack propagating away from a constant source of buoyant fluid consists of a narrow conduit that feeds a crack head of slowly changing size, shape, and velocity. The condition of constant crack head size requires that the crack head velocity is matched by the average fluid velocity in the conduit. In this section we add the effects of the fluid viscosity to the pressure gradient in the tail of the propagating fracture. Figure 7b shows the (idealized) geometry considered. The analysis is similar to that given by Lister (1991) for a density stratified lithosphere. The crack conduit is of uniform breadth ($2a_\infty$) and is smoothly attached to the circular leading edge of the crack. It is convenient to place the origin of coordinates ($z = 0$) at the tip of the crack. We assume that the horizontal pressure gradient is negligible so that horizontal cross sections of the fracture are ellipses. The gradient in overpressure for the laminar flow of buoyant fluid in a planar crack arises from the sum of the buoyant force of the fluid and the viscous stresses transmitted into the fluid from the sides of the crack:

$$\frac{dP}{ds} = -\rho g' + \frac{3\eta Q_s}{2a^3 r}, \quad (16a)$$

where

$$Q_s = u\pi a r \quad (16b)$$

and u is the average fluid velocity and η is the dynamic shear viscosity of the fluid in the crack.

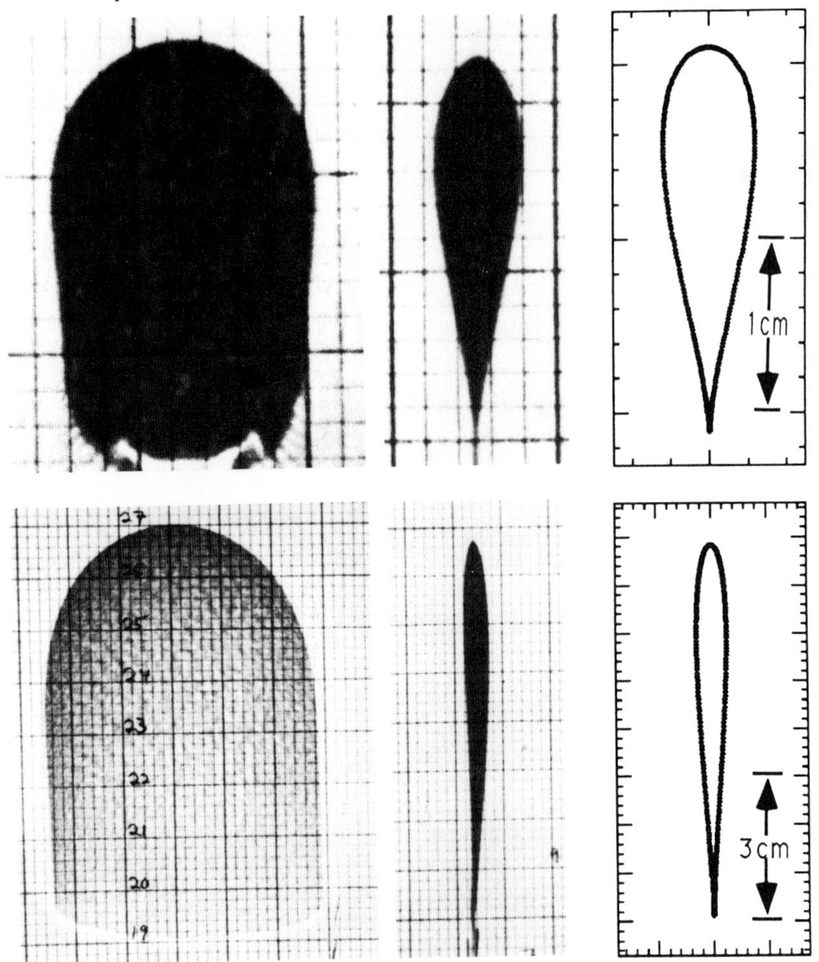

Figure 8 Two comparisons of the experimental versus theoretical elastostatic crack shape, showing the in-plane shadow photograph in the first column, the cross-sectional photograph in the second column, and a plot of the theoretical shape given by Eq. (14) in the third column. The upper row shows 0.8 ml air in 1.4% gelatin. The lower row shows 10 ml air in 4.0% gelatin.

Since horizontal pressure gradients are assumed to be negligible, the elastic normal stress acting on the fluid is

$$P(s) = \Theta \frac{\mu a(s)}{(1 - \nu) r(s)}, \tag{17}$$

where Θ is a dimensionless parameter that ranges in value from $\Theta = \pi/2$ for the section of the crack near the crack head (where the crack contour is circular) to $\Theta = 1$ for the section of the crack far downstream of the crack tip (Rubin and Pollard, 1987). Since we are interested in the details of the shape near the head of the crack we will take $\Theta = \pi/2$. The average vertical fluid velocity is assumed to be constant so that the pressure gradient in the crack head is approximately hydrostatic when the maximum thickness of the crack head

---

Figure 9 Comparison of the experimental versus theoretical crack shape with constant source flux, showing the in-plane shadow photograph in the first column, the cross-sectional photograph in the second column, and a plot of the theoretical shape given by Eqs. (18)–(20c) in the third column, all from the same experiment using a corn syrup, cadmium chloride, and water solution of specific gravity 1.41 in 1.8% concentration gelatin.

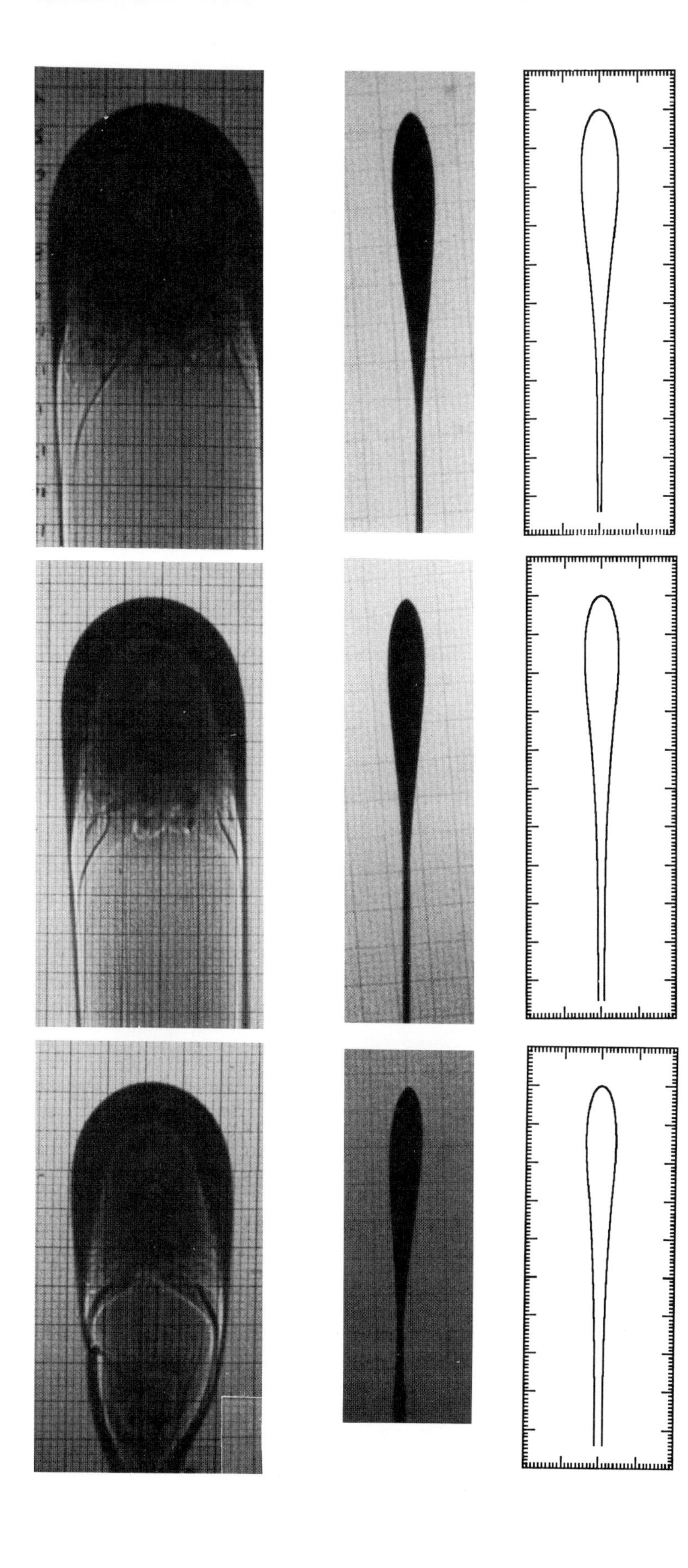

$2a_{max}$ is a few times greater than the crack conduit thickness $2a_\infty$. Reasoning thus, we approximate the thickness $2a(s)$ in the crack head by Eq. (14), rewritten in terms of s,

$$2a(s) \cong \frac{2}{\pi}\rho g'\left(\frac{1-\nu}{\mu}\right) \cdot (2r - s)(2rs - s^2)^{1/2} \tag{18}$$

for $0 < s < r$ and $a_\infty/a_{max} < 1/3$. For a crack head of slowly changing volume, the breadth $2r$ of the crack is nearly constant for $s > r$. Differentiating Eq. (17) with respect to s for constant r and combining the result with Eqs. (16) we obtain for $s > r$,

$$\frac{da}{ds} = \frac{2r}{\pi}\left(\frac{1-\nu}{\mu}\right)\left(\frac{3\eta Q_s}{2a^3 r} - \rho g'\right). \tag{19}$$

Note that $da/ds = 0$ when $3\eta Q_s/2a^3r = \rho g'$. That is, for constant flux conditions the width of the conduit is constant as s tends to ∞ where the fluid buoyancy is balanced by the viscous stress on the crack walls. Equation (19) is easily integrated to yield an analytical expression relating s and a. The constant of integration is evaluated by imposing the condition of continuous width of the crack at $s = r$. The result for $s > r$ is

$$s = 2r\left\{1 - \frac{a}{2a_1} + \frac{a_\infty}{a_1} \cdot \ln\left[\frac{(a + a_\infty)(a_1 - a_\infty)}{(a - a_\infty)(a_1 + a_\infty)}\right]\right\}, \tag{20a}$$

where

$$a_\infty = \left(\frac{3\eta Q_s}{2\rho g' r}\right)^{1/3} = \sqrt{\frac{3\pi\eta u}{2\rho g'}} \tag{20b}$$

and

$$a_1 = \frac{2}{\pi}\left(\frac{1-\nu}{\mu}\right)\rho g' r^2 \tag{20c}$$

is the half-width of the crack at $s = r$ determined from Eq. (18).

Figure 9 shows an example of the constant source flux experiments at three different times in the development of the buoyancy-driven fluid-filled fracture and compares the experimental cross-sectional shape of the propagating fluid-filled fractures near the crack head and the theoretical shape given by Eqs. (18)–(20c).

Application to Magma Transport

The relevance of fluid-filled fracture models to magma fracture depends strongly on both the fracture properties of the country rock and the buoyancy and viscosity of the ascending magma. Buoyancy-driven magma fracture is important at depths in the lithosphere ranging from the partially molten source region where dike nucleation most likely occurs to the level of neutral buoyancy (LNB) where magma accumulates to form crustal magma chambers (Ryan, 1987). We are therefore particularly interested in the country rock and the magma properties at depths ranging from approximately 100 to 2 km, and thus in confining pressures from 3500 to 60 MPa.

Fracture Properties of Country Rock

The theoretical yield strength of a solid with no flaws is based on the bond strength between atoms and is approximately $E/10$, where E is the Young's modulus (Lawn and Wilshaw, 1975). This gives an upper limit of $\sigma_y \approx 5000$ MPa for the extensional yield strength in the lithosphere at subcrustal depths.

The experimental value of σ_y under high confining pressures is difficult to test directly and estimates are thus generally obtained from triaxial tests. A yield criterion that relates the shear failure of the triaxial test to the extensional yield strength is then used (Paterson, 1978). In the following sections we use values of $\sigma_y = 100-1000$ MPa as representative of igneous rocks under confining pressures of 60 to 1000 MPa. For active magmatic systems, where magma rises frequently (with respect to the magma-channel solidification time) along preexisting fracture planes, the extensional strength will be negligible and will probably not exceed 10 MPa (Ryan, 1988).

The estimation of the fracture toughness of rocks appropriate to magma fracture in the lithosphere has been a subject of debate in the rock mechanics and magma transport literature. The theoretical value of the fracture toughness is obtained by considering a crack in an elastic solid as a pressurized slit of half-length b. The mode I stress intensity factor is given by

$$K_I = P\sqrt{\pi b}, \tag{21}$$

where P is the uniform remote tensional load or the uniform internal overpressure on the crack. Lawn and Wilshaw (1975) give an estimate of the stress concentration factor at the tips of an elliptical "crack" ($a << b$) of thickness a and crack tip radius of curvature r:

$$\frac{\sigma_b}{P} = 2\sqrt{\left(\frac{b}{r}\right)}, \tag{22}$$

where σ_b is the extensional stress at the crack tip. The fracture toughness K_{Ic} is defined as the critical value of the stress intensity required for crack propagation. Letting $\sigma_b = \sigma_y \approx E/10$ when $K_I = K_{Ic}$ and combining Eqs. (21) and (22) result in the theoretical estimate of the fracture toughness for an ideal solid with an atomically sharp crack:

$$K_{Ic} \approx \frac{E\sqrt{r}}{20}. \tag{23}$$

For $r = 10^{-9}$ m and $E = 10^{11}$ Pa, the resulting estimate for the ideal (mode I) fracture toughness is $K_{Ic} \approx 10^5$ Pa $\cdot$ m$^{1/2}$.

As defined above, the fracture toughness is a material property. However, microcrack formation in the region of extensional stress ahead of the crack tip may complicate the crack-tip energy balances. Off-axis microcracking increases the fracture energy and hence the fracture toughness and has been shown to accompany the propagation of laboratory scale cracks (Kobayashi and Fourney, 1978). In a field study of mafic dikes, Delaney *et al.* (1986) described the similarity between microcracking in laboratory experiments and dike-parallel joints in the field. They showed that the difference in magnitude of the high stress region or the "process zone" ahead of the crack tip between laboratory cracks and dikes is consistent with the difference in scale between the microcrack zone and the jointing zone, respectively. Comparison of laboratory fracture toughness test results and field studies of dikes indiates that the magnitude of microcracking and jointing is dependent on the size of the crack. High-temperature (diffusion-driven or dislocation-driven) crack-tip plasticity is likely to be another important scale-dependent mechanism in mantle magma fracture.

As a result of one or both of the above processes, the fracture toughness of country rock obtained from the dimensions of dikes is generally two or three orders of magnitude greater than those obtained from laboratory experiments. This suggests that, in addition to being a material property, the fracture toughness is a parameter that depends upon the length scale of the crack and the crack-tip process zone size. Thus, fracture toughness values obtained from laboratory tests on centimeter-scale samples are probably not appropriate as a propagation criterion for kilometer-scale magma fractures.

Because of the great lateral extent of dikes (10 km is typical) few can be traced along their entire length and it is not surprising that little data exist on the dimensions of dikes in the field. Table 3 is a compilation of data collected by various authors of the dimensions of dikes. The value of the apparent fracture toughness is estimated as

$$K_{Ic} = \frac{\mu}{(1 - \nu)}\frac{a}{\sqrt{r}}, \tag{24}$$

where $\mu/(1 - \nu)$ is the rock rigidity, a is the average dike half-thickness, and r is the dike half-

Table 3

Country Rock Fracture Toughness Estimates from Dike Dimensions

Host rock type	Magma type	$\mu/(1-\nu)$ (GPa)	2a (m)	2r (m)	K_{Ic} (MPa$\sqrt{m}$)	Reference
Granite	Rhyolite	22	7	11,000	**1,000**	Reches and Fink (1988)
Igneous	Basalt/andesite	4	25	10,000	**700**	Macdonald *et al.* (1988)
Shale	Minette	1	2	346	**~100**	Delaney and Pollard (1981)
Sedimentary	Lamprophere	~2	12	11,550	**1,300**	Pollard and Muller (1976)
Basalt	Tholeiite	10	12	4,000	**1,300**	Gudmundsson (1983)
Pyroclastic flow	Basalt/andesite	5	1	480	**200**	Fedatov *et al.* (1978)

breadth. The form of the dike or fissure is assumed to be elliptical. The resulting laboratory and field estimates of K_{Ic} lie in the range $1 < K_{Ic} < 1000$ MPa · m$^{1/2}$. We consider $100 < K_{Ic} < 1000$ MPa · m$^{1/2}$ an appropriate estimate for magma fracture in the lithosphere.

Melt Properties

The driving force for magma fracture is controlled by the gradient in overpressure of the magma. Flow is driven by the density difference between the country rock and magma in the presence of gravity (buoyancy forces) and is resisted by fracture near the crack tip and by viscous shear transmitted into the flowing magma from the crack faces.

The density of basaltic magma (Kilauea 1921 olivine tholeiite) at confining pressures corresponding to depths as great as 50 km in the lithosphere has been studied by Fujii and Kushiro (1977). They found an approximately linear increase in the melt density as a function of pressure from 2600 kg · m^{-3} at atmospheric pressure to 2900 kg · m^{-3} at 1.5 GPa. The density of lithospheric rocks from the Moho to 100 km depth varies from 2900 to 3300 kg · m^{-3}. Hence a typical value for the magma buoyancy $g' = g(\rho - \rho_l)/\rho$ is 1 m · s^{-2}.

Magma viscosities for a variety of melts have been compiled by Ryan and Blevins (1987). The viscosity is sensitive to the absolute temperature. Since the basalt liquidus increases with pressure from about 1230°C at the surface to about 1500°C at 100 km depth, the melt viscosity ranges from about 100 Pa · s at the surface to 1 Pa · s at the base of the lithosphere, respectively, assuming that the magma temperature is near the liquidus temperature during ascent.

Implications for Magma Ascent Velocities

Our experiments indicate a new regime for propagating fluid-filled cracks, characterized by a fracture resistance velocity U_F. Does U_F apply to magma fracture propagation velocities in the lithosphere as well? To answer this question we must first assess the relevance of our experiments to magma fracture propagation in the lithosphere. In our constant flux experiments the low fluid viscosities ensured that the fracture resistance of the solid, rather than the fluid viscosity, determined the crack propagation velocity. However, even when the fluid viscosity is sufficient to allow a fracture conduit to form, $U_F < U_P$ for a fracture with head width significantly greater than conduit width. In this case the fracture resistance must still control the crack propagation velocity. Hence the results of our experiments apply to the elastically deforming head of a magma-filled fracture that is propagating in the lithosphere when $U_F < U_P$ in the dike head even though U_P defines the fluid velocity in the magma fracture conduit. Since U_F is proportional to the process zone size and inversely proportional to the elastic wave travel time down the length of the crack, a process zone size of 1 m with a magma-filled fracture head length of several kilometers implies a propagation velocity of about 1 m · s^{-1}. This velocity is of the same order of magnitude as that inferred from geophysical observations of magma fracture (Spera, 1980; Aki *et al.*, 1977). It is also similar to estimates of the Poiseuille velocity for the same fracture size (Lister and Kerr, 1991). Figure 10 compares the fracture resistance velocity and the Poiseuille velocity for a range of rock fracture toughnesses and magma viscosities. For high-viscosity magma or

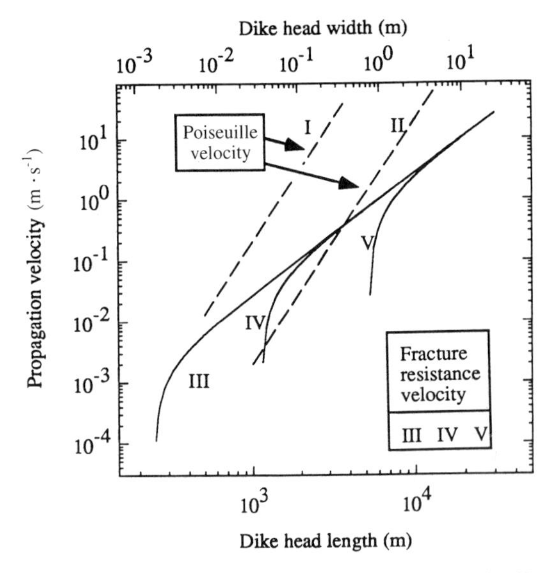

Figure 10 Comparison of the fracture resistance velocity U_F to the Poiseuille velocity U_P applied to magma fracture in the lithosphere. Numerals I through V refer to the following: (I and II) U_P with viscosities 100 and 1 Pa · s, respectively. (III–V) U_F with fracture toughnesses 10^7, 10^8, and 10^9 Pa · m$^{1/2}$, respectively. In all cases $\mu/(1 - \nu) = 5 \times 10^{10}$ Pa, $\sigma_y = 10^9$ Pa, $g' = 1.0$ m · s^{-2}, and $U_E = 4.0$ km · s^{-1} (the value for U_E is chosen as a typical value of the shear velocity in the elastic lithosphere).

for weak, near-surface crustal rocks, U_F can be greater than the Poiseuille velocity in the head of the forming dike. Hence, very near the surface, the magma viscosity probably limits the crack propagation velocity. On the other hand, for low-viscosity basaltic magmas and for elastic and fracture properties representative of the lithosphere, the Poiseuille velocity U_P is typically greater than U_F. This suggests that the fracture resistance velocity U_F limits the magma fracture velocity at depth in the lithosphere.

How is the fracture resistance model of buoyancy-driven magma fracture propagation distinguishable from models that assume that the fracture resistance is negligible once the fracture toughness has been exceeded? Models in which only the Poiseuille velocity limits the propagation speed imply that all of the gravitational potential energy lost during magma ascent goes into frictional (viscous) heating of the magma. The energetics of these models do not allow for seismic wave radiation accompanying crack propagation, except by postulating the existence of a vapor phase present in the dike tip (Anderson, 1978; Sammis and Julian, 1987). In the fracture resistance model for magma fracture propagation presented here, the shape of the dike head and the propagation velocity are coupled by seismic waves. Hence seismicity is an integral element of our model.

The average fluid velocity for a fracture with a constant shape and constant source flux is uniform over the entire fracture length and equal to the propagation velocity. In the magma fracture conduit, the fluid buoyancy is balanced by the viscous stresses (see Eqs. (19)–(20c). However, the viscous stress contribution to the total pressure gradient ΔP_v is strongly dependent on the local crack width ($\Delta P_v \propto 1/a^3$) whereas the hydrostatic component of the pressure gradient ΔP_h is constant ($\Delta P_h = \rho g' \propto 1/a_\infty^3$; see Eq. (16a)). This means that where the magma fracture head thickness $2a$ is a few times greater than the conduit thickness $2a_\infty$, the pressure gradient in the crack head is nearly hydrostatic ($a_\infty/a = 0.3$ gives $\Delta P_v/\Delta P_h = 0.03$). We can use this to check the applicability of our model to magma fracture propagation (and dike formation) in the lithosphere by determining the conditions under which the ratio of the conduit thickness to the maximum dike head thickness a_∞/a_m is less than 0.3. When the head of an ascending buoyant fracture changes volume and shape slowly, the average fluid velocity in the conduit approximates the fracture propagation velocity. The average fluid velocity is related to the source flux and the conduit dimensions by

$$u = \frac{Q_s}{\pi a_\infty r}. \tag{25}$$

Dividing Eq. (20b) by Eq. (15) yields an expression for a_∞/a_m,

$$\frac{a_\infty}{a_m} = \frac{2\pi\sqrt{3}}{9}\left(\frac{\mu}{1-\nu}\right)\left[\frac{3\eta Q_s}{2(\rho g')^4 r^7}\right]^{1/3}. \tag{26}$$

An approximate expression for the newly formed dike breadth $2r$ is obtained by combining Eqs. (12) and (25) and rearranging,

$$2r \cong \left[0.08\left(\frac{Q_s^2\sigma_y^6}{\eta U_E^3(\rho g')^5}\right)^{3/8} + \frac{\pi K_{Ic}^2}{16(\rho g')^2}\right]^{1/3}. \tag{27}$$

Figures 11 and 12 are plots of Eqs. (27) and (26), respectively. Both a_∞/a_m and $2r$ are plotted versus the source flux Q_s for three viscosities representative of basaltic magma and for two values of the fracture toughness. Figure 12 shows that for a

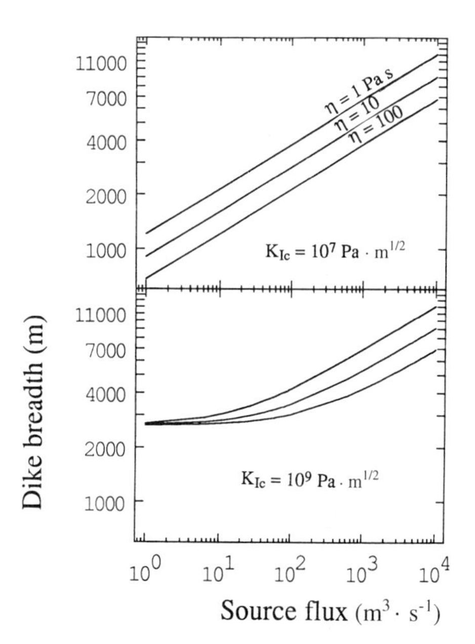

Figure 11 The magma fracture head radius is plotted as a function the source flux Q_s for $1 < Q_s < 10^4$ m$^3 \cdot$ s^{-1}, for η = 1, 10, and 100 Pa · s, and for $K_c = 10^7$ and 10^9 Pa · m$^{1/2}$ (see Eq. (27)). All other material properties are constants chosen to be representative of the lithosphere; $\mu/(1-\nu) = 5 \times 10^{10}$ Pa, $\sigma_y = 10^9$ Pa, $\rho = 3300$ kg · m^{-3}, $g' = 1$ m · s^{-2}.

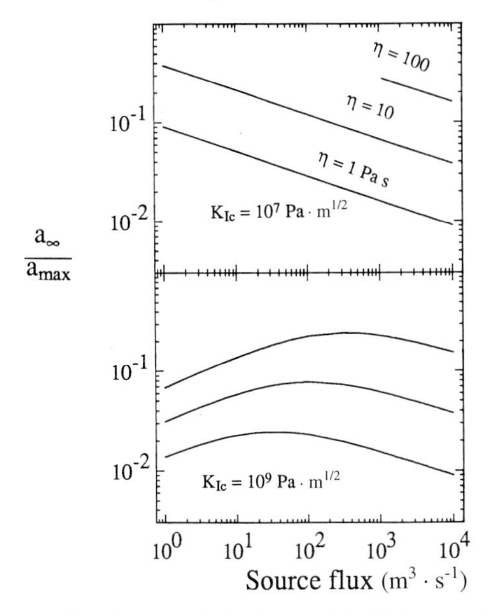

Figure 12 The ratio of the dike conduit thickness to the maximum dike head thickness $a_{\infty}/a_{\mathrm{m}}$ is plotted as a function the source flux Q_s for $1 < Q_s < 10^4\ m^3 \cdot s^{-1}$, for $\eta = 1$, 10, and 100 Pa · s, and for $K_c = 10^7$ *and* 10^9 *Pa* · $m^{1/2}$ (see Eq. (26)). All other material properties are constants chosen to be representative of the lithosphere; $\mu/(1 - \nu) = 5 \times 10^{10}$ Pa, $\sigma_y = 10^9$ Pa, $\rho = 3300$ kg · m^{-3}, $g' = 1$ m · s^{-2}.

fracture toughness value of $K_{\mathrm{Ic}} = 10^3$ MPa · m$^{1/2}$ (representative of estimates from the dimensions of dikes in the field), $a_{\infty}/a_{\mathrm{m}}$ is always less than 0.3. This implies that the fracture resistance velocity U_{F} will generally control the buoyancy-driven magma fracture propagation speed if the fracture toughness of the lithosphere is about 10^3 MPa · m$^{1/2}$. If the fracture toughness is much lower than 10^3 MPa · m$^{1/2}$ then Fig. 12 shows that the Poiseuille velocity is probably appropriate for low-source-flux and high-viscosity magma and the fracture resistance velocity is appropriate for relatively high-source-flux and low-viscosity magma.

Additional Processes

In this chapter we have presented experimental work on isolated, vertically propagating fluid-filled fractures and applied the results to magma fracture in the lithosphere. There are several important processes associated with magma transport in the lithosphere that were not addressed by these experiments.

In most igneous environments the near-surface crustal density falls below the magma density. The horizon at which the crustal density equals the magma density is known as the *level of neutral buoyancy* (LNB; Ryan, 1987). As a parcel of magma approaches the LNB from below, the driving force for vertical propagation becomes depleted and further fracture propagation proceeds in blade-like magma fractures that spread laterally away from the magma source and may result in volcanic rift zones (Rubin and Pollard, 1987; Ryan, 1987; Lister and Kerr, 1991).

When two or more magma fractures ascend in close proximity to each other the stress fields induced by the individual parcels will mechanically perturb each other depending on their crack-tip offsets, overlaps, and mutual distances (Ryan, 1990; Takada, 1994). Acceleration associated with crack-tip splitting and coalescence is also a likely source of seismic radiation (Sammis and Julian, 1987).

Heat exchange between flowing magma and the wallrock is another important process in magma transport. Whether magma tends to solidify in a magma conduit or whether the wallrock melts depends on the relative contributions of conductive heat transfer by the wallrock and advective heat transfer by the flowing magma, respectively (Bruce and Huppert, 1990; Delaney and Pollard 1982).

Summary

Crack propagation (in our experiments) is found to be controlled by the fracture resistance of the solid. We define the fracture toughness of the gelatin gels to be the critical stress intensity associated with the transition between two observed crack propagation regimes; in regime 1 slow crack propagation is characterized by a strong velocity dependence on K_{I}; in regime 2 fast propagation is characterized by velocity $U \propto K_{\mathrm{I}}^2$. Dimensional analysis based on the experimental results yields an expression (Eq. (9)) for the propagation velocity, in regime 2, that collapses the data (Fig. 6).

To provide physical justification of the dimensional analysis we introduce a model for buoyancy-driven fracture in which the loading configuration provides a dynamical constraint on the crack propagation velocity. The idea behind the model is that the crack tip of a buoyancy-driven fracture with stress intensity exceeding the fracture toughness cannot undergo unlimited unstable (dynamical) propagation since the rest of the crack takes a finite time to respond to changes in the crack geometry.

The observed shape of the propagating constant volume fractures with low-viscosity fluids closely approximates the theoretical elastostatic shape of an internally pressurized crack. This agreement indicates (*a*) the pressure gradient in the experimental cracks is nearly hydrostatic and propagation is independent of the fluid viscosity, and (*b*) propagation in regime 2 is "quasi-dynamic" in the sense that the crack shape adjusts dynamically to the elastostatic shape in order to maintain crack-tip loading.

The observed in-plane shape of the propagating fractures resulting from the experiments using a constant flux of viscous fluid is used in conjunction with elasticity theory to derive a semi-empirical formula for the cross-sectional shape. The cross-sectional shape consists of a crack head that tapers at depth to a thin conduit.

Application of U_F to magma fracture implies propagation velocities of order $1 \text{ m} \cdot \text{s}^{-1}$ in the lithosphere. Scaling of the relative contribution of the buoyancy and viscous stress to the pressure gradient in a fluid-filled fracture shows that when the fracture head is a few times thicker than the fracture conduit, the pressure gradient in fracture head is nearly hydrostatic. This indicates that even when the Poiseuille velocity controls flow in the conduit, the fracture resistance velocity U_F can still control the crack propagation velocity. Applying this scaling to magma fracture in the lithosphere we show that U_F is likely to control the propagation of low-viscosity magma fractures at depth.

Acknowledgments

This research has been supported by NSF Grant EAR8916152. M. Heimpel thanks Paul T. Delaney for a thorough review and Dr. Michael Ryan for all his advice and support. Thanks also to Allan Rubin for some lively discussions on fracture mechanics.

References

Aki, K., Fehler, M., and Das, S. (1977). Source mechanisms of volcanic tremor: Fluid-driven crack models and their application to the 1963 Kilauea eruption, *J. Volcanol. Geotherm. Res.* **2,** 259–287.

Anderson, O. L. (1978). The role of magma vapors in volcanic tremors and rapid eruptions, *Bull. Volcanol.* **41,** 341–353.

Anderson, O. L., and Grew, P. C. (1977). Stress corrosion theory of crack propagation with applications to geophysics, *Rev. Geophys.* **15,** 77–103.

Atkinson, B. K., and Meredith, P. G. (1987). The theory of subcritical crack growth with applications to minerals and rocks, *in* "Fracture Mechanics of Rock" (B. K. Atkinson, ed.), pp. 111–166, Academic Press, London, England.

Barenblatt, G. I. (1962). The mathematical theory of equilibrium cracks in brittle fracture, *Adv. Appl. Mech.* **7,** 55–129.

Bruce, P. M., and Huppert, H. E. (1990). Solidification and melting along dykes by the laminar flow of basaltic magma, *in* "Magma Transport and Storage" (M. P. Ryan, ed.), pp. 87–101, Wiley, Chichester/Sussex, England.

Delaney, P. T., and Pollard, D. D. (1981). Deformation of host rocks and flow of magma during growth of minette dikes and breccia-bearing intrusions near Ship Rock, New Mexico, "U.S. Geol. Surv. Prof. Paper," 1202.

Delaney, P. T., and Pollard, D. D. (1982). Solidification of basaltic magma during flow in a dike, *Am. J. Sci.* **282,** 856–865.

Delaney, P. T., Pollard, D. D., Ziony, J. I., and McKee, E. H. (1986). Field relations between dikes and joints: Emplacement processes and paleostress analysis, *J. Geophys. Res.* **91,** 4920–4938.

Evans, A. G., and Langdon, T. G. (1976). "Structural Ceramics," "Progress in Materials Science," Vol. 21, pp. 174–441, Pergamon Press, Oxford, England.

Fedotov, S. A., Enman, V. B., Magus'kin, M. A., Levin, V. Y., Zharinov, N. A., and Enman, S. V. (1978). Deformations of the Earth's surface in the vicinity of the New Tolbachik Volcanoes (1975–1976), *in* "The Great Tolbachik Fissure Eruption, Geological and Geophysical Data 1975–1976" (S. A. Fedotov and Ye. K. Markhinin, eds.), pp. 267–282, Cambridge Univ. Press, Cambridge.

Fiske, R. S., and Jackson, E. D. (1972). Orientation and growth of Hawaiian volcanic rifts: The effect of regional structure and gravitational stresses, *Proc. R. Soc. London Ser. A* **329,** 299–326.

Fowler, A. C. (1990). A compaction model for melt transport in the Earth's asthenosphere. Part II. Applications, *in* "Magma Transport and Storage" (M. P. Ryan, ed.), pp. 16–32, Wiley, Chichester/Sussex, England.

Freund, L. B. (1990). "Dynamic Fracture Mechanics," Cambridge Univ. Press, Cambridge.

Fujii, T., and Kushiro. (1977). Density, viscosity, and compressibility of basaltic liquid at high pressures, *Carnegie Inst. Washington Yearbook,* 419–424.

Gudmundsson, A. (1983). Form and dimensions of dykes in eastern Iceland, *Tectonophysics* **95,** 295–307.

Irwin, G. R. (1958). Fracture, *in* "Handbuch der Physik" (S. Flugge, ed.), Vol. 6, pp. 551–590, Springer-Verlag, New York.

Kobayashi, T., and Fourney, W. L. (1978). Experimental characterization of the development of the microcrack process zone at a crack tip in rock under load, "Proceedings, 19th Natl. Symp. Rock Mech.," pp. 243–246.

Lawn, B. R., and Wilshaw, T. R. (1975). "Fracture of Brittle Solids," Cambridge Univ. Press, Cambridge.

Lister, J. R. (1990). Buoyancy-driven fluid fracture: The effects of material toughness and of low-viscosity precursors, *J. Fluid Mech.* **210,** 263–280.

Lister, J. R. (1991). Steady solutions for feeder dykes in a density stratified lithosphere, *Earth Planet. Sci. Lett.* **107,** 233–242.

Lister, J. R., and Kerr, R. C. (1991). Fluid-mechanical models of crack propagation and their application to magma transport in dykes, *J. Geophys. Res.* **96,** 10049–10077.

Macdonald, R., Wilson, L., Thorpe, R. S., and Martin, A. (1988). Emplacement of the Cleveland Dyke: Evidence from geochemistry, mineralogy, and physical modelling, *J. Petrol.* **29,** 559–583.

Marsh, B. D. (1982). On the mechanics of igneous diapirism, stoping, and zone melting, *Am. J. Sci.* **282,** 808–885.

Maaloe, S. (1987). The generation and shape of feeder dykes from mantle sources, *Contrib. Mineral. Petrol.* **96,** 47–55.

McKenzie, D. P. (1984). The generation and compaction of partially molten rock, *J. Petrol.* **25,** 713–765.

Morita, N. (1993). Personal communication.

Nicolas, A. (1986). A melt extraction model based on structural studies in mantle peridotites, *J. Petrol.* **27,** 999–1022.

Nicolas, A. (1990). Melt extraction from mantle peridotites: Hydrofracturing and porous flow, with consequences for oceanic ridge activity, *in* "Magma Transport and Storage" (M. P. Ryan, ed.), pp. 159–173, Wiley, Chichester/Sussex, England.

Olson, P., and Christensen, U. (1986). Solitary wave propagation in a fluid conduit within a viscous matrix, *J. Geophys. Res.* **91,** 6367–6374.

Paterson, M. S. (1978). "Experimental Rock Deformation—The Brittle Field," Minerals and Rocks, Vol. 13, Springer-Verlag, New York.

Pollard, D. D. (1987). Elementary fracture mechanics applied to the structural interpretation of dykes, *in* "Mafic Dyke Swarms" (H. C. Halls and W. H. Fahrig, eds.), Geol. Assoc. Canada Spec. Paper 34, pp. 5–24.

Pollard, D. D., and Muller, O. H. (1976). The effect of gradients in regional stress and magma pressure on the form of sheet intrusions in cross section, *J. Geophys. Res.* **81,** 975–984.

Reches, Z., and Fink, J. (1988). The mechanism of intrusion of the Inyo Dike, Long Valley caldera, California, *J. Geophys. Res.* **93,** 4321–4334.

Richards, R., and Mark, R. (1966). Gelatin models for photoelastic analysis of gravity structures, *Exp. Mech.* **6,** 30–38.

Rubin, A. M., and Pollard, D. D. (1987). Origins of blade-like dikes in volcanic rift zones, "U.S. Geol. Surv. Prof. Paper," **1350,** pp. 1449–1470.

Rubin, A. M. (1993). Tensile fracture of rock at high confining pressure: Implications for dike propagation, *J. Geophys. Res.,* **98,** 15,919–15,935.

Ryan, M. P. (1987). Neutral buoyancy and the mechanical evolution of magmatic systems, *in* "Magmatic Processes: Physicochemical Principles," (B. O. Mysen, ed.), Spec. Publ. 1, pp. 259–287, The Geochemical Society, University Park, PA.

Ryan, M. P. (1988). The mechanics and three-dimensional internal structure of active magmatic systems: Kilauea Volcano, Hawaii, *J. Geophys. Res.* **93,** 4213–4248.

Ryan, M. P. (1990). The physical nature of the Icelandic magma transport system, *in* "Magma Transport and Storage" (M. P. Ryan, ed.), Wiley, Chichester/Sussex, England.

Ryan, M. P., and Blevins, J. Y. K. (1987). The viscosity of synthetic and natural silicate melts and glasses at high temperatures and 1 bar (10^5 Pascals) pressure and at higher pressures, "U.S. Geol. Surv. Bull.," **1764.**

Sammis, C. G., and Julian, B. R. (1987). Fracture instabilities accompanying dike intrusion, *J. Geophys. Res.* **92,** 2597–2605.

Scott, D. R., and Stevenson, D. J. (1984). Magma solitons, *Geophys. Res. Lett.* **11,** 1161–1164.

Sleep, N. H. (1988). Tapping of melt by veins and dikes, *J. Geophys. Res.* **93,** 10255–10272.

Sneddon, I. N. (1946). The distribution of stress in the neighbourhood of a crack in an elastic solid, *Proc. R. Soc. London* **187,** 229–260.

Spence, D. A., Sharp, P. W., and Turcotte, D. L. (1987). Buoyancy-driven crack propagation: A mechanism for magma migration, *J. Fluid Mech.* **174,** 135–153.

Spera, F. (1980). Aspects of magma transport, *in* "Physics of Magmatic Processes" (R. B. Hargraves, ed.), pp. 265–323, Princeton Univ. Press, Princeton, NJ.

Takada, A. (1994). Accumulation of magma in space and time by crack interaction, *in* "Magmatic Systems" (M. P. Ryan, ed.), Academic Press, San Diego.

Takada, A. (1990). Experimental study on propagation of liquid-filled crack in gelatin: Shape and velocity in hydrostatic stress condition, *J. Geophys. Res.* **95,** 8471–8481.

Turcotte, D. L. (1990). On the role of laminar and turbulent flow in buoyancy driven magma fractures, *in* "Magma Transport and Storage" (M. P. Ryan, ed.), pp. 103–111, Wiley, Chichester/Sussex, England.

Weertman, J. (1971). Theory of water-filled crevasses in glaciers applied to vertical magma transport beneath oceanic ridges, *J. Geophys. Res.* **76,** 1171–1183.

Whitehead, J. A., and Luther, D. S. (1975). Dynamics of laboratory diapir and plume models, *J. Geophys. Res.* **80,** 705–717.

Chapter 11 Accumulation of Magma in Space and Time by Crack Interaction

Akira Takada

Overview

Magma accumulates in space and time during its ascent. Crack interactions play a significant role in the process of magma accumulation. Gelatin experiments on crack interactions with a range of initial geometries are summarized. For example, two vertical, collinear, buoyancy-driven cracks; two vertical, buoyancy-driven cracks whose planes are oriented through a variety of angles to each other; and two vertical, parallel, offset, buoyancy-driven cracks can coalesce. Buoyancy-driven cracks coalesce with relative ease, compared with non-buoyant internally pressurized cracks. The stress fields near two collinear, pressurized cracks under hydrostatic conditions are evaluated, using the stress intensity factors at a crack tip. The stress fields near two parallel, offset, pressurized cracks under various regional stress conditions are evaluated, using the expansions of Westergaard stress functions. On the basis of the results of gelatin experiments and stress analyses, a qualitative diagram of crack height vs horizontal distance between two cracks, indicating the conditions for coalescence of vertical, buoyancy-driven cracks, is proposed. When the height of the upper crack is smaller than that of the lower main crack, the main crack can catch up with the upper crack, and there is a minimum coalescence height of the upper crack at a certain horizontal separation distance. When the height of the lower crack is larger than that of the upper main crack, the lower crack can catch up with the main crack, and there is a maximum coalescence height of the lower crack. When the horizontal separation distance is reduced, the range of crack heights for coalescence increases. An increase of the applied differential stress on remote boundaries prevents buoyancy-driven cracks from interacting.

The efficiency of magma-filled crack coalescence in the Earth's mantle and in the crust is discussed. Magma-filled cracks can coalesce and increase in volume during their ascent as long as a range of crack sizes and ascent velocities exists. One simple model for crack coalescence is that cracks of nearly the same size coalesce n times, one after another. The volume of a magma-filled crack after the nth coalescence amounts to (the initial volume of a stationary crack) $\times$ 2^n. If a magma volume is accumulated of the order of 10^8 m^3 (0.1 km^3), 23–10 coalescences are needed.

When the differential stress in the horizontal plane is small at a large supply rate of magma, cracks may coalesce with relative ease. Magma concentrates in space in its ascent process, so that a magma path may be formed. Such a case plays a role in the formation of a polygenetic volcano. When the differential stress in the horizontal plane is large, cracks coalesce with difficulty. Such a case plays a role in the formation of a monogenetic volcano. On the other hand, at a small supply rate of magma, the possibility of crack interactions in space and time decreases. A monogenetic volcano is formed whether the differential stress is large or small.

Notation

		Units
$F_1(\omega)$	complete elliptic integral of the first kind	
$F_2(\omega)$	complete elliptic integral of the second kind	
Im	imaginary part	
K_c	fracture toughness	$Pa \cdot m^{1/2}$
K_I	stress intensity factor (mode I) at a crack tip	$Pa \cdot m^{1/2}$
M	volume of the stationary crack	m^3
$H(h)$	relation between crack width and height	m
$P(x)$	excess pressure in a crack	Pa
$P_j(x)$	excess pressure in crack j ($j = 1, 2$)	Pa
P_{max}	maximum excess pressure	Pa
P_0	excess liquid pressure in a crack	Pa
P_{0j}	excess liquid pressure in crack j ($j = 1, 2$)	Pa

		Units
Re	real part	
V	ascending velocity	$m \cdot s^{-1}$
$Z_I(z)$	complex function	
$Z_{II}(z)$	complex function	
a	half-height of a crack	m
a_j	half-height of crack j ($j = 1, 2$)	m
a_s	half length of the smaller crack	m
d	horizontal distance between two cracks	m
e	distance between two cracks normalized by a_s	dimensionless
g	acceleration of gravity	$m \cdot s^{-2}$
h	crack height	m
l_0	crack height	m
n	half-length of the larger crack normalized by a_s	dimensionless
n_0	number	dimensionless
r_j	distance from crack j ($j = 1, 2$)	m
s_u	u component of the regional stress ($u = x, y, xy$)	Pa, MPa
s_{uj}	u component of the stress from crack j ($u = x, y, xy; j = 1, 2$)	Pa, MPa
t	time	s
w	crack width	m
x	Cartesian coordinate	m
x_{0j}	x coordinate of the center of the crack j ($j = 1, 2$)	m
y	Cartesian coordinate	m
y_{0j}	y coordinate of the center of the crack j ($j = 1, 2$)	m
z	complex number	
$\Delta\rho$	density difference	$kg \cdot m^{-3}$
ξ	Cartesian coordinate	m
η	viscosity	$Pa \cdot s$
μ	shear modulus	Pa, MPa, GPa
ν	Poisson's ratio	dimensionless
π	Pi = 3.1415	dimensionless
ρ	density	$kg \cdot m^{-3}$
σ_{max}	maximum principal stress	Pa, MPa
σ_{min}	minimum principal stress	Pa, MPa
$\sigma_u(x, y)$	u component of stress around a crack ($u = x, y, xy$)	Pa, MPa
$\sigma_{uj}(x, y)$	u component of stress originated by crack j ($u = x, y, xy; j = 1, 2$)	Pa, MPa
$\sigma_{u12}(x, y)$	u component of stress originated by crack 1 and crack 2 ($u = x, y, xy$)	Pa, MPa
σ_j	principal stress ($j = 1, 2, 3$)	Pa, MPa
ϕ	rotation angle	radians
ω	number	dimensionless

Introduction

Beneath a polygenetic volcano, magma concentrates in space and through time so that an approximately steady-state magma path may be formed. To supply discrete volcanic centers, magma must accumulate in space when it is otherwise scattered in the partially melted mantle (Fig. 1). On the other hand, a volcano erupts intermittently. Thus, magma must accumulate in time when it is otherwise believed to be produced at a constant rate in the partially melted mantle. The role of magma accumulation is therefore vitally important for magma ascent and eruption.

There is a well-understood currently active magmatic system in Hawaii where it has been demonstrated that magma ascends great distances by a magma fracture mechanism (e.g., Shaw, 1980; Ryan *et al.*, 1981, 1987a, 1988). Seismic data with well-located hypocenters, for example, define a primary conduit region for Kilauea that is most consistent with a side-by-side ascent mode of buoyant fractures (Ryan, 1988). This chapter deals with magma transport through a crack. Magma ascent by a buoyancy-driven crack in the

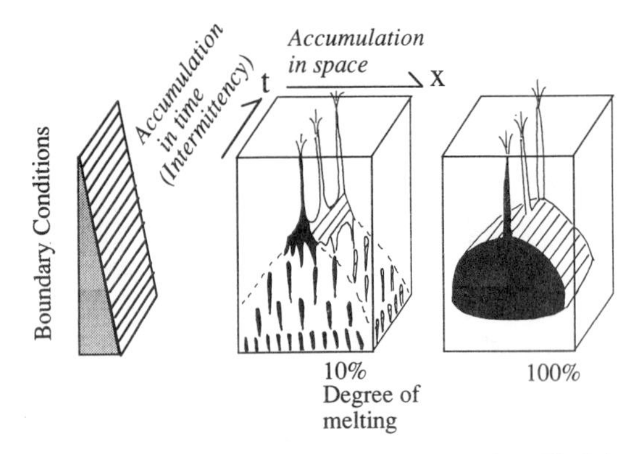

Figure 1 Magma accumulation in space and time. The left diagram shows magma accumulated from the partial melting rock. The right diagram shows a large magma body produced from the total melting rock. t and x represent time and space, respectively. The boundary conditions are composed of the magma production rate as input, the density of the host rocks, the stress condition, and the geotherm.

Earth's lithosphere has been studied by Weertman (1971a, 1971b), Spence and Turcotte (1990), Takada (1989), Lister and Kerr (1991), and others. Takada (1989) proposed a magma transport model by considering an extensive crack propagation system. Unlike soliton wave transport (e.g., Scott and Stevenson, 1984; Scott *et al.*, 1986; Whitehead, 1986; Whitehead and Helfrich, 1990), crack interactions were found to play a significant role in the process of magma accumulation in space and through time.

Fundamental theories of crack interaction have been studied with reference to the mechanics of magma transport and the earthquake process. Two-dimensional open crack interactions under a tensile stress field have been developed by using dislocation theory and fracture mechanics (e.g., Yokobori *et al.*, 1965, 1971; Yokobori and Ichikawa, 1967). Analog experiments of two-dimensional crack interaction problems have been carried out by using a photoelastic material (Lange, 1968) and a soda-lime glass (Swain and Hagen, 1978). Delaney and Pollard (1981), Pollard *et al.* (1982), and Pollard and Aydin (1984) applied the two-dimensional interaction of parallel, offset, internally pressurized cracks to dike intrusions and spreading mid-ocean ridge segments. A pressurized crack is one with no buoyancy of the fluid in the crack relative to the surrounding medium; a buoyancy-driven crack is one with buoyancy of the fluid relative to the surrounding medium. Stoyanov and Dabovski (1986) studied the development of echelon cracks under oblique extension. Olson and Pollard (1989) estimated the crack paths by the interactions of the two pressurized cracks under a remote differential stress. Ryan (1990) evaluated the displacement fields near three parallel, offset, pressurized cracks to study the internal mechanics of the Icelandic magma transport system. Takada (in press) performed gelatin experiments on the three-dimensional crack interactions between buoyancy-driven cracks and applied the results to the development of the subvolcanic structure including a dike and sheet complex. In contrast to previous studies, the crack interactions were considered from the viewpoint of magma accumulation in space and through time.

This chapter summarizes gelatin experiments on fluid-filled crack interactions and then evaluates the mechanics of crack interactions between buoyancy-driven cracks, using stress analyses. In addition, the efficiency of crack coalescence in the mantle and in the crust, and the implications of crack interactions on volcano evolution, are discussed.

Crack Interactions

Approach to Crack Interaction

It is difficult to calculate the detailed patterns of three-dimensional crack interactions. The effect of positive fluid buoyancy on crack interactions, for example, may be inferred from analog experiments on the interactions of liquid-filled cracks in gelatin. The two-dimensional interaction of buoyancy-driven cracks is evaluated using stress analyses. There are two approaches to evaluate the two-dimensional crack interaction: one is an approach utilizing stress intensity factors at a crack tip and the other is an approach that examines stress trajectories near cracks. The first is a fracture mechanics approach and the latter employs classical continuum mechanics.

The former approach is applied to the interaction of collinear cracks. *Whether or not* a crack will propagate (the crack propagation criteria) depends on the crack extension force; the *direction* in which a crack will propagate depends on the local stress field at the crack tip as measured by the ratio of stress intensity factors, for example, mode II/mode I (Lawn and Wilshaw, 1975; Pollard, 1987). Delaney and Pollard (1981), Pollard *et al.* (1982), and Pollard and Aydin (1984) studied the interaction of internally pressurized cracks using linear fracture mechanics. Erdogan and Sih (1963) evaluated the stress fields represented by stress intensity factors at a crack tip and estimated the direction of crack extension under plane loading and traverse shear.

The crack extension paths near two parallel offset cracks under various stress boundary conditions are inferred from the latter approach. It is difficult to estimate the interaction between buoyancy-driven cracks. However, the stress trajectories approach has an advantage in knowing the approximate crack extension pathways. The maximum principal stress (σ_{max}) trajectories control the approximate direction of crack extension. The gradient of the minimum principal stress (σ_{min}) along a σ_{max} stress trajectory governs whether the crack will extend or not.

There are two types of crack interactions: the interaction of liquid-filled cracks and the interaction between a liquid-filled crack and a solid-filled crack (Takada, in press). The latter type of interaction will not induce two cracks to coalesce. This chapter deals with the interaction of liquid-filled cracks.

Takada (1990) reported the existence of both a growing crack and an isolated crack in an elastic medium. Lister and Kerr (1991) concluded that a growing crack driven by positive buoyancy with a constant flow rate has a bulbous nose at the crack tip and a growing feeder tail connected with the source "feeder dike." However, the source must have sufficient magma volume to maintain a relatively constant flow rate. Prior to crack ascent, magma accumulation has already occurred. It is thus important to study the interactions of isolated cracks in the process of transporting the magma accumulation from the partially melted mantle. This chapter deals with the spatial accumulation of otherwise isolated cracks and adds some considerations toward understanding the accumulation of buoyancy-driven, growing cracks with a source operating at a constant flow rate.

A quasi-static isolated crack is approximately equivalent to one end member of cracks such that $K_I = K_c$ at the upper dike tip and $K \sim 0$ at the lower tip. For example, the additional length scale of a quasi-static, isolated crack in gelatin is a few to several centimeters (Takada, 1990). As crack coalescences repeat, an isolated crack increases in volume and in length so that it departs from the quasi-static case. It is difficult to evaluate the stress field near a long crack that is not quasi-static. In the stress analyses and in the discussions of this chapter, it is assumed that isolated cracks with various lengths are quasi-static.

Summary of Gelatin Experiments

Analog experiments employing gelatin suggest that buoyancy-driven cracks are relatively easy to coalesce with each other; that is, buoyancy is effective for crack coalescence (Takada, in press). In gelatin experiments, one can observe the three-dimensional interactions of buoyancy-driven, liquid-filled cracks. The fracture resistance of a crack tip is relatively large in gelatin (Takada, 1990). Lister and Kerr (1991) pointed out that the fracture resistance effect by a magma-filled crack in rocks is far less important in controlling the fracture dynamics than that of buoyancy, once a magma-filled crack begins to extend. However, when stress fields are discussed, the effects of the fracture resistance at a crack tip is very small, in comparison with that of the velocity of crack propagation. Thus, gelatin experiments are valid for constructing an approximate view of the complex pattern of three-dimensional crack interaction.

Several experiments on crack growth under nonhydrostatic conditions, using gelatin or soft organic polymers, have been performed: the interaction between parallel offset cracks in a photoelastic material (Lange, 1968), crack growth in gelatin under the gravitational stress field (Fiske and Jackson, 1972), crack growth in acethylcellulose jelly (Dabovski and Stojanov, 1981), and the development of echelon cracks (Stoyanov and Dabovski, 1986). On the other hand, a few, quantitative gelatin or agar experiments on the propagation of a buoyancy-driven crack under hydrostatic conditions were carried out: on crack shape (Maaløe, 1989), on crack shape and crack propagation (Takada, 1990), and on the lateral growth of a crack at the neutral buoyancy level (Lister and Kerr, 1991). Takada (in press) studied the interaction of two buoyancy-driven, liquid-filled cracks.

The methods of experiments and physical properties of gelatin were reported in detail in Takada (1990; in press). E-290 gelatin powder prepared by Miyagi Chemical Industrial Company Ltd., Sendai, Japan, was used. A 1.25% gelatin solution was solidified in a rectangular container 3 cm in length, 30 cm in width, and 50 cm in height. The density of gelatin is 1007 ± 1 kg $\cdot$ m^{-3}, the Young's modulus is about $0.6–1.3 \times 10^3$ Pa, and the Poisson's ratio is nearly 0.5. The stress state in gelatin is nearly hydrostatic before the injection of liquids. As a working fluid, 1 cSt-silicon oil (density 822 kg $\cdot$ m^{-3}, viscosity 0.8×10^{-3} Pa $\cdot$ s) was injected into the tank gelatin from the bottom and the wall of the container by a syringe or a micropump, to generate two vertical cracks (Fig. 2). Both the inclination and the trend of a crack in gelatin can be controlled by the obliquely cut tip of a needle. If the lower crack is larger than the upper crack, the lower crack ascends at a faster speed to catch up with and coalesce with the upper crack that is ascending at a lower velocity. The ascent velocity is proportional to (the crack height)^4 in gelatin (Takada, 1990). The results of the gela-

GELATIN EXPERIMENTAL METHODS

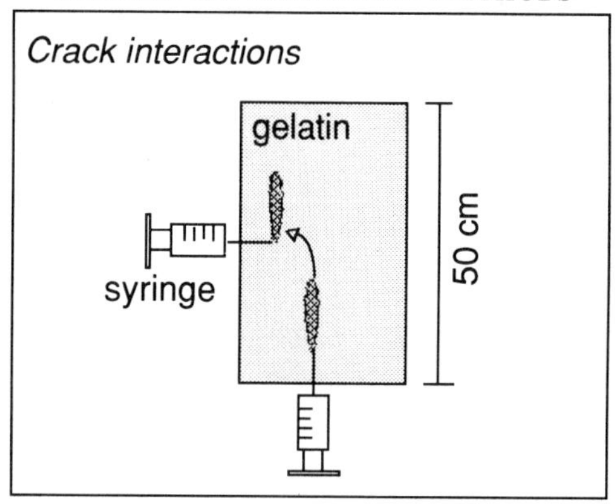

SUMMARY OF THE INTERACTION OF BUOYANCY- DRIVEN CRACKS

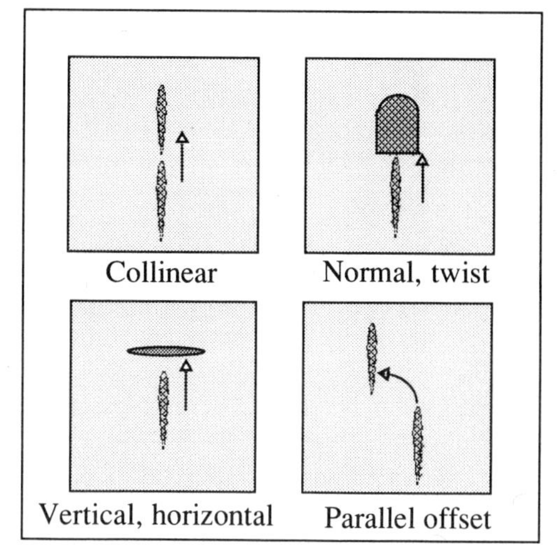

Figure 2 Gelatin tank experimental methods for the study of crack interactions (upper), and a summary of the types of interactions of buoyancy-driven cracks (lower).

tin experiments are summarized in Fig. 2. This chapter adds two new cases shown in Fig. 3 to the results of Takada (in press). The additional length scales of cracks in Figs. 2 and 3 are in the range of the quasi-static case.

Two vertical, collinear, buoyancy-driven cracks can coalesce as illustrated in Fig. 3A. Two vertical cracks whose planes are oriented through a variety of angles to each other can also coalesce. A lower vertical buoyancy-driven crack can coalesce with an upper horizontal crack, inducing the lower crack to close as its fluid is discharged into the upper crack and inducing the upper crack to start to grow horizontally; however, some liquid is retained in the lower crack. On the other hand, two cracks subjected to a set of external boundary loads do not appear to coalesce without a prior increase in the liquid volume within the cracks. Two vertical, parallel, offset, buoyancy-driven cracks can coalesce if the horizontal distance between two cracks is not large: for example, about a separation distance of one-half the vertical height of the upper crack. On the other hand, two externally loaded cracks have never been observed to coalesce.

In Fig. 3B, the lower crack was initiated vertically, and the upper crack was inclined and dipping to the right. After injection, the lower crack propagated upward with positive buoyancy and was unperturbed by the stress field from the the upper buoyancy-driven crack (Fig. 3B, 1). Under the influence of the stress field of the upper crack, the lower crack began to interact with the lower tail of the upper crack, turned to the left, and finally coalesced (Fig. 3B, 2–4).

In Fig. 3C, liquid was supplied from the bottom of the gelatin block. Cracks of the same size ascended one after the other at a fairly regular interval through the preexisting fracture. Once a substantially larger crack was formed, the two cracks coalesced. The lower larger crack could thus catch up with the upper smaller crack and coalesce.

Interaction of Two Collinear Cracks

Delaney and Pollard (1981) evaluated stress intensity factors at the tips of two collinear, internally pressurized cracks of the same size. They reported that collinear cracks are attracted to each other. Yokobori *et al.* (1965) studied the stress intensity factors at the tips of collinear open cracks of different sizes under a uniaxial stress field. To study two collinear pressurized cracks under hydrostatic conditions, this study expands the stress intensity factors (mode I) of Yokobori *et al.* (1965):

$$K_{\mathrm{I}} = P_0 \sqrt{\frac{\pi(2n+e)}{4ea_s}} \cdot \left(2a_s - (2a_s + ea_s) \cdot \left(1 - \frac{F_1(\omega)}{F_2(\omega)}\right)\right) \tag{1a}$$

$$\omega = \sqrt{\frac{4n}{(2n+e)(e+a_s)}}. \tag{1b}$$

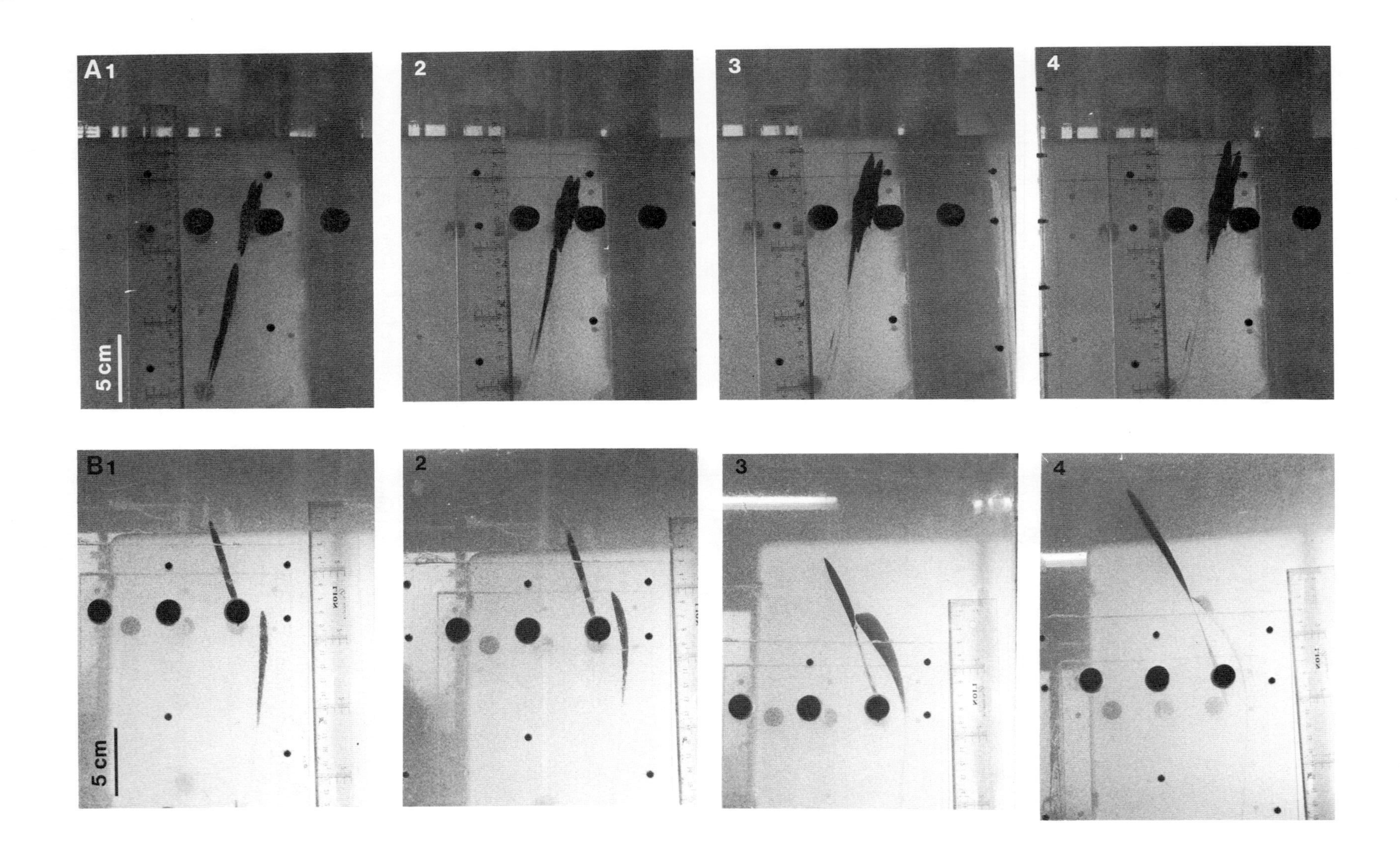
A1
2
3
4
5 cm
B1
2
3
4
5 cm

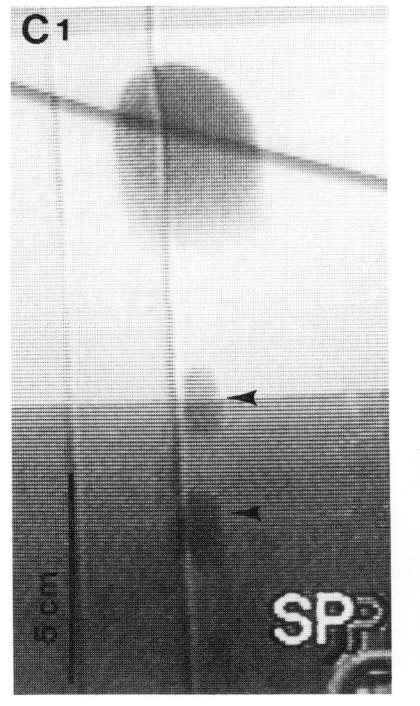

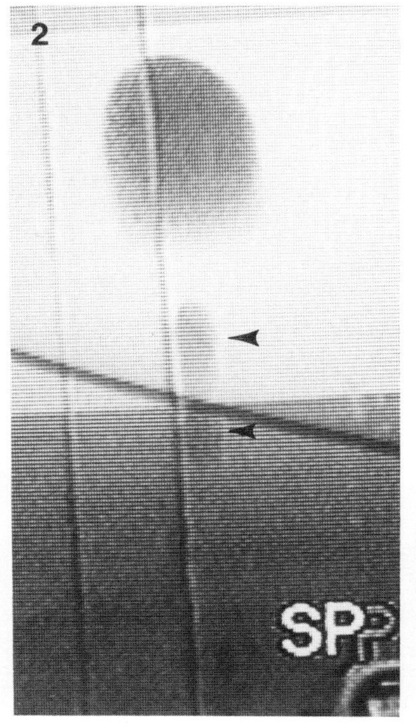

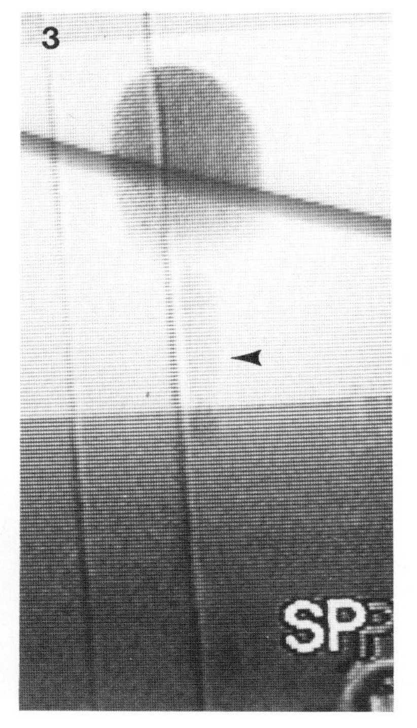

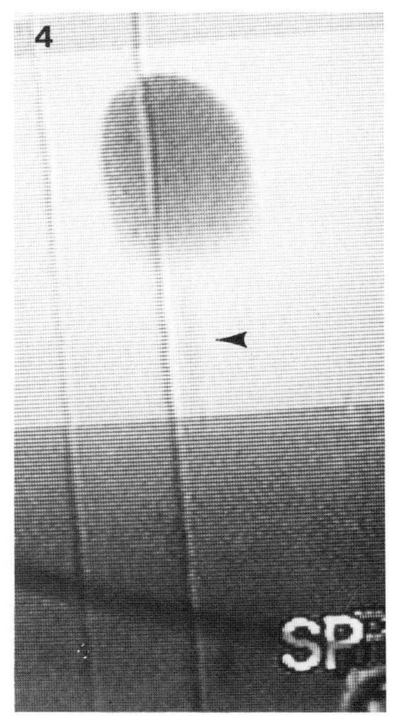

Figure 3 Photographs showing gelatin experiments on the interactions of collinear buoyancy-driven cracks (A), the interaction of two buoyancy-driven cracks that approach in an oblique position (B), the coalescence involving fluid exchanges and changes in the crack size (C), respectively. The larger crack at the lower position catches up with the smaller one at the upper position, and they coalesce (A, 1 $\rightarrow$ 4; B, 1 $\rightarrow$ 4). As the lower crack approaches the upper crack, it begins to turn toward the upper crack. The lower crack develops a C-shaped, curved structure in the horizontal section at its tip, which appears as differential thickness in the photograph (B, 3). The lower small cracks, marked with open triangles, are ascending rapidly one after the other through the preexisting fracture just below the upper slowly ascending large crack (C, 1), and begin to coalesce by fluid exchanges and changes in crack size (C, 2–4).

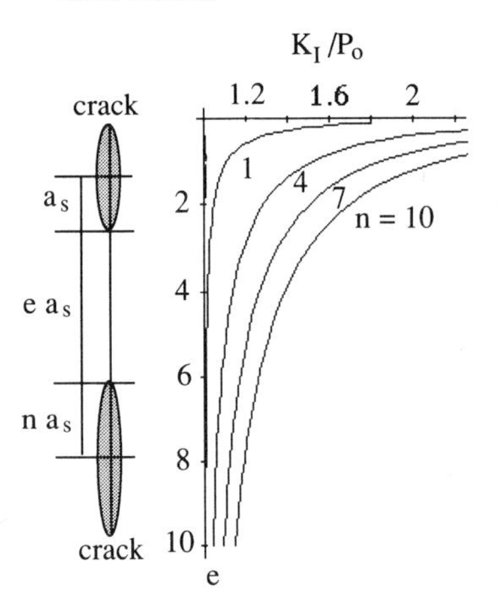

Figure 4 The stress intensity factor K_I (mode I) at the tip of the smaller crack due to the interaction of two collinear pressured cracks (after Yokobori *et al.*, 1965). P_0 is the excess pressure. The distance between two cracks, and the half-length of the larger crack, are normalized by the half-length of the smaller crack, a_s, respectively.

$F_1(\omega)$, $F_2(\omega)$ are complete elliptic integrals of the first and second kind, respectively. P_0 is the excess liquid pressure against the crack wall, a_s, e, and n are the half-height of the smaller crack, the distance between two cracks normalized by a_s, and the half-height of the larger crack normalized by a_s, respectively. Figure 4 shows that the smaller crack is attracted by the larger crack and that two cracks with shorter distances are attracted more strongly. The attraction force is the stress gradient induced by the larger crack. The gelatin experiments suggest that collinear, buoyancy-driven, liquid-filled cracks are easy to coalesce compared with two remotely loaded cracks (Figs. 3A, and 3C).

Interaction of Two Parallel Offset Cracks

Pollard (1973) estimated the stress fields around a two-dimensional, internally pressurized crack under hydrostatic conditions. Roberts (1970) and Dabovski (1979) studied the stress fields around an elliptical magma body under regional stress conditions. Delaney and Pollard (1981), Pollard *et al.* (1982), and Ryan (1990) calculated the displacement fields near two offset cracks. Pollard and Aydin (1984) estimated the stress field near two internally pressurized cracks under extensional stress conditions. Olson and Pollard (1989) evaluated the crack extension paths by the interactions of two parallel offset pressurized cracks under a remote differential stress, using the boundary element method. For a small remote differential stress, the closely spaced, two parallel offset cracks exhibit an asymptotic approach to each other. A large remote differential stress produces a nearly planar path. Takada (in press) evaluated the stress fields near an internally pressurized crack under various regional stress fields, and the stress fields near a buoyancy-driven crack under hydrostatic conditions.

This chapter evaluates the stress fields near two internally pressurized cracks under various regional stress conditions to determine the approximate patterns of the stress trajectories under various remote stress conditions and to discuss the interactions of buoyancy-driven cracks. Westergaard stress functions (Westergaard, 1939) were adopted to expand the two-dimensional stress analysis. Complex stress functions, $Z_I(z)$ and $Z_{II}(z)$, represent the stress field near a crack with the symmetrical excess pressure distribution $P(x)$ along the long axis (Tada *et al.*, 1973),

$$z = z + yi \tag{2}$$

$$Z_I(z) = \frac{1}{\pi\sqrt{z^2 - a^2}} \cdot \int_a^{-a} \frac{P(\xi)\sqrt{a^2 - \xi^2}}{z - \xi}\, d\xi \tag{3}$$

$$Z_{II}(z) = 0 \tag{4}$$

$$P(x) = P_0 \text{ (internally pressurized crack)} \tag{5}$$

$$P(x) = P_0 + \Delta\rho g(x + a) \text{ (fluid-buoyancy-driven crack),} \tag{6}$$

where z is a complex number, a is the half-height of the crack, ξ is a variable in the range $-a$ to a, and x and y are the coordinates on the long axis and the short axis of a crack. The x component of stress, $\sigma_x(x, y)$, the y component, $\sigma_y(x, y)$, and the shear (xy) component, $\sigma_{xy}(x, y)$, are

$$\sigma_x(x, y) = \mathrm{Re}[Z_I(z)] - y\, \mathrm{Im}[dZ_I(z)/dz] \tag{7}$$

$$\sigma_y(x, y) = \mathrm{Re}[Z_I(z)] + y\, \mathrm{Im}[dZ_I(z)/dz] \tag{8}$$

$$\sigma_{xy}(x, y) = -y\, \mathrm{Re}[dZ_I(z)/dz], \tag{9}$$

where Re$[Z(z)]$, Im$[Z(z)]$ are the real and imaginary parts of the complex function $Z(z)$, respec-

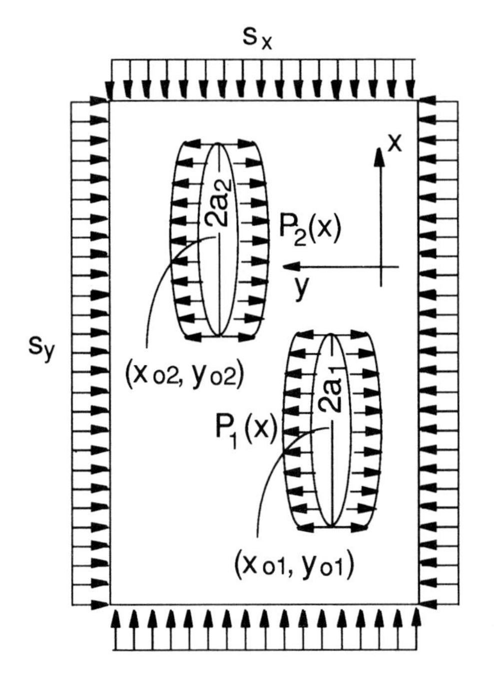

Figure 5 The two-dimensional stress field near two liquid-filled cracks subject to the regional stress field. $P_1(x)$, $P_2(x)$, s_x, and s_y are the excess pressures of crack 1 and crack 2, and the regional stresses, respectively.

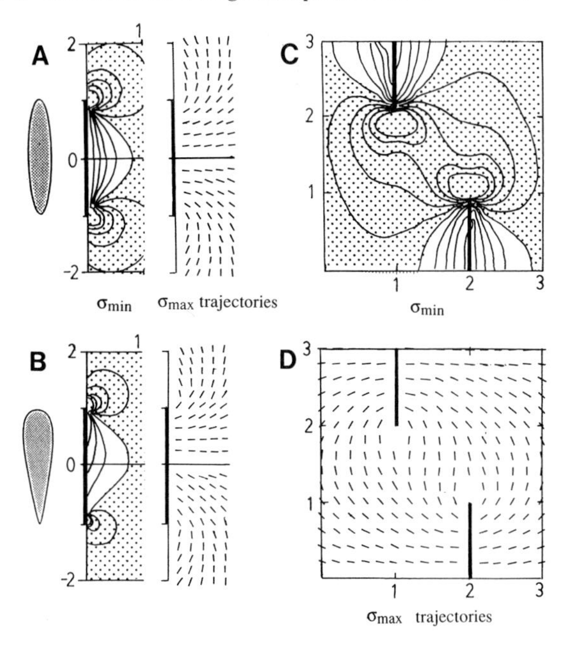

Figure 6 Contour map of the minimum principal stress (σ_{min}) and the maximum principal stress (σ_{max}) trajectories near a remotely loaded internally pressurized crack (A), those near a buoyancy-driven crack (B), the contour map of σ_{min} (C), and the σ_{max} trajectories near two internally pressurized cracks (D). The excess internal pressure of an equally pressurized crack, and the maximum excess pressure of a buoyancy-driven crack, are assumed to be 1. The compressional stress is positive. The stress in the stippled area represents extensional components. The compressive component fields are unstippled. The contour interval of the stress is 0.2. A short bar represents the calculated σ_{max} trajectory at each mesh point.

tively. $\sigma_{xj}(x - x_{0j}, y - y_{0j})$, $\sigma_{yj}(x - x_{0j}, y - y_{0j})$, $\sigma_{xyj}(x - x_{0j}, y - y_{0j})$ are the stresses near crack j centered at the coordinate (x_{0j}, y_{0j}) $(j = 1, 2)$ (Fig. 5). The stresses near two offset cracks (crack 1 and crack 2) under the remote regional stress fields, σ_{x12}, σ_{y12}, and σ_{xy12}, are, based on the method of superposition,

$$\sigma_{x12} = \sigma_{x1}(x - x_{01}, y - y_{01}) + \sigma_{x2}(x - x_{02}, y - y_{02}) + s_x \tag{10}$$

$$\sigma_{y12} = \sigma_{y1}(x - x_{01}, y - y_{01}) + \sigma_{y2}(x - x_{02}, y - y_{02}) + s_y \tag{11}$$

$$\sigma_{xy12} = \sigma_{xy1}(x - x_{01}, y - y_{01}) + \sigma_{xy2}(x - x_{02}, y - y_{02}) + s_{xy}, \tag{12}$$

where s_x, s_y, s_{xy} are x, y, and xy components of the regional stress, respectively.

Solving Eqs. (2)–(12), the σ_{max} trajectories and the values of the σ_{min} near an internally pressurized crack, near a buoyancy-driven crack, and near two offset, internally pressurized cracks are obtained as shown in Figs. 6 and 7. The results are approximate, but they are useful for knowing the crack interactions of two buoyancy-driven cracks. The compressional stress is positive. The principal stress axes are given, rotating the x–y coordinate through the angle ϕ:

$$\phi = 0.5 \text{ Arctan}[2\sigma_{xy12}/(\sigma_{x12} - \sigma_{y12})]. \tag{13}$$

ϕ depends on the liquid pressure and the differential regional stress defined as "the maximum principal stress—the minimum principal stress." The pattern of stress trajectories depends only on the differential regional stress, if the differential stress is normalized by the constant excess liquid pressure (Takada, in press).

Parallel, offset, internally pressurized cracks may interact under hydrostatic conditions. The detailed paths of two parallel offset pressurized cracks under an isotropic remote stress were simulated by Olson and Pollard (1989). Such paths are

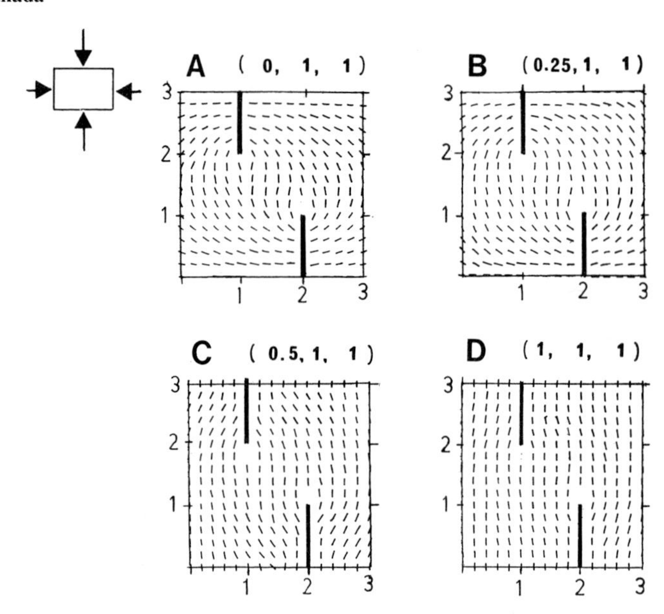

Figure 7 The σ_{max} trajectories near two internally pressurized cracks. Numerals in parentheses are the differential stress normalized by liquid pressure, the crack height contrast, and the distance between two cracks normalized by the half-height of the left crack. (A) Hydrostatic conditions; (B), (C), and (D) differential stresses of 0.25, 0.5, and 1, respectively. The pattern is the same as in Figure 6.

observed at various scales in the field (Fig. 13 in Pollard *et al.,* 1982; Fig. 8). Figures 8A and 8B may be equivalent to the cross sections of parallel, offset, internally pressurized cracks. For example, the offset dikes in Fig. 8A outcrop on a river bed as the horizontal cross section, so that the effect of buoyancy on crack pattern cannot be found. Moreover, if the density difference between the host rocks and magma is small, the effect cannot be found. For small cracks such as the veins in Fig. 8B, this effect also cannot be detected.

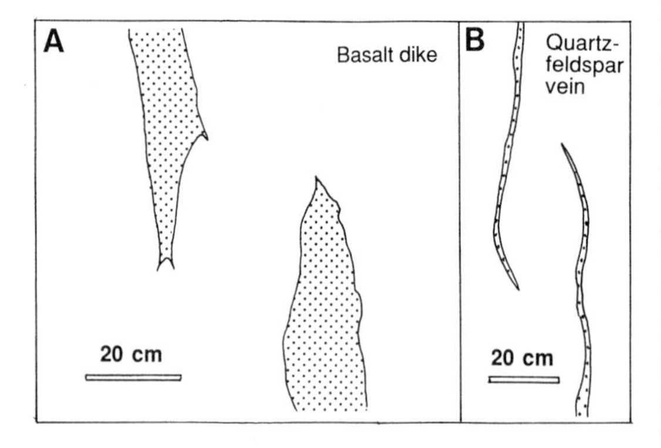

Figure 8 Plan views of interacting dike tips. Two basalt dikes in the welded tuff of the Shitara igneous complex, in the Shitara district, central Japan (A), and two quartz-feldspar veins in the granitic rocks of the Cottonwood stock, Utah (after Pollard *et al.*, 1982) (B).

The σ_{max} trajectories near two cracks curve with a sigmoidal pattern as shown in Fig. 6D. The contour map of the σ_{min} value near two internally pressurized cracks in Fig. 6C indicates that the contour interval on a σ_{max} trajectory decreases toward the cracks, that is, the stress gradient on the σ_{max} trajectory increases. The stress near the crack becomes more compressional toward the crack. Thus, it is difficult for a crack tip to extend from the right crack to the left crack in this illustration. This tendency is supported by the crack pattern in Fig. 8. However, the contour interval on a σ_{max} trajectory near the tail of a buoyancy-driven crack (Fig. 6B) is larger than that near an internally pressurized crack (Fig. 6A). The compressional stress near the tail of a buoyancy-driven crack does not increase. If it is assumed that another buoyancy-driven crack propagates from the right side toward the crack tail of Fig. 6B, the two cracks can coalesce. The buoyancy of the buoyancy-driven crack prevails over the stress gradient, in that σ_{min} becomes more compressional toward the crack tail of Fig. 6B. If we exchange internally pressurized cracks for buoyancy-driven cracks in Figs. 6C and 6D, it is possible that the buoyancy-driven crack on the right extends approximately on the σ_{max} trajectory to coalesce with the crack on the left. This supports the results of gelatin experiments on the interactions of two parallel, offset, buoyancy-driven cracks. Parallel, offset, buoyancy-driven cracks are relatively easy to coalesce under hydrostatic conditions.

As the differential stress increases from 0 to 1, the contour patterns of the σ_{max} trajectories become parallel (Figs. 7A and 7B). This suggests that it is more difficult for cracks to coalesce at larger differential stresses. The detailed paths of two parallel offset pressurized cracks under several remote stresses were reported by Olson and Pollard (1989). This tendency is valid for the case near two buoyancy-driven cracks. The critical value of the differential stress for coalescence is around 0.5 (Fig. 7C).

On the other hand, the possibility of coalescence decreases with the increasing size contrast of two cracks. The far-field stress perturbations from a crack are in an inverse relationship to (the crack separation distance normalized by the crack size)^2 (Pollard and Segall, 1987). The far-field stress from crack j, s_{uj}, is

$$s_{uj} \propto P_{0j}(a_j/r_j)^2, \tag{14}$$

where P_{0j}, a_j, and r_j are the maximum excess pressure of crack j, the height of crack j, and the distance from crack j, respectively ($j = 1, 2$). It is assumed that the maximum excess pressure in a crack is in proportion to the crack height. For $a_2 = (1/n_0)a_1$, $P_{02} = (1/n_0)P_{01}$, the stress from crack 2 at the far stress field is given by Eq. (14):

$$\begin{aligned} s_{u2} \propto P_{02}(a_2/r_2)^2 &= (1/n_0)P_{01}(a_1/n_0 r_2)^2 \\ &= (1/n_0)^3 P_{01}(a_1/r_2)^2. \end{aligned} \tag{15}$$

As the size of crack 2 becomes smaller than crack 1, that is, n_0 becomes larger, the stress derived from crack 2 decreases abruptly (Eq. (15)). Moreover, Eq. (15) indicates that the interaction between two cracks decreases as the distance between the two cracks increases, as intuitively expected. If the horizontal separation distance between two cracks is n_0 times as large as the case in Fig. 7A, the stress from crack 2 is $(1/n_0)^2$ times as large as that case.

Conditions for Crack Coalescence

The gelatin experiments combined with the stress analyses indicate that the conditions for offset crack coalescence depend on the crack heights, the separation distance between the two cracks, and the differential stress. The maximum excess pressure caused by buoyancy in a vertical crack, P_{max}, is proportional to the crack height, h, if $P_0 = 0$ in Eq. (6):

$$P_{max} = \Delta\rho g h. \tag{16}$$

Thus, a parameter, the crack height, includes the excess pressure. The condition shown in Fig. 9 is inferred from Figs. 2–4, 6, and 7.

The ascent velocity of an isolated crack, V, in the case of laminar flow, is presented in several previous works (e.g., Weertman, 1971b; Takeuchi *et al.*, 1972; Takada, 1990),

$$V = \frac{\Delta\rho g w^2}{3\eta}, \tag{17}$$

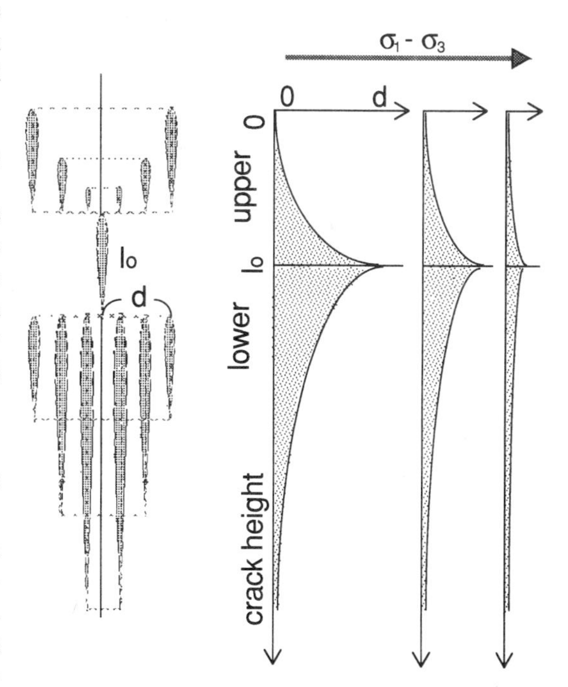

Figure 9 Crack height vs the horizontal separation distance between two crack diagrams showing conditions promoting crack coalescence. The height of the main crack is l_0. In the stippled regions of the diagrams, two cracks may coalesce. The upper region ($<l_0$) and the lower region ($>l_0$) represent coalescence with the upper and lower cracks, respectively.

where η and w are the viscosity of injected liquid and the crack width, respectively. The crack width is given by the function $H(h)$. Substituting into Eq. (17), the ascent velocity is

$$V \propto H(h)^2. \tag{18}$$

For example, the ascent velocity of an isolated crack in gelatin was obtained by Takada (1990):

$$H(h) = h^2, \qquad V \propto h^4. \tag{19}$$

It is assumed that the ascent velocity becomes larger with increases in the height of an isolated crack. Thus, when the height of the upper crack is smaller than that of the lower main crack, the main crack can catch up with the upper crack. When the height of the lower crack is larger than that of the main crack, the lower crack can catch up with the main crack. When the horizontal separation distance is smaller, the range of crack heights for coalescence becomes wider, which is inferred from Figs. 3 and 6, and Eqs. (14) and (15). The concave curve of Fig. 9 represents the minimum coalescence height of the upper crack at a certain horizontal separation distance, d, when interactions occur between the main crack and the upper crack. The convex curve represents the maximum coalescence height when interactions occur between the main crack and the lower crack. In the stippled regions between two curves, two cracks can coalesce. l_0 is the height of the main crack. The upper ($<l_0$) and lower regions ($>l_0$) represent the coalescence with the upper and lower cracks, respectively. In the case of collinear cracks, the main crack can coalesce with the upper crack with a height in the range 0–l_0, or with the lower crack with a height greater than l_0 (Figs. 3 and 4). If the differential stress increases, the region where the two cracks can coalesce is reduced (Fig. 7).

The elastic stress perturbations from a buoyancy-driven, growing crack of Lister and Kerr (1991) originate mainly from a bulbous nose at the crack tip and a feeder tail. The height of the growing crack from the source may be equivalent to the height of an isolated crack. The crack height from the source should be used for evaluating the crack interactions. The feeder tail with constant width generates a local stress perturbation, in contrast with the closed tail of an isolated crack (Fig. 6). It prevents coalescence from the lower isolated crack to the upper growing crack. The condition for crack coalescence focuses primarily on the conditions of the coalescence from the lower growing crack into the upper isolated crack.

Applications to Magma Accumulation in Space and Time

Magma Accumulation

If a magma reservoir is not initiated in the partially melted mantle, magma may accumulate through magma-filled cracks. It is possible that a magma-filled crack is generated in the partially melted mantle or around this region (e.g., Nicolas, 1986, 1990; Sleep, 1988). According to Lister and Kerr (1991), the maximum thickness and the maximum height of an isolated stationary crack are

$$h_{\max} \sim \left(\frac{K_c}{\Delta\rho g}\right)^{2/3} \tag{20}$$

$$w_{\max} \sim \left(\frac{K_c^4}{\Delta\rho g m}\right)^{1/3} \tag{21}$$

$$m = \mu/(1 - \nu),$$

where K_c, μ, and ν are the fracture toughness, the shear modulus, and the Poisson's ratio, respectively. The volume of the crack (assuming equal height and lengths) is given by Eqs. (20)–(21):

$$M \sim (h_{\max})^2 w_{\max}. \tag{22}$$

For $K_c = 4$ MPa $\cdot$ m$^{1/2}$ (e.g., Atkinson, 1984), $\Delta\rho = 400$ kg $\cdot$ m^{-3}, $\mu = 20$ GPa, and $\nu = 0.3$, $h_{\max} = 100$ m, $w_{\max} = 1.4$ mm, and $M = 14$ m^3 are obtained. The apparent resistance tends to increase due to the cooling of magma, or due to plastic deformation at a crack tip. The estimated fracture toughness at very low confining pressures from the shape of a dike in the field (e.g., Delaney and Pollard, 1981) may be useful. For example, $h_{\max} = 8 \times 10^2$ m, $w_{\max} = 0.1$ m, and $M = 8 \times 10^4$ m^3 are obtained for $K_c = 100$ MPa $\cdot$ m$^{1/2}$.

If a magma volume of 10^8 m^3 (0.1 km^3) is accumulated, 7×10^6 isolated stationary cracks for $K_c = 4$ MPa $\cdot$ m$^{1/2}$ or 1.2×10^3 stationary isolated cracks for $K_c = 100$ MPa $\cdot$ m$^{1/2}$ must coalesce. One simple model for crack coalescence is adopted here as an example. Cracks of nearly the same size coalesce n times one after another (Fig. 10). Two cracks of a volume M coalesce with each other so that the volume of a new crack becomes $M \times 2$ (first coalescence). Two cracks of a volume $M \times 2$

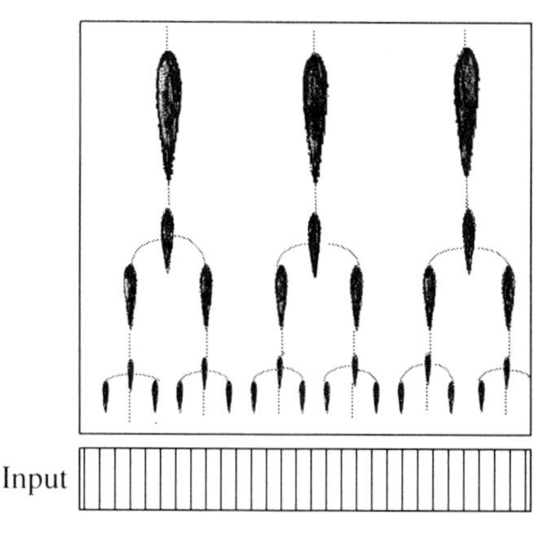

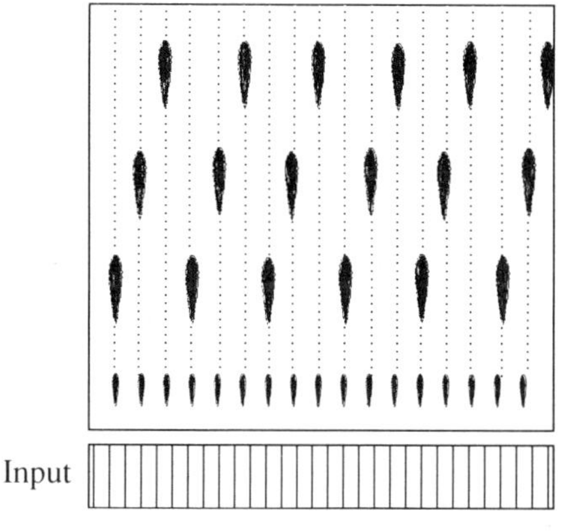

Figure 10 Schematic diagram showing magma accumulation in space. The upper diagram (A) represents the case of a small differential stress; the lower diagram (B) represents the case of a large differential stress. In (A and B) the production rate of magma is large enough for magma-filled cracks to interact. Magma-filled cracks are generated in the input block. In the case of basalt magma, the input block means the asthenosphere. At a small magma production rate, the possibility of crack interactions in space and time decreases.

coalesce so that the volume of a new crack becomes $M \times 2^2$ (second coalescence). The volume of a magma-filled crack after the nth coalescence amounts to

$$M \times 2^n. \tag{23}$$

The stationary crack increases in volume so that the crack becomes unstable to propagate. Crack coalescence occurs during crack propagation under the conditions envisioned in Fig. 9. If a magma volume is accumulated of the order of 10^8 m^3 (0.1 km^3), 23 coalescences (2^{23} times) for K_c = 4 MPa · $m^{1/2}$ or 10 coalescences (2^{10} times) for K_c = 100 MPa · $m^{1/2}$ are needed (Eqs. (22) and (23)). Moreover, fluctuations of the crack size can produce significant magma accumulations during their ascent (Fig. 3C). It is more efficient that cracks with various orientations (Fig. 2) or those with a network of cracks coalesce.

Magma-filled cracks can coalesce and increase in volume during their ascent (Takada, 1989) as long as a range of crack sizes and ascent velocities exists. Both the size and ascent speed are controlled by the physical properties of the crust, such as the density difference between the magma and the host rocks, and the stress conditions. A larger crack at a lower velocity is formed as the density difference decreases. As basalt magma approaches the depth where the density difference is zero (the level of neutral buoyancy, e.g., Ryan, 1987a, 1987b; Rubin and Pollard, 1987; Takada, 1989; Lister and Kerr, 1991, and Ryan, 1994 (this volume)), the ascent velocity approaches zero. The stress gradient with depth is proportional to the apparent density. The stress gradient adjusts the crack propagation driven by buoyancy (Takada, 1989), especially the level of neutral buoyancy (Lister and Kerr, 1991). A constant supply of magma-filled cracks promotes the coalescence of magma-filled cracks, that is, the magma accumulation in time at this level.

Lister and Kerr (1991) proposed that, once a magma-filled crack starts to rise, it never stops until it reaches the neutral buoyancy level. It is an ideal liquid-filled crack. The apparent resistance is generated due to the solidification of magma at a crack tip or due to the ductile deformation at a crack tip. This effect promotes the efficiency of the magma accumulation during magma ascent in the crust. It is possible that magma-filled cracks coalesce to increase in volume during magma ascent (Takada, 1989).

Variation of a Volcano

The arrangement of the principal stress axes and the value of the regional stress govern the magma accumulation in space. If σ_3 is in the horizontal plane, a vertical crack is initiated (Anderson, 1951). In this case, there are two types in the arrangement of the principal stress axes: One is that σ_1 is in the vertical plane; the other is that σ_1 is in the horizontal plane (Fig. 11). The former includes the case in which a normal fault occurs; the

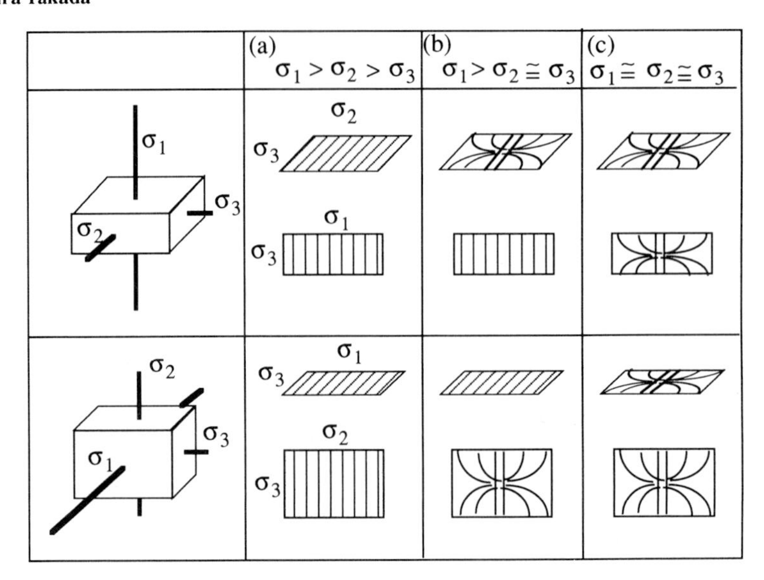

Figure 11 Schematic three-dimensional stress trajectories under various arrangements of the magnitudes of the principal stress axes. The upper row shows σ_1 is in the vertical plane and σ_2 and σ_3 lie in the horizontal plane. The lower row shows σ_2 is in the vertical plane and σ_1 and σ_3 are in the horizontal plane. (a–c) $\sigma_1 - \sigma_3$ is large (large differential stress), $\sigma_2 - \sigma_3$ is small, and $\sigma_1 - \sigma_3$ is small (small differential stress), respectively.

latter includes the case in which a strike-slip fault occurs. Nakamura (1986) proposed that a monogenetic volcano occurs in the region with crustal extension. The former is equivalent to the condition proposed by Nakamura (1986), that is, σ_3 lies in the horizontal plane. On the other hand, Fedotov (1981) proposed that the activity of monogenetic volcanoes such as cinder cones occurs at a small magma supply rate. First, the case of a magma supply rate sufficient for cracks to interact is discussed.

Column (a) in Fig. 11 illustrates the case $\sigma_1 - \sigma_3$, that is, the differential stress is large and both $\sigma_1 - \sigma_2$ and $\sigma_2 - \sigma_3$ are also large. According to crack interaction theory, parallel stress trajectories develop in the horizontal plane and in the vertical plane. This condition prevents cracks from coalescing. This condition also promotes the formation of a monogenetic volcano (Fig. 10).

Column (b) in Fig. 11 illustrates the case in which $\sigma_1 - \sigma_3$ is large and $\sigma_2 - \sigma_3$ is small. When σ_1 is in the vertical plane, vertical stress trajectories parallel to σ_1 develop. This condition lends itself to the formation of a monogenetic volcano (Fig. 10A). When σ_1 is in the horizontal plane, the radial stress trajectories develop in the vertical plane, and the parallel σ_1 trajectories develop in the horizontal plane. This case has an advantage in the formation of a polygenetic volcano with parallel fissure eruptions (Fig. 10B).

Column (c) in Fig. 11 illustrates the case in which $\sigma_1 - \sigma_3$ is small. The differential stresses in the horizontal plane and in the vertical plane are smaller than $\sigma_1 - \sigma_3$. The radial σ_1 trajectories are formed in both planes. The coalescence of magma-filled cracks is accelerated in the crust. This condition advances the formation of a polygenetic volcano with radial fissure eruptions (Fig. 10).

Two parallel cracks are difficult to coalesce under regional stress with the normalized differential stress more than around 0.5. If the production or supply rate of magma-filled cracks increases, several parallel cracks will interact with one another. Such multicrack interactions tend to cancel the effects on the regional extensional stress, that is, the differential stress. The cracks can coalesce with one another. The increase in magma production or the supply rate of magma-filled crack promotes the concentration of magma in space, that is, the formation of a magma path (Takada, in press). This condition tends to promote the formation of a polygenetic volcano. If the crustal extension is large enough at a large magma production rate, such as an oceanic ridge, a parallel dike swarm develops and may grow into a magma reservoir in the lower crust (Takada, 1989). On the other hand, for the formation of a monogenetic volcano with primary magma, special conditions such as the increase in buoyancy by volatile components, or the imposition of a stress gradient such that the tensile stress increases upward, may be needed (Takada, 1989). At a small production or supply rate of magma, the possibility of crack interactions in space and time decreases. A monogenetic volcano will be formed whether the differential stress is large or small. The theory of crack interactions can explain the case proposed by Fedotov (1981).

The basic crack interaction concept has been discussed. However, these are ideal cases. A magma-filled crack with a small volume cannot reach the surface because of heat losses and solidification. For a monogenetic volcano, magma-filled cracks must coalesce with one another to increase in volume. In general, the differential stress in the mantle and the lower crust is not large compared with that of the upper crust. For a monogenetic volcano, magma-filled cracks may coalesce in the mantle and the lower crust.

Conclusions

Crack interactions play a significant role in the process of magma accumulation in space and time. Gelatin experiments on crack interactions with various geometric initial conditions suggest that positive fluid buoyancy is effective for crack coalescence. First, the interaction of collinear, remotely loaded cracks was evaluated by the stress intensity factors at a crack tip. Second, the stress fields near two parallel, offset, remotely loaded cracks subjected to various regional stress fields were calculated, using the the expansion of Westergaard stress functions. The coalescence effects of the height contrasts of cracks and of that on the separation distance between two cracks are also considered. Whether two cracks can coalesce or not depends on the height contrast of two neighboring cracks, the horizontal separation distance between two cracks, and the differential stress field. The increase in the differential stress level prevents cracks from interacting. A qualitative diagram of the crack height vs the horizontal distance between two cracks, indicating the conditions of crack coalescence, was proposed on the basis of the results of gelatin experiments and stress analyses.

The interactions of isolated cracks may apply to magma accumulation in the Earth's mantle and in the crust. The generation of apparent fracture toughness, crack coalescences with various arrangements, and fluctuations on crack size causes an increase in the efficiency of the magma accumulation process.

Magma accumulation due to the three-dimensional crack interactions on various arrangements of the principal stress axes are considered, if the production rate of magma is large enough for magma-filled cracks to interact. When the differential stress in the vertical plane is small, cracks tend to coalesce relatively easily with one another. Magma concentrates by crack interactions in space in its ascent process so that a magma path may be formed. This case has an advantage in the formation of a polygenetic volcano. When the differential stress in the vertical plane is large, cracks are difficult to coalesce. This case advances the formation of a monogenetic volcano. On the other hand, at a small production rate of magma-filled cracks, a monogenetic volcano will be formed whether the differential stress is large or small.

Acknowledgments

The author is grateful to H. Koide of the Geological Survey of Japan for valuable comments and critical readings of this manuscript. The author has benefited from discussions with K. Kurita and A. Shinya of Tsukuba University. The author appreciates critical reviews by L. Wilson of University of Lancaster, A. M. Rubin of Princeton University, and M. P. Ryan of the U.S. Geological Survey.

References

Anderson, E. M. (1951). "The Dynamics of Faulting and Dyke Formation," 2nd ed., Oliver and Boyd, London, England.

Atkinson, B. K. (1984). Subcritical crack growth in geologic materials, *J. Geophys. Res.* **89,** 4077–4114.

Dabovski, C. (1979). Stress and faulting around sheet-like magmatic chambers, *Geotect. Tectonophys. Geodyn.* **8–9,** 17–38 (in Russian with English abstract).

Dabovski, C., and Stojanov, S. (1981). Fracturing around magmatic chambers as observed in laboratory models, *Geotect. Tectonophys. Geodyn.* **13,** 3–18 (in Russian with English abstract).

Delaney, P. T., and Pollard, D. D. (1981). Deformation of host rocks and flow of magma during growth of minette dikes and breccia-bearing intrusion near Ship Rock, New Mexico, "U.S. Geol. Surv. Prof. Paper," 1202.

Erdogan, F., and Sih, G. C. (1963). On the crack extension in plates under plane loading and transverse shear, *J. Basic Eng. Trans. ASME Ser. D* **35,** 519–527.

Fedotov, S. A. (1981). Magma rate in feeding conduits of different volcanic centers. *J. Volcanol. Geotherm. Res.* **9,** 379–394.

Fiske, R., and Jackson, E. D. (1972). Orientation and growth of Hawaiian volcanic rifts: The effect of regional structure and gravitational stresses, *Proc. R. Soc. London. Ser. A* **329,** 299–326.

Lange, F. F. (1968). Interaction between overlapping parallel cracks: A photoelastic study, *Int. J. Fracture Mech.* **4,** 287–294.

Lawn, B. R., and Wilshaw, T. R. (1975). "Fracture of Brittle Solids," Cambridge Univ. Press, New York.

Lister, J. R., and Kerr, R. C. (1991). Fluid-mechanical model of crack propagation and their application to magma transport in dykes, *J. Geophys. Res.* **96,** 10049–10077.

Maaløe, S. (1989). The generation and shape of feeder dykes from mantle sources, *Contrib. Mineral. Petrol.* **96,** 47–55.

Nakamura, N. (1986). Volcanology and the plate tectonics, *Bull. Volcanol. Soc. Japan* Special number "What is a volcano," S1–S16 (Japanese with English abstract).

Nicolas, A. (1986). A melt extraction model based on structural studies in mantle peridotites, *J. Petrol.* **23,** 568–582.

Nicolas, A. (1990). Melt extraction from mantle peridotites: hydrofracturing and porous flow, with consequences for oceanic ridge activity, *in* "Magma Transport and Storage" (M. P. Ryan, ed.), pp. 159–173, Wiley, Chichester/Sussex, England.

Olson, J., and Pollard, D. D. (1989). Inferring paleostress from natural fracture pattern: A new method, *Geology* **17,** 345–348.

Pollard, D. D. (1973). Derivation and evaluation of a mechanical model for sheet intrusion, *Tectonophysics* **19,** 233–269.

Pollard, D. D. (1987). Elementary fracture mechanics applied to the structural interpretation of dykes, *in* "Mafic Dyke Swarm" (H. C. Hall and W. F. Fahrig, eds.), Geol. Assn. Canada Spec. Paper, 34, pp. 5–24.

Pollard, D. D., and Aydin, A. (1984). Propagation and linkage of oceanic ridge segments, *J. Geophys. Res.* **89,** 10017–10028.

Pollard, D. D., and Segall, P. (1987). The theoretical displacements and stresses near fractures in rock: With applications to faults, joints, veins, dikes, and solution surfaces, *in* (B. K. Atkinson, ed.), Fracture Mechanics of Rock, pp. 277–349. Academic Press, London.

Pollard, D. D., Segall, P., and Delaney, P. T. (1982). Formation and interpretation of dilatant echelon cracks, *Geol. Soc. Am. Bull.* **93,** 1291–1303.

Roberts, J. L. (1970). The intrusion of magma into brittle rocks, *in* "Mechanism of Igneous Intrusion" (G. Newall, and N. Rast, eds.), Geol. J. Spec. Issue, **2,** 287–338.

Rubin, A. M., and Pollard, D. D. (1987). Origins of blade-like dikes in volcanic rift zones, *in* (R. W. Decker, T. L. Wright, and P. Stauffer, eds.), "Volcanism in Hawaii," U.S. Geol. Surv. Prof. Paper, 1350, pp. 1449–1470.

Ryan, M. P. (1987a). The elasticity and contractancy of Hawaiian olivine tholeiite, and its role in the stability and structural evolution of sub-caldera magma reservoirs and rift systems, *in* "Volcanism in Hawaii" (R. W. Decker, T. L. Wright, and P. Stauffer, eds.), U.S. Geol. Surv. Prof. Paper, 1350, pp. 1395–1447.

Ryan, M. P. (1987b). Neutral buoyancy and the mechanical evolution of magmatic systems, *in* "Magmatic Process: Physicochemical Principles" (B. O. Mysen, ed.), Geochemical Soc. Spec. Publ. No. 1, pp. 259–287.

Ryan, M. P. (1988). Mechanics and three-dimensional internal structure of active magmatic systems: Kilauea volcano, Hawaii, *J. Geophys. Res.* **93,** 4213–4238.

Ryan, M. P. (1990). The physical nature of the Icelandic magma transport system, *in* "Magma Transport and Storage" (M. P. Ryan, ed.), pp. 176–224, Wiley, Chichester/Sussex, England.

Ryan, M. P. (1994). Neutral buoyancy controlled magma transport and storage: A summary of basic relationships, *in* "Magmatic Systems" (M. P. Ryan, ed.), Academic Press, San Diego.

Ryan, M. P., Koyanagi, R. Y., and Fiske, R. S. (1981) Modeling the three-dimensional structure of macroscopic magma transport systems: Application to Kilauea volcano, Hawaii, *J. Geophys. Res.* **86,** 7111–7129.

Scott, D. R., and Stevenson, D. J. (1984). Magma solitons, *Geophys. Res. Lett.* **11,** 1161–1164.

Scott, D. R., Stevenson, D. J., and Whitehead, J. A., Jr. (1986). Observation of solitary waves in a viscously deformable pipe, *Nature* **319,** 759–761.

Shaw, H. R. (1980). The fracture mechanism of magma transport from the mantle to the surface, *in* "Physics of Magmatic Processes" (R. B. Hargraves, ed.), pp. 201–264, Princeton Univ. Press, Princeton, NJ.

Sleep, N. H. (1988). Tapping of melt by veins and dikes, *J. Geophys. Res.* **93,** 10255–10272.

Spence, D. A., and Turcotte, D. L. (1990). Buoyancy-driven magma fracture: A mechanism for ascent through the lithosphere and the emplacement of diamonds, *J. Geophys. Res.* **95,** 5133–5139.

Stoyanov, S., and Dabovski, C. (1986). Morphology of fracturing in zones of oblique extension: Experimental results and geological implications, *Geotect. Tectonophys. Geodyn.* **19,** 3–22 (in Russian with English abstract).

Swain, M. V., and Hagan, J. T. (1978). Some observations of overlapping interacting cracks, *Eng. Fracture Mech.* **10,** 299–304.

Tada, H., Paris, P. C., and Irwin, G. R. (1973). "The Stress Analysis of Cracks Handbook," Del Research Corp., Hellertown, PA.

Takada, A. (1989). Magma transport and reservoir formation by a system of propagating cracks, *Bull. Volcanol.* **52,** 118–126.

Takada, A. (1990). Experimental study on propagation system of liquid-filled crack in gelatin: Shape and velocity in hydrostatic stress condition, *J. Geophys. Res.* **95,** 8471–8481.

Takada, A. (1994). Development of a subvolcanic structure by the interaction of liquid-filled cracks, *J. Volcanol. Geotherm. Res.* (in press).

Takeuchi, H., Fujii, N., and Kikuchi, M. (1972). How magma goes up? *J. Seism. Soc. Japan* **25,** 266–268 (in Japanese).

Weertman, J. (1971a). Theory of water-filled crevasses in glaciers applied to vertical magma transport beneath oceanic ridge, *J Geophys Res.* **76,** 1171–1183.

Weertman, J. (1971b). Velocity at which liquid-filled cracks move in the Earth's crust or in glaciers, *J. Geophys. Res.* **76,** 8544–8553.

Westergaard, H. M. (1939). Bearing pressures and cracks, *Trans. ASME J. Appl. Mech.* **6,** 49–53.

Whitehead, J. A. (1986). Buoyancy-driven instabilities of low-viscosity zone as models of magma-rich zone, *J. Geophys. Res.* **91,** 9303–9314.

Whitehead, J. A., and Helfrich, K. R. (1990). Magma waves and diapiric dynamics, *in* "Magma Transport and Storage" (M. P. Ryan, ed.), pp. 53–76, Wiley, Chichester/Sussex, England.

Yokobori, T., and Ichikawa, M. (1967). The interaction of parallel elastic cracks and parallel slip bands respectively based on the concept of continuous distribution of dislocations. II. *Rep. Res. Inst. Strength Fracture Materials Tohoku Univ.* **3,** 15–37.

Yokobori, T., Ohashi, M., and Ichikawa, M. (1965). The interaction of two collinear asymmetrical elastic cracks, *Rep. Res. Inst. Strength Fracture Materials Tohoku Univ.* **1,** 33–39.

Yokobori, T., Uozumi, M., and Ichikawa, M. (1971). Interaction between non-coplanar parallel staggered elastic cracks, *Rep. Res. Inst. Strength Fracture Materials Tohoku Univ.* **7,** 25–47.

Chapter 12 Generalized Upper Mantle Thermal Structure of the Western United States and Its Relationship to Seismic Attenuation, Heat Flow, Partial Melt, and Magma Ascent and Emplacement

Hiroki Sato and Michael P. Ryan

Overview

The crust and upper mantle of the western United States are characterized by low seismic velocities and high seismic attenuation values, as well as high heat flow and high electrical conductivity values. Generalized temperature distributions inferred for the region provide partial constraints for understanding these anomalous and apparently fundamental features. Laboratory seismic measurements of rocks from the upper mantle at high pressures and high temperatures are compared with seismic observations in an effort to estimate the overall thermal structure of this portion of the upper mantle. We report these estimated temperature distributions beneath the western United States and further estimate the laterally averaged degree of melting as a function of depth. Beneath the eastern Rockies, the resulting upper mantle temperature is below the dry solidus, and no melting is generally expected. From the Rocky Mountains to the far west, anomalously low seismic velocities and low Q—in regions such as the Rio Grande rift, the Basin and Range province and the Cascade volcanoes—yield temperatures higher than the dry solidus, and partial melting (to ≈10 vol.%) is inferred. The relatively high gradient temperature–depth profiles of the western United States are higher than normal mantle geotherms and are comparable to the oceanic and hot continental geotherms. We summarize available heat flow data in the western United States and calculate conductive temperatures within the lithosphere. The temperatures inferred from surface heat flow values are consistent with those from laboratory seismic data for a dry peridotite.

Seismic velocity structures have permitted estimations of anomalously high temperatures in magmatic regions. Beneath the Rio Grande rift, the Basin and Range province, and the Salton Trough, velocity perturbations below −6% indicate the presence of partial melt and temperatures above the dry solidus. Relatively high heat flow regions coincide with these low velocity regions, and have yielded conductive temperatures consistent with the temperatures estimated from the velocity data. The thermal structures thus inferred from seismic velocities are generally consistent with those from seismic anelasticity studies. However, fine velocity structures may determine small-scale low velocity anomalies (and therefore small-scale, high-temperature anomalies) that are not always resolved from the more generalized seismic anelasticity analysis. Such regional anomalies may correspond to discrete magma generation sites and to local concentrations of upper mantle magma, ascending diapirs, and local areas of asthenospheric upwelling. Peaks in partial melt content occur at ≈165 km (≈3%) and at ≈145 km (≈10%) beneath the Intermountain (IM) and Western Margin (WM) regions, respectively. Elastic dislocation treatments of veins and deep dikes yield magma-filled fracture heights that range from 500 m to 11 km, suggesting that the magma ascent pathway is highly disconnected. The ascent of positively buoyant melt batches across the Mohorovičić discontinuity is treated as a composite material interface characterized by paired Young's modulus ratios ($^{E_2}/_{E_1}$). The modulus defect induced by melt in low aspect ratios in the semi-consolidated gabbro complexes near the Moho produces pronounced crack-tip enlargements for the magma-charged fractures that penetrate this interface. These bulbous enlargements may thus combine with the interface rheological contrasts to locally inhibit further crack advance, modify the transport pathway, and serve as nuclei for deep sills.

Notation

		Units
B	Medium 1 crack flank height	m
E	Isotropic Young's modulus	MPa, GPa
E^*	Effective aggregate Young's modulus of microporous solid	MPa, GPa
E_1	Young's modulus surrounding the crack flanks (medium 1)	MPa, GPa
E_2	Young's modulus surrounding the crack-tip (medium 2)	MPa, GPa

Magmatic Systems
Edited by M. P. Ryan

		Units
E_2/E_1	Bimodulus contrast	dimensionless
ΔG	Difference in the *in situ* stress gradient and the magma pressure gradient	$MPa \cdot km^{-1}$
g	Gravitational acceleration	$m \cdot s^{-2}$
h	Height above subjacent magma along a completely interconnected fluid pathway	m
h^*	Fracture semiheight	m
H_c	Buoyancy-driven crack critical height	m
K^*	Effective aggregate bulk modulus of microporous solid	MPa, GPa
P	Uniform fracture magma pressure	MPa
Q	Seismic quality factor	dimensionless
Q_p	Seismic quality factor of compressional waves	dimensionless
Q_{pm}	Seismic quality factor of compressional waves at solidus temperature	dimensionless
Q_s	Seismic quality factor of shear waves	dimensionless
T	Temperature	°C, K
T_m	Solidus temperature of dry peridotite	°C, K
T/T_m	Homologous temperature	dimensionless
T_p	Potential temperature	°C
V	Volume of melt per unit crack height	m^3/m
V_a	Seismic velocity of the asthenosphere	$km \cdot s^{-1}$
V_L	Seismic velocity of the lithosphere	$km \cdot s^{-1}$
V_m	Seismic velocity at solidus temperature	$km \cdot s^{-1}$
V_p	Elastic compressional wave velocity	$km \cdot s^{-1}$
V_s	Elastic shear wave velocity	$km \cdot s^{-1}$
W	Magma fracture width	m
Z	Vertical Cartesian coordinate	m
α	$= B/h^*$	dimensionless
β	$= \sqrt{(1-\eta)^2/(1-\alpha^2)}$	
η	$= Z/h^*$	dimensionless
μ^*	Effective aggregate shear modulus of a microporous solid	MPa, GPa
μ	Elastic shear modulus	MPa, GPa
ν	Poisson's ratio	dimensionless
ϕ	Porosity	dimensionless
ρ_{is}	*In situ* country rock density	$g \cdot cm^{-3}$
ρ_m	Melt density	$g \cdot cm^{-3}$
$\Delta\sigma_H$	Difference in the horizontal components of boundary loads along the crack semiheight	MPa
θ	Dihedral angle	degrees
γ_{ss}	Grain boundary energy between matrix grains	Nm^{-1}
γ_{sF}	Grain boundary energy between matrix and melt	Nm^{-1}

Introduction

Many geophysical surveys have indicated anomalous features in the crust and upper mantle beneath the western United States: low seismic velocities (e.g., Holbrook, 1990; Hearn *et al.*, 1991), high seismic wave attenuation (e.g., Patton and Taylor, 1984; Al-Khatib and Mitchell, 1991), relatively high electrical conductivities (e.g., Wannamaker *et al.*, 1989; Klein, 1991), high heat flow values (e.g., Sass *et al.*, 1981; Blackwell *et al.*, 1990), and a relatively thin crustal and lithospheric thickness (e.g., Soller *et al.*, 1982; Bechtel *et al.*, 1990). Seismic studies have further indicated the deep presence of regional magmatism as inferred from observed low velocities, high wave attenuation, and the occurrence of wave reflections from magma reservoir margins (e.g., Rinehart and Sanford, 1981; Kissling, 1988; Iyer *et al.*, 1990; Holbrook, 1990). From a geodynamic and regional tectonic perspective, in the northern portion of the western United States, the Juan de Fuca and the Gorda plates are being subducted beneath the Cascade volcanic system, whereas in the southwestern portion, the subduction of the oceanic crust from the East Pacific Rise tectonically loads the San Andreas fault. In addition, there is a region of continental extension and shallow igneous activity: the Basin and Range province. These fundamental structural relationships are well-known characteristics of the region and provide geologic contrasts with other cooler and more stable areas of North America (e.g., Robertson, 1972; Der *et al.*, 1982; Gough, 1984; Nathenson and Guffanti, 1988).

Temperature distributions determined in the upper mantle beneath the western United States may provide important constraints for increasing our understanding of the anomalous features outlined previously. Peridotites are considered the dominant lithology of the upper mantle. The laboratory seismic (acoustic) data for a dry peridotite at high pressures and high temperatures have been compared with *in situ* seismic observations, and the generalized temperatures of the upper mantle have thus been estimated. In previous papers (Sato and Sacks, 1989, 1990; Sato *et al.*, 1989a), we have investigated the thermal structure of, for example, the oceanic upper mantle from seismic anelasticity and velocity data. Temperatures so derived were consistent with the temperatures inferred from surface heat flow data; therefore, laboratory seismic results have been demonstrated to be applicable to the Earth.

In this chapter we estimate the generalized temperature profile and the degree of partial melt in the upper mantle beneath the western United States from published seismic velocity and anelasticity data. These inferred temperatures are then compared with conductive geotherms derived from heat flow. Our procedure thus yields temperature profiles from the surface to ~300 km depth beneath the western United States. In addition, there are many small-scale anomalies (low seismic velocity, low Q, and high heat flow) in regions such as the Cascade volcanoes (e.g., Hearn *et al.*, 1991; Harris *et al.*, 1991), the Rio Grande rift (e.g., Carpenter and Sanford, 1985; Halderman and Davis, 1991), the Long Valley caldera (e.g., Ryall and Ryall, 1981; Dawson *et al.*, 1990), the Coso hot springs area (Reasenberg *et al.*, 1980; Young and Ward, 1980), and the Salton Trough (Humphreys and Clayton, 1990). Abnormal thermal structures are to be expected beneath these regions. We have also estimated temperature distributions in these anomalous regions. The temperatures so derived are then discussed with previous tectonic and magmatic models derived from seismic, petrologic, and geothermal studies.

Laboratory Seismic Data for Mantle Peridotite

Seismic Velocity of Mantle Peridotite

In the laboratory, the seismic properties of a dry mantle peridotite have been determined at high pressures and temperatures in order to provide a comparative basis with seismic observations from regional surveys and to enable the estimation of temperatures at depth (e.g., Murase and Fukuyama, 1980; Kampfmann and Berckhemer, 1985; Sato *et al.*, 1989b). Detailed descriptions of the experimental techniques and laboratory measurements have been reported by Murase and Kushiro (1979) and Sato *et al.* (1989b). Here we briefly describe the laboratory results.

Seismic velocities in a dry peridotite (a spinel lherzolite) have been determined by Murase and Kushiro (1979) and Murase and Fukuyama (1980) as a function of temperature (1000–1300°C) at confining pressures to 1 GPa (10 kbar). Since the temperature range of their measurements extended above the solidus of dry peridotite, the effect of partial melting on seismic velocities was examined. The partial melt fraction of the sample was also determined over the same pressure and temperature ranges by Murase and Fukuyama (1980), using the β-track method described by Mysen and Kushiro (1977) and thin-section-based petrographic observations.

The experimental results have shown that the seismic velocities in partially molten peridotite decrease sharply with increasing temperature; i.e., the velocity at hypersolidus temperatures may be used to constrain reliable mantle temperatures. Importantly, the pressure dependence of both the velocity and the partial melt fraction is essentially accounted for by that of the solidus temperature of peridotite, i.e., *the homologous temperature dependence* (Sato *et al.*, 1989b). The *homologous* temperature is a dimensionless number that is formed from the ratio T/T_m, where T_m is the solidus temperature at a given composition and pressure. It expresses the dimensionless thermal relationship of a rock to its melting point. The laboratory velocities, V, normalized by the velocity, V_m, at the solidus temperature, T_m, for both compressional and shear waves determined at 0.5 and 1 GPa plot as a single trend as a function of the homologous temperature T/T_m (T is absolute temperature in kelvins), as shown in Fig. 1A. The melt fractions determined at 0.5 and 1 GPa also depend on a single homologous temperature (Fig. 1B). These observations, therefore, allow one to extrapolate the experimental results to higher pressures (greater depths), by simply knowing the solidus as a function of pressure. The solidus of dry peridotite has been determined up to 14 GPa (140 kbar and ≃400 km depth) by Takahashi (1986). The solidi, T_m, used here were

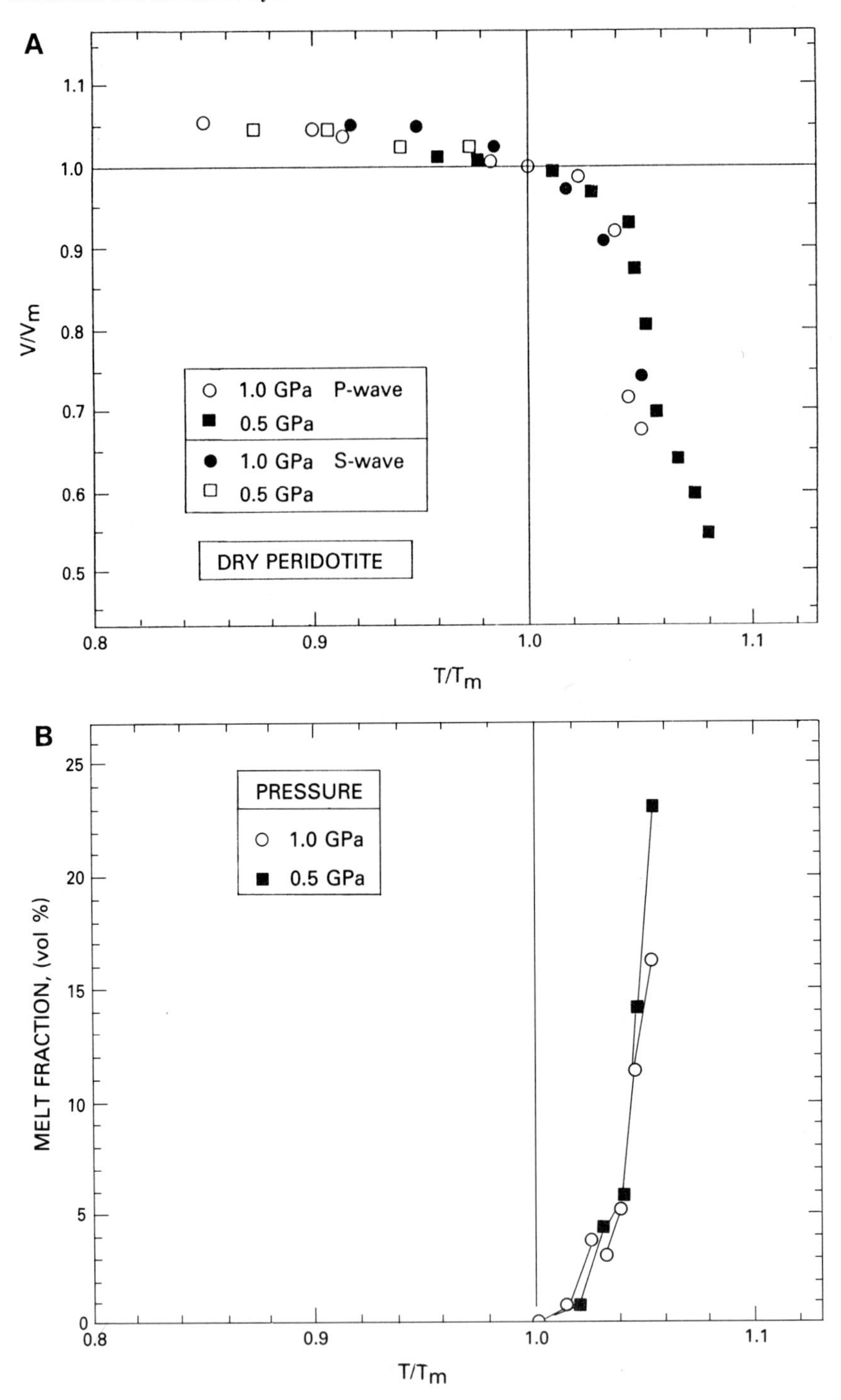

Figure 1 Diagrams for determining temperature and degree of partial melt from seismic velocity data. (A) Normalized velocity as a function of the homologous temperature (T/T_m) in dry peridotite; solid (open) squares and open (solid) circles at 0.5 and 1.0 GPa, respectively, for compressional (shear) waves (original velocity data from Murase and Kushiro (1979) and Murase and Fukuyama (1980)). Note that T and T_m are given in kelvins in this figure. Within the experimental errors in the velocity and the solidus determinations, both compressional and shear velocities have a single trend in this diagram. (B) Fraction of partial melt as a function of the homologous temperature (T/T_m). The degree of partial melt determined at 0.5 (squares) and 1.0 GPa (circles) again depends on the single homologous temperature (after Sato *et al.*, 1988).

1180 and 1220°C (1450 and 1490 K) at 0.5 and 1 GPa, respectively.

The normalized velocity in Fig. 1A has been compared with the seismic velocity structure, i.e., the ratio of asthenospheric velocity, V_a, to the lithospheric velocity, V_L. Fig. 1A shows that the temperature dependence of velocity becomes smaller at lower temperatures and that the velocity ratio approaches 1.05. The velocity increase at even lower temperatures seems to be fairly small. We may therefore define lithospheric velocity, V_L, by (Sato *et al.*, 1989a)

$$V_L/V_m = 1.05. \quad (1)$$

This equation is equivalent to

$$V_a/V_m = 1.05\ V_a/V_L. \quad (2)$$

We calculate V_a/V_m from Eq. (2) and obtain an estimate of T/T_m by using Fig. 1A. T_m as a function of depth is obtained from the solidus determined by Takahashi (1986) (cf. Fig. 8A). Therefore the temperature T is calculated as a function of depth. For $T > T_m$, we employ Fig. 1B to estimate the volume fraction of partial melt.

Figure 1A indicates that the velocity drops by about 6% at subsolidus temperatures ($0.85 < T/T_m < 1$). This implies that some low velocities in the asthenosphere could be explained by subsolidus temperature increases *without* invoking partial melting (Sato *et al.*, 1989a). The more gradual drop of the velocity at $T < T_m$ (as contrasted with $T > T_m$) does not allow one to estimate accurate temperatures within the subsolidus region. Only in the region where the melt fraction is not less than 2 vol%, may we constrain a thermal structure from Fig. 1.

We illustrate with some examples of thermal structure determined from laboratory and seismic velocities (Fig. 1). For the Iceland Plateau, Evans and Sacks (1979) determined the seismic velocity structure ($V_L = 4.5\ \text{km} \cdot \text{s}^{-1}$) from the inversion of both Rayleigh and Love wave data. In the region of 0–5 Ma asthenosphere, an anomalously low velocity ($V_a = 3.9\ \text{km} \cdot \text{s}^{-1}$) is found at a depth of 25–65 km. Using Eq. (2), $V_a/V_L = 0.87$ ($=3.9\ \text{km} \cdot \text{s}^{-1}/4.5\ \text{km} \cdot \text{s}^{-1}$) and gives $V_a/V_m = 0.91$, which then yields $T/T_m = 1.04$ (Fig. 1A) and the melt fraction 7 vol% (Fig. 1B). An average depth for this low velocity zone is 45 km where T_m is 1520 K from high pressure experiments by Takahashi (1986). Therefore the temperature at this depth is estimated to be 1580 K (1310°C). We also estimate temperatures at various depths using the known T/T_m ($=1.04$) and the T_m from Takahashi (1986). The thermal structure thus obtained is summarized in Table 1. The actual temperature and melt fraction could both be higher at 0 Ma than at 5 Ma in the asthenosphere. Because of the limited resolution in seismic surface wave studies, the temperature and partial melt fraction in Table 1 show bulk average values. For asthenosphere older than 5 Ma, the relative velocity drop in the low velocity zone is less than 6% and no melting is expected in the region. These results indicate that extensive partial melting exists in the asthenosphere younger than 5 Ma, but that melting is not necessarily required in asthenosphere older than 5 Ma under the slowly spreading ($\approx 20\ \text{mm} \cdot \text{yr}^{-1}$) Iceland Plateau. More detailed descriptions of the temperature estimates from seismic velocities have been reported by Sato *et al.* (1989a).

Table 1

Example Thermal Structure of a Low Velocity Zone[a]

Age (Ma)	Depth (km)	Temperature (°C)	Fraction of partial melt (vol%)
0–5	30	1260	
	45	1310	7[b]
	60	1370	
	70	1410	
	85	1470	5[b]
	100	1530	

[a] Constructed for the Iceland Plateau, after Sato *et al.* (1989a)

[b] Because of the limited resolution in seismic surface wave studies, the melt fraction is determined only for the average depth of the low velocity zone.

Seismic Anelasticity of Mantle Peridotite

The thermal structure of the upper mantle has also been estimated from seismic anelasticity (Q^{-1}) data of mantle peridotite by Sato and Sacks (1989). We briefly describe the laboratory results and the method of constructing a temperature–depth profile.

Laboratory measurements show, importantly, that the seismic attenuation of compressional waves (Q_p^{-1}) again depends on the ratio of rock temperature to the solidus temperature, that is, at-

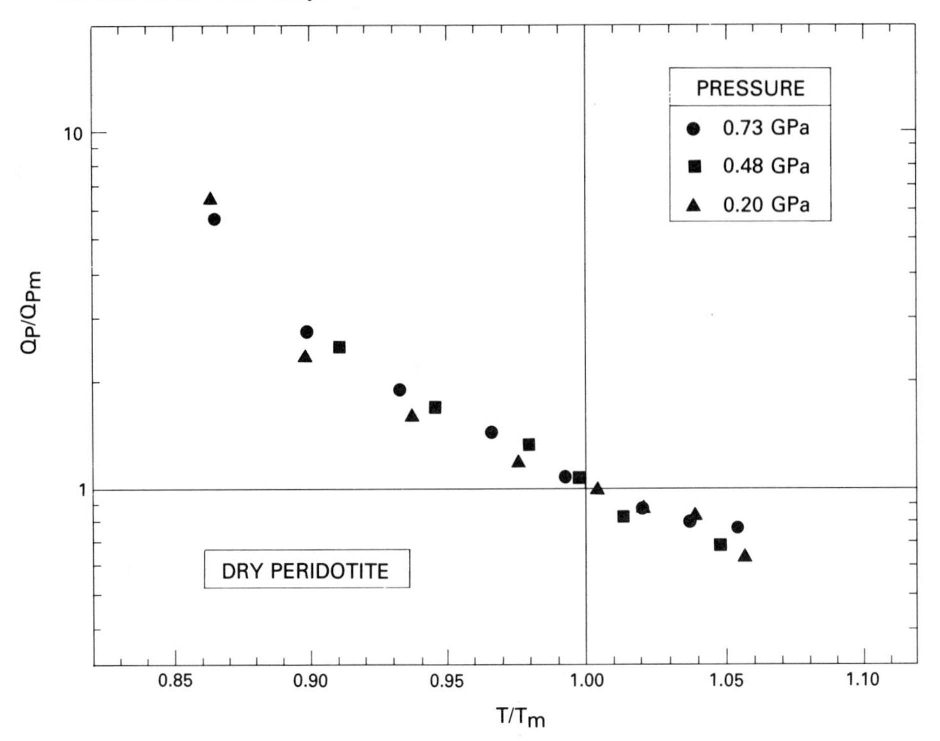

Figure 2 Normalized Q_p as a function of the homologous temperature (T/T_m) in dry peridotite. T and T_m are given in kelvins in this figure. The data taken at three different pressures have a single trend within the experimental errors in the Q_p and the solidus determinations. This diagram is used to estimate the temperature from seismic anelasticity data. Data from Sato *et al.* (1989b).

tenuation shows a homologous temperature dependence (Fig. 2). We may thus extrapolate the laboratory results to higher pressures by simply using the known solidus as a function of pressure. The parameter Q_{pm} (the Q_p value at the solidus temperature) is expressed as (Sato and Sacks, 1989)

$$Q_{pm} = Q_\phi + P/P_\phi, \tag{3}$$

where P is pressure and Q_ϕ and P_ϕ are 3.5 and 0.073 GPa, respectively. From the seismic Q structure, therefore, both Q_p and Q_{pm} are known. The ratio Q_p/Q_{pm} then yields an estimate of T/T_m from Fig. 2. The value T_m is obtained from melting experiments by Takahashi (1986) (cf. Fig. 8A), and thus the temperature T may be determined. If the quality factor of shear waves, Q_s, is determined from seismic studies, we employ the assumption $Q_p/Q_s = 2.25$ (assuming that the fundamental loss of seismic energy occurs not in compression but in shear; e.g., Anderson *et al.*, 1965) to make comparisons with the laboratory Q_p. If the temperature so derived is higher than T_m, we may estimate the melt fraction by employing Fig. 1B.

The laboratory results (Fig. 2)—determined at sufficiently high temperatures ($0.86 \leqslant T/T_m \leqslant 1.05$)—are used to estimate the thermal structure of the asthenosphere. In the lithosphere where temperatures are low ($T/T_m < 0.7$–0.8), the physical mechanism of attenuation may differ from that at higher temperatures. We therefore determine the temperature from seismic Q at asthenospheric depths. Observations of heat flow have yielded conductive geotherms in the lithosphere (e.g., Chapman and Pollack, 1977). We discuss later the generalized lithospheric geotherm from heat flow data.

The comparison of the laboratory Q_p results (Fig. 2) with the *in situ* seismic Q structure has provided estimates of mantle temperatures (Sato and Sacks, 1989). In the asthenosphere beneath the Iceland Plateau, for example, Chan *et al.* (1989) have determined $Q_s = 7$ at 45 km depth in the 0–5 Ma region and $Q_s = 50$ at 65 km depth in the 5–10 Ma region. Employing $Q_p/Q_s = 2.25$ and then comparing with Fig. 2, we obtain a temperature of 1290°C and 3 vol% melt in the 0–5 Ma asthenosphere (Sato and Sacks, 1989). In 5–10 Ma asthenosphere, however, the temperature is below the dry solidus (1140°C), and no melting is expected. It has been shown (Sato and Sacks, 1989) that the temperature and the fraction of melt determined from seismic anelasticity studies are consistent with those from seismic velocity and heat flow data. More detailed descriptions of the temperature estimates from seismic Q have been reported by Sato and Sacks (1989).

Generalized Thermal Structure beneath the Western United States from Seismic Anelasticity Data

Upper Mantle Anelasticity Structure

We first describe generalized upper mantle temperature profiles inferred from seismic Q data. Beneath the western United States, detailed Q structures have been reported (e.g., Patton and Taylor, 1984; Al-Khatib and Mitchell, 1991; Halderman and Davis, 1991). Patton and Taylor (1984), for example, showed that both the Rayleigh-wave and the short-period Love-wave attenuation data are consistent with a frequency-independent Q model, with low Q_s ($\simeq$100) in the lower crust and with the lowest values ($Q_s \simeq 30$) in the upper mantle at a depth of 60 km beneath the Basin and Range province. In this region, Q is generally low and extensive partial melting in the upper mantle has been indicated (e.g., Patton and Taylor, 1984; Lay and Wallace, 1988). Al-Khatib and Mitchell (1991) have also obtained low Q values in the upper mantle starting at about 50–70 km depth beneath the western United States (Fig. 3). The studies by Al-Khatib and Mitchell (1991) have de-

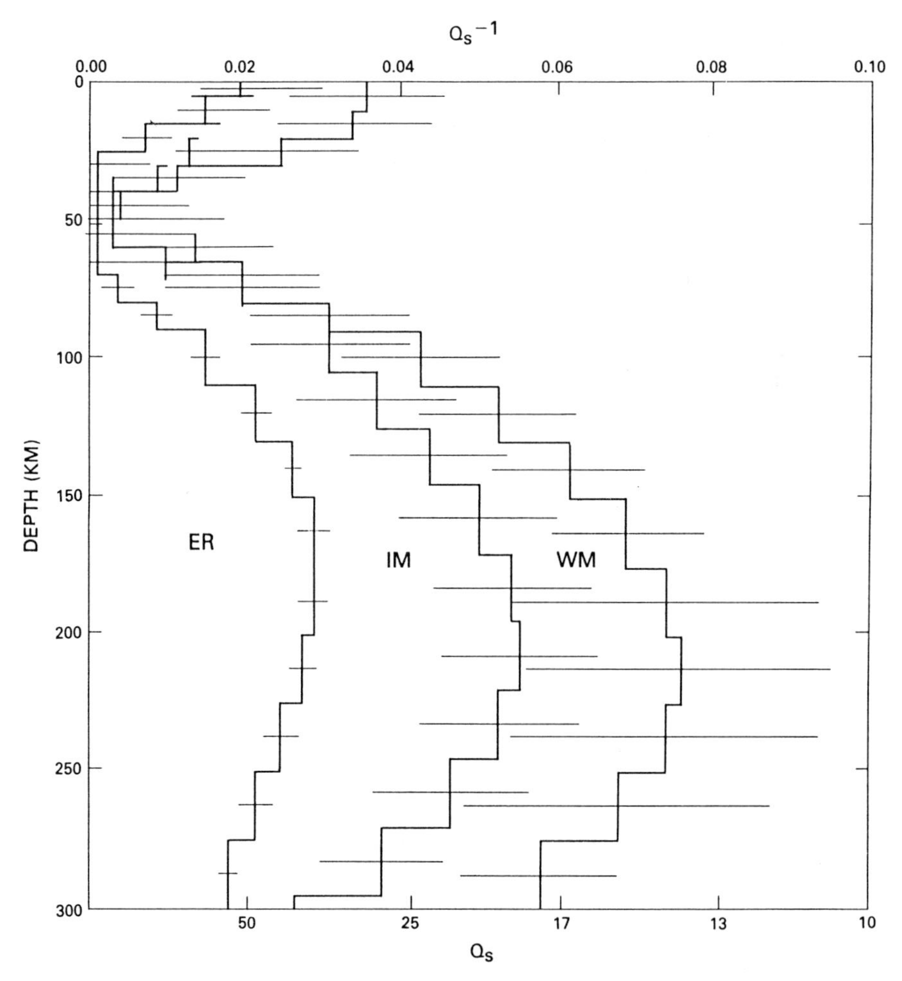

Figure 3 The Q_s structure for the Eastern Rockies (ER), the Intermountain (IM) region, and the Western Margin (WM) region. Horizontal bars denote one standard deviation of the Q value (from Al-Khatib and Mitchell, 1991).

termined the lateral variations of the upper mantle anelasticity structure in some detail from recorded surface waves. They divided the western United States into three regions: the Eastern Rockies (ER), the Intermountain (IM) region, and the Western Margin (WM) region. In each region, they have determined the Q_s structure to 300 km depth. Their results are consistent with previous studies and show a good overall correlation with the broad-scale tectonic and magmatic activity in the western United States. The highest Q_s values are obtained in the Rocky Mountain region where tectonic activity has died out, whereas the lowest upper mantle Q_s values are obtained beneath the western margin of the United States where the occurrence of tectonic activity is more recent. We here employ the Q structure determined by Al-Khatib and Mitchell (1991) to estimate temperature–depth profiles.

Volcanic Setting

Seismic transect CO-BKS of Al-Khatib and Mitchell (1991) passes through southwestern Oregon and northern California, just west of the Cascades and then on beneath the Geysers–Clear Lake volcanic center. Transect CO-GSC passes through that portion of the Cascades that includes Crater Lake and the Medicine Lake volcano, then on southward beneath the Long Valley, Big Pine, Coso, and Lava Mountains volcanic regions of California. Both transects enclose a slender wedge-shaped region that includes the southern Cascades centers of Mt. Shasta and Mt. Lassen. The relationship of these transects to the distribution of upper Cenozoic volcanic centers is given in Fig. 4, and their relationship to Quaternary and active volcanic centers is given in Fig. 5 and in Table 2. The physiographic nature of the crust above the transects is illustrated in Fig. 6. Collectively these traverses include portions of the Cascades and Pacific Coast Ranges, the Sierra Nevada, and a westernmost sliver of the Basin and Range province. They are here referred to as the Western Margin (WM) region.

The IM region seismic transects of Al-Khatib and Mitchell (1991) have great cumulative length and diagonally cut across the Cascades and then cut the grain of the Basin and Range province. Seismic transect CO-DUG passes beneath the Cascade centers of Three Sisters and the Newberry volcano and then through the High Lava Plains of south central Oregon and north of the Railroad Point and Fish Creek volcanic centers of Nevada. Transect LON-DUG begins just north of Mt. St. Helens and then passes beneath the Cascade volcanic center of Mt. Adams and on to the western Snake River Plain. Transect LON-ALQ effectively resamples the same volume as LON-DUG, but orthogonally cuts the Jemez volcanic zone as well. The Jemez volcanic zone is defined by the NE–SW trending San Carlos – Springerville – Quemado – Zuni Bandera – Mt. Taylor – Jemez – Brazos – Ocate and Raton volcanic centers of Arizona, New Mexico, and southern Colorado (Figs. 4–6; Table 2). Transect TUC-LUB cuts the Rio Grande rift at a high angle near the Jor-

Table 2

Centers of Quaternary Volcanism[a] in Relation to Their State, Their Location in the Index Map of Figure 5, and the Principal Seismic Transects for Anelasticity Determinations

Arizona	
GC	Grand Canyon
P	Pinnacate
S	Springerville
SA	Sentinel-Arlington
SB	San Bernardino
SC	San Carlos
SF	San Francisco
California	
AP	Amboy-Pisgah
BP	Big Pine
CA	Cima
CH	Cinder Hill
CL	Clear Lake
CM	Cargo Muchacho Mtns.
CO	Coso
EM	Eagle Mountains
G	Goffs
K	Kearn
LM	Lava Mountains
LV	Long Valley
SP	Sonora Pass
SS	Salton Sea
T	Truckee-Donner Pass
U	Ubehebe
Colorado	
AS	Aspen
E	Eagle
Idaho	
B	Blackfoot
Nevada	
CD	Carson Desert
FC	Fish Creek
LC	Lunar Crater
RP	Railroad Point
TM	Timber Mountain
New Mexico	
A	Albuquerque
BR	Brazos
CV	Cerro Verde
CZ	Carrizozo
J	Jemez Mountains
JM	Jornado del Muerto
MT	Mount Taylor
O	Ocate
PO	Portrillo
Q	Quemado
R	Raton
Z	Zuni-Bandera
Utah	
BD	Black Rock Desert
KA	Kanab
KO	Kolob
SG	St. George
Wyoming	
LH	Leucite Hills
YP	Yellowstone Plateau

[a] After Luedke and Smith (1991).

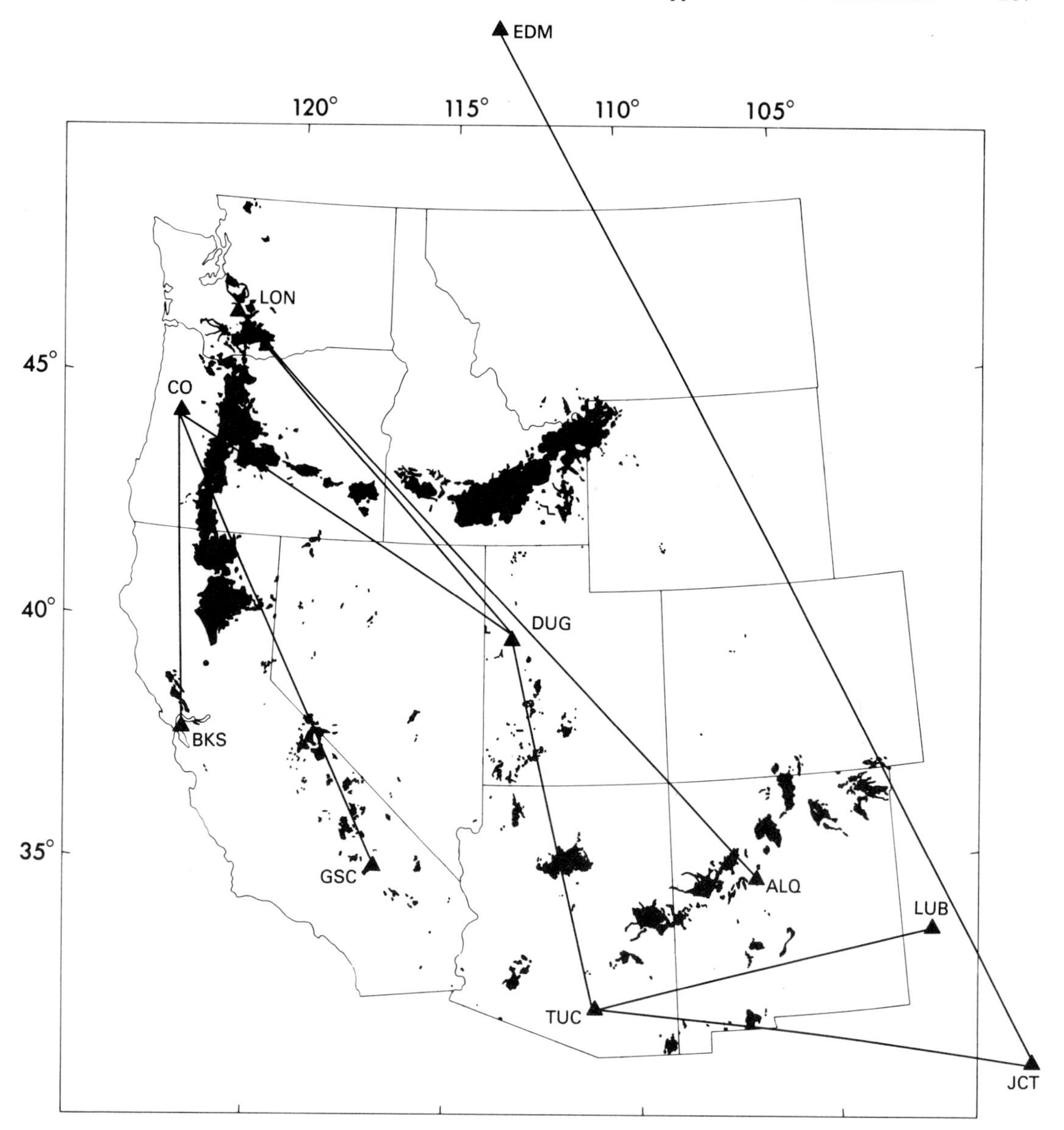

Figure 4 Seismic ray path distributions used for anelasticity evaluations in relationship to the distribution of upper Cenozoic volcanic rocks erupted over the interval 0—5 Ma. Endpoints on the seismic transects are: EDM, Edmonton, Alberta, Canada; JCT, Johnson City, Texas; LUB, Lubbock, Texas; TUC, Tuscon, Arizona; ALQ, Albuquerque, New Mexico; LON, Longview, Washington; DUG, Dugdale, Utah; CO, Corvallis, Oregon; BKS, Berkeley, California; GSC, Goldstone, California. Seismic station locations from Al-Khatib and Mitchell (1991). Volcanic eruptive distribution from Luedke and Smith (1991) as modified from Smith and Luedke (1984).

nado del Muerto center north of Truth or Consequences, New Mexico (Fig. 5). Finally, along transect DUG-TUC, the Black Rock Desert, Kolob, and Kanab volcanic centers of southwestern Utah are traversed, in addition to the San Francisco volcanic field of central Arizona (Figs. 4 and 5). Figure 7 illustrates the major physiographic provinces of the western United States in relation to the digital shaded relief topography of Thelin and Pike (1991).

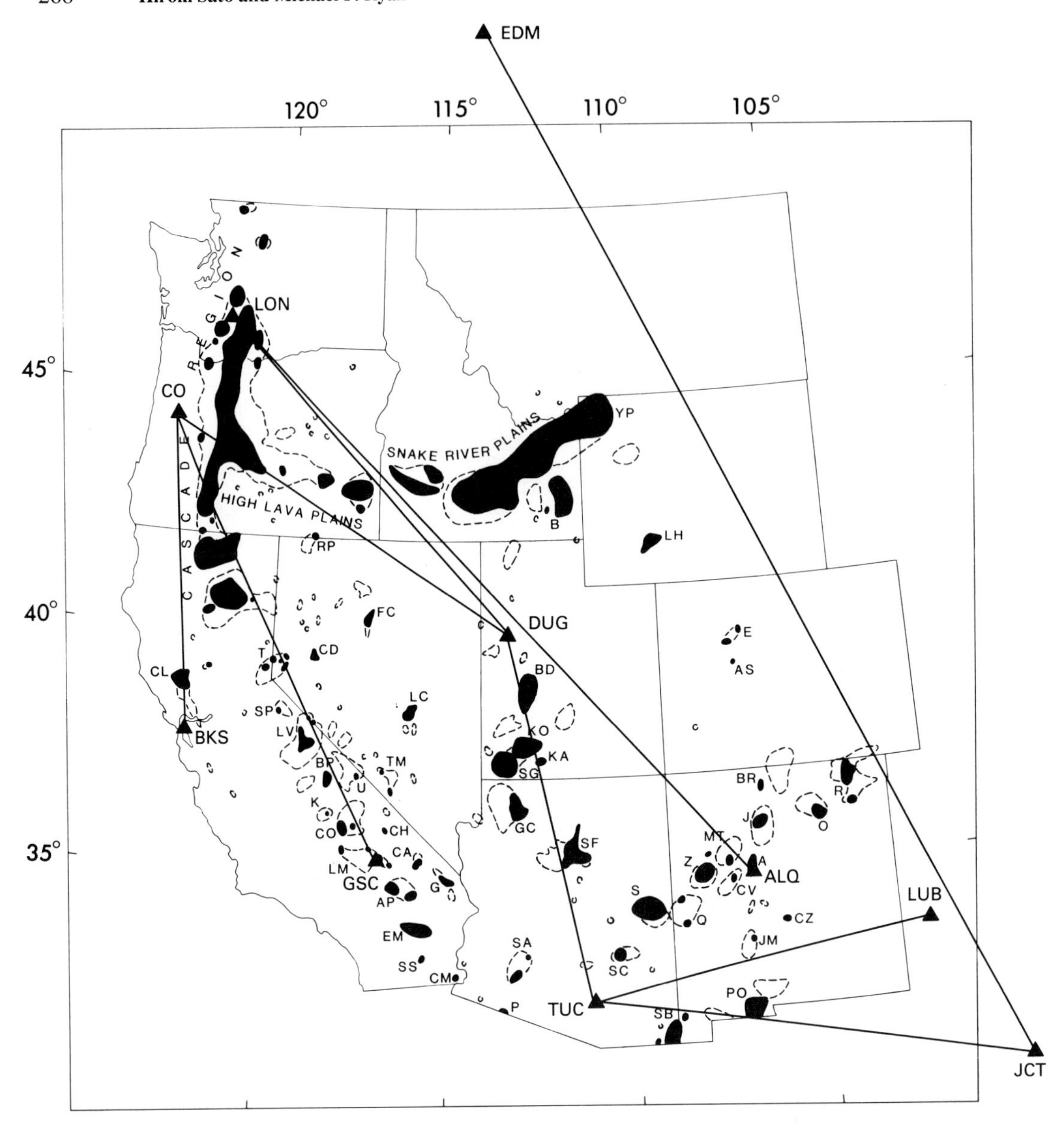

Figure 5 Distribution of volcanic centers active over the interval 0–1.6 Ma (solid pattern) in relationship to (*a*) the distribution of volcanic activity over the period 0–5 Ma (dashed lines); and (*b*) the seismic transects used to infer the generalized anelasticity structure of the deep crust and upper asthenosphere. Quaternary volcanic centers are listed by state in Table 2 and are after Luedke and Smith (1991). The seismic transects are based on the study of Al-Khatib and Mitchell (1991). Endpoint seismic station codes are provided in Fig. 4.

Upper Mantle Temperature and Partial Melt Fraction

We compare the Q_s structure of Fig. 3 with the laboratory data of Fig. 2 by using the Q_p/Q_s ratio (=2.25) described earlier, and obtain the generalized geotherms as shown in Fig. 8A. The lithosphere–asthenosphere boundary from the Q structure by Al-Khatib and Mitchell (1991) is at the depth 50–70 km. Therefore, we determine temperatures beneath that depth. Temperature uncertainties are also shown in Fig. 8A, allowing for a ±15% error in the experimental measurements. If one considers errors in the seismic Q structure, the temperature uncertainty would be larger. However, some of these errors are removed in the cali-

Figure 6 Digital shaded relief topography of the western United States in relation to the principal seismic transects for the anelasticity study of the upper asthenosphere. Seismic station identities are given in the legend to Fig. 4. Relief topography from Thelin and Pike (1991).

bration process by using temperatures from other data sources such as heat flow. This is discussed later (see also Sato and Sacks, 1989). A complete assessment of the uncertainty in the seismic Q is beyond the scope of this study. Errors in T_m may also cause errors in estimated temperature. The

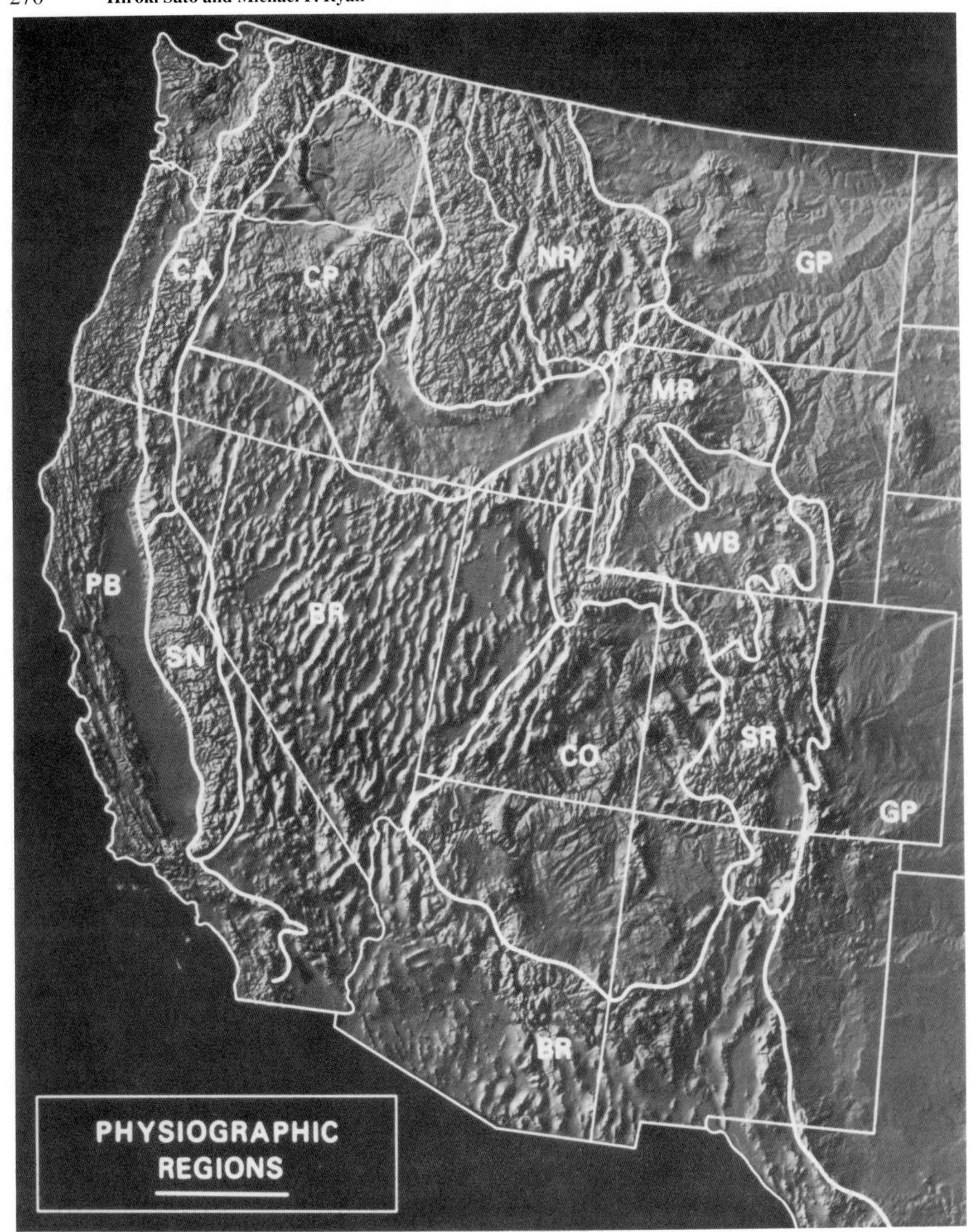

Figure 7 Digital shaded relief topography of the western United States in relation to physiographic regions discussed in the text. Region codings are: BR, Basin and Range; CA, Cascade Mountains; CO, Colorado Plateau; CP, Columbia Plateau; GP, Great Plains; MR, Middle Rocky Mountains; NR, Northern Rocky Mountains; PB, Pacific Border; SN, Sierra Nevada; SR, Southern Rocky Mountains; WB, Wyoming Basin. Physiographic provinces from Fenneman (1928). Relief topography from Thelin and Pike (1991).

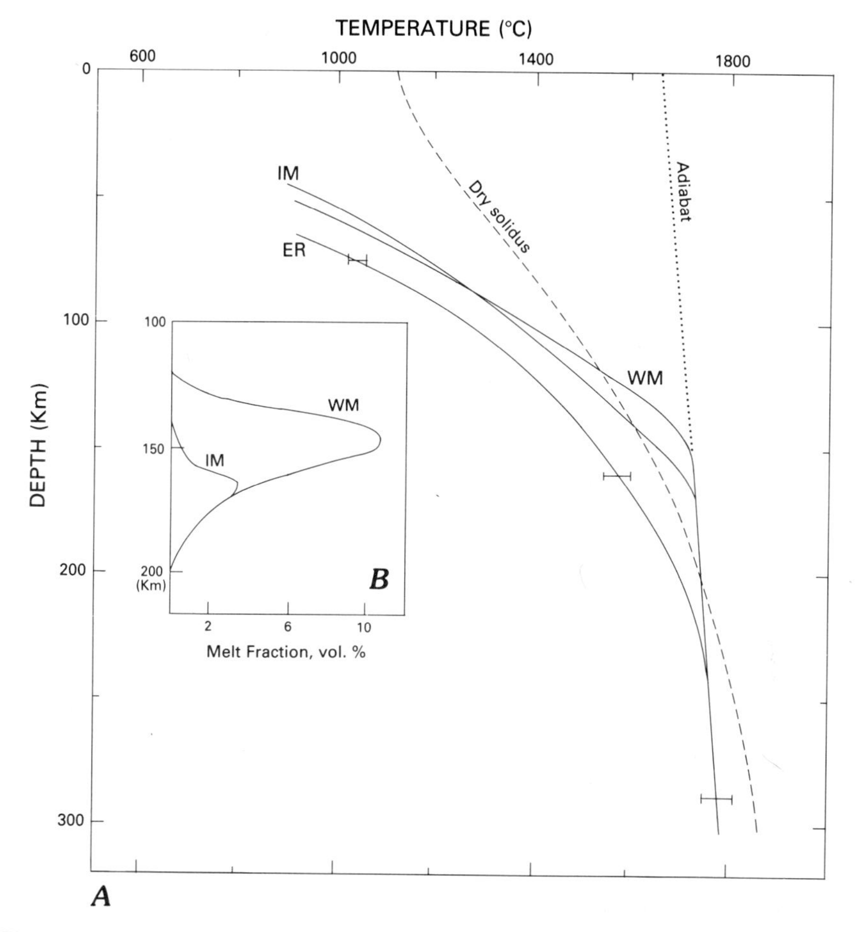

Figure 8 Thermal structure of the upper mantle beneath the western United States from the seismic anelasticity structure. (A) Temperature profiles (solid lines) for the Eastern Rockies (ER), the Intermountain (IM), and the Western Margin (WM) regions. Horizontal bars denote uncertainties in temperature from $\pm 15\%$ errors in the laboratory Q data. Dashed and dotted lines are dry peridotite solidus from Takahashi (1986) and mantle adiabat (0.4°C · km^{-1}), respectively. Although the temperatures in the ER region are slightly less than the dry solidus, those in the IM and WM regions cross the solidus and partial melt is expected. (B) Partial melt fraction as a function of depth. A relatively large amount of melt (to ~10 vol%) is expected in the WM region, where the temperature is above the solidus at ~120–200 km depth.

dry peridotite solidi determined by Takahashi and Kushiro (1983), Scarfe and Takahashi (1986), and Takahashi (1986), however, are found to be consistent (within ± 20°C or so) despite some compositional variations between their studies.

An extrapolation of the conductive geotherms upward in temperature until they intersect the mantle adiabat provides an estimator of lithospheric thickness (e.g., McKenzie and Bickle, 1988). Such estimators applied to the WM and IM geotherms of Fig. 8 yield lithospheric thicknesses of ≈140 km and ≈150 km, respectively. These transitions, from lithosphere to asthenosphere, fall within the thermal boundary layer that separates the convective adiabatic interior of the upper mantle from the conductive lid above. The mechanical boundary layer of the upper lithosphere contains the crust–mantle interface that may experience brittle elastic fracture and creep rupture during the dike and sill emplacement episodes that underplate the crust.

The Q structure by Al-Khatib and Mitchell (1991) is noted to be an *average* over a wide area of the western United States (Figs. 3 and 7). Figure 7 illustrates the major physiographic provinces of the western United States in relation to the digital shaded relief topography of Thelin and Pike (1991). The IM region, for example, includes the Basin and Range province and the Colorado and Columbia Plateaus, where the local geology and the geophysical characteristics differ. The Q structure in the ER region is averaged along a seismic ray path as long as ~3000 km. Therefore, the temperatures in this study represent only broad and *generalized* averages.

Beneath about 120 km depth, the temperatures derived are well above the dry peridotite solidus in the IM and WM regions, and therefore partial melting is inferred. By using Fig. 1B, the degree of this melting for the dry upper mantle is determined and is shown in Fig. 8B. A relatively large amount of melt (up to ≈10 vol%) is obtained beneath the WM region. The partial melting is consistent with high heat flow values, high electrical conductivities and the surface manifestation of magmatic activity in the IM and WM regions as discussed later. On the other hand, beneath the ER region, temperatures are slightly lower than the solidus and no melting is generally required.

The lateral variations of Q yield lateral temperature changes, indicating horizontal gradients in T and an increasing upper mantle temperature and melt fraction from the east to west between the Rocky Mountains and the Pacific coast. At 65- and 160-km depths, there are about 190 and 130°C differences, respectively, between the ER and IM regions (Fig. 8A), which correspond to a lateral temperature gradient of about 0.3°C · km^{-1}. At 125 km depth, the temperature difference is about 90°C between the ER and IM regions and between the IM and WM regions. These lateral temperature variations are comparable to the adiabatic gradient. The vertical temperature variation at shallower (50–80 km) depths is, however, much larger (10–15°C · km^{-1}), and is comparable to the gradient of the conductive geotherm derived from surface heat flow, as described later (cf. Fig. 11).

At greater depths, the geothermal gradient approaches the mantle adiabat (~0.4°C · km^{-1}, Fig. 8A). This adiabat coincides with the temperature profile determined from the Q structure of the ER region. However, some of the low Q values at depths in the IM and WM regions yield higher temperatures (i.e., $T/T_m > 1.05$) than the adiabat, by simply extrapolating the laboratory data to higher temperatures. The laboratory Q measurements, however, have been carried out at temperatures to $T \simeq 1.05T_m$ (Fig. 2). At higher temperatures, the melt fraction increases and the seismic velocity then drops rapidly with increasing temperature (cf. Fig. 1). This may also cause a rapid decrease of Q. We may not, therefore, be able to extrapolate our Q data to temperatures above $1.05T_m$. A rapid Q decrease (that is expected at $T/T_m >$ 1.05) gives lower temperature estimates (as the adiabat in the IM and WM regions; Fig. 8A) than the temperatures determined from the extrapolation of experimental data. The scatter of Q values in the IM and WM regions becomes larger beneath about 150 km depth (Fig. 3). Fairly small error bars in the ER region should give much more reliable temperature estimates than those in the IM and WM regions. Here we simply employ the mantle adiabat at the greater depths derived from the temperature profile of the ER region (Fig. 8A).

Comparison with Previous Temperature Estimates

Figure 9 compares the geotherms in this study with those from other geophysical or geological data sources. Ito and Sato (1992) have given a model mantle geotherm that is determined by combining heat flow data, seismic anelasticity data and detailed phase diagrams of mantle minerals. Sclater *et al.* (1980) have used estimates of the mean heat flux and the radioactive heat production in the crust to compute the temperature as a function of depth beneath the continents and oceans. A thermal history of the Rio Grande rift has been numerically calculated by assuming a conductive cooling of the lithosphere after the cessation of magmatic activity and rift extension at about 25 Ma (Morgan *et al.*, 1986). Phase equilibria and compositions of mineral assemblages have been used to estimate equilibration temperatures and pressures in the upper mantle (e.g., Mercier and Carter, 1975; Finnerty and Boyd, 1984). In addition, Herrin (1972) and Griggs (1972) have used constraints from surface heat flux, seismic data, melting curves and thermal conduction models to estimate mantle geotherms. These previous temperature estimates for the oceans and the hot continent are comparable to the high temperatures found beneath the western United States as shown in Fig. 9. The geotherms inferred in this study (as well as others compared in Fig. 9) suggest higher potential temperatures, T_p, than have

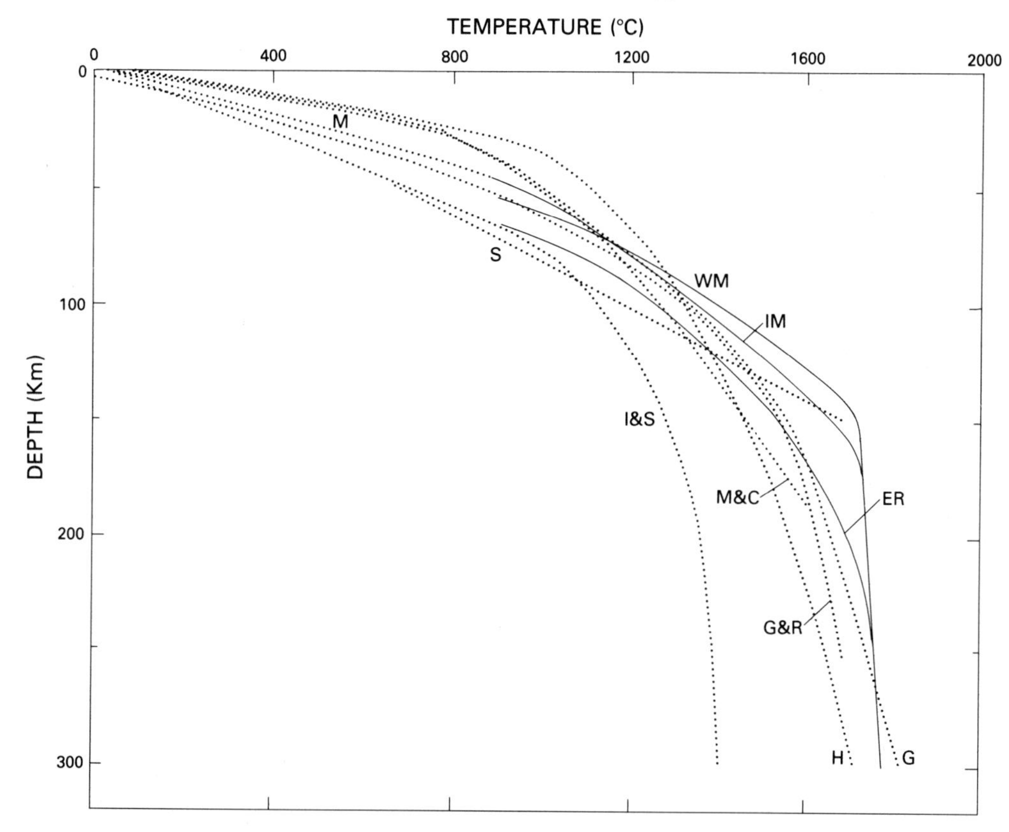

Figure 9 Geotherms of the upper mantle. Solid lines are temperatures in this study for the ER, IM, and WM regions. Dashed lines are a model mantle geotherm (I&S) by Ito and Sato (1992), a Rio Grande rift geotherm (M) by Morgan *et al.* (1986), a hot continental geotherm (S) by Sclater *et al.* (1980), a high temperature oceanic geotherm (M&C) by Mercier and Carter (1975), an oceanic geotherm (G&R) by Green and Ringwood (1967), a Basin and Range geotherm (H) by Herrin (1972), and a theoretical steady-state geotherm (G) by Griggs (1972). The temperatures beneath the western United States in this study are comparable with oceanic or hot continental geotherms.

been suggested by McKenzie (1984) and by McKenzie and Bickle (1988). At the higher temperature end, for example, McKenzie (1984) suggests convective diapirs with core temperatures of ≈1550°C rising through ambient surrounding mantle with a temperature of ≈1350°C and producing melt at P ≅ 3.5 GPa, or about 110-km depth. The physical context considered is suboceanic mantle melting, a dry peridotite solidus (Takahashi and Kushiro, 1983), and the very low mantle retention of the melts produced, i.e., volumes of basalt observed at the surface are used as a constraint on the melting process.

Surface Heat Flow in the Western United States

We here briefly describe the heat flow measurements reported in the western United States and estimate the average heat flow values in the ER, IM, and WM regions. The presence of Cascade magmatism as well as many geothermal areas have produced a high heat flux in the region. The heat flow map of the United States of Sass *et al.* (1981) has shown values above 100 mW · m^{-2} in the Cascade Range, the Rio Grande rift, and the northern part of the Basin and Range province. The high heat flow in the Rio Grande rift averages ~95 mW · m^{-2} south of Socorro and is 77 mW · m^{-2} to the north of Socorro (Reiter *et al.*, 1986). Blackwell *et al.* (1982, 1990) have measured the heat flow of the Cascade Range and show a major change from 40 mW · m^{-2} in the Western Cascades to 100 mW · m^{-2} in the High Cascades. The heat flow averages 88 mW · m^{-2} in the High Cascades and 100 mW · m^{-2} for the Western Cascade-to-High Cascade anomaly. Mean heat flow values in the Basin and Range and in the Sierra Nevada provinces are 92 and 39 mW · m^{-2}, respectively (Sass and Lachenbruch, 1978). Extension may have been an important factor in producing or maintaining the high heat flow in the Basin and Range province (Morgan and Gosnold, 1989). On the other hand, a deep heat sink caused by downward convection of heat associated with subduction has been indicated for the relatively low heat flow in the Sierra Nevada (Roy *et al.*, 1972). In the California Coast Ranges, the subducting slab has been cut off east of the San Andreas fault, allowing the growth of a "slabless window" (e.g., Dickinson and Snyder, 1979; Lachenbruch and Sass, 1980). The relatively high heat flow (~80 mW · m^{-2}) in this region is therefore thought to be related to a "hole" in the lithosphere beneath which hot asthenosphere rises, resulting in an increase in surface heat flow by a factor of 2 within 4 Myr of the passage of the Mendocino triple junction (Lachenbruch and Sass, 1980; Zandt and Furlong, 1982). Although heat flow is generally low (49 mW · m^{-2}) in the Precambrian platforms, the Mesozoic orogenic belts including the Pacific coast and the Rocky Mountains show a moderate value of 65 mW · m^{-2} (Sclater and Francheteau, 1970). A mean heat flow of 76 mW · m^{-2} has been reported in the western United States by Sass *et al.* (1971) and by Nathenson and Guffanti (1988). A comprehensive summary of heat flow data of the United States by Morgan and Gosnold (1989) has reported 68 mW · m^{-2} in the Colorado Plateau, 86 mW · m^{-2} in the Basin and Range province, 101 mW · m^{-2} in the Cascade Range, 54 mW · m^{-2} in the Sierra Nevada, and 141 mW · m^{-2} in the Salton Trough. The preceding heat flow values have also been described clearly in a heat flow map of the western United States constructed by Blackwell *et al.* (1991). Their map shows values below 60 mW · m^{-2} in the Sierra Nevada, the northern Pacific coast, the Great Plains, the Wyoming Basin, and parts of the Columbia and Colorado Plateaus. A relatively high heat flow region (80–100 mW · m^{-2}) covers the Basin and Range province, the Rio Grande rift, and the southern Rockies. Heat flow values above 100 mW · m^{-2} occur in the High Cascades, the Snake River Plain, the Yellowstone area, the Battle Mountain high, the Salton Trough, and the southern Rockies. It has been pointed out (Blackwell *et al.*, 1991) that about 75% of the area of the Cordillera is characterized by heat flow values in the range 70 to 90 mW · m^{-2}.

The regional heat flow of the western United States in relation to the digital shaded relief topography is shown in Fig. 10. The ER region consists mainly of Mesozoic orogenic belts, where an average heat flow is 65 mW · m^{-2}. The heat flow is high in the Basin and Range province (86–92 mW · m^{-2}: Sass and Lachenbruch, 1978; Blackwell *et al.*, 1978) and in the Rio Grande rift zone (95 mW · m^{-2}: Reiter *et al.*, 1986). Low heat flows occur, however, in the Columbia Plateau (62 mW · m^{-2}: Blackwell *et al.*, 1990) and in the Colorado Plateau (~60 mW · m^{-2}: Gosnold, 1990; and ~68 mW · m^{-2}: Morgan and Gosnold, 1989). An average heat flux could be thus about 80 mW · m^{-2} for the seismic ray path distributions in the Basin and Range, the Columbia Plateau, and the Colorado Plateau of the IM region, where many ray paths pass through the Basin and Range province (Figs. 6 and 7). The WM region includes the Pacific coast, the Cascades, the Sierra Nevada, and the western part of the Basin and Range province. Although heat flow values are only about 40–54 mW · m^{-2} in the Sierra Nevada (e.g., Sass and Lachenbruch, 1978; Sass *et al.*, 1981; Morgan and Gosnold, 1989) and in the Coast Range west of the Cascades (Blackwell *et al.*, 1982, 1990), other areas produce high heat flows. Examples of such high thermal fluxes are an average 75 mW · m^{-2} in the southern Washington Cascades, and range from ~80 to >100 mW · m^{-2} in the High Cascades (Blackwell *et al.*, 1982, 1990). Heat flows are generally higher than 60 mW · m^{-2} along the Pacific coast of California (Sass *et al.*, 1971, 1981). An average heat flow could be thus about 75 mW · m^{-2} for the seismic ray path distributions in the WM region.

Comparison with Heat Flow Temperature–Depth Profiles

We next estimate temperatures from the surface heat flow and compare them with the geotherm of this study. Even though there are some uncertainties in the seismic anelasticity structure, experimental data, or the chemical composition of the upper mantle, geotherms may be well constrained by such a comparison (cf. Sato and Sacks, 1989).

Figure 10 Regional heat flow in relation to the digital shaded relief topography of the western United States. Heat flow contours are from Morgan and Gosnold (1989). Relief topography from Thelin and Pike (1991).

Conductive geotherms for 65, 75, and 80 mW · m^{-2} are shown in Fig. 11, following the calculations by Chapman and Pollack (1977). They have used a thermal conductivity of 2.5 W · m^{-1} · k^{-1} throughout the crust and a temperature-dependent conductivity above 500°C based on experimental data by Schatz and Simmons (1972). Their model comprises an upper crustal region en-

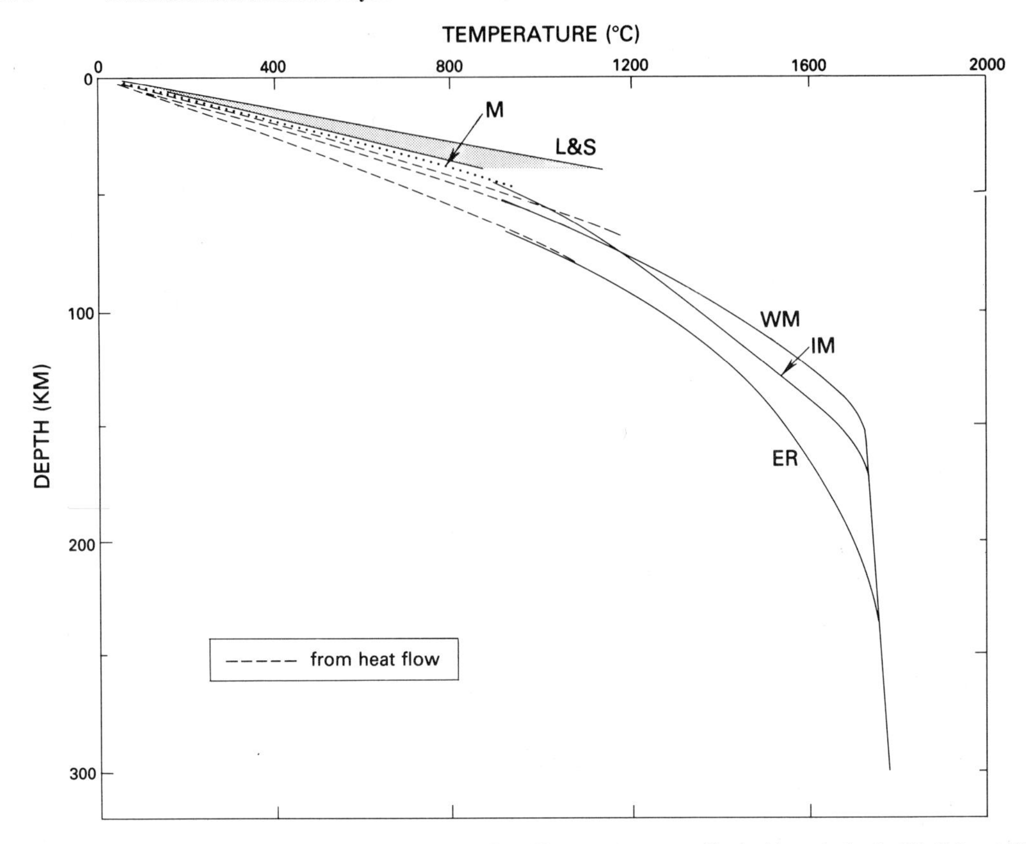

Figure 11 Comparison with geotherms derived from heat flow. The temperature profiles in this study for the ER, IM, and WM regions (solid lines) are consistent with the conductive geotherms of 65, 80, and 75 mW · m^{-2} (dashed lines), respectively. Since laboratory anelasticity measurements were carried out for a dry peridotite, and a dry solidus was employed to estimate the temperatures, this consistency suggests a generally dry upper mantle in the western United States (within the resolution of the seismic anelasticity structures, i.e., excluding regions of local enrichment of volatiles). The temperature distributions indicate a gradual change from a conductive to an adiabatic regime in the upper mantle. A dotted line (M) is a Rio Grande rift geotherm by Morgan *et al.* (1986). A hatched area (L&S) is a Basin and Range geotherm by Lachenbruch and Sass (1977). (See also the curves accounting for crustal extension labelled C and D for "characteristic Basin and Range" in Fig. 9-18 of Lachenbruch and Sass, 1978).

riched in radioactive sources, a granulite-facies lower crust (heat production A of 0.25 μW/m^3), and a depleted ultrabasic zone (A = 0.01 μW/m^3) overlying pyrolite mantle (A = 0.084 μW/m^3). Because the laboratory Q data measured at high homologous temperatures (Fig. 2) are used to estimate the temperature of the asthenosphere, we compare it with the conductive geotherm at the top of the asthenosphere. The anelasticity structure by Al-Khatib and Mitchell (1991) has shown an increase of Q^{-1} with depth in the lithosphere and a decrease of Q^{-1} at the lithosphere–asthenosphere boundary at 50–70 km depth. We therefore make temperature comparisons within this depth range. As shown in Fig. 11, the temperatures inferred from heat flow are consistent with the geotherms of this study. The temperature profile of the ER region also coincides with the geotherm for the mean heat flow of 66 mW · m^{-2} in the Great Plains calculated by Morgan and Gosnold (1989). A similar geotherm is suggested for the heat flow of 68 mW · m^{-2} in the Colorado Plateau region. Since experimental anelasticity measurements were carried out for a dry peridotite, and a dry solidus was employed to estimate the temperatures, this consistency suggests that the upper mantle beneath the western United States may be generally dry (within the resolution of the seismic anelasticity structures). However, because of the limited resolution of seismic anelasticity studies, volatiles localized in regions less than several tens of kilometers may not be resolved. Indeed, a local enrichment of volatiles may certainly occur (e.g., beneath regional geothermal areas such as the Long Valley caldera, the Coso hot springs, and the Cascade volcanoes).

Large vertical temperature gradients at shallow depths are also consistent with the steep temperature increases with depth as inferred by conduction heat transfer (Fig. 11). The temperature gradient becomes comparable to the mantle adiabat beneath ~150–220 km depth. Thus the geotherm derived from seismic Q data indicates a generally conductive regime of heat transfer at the top of the asthenosphere and a gradual change from a conductive to an adiabatic regime with depth.

Generalized Thermal Structure beneath the Western United States from Seismic Velocity Data

General Temperature Distributions

We now compare the laboratory velocity data (Fig. 1) with the seismic velocity structure, and estimate temperatures. Overall, the P_n velocity structure of the western United States has shown a velocity variation from 7.6 to 8.1 km · s^{-1} in the uppermost mantle (Hearn *et al.*, 1991). Low P_n velocities beneath the Yellowstone region and the Basin and Range province have been interpreted as resulting from anomalously hot upper mantle, but those beneath the Sierra Nevada have been ascribed to a deep crustal root. A normal continental lithosphere has an average P_n velocity of 8.1 km · s^{-1} (e.g., Mooney and Braile, 1989). We therefore employ V_L = 8.1 km · s^{-1} in Eq. (2). The lowest P_n velocity of 7.6 km · s^{-1} (nearly a 6% velocity drop) yields V_a/V_m = 0.99 from Eq. (2). This indicates that only a minor amount of melt may exist in the region, and thus, the temperature is expected to be slightly above the solidus (about 1250°C at 40 km depth). In the Juan de Fuca plate subduction zone, Harris *et al.* (1991) have determined a dipping high velocity anomaly that corresponds to the subducting plate, and low velocity anomalies to 6% in the mantle wedge beneath the Cascade volcanoes. The 6% velocity drop again suggests the presence of minor amounts of melt. The amount of melt, however, could be larger in the *cores* of these low velocity regions, where the velocity may be less than 7.6 km · s^{-1}. Because of the limited resolution of the seismic tomographic method, melting localized in less than a few tens of kilometers may not be resolved from the velocity structures of Hearn *et al.* (1991) and Harris *et al.* (1991). We here report, therefore, bulk and *averaged* values of the partial melt fraction and the temperature. Local enrichments of melt, however, are certainly expected to occur. A fairly high heat flow region (>100 mW · m^{-2}) is in fact a relatively localized occurrence (e.g., in the Cascade volcanoes, the Basin and Range province, and the Rio Grande rift; Blackwell *et al.*, 1982, 1990; Lachenbruch *et al.*, 1985).

Al-Khatib and Mitchell (1991) have determined the shear velocity structure of the ER, IM, and WM regions in the western United States, as well as the anelasticity structure described earlier (cf. Fig. 3). In the ER region, the velocity (V_a) decreases from 4.6 km · s^{-1} at 140 km depth to the minimum 4.3 km· s^{-1} at 240 km depth. Using V_L = 4.55 km · s^{-1} in Eq. (2), V_a = 4.3 km · s^{-1} gives near-solidus temperatures, and a state of widely distributed partial melt may not exist in the region. In the IM and WM regions, however, the low velocity zones show V_a = 4.2 to 4.0 km · s^{-1}, and 1 to 6 vol% melt is expected (T/T_m = 1.02–1.04). The low velocity zone is located in the 90- to 150-km-depth interval in the WM region, and the temperature is estimated to be about 1580°C at 120 km depth. The melt fractions and temperatures determined here from the velocity structure by Al-Khatib and Mitchell (1991) are comparable with those determined from the anelasticity structure (Fig. 8).

Regional Temperature Distributions

A detailed model of the Basin and Range velocity structure by seismic array studies shows a small-scale low velocity zone (less than a 5-km-thick V_p = 7.5 km · s^{-1} zone) in the uppermost mantle beneath northwestern Nevada (Holbrook, 1990). Using V_L = 8.1 km · s^{-1} and Eq. (2), this low velocity suggests ≈1 vol% melt and a temperature of 1260°C. A fairly low P-wave velocity (7.4 km · s^{-1}) has also been determined beneath the Salton Trough from a relatively high resolution velocity structure (a block size of 15 km width and 30 km depth) by using the southern California seismic array (Humphreys and Clayton, 1990). We infer up to 3–4 vol% melt in the core of this low velocity region (less than 50 km width), and the corresponding temperatures are up to 1300, 1410, and 1520°C at 45-, 75-, and 105-km depths, respectively. A similar amount of partial melt has been estimated by Humphreys and Hager (1990), although they have used the numerical calculations of Mavko (1980) to provide this estimate.

In the asthenosphere beneath the Coast Ranges of northern California, shear velocity studies by Levander and Kovach (1990) have determined $V_L = 4.55 \text{ km} \cdot \text{s}^{-1}$ and $V_a = 4.1 \text{ km} \cdot \text{s}^{-1}$. This yields 3 vol% melt, and $T = 1280$ and 1360°C at 40- and 60-km depths, respectively. Lateral P-wave velocity variations determined in detail in northern California from 9383 travel time residuals by Benz *et al.* (1992) showed (-6%) low velocity regions, indicative of magma bodies within the crust beneath the active volcanoes. They also imaged the steep 70° east-dipping Gorda plate to 270 km depth as having a +5% high velocity anomaly, and the shallow asthenosphere at 30–100 km depth beneath the northern Coast Ranges as having an average of -4% low velocity. In the core of these low velocity regions (less than ~40 km wide), one may expect melting (up to 4 vol%) and temperatures to 1320°C (T/T_m = 1.03) at a depth of 50 km. A few percent partial melt and temperatures higher than the mantle solidus are also estimated from the 8% velocity drop and the low Q ($Q_p = 57$) in the asthenosphere beneath the Rio Grande rift (Halderman and Davis, 1991). From Eq. (2), our estimates for the 8% drop ($V_a/V_L = 0.92$) are $T/T_m = 1.02$ and 2 vol% melting. The asthenospheric Q_p of 57 yields comparable values as $T/T_m = 1.01$ and 0.5 vol% melting from Fig. 2. Beneath the east African rift zone, by comparison, Halderman and Davis (1991) determined a 12% velocity drop, yielding T/T_m = 1.04 and about 5 vol% melt. The corresponding temperatures are 1500 and 1530°C at 100 km depth beneath the Rio Grande and east African rift zones, respectively. In every geothermal area, one expects partial melting whatever the details of its three-dimensional distribution.

The temperatures derived here from small-scale low velocity anomalies at relatively shallow depths (<100 km) are generally higher than the geotherms from seismic Q as shown in Fig. 8. This is consistent with the idea that small-scale magma generation sites, magma reservoirs, small diapirs, or geothermal root regions as resolved by fine-scale velocity inversions may have higher core temperatures than the bulk averages resolved by anelasticity studies. The equilibration temperatures of mantle xenoliths as calculated from element partitioning geothermometers have also shown relatively high temperatures (e.g., 860–980°C at 30–40 km depth and about 1100°C at ~60 km depth beneath southern British Columbia; Brearley *et al.*, 1984; Canil *et al.*, 1987; compare with Fig. 8A.).

A Remark on Melt Retention and the Effective Melt Content of a *Source Region*

The suggestion that melt fractions may approach ≈10% beneath the WM region and may be near ≈3% below the IM region deserves comment in light of the tendency for low viscosity basaltic melts to wet grain-edge intersections and to thus escape the deforming matrix. Using the liquid-phase sintering results of Beeré (1975) and the two-phase flow modeling approach of Drew (e.g., 1983) and Didwania and Homsy (e.g., 1981), McKenzie (1984, 1985) has treated the matrix compaction—melt expulsion problem with application to decompression melting. From the perspective of matrix melt retention, a critical relationship is

$$\cos(\theta/2) = \gamma_{ss}/2\gamma_{sF} \tag{4}$$

where θ is the dihedral angle defined by the melt-crystal interfaces, γ_{ss} is the interfacial free energy of the crystal–crystal contacts and γ_{sF} is the interfacial free energy of the melt–crystal contacts. Importantly, the value of θ controls the geometry and stability of the pores and *if $\theta < 60$, the pore space at all grain edge intersections remains interconnected for all values of the porosity ϕ*. If, however, due to mineralogical and pore fluid heterogeneities, $\theta > 60°$, then there is a threshold ϕ required for the establishment of interconnected melt tubules in three dimensions. Results of experiments on texture-equilibrated olivine-rich basaltic partial melts show typically $20° \leq \theta \leq 47°$ (Waff and Bulau, 1979, 1982, but Cheadle, 1989, presents a more complete compilation) however dry melts in an OPX-rich environment may lead to values of θ in excess of 60°.

Fujii, Osamura, and Takahashi (1986) have

carried out experiments on sintered orthopyroxene + basalt mixtures in efforts to determine the dihedral angle θ, and make inferences about the relative interfacial free energies and the extent of interconnected permeable pathways at the grain scale. In melts of low water content, the high dihedral angles between dry melts and orthopyroxene (median $\theta = 70°$) suggest low degrees of grain edge wetting. This implies isolated melt pockets and a local disruption of the melt network, thus inhibiting permeability. For water-saturated melts, however, the pyroxene-melt-pyroxene θ is $\approx 52°$, suggesting the restoration of complete fluid connectivity (see also von Bargen and Waff, 1988). Collectively, the dry melt results suggest that pyroxene-rich microdomains may contain melt fractions that are notably in excess of those expected from applications of the McKenzie (1984, 1985) treatment.

The magma generation regions suggested in Fig. 8 are expected to contain swarms of melt-filled veins as well as deep dikes. These are the structures that play an important role in transporting magma out of the generation volume and upward toward the Moho. In this sense they form a bridgework or set of connecting links between the grain-scale melt flow regimes in the heart of the generation volume and the igneous structures that underplate the continental keel. They are not incorporated in the McKenzie (1984, 1985) treatment but have been discussed by Sleep (1988) and have a fundamental role to play in upper mantle magmatism. As briefly sketched below, the stability criteria for melt-pressurized cracks include considerations of the elastic moduli of the host rock and the *in situ* density contrasts between the melt and matrix. Stability is also a sensitive function of melt volume and the correlative crack widths. For example, basaltic melt-filled cracks embedded in peridotite that form veins less than roughly 20 mm in width tend to be gravitationally stable when their heights are less than about 480 m, and they will thus tend to remain in the vicinity of the source region. If they do remain at depth, they will therefore contribute to the regionally averaged low Q values inferred seismically. These deep veins will thus add to the Q reductions induced by trapped grain-scale melt. It is then expected that *low Q values associated with regions of melting upper mantle may be the volume-averaged result of both grain-scale and vein + dike-scale melt distributions.*

Comparison with Heat Flow-Derived Temperatures

Temperatures determined from seismic velocity data are again generally consistent with the temperatures derived from surface heat flow studies. The low velocity regions in fact coincide spatially with the high heat flow regions. The low velocity zone beneath the Salton Trough, for example, is located beneath the Imperial Valley (Humphreys and Clayton, 1990), where the heat flow is above 100 mW $\cdot$ m^{-2} and averages about 140 mW $\cdot$ m^{-2} (Lachenbruch *et al.*, 1985). Such a high heat flow yields the notably high temperature of about 1300°C at 45 km depth (Chapman and Pollack, 1977), which is consistent with our estimate. The northwestern Nevada region studies by Holbrook (1990) and the Rio Grande rift area studied by Halderman and Davis (1991) are also characterized by high regional heat flow levels above 105 mW $\cdot$ m^{-2} (e.g., Sass *et al.*, 1981; Blackwell *et al.*, 1991), which again give conductive temperatures (Chapman and Pollack, 1977) comparable to this study. These results indicate that the seismic velocity determined in the laboratory for a dry peridotite is appropriate to the upper mantle lithology beneath.

If, however, local enrichments of water in volumes that are less than a few tens of kilometers occur in these regions (not resolved from seismic velocity studies), the inferred temperature should be lower than those determined from the dry solidus, since water depresses the solidus of mantle rocks. In the case of water-saturated (or water-undersaturated) upper mantle, the derived temperatures would then be reduced by an amount up to (or less than) about 250°C (Sato, 1992).

On Discontinuous Magma Ascent and the Mechanics of Basaltic Underplating

Establishing the linkage between the generalized depth intervals for partial melt in the WM and IM regions (as illustrated in Fig. 8) and the basaltic and then the andesitic, dacitic, and rhyolitic volcanic centers of Figs. 4 and 5 is fundamental for understanding the magmatism of the western United States. In turn, this requires constraining the likely connections between the ascent of basic magmas and the mechanics of basaltic underplating, since basic melts pond at or near the base of the continental crust prior to migration into the mid-crust or eventual eruption (e.g., Rivalenti *et al.,* 1981; Shervais, 1979; Hamilton, 1989), and play a substantial role in providing the heat sources for subsequent silicic magmatism. Bergantz and Dawes (this volume) discuss aspects of the fluid dynamics and the relative roles of convection and conduction in lower crustal melt generation events induced through basaltic underplating. This section reviews aspects of the discontinuous magma ascent pathways and the elastic fracture morphology as buoyancy-driven fractures rise from the upper mantle and penetrate the crust–mantle interface, nucleating melt-ponding episodes at the base of the crust. We leave for later, the important problems of andesitic, dacitic, and rhyolitic melt migration in higher portions of the continental crust—but remark that looking at aspects of the mechanics of magma emplacement across the Mohorovičić Discontinuity is a logical first step in developing a physical process framework for the generation of higher level magmas.

The process of upward melt migration begins within the depth intervals indicated in Fig. 8B. For the WM region this interval is centered at ≈145-km depth, whereas for the IM region it is centered at ≈165-km depth. The generalized melt fractions indicated for the WM and IM regions are up to about 10% and ≈3%, respectively, as indicated in Fig. 8B. Considerable local variations are expected, as discussed below, both in terms of melt fractions and depth intervals. The mechanics of vein and dike *nucleation* within the region of partial melting is discussed by Sleep (1988).

Inherently Discontinuous Nature of the Ascent Path

One of the most fundamental attributes of the magma ascent pathway is its disconnected nature. For basaltic and ultrabasic melts there is a second fundamental attribute: it is a fluid-filled fracture. This section briefly reviews the basis for these assertions and outlines the process of magma migration from the regions of partial melt shown in Fig. 8. We anticipate the results of this section by recognizing that it is the *finite strength* of fluid-weakened mantle rocks at high temperatures that restricts the fully fluid-connected magma column segments to *finite heights*.

Employing the mixed boundary value solutions of Muskhelishvilli (1953) and Mikhlin (1957), Weertman (1971a, 1971b) has used elastic dislocation theory to determine the outward displacements of the fracture's walls as functions of the internal liquid pressure and the relevant material properties. These studies have produced relationships that relate the heights of individual crack segments to the rock strength—as reflected in the shear modulus. The critical crack height, H_c, may be defined as the maximum vertical extent of a buoyancy-driven fluid-filled crack that is embedded in an elastic solid. In the Weertman (1971a, 1971b) relations,

$$H_c = 2\,[2\,V\mu\,/\,\pi\,(1 - \nu) \cdot g(\rho_{is} - \rho_m)]^{1/3}, \tag{5}$$

where μ is the shear modulus, ν is the Poisson's ratio, V is the volume of magma per unit crack height, g is the gravitational acceleration constant, and ρ_{is} and ρ_m are the *in situ* and magma densities, respectively. As is evident in (5), low shear moduli and sizable *in situ* density contrasts promote the development of modest crack heights, as do modest or narrow dike and vein widths. When $H \geq H_c$, the crack tip becomes unstable, and fracture at the leading (upper) edge induces upward melt flow into the newly created fracture opening. Finite rock strength and conservation of mass within the fluid then promote the withdrawal of magma from the crack tail, and the lower fracture surfaces move toward each other and close. This process promotes an increment of magma ascent.

Vein and dike widths relevant to upper mantle melt migration are only roughly constrained at present, but scale with the overall fracture height and tend to be maximized as $H \rightarrow H_c$. Vein widths observed in the clinopyroxenite and basaltic glass infillings of fractures in the xenoliths of the Cima volcanic field, California (Fig. 5), are variable (Wilshire, 1990), but about 20 mm may be taken as their approximate upper limit. Toward the other end of the width range are feldspathic and gabbroic veins and dikes observed in upper mantle peridotites. Typically these are in the 10- to 20-cm range, but widths may also range up to ≈100 cm (e.g., Nicolas, 1986, 1990). Within a reconstructed subcontinental setting, observations of feldspathic veins and gashes in the spinel lherzolite massif of Liguria are $\approx 1 - \approx 3$ cm in width, for example, while the Lanzo massif of northwestern Italy has clustered feldspathic veins of comparable width (Nicolas, 1986). The critical fracture height for average vein widths of 20 mm is $H_c \approx 480$ m [as estimated by Weertman's (1971a, 1971b) dislocation theory], whereas the upper end of the dike widths ($\approx$95 cm) produces a critical fracture height of $H_c \approx 10.6$ km. Hence, about 11 km appears to be a rough upper limit to the (completely fluid-connected) vertical extent of fracture-bound upward-migrating basaltic magma beneath the western United States. Aspects of the positively buoyant fluid ascent process have been outlined by Secor and Pollard (1975) and by Pollard (1976) with application to geothermal fluids. Lister (1990) and Lister and Kerr (1991) have used similarity solutions to discuss the propagation of positively buoyant and neutrally buoyant fractures, and Takada (1990; and this volume) and Heimpel and Olson (this volume) have used gelatin modeling to study the mechanics of ascending fluid-filled cracks with application to the process of magma ascent. Nakashima (1993) provides a recent application of Weertman's (1971a, 1971b) approach to the fracture-assisted migration of metamorphic fluids.

Considering the region of primary underplating to be a laterally extensive layered gabbro complex that surmounts the peridotites of the uppermost mantle and is incrementally replenished by the rising buoyancy-driven fractures discussed previously, permits a consideration of this region as a bimodulus elasticity transition zone. In this section we ask if the concept is plausible. In the following sections we ask how it works mechanically, and if the information gained can lead to greater insights into how regions of underplating can nucleate, and why rising fractures may have difficulty penetrating the Moho. These sections make use of a review and reapplication of the work of Morita *et al.* (1988). Note in particular that the buoyancy contrasts imparted by vertical gradients in *in situ* density are a fundamental part of the Moho penetration story, but are reserved for a separate discussion and not treated here.

Contributors to contrasting aggregate elastic moduli on either side of the transition zone include changes in mineralogy, gradients in melt fraction, changes in the melt + crystal microtexture in partially consolidated portions of the gabbro complex (=the residual melt aspect ratios), and gradients in temperature. The transition zone itself will be defined by the length and steepness of these gradients, and by the dominant source of the modulus defect.

Fountain (1976) and Fountain and Christensen (1989) review the compressional wave velocity structure of the upper mantle–lower crust transition of the Ivrea-Verbano zone of northern Italy and beneath the western United States. In addition, Furlong and Fountain (1986) have computed, for example, the Vp (Z) for olivine gabbro and quartz tholeiite compositions through the gabbro-garnet eclogite transition region. For all rock types considered, an appreciable (≈ 7.4 km $\cdot$ s$^{-1} \leq$ Vp $\leq$ ≈ 8.1 km $\cdot$ s^{-1}) range in velocity occurs. Recollecting that the compressional wave velocity of basaltic melt is 2.6 km $\cdot$ s^{-1}, and the Voigt-Reuss-Hill (VRH)-averaged Vp of olivine (Fo_{100}) is 8.59 km $\cdot$ s^{-1} (Graham and Barsh, 1969), while clinopyroxene is 7.22 km $\cdot$ s^{-1} (Aleksandrov *et al.*, 1964) and Plagioclase (An_{56}) is 6.70 km $\cdot$ s^{-1} (Aleksandrov and Ryzhova, 1962; Ryzhova, 1964), a consideration of weighted-average mixing model approaches to constructing polycrystalline–polyphase bulk estimates of Vp for the partially molten margins of deep gabbroic intrusions will not produce unique inversions of melt content. In addition, the aggregate bulk (K^*) and shear (μ^*) moduli (and thus the bulk Vp and Vs) are known to be sensitively dependent on melt aspect ratios (Walsh, 1969; Ryan, 1980). Despite these general levels of uncertainty, dramatic reductions in the aggregate Young's Modulus, E^*, as a function of the melt percent and melt aspect ratio have been shown for melt-weakened basalt (Ryan, 1980), and broadly comparable weakening is expected for semi-consolidated gabbro near the Mohorovičić discontinuity. Therefore both composition and melt fraction conspire with melt geometry to make the transition from the peridotites of the upper mantle to the gabbroic cumulates of the lower crust a composite material interface.

Penetration of the Mantle–Crust Transition Zone by a Buoyancy-Driven Crack

A rising buoyancy-driven crack does work on its environment in an effort to reduce the gravi-

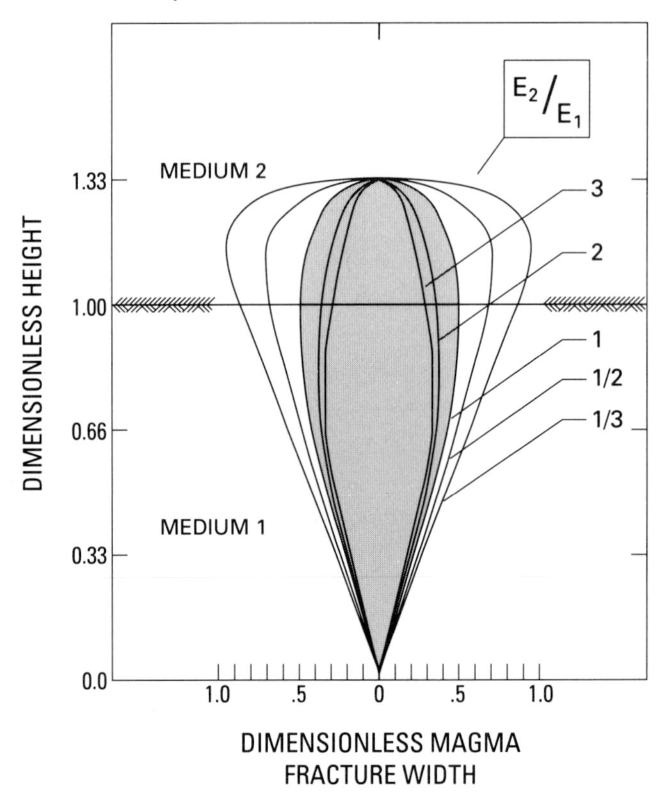

Figure 12 Dimensionless magma fracture width vs dimensionless fracture height (B/h^*) for a buoyancy-driven crack penetrating a bimodulus interface. The interface is defined by the ratios of the Young's moduli (E_2/E_1) above and below the transition region (modified after Morita *et al.*, 1988, and Ryan 1993, published by the American Geophysical Union).

tational potential energy induced through the magma/country rock density contrasts. This work occurs through crack opening displacements, and in geologic sections, it correlates with the resulting dike widths. For rising fractures that cross the mantle–crust transition zone, the bielastic interface of Fig. 12 will be approximately horizontal and normal to the crack centerline. The outward displacements along the fracture length determine the cross-sectional crack profile, and thus the overall geometry. For given levels of magma pressure, the magnitudes of the displacements are a sensitive function of local aggregate elastic moduli—for fixed remote boundary conditions. For the case of elastic isotropy ($E_2/E_1 = 1$) the fracture width is given by

$$W = \frac{4\,(1 - \nu^2)\,h^*}{E} \cdot (\Delta G h^*)\,\frac{\eta}{2}\sqrt{1 - \eta^2}. \qquad (6)$$

In (6), E is the (isotropic) Young's modulus, ΔG is the difference between the magma fracture pressure gradient and the *in situ* stress gradient, h^* is the fracture semiheight, $\eta = Z/h^*$ is the *dimensionless* fracture semiheight, and ν is the Poisson's ratio. In Fig. 12, the isotropic case is shaded, and the nonisotropic cases are given by the nonshaded profiles. These profiles are determined by the ratios of the transition zone elastic anisotropy (e.g., $E_2/E_1 = 3$). The end members of this grouping are profiles $\frac{1}{3}$ and 3. Profile $\frac{1}{3}$ is produced by the crack-tip widening induced by softer material above the transition. Profile 3, however, shows relative crack-tip constriction, induced by the now higher modulus rock above. In Figs. 12, 13, and 14, the expressions for the isotropic cases are given by expressions (6), (7), and (8), respectively, whereas the nonisotropic curves have been computed by a finite element approach (Morita *et al.*, 1988).

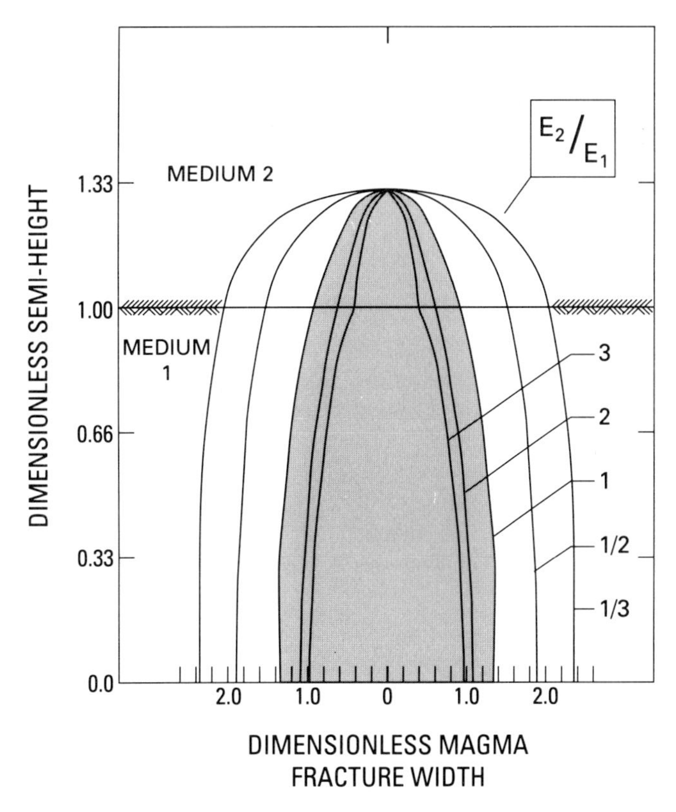

Figure 13 Dimensionless magma fracture width vs dimensionless semiheight (B/h^*) for a constant fluid pressure-driven crack expansion across a bimodulus interface. Curves describe the cross-sectional profile over the range of Young's modulus ratios: $\frac{1}{3} \leq (E_2/E_1) \leq 3$ (modified after Morita *et al.*, 1988).

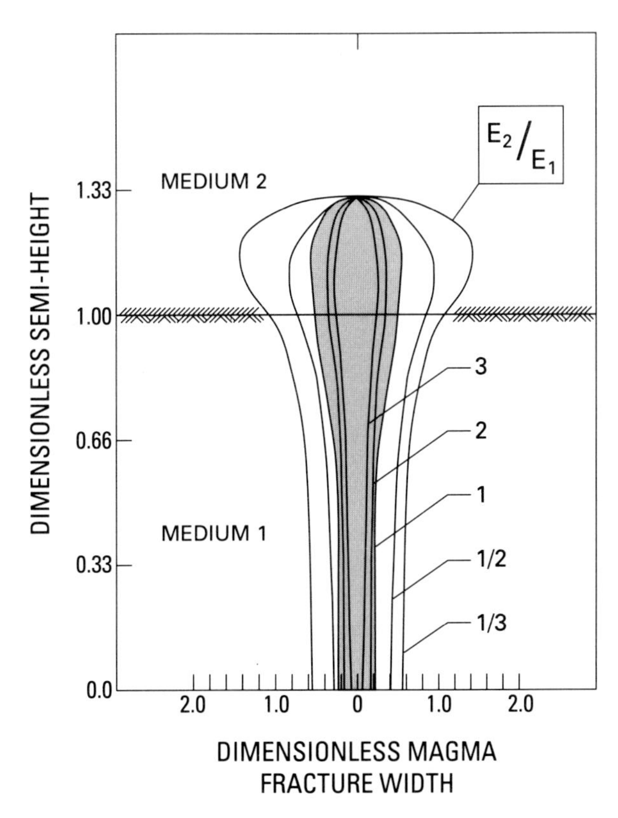

Figure 14 Dimensionless magma fracture width vs dimensionless semiheight (B/h^*) for a constant fluid pressure crack with differences in boundary loads in media 1 and 2 (modified after Morita *et al.*, 1988).

Penetration of the Mantle–Crust Transition Zone by a Constant Fluid Pressure Crack

Magma pressure is a function of depth, h, beneath the Earth's surface, $P = \rho g h$, and increases with depth h, in surroundings of higher density. Over *restricted* vertical intervals, however, the magma pressure within a fracture may be idealized as approximately constant, and the parametric response of the crack wall displacements (the crack profile) can be viewed as a function of the extent of elastic anisotropy. Thus, if one focuses on cracks that penetrate and straddle the Moho, but have a yet rather restricted vertical extent, the ratios of the elastic moduli above and below the crust–mantle interface can be examined for their contributions to crack-tip swelling.

The crack tip itself is embedded in rock with an aggregate Young's modulus E_2, whereas the crack flanks are contained in material of modulus E_1 and have a semiheight B. The fracture width for the special case of elastic isotropy ($E_2/E_1 = 1$) is given by

$$W = \frac{4\,(1 - \nu^2)\, h^* P}{E} \sqrt{1 - \eta^2}, \tag{7}$$

where the approximately uniform magma pressure is denoted by P, and the other symbols have been given previously. In Fig. 13, the isotropic profile is shaded, whereas the anisotropic profiles are not. For a given set of far-field boundary conditions, fluid pressure and elastic moduli have competing effects, as expected. Higher elastic moduli favor crack profile constriction, whereas higher fluid pressures produce crack enlargements. For profile $\frac{1}{3}$, magma rising from the higher modulus upper mantle has penetrated the lower modulus fluid-weakened rocks above, and significant crack-tip widening has occurred. The widening at the crack-tip region, if substantial, may promote the further enlargement of the fracture in the higher modulus rocks below, but this will occur only when there is complete mechanical bonding across the interface, that is, no lateral slippage. In profile 3, the extreme crack-tip position is relatively pinched, whereas the crack flanks below fail to achieve the dilation of profile $\frac{1}{3}$ due to the assumed material bonding across the transition zone. Note that the requirement of complete material bonding across the interface prevents discontinuities in lateral displacement (horizontal slippage). Rubin (1993) computes dike aspect ratios for fluid-pressured cracks embedded in isotropic Maxwell viscoelastic media.

Penetration of a Material Transition Zone by a Fluid-Pressurized Crack with Differential Boundary Loading

As magma-driven fractures rise through the uppermost mantle and penetrate the base of the crust, they may enter environments where substantially different gradients in the horizontal component of confining pressure, $\Delta\sigma_H$, exist. In many ways this is expected, since the transition separates an environment dominated by the density of harzburgites and less depleted peridotites (densities: $\rho \approx 3.2$–$3.3\ \mathrm{g \cdot cm^{-3}}$) from layered gabbros (densities: $\rho \approx 2.8$–$2.9\ \mathrm{g \cdot cm^{-3}}$) and their associated cumulates. If we idealize this transition as a step change (a reduction) in lateral confining pressure that is applied normal to the crack walls as one moves vertically upward across the transition, the relative roles of elastic anisotropy can be explored in terms of influences on the crack-tip profiles. For the special case of elastic isotropy, the fracture width is given by

$$\begin{aligned} W = \frac{4\,(1 - \nu^2)\, h^* \Delta\sigma_H}{E} \Big[& \sqrt{1 - \eta^2} \\ & - \frac{2}{\pi} \sqrt{1 - \eta^2}\, \sin^{-1}\alpha \\ & + \frac{\alpha}{\pi} \Big(\mathrm{Log} \left| \frac{\beta - 1}{\beta + 1} \right| \\ & - \frac{\eta}{\alpha}\, \mathrm{Log} \left| \frac{\beta - \eta/\alpha}{\beta + \eta/\alpha} \right| \Big) \Big], \end{aligned} \tag{8}$$

where $\alpha = B/h^*$ is the ratio of medium 1 (crack-flank) height to the total fracture semiheight, and $\beta = \sqrt{(1 - \eta)^2 / (1 - \alpha^2)}$. Expressed in other terms, $\Delta\sigma_H$ is the difference in the horizontal components of boundary loads in the crack-tip and crack-flank regions. In Fig. 14, the isotropic case has been shaded. For *all* profiles, the sudden reduction to lower values of σ_H in the crack-tip region promotes some level of swelling just behind the crack terminus. When these lower levels of crack-tip confining stress are coupled with appreciably lower crack-tip elastic moduli (as illustrated by profile $\frac{1}{3}$), the crack-tip region experiences considerable swelling throughout "medium 2." However, as also illustrated by profile $\frac{1}{3}$, the crack flanks beneath (in medium 1) experience significant reductions in width: a result of the higher modulus ultramafic lithologies beneath.

The three crack-tip deformation modes illustrated above suggest that the penetration of the semi-consolidated margins of a gabbroic body may promote significant crack-tip widening. This, in turn, suggests potential sets of (*i*) loading conditions; (*ii*) boundary conditions, and (*iii*) material property mixtures that may ultimately stimulate still nucleation. It has been long appreciated that the *in situ* density reductions of the upper mantle–lower crust transitions promote the stagnation of rising magmas by reducing the rock-melt density contrast. The crack-tip swelling evident in this discussion suggests an additional mechanism for modulating ascent.

Closing Discussion

Geophysical and petrological data have suggested the existence of partial melting beneath the western United States. Shankland and Waff (1977), for example, have estimated 5–8 vol% melt for the relatively high electrical conductivity of 0.2–0.4 $S \cdot m^{-1}$ beneath the western United States, using numerical conductivity calculations and the effective medium theory for a model rock consisting of a basaltic partial melt within a dominantly olivine matrix. Their model melt fractions are comparable to our present results. The high electrical conductivities in the crust beneath the Basin and Range province (Klein, 1991) and the Western and High Cascades (Jiracek *et al.*, 1989; Stanley *et al.*, 1990) have also suggested the presence of melts and/or hydrous solutions. From the observations of topography, regional seismology, heat flow, electromagnetic structure, petrochemistry, and the stress orientations within the western United States, Gough (1984) concluded that there is widespread partial melting in the upper mantle (at a depth of roughly 150 km) with mantle upflow inferred beneath. Wilshire (1990) has reported extensive *in situ* partial melting in xenoliths brought to the surface of the Basin and Range province. Partial melting has been commonly observed along grain boundaries in mantle xenoliths from southern British Columbia (e.g., Brearley *et al.*, 1984; Canil *et al.*, 1987).

Relatively high heat flow in the Basin and Range province suggests Moho temperatures close to or exceeding the lower crustal solidus (Morgan and Gosnold, 1989). High heat flow (>100 $mW \cdot m^{-2}$) in the Battle Mountain high in northern Nevada and extending into the Snake River Plain of Idaho implies hypersolidus temperatures in the lower crust (e.g., Lachenbruch and Sass, 1978). Mantle-derived magmas appear to be the source of the high heat flow in the active arc volcanism of the Cascade Range (e.g., Morgan and Gosnold, 1989). From gravity gradient data and the width of the heat flow transition zone in the Cascades, Blackwell *et al.* (1982) have postulated the presence of a hot, low density region of about 60 km in width in the crust below a depth of 7 to 10 km, suggesting, in turn, the potential for a zone of temporary residence for magmas derived from the subduction zone. This inference has been challenged by Blakely (1994) who shows that the source depth of the western Cascades gravity gradient can not be deeper than 2.5 km. Magmatic additions to the crust due to extension are suggested for the source of the high heat flow values in the Salton Trough (Lachenbruch *et al.*, 1985; Morgan and Gosnold, 1989). High heat flow values in young calderas such as Yellowstone, Long Valley, and the Valles caldera have been thought to be the consequences of subjacent mantle magmatism. Upward convective heat transfer into the crust by mantle-derived magmas has commonly been indicated from high heat flow geotherms in the western United States (e.g., Blackwell, 1978; Morgan and Gosnold, 1989).

A high seismic attenuation (Q_p = 40) in the upper crust beneath northeastern Yellowstone indicates unusual hydrothermal activity, a shallow magmatic presence, and/or a relatively porous steam-saturated body (Clawson *et al.*, 1989). The presence of a magma chamber has been indicated in the crust beneath Long Valley caldera, California, from, among other evidence, the lack of S-wave transmission by shallow earthquakes (Ryall and Ryall, 1981) and from the relatively low P-wave velocities (Dawson *et al.*, 1990). Beneath the Valles caldera, New Mexico, a region of magma storage is indicated from relatively low P-wave velocities over the depth range 6 to 14 km (Roberts *et al.*, 1991). Detailed P-wave velocity studies by Iyer *et al.* (1990) have shown a magma chamber (7% below normal velocities) of 10 km width extending from 7 to 28 km depth beneath the western half of the Long Valley caldera, and a small magma chamber of a few cubic kilometers in the upper crust beneath the Newberry volcano, Oregon. The lower crust of the southwestern Basin and Range province has low velocities, despite the observation that lower crustal xenoliths have mafic compositions and high densities (Wilshire, 1990). This low velocity again indicates the presence of partial melt, high temperatures, and (fluid-filled) fracture systems. Low Q values indicating magmatic activity in the crust have also been reported beneath the Rio Grande rift (Carpenter and Sanford, 1985) and the Coso hot springs region (Young and Ward, 1980).

Beneath the Cascade volcanoes, a high conductivity layer on the top of the descending Juan de Fuca plate has been observed and lends itself to the inference of water-charged sediments that are partially subducted with the oceanic crust (e.g., Wannamaker *et al.,* 1989; Young and Kitchen, 1989; Kurtz *et al.,* 1990). Hydrous emanations from the subducted sediments have been found to be consistent with a vertical conductive zone from the slab to the mantle wedge at ~15–40 km depth beneath the Willamette Basin (Young and Kitchen, 1989), and by a subhorizontal conductor from the Willamette Basin to the Cascade Range at about 30 km depth (Wannamaker *et al.,* 1989).

Materials that ascend adiabatically in the Earth's mantle have a potential temperature T_p at the Earth's surface of about 1300°C for the normal mantle (e.g., McKenzie and Bickle, 1988). T_p of an upper mantle diapir is 200–300°C higher than this normal value (Courtney and White, 1986; Wyllie, 1988). This is compatible with our estimate from Fig. 8A, which is T_p = 1650°C, and is 1590°C if one uses an adiabatic gradient of 0.6°C · km^{-1}. Hot mantle upflow may be a possible cause of the high temperatures and partial melt beneath the WM and IM regions. The melt fractions to ~10 vol% are also compatible with those of an ascending diapir experiencing decompression melting. The basaltic products of the supersolids intervals resolved in this study migrate toward the Moho as positively buoyant magma-pressurized fractures. In geologic sections, they form veins and deep dikes. Ascending magma fractures penetrate the Moho and enter the basal sections of semi-consolidated gabbroic intrusives that may be mechanically treated as a composite material interface. Significant crack-tip swelling for dikes that penetrate this interface suggests that dike-tip inflation processes may act as nuclei for sill initiation in regions of basaltic underplating.

Acknowledgments

The authors thank C. Herzig, I.S. Sacks, T. Watanabe, E. Ito, and E. Nakamura for valuable discussions on the subject of this chapter. The chapter benefited greatly from careful and thoughtful reviews by David D. Blackwell (Southern Methodist University), Arthur H. Lachenbruch, and John H. Sass (U.S. Geological Survey). This work was partially supported by Grant 05231221 from the Ministry of Education, Science and Culture, Japan, and by the U.S. Geological Survey's Geothermal Research Program. Robert Luedke helped with original photobases for select figures. Mary Woodruff prepared the typescript.

References

Aleksandrov, K. S., and Ryzhova, T. V. (1962). Elastic properties of rock-forming minerals, 3, Feldspars, Bull. Acad. Sci., USSR Geophys. Ser., 2, 1129.

Aleksandrov, K. S., Ryzhova, T. V., and Belikov, B. P. (1964). The elastic properties of pyroxenes, *Sov. Phys.-Crystallog.,* **8,** (5), 589–591.

Al-Khatib, H. H., and Mitchell, B. J. (1991). Upper mantle anelasticity and tectonic evolution of the western United States from surface wave attenuation, *J. Geophys. Res.* **96,** 18129–18146.

Anderson, D. L., Menahem, A. B., and Archambeau, C. B. (1965). Attenuation of seismic energy in the upper mantle, *J. Geophys. Res.* **70,** 1441–1448.

Bechtel, T. D., Forsyth, D. W., Sharpton, V. L., and Grieve, R. A. F. (1990). Variations in effective elastic thickness of the North American lithosphere, *Nature* **343,** 636–638.

Beeré, W. (1975). A unifying theory of the stability of penetrating liquid phases and sintering pores, *Acta Metall.,* **23,** 131–138.

Benz, H. M., Zandt, G., and Oppenheimer, D. H. (1992). Lithospheric structure of northern California from teleseismic images of the upper mantle, *J. Geophys. Res.* **97**, 4791–4807.

Bergantz, G. W., and Dawes, R. (1994). Aspects of magma generation and ascent in continental lithosphere, *in* "Magmatic Systems" (M. P. Ryan, ed.), Academic Press, San Diego.

Blackwell, D. D. (1978). Heat flow and energy loss in the western United States, *in* "Cenozoic Tectonics and Regional Geophysics of the Western Cordillera" (R. B. Smith and G. P. Eaton, eds.), Memoir 152, pp. 175–208, Geol. Soc. Am., Boulder, Co.

Blackwell, D. D., Steele, J. L., and Carter, L. S. (1991). Heat-flow patterns of the North American continent: A discussion of the geothermal map of North America, *in* "Neotectonics of North America" (D. B. Slemmons, E. R. Engdahl, M. D. Zoback, and D. D. Blackwell, eds.), Decade Map Vol. 1, pp. 423–436, Geol. Soc. Am., Boulder, Co.

Blackwell, D. D., Hull, D. A., Bowen, R. G., and Steele, J. L. (1978). "Heat flow in Oregon," Oreg. Dep. Geol. Miner. Ind. Spec. Paper 4.

Blackwell, D. D., Steele, J. L., Kelly, S., and Korosec, M. A. (1990). Heat flow in the State of Washington and thermal conditions in the Cascade Range, *J. Geophys. Res.* **95,** 19495–19516.

Blackwell, D. D., Bowen, R. G., Hull, D. A., Riccio, J., and Steel, J. L. (1982). Heat flow, arc volcanism, and subduction in northern Oregon, *J. Geophys. Res.* **87,** 8735–8754.

Blakely, R. J. (1994) Extent of partial melting beneath the Cascade Range, Oregon: Constraints from gravity anomalies and ideal-body theory. *J. Geophys. Res.* **99,** 2757–2773.

Brearley, M., Scarfe, C. M., and Fujii, T. (1984). The petrology of ultramafic xenoliths from Summit Lake, near Prince George, British Columbia, *Contrib. Mineral. Petrol.* **88,** 53–63.

Canil, D., Brearley, M., and Scarfe, C. M. (1987). Petrology of ultramafic xenoliths from Rayfield River, south-central British Columbia, *Can. J. Earth Sci.* **24,** 1679–1687.

Carpenter, P. J., and Sanford, A. R. (1985). Apparent Q for upper crustal rocks of the central Rio Grande rift, *J. Geophys. Res.* **90,** 8661–8674.

Chan, W. W., Sacks, I. S., and Morrow, R. J. (1989). Anelasticity of the Iceland Plateau from surface wave analysis, *J. Geophys. Res.* **94,** 5675–5688.

Chapman, D. S., and Pollack, H. N. (1977). Regional geotherms and lithospheric thickness, *Geology* **5,** 265–268.

Cheadle, M. J. (1989). Properties of texturally equilibrated two-phase aggrevates. Ph.D. dissertation, University of Cambridge.

Clawson, S. R., Smith, R. B., and Benz, H. M. (1989). P wave attenuation of the Yellowstone caldera from three-dimensional inversion of spectral decay using explosion source seismic data, *J. Geophys. Res.* **94,** 7205–7222.

Courtney, R. C., and White, R. S. (1986). Anomalous heat flow and geoid across the Cape Verde Rise: Evidence for dynamic support from a thermal plume in the mantle, *Geophys. J.R. Astron. Soc.* **87,** 815–867.

Dawson, P. B., Evans, J. R., and Iyer, H. M. (1990). Teleseismic tomography of the compressional wave velocity structure beneath the Long Valley region, California, *J. Geophys. Res.* **95,** 11021–11050.

Der, Z. A., McElfresh, T. W., and O'Donnell, A. (1982). An investigation of the regional variations and frequency dependence of anelastic attenuation in the mantle under the United States in the 0.5–4 Hz band, *Geophys. J.R. Astron. Soc.* **69,** 67–99.

Dickinson, W. R., and Snyder, W. S. (1979). Geometry of subducted slabs related to San Andreas transform, *J. Geology* **87,** 609–627.

Didwania, A. K., and Homsy, G. M. (1981). Flow regimes and flow transitions in liquid fluidized beds. *Int. J. Multiphase Flow* **7,** 563–580.

Drew, D. A. (1983). Mathematical Modeling of two-phase flow. *Annu. Rev. Fluid Mech.,* **15,** 261–291.

Evans, J. R., and Sacks, I. S. (1979). Deep structure of the Iceland Plateau, *J. Geophys. Res.* **84,** 6859–6866.

Fenneman, N.M. (1928). "Physical divisions of the United States," national atlas map, sheet 59, scale 1:17,000,000, U.S. Geol. Surv., Washington, DC.

Finnerty, A. A., and Boyd, F. R. (1984). Evaluaton of thermobarometers for garnet peridotites, *Geochim. Cosmochim. Acta* **48,** 15–27.

Fountain, D. M. (1976). The Ivrea–Verbano and Strona–Ceneri zones, Northern Italy: A cross-section of the continental crust—New evidence from seismic velocities of rock samples, *Tectonophysics,* **33,** 145–165.

Fountain, D. M., and Christensen, N. I. (1989). Composition of the continental crust and upper mantle: A review, *in* "Geophysical Framework of the Continental United States" (L. C. Pakiser and W. D. Mooney, eds.), Memoir 172, Geological Society of America, Boulder, Co.

Fujii, N., Osamura, K., and Takahashi, E. (1986). Effect of water saturation on the distribution of partial melt in the olivine-pyroxene-plagioclase system. *J. Geophys. Res.,* **91,** 9253–9259.

Furlong, K. P., and Fountain, D.M. (1986). Continental crustal underplating: Thermal considerations and seismic–petrologic consequences, *J. Geophys. Res.* **91,** 8285–8294.

Gosnold, W. D. (1990). Heat flow in the Great Plains of the United States, *J. Geophys. Res.* **95,** 353–374.

Gough, D. I. (1984). Mantle upflow under North America and plate dynamics, *Nature* **311,** 428–433.

Graham, E. K., and Barsch, G. (1969). Elastic constants of single-crystal forsterite as a function of temperature and pressure, *J. Geophys. Res.* **74,** 5949–5960.

Green, D. H., and Ringwood, A. E. (1967). The stability fields of aluminous pyroxene peridotite and garnet peridotite and their relevance in upper mantle structure, *Earth Planet. Sci. Lett.* **3,** 151–160.

Griggs, D. T. (1972). The sinking lithosphere and the focal mechanism of deep earthquakes, *in* "The Nature of the Solid Earth" (E. C. Robertson, ed.), pp. 361–384, McGraw–Hill, New York.

Halderman, T. P., and Davis, P. M. (1991). Q_p beneath the Rio Grande and east African rift zones, *J. Geophys. Res.* **96,** 10113–10128.

Hamilton, W. B. (1989). Crustal geologic processes of the United States, *in* "Geophysical framework of the Continental United States" (L. C. Pakiser and W. D. Mooney, eds.), Memoir 172, Geological Society of America, Boulder, Co.

Harris, R. A., Iyer, H. M., and Dawson, P. B. (1991). Imaging the Juan de Fuca plate beneath southern Oregon using teleseismic P wave residuals, *J. Geophys. Res.* **96,** 19879–19889.

Hearn, T., Beghoul, N., and Barazangi, M. (1991). Tomography of the western United States from regional arrival times, *J. Geophys. Res.* **96,** 16369–16381.

Heimpel, M., and Olson, P. (1994). Buoyancy-driven fracture and magma transport through the lithosphere: Models and experiments, *in* "Magmatic Systems" (M. P. Ryan, ed.), Academic Press, San Diego.

Herrin, E. (1972). A comparative study of upper mantle models: Canadian shield and Basin and Range provinces, *in*

"The Nature of the Solid Earth" (E. C. Robertson, ed.), pp. 216–231, McGraw–Hill, New York.

Holbrook, W. S. (1990). The crustal structure of the northwestern Basin and Range province, Nevada, from wide-angle seismic data, *J. Geophys. Res.* **95,** 21843–21869.

Hughes, S. S. (1990). Mafic magmatism and associated tectonism of the central High Cascade Range, Oregon, *J. Geophys. Res.* **95,** 19623–19638.

Humphreys, E. D., and Clayton, R. W. (1990). Tomographic image of the southern California mantle, *J. Geophys. Res.* **95,** 19725–19746.

Humphreys, E. D., and Hager, B. H. (1990). A kinematic model for the late Cenozoic development of southern California crust and upper mantle, *J. Geophys. Res.* **95,** 19747–19762.

Ito, E., and Sato, H. (1992). Effect of phase transformations on the dynamics of the descending slab, *in* "High-Pressure Research: Application to Earth and Planetary Sciences" (Y. Syono, and M. H. Manghnani, eds.), pp. 257–262. Terra Sci. Publ. Co., Tokyo.

Iyer, H. M., Evans, J. R., Dawson, P. B., Stauber, D. A., and Achauer, U. (1990). Differences in magma storage in different volcanic environments as revealed by seismic tomography: silicic volcanic centers and subduction-related volcanoes, *in* "Magma Transport and Storage" (M. P. Ryan, ed.), pp. 293–316, Wiley, Chichester/Sussex, England.

Jiracek, G. R., Curtis, J. H., Ramirez, J., Martinez, M., and Romo, J. (1989). Two-dimensional magnetotelluric inversion of the EMSLAB Lincoln line, *J. Geophys. Res.* **94,** 14145–14151.

Kampfmann, W., and Berckhemer, H. (1985). High temperature experiments on the elastic and anelastic behaviour of magmatic rocks, *Phys. Earth Planet. Inter.* **40,** 223–247.

Kissling, E. (1988). Geotomography with local earthquake data, *Rev. Geophys.* **26,** 659–698.

Klein, D. P. (1991). Crustal resistivity structure from magnetotelluric soundings in the Colorado Plateau–Basin and Range provinces, central and western Arizona, *J. Geophys. Res.* **96,** 12313–12331.

Kurtz, R. D., DeLaurier, J. M., and Gupta, J. C. (1990). The electrical conductivity distribution beneath Vancouver Island: A region of active plate subduction, *J. Geophys. Res.* **95,** 10929–10946.

Lachenbruch, A. H., and Sass, J. H. (1977). Heat flow in the United States and the thermal regime of the crust, *in* "The Earth's Crust" (J. G. Heacock, ed.), pp. 626–675, Geophys. Monogr. 20, Am. Geophys. Union, Washington, DC.

Lachenbruch, A. H., and Sass, J. H. (1978). Models of an extending lithosphere and heat flow in the Basin and Range province, *in* "Cenozoic Tectonics and Regional Geophysics of the Western Cordillera"(R. B. Smith, and G. P. Eaton, eds.), pp. 209–250, Memoir 152, Geol. Soc. Am., Boulder, Co.

Lachenbruch, A. H., and Sass, J. H. (1980). Heat flow and energetics of the San Andreas fault zone, *J. Geophys. Res.* **85,** 6185–6222.

Lachenbruch, A. H., Sass, J. H., and Galanis, S. P., Jr. (1985). Heat flow in southernmost California and the origin of the Salton Trough, *J. Geophys. Res.* **90,** 6709–6736.

Lay, T., and Wallace, T. C. (1988). Multiple ScS attenuation and travel times beneath western North America, *Bull. Seismol. Soc. Am.* **78,** 2041–2061.

Levander, A. R., and Kovach, R. L. (1990). Shear velocity structure of the northern California lithosphere, *J. Geophys. Res.* **95,** 19773–19784.

Lister, J. R. (1990). Buoyancy-driven fluid fracture: Similarity solutions for the horizontal and vertical propagation of fluid-filled cracks, *J. Fluid Mech.* **217,** 213–239.

Lister, J. R., and Kerr, R. C. (1991). Fluid-mechanical models of crack propagation and their application to magma transport in dikes. *J. Geophys. Res.* **96,** 10049—10077.

Luedke, R. G., and Smith, R. L. (1984). Map showing distribution, composition, and age of late Cenozoic volcanic centers in the Western conterminous United States, U.S. Geological Survey, MI Series, MAP I-1523.

Luedke, R. G., and Smith, R. L. (1991). Quaternary volcanism in the western conterminous United States in the Geology of North America, vol. K-2, "Quaternary Nonglacial Geology: Conterminous U.S." (R. B. Morrison, ed.), pp. 75–92, The Geological Society of America.

Mauko, G. M. (1980). Velocity and attenuation in partially molten rocks, *J. Geophys. Res.* **85,** 5173–5189.

McKenzie, D. P. (1984). The generation and compaction of partially molten rock, *J. Petrol.* **25,** 713–765.

McKenzie, D. P. (1985). The extraction of magma from the crust and mantle, Earth and Planet. Sci. Lett. **74,** 81–91.

McKenzie, D., and Bickle, M. J. (1988). The volume and composition of melt generated by extension of the lithosphere, *J. Petrol.* **29,** 625–679.

Mercier, J.-C., and Carter, N. L. (1975). Pyroxene geotherms, *J. Geophys. Res.* **80,** 3349–3362.

Mikhlin, S. G. (1957). "Integral equations" (A. H. Armstrong, transl.), Pergamon Press. New York.

Mooney, W. D., and Braile, L. W. (1989). The seismic structure of the continental crust and upper mantle of North America, *in* "The Geology of North America—An Overview" (A. W. Bally and A. R. Palmer, eds.), pp. 39–52, Geol. Soc. Am., Boulder, Co.

Morgan, P., and Gosnold, W. D. (1989). Heat flow and thermal regimes in the continental United States, *in* "Geophysical Framework of the Continental United States" (L. C. Pakiser and W. D. Mooney, eds.), pp. 493–522, Memoir 172, Geol. Soc. Am., Boulder, Co.

Morgan, P., Seager, W. R., and Golombek, M. P. (1986). Cenozoic thermal, mechanical and tectonic evolution of the Rio Grande rift, *J. Geophys. Res.* **91,** 6263–6276.

Morita, N., Whitfill, D. L. and Wahl, H. A. (1988). Stress-intensity factor and fracture cross-sectional shape predictions from a three-dimensional model for hydraulically induced fractures, *J. Petrol. Technol.* **Oct.,** 1329–1342.

Murase, T., and Fukuyama, H. (1980). Shear wave velocity in partially molten peridotite at high pressures, *Carnegie Inst. Wash. Yrbk.* **79,** 307–310.

Murase, T., and Kushiro, I. (1979). Compressional wave velocity in partially molten peridotite at high pressures, *Carnegie Inst. Wash. Yrbk.* **78,** 559–562.

Muskhelishvilli, N.I. (1953). "Singular Integral Equations" (J. R. M. Radok, transl.), Noordhoff, Groningen, The Netherlands.

Mysen, B. O., and Kushiro, I. (1977). Compositional variations of coexisting phases with degree of melting of peridotite in the upper mantle, *Am. Mineral.* **62,** 843–865.

Nakashima, Y. (1993). Buoyancy-driven propagation of an isolated fluid-filled crack in rock: Implication for fluid transport in metamorphism, *Contrib. Mineral. Petrol.* **114,** 289–295.

Nathenson, M., and Guffanti, M. (1988). Geothermal gradients in the conterminous United States, *J. Geophys. Res.* **93,** 6437–6450.

Nicolas, A. (1986). A melt extraction model based on structural studies in mantle peridotites, *J. Petrol.* **27,** 999–1022.

Nicolas, A. (1990). Melt extraction from mantle peridotites: Hydrofracturing and porous flow with consequences for oceanic ridge activity, *in* "Magma Transport and Storage" (M. P. Ryan, ed.), pp. 159–173, Wiley, Chichester/Sussex, England.

Patton, H. J., and Taylor, S. R. (1984). *Q* structure of the Basin and Range from surface waves, *J. Geophys. Res.* **89,** 6929–6940.

Pollack, H. N., and Chapman, D. S. (1977). On the regional variation of heat flow, geotherms, and lithospheric thickness, *Tectonophysics* **38,** 279–296.

Pollard, D. D. (1976). On the form and stability of open hydraulic fractures in the Earth's crust, *Geophys. Res. Lett.* **3,** 513–516.

Reasenberg, P., Ellsworth, W., and Walter, A. (1980). Teleseismic evidence for a low-velocity body under the Coso geothermal area, *J. Geophys. Res.* **85,** 2471–2483.

Reiter, M., Eggleston, R. E., Broadwell, B. R., and Minier, J. (1986). Estimates of terrestrial heat flow from deep petroleum tests along the Rio Grande rift in central and southern New Mexico, *J. Geophys. Res.* **91,** 6225–6245.

Rinehart, E. J., and Sanford, A. R. (1981). Upper crustal structure of the Rio Grande rift near Socorro, New Mexico, from inversion of microearthquake S-wave reflections, *Bull. Seismol. Soc. Am.* **71,** 437–450.

Rivalenti, G., Garuti, G., Rossi, A., Siena, F. and Sinigoi, S. (1981). Existence of different peridotite types and of a layered igneous complex in the Ivrea zone of the Western Alps, *J. Petrol.* **22,** 127–153.

Roberts, P. M., Aki, K., and Fehler, M. C. (1991). A low-velocity zone in the basement beneath the Valles caldera, New Mexico, *J. Geophys. Res.* **96,** 21583–21596.

Robertson, E. C. (1972). "The Nature of the Solid Earth," McGraw–Hill, New York.

Roy, R. F., Blackwell, D. D., and Decker, E. R. (1972). Continental heat flow, *in* "The Nature of the Solid Earth" (E. C. Robertson, ed.), pp. 506–543, McGraw–Hill, New York.

Rubin, A. M., (1993). Dikes vs. Diapirs in viscoelastic rock. *Earth and Planet. Sci. Lett.* **119,** 641–659.

Ryall, F., and Ryall, A. (1981). Attenuation of P and S waves in a magma chamber in Long Valley caldera, California, *Geophys. Res. Lett.* **8,** 557–560.

Ryan, M. P. (1980). Mechanical behavior of magma reservoir envelopes: Elasticity of the olivine tholeiite solidus, *Bull. Volcanol.* **43,** 743–772.

Ryan, M. P. (1993). Neutral buoyancy and the structure of mid-ocean ridge magma reservoirs, *J. Geophys. Res.* **98,** 22321–322338.

Ryzhova, T.V. (1964). The elastic properties of plagioclase, *Izv. Geophys. Ser.* **7,** 1049–1051.

Sass, J. H., and Lachenbruch, A. H. (1978). Thermal regime of the Australian continental crust, *in* "The Earth, Its Origin, Structure and Evolution" (M. W. McElhinny, ed.), pp. 301–351, Academic Press, New York.

Sass, J. H., Lachenbruch, A. H., Munroe, R. J., Greene, G. W., and Moses, T. H., Jr. (1971). Heat flow in the western United States, *J. Geophys. Res.* **76,** 6376–6413.

Sass, J. H., Blackwell, D. D., Chapman, D. S., Costain, J. K., Decker, E. R., Lawver, L. A., and Swanberg, C. A. (1981). Heat flow from the crust of the United States, *in* "Physical Properties of Rocks and Minerals" (Y. S. Touloukian, W. R. Judd, and R. F. Roy, eds.), pp. 503–548, McGraw–Hill, New York.

Sato, H. (1992). Thermal structure of the mantle wedge beneath northeastern Japan: Magmatism in an island arc from the combined data of seismic anelasticity and velocity and heat flow, *J. Volcanol. Geotherm. Res.* **51,** 237–252.

Sato, H., and Sacks, I. S. (1989). Anelasticity and thermal structure of the oceanic upper mantle: Temperature calibration with heat flow data, *J. Geophys. Res.* **94,** 5705–5715.

Sato, H., and Sacks, I. S. (1990). Magma generation in the upper mantle inferred from seismic measurements in peridotite at high pressure and temperature, *in* "Magma Transport and Storage" (M. P. Ryan, ed.), p. 277–292, Wiley, Chichester/Sussex, England.

Sato, H., Sacks, I. S., and Murase, T. (1989a). The use of laboratory velocity data for estimating temperature and partial melt fraction in the low velocity zone: Comparison with heat flow and electrical conductivity studies, *J. Geophys. Res.* **94,** 5689–5704.

Sato, H., Sacks, I. S., Murase, T., and Scarfe, C. M. (1988). Thermal structure of the low velocity zone derived from laboratory and seismic investigations, *Geophys. Res. Lett.* **15,** 1227–1230.

Sato, H., Sacks, I. S., Murase, T., Muncill, G., and Fukuyama, H. (1989b). Q_p-melting temperature relation in peridotite at high pressure and temperature: Attenuation mechanism and implications for the mechanical properties of the upper mantle, *J. Geophys. Res.* **94,** 10647–10661.

Scarfe, C. M., and Takahashi, E. (1986). Melting of garnet peridotite to 13 GPa and the early history of the upper mantle, *Nature* **322,** 354–356.

Schatz, J. F., and Simmons, G. (1972). Thermal conductivity of earth materials at high temperatures, *J. Geophys. Res.* **77,** 6966–6983.

Sclater, J. G., and Francheteau, J. (1970). The implication of terrestrial heat flow observations on current tectonic and geochemical models of the crust and upper mantle of the earth, *Geophys. J.R. Astron. Soc.* **20,** 509–542.

Sclater, J. G., Jaupart, C., and Galson, D. (1980). The heat flow through oceanic and continental crust and the heat loss of the earth, *Rev. Geophys. Space Phys.* **18,** 269–311.

Secor, D. T., and Pollard, D. D. (1975). On the stability of open hydraulic fractures in the Earth's crust, *Geophys. Res. Lett.* **2,** 510–513.

Shankland, T. J., and Waff, H. S. (1977). Partial melting and electrical conductivity anomalies in the upper mantle, *J. Geophys. Res.* **82,** 5409–5417.

Shervais, J. W. (1979). Thermal emplacement model for the Alpine lherzolite massif at Balmuccia, Italy, *J. Petrol.* **20,** 795–820.

Sleep, N. H. (1988). Tapping of melt by veins and dikes, *J. Geophys. Res.* **93,** 10255–10272.

Smith, R.L., and Luedke, R.G. (1984). Potentially active volcanic lineaments and loci in western conterminous United States, *in* "Explosive Volcanism, Inception, Evolution and Hazards" (F. R. Boyd, ed.), pp. 47–66, National Academy Press, Washington, DC.

Soller, D. R., Ray, R. D., and Brown, R. D. (1982). A new global thickness map, *Tectonics* **1,** 125–149.

Stanley, W. D., Mooney, W. D., and Fuis, G. S. (1990). Deep crustal structure of the Cascade Range and surrounding regions from seismic refraction and magnetotellurgic data, *J. Geophys. Res.* **95,** 19419–19438.

Takada, A. (1990). Experimental study on propagation of liquid-filled crack in gelatin: Shape and velocity in hydrostatic stress condition, *J. Geophys. Res.* **95,** 8471–8481.

Takada, A. (1994). Accumulation of magma in space and time by crack interaction, *in* "Magmatic Systems" (M. P. Ryan, ed.), Academic Press, San Diego.

Takahashi, E. (1986). Melting of a dry peridotite KLB-1 up to 14 GPa: Implications on the origin of peridotitic upper mantle, *J. Geophys. Res.* **91,** 9367–9382.

Takahashi, E., and Kushiro, I. (1983). Melting of a dry peridotite at high pressures and basalt magma genesis, *Am. Mineral.* **68,** 859–879.

Thelin, G. P., and Pike, R. J. (1991). "Landforms of the Conterminous United States," digital shaded relief image, Albers equal area conic projection, scale: 1:3,500,000, MI-MAP I-2206, U.S. Geological Survey.

von Bargen, N., and Waff, H. S. (1988). Wetting of enstatite by basaltic melt at 1350°C and 1.0–2.5 GP_a pressure, *J. Geophys. Res.* **93,** 1153–1158.

Waff, H. S., and Bulau, J. R. (1979). Equilibrium fluid distribution in an ultramatic partial melt under hydrostatic stress conditions, *J. Geophys. Res.* **84,** 6109–6114.

Waff, H. S., and Bulau, J. R. (1982). Experimental determination of near-equilibrium textures in partially molten silicates at high pressures, *in* "High Pressure Research in Geophysics, Vol. 12" (S. Akimoto, and M. H. Manghnani, eds.), pp. 229–236, Advances in Earth and Planetary Sciences, Center for Academic Publications, Tokyo.

Walsh, J. B. (1969). New analysis of attenuation in partially melted rock, *J. Geophys. Res.* **74,** 4333—4337.

Wannamaker, P. E., Booker, J. R., Jones, A. G., Chave, A. D., Filloux, J. H., Waff, H. S., and Law, L. K. (1989). Resistivity cross section through the Juan de Fuca subduction system and its tectonic implications, *J. Geophys. Res.* **94,** 14127–14144.

Weertman, J. (1971a). Theory of water-filled crevasses in glaciers applied to vertical magma transport beneath oceanic ridges, *J. Geophys. Res.* **76,** 1171–1183.

Weertman, J. (1971b). Velocity at which liquid-filled cracks move in the Earth's crust or in glaciers, *J. Geophys. Res.* **76,** 8544–8553.

Wilshire, H. G. (1990). Lithology and evolution of the crust–mantle boundary region in the southwestern Basin and Range province, *J. Geophys. Res.* **95,** 649–665.

Wyllie, P. J. (1988). Solidus curves, mantle plumes, and magma generation beneath Hawaii, *J. Geophys. Res.* **93,** 4171–4181.

Young, C. T., and Kitchen, M. R. (1989). A magnetotelluric transect in the Oregon coast Range, *J. Geophys. Res.* **94,** 14185–14193.

Young, C.-Y., and Ward, R. W. (1980). Three-dimensional Q^{-1} model of the Coso hot springs known geothermal resource area, *J. Geophys. Res.* **85,** 2459–2470.

Zandt, G., and Furlong, K. P. (1982). Evolution and thickness of the lithosphere beneath coastal California, *Geology* **10,** 376–381.

Chapter 13 Aspects of Magma Generation and Ascent in Continental Lithosphere

George W. Bergantz and Ralph Dawes

Overview

The variety of melt-producing reactions and rheological conditions in the crust make it difficult to generalize many aspects of crustal magmatism. Basalt provides both the material and the heat to initiate and sustain magmatism as demonstrated by unequivocal geophysical and geological evidence. Following intrusion of basaltic magma in the deep crust, the quantity of melt produced and its composition are controlled by the presence of hydrous phases; however, it is inappropriate to parameterize the amounts of melt produced by consideration of the modal percentage of the hydrous phases alone. Laboratory and numerical experiments on the heat transfer following hypothetical basaltic underplating indicate that the thermal exchange is largely conductive, in agreement with geological examples. Aggregrate viscosity is the most important physical parameter in controlling melt homogenization and ascent, although there is some ambiguity as to the utility of existing models for magma rheology. Compaction, diapirism, and diking do not appear to be viable ascent mechanisms when considered alone. The observations that midcrustal plutons often occupy crustal scale shear zones and that melt extraction occurs at a variety of scales suggest that models of magmatism require an approach that is not simply cast in terms of simple end member mechanical models. Crustal magmatism might be usefully considered in a geodynamical context where magma segregation is explicitly keyed to tectonic conditions and petrologic diversity is generated and properly understood on a crustal scale.

Notation

		Units
H	Height of underplating model	m
I	Thickness of basalt layer	m
L	Specific latent heat	$J \cdot kg^{-1} \cdot K^{-1}$
Nu	Nusselt number	dimensionless
P	Pressure	$kg \cdot m^{-1} \cdot s^{-2}$
Ste	Stefan number	dimensionless
T	Temperature	K, °C
T_{CR}	Initial country rock temperature	K, °C
T_M	Initial basalt temperature	K, °C
W	Width of underplating model	m
X	Thickness of melt region	m
b	Grain diameter	m, cm
c_p	Specific heat at constant pressure	$J \cdot kg^{-1}$
g	Scalar acceleration of gravity	$m \cdot s^{-2}$
k	Thermal conductivity	$J \cdot s^{-1} \cdot m^{-1} \cdot K^{-1}$
t	Time	s
u	Horizontal component of the velocity	$m \cdot s^{-1}$
w	Vertical component of the velocity	$m \cdot s^{-1}$
x	Horizontal independent variable	m
z	Vertical independent variable	m
β	Thermal expansion coefficient	K^{-1}
γ	Power-law coefficient in viscosity relation	dimensionless
κ	Molecular thermal diffusivity	$m^2 \cdot s^{-1}$
μ	Dynamic viscosity	$kg \cdot m^{-1} \cdot s^{-1}$
μ_r	Reduced dynamic viscosity	dimensionless
ν	Kinematic viscosity	$m^2 \cdot s^{-1}$
ρ_0	Reference density	$kg \cdot m^{-3}$
ϕ	Volume fraction solids	dimensionless
ϕ_M	Limiting volume fraction solids	dimensionless
χ	Porosity	dimensionless

Introduction

A number of models for melt migration and compositional diversity have been constructed for the mantle (McKenzie and Bickle, 1988). Less well understood are the processes that drive petrologic diversity in the crust. The differences lie in that the crustal composition is not readily generalized, nor are the changes in chemical potential that drive melt production well understood. The movement of melt is also more complex, due to the extreme changes in temperature and rheological state in the crust. Thus, the chemical and dynamic generalizations that have permitted some degree of progress in the modeling of processes in the mantle are not widely applicable. Despite the accessibility of crustal geologic features, even first-order answers to classical questions such as the origin and meaning of migmatites or the "room problem" in granite magmatism are not yet resolved. The crustal sections that have been described by metamorphic petrologists and structural geologists are often not like those that would seem to be required from the mass balance and transport calculations of igneous petrologists. The goal of this study is to inventory many aspects of crustal magmatism around a central theme of basaltic injection and underplating. The *geological* expression of magmatism and transport is emphasized. The purpose is to present and discuss the kinds of data that are potentially relevant to the construction of transport models.

The face of magmatism can be complex, and temporal and spatial variability is evident: at Arenal volcano, significant chemical variability is manifested on time scales of 10^1 to 10^3 yr (Reagan *et al.*, 1987), which is in contrast to compositionally monotonous systems such as Mt. Hood, and to Katmai where spatially overlapping vents erupt a variety of magma types with diverse plumbing systems and apparently unrelated eruptive styles (Hildreth, 1987). There is also some question as to the characteristic rates of intrusion and the lifetimes of chemical processes in magmatic systems. Halliday *et al.* (1989), for example, estimate that the silicic magma at Glass Mountain may have had a residence time of 0.7 Myr (also see Mahood, 1990), whereas the models of Reagan *et al.* (1991) and Gill and Condomines (1992) posit transit times from 8,000 to 50,000 yr for melt generation, transport, fractionation, and eruption. Intrusive to extrusive ratios vary as well: from 5:1 in oceanic settings to 10:1 for continents (Crisp, 1984). The physical controls on the variability and evolution of magmatic systems remain poorly characterized: it is not clear what generic feature or index of chemical and physical change is appropriate when trying to compare magmatic systems or generalize the conditions of petrologic diversity. Clearly, broad patterns emerge when the style of intrusion and eruption is indexed to tectonic regime, but within individual centers a diversity of processes is evident.

Petrologic diversity is usually considered to involve crustal melting with variable removal of refractory phases, or crystal fractionation of a mantle-derived melt that may mix with crustally derived components. The importance of crustal sources and the potential contribution of the partial retention of restite in the variations in granitoids are indicated by the presence of complexly zoned zircons (Miller *et al.*, 1992; Paterson *et al.*, 1992), the geochemistry of some enclaves (Chen *et al.*, 1990), and rather remarkable agreement between calculated magma temperatures and plagioclase compositions (Burnham, 1992). These observations indicate that the restite unmixing relations proposed by Chappell and White (1992) and Chappell *et al.* (1987) have a place in petrogenetic schemes. However, major and trace element systematics often require a more complex interpretation for the origin of many plutons and volcanic suites, involving the presence of a mantle component that yields petrologic diversity by fractionation and by providing a mixing component for anatectic melts (DePaolo *et al.*, 1992). The geochemical evidence for a mantle component is discussed in more detail below.

Both scenarios of magma generation require a change in thermal conditions in the crust. One potentially unifying element in the characterization of magmatism comes from the suggestion of Hildreth (1981) that crustal magmatism is fundamentally basaltic. In keeping with the broad interpretation suggested by Hildreth (1981), we do not mean simply a Bowen-style fractional crystallization relation. Instead basalt acts as a heat source for crustal melting and as a chemical reservoir for magma mingling/mixing. It also provides the necessary changes in rheological state to permit extension and large tectonic strains. We explore some elements of the premise that much of the petrologic diversity in magmas originates in the mid- to deep crust where the thermal conditions enhance the likelihood of melt generation and mingling between the basalt and crustal melts.

The term "underplating" has been used to describe the accumulation of basaltic magma in the mid- to deep crust. According to the underplating scenario magma is, initially at least, overlain by upward propagating regions of partial melt and also subject to continued intrusion from below. The Moho is often suggested as the site of basaltic magma accumulation and indeed may be defined by such a process. Underplating has also been used to describe the tectonic accretion of material from a subducting slab, or even the attachment of the slab itself onto the bottom of the overriding plate. Admittedly, this "underplating" nomenclature is potentially confusing; in the rest of our discussion we use underplating only to refer to the interaction of basaltic magma and the crust.

The geological and geophysical evidence for the presence of basaltic magma is as follows:

(*a*) The petrologic diversity of crustal igneous rocks requires basaltic magma as a material source, and just as importantly, as a means of providing the thermal input to generate partial melting and subsequent hybridization of the adjacent crust. In this model basalt is not simply a parent magma, but rather just one of several possible geochemical reservoirs. Numerous petrologic studies of both intrusive and extrusive rocks (Davidson *et al.*, 1990; Feeley and Grunder, 1991; Hildreth *et al.*, 1986; Manduca *et al.*, 1992) have documented the presence of basalt and other geochemical reservoirs that have interacted with the basalt. Repeated intrusion of basalt may lead to partial melting of previous underplates, which would comprise a young and mafic lower crust. Repeated melting of this material could give rise to evolved suites that would be difficult to geochemically distinguish from simple basalt fractionation (Kay *et al.*, 1990; Tepper *et al.*, 1993) if the lower crust is isotopically young. Among the most dramatic examples of the type of interaction of basaltic magma and crust is the basalt–rhyolite association at the Yellowstone Plateau volcanic field (6000 km^3 rhyolite, 100 km^3 basalt). The isotopic character of the eruptive rocks and the absence of intermediate compositions require a large-scale, deep-crustal hybridization between basalt and Archaean crust (Hildreth *et al.*, 1991). The origin of the rhyolites of the Snake River Plain–Yellowstone Plateau province is also consistent with an origin by deep-crustal melting following the intrusion of basalt (Leeman, 1982). Observations of this type provide the key elements of the MASH hypothesis of Hildreth and Moorbath (1988) (melting–assimilation–storage–homogenization), which holds that the "base level" chemical signature in some magmatic systems may be due to intrusion, partial melting, and hybridization of the deep crust by mantle-derived melts.

(*b*) Taken together, the geophysical data such as the high heat flow, uplift, the seismic reflectivity of the lower crust and a well-defined Moho are consistent with an origin by the accumulation of basaltic magma in the mid- and lower crust. The rapid rise time of the very high heat flow in the Basin and Range can be attributed only to a combination of tectonic extension and intrusion of basalt (Lachenbruch and Sass, 1978; Mareschal and Bergantz, 1990). The presence of well-defined subhorizontal reflections and downward-increasing seismic velocities, from 6.8 to 7.8 $km \cdot s^{-1}$, in the lower crust of the Basin and Range is consistent with magmatically underplated rocks (Klemperer *et al.*, 1986; Valasek *et al.*, 1987). Seismic velocity models for the deep-crustal structure in southern Alaska suggest that magmatically emplaced rocks may form part of the "crustal" root (Fuis and Plafker, 1991). Variations in seismically determined crustal thickness in Australia can be attributed to crustal underplating by mantle-derived magmas (Drummond and Collins, 1986). A number of other deep-reflection profiles have features that would permit an interpretation involving basaltic magmatism in the deep crust (Mereu *et al.*, 1989). In one sense, the Moho is a "Stefan" boundary in that its spatial variation and definition represents a degree of freedom in the response of the crust to thermal throughput on a planetary scale (Nelson, 1991).

(*c*) Exposures of the deep crust provide direct examples of magmatic additions to the lower crust with the subsequent generation of zones of partial melting and granulite formation. The best studied of these is the Ivrea–Verbano zone of northern Italy where a crustal column containing mantle peridotite, overlain by layered and homogeneous gabbro and amphibolite facies metasediments, is exposed (Handy and Zingg, 1991; Voshage *et al.*, 1990; Zingg, 1990). The isotopic similarity between the magmatic assemblage and the overlying granitoids, and the positive Eu anomaly in the mafic portions point to a genetic relationship between the two and may be an example of the lower crustal regions of partial melting and mixing.

Quick *et al.* (1992) have mapped pervasive deformation features in the Ivrea–Verbano zone that must have formed under near solidus conditions, yielding fabrics that indicate substantial lateral mass transfer while the melt was in a rheologically "mushy" state, much like textures observed in ophiolite complexes. The compositional variety and volumes of mafic magma suggest that the Ivrea–Verbano zone may represent a very thermally mature system with sustained thermal input. The other end member, perhaps representing the incipient generation of a MASH zone, where the melting of surroundings is minor, may be the Fiambala gabbronorite in Argentina (Grissom *et al.*, 1991). This sill-like body was emplaced at a depth of 21–24 km, synchronous with the local development of granulite facies conditions. The amount of crustal partial melting as a result of the thermal perturbation during intrusion is much less than that observed in the Ivrea–Verbano zone, and like that estimated by Bergantz (1989b) for a single episode of underplating.

(*d*) The occurrence of counterclockwise P–T paths and the coincidence of peak metamorphism and peak temperatures in granulite terrains are consistent with a metamorphic history where crustal thickening occurs at least partly as a result of the addition of magma to the lower crust (Bohlen and Mezger, 1989). An in-depth inventory of the granulite "controversy" is beyond the scope of this paper (the interested reader is directed to the comprehensive volume edited by Vielzeuf and Vidal, 1990). We note however that some granulites clearly have mineral phases consistent with an origin by the generation and removal of partial melt, although there are few examples where the evidence is unequivocal on a regional basis. Incomplete melt removal (Rudnick, 1992) and the uncertainty in the assumptions of trace element partitioning during melting (Bea, 1991; Sawyer, 1991) may render the standard geochemical arguments of little use in discriminating between a restitic and nonrestitic granulite. Nonetheless, the abundance of mafic xenoliths recording pressures of 0.5–1.0 GPa and magmatic or near magmatic temperatures and textures suggest that one component of crustal growth is the ponding of basic magmas at or near the Moho (Cull, 1990; Rudnick, 1990; Rudnick and Taylor, 1987). The isotopic and trace element character of lower crustal xenoliths requires an origin by basaltic intrusion of the lower crust, followed by simultaneous partial melting of overlying crust, which may include the residue of previous episodes of underplating, fractionation of the mafic melt, and the production of melts representing varying degrees of hybridization (Kempton and Harmon, 1992; Rudnick, 1992). Continued magmatic underplating leads to the cratonization of the crust, yielding a crust that has grown from below, perhaps with episodes of delamination of hybridized and cumulate regions into the mantle (Arndt and Goldstein, 1989; Kay and Kay, 1991).

Melt Generation

The primary factors that control melting of the lower crust are (*a*) bulk composition and mineral mode, (*b*) temperature, (*c*) pressure, (*d*) volatiles (primarily H_2O and CO_2), and (*e*) the physical dynamics of the melting process. The first four factors control the chemical potential of the melt phase, and thus directly influence the extent of melting and the melt composition. The fifth factor includes kinetic factors and transport mechanisms that may overlap with the melt segregation process. This section outlines recent progress in understanding the phase equilibria that are likely to be involved in lower crustal melting. We suggest that what is currently known about lower crustal melting is consistent with the following: (*a*) the generation of many common magmatic suites is initiated by intrusion of the crust by mantle-derived mafic magma; (*b*) heat and volatiles from the mafic magmas partially melt proximal lower crust, yielding a major component of resulting magmatic suites.

A good experimental understanding of the haplogranite system is fairly well in hand, but the experiments are only indirectly applicable to melting of natural rocks besides acid granites. As stated by Tuttle and Bowen (1958), lower-crustal melting is likely to be water-undersaturated and involve mica and amphibole in the phase equilibria that control the melting reactions, the water activities in the melts, and melt productivity. It is now generally accepted that crustal melts that mobilize to form voluminous granitoids are water-undersaturated (Burnham, 1967; Clemens, 1984; Clemens and

Vielzeuf, 1987; Powell, 1983; Thompson, 1983). This has led to a focus on so-called "damp melting," in which the only water in the system is that contained in hydrous metamorphic minerals present before melting began.

Free water is unlikely to be present in pore spaces at >15–20 km depths (Yardley, 1986). It may be possible for water-rich fluids to enter the lower crust by exsolution from crystallizing, mantle-derived magmas (Lange and Carmichael, 1990; Wickham and Peters, 1992). Similarly, CO_2 or mixed H_2O–CO_2 fluids released from mafic magmas may play a role in lower-crustal melting although the source of the CO_2-dominated fluids remains to be resolved (Peterson and Newton, 1990). Although appinites and lamprophyres, common in calc-alkaline intrusions, suggest a possible role for H_2O or H_2O–CO_2 fluid-saturated mafic magmas in the generation of calc-alkaline granitoids, little research has been devoted to this topic. The production of plausible granitoid compositions in fluid-absent, lower-crustal melting experiments, and the discovery by Beard and Lofgren (1989) that water-saturated melts of amphibolite, generated at midcrustal pressures, are unlike natural granitoids, are oblique evidence in support of fluid-absent melting as the dominant process in the lower crust.

Recent chemographic and experimental work on lower-crustal melting has focused on four types of systems: model systems (i.e., specific reactions involving a few components and phases), amphibolites (of more or less basaltic composition), common granitoids (e.g., tonalites), and metapelites. Theoretical phase equilibria of model systems provide a basis for understanding the melting behavior of rocks containing hydrous minerals (Brown and Fyfe, 1970; Burnham, 1979; Thompson, 1982; Thompson and Algor, 1977; Thompson and Tracy, 1979; Wyllie, 1977). Reactions in natural rocks generally have greater thermodynamic variance than those in model systems. Melting is likely to take place through both continuous and discontinuous reactions (Thompson, 1988), making it difficult to generalize melt production.

Melting experiments that used natural rocks as starting materials, evaluated melt fraction through the melting interval, and applied pressures equivalent to the middle to deep crust, have been performed by Vielzeuf and Holloway (1988), Rutter and Wyllie (1988), Wolf and Wyllie (1989), Patiño-Douce and Johnston (1991), Beard and Lofgren (1991), and Rushmer (1991). The following points derive from these studies. (*a*) Closed-system melting of metapelites produces peraluminous melt compositions similiar to S-type granites (Patiño-Douce and Johnston, 1991; Vielzeuf and Holloway, 1988). The main melt producing reaction in both studies is biotite + plagioclase + aluminosilicate + quartz = garnet + melt. At higher temperatures garnet is also consumed to form melt, with spinel and aluminosilicate the most refractory phases. (*b*) Amphibolite melting at lower pressures (7–10 kbar) occurs by "damp" melting mainly via breakdown of hornblende, producing clinopyroxene ± orthopyroxene (Rushmer, 1991; Wolf and Wyllie, 1989). Melts range from granite or trondhjemite to tonalite with increasing temperature and cross the boundary from slightly peraluminous to metaluminous. (*c*) The importance of hydrous minerals in controlling melt productivity and composition is consistently highlighted in all of these studies; the linked parameters mineral mode and bulk rock composition exert a strong control on the amount of melt produced in fluid-absent melting.

This third point, on the relation between mineral mode and melt production, is illustrated by the very different melt fraction vs temperature results for melting of metapelite obtained by Vielzeuf and Holloway (1988) and Patiño-Douce and Johnston (1991). This difference, and the more potassic and felsic nature of the melts produced in the experiments of Vielzeuf and Holloway, is ascribed primarily to the different bulk compositions and mineral modes of the two metapelites, and how well or poorly the initial mineral mode corresponds to the stoichiometry of the major melt-producing reaction, which in this case involves biotite dehydration (Patiño-Douce and Johnston, 1991).

The results of a variety of partial melting experiments (Fig. 1) suggest that melt production vs temperature of natural, hydrous mineral-bearing metamorphic rocks covers a spectrum. At one end of the spectrum are rocks in which the relation between modes of hydrous minerals, quartz, and feldspar, and bulk minimum-melt components, fits well enough with reaction stoichiometry that most melting occurs at the reaction that consumes the hydrous mineral and releases H_2O fluid to dis-

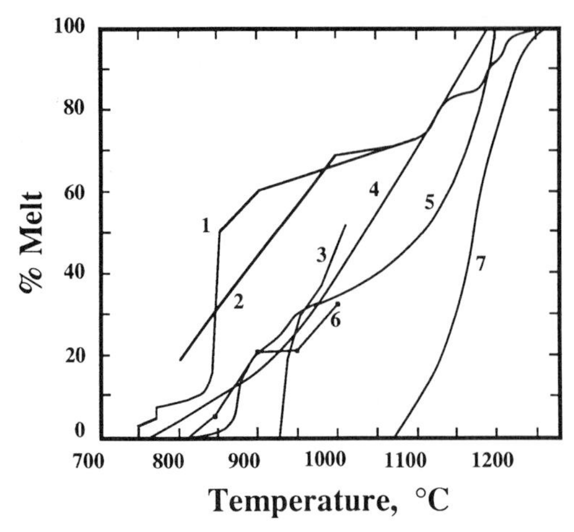

Figure 1 Percentage melt as a function of temperature for a number of lithologies. Pressures range between 7 and 10 kbar. Compositions are metapelites (1,2), silicic plutons (4,5), and amphibolite (3,6). All are damp melting: no free water is present at the start of the experiment. Note rapid and nonmonotonic changes in percentage melt as temperature increases. Curve 7 is the crystallization of a basalt. References for individual curves: (1) Vielzeuf and Holloway (1988); (2) Patiño-Douce and Johnston (1991); (3) Rushmer (1991); (4) Wyllie (1977); (5) Rutter and Wyllie (1988); (6) Beard and Lofgren (1989); (7) Marsh (1981).

solve in the melt. Such rocks will produce melt in step-like fashion, with large volumes produced over a narrow temperature interval. At the other end of the spectrum are rocks that do not contain the components in proportions suitable to maintaining quartz- and feldspar-saturated melting via dehydration reactions. These rocks will not produce major volumes of melt over narrow temperature intervals within the melting range. Because the total volume of rock undergoing partial melting will probably vary over some range of heterogeneity, it seems likely that in many real situations the intermediate pattern of the temperature vs melt fraction spectrum holds, with a dampened correlation between breakdown reactions of hydrous minerals and pulses of increased melt production.

In sum, recent work on lower crustal melting suggests a dominant role for fluid-absent, "damp" melting. Dehydration of biotite occurs at 800–900°C, producing melt fractions of >35% with granite or felsic granodiorite composition. Dehydration of hornblende occurs at 900–1000°C, producing melt fractions of >35% with tonalite or intermediate granodiorite composition. The residual, anhydrous mineral assemblages are similar to those found in granulites (Rudnick, 1992). Metapelites and mica-rich quartzofeldspathic rocks (e.g., some types of metagraywackes) produce the largest volume proportions of melt. What rock types are present in the crust is uncertain, but it is not necessarily limited to direct products of mantle-derived magmas. Metasupracrustal rocks may also be emplaced in the lower crust, particularly in convergent margin orogens (Kempton and Harmon, 1992).

In generic terms, the experimental evidence accords with the geological evidence, in that both are consistent with the initiation of crustal-scale magmatism by the intrusion of basaltic magma at the base of the crust, and the involvement of lower-crustal melting as an important part of the process. Major episodes of lower-crustal melting are associated with temperatures achievable only if basalt intrudes the lower crust, based on the temperatures required for the generation of intermediate magmas. Similarly high temperatures are indicated for many incompatible-element-depleted granulites (Bohlen and Mezger, 1989; Clemens, 1990). The compositions of magmatic rocks that may have been generated in the lower crust are very similar to the experimentally produced melts from vapor-absent runs, with an important exception.

The major point of disagreement between experimentally produced melts and natural magmatic rocks is that the Mg numbers of experimental melts are consistently lower than those of most natural rocks interpreted as probable lower-crustal melts. This is particularly true at intermediate SiO_2 values. Assuming that crustal melting was involved, this discrepancy may be accounted for by partial mingling or mixing of basaltic magmas with lower-crustal melts that were generated by the intrusion of the basaltic magmas, or by restite retention, which assumes that Ca- and Mg-rich refractory minerals from the crustal melting source are entrained in the melt when it leaves the source area. Even if restite unmixing (Chappell and Stephens, 1988) is an important process, intrusion of the lower crust by basalts must still be invoked (Burnham, 1992). The question is then whether magmas are mobilized en masse as melt plus residual solids or whether the melt segregates efficiently from the residual solids, although the answer is undoubtedly somewhere in between and will vary on a case-by-case basis.

Modeling the Physical Interaction of Basalt and Crust

The effectiveness of melt generation, segregation, and homogenization will depend on the rate at which melt is generated, the differences between the viscosity of the crystal–liquid *mixture,* the melt-only viscosity, and the bulk viscosity of the solid matrix. Hence, one of the most important elements in characterizing the dynamics is the rheological state of the partial melt zone (Bergantz, 1990; Wickham, 1987). In addition, there is a growing body of geological evidence that there is a feedback between the stress field in the crust and the style and timing of ascent. Before we consider models of underplating and segregation mechanisms, we will consider the rheological characterization of crystal–melt mixtures.

Rheology of Suspensions

Models for the viscosity of magmatic *liquids* are given by Bottinga and Weill (1972) and Shaw (1972) and a comprehensive summary of the available data on the viscosity of melts of geologic interest can be found in Ryan and Blevins (1987). Obviously, estimates of viscosity based on consideration of only the changing liquid composition have little application to dynamical models of crystal–melt systems; the viscosity can range from nearly infinite under near solidus conditions to 1 Pa · s under near liquidus conditions. Developing robust quantitative models for the influence of increasing crystallinity on the viscosity is difficult; some scaling relations are reviewed by Jomha *et al.* (1991). Even making reliable measurements of the viscosity of suspensions has proven challenging: the same crystal–liquid suspensions yield different viscosities when measured in different viscometers and the measurements are often difficult to reproduce even when repeated evaluations are made on the same viscometer. These results cast doubt on the notion that a single expression can adequately characterize the viscosity of dense suspensions (Cheng, 1984; Frith *et al.,* 1987). The origin of the irreproducibility may reside in variations of the solid packing structure, an organization of the crystals that is an inevitable consequence of flow. This will vary depending on the flow regime and crystal shape and induce a feedback that is particular to a given geometry, shear rate, etc. Despite these difficulties, various empirical relationships can be taken from the engineering literature, although it is important that these results be applied to geologic conditions with caution.

To the viscosity calculated for the crystal-free liquid, a "correction" for the influence of crystals can be added. One such expression often cited is a form of the Krieger–Dougherty expression given by Wildemuth and Williams (1984),

$$\mu_r = \frac{1}{(1 - \phi/\phi_M)^\gamma}, \quad (1)$$

where μ_r is the reduced viscosity, which is the actual viscosity divided by the viscosity of the crystal-free liquid at the same temperature, pressure, and composition; ϕ is the volume particle fraction; and ϕ_M is the volume fraction of particles at which the three-dimensional contact between the particles causes strain to cease at a given shear stress. It should be appreciated that ϕ_M is shear stress dependent and using the absolute maximum packing fraction for ϕ_M may overestimate ϕ_M for many magmatic conditions where strain rates associated with convection may be low. ϕ_M ranges from about 0.5 at low shear stress to a value of 0.75 for infinite shear stress, and γ is a coefficient with experimentally determined values that vary from about 1.3 to almost 3 with a value of 2 being typical of many systems (Barnes *et al.*, 1989; Sengun and Probstein, 1989; Wildemuth and Williams, 1984). Although weakly shear stress dependent, γ is nearly invariant with variations in stress over two orders of magnitude for a wide variety of particle sizes and mixture properties (Wildemuth and Williams, 1984).

The MELTS program (a substantial refinement of the SILMIN algorithm of Ghiorso (1985)) was used to calculate the viscosity as cooling proceeds for a given initial composition and pressure using the method of Shaw (1972). The MELTS program is robust in that it allows for the crystallization of practically all the relevant phases: pyroxenes, common feldspars, quartz, biotite, olivine, the Fe–Ti oxides, and others. In addition, the oxygen fugacity can be controlled. Using MELTS, changes in melt composition and volume percent crystallinity can be calculated as a function of temperature, including the effects of water in the melt.

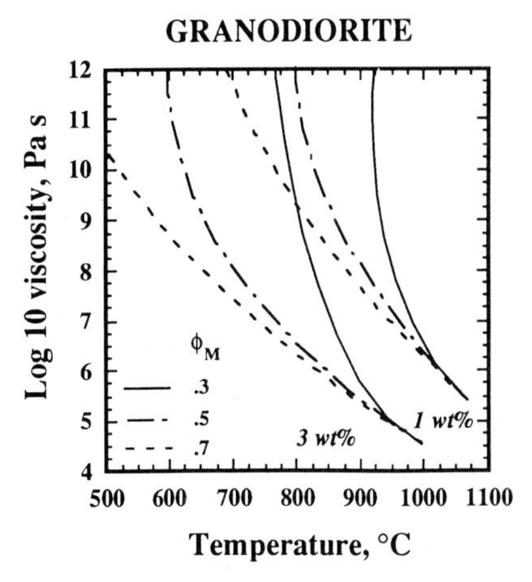

Figure 2 Calculated variation of viscosity with temperature for the Red Lake granodiorite (Noyes *et al.*, 1983) at two different initial water contents, 1 and 3 wt% H_2O. The calculations were done with the MELTS algorithm assuming the oxygen was buffered at Ni–NiO and a total pressure of 2 kbar. Water content will increase as crystallization of anhydrous phases continues. Very high water contents yield low solidus temperatures; given the absence of experimental data under these conditions, viscosity values for temperatures under 600°C should be used with caution.

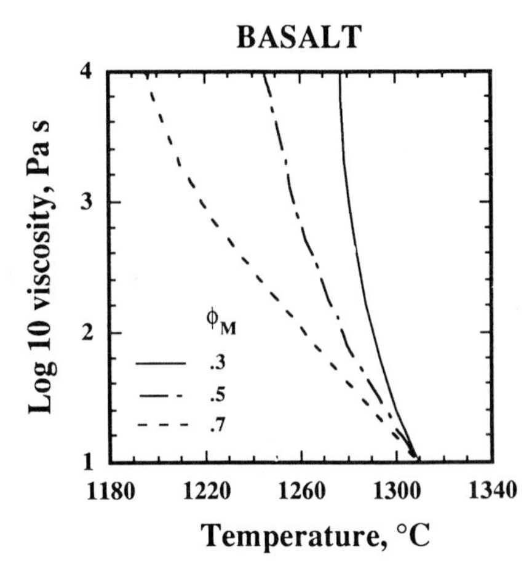

Figure 3 Calculated variation of viscosity with temperature for the mafic inclusions (basaltic composition) in the Burnt lava flow of Grove *et al.* (1988). The calculations were done with the MELTS algorithm assuming the oxygen was buffered at Q–F–M and a total pressure of 8 kbar. These curves are typical of a variety of basalts. Adding small amounts of water to the system (1–2 wt%) had a negligible effect on the calculated viscosity.

Thus at any temperature, the calculated melt composition, water content, and volume fraction crystals can be used in Eq. (1) to yield the viscosity.

Two examples of the calculated viscosity are given here, a granodiorite in Fig. 2 and a basalt in Fig. 3. Three curves are shown in each of the figures; each curve corresponds to a different assumed value of the critical volume fraction, ϕ_M. There is little unambiguous geological evidence of the value of the critical volume fraction as discussed by Bergantz (1990) but a value around 0.5 is suggested from observations on extrusive and plutonic rocks (Marsh, 1981; Miller *et al.*, 1988). For each curve in Figs. 2 and 3, all values of melt viscosity and crystallinity are the same at a given temperature, the only difference among the curves is the value of ϕ_M. The curves for a critical volume fraction of 0.3 and 0.7 delimit an envelope of reasonable values. It is apparent that increasing crystallinity dominates the variation in viscosity. Two initial water contents are given for the granodiorite to demonstrate the effect that increasing water content in the melt will have on the viscosity.

This is important when one considers that the volume fraction of solids as a function of temperature can be very different depending on whether temperature is falling, such as in the case of a magma, or temperature is rising, which is the case in partial melting. Volume fraction curves tend to be smoother for solidification as the nucleation is typically heterogeneous and the undercoolings are demonstrably small (Cashman, 1993). In compositionally similar, but "dirtier" rocks, retrograde assemblages can profoundly affect the melt fraction distribution during partial melting; see Fig. 1. Hence, the disposition of the partial melt may not be simply related to temperature as discussed in Bergantz (1992). This has been demonstrated by Wolf and Wyllie (1991) where it was observed that melt distribution in anisotropic partial melts was not controlled by dihedral angles. This raises the question as to the degree to which equilibrium is maintained during partial melting. The agreement between experimentally produced and calculated major element melt compositions often agrees well with those seen in the rock record (Ashworth and Brown, 1990; Burnham, 1992). Direct natural evidence is uncommon; however, in a remarkable study of the partial melting of a magma chamber's walls that was interrupted by eruption, Bacon (1992) documents that equilibrium was not achieved between plagioclase and melt on time scales of 10^2–10^4 yr. The implica-

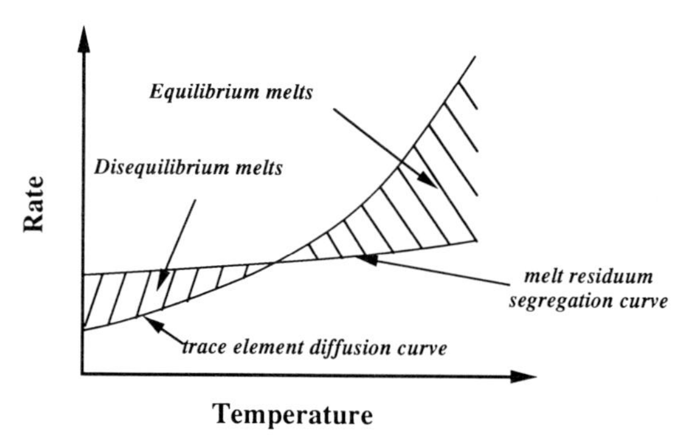

Figure 4 Schematic of relationship between melt segregation and rate of trace element equilibration between melt and residuum. After Sawyer (1991).

tions of disequilibrium melting during melt generation and the geochemical interpretation of migmatites have been addressed by Bea (1991) and Sawyer (1991). The approach to equilibrium can be considered a competition between diffusion and segregation and is shown schematically in Fig. 4. Rubie and Brearley (1990) explore a number of simple models of disequilibrium melting to examine the sensitivity of the melt fraction distribution to the overstepping of the solidus temperature and the effects of variable latent heats of melting, the volume fraction of the melting species, grain sizes, and the diffusion activation energy. For the conditions of magmatic underplating, the initially large thermal overstepping rapidly yields the appropriate melt fraction. However, as the thermal gradient extends outward, the melt fraction may not be simply related to temperature as the *rates* of heating and cooling will be less. Unfortunately, no complete kinetic model for melt-forming reactions exists. This is due in part to the experimental difficulty in establishing the reaction stoichiometry and developing the appropriate kinetic model; the most recent summary of metamorphic kinetics can be found in Kerrick (1991). Hence, estimating the rheological state during partial melting is subject to considerable uncertainty. The available experiments are few and reveal a strong sensitivity of deformation style to melt fraction, particularly near the solidus (Dell'Angelo and Tullis, 1988).

Numerical Model of Basaltic Underplating

We begin our discussion by invoking the simplest imaginable process, as illustrated in Fig. 5. Basaltic magma at the liquidus intrudes some region of the crust that is at ambient temperature T_{CR}. In response to the intrusion, melting begins in the country rock and, subsequently, a region of solid and liquid will form in the region once occupied by the magma. This simple picture has been examined using both numerical and laboratory experiments by a number of workers (Bergantz, 1989a, 1989b; Campbell and Turner, 1987; Fountain *et al.*, 1989; Hodge, 1974; Huppert and

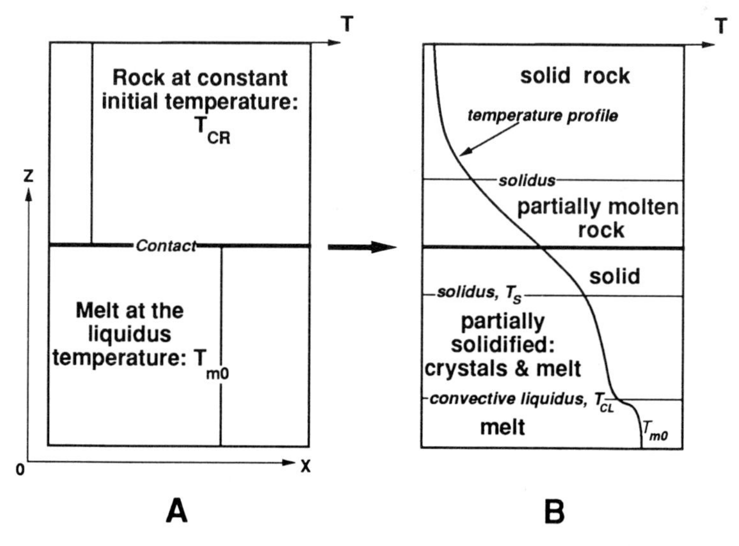

Figure 5 The initial conditions at the time of basaltic underplating, with subsequent evolution. Well-mixed basalt at the liquidus temperature forms a planar interface contact against country rock, which is at some ambient temperature T_{CR}. (A) The temperature profile initially has a step profile that decays, yielding (B) regions of crystal–melt "mush."

Sparks, 1988; Wells, 1980; Yoder, 1990; Younker and Vogel, 1976). Each of these studies has emphasized different physical aspects of basaltic underplating and yielded results that are not all in agreement. The laboratory experiments often demonstrate a wide variety of convective phenomena; however, the solid fraction–enthalpy, and hence transport property, relationships between the model basalt and country rock do not match those found in natural systems. The analytical models usually *assume* that the enthalpy transfer is only by conduction, and so there is always some question as to the role of convection in the magma in enhancing the heat transfer and partial melting of the country rock.

The style and vigor of convection in magma chambers has been subject to uncertainty, and even controversy (Huppert and Sparks, 1991; Marsh, 1991). Much of the uncertainty lies in the difficulty in forming *testable* hypotheses. The simple fluids available for table-top experiments rarely demonstrate the strong rheological contrasts and buoyancy relationships present in magmatic systems. On the other hand, the products of magmatism rarely provide unambiguous evidence of the processes involved in their generation and transport. Buoyancy in magmas is also generated in complex ways, with thermal effects being important near the liquidus and compositional ones important near the solidus. There is no question that magmas convect, although time dependence and the presence of multiple length scales often make the fluid structures and thermal history difficult to generalize with the conventional parameterizations developed for simple fluids. The emphasis of the model presented here is convection and partial melting under one set of arguably geologically relevant conditions. We note that our model assumptions do preclude some aspects of the heat and mass transfer in magmas, such as volatile release and the capacity to model individual crystal growth and settling. However, it is hoped that calculations presented here, which emphasize some of the known rate-limiting features present in geological systems, will encourage a discussion of magmatic convection that directly addresses geological conditions.

One model of the conjugate enthalpy transfer between crystallizing basalt and simultaneous partial melting of the overlying country rock is presented. The important new features of this model are that we are using a set of "best guess" calculated physical property variations: viscosity, density, and all the thermophysical properties vary temporally and spatially as crystallization and concomitant partial melting proceed. In particular, the model attempts to retain the appropriate relationships for enthalpy content and transport properties in the combined basalt–country rock system. Convection in the basalt occurs, and the time dependence, vigor, and form of the convection can be directly related to the crystallization history, which itself is constrained by the phase relations determined under geologically relevant conditions and the progress of melting in the country rock. Thus, the questions of interest are: How might magmatic convection influence the timing of partial melting in the overlying country rock and what is the style of this convection? How does variable viscosity generated by the changing crystallinity and composition of the melt influence the convection? What might the characteristic convective velocities be, and what is the form of the instabilities? What are the overall heat transfer rates from the system relative to those where the transfer is by conduction only?

Oldenburg and Spera (1992) have developed a clever hybrid model for the numerical simulation of crystallizing systems. Their model is a combination of two end-member states: the relative motion condition where interdendritic melt can flow through the crystal pile, and the no relative motion condition where the fluid that is moving through any control volume includes the crystals and equilibrium is assumed. The model considered here uses the no relative motion assumption. In practice this means that at any point in the flow field, the crystallinity, composition, density, viscosity, and all thermophysical properties are determined from the phase constraints directly. This is essentially just application of the phase rule to the control volume over which the local average temperature is determined.

The advantage of this approach is that all the transport property variations can be indexed to temperature. What is required is a means of calculating how density, specific heat, crystallinity, viscosity, and conductivity vary as a function of temperature, including the effects of melt composition as crystallization proceeds. The MELTS algorithm of Ghiorso (discussed previously) provides these quantities. The disadvantage of this approach is that it does not allow for nonequilibrium crystal–liquid segregation. For example, if crystals begin to settle out individually or in the form of a plume, it is possible that interstitial melt could be removed as the plume falls. In our model, any time magma convects to a new region where temperature is different, it is assumed that the crystallinity changes accordingly; temperature and crystal fraction are linked. Thus, convection driven strictly by crystals falling cannot be accommodated. We do not consider this a restrictive condition as it is not clear what role settling crystals may have in driving *bulk* convection.

The combined basalt–country rock system shown in Fig. 5 was represented on a two-dimensional 75 × 75 grid of total width W of 15 m and total height H of 25 m. The depth of the initial basalt layer is l and was 20 m thick. The origin was placed at the lower left corner and Cartesian coordinate axes were introduced: x being the horizontal coordinate axis and z the vertical axis, where z is taken as positive in the upward direction. Fifty-five grid points in the vertical direction were assigned to the fluid layer and twenty to the overlying country rock. Variable grid spacing was used in both the vertical and horizontal directions to ensure that the boundary layers near the intrusive contact could be adequately characterized. Using fewer grid points (40 × 40) yielded the same Nusselt number relationships discussed here, but some details of the thermal structure of the plumes were lost so we elected to use the finer grid.

A comprehensive discussion of the continuum equations describing the heat and mass transfer in systems undergoing crystallization, with possible double-diffusive convection is available (Beckermann and Viskanta, 1988, 1993; Bennon and Incropera, 1987; Ni and Beckermann, 1991; Oldenburg and Spera, 1991, 1992). We will not revisit these derivations, but direct the reader to these sources for clarification of the continuum expression of the model.

The equations required to describe the velocity and temperature fields following underplating, invoking the Boussinesq approximation, are given for

conservation of mass (continuity),

$$\frac{\partial u}{\partial x} + \frac{\partial w}{\partial z} = 0, \qquad (2)$$

conservation of momentum,

$$\rho_0\left[\frac{\partial u}{\partial t} + u\frac{\partial u}{\partial x} + w\frac{\partial u}{\partial z}\right] = -\frac{\partial P}{\partial x} + \frac{\partial}{\partial x}\left(\mu\frac{\partial u}{\partial x}\right) + \frac{\partial}{\partial z}\left(\mu\frac{\partial u}{\partial z}\right) + \frac{\partial \mu}{\partial x}\frac{\partial u}{\partial x} + \frac{\partial \mu}{\partial z}\frac{\partial w}{\partial x} \qquad (3)$$

$$\rho_0\left[\frac{\partial w}{\partial t} + u\frac{\partial w}{\partial x} + w\frac{\partial w}{\partial z}\right] = -\frac{\partial P}{\partial z} + \rho g + \frac{\partial}{\partial x}\left(\mu\frac{\partial w}{\partial x}\right) + \frac{\partial}{\partial z}\left(\mu\frac{\partial w}{\partial z}\right) + \frac{\partial \mu}{\partial x}\frac{\partial u}{\partial z} + \frac{\partial \mu}{\partial z}\frac{\partial w}{\partial z}, \qquad (4)$$

and conservation of energy,

$$\rho_0 c_P\left[\frac{\partial T}{\partial t} + u\frac{\partial T}{\partial x} + w\frac{\partial T}{\partial z}\right] = \frac{\partial}{\partial x}\left(k\frac{\partial T}{\partial x}\right) + \frac{\partial}{\partial z}\left(k\frac{\partial T}{\partial z}\right), \qquad (5)$$

with initial and boundary conditions on region and boundary temperatures,

$$T(x,z,0) = T_M, \quad 0 \le z \le l; \qquad T(x,z,0) = T_{CR}, \quad l \le z \le T_{CR}, \qquad (6)$$

$$\left.\frac{\partial T}{\partial x}\right|_{x=0} = 0, \quad \left.\frac{\partial T}{\partial x}\right|_{x=W} = 0, \quad \left.\frac{\partial T}{\partial z}\right|_{z=0} = 0, \quad T(x,H,t) = T_{CR}, \qquad (7)$$

and on mass flux (motion components) at the boundaries,

$$\left.\frac{\partial u}{\partial z}\right|_{x=0} = 0, \quad \left.\frac{\partial u}{\partial z}\right|_{x=W} = 0, \quad u(x,0,t) = 0, \quad u(x,l,t) = 0 \qquad (8)$$

$$\left.\frac{\partial w}{\partial x}\right|_{x=0} = 0, \quad \left.\frac{\partial w}{\partial x}\right|_{x=W} = 0, \quad u(x,0,t) = 0, \quad w(x,l,t) = 0 \qquad (9)$$

where l is the thickness of the fluid layer. The initial conditions are constant temperature in the country rock T_{CR}, and the basalt is assumed to have intruded at its liquidus, T_M. The horizontal boundary conditions on temperature and velocity are reflecting conditions where it is assumed that the magma layer is infinite in the horizontal dimensions. Thus, effects due to the presence of side walls as sources of drag or heat are not considered. The vertical conditions are that the magma layer is effectively insulated at the bottom. This means that the convection is driven only by cooling from above and is fundamentally different from systems that are heated from below and cooled from above, and buoyancy is generated at *both* boundaries. This seems reasonable as it is difficult to imagine what heat source would be hotter than the basalt itself.

The numerical algorithm used to solve Eqs. (2–5) is a modified version of the TEMPEST code developed at Pacific Northwest Laboratories (Trent and Eyler, 1991). It uses a semi-implicit, finite volume method. We have established code accuracy and reliability by doing benchmark runs on a number of classical natural convection problems, and we have found excellent agreement with published numerical and experimental results, including conditions of variable transport properties.

We consider only one set of calculations here, but one that reveals the important features observed in a variety of simulations. The model basalt composition used was that of the Burnt Lava flow (Grove *et al.,* 1988). The crystallization was assumed to occur under oxygen-buffered conditions of quartz–fayalite–magnetite. It was also assumed that crystallization occurred in the fractionation mode. This does not mean that the crystals were physically removed as temperature dropped, but rather that as crystals grow, they do not back-react and the melt composition evolves accordingly. This is really a statement that thermal equilibrium might be maintained, but the crystals will remain zoned. If resorption is required by advection of crystals to a region of higher temperature, it will initially involve the rim of the crystal. This decision was based on the fact that plagioclase was the dominant near liquidus phase. Very little difference in the convective state or the progress of melting of the country rock was found in simulations where the physical properties were generated assuming complete equilibrium between melt and solid phases.

The MELTS algorithm was used to calculate the required physical property data. The numerical prescription of the enthalpy changes associated with phase changes was done by defining a temperature-dependent effective specific heat capacity that includes changes in both sensible and latent heats. Figure 6 is a plot of the calculated variation in effective specific heat as a function of temperature for the basalt. The nonmonotonic behavior of the effective specific heat as a function of temperature is the result of the magma becoming saturated with new phases as cooling proceeds. Once the melt has become saturated with a phase, with a concomitant "burst" of crystal growth, the effective specific heat will diminish as the increase in crystallinity per degree of temperature drop diminishes. Although the liquidus phase was olivine, the increase in crystallinity with temperature at the liquidus is modest; the two dominant peaks in Fig. 6 are due to the saturation of the magma with plagioclase at about 1275°C and with spinel at 1085°C. Pyroxene appears at about 1260°C. Sensitivity analysis of the model results indicated that geologically reasonable variations in the absolute values, or reasonable variations in the functional form of the effective specific heat function, had little impact on the results. Calculated density variations are shown in Fig. 7. The system density was used in the calculations, in keeping with our no relative motion postulate. This system density reflects both changes in the melt density as cooling and composition of the melt changes during crystallization and increases in density associated with the appearance of the solid phases. The viscosity–crystallinity–temperature relationships for the crystallizing basalt are shown in Fig. 3.

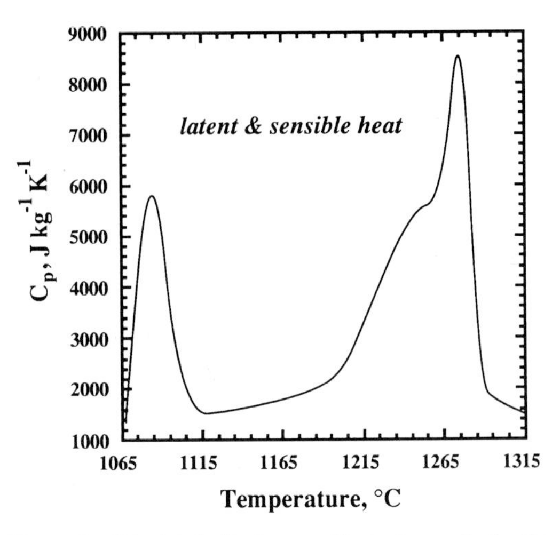

Figure 6 Calculated effective specific heat capacity for the model basalt. This effective specific heat includes both the effects of phase change and an overall drop in system temperature. It also includes the temperature-dependent heat capacities of the minerals and melt. The peak in effective heat capacity at about 1275°C is due to the onset of feldspar crystallization; the peak at about 1085°C is due to the appearance of spinel.

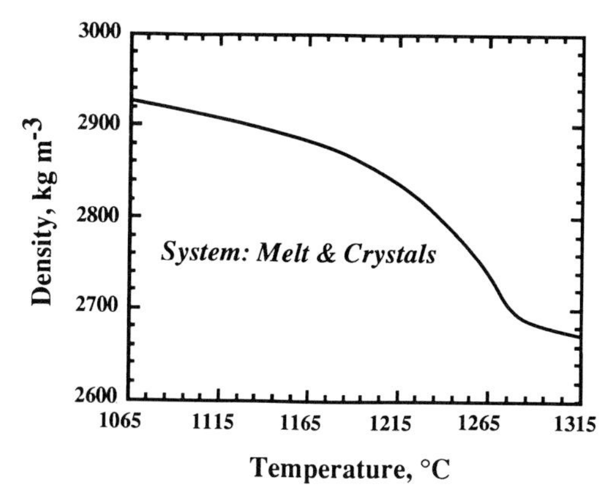

Figure 7 Calculated density of the model basalt. This density includes all the phases present as both crystals and liquid.

The melt fraction–temperature relationship used for the model country rock was that of Rutter and Wyllie (1988) who melted a tonalite under vapor-absent conditions (Fig. 1). They do not report enthalpy changes associated with the melting and so a constant value of 1.5×10^5 J · kg^{-1} was used. This value is consistent with those calculated by Hon and Weill (1982). Temperature-dependent thermal conductivity curves for both the basalt and the tonalite protolith were taken from Touloukian *et al.* (1981).

We initialize the model by assuming a pressure of 8 kbar with an ambient temperature of T_{CR} equal to 450°C and a basalt liquidus, T_M equal to 1315°C. After a short period following intrusion to allow the cold boundary layer at the roof to develop, the initial instability was generated by introducing a random thermal perturbation of 1 to 3°C at each grid point near the basalt–country rock boundary. These rather high perturbation values were required as the presence of variable viscosity rendered the cooling boundary layer very stable. Once convection started, the form and disposition of the instabilities were independent of the location and magnitude of the initial perturbations.

Many of the questions related to the nature of the coupled crystallizing basalt and melting of the country rock can be resolved by considering the time-dependent Nusselt number for the system. In this instance the Nusselt number is defined to be the ratio of the total heat transfer in the presence of convection to that which would occur under conditions of conduction only. This ratio was calculated using the thermal flux determined at the contact between the magma layer and the country rock. The maximum Nusselt number under these conditions is 2 and would occur if the magma were perfectly well mixed by some unspecified stirring agent. To obtain the conduction results for calculation of the Nusselt number, we did a numerical simulation identical to that with convection, except that no fluid motion was permitted. The Nusselt number as a function of the dimensionless conduction time is shown in Fig. 8; four curves are shown and we consider each in turn.

The first is the case of constant viscosity. In this case, crystallization is occurring, but the viscosity is held constant at the value the magma had at the time of intrusion; hence, there is no effect on the viscosity due to increasing crystallinity or changing composition and temperature. Although this is clearly unrealistic for magmas, it provides a useful end-member result. The Nusselt number rises rapidly to a maximum of about 1.77 and then decreases as the magma reservoir temperature decreases and the country rock heats up. The other three curves correspond to cases where the viscosity varied as a function of critical solid fraction, ϕ_M. The curve for a critical volume fraction of 0.7 is queried as there appeared to be abrupt and unreproducible oscillation in the thermal flux,

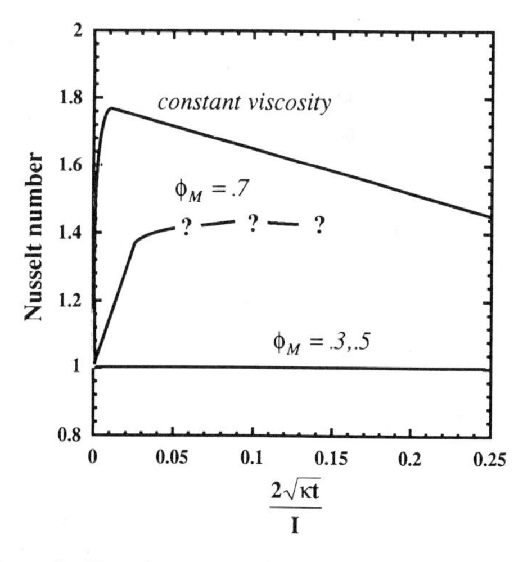

Figure 8 Nusselt number as a function of the dimensionless conductive cooling time. The thickness of the melt layer is l, the thermal diffusivity is κ, and time is t. Four curves are shown: one for the case where viscosity is held constant and three that correspond to the viscosity–temperature relations given in Figure 3. The curves for critical volume fractions, ϕ_M, of 0.3 and 0.5 are practically indistinguishable from each other. See text for a discussion of the query on the curve for ϕ_M equal to 0.7.

perhaps indicating the abrupt destruction of the stagnant layer at the roof (see discussion following); because this was not reproducible, that curve should be interpreted with caution. The cause of this behavior is still under investigation; we will not discuss it further.

The curves for critical volume fractions of 0.3 and 0.5 fall practically on top of each other in Fig. 8. Both have Nusselt number values so close to 1 that it is not apparent from Fig. 8 that there is any difference between them. Under these conditions, even though convection is occurring, the heat transfer to the country rock, and hence the progress of melting, is as if the magma were cooling by conduction only. This is despite the fact that the initial temperature difference between the magma and the country rock was 865°C. For these model runs the region of the magma in which convection is initiated is where the viscosity is varying by no more than an order of magnitude, and more typically by no more than a factor of 2, and where the crystallinity is quite low. This indicates that convection is occurring in a region where the temperature differs by no more than a few (one to five) degrees. Another important observation was that the Nusselt number decayed monotonically; there was no melting back or resorption of the high-viscosity sublayer.

The disposition of the isotherms and plume morphology was similiar to that seen in the study of variable viscosity convection by Olson *et al.* (1988); see their Figs. 17, 18, and 21. For all the cases considered in this study, the convection formed plumes that fell from the leading edge of the crystallizing front. There was no regularity in the positions from which the plumes originated or in the numbers of plumes that might exist in a fully developed state at any time; the plume morphology was broadly like that of classic thermals. The flow field was always irregular but not turbulent; the maximum vertical velocity associated with this convection was of order 10^{-4} m · s^{-1}, and typically much smaller, about 10^{-5} m · s^{-1}. Despite these low velocities the fluid motion was able to keep the interior largely isothermal, probably due to the small temperature contrast associated with convection. Although the flow would on occasion be organized into cells with high aspect ratios (length/width), which did not penetrate to the floor, it was more common for the return flow to be dispersed and irregular.

The partitioning of the fluid into a region with a rigid lid where high viscosity inhibits convection, and an adjacent region where convection can occur, has been noted in other studies of variable-viscosity convection (Chen and Pearlstein, 1988; Chu and Hickox, 1990; Ogawa *et al.*, 1991; Olson *et al.*, 1988; Richter *et al.*, 1983). Of particular relevance are the studies of Jaupart and Parsons (1985) who examine the case where a variable-viscosity fluid layer is cooled only from above, and Brandeis and Jaupart (1986) and Smith (1988) who also add the complication of simultaneous crystallization. Smith (1988) uses linear stability analysis to determine the critical Rayleigh number and demonstrates that, in the near-critical regime, high aspect ratio cells that may not penetrate to the bottom of the chamber form and that the appropriate length scale represents a balance between thermal diffusion and propagation rate of the solidification front. For supercritical conditions where the critical Rayleigh number is substantially exceeded, it is estimated that the convection will still be gentle, as the convective vigor is modulated by variable viscosity and the rate-limiting steps involved with heat transfer out of the roof by conduction.

The results obtained from our modeling validate the suggestions of Smith (1988) and are in agreement with the laboratory experiments of Bergantz (1989a). The convection is never vigorous, although it certainly does occur. The high aspect ratio cells predicted by Smith (1988) were observed, but they appear on an intermittent basis, which is to be expected in the supercritical regime. This type of behavior has been seen in geophysical models of lithospheric convection below the oceanic crust where the mantle has been modeled as a variable-viscosity fluid (Buck and Parmentier, 1986). These results are also consistent with the models of magma chamber convection proposed by Marsh (1989), who addresses some of the petrologic implications and details of the parameterization of convection.

The calculated progress of melting of the country rock due to the presence of a convecting basalt layer follows directly from the low values of the Nusselt number. For the cases where the critical melt fraction ϕ_{M} was 0.3 and 0.5, the amount of melt generated at the intrusive contact and the way it changed as melting proceeded were indistinguishable from that calculated in the conduction-only models (Bergantz, 1989b). It thus appears that within the framework of the assumptions invoked in this model, calculating the progress of melting following a single episode of underplating can be facilitated by assuming a simple, but still nonlinear, conductive cooling model. These results clearly indicate that multiple intrusions are required to thermally mature a deep crustal section to incite melt generation on a regional scale. However, it should be remembered that most geochemical models of partial melting (see preceding discussion) require melting in the range 10–35%, which is like that estimated here and in Bergantz (1989b). The compositions of melts produced by wholesale melting of the lower crust deviate from naturally occurring compositions (Beard and Lofgren, 1991) and models that would predict melting in excess of 50% should be invoked with caution. Thus, there is no justification for models that suggest that the incipient stages of partial melting involve *bulk* melting of the lower crust.

Our results are also in keeping with observations made during large-scale melting experiments. An artificial magma chamber, or melt layer, 3 × 1.5 m was generated by melting soil at the Oak Ridge National Laboratory (Jacobs *et al.*, 1992). The melting was initiated and sustained by applying electrical power to the ground through four graphite electrodes. The melt pool convected vigorously as it was thermally "pumped" and superheated conditions were attained: melt temperatures reached 1500°C. However, once the power was shut off, convection could be driven only by heat losses to the environment under conditions of thermal decay; convection ceased once the liquidus of the melt was attained (Dunbar, 1993; oral communication). This is an important demonstration of the fact that the rheological conditions associated with the onset of crystallization control the subsequent dynamic evolution of the body.

It should be noted that a number of conditions that could alter the broad conclusions of our model might exist. If the magma chamber experienced reintrusion and subsequent disruption of the upper viscous boundary layer, then a pulse of melting might follow. If the roof were not infinitely rigid, as assumed here, it could buckle under its own weight and delamination of the roof could occur. This would abruptly bring together partially melted country rock with near liquidus temperature magma, yielding a condition that could lead to magma mixing. Any volatile release will aid melt production in the country rock, which we have not considered. Other geometries, such as the multiple dikes depicted by Grove *et al.* (1988), may yield regions of overlapping thermal gradients and enhanced melting. It was also found in the course of a sensitivity analysis that some variable viscosity–temperature relationships *did not* yield a steady-state viscous lid and whole-layer convection occurred. Although these viscosity–temperature relationships are not like those calculated by the MELTS code for use in this example, it is not unreasonable to speculate that there is more to be learned about variable-viscosity convection. Similar results, although on a system that was heated from below, were found by Ogawa *et al.* (1991). A complete parametric treatment of our work on variable-viscosity convection, crystallization, and partial melting will appear elsewhere.

The results of our numerical experiments differ from the laboratory experiments of Huppert and Sparks (1988) and Campbell and Turner (1987). These deservedly influential papers evaluate a number of the fluid dynamical and petrological consequences of basaltic underplating. We agree with many of the geological implications for melt production and the origin of petrologic diversity discussed in these works. However, our continuum models yield estimates of time and length scales of magma generation and cooling that are quite different from theirs. The important differences are that Huppert and Sparks (1988) predict very turbulent conditions in the magma, leading to rapid cooling and concomitant bulk melting of the overlying country rock. The differences can be ascribed to the following: (*a*) we explicitly model strongly variable-viscosity flow, and as importantly, (*b*) the melting and solidification relations between the country rock and the model basalt in our work are like those in natural systems. We preface our discussion by noting that magmas are rarely superheated and that the melt fraction–temperature curves used in models must retain the relationships between melting intervals between basalt and country rock as shown in Fig. 1: the underplated material must have a higher liquidus temperature and a narrower and higher solidus-to-liquidus temperature range than the country rock.

The laboratory experiments of underplating of Huppert and Sparks (1988) use a superheated aqueous solution of $NaNO_3$, underplated beneath a wax (PEG 1000). We were unable to find any reference to the the actual weight percent of $NaNO_3$ for the experiments of Huppert and Sparks (1988); however, the liquidus slope of the $NaNO_3$ solution varies between 0 and $-18.1°C$ (Kirk-Othmer, 1978). The $NaNO_3$ was substantially superheated to 70°C before underplating. The PEG wax used as the country rock has a melting interval between 37 and 40°C and at the start of the experiment was isothermal with a temperature of 20°C. Thus, in no circumstances could the underplated "basalt" solution crystallize in the models, and hence no rheological penalty for crystallization is possible. This is unlike the geological case in many important respects, where basalt crystallizes at temperatures above the ambient temperatures of the country rock. The laboratory roof melting experiments of Campbell and Turner (1987) show more complex behavior as simultaneous crystallization and melting occur in some of their experiments. However, the model basalt in their experiments is also superheated, and hence the same cautions detailed above apply. We hasten to add that the experiments of Huppert and Sparks (1988) and Campbell and Turner (1987) are well motivated and that it should be appreciated that getting the right combination of material in any analog model of underplating is very difficult.

In summary, numerical experiments using geologically relevant thermophysical properties indicate that partial melting from underplated basalt will be largely conductive, *in the early stages of underplating*. This is in agreement with many features from the rock record, including measured amounts of partial melting at plutonic contacts and the compositions of naturally occurring granitoids. We concur with the important conclusions from Huppert and Sparks (1988) regarding the thermal and compositional evolution of regions of underplating: if further underplating occurs, the region will thermally mature and reach a condition where large amounts of melt can be generated by a modest increase in temperature caused by incremental addition of magma to the region of partial melting. Neither the model presented here nor those in Huppert and Sparks (1988) address these conditions in a rigorous way. Without some geological evidence, constructing a model that could be generalized would be difficult.

Melt Collection and Ascent

Once partial melting has been initiated, a gradient in melt fraction will exist away from the underplating interface. Wickham (1987) has considered a number of the controls on the segregation and transport of magma during partial melting; we will focus largely on refinements and new paradigms. A number of processes of homogenization, segregation, and transport can occur during partial melting: convective overturn in the style of porous medium convection; magma mixing involving underlying partially molten basalt; bulk convection of the partially molten region; compaction of the matrix, yielding discrete melt bodies; and, finally, melt (and matrix) transfer by virtue of diking or diapirism.

Melt Convection in a Porous Medium

Once a partly melted region is generated, a variety of dynamical states are possible: simple convection of the melt in a rigid matrix, movement of melt and matrix due to compaction, or bulk convective motion of the region of partial melting. The first is the most simple: it is assumed that the unmelted residuum forms a rigid matrix. This may have application to magma chambers that are composed of regions of largely crystal–liquid mush, such as that now proposed for mid-ocean ridges (Nicolas *et al.*, 1993). This is appropriate for conditions where the melt fraction is below about 50%, although this parameter is very poorly constrained (Bergantz, 1990). Assuming that an interconnected network forms, which can be for as little as 2 wt% melt for a mixture of olivine and basalt powder (Daines and Richter, 1988), the motion of the melt may be considered flow in a porous medium, and the formalism for convection in a porous medium may be applied as noted by Lowell (1982). For the underplating conditions described previously, where the region undergoing partial melting is being heated from below and growing upward, the permeability will be anisotropic and heterogeneous and may well produce local reversals in density due to the presence of water, which will also affect the density. There is no theoretical formalism to predict the onset of convection for these geologically relevant conditions. However, if one assumes that permeability is homogeneous and isotropic, the model relationships of Lowell (1982) can be used. These are summarized in the following expression, which specifies the dependence of the required thickness of the melt region, $X(t)$, on other system properties,

$$X(t) \geq \frac{288\pi^3 \kappa \nu c_p}{\beta g \chi^3 b^2 L \cdot \mathrm{Ste}}, \quad (10)$$

where κ is the thermal diffusivity, ν the kinematic viscosity, c_p the specific heat, β the thermal expansion coefficient, g the scalar acceleration of gravity, b the average grain diameter, L the latent heat of melting, and Ste the Stefan number modified by multiplying the latent heat term by the melt fraction. For typical values of these quantities and a constant melt fraction of 0.3, Eq. (10) is plotted in Fig. 9. Low-viscosity melts may become homogenized over scales of a few meters to tens of meters, which is the length scale over which the model assumptions may be relevant. For more viscous melts, the layer thickness will exceed any reasonable length scale of lithologic homogeneity and convection cannot be simply characterized by an expression like (10).

Expression (10) is based on model assumptions that severely limit its application to crustal melting except as an end-member estimate. A number of additional, potentially relevant complexities have been addressed in the engineering literature: Kaviany (1984) considered the case where the lower boundary temperature is increased linearly, variable porosity was considered by David *et al.* (1991), and variable viscosity of the melt by Blythe and Simpkins (1981). Given the nonlinear nature of these effects, it is not possible to combine the results of these different studies and to assess their importance for convection in partial melts. None of the current studies considers the conditions for convection in a growing melt layer where the permeability and melt viscosity are changing as a function of temperature. In summary, it is unclear whether a distributed melt will be homogenized on anything but small scales. Whether melt homogenization occurs depends most importantly on melt viscosity, which will vary dramatically as a function of volatile content and hence melt progress.

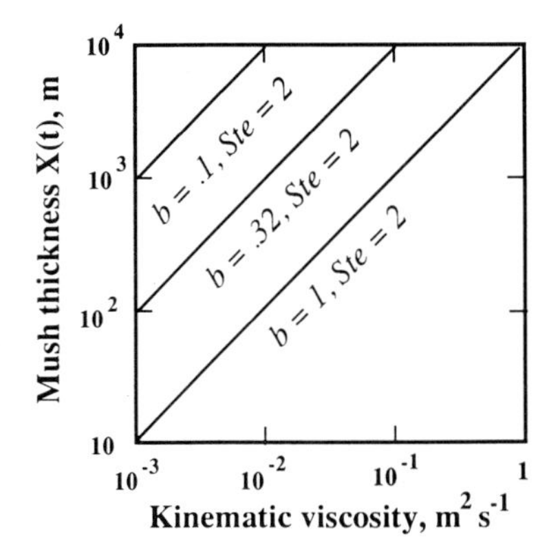

Figure 9 Critical mush thickness for porous media convection to begin as a function of kinematic viscosity. For these curves the grain diameter, *b*, is in centimeters. Ste is the Stefan number. After Lowell (1982).

Compaction

If the unmelted matrix is not infinitely rigid, compaction and subsequent reorganization of the partial melt region can occur. Compaction refers to the change in melt fraction that occurs as buoyant interstitial fluid moves upward and solid material is displaced downward. The governing equations that describe compaction are developed in the context of mixture theory or averaging techniques and hence the calculated values of the dependent variables represent averages at scales that are large relative to individual grains but small relative to the scale of the gradients in the dependent variables. This is the usual approach in the continuum description of porous media and mixed phase processes and does not limit the utility of the model for most problems of geologic interest. The physical processes and petrologic implications of this form of melt segregation have been considered in the context of melt migration in the mantle (Fowler, 1985, 1990a, 1990b; McKenzie, 1984, 1985; Ribe, 1985; Scott and Stevenson, 1986; Sleep, 1974), in magma chambers (Shirley, 1986, 1987), and in the crust (Fountain *et al.*, 1989; Lowell and Bergantz, 1987; McKenzie, 1985; Wickham, 1987).

A succinct summary of the equations describing compaction is given by Ribe (1987). Of interest here is the case where compaction is occurring in a region of variable partial melt and with gradients where the partial melt fraction goes to zero. Under these conditions, melt can collect into "solitary waves," dubbed solitons or magmons (Scott and Stevenson, 1984; Stevenson and Scott, 1987), where under certain conditions the melt fraction can go to 100% and fully liquid magma chambers can ostensibly form. The model of Fountain *et al.* (1989) demonstrates this in the context of melt migration during crustal anatexis following underplating by mafic magma. Provided the melt fraction at the contact exceeds 25%, it is demonstrated that compaction can yield melt bodies with dimensions on the order of kilometers extending from the contact to a few kilometers above the contact. These length scales are in rough agreement with some geological examples; however, the upward migration is limited to regions that have undergone partial melting. Figure 10 shows the calculated distribution of partial melt. Features not explicitly considered in the model of Fountain *et al.* (1989) are the specifics of thermal or compositional magmatic convection; e.g., the model is spatially one-dimensional. These authors argue that the density will be dominated by the influence of crystals falling off the roof, an ad hoc but not unreasonable assumption once the growing melt layer is itself largely crystal free. Lowell and Bergantz (1987) also consider compaction of a growing partial melt layer and conclude that the compositional effects on density may yield instabilities in the growing melt layer (see discussion on porous media convection above) and hence yield conditions in the region of compaction that would be hard to generalize and would require a multidimensional model to adequately represent the structure of the fluid and solid flow.

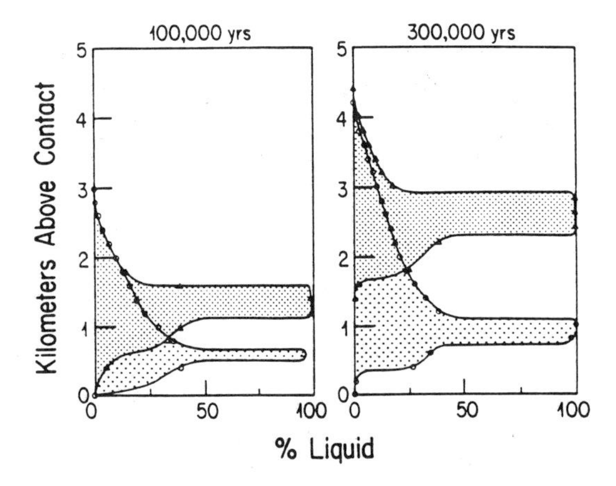

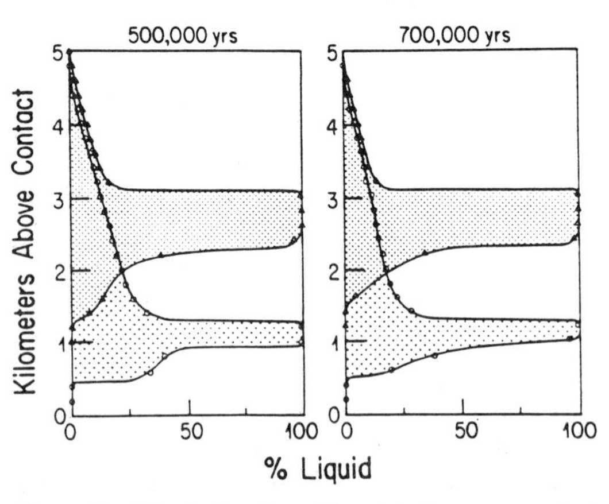

Figure 10 Distribution of granitic melt in the country rock at four different times, 100,000, 300,000, 500,000, and 700,000 yr after intrusion. The stippled area is defined by the curve of percentage liquid versus distance from the intrusion. From Fountain *et al.* (1989).

Diapirism and Diking

The crustal scale transport of magma is often considered in terms of two end members: diapirism and diking. This is somewhat surprising as there is little theoretical or unambiguous geological evidence to suggest that either of these mechanisms alone is dominant in the *crustal* transfer of silicic melts. Diapirism is a type of Rayleigh–Taylor instability and has traditionally been invoked to explain the growth and disposition of salt domes, the spacing of volcanoes (Marsh, 1979), and the ascent of granitic magma in the crust (Miller *et al.*, 1988). The conditions leading to the onset of the instability are developed by Turcotte and Schubert (1982) and more recently by Lister and Kerr (1989) who caution that variable physical properties, geometries, and the presence of a deformable lower layer render many of the more classical estimates of the wavelength and rise time of the fastest growing instability inappropriate. These conditions may be present in partial melt regions that have vertical lithological and rheological variations where a region of greater partial melt is sandwiched between regions with smaller amounts of partial melt and the viscosity contrasts are not large between the layers.

The important rate-limiting element in magmatic diapirism lies in the thermal coupling of the diapir and the surrounding rock. It is thought that heat loss from the magma to the surrounding rock lowers the viscosity of the country rock allowing it to "flow" around the magma body. Implicit in this model is that there is a finite distance that the magma can rise before the enthalpy difference between the diapir and the surroundings diminishes to the point where the density and viscosity contrasts are no longer sufficient to permit continued ascent. This is the basis of what has become known as the hot Stokes model of diapirism. The quantitative description of the model has been developed by Marsh (1982), Morris (1982), Daly and Raefsky (1985), and Mahon *et al.* (1988). It should be noted that when the hot Stokes model of diapirism was first proposed by Marsh and others, the intended application was melt transfer in the mantle.

The hot Stokes model of diapiric transfer of magma contains features that, in principle, should be testable by geologic observation (Bateman, 1984; England, 1990; Paterson *et al.*, 1991). However, the ambiguity in uniquely identifying strain fields that record diapiric ascent as opposed to the fabric generated during (final) emplacement precludes a direct interpretation in many cases. Paterson *et al.* (1991) have considered a wide variety of natural examples and conclude that the width of aureoles and the intensity of wall rock strain around granitoids are generally not as great as those required by laboratory or numerical models for diapirism. England (1990; 1992) argues that the fabrics generated during an emplacement process, such as "ballooning," may make it difficult to recognize the signature of diapirism. This condition is exacerbated if the diapir is compositionally nonuniform, because a number of mixing conditions can arise if there are multiple magmatic elements in the diapir, leading to strain localization and nonuniformity in the thermal and compositional fields. Weinberg (1992) notes that reverse zoning would be the most common pluton geometry if the zoning were controlled only by internal circulation. However, most zoned systems have a normal zonation pattern that progresses from more mafic to felsic, and it appears that an internal circulation driven by internal shear coupling may be of limited importance in generating zonation patterns. Isothermal experiments of diapirism by Cruden (1990) reveal some of the fabrics that can occur within an upward moving melt body by virtue of the viscous coupling to the country rock. The deformation is strongly time dependent and a variety of fabrics can form with penetrative foliations at the margins and a possibly isotropic fabric internally. An element not included in this or other models is the inevitable crystallization that must occur at the boundary as diapirism proceeds. This crystallization will effectively act to move the isotherm associated with internal slip inward, yielding a telescoping system where fractionation can occur by virtue of ascent, and the time history of fabrics will be even more complex than that given by Cruden (1990). These considerations cast some doubt on the notion that kinematic indicators taken from intrusions and the surrounding rock can discriminate between diapirism, as understood in the classic sense, and magma chamber development following initial transport in a dike.

Mechanical and thermal models of magma transport in dikes are considered in this text. It is important to note, however, that some models of melt transfer in dikes suggest that it is a very efficient means of magma transfer. Clemens and Mawer (1992) present a model where it is estimated that a 2000-km^3 batholith can be filled in less than 900 yr. Although this model requires some end-member assumptions, it demonstrates the kinds of time scales that may be possible if magma is transported in dikes.

Tectonic Regimes and Magma Ascent

The previous discussions considered simple systems where the material surrounding the melt was considered to be a mechanical continuum, and the role of tectonism and the rheological partitioning of the crust were not explicitly included. Conflicting and ambiguous geological evidence indicate that this approach, while amenable to modeling, is not sufficient to account for the variety of scales of melt extraction and ascent that are observed in the rock record. For example, on the basis of the compositions of migmatites and partial melts of the midcrust, Sawyer (1991) provides evidence that the rates of melting and melt extraction often are greater than the rates at which chemical equilibrium can be achieved in the melt. Thus, it may be inappropriate to model many important melt-forming reactions with an assumption of thermodynamic equilibrium. There are numerous studies that demonstrate the broadly syntectonic nature of magma ascent. Some studies even infer that a feedback exists between magma generation and tectonism. It is apparent that the material transfer required to accommodate the ascent of magma occurs simultaneously at a variety of scales; therefore, models that attempt to describe the ascent and magma chamber assembly process only on the basis of the near field deformation will necessarily be incomplete.

The possible role of crustal scale structures controlling the ascent and emplacement of magmas is typically determined by the use of kinematic indicators measured in and at the margins of plutons and the disposition of magmatic centers relative to the positions of structural features. A hypothesis linking the ascent of magmas and the development of crustal scale structural features where magmas can possibly follow up and into any zones of weakness created by faulting is attractive in that it appears, on the face of it at least, to provide a solution to the "room problem" in the near field. We are not suggesting that fault bends produce "holes" in the crust that magmas then pour into, but rather that magmas can intrude along shear zones while they are active. With this type of intrusion mechanism the need for a crustal scale ductile halo as required for diapiric rise is obviated. Gravity studies indicate that the form of granitic plutons can be correlated with the tectonic style and occurrence of shear zones: intrusion during extension yields thin plutons composed of a number of subunits and intrusion during transcurrent shear yields a few feeder zones that lie off the shear zone (Vigneresse, 1994).

The structural control of the intrusion of plutons is further suggested by the frequent occurrence of plutons in linear belts as noted by Pitcher and Bussell (1977) in the Andes. This tectonic control on the character of magmatism at convergent margins has been considered in the broadest sense by Glazner (1991) who suggests that plutonism is favored when the tangential component of convergence is large, and volcanism is favored when the normal component of the convergence vector is dominant. However, no specific mechanism for this correlation is proposed. Considering tectonic elements at a similar scale, Tikoff and Teyssier (1992) provide a model where plutonism and batholith assembly occurs in zones of dilatation that accompany the development of en echelon P-shear arrays, which result from a transpressional strain regime created by oblique convergence. The evidence used to support this is the linear distribution of intrusions that lie just off the axis of the strike-slip faults, and the occurrence of syntectonic fabrics whose orientation is coherent on a pluton scale and consistent with an origin by dextral shearing.

Mineral fabrics provide the only direct means for the determination of the kinematic conditions during ascent, emplacement, and subsequent deformation. These fabrics are defined by mineral orientations and their mechanical condition. In general, a plutonic rock can have a fabric that is the result of a sequence of stress fields, from those generated during flow in the magmatic state, to near solidus and subsolidus brittle fabrics due to regional stresses. The use of kinematics to determine the flow in magmatic rocks has been reviewed by Nicolas (1992). The criteria for the determination of those elements of a rock fabric that have an origin by magmatic flow, as opposed to solid-state deformation, are reviewed by Paterson *et al.* (1989). The criteria involve observations at the outcrop and thin section scale: alignment of near liquidus phases indicates flow where crystals are dispersed in a fluid medium. Tiling of crystals can also indicate magmatic flow with a high solid fraction. The presence of plastic strain such as kinking, undulatory extinction, and crystal regrowth indicates deformation in the solid state. When deformation is taking place near the critical melt fraction, the textures can be ambiguous as a transition from magmatic to solid-state flow may occur. Establishing the sense of shear under these conditions, which are of particular interest in mid- to deep-crustal exposures where magma generation and extraction may be occurring, is complex and the interested reader is referred to the study of Blumenfeld and Bouchez (1988) who consider deformation in migmatites and partial melts. The magnetic susceptibility anisotropy of plutonic rocks has also been used to determine flow direction when other means of determining the fabric are inadequate or ambiguous. General agreement is found between the magmatic mineral fabric and the orientation of the magnetic susceptibility anisotropy (Bouchez *et al.*, 1990).

Syntectonic mineral fabrics, that is, mineral fabrics with a component that can be ascribed to both magmatic and regional shear during solidification, provide compelling evidence that magma is, in some instances, intruding *during* tectonic movement. Crustal "openings" and bends, which are the inevitable consequence of shearing, provide access for magmas that reside at deeper crustal levels. Hutton (1988, 1992) and Hutton *et al.* (1990) review a few case histories where magma has intruded along shear zones both vertically and nearly horizontally. Antonellini and Cambray (1992) document magma transport by the process of stepping up along bedding-parallel shear zones in rift systems. Although these models of magma transfer along shear zones appear to provide a resolution to the space problem in the near field, the space must be accommodated elsewhere, ostensibly by thrusting or extension in the far field. Thus, the rate-limiting steps in magmatic ascent will involve some combination of regional stresses and local buoyancy forces related to the magma density contrast.

Among the more provocative corollaries of such models are the suggestions that shearing itself may act as an agent of melt production, or conversely, the presence of melt may act to localize shear. Karlstrom *et al.* (1993) document melt extraction in conjunction with thrusting in a small midcrustal pluton: tectonic movement was instrumental in driving melt segregation within the pluton. Hutton and Reavy (1992) propose that strike-slip transpression thickens the crust, yielding anatectic granitic magmas that will have the usual geochemical signature of crustal melts. This thickening may also yield an undulating Moho, which itself may undergo melting and mixing with mantle diapirs, yielding magmas with geochemical attributes that indicate interaction with mantle-derived materials. Thus, thrusting and transpression provide a means to generate magmas whose geochemical indicators suggest diverse origins, although additional geologic tests need to be articulated to fully test this hypothesis. The presence of melt in shear zones may also act to enhance and/or localize deformation as argued by Davidson *et al.* (1992), who combine a thermal model with kinematic indicators of shearing to demonstrate that large amounts of crustal strain can be accommodated by regions of melt. One key element in this mechanism is that the crust is at near solidus temperatures and hence the melt layer has a long thermal lifetime.

It should be appreciated that kinematic indicators interpreted from the mineral fabric record the last episode of the crystal–melt organization in magma. They record the end stages of the process, whether one of ascent or emplacement, and as noted in the discussion on diapirism, they may record mechanical conditions quite different than those that existed at the time of melt extraction and ascent. Nonetheless, these fabrics are the only structural information directly available.

The tectonic control of the distribution of volcanic vents at both regional and local scales is evident, although complex, perhaps due to the redirection of regional stress fields in the near surface. This is well documented in the study of Bacon (1985) who also notes the correlation between earthquakes and the timing of eruption. More complex relationships between the clustering of cinder cones and the presence of fault zones are recognized in Mexico (Conner, 1990). Vent clustering was found to be pervasive and the azimuths of the clusters appear to be dominantly controlled by the convergence direction, and less so by the orientation of faults. The long-standing conundrum in this regard is the spacing and linear arrangements of volcanic centers in arcs (Jarrard, 1986; Marsh, 1979; Sherrod and Smith, 1990). The positions of these centers often persist for millions of years, erupting a variety of magma types with no clear association with faulting, suggesting instead a control due to processes in the magma supply region. Or it may be that the crust is sufficiently fractured that magmas can at any time exploit an available shear zone, and so the connection between magmatism and shear zones potentially tells us little about the rate-limiting elements in crustal magmatism.

Thus, the distribution and disposition of plutons often indicate some form of structural control, while volcanism reflects structural control in some cases and not in others. One feature that both plutonic and volcanic systems do share is some degree of open system behavior. This is expressed in plutons as repeated intrusion and pluton assembly in an incremental fashion (Harry and Richey, 1963) which can yield a wide variety of zoning patterns (Bergantz, 1990) as well as chemical evidence of magma mixing. In the volcanic record it is expressed by the diversity of magmas erupting from a single center: as individual units often unrelated geochemically in any simple way, or mingling with adjacent magma bodies as demonstrated in the spectacularly exposed Quaternary arc volcano, Tatara–San Pedro in the Chilean Andes (Dungan, 1992).

It appears then that the time-integrated geodynamic and petrologic expression of magmatism might be that of a crustal scale conduit rather than a single chamber (Singer *et al.*, 1989). Implicit in this conduit model is that magmatic systems are assembled from the bottom up, which yields a downward and outward flaring thermal anomaly. The region from which magmatic contributions can be generated by partial melting is potentially much larger at the bottom than at the top and petrologic diversity is generated in this case by crystal–liquid separation by virtue of ascent and open system processes. The style of ascent will vary from processes that are ductile in the deep crust to diking and intrusion along fault-generated bends in the upper crust as rheological conditions will migrate upward with the thermal anomalies. As the thermal anomaly that is driving magmatism in the deep crust migrates upward, it will "consume" the geological evidence of the early stages of magma generation, ascent, and assembly. The end result may be a thermal anomaly on a crustal scale like that interpreted from compressional wave velocity studies at Long Valley, California (Dawson *et al.*, 1990). Geological evidence of the early stages of this process would have to be found in regions where this process has failed, for example, an immature episode of underplating as discussed previously. If melt is generated by anatexis during transpressive orogenesis, a similar style of melt generation and ascent may result. D'Lemos *et al.* (1992) document these transitions in intrusive style with rheological state on a crustal scale where the important physical elements linking regions of melt generation and final pluton formation are "megadikes."

Given the temporal complexity in the spectrum of chemical compositions within single centers, and the presence of intermediate magmas in the midcrust, it is clear that we need to move beyond paradigms of chemical petrology that rely solely on the concept of parental magma chambers as sites for the generation of petrologic diversity. Nor is it clear that simple mechanical models adequately express the many scales at which magma generation, extraction, and ascent occur simultaneously. As discussed previously, basaltic magmatism is a ubiquitous and seemingly necessary element in the generation of crustal melts in some tectonic regimes and represents the first of the likely connections between the mantle and crustal magma chambers. As this "master" perturbation propagates upward in the crust, the conditions and style of magmatic ascent will change as the temperature field changes both spatially and temporally. The success of the methods taken from structural geology in illustrating some of the kinematic aspects of magma ascent and the connection between rheological conditions and temperature suggest that a geodynamic approach to the quantitative description of magmatic processes may provide the missing link in developing scientifically sound models of crustal scale magma genesis.

Acknowledgments

This work was supported by National Science Foundation Grants OCE-90009993 and EAR-9019217 to G. W. B. We are grateful for helpful reviews by C. Wayne Burnham, Juliet McKenna, Peter Reiners, Donna Whitney, and in particular, Steve Wickham. These reviews sharpened our discussion considerably, and we acknowledge that our interpretations are not shared by all the reviewers. The patient and detailed editorial assistance of Michael P. Ryan is gratefully acknowledged.

References

Antonellini, M. A., and Cambray, F. W. (1992). Relations between sill intrusions and bedding-parallel extensional shear zones in the Mid-continent Rift System of the Lake Superior region, *Tectonophysics* **212,** 331–349.

Arndt, N. T., and Goldstein, S. L. (1989). An open boundary between lower continental crust and mantle: Its role in crust formation and crustal recycling, *Tectonophysics* **161,** 201–212.

Ashworth, J. R., and Brown, M. (1990). An overview of diverse responses to diverse processes at high crustal temperatures, *in* "High-Temperature Metamorphism and Crustal Anatexis" (J. R. Ashworth and M. Brown, eds.), pp. 1–18. Unwin Hyman, London.

Bacon, C. R. (1985). Implications of silicic vent patterns for the presence of large crustal magma chambers, *J. Geophys. Res.* **90,** 11243–11252.

Bacon, C. R. (1992). Partially melted granodiorite and related rocks ejected from Crater Lake caldera, Oregon, *Trans. Roy. Soc. Edinburgh* **83,** 27–47.

Barnes, H. A., Hutton, J. F., and Walters, K. (1989). "An Introduction to Rheology," Elsevier, New York.

Bateman, R. (1984). On the role of diapirism in the segregation, ascent and final emplacement of granitoid magma, *Tectonophysics* **110,** 211–231.

Bea, F. (1991). Geochemical modeling of low melt-fraction anatexis in a peraluminous system: The Pena Negra Complex (central Spain), *Geochem. Cosmochim. Acta* **55,** 1859–1874.

Beard, J. S., and Lofgren, G. E. (1989). Effects of water on the composition of partial melts of greenstone and amphibolite, *Science* **244,** 195–197.

Beard, J. S., and Lofgren, G. E. (1991). Dehydration melting and water-saturated melting of basaltic and andesitic greenstones and amphibolites at 1, 3, and 6.9 kb, *J. Petrol.* **32,** 365–401.

Beckermann, C., and Viskanta, R. (1988). Double-diffusive convection during dendritic solidification of a binary mixture, *Physicochem. Hydrodyn.* **10,** 195–213.

Beckermann, C., and Viskanta, R. (1993). Mathematical modeling of transport phenomena during alloy solidification, *Appl. Mech. Rev.* **46,** 1–27.

Bennon, W. D., and Incropera, F. P. (1987). A continuum model for momentum, heat and species transport in binary solid–liquid phase change systems. I. Model formulation, *Int. J. Heat Mass Trans.* **30,** 2161–2170.

Bergantz, G. W. (1989a). Thermal and dynamical state of the crust following underplating: Implications for melt generation and petrologic diversity, *EOS Trans. Am. Geophys. Union* **70,** 1320.

Bergantz, G. W. (1989b). Underplating and partial melting: Implications for melt generation and extraction, *Science* **245,** 1093–1095.

Bergantz, G. W. (1990). Melt fraction diagrams: The link between chemical and transport models, *in* "Modern Methods of Igneous Petrology: Understanding Magmatic Processes" (J. Nicholls and J. K. Russell, eds.), pp. 240–257, Mineralogical Society of America.

Bergantz, G. W. (1992). Conjugate solidification and melting in multicomponent open and closed systems, *Int. J. Heat Mass Trans.* **35,** 533–543.

Blumenfeld, P., and Bouchez, J. L. (1988). Shear criteria in granite and migmatite deformed in the magmatic and solid stages, *J. Struct. Geol.* **10,** 361–371.

Blythe, P. A., and Simpkins, P. G. (1981). Convection in a porous layer for a temperature dependent viscosity, *Int. J. Heat Mass Trans.* **24,** 497–506.

Bohlen, S. R., and Mezger, K. (1989). Origin of granulite terranes and the formation of the lowermost continental crust, *Science* **244,** 326–329.

Bottinga, Y., and Weill, D. F. (1972). The viscosity of magmatic silicate liquids: A model for calculation, *Am. J. Sci.* **272,** 438–475.

Bouchez, J.-L., Gleizes, G., Djouadi, T., and Rochette, P. (1990). Microstructure and magnetic susceptibility applied to emplacement kinematics of granites: The example of the Foix pluton (French Pyrenees), *Tectonophysics* **184,** 157–171.

Brandeis, G., and Jaupart, C. (1986). On the interaction between convection and crystallization in cooling magma chambers, *Earth Planet. Sci. Lett.* **77,** 345–361.

Brown, G. C., and Fyfe, W. S. (1970). The production of granitic melts during ultrametamorphism, *Contrib. Mineral. Petrol.* **28,** 310–318.

Buck, W. R., and Parmentier, E. M. (1986). Convection beneath young oceanic lithosphere: implications for thermal structure and gravity, *J. Geophys. Res.* **91,** 1961–1974.

Burnham, C. W. (1967). Hydrothermal fluids at the magmatic stage, *in* "Geochemistry of Hydrothermal Ore Deposits" (H. L. Barnes, ed.), pp. 34–76, Holt, Rinehart & Winston, New York.

Burnham, C. W. (1979). Magmas and hydrothermal fluids, *in* "Geochemistry of Hydrothermal Ore Deposits" (H. L. Barnes, ed.), 2nd ed., pp. 71–136, Wiley, New York.

Burnham, C. W. (1992). Calculated melt and restite compositions of some Australian granites, *Trans. Roy. Soc. Edinburgh* **83,** 387–397.

Campbell, I. H., and Turner, J. S. (1987). A laboratory investigation of assimilation at the top of a basaltic magma chamber, *J. Geol.* **95,** 155–172.

Cashman, K. V. (1993). Relationship between plagioclase crystallization and cooling rate in basaltic melts, *Contrib. Mineral. Petrol.* **113,** 126–142.

Chappell, B. W., and Stephens, W. E. (1988). Origin of infracrustal (I-type) granite magmas, *Trans. Roy. Soc. Edinburgh* **79,** 71–86.

Chappell, B. W., and White, A. J. R. (1992). I- and S-type granites of the Lachlan Fold Belt, *Trans. Roy. Soc. Edinburgh* **83,** 1–26.

Chappell, B. W., White, A. J. R., and Wyborn, D. (1987). The importance of residual source material (restite) in granite petrogenesis, *J. Petrol.* **28,** 1111–1138.

Chen, Y.-M., and Pearlstein, A. J. (1988). Onset of convection in variable viscosity fluids: Assessment of approximate viscosity–temperature relations, *Phys. Fluids* **31,** 1380–1385.

Chen, Y. D., Price, R. C., and White, A. J. R. (1990). Mafic inclusions from the Glenbog and Bluegum granite suites, southeastern Australia, *J. Geophys. Res.* **95,** 17757–17785.

Cheng, D. C.-H. (1984). Further observations on the rheological behavior of dense suspensions, *Powder Tech.* **37,** 255–273.

Chu, T. Y., and Hickox, C. E. (1990). Thermal convection with large viscosity variation in an enclosure with localized heating, *J. Heat Trans.* **112,** 388–395.

Clemens, J. D. (1984). Water contents of silicic to intermediate magmas, *Lithos* **17,** 273–287.

Clemens, J. D. (1990). The granulite–granite connection, *in* "Granulites and Crustal Evolution" (D. Vielzeuf and P. Vidal, eds.), pp. 25–36, Kluwer, Dordrecht.

Clemens, J. D., and Mawer, C. K. (1992). Granitic magma transport by fracture propagation, *Tectonophysics* **204,** 339–360.

Clemens, J. D., and Vielzeuf, D. (1987). Constraints on melting and magma production in the crust, *Earth Planet. Sci. Lett.* **86,** 287–306.

Conner, C. B. (1990). Cinder cone clustering in the Trans-Mexican volcanic belt: Implications for structural and petrologic models, *J. Geophys. Res.* **95,** 19395–19405.

Crisp, J. A. (1984). Rates of magma emplacement and volcanic output, *J. Volc. Geotherm. Res.* **20,** 177–211.

Cruden, A. R. (1990). Flow and fabric development during the diapiric rise of magma, *J. Geol.* **98,** 681–698.

Cull, J. P. (1990). Underplating of the crust and xenolith geotherms in Australia, *Geophys. Res. Lett.* **17,** 1133–1136.

D'Lemos, R. S., Brown, M., and Strachan, R. A. (1992). Granite magma generation, ascent and emplacement within a transpressional orogen, *J. Geol. Soc. London* **149,** 487–490.

Daines, M. J., and Richter, F. M. (1988). An experimental method for directly determining the interconnectivity of melt in a partially molten system, *Geophys. Res. Lett.* **15,** 1459–1462.

Daly, S. F., and Raefsky, A. (1985). On the penetration of a hot diapir through a strongly temperature dependent medium, *Geol. J. Roy. Astron. Soc.* **83,** 657–682.

David, E., Lauriat, G., and Cheng, P. (1991). A numerical solution of variable porosity effects on natural convection in a packed-sphere cavity, *J. Heat Trans.* **113,** 391–399.

Davidson, C., Hollister, L., and Schmid, S. M. (1992). Role of melt in the formation of a deep crustal compressive shear zone: The Maclaren Glacier metamorphic belt, south central Alaska, *Tectonics* **11,** 348–359.

Davidson, J. P., McMillian, N. J., Moorbath, S., Worner, G., Harmon, R. S., and Lopez-Escobar, L. (1990). The Nevados de Payachata volcanic region. II. Evidence for widespread crustal involvement in Andean magmatism, *Contrib. Mineral. Petrol.* **105,** 412–432.

Dawson, P. B., Evans, J. R., and Iyer, H. M. (1990). Teleseimic tomography of the compressional wave velocity structure beneath the Long Valley region, California, *J. Geophys. Res.* **95,** 11021–11050.

Dell'Angelo, L. N., and Tullis, J. (1988). Experimental deformation of partially melted granitic aggregates, *J. Meta. Geol.* **6,** 495–515.

DePaolo, D. J., Perry, F. V., and Baldridge, W. S. (1992). Crustal versus mantle sources of granitic magmas: A two-parameter model based on Nd isotopic studies, *Trans. Roy. Soc. Edinburgh* **83,** 439–446.

Drummond, B. J., and Collins, C. D. N. (1986). Seismic evidence for underplating of the lower continental crust in Australia, *Earth Planet. Sci. Lett.* **79,** 361–372.

Dunbar, N. W., Riciputi, L. R., Jacobs, G. K., and Christie, W. (1993). Generation of rhyolitic melt in an artificial magma: Implications for fractional crystallization processes in natural magmas, *J. Volc. Geotherm. Res.,* **57,** 157–166.

Dungan, M. (1992). The life history of an Andean volcano, *EOS Trans. Am. Geophys. Union* **73,** 406–407.

England, R. W. (1990). The identification of granitic diapirs, *J. Geol. Soc. London* **147,** 931–933.

England, R. W. (1992). The genesis, ascent, and emplacement of the Northern Arran Granite, Scotland: Implications for granite diapirism, *Geol. Soc. Am. Bull.* **104,** 606–614.

Feeley, T. C., and Grunder, A. L. (1991). Mantle contribution to the evolution of middle tertiary silicic magmatism during early stages of extension: The Egan Range volcanic complex, east-central Nevada, *Contrib. Mineral. Petrol.* **106,** 154–169.

Fountain, J. C., Hodge, D. S., and Shaw, R. P. (1989). Melt segregation in anatectic granites: A thermomechanical model, *J. Volc. Geotherm. Res.* **39,** 279–296.

Fowler, A. C. (1985). A mathematical model of magma transport in the asthenosphere, *Geophys. Astrophys. Fluid Dyn.* **33,** 63–96.

Fowler, A. C. (1990a). A compaction model for melt transport in the Earth's asthenosphere. Part I. The basic model, *in* "Magma Transport and Storage" (M. P. Ryan, ed.), pp. 3–14, Wiley, Chichester/Sussex, England.

Fowler, A. C. (1990b). A compaction model for melt transport in the Earth's asthenosphere. Part II. Applications, *in* "Magma Transport and Storage" (M. P. Ryan, ed.), pp. 15–32, Wiley, Chichester/Sussex, England.

Frith, W. J., Mewis, J., and Strivens, T. A. (1987). Rheology of concentrated suspensions: Experimental investigations, *Powder Tech.* **51,** 27–34.

Fuis, G. S., and Plafker, G. (1991). Evolution of deep structure along the trans-Alaska crustal transect, Chugach Mountains and Copper River basin, Southern Alaska, *J. Geophys. Res.* **96,** 4229–4253.

Ghiorso, M. S. (1985). Chemical mass transfer in magmatic systems. I. Thermodynamic relations and numerical algorithms, Contrib. Mineral. Petrol. **90,** 107–120.

Gill, J., and Condomines, M. (1992). Short-lived radioactivity and magma genesis, *Science* **257,** 1368–1376.

Glazner, A. F. (1991). Plutonism, oblique subduction, and continental growth: An example from the Mesozoic of California, *Geology* **19,** 784–786.

Grissom, G. C., DeBari, S. M., Page, S. P., Page, R. F. N., Villar, L. M., Coleman, R. C., and Ramirez, M. V. (1991). The deep crust of an early Paleozoic arc: The Sierra de Fimbala, northwestern Argentina, *in* "Andean Magmatism and Its Tectonic Setting" (R. S. Harmon and C. W. Rapela, eds.), Special Paper 265, pp. 189–200, Geol. Soc. Am., Boulder, CO.

Grove, T. L., Kinzler, R. J., Baker, M. B., Donnelly-Nolan, J. M., and Lesher, C. E. (1988). Assimilation of granite by basaltic magma at Burnt Lava flow, Medicine Lake volcano, northern California: Decoupling of heat and mass transfer, *Contrib. Mineral. Petrol.* **99,** 320–343.

Halliday, A. N., Mahood, G. A., Holden, P., Metz, J. M., Dempster, T. J., and Davidson, J. P. (1989). Evidence for long residence times of rhyolitic magma in the Long Valley magmatic system: The isotope record in precaldera lavas of Glass Mountain, *Earth Planet. Sci. Lett.* **94,** 274–290.

Handy, M. R., and Zingg, A. (1991). The tectonic and rheological evolution of an attenuated section of the continental crust: Ivrea crustal section, southern Alps, northwestern Italy and southern Switzerland, *Geol. Soc. Am. Bull.* **103,** 236–253.

Harry, W. T., and Richey, J. E. (1963). Magmatic pulses in the emplacement of plutons, *Liverpool Manchester Geol. J.* **3,** 254–268.

Hildreth, W. (1981). Gradients in silicic magma chambers: Implications for lithospheric magmatism, *J. Geophys. Res.* **86,** 10153–10192.

Hildreth, W. (1987). New perspectives on the eruption of 1912 in the Valley of Ten Thousand Smokes, Katmai National Park, Alaska, *Bull. Volc.* **49,** 680–693.

Hildreth, W., Grove, T. L., and Dungan, M. A. (1986). Introduction to the special section on open magmatic systems, *J. Geophys. Res.* **91,** 5887–5889.

Hildreth, W., Halliday, A. N., and Christiansen, R. L. (1991). Isotopic and chemical evidence concerning the genesis and contamination of basaltic and rhyolitic magma beneath the Yellowstone Plateau volcanic field, *J. Petrol.* **32,** 63–138.

Hildreth, W., and Moorbath, S. (1988). Crustal contributions to arc magmatism in the Andes of Central Chile, *Contrib. Mineral. Petrol.* **98,** 455–489.

Hodge, D. S. (1974). Thermal model for the origin of granitic batholiths, *Nature* **251,** 297–299.

Hon, R., and Weill, D. (1982). Heat balance of basaltic intrusion vs granitic fusion in the lower crust, *EOS Trans. Am. Geophys. Union* **63,** 470.

Huppert, H., and Sparks, R. S. J. (1991). Comments on "On convective style and vigor in sheet-like magma chambers" by Bruce D. Marsh, *J. Petrol.* **32,** 851–854.

Huppert, H. E., and Sparks, R. S. J. (1988). The generation of granitic magmas by intrusion of basalt into continental crust, *J. Petrol.* **29,** 599–624.

Hutton, D. H. W. (1988). Granite emplacement mechanisms and tectonic controls: Inferences from deformation studies, *Trans. Roy. Soc. Edinburgh* **79,** 245–255.

Hutton, D. H. W. (1992). Granite sheeted complexes: Evidence for the dyking ascent mechanism, *Trans. Roy. Soc. Edinburgh* **83,** 377–382.

Hutton, D. H. W., Dempster, T. J., Brown, P. E., and Becker, S. D. (1990). A new mechanism of granite emplacement: Intrusion in active extensional shear zones, *Nature* **343,** 452–455.

Hutton, D. H. W., and Reavy, R. J. (1992). Strike-slip tectonics and granite petrogenesis, *Tectonics* **11,** 960–967.

Jacobs, G. K., Dunbar, N. W., Naney, M. T., and Williams, R. T. (1992). Petrologic and geophysical studies of an artificial magma, *EOS Trans. Am. Geophys. Union* **73,** 411–412.

Jarrard, R. D. (1986). Relations among subduction parameters, *Rev. Geophys.* **24,** 217–284.

Jaupart, C., and Parsons, B. (1985). Convective instabilities in a variable viscosity fluid cooled from above, *Phys. Earth Planet. Inter.* **39,** 14–32.

Jomha, A. I., Merrington, A., Woodcock, L. V., Barnes, H. A., and Lips, A. (1991). Recent developments in dense suspension rheology, *Powder Tech.* **65,** 343–370.

Karlstrom, K. E., Miller, C. F., Kingsbury, J. A., and Wooden, J. L. (1993). Pluton emplacement along an active ductile thrust zone, Piute Mountains, southeastern California: Interaction between deformational and solidification processes, *Geol. Soc. Am. Bull.* **105,** 213–230.

Kaviany, M. (1984). Onset of thermal convection in a saturated porous medium: Experiment and analysis, *Int. J. Heat Mass Trans.* **27,** 2101–2110.

Kay, R. W., and Kay, S. M. (1991). Creation and destruction of lower continental crust, *Geol. Rund.* **80,** 259–278.

Kay, S. M., Kay, R. W., Citron, G. P., and Perfit, M. R. (1990). Calc-alkaline plutonism in the intra-oceanic Aleutian arc, Alaska, *in* "Plutonism from Antarctica to Alaska" (S. M. Kay and C. W. Rapella, eds.), Special Paper 241, pp. 233–255, Geol. Soc. Am., Boulder, CO.

Kempton, P. D., and Harmon, R. S. (1992). Oxygen isotope evidence for large-scale hybridization of the lower crust during magmatic underplating, *Geochem. Cosmochim. Acta* **56,** 971–986.

Kerrick, D. M. (1991). Contact metamorphism, *Rev. Mineral.* **26,** 847.

Kirk-Othmer (1978). "Encyclopedia of Chemical Technology," Vol. 21, 3rd ed., Wiley, New York.

Klemperer, S. L., Hauge, T. A., Hauser, E. C., Oliver, J. E., and Potter, C. J. (1986). The Moho in the northern Basin and Range province, Nevada, along the COCORP 40°N seismic-reflection transect, *Geol. Soc. Am. Bull.* **97,** 603–618.

Lachenbruch, A. H., and Sass, J. H. (1978). Models of an extending lithosphere and heat flow in the Basin and Range province, *Geol. Soc. Am. Bull. Mem.* **152,** 209–250.

Lange, R. A., and Carmichael, I. S. E. (1990). Hydrous basaltic andesites associated with minette and related lavas in western Mexico, *J. Petrol.* **31,** 1225–1259.

Leeman, W. P. (1982). Rhyolites of the Snake River Plain–Yellowstone Plateau province, Idaho and Wyoming: A summary of petrogenetic models, *in* "Cenozic Geology of Idaho" (B. Bonnichsen and R. M. Breckenridge, eds.), pp. 203–212, Idaho Geological Survey.

Lister, J. R., and Kerr, R. C. (1989). The effect of geometry on the gravitational instability of a buoyant region of viscous fluid, *J. Fluid Mech.* **202,** 577–594.

Lowell, R. P. (1982). Thermal convection in magmas generated by hot-plate heating, *Nature* **300,** 253–254.

Lowell, R. P., and Bergantz, G. W. (1987). Melt stability and compaction in a partially molten silicate layer heated from below, *in* "Structure and Dynamics of Partially Solidified Systems" (D. E. Loper, ed.), pp. 383–400, Martinus Nijhoff, Dordrecht.

Mahon, K. I., Harrison, T. M., and Drew, D. A. (1988). Ascent of a granitoid diapir in a temperature varying medium, *J. Geophys. Res.* **93,** 1174–1188.

Mahood, G. A. (1990). Second reply to comment of R. S. J. Sparks, H. E. Huppert, and C. J. N. Wilson on "Evidence for long residence times of rhyolitic magma in the Long Valley magmatic system: The isotopic record in the precaldera lavas of Glass Mountain," *Earth Planet. Sci. Lett.* **99,** 395–399.

Manduca, C. A., Silver, L. T., and Taylor, H. P. (1992). $^{87}Sr/^{86}Sr$ and $^{18}O/^{16}O$ isotopic systematics and geochemistry of granitoid plutons across a steeply-dipping boundary between contrasting lithospheric blocks in western Idaho, *Contrib. Mineral. Petrol.* **109,** 355–372.

Mareschal, J.-C., and Bergantz, G. B. (1990). Constraints on thermal models of the Basin and Range province, *Tectonophysics* **174,** 137–146.

Marsh, B. D. (1979). Island arc development: Some observations, experiments and speculations, *J. Geol.* **87,** 687–713.

Marsh, B.D. (1981). On the crystallinity, probability of occurrence, and rheology of lava and magma, *Contrib. Mineral. Petrol.* **78,** 85–98.

Marsh, B. D. (1982). On the mechanics of igneous diapirism, stoping and zone melting, *Am. J. Sci.* **282,** 808–855.

Marsh, B. D. (1989). On convective style and vigor in sheet-like magma bodies, *J. Petrol.* **30,** 479–530.

Marsh, B. D. (1991). Reply to comments of Huppert and Sparks, *J. Petrol.* **32,** 855–860.

McKenzie, D. (1984). The generation and compaction of partially molten rock, *J. Petrol.* **25,** 713–765.

McKenzie, D. (1985). The extraction of magma from crust and mantle, *Earth Planet. Sci. Lett.* **74,** 81–91.

McKenzie, D., and Bickle, M. J. (1988). The volume and composition of melt generated by extension of the lithosphere, *J. Petrol.* **29,** 625–679.

Mereu, R. F., Mueller, S., and Fountain, D. M. (1989). "Properties and Processes of Earth's Lower Crust," Geophysical Monograph 51, Amer. Geophys. Union.

Miller, C. F., Hanchar, J. M., Wooden, J. L., Bennett, V. C., Harrison, T. M., Wark, D. A., and Foster, D. A. (1992). Source region of a granite batholith: Evidence from lower crustal xenoliths and inherited accessory minerals, *Trans. Roy. Soc. Edinburgh* **83,** 49–62.

Miller, C. F., Watson, E. B., and Harrison, T. M. (1988). Perspectives on the source, segregation and transport of granitoid magmas, *Trans. Roy. Soc. Edinburgh* **79,** 135–156.

Morris, S. (1982). The effects of strongly temperature dependent viscosity on slow flow past a hot sphere, *J. Fluid Mech.* **124,** 1–26.

Nelson, K. D. (1991). A unified view of craton evolution motivated by recent deep seismic reflection and refraction results, *Geophys. J. Int.* **105,** 25–35.

Ni, J., and Beckermann, C. (1991). A volume-averaged two-phase model for transport phenomena during solidification, *Metall. Trans. B* **22,** 349–361.

Nicolas, A. (1992). Kinematics in magmatic rocks with special reference to gabbros, *J. Petrol.* **33,** 891–915.

Nicolas, A., Freydier, C., Godard, M., and Vauchez, A. (1993). Magma chambers at oceanic ridges: How large? *Geology* **21,** 53–56.

Noyes, H. J., Wones, D. R., and Frey, F. (1983). A tale of two plutons: Petrographic and mineralogic constraints on the petrogenesis of the Red Lake and Eagle Peak plutons, Central Sierra Nevada, California, *J. Geol.* **91,** 353–379.

Ogawa, M., Schubert, G., and Zebib, A. (1991). Numerical simulations of three-dimensional thermal convection with strongly temperature-dependent viscosity, *J. Fluid Mech.* **233,** 299–328.

Oldenburg, C. M., and Spera, F. J. (1991). Numerical modeling of solidification and convection in a viscous pure binary eutectic system, *Int. J. Heat Mass Trans.* **34,** 2107–2121.

Oldenburg, C. M., and Spera, F. J. (1992). Hybrid model for solidification and convection, *Numer. Heat Trans. B* **21,** 217–229.

Olson, P., Schubert, G., Anderson, C., and Goldman, P. (1988). Plume formation and lithosphere erosion: A comparison of laboratory and numerical experiments, *J. Geophys. Res.* **93,** 15065–15084.

Paterson, B. A., Stephens, W. E., Rogers, G., Williams, I. S., Hinton, R. W., and Herd, D. A. (1992). The nature of zircon inheritance in two granite plutons, *Trans. Roy. Soc. Edinburgh* **83,** 459–472.

Paterson, S. R., Vernon, R. H., and Fowler, T. K. (1991). Aureole tectonics, *in* "Contact Metamorphism" (D. M. Kerrick, ed.), Reviews in Mineralogy, Vol. 26, pp. 673–722.

Paterson, S. R., Vernon, R. H., and Tobisch, O. T. (1989). A review of the criteria for the identification of magmatic and tectonic foliations in granitoids, *J. Struct. Geol.* **11,** 349–363.

Patiño-Douce, A. E., and Johnston, A. D. (1991). Phase equilibria and melt productivity in the pelitic system: Implications for the origin of peraluminous granitoids and aluminous granulites, *Contrib. Mineral. Petrol.* **107,** 202–218.

Peterson, J. W., and Newton, R. C. (1990). Experimental biotite-quartz melting in KMASH-CO_2 system and the role of CO_2 in the petrogenesis of granites and related rocks, *Am. Min.* **75,** 1029–1042.

Pitcher, W. S., and Bussell, M. A. (1977). Structural control of batholith emplacement in Peru: A review, *J. Geol. Soc. London* **133,** 249–256.

Powell, R. (1983). Processes in granulite-facies metamorphism, *in* "Migmatites, Melting, and Metamorphism" (M. P. Atherton and C. D. Gribbs, eds.), pp. 127–139, Shiva, Cheshire.

Quick, J. E., Sinigoi, S., Negrini, L., Demarchi, G., and Mayer, A. (1992). Synmagmatic deformation in the underplated igneous complex of the Ivrea–Verbano zone, *Geology* **20,** 613–616.

Reagan, M. K., Gill, J. B., Malavassi, E., and Garcia, M. O. (1987). Changes in magma composition at Arenal volcano, Costa Rica, 1968–1985: Real time monitoring of open system differentiation, *Bull. Volc.* **49,** 415–434.

Reagan, M. K., Herrstrom, E. A., and Murrell, M. T. (1991). The time scale of magma generation in Nicaragua from ^{238}U-series nuclide abundances, *Geol. Soc. Am. Abst.* A114.

Ribe, N. M. (1985). The generation and composition of partial melts in the Earth's mantle, *Earth Planet. Sci. Lett.* **73,** 361–376.

Ribe, N. M. (1987). Theory of melt segregation—A review, *J. Volc. Geotherm. Res.* **33,** 241–253.

Richter, F. M., Nataf, H.-C., and Daly, S. F. (1983). Heat transfer and horizontally averaged temperature of convection with large viscosity variation, *J. Fluid Mech.* **129,** 173–192.

Rubie, D. C., and Brearley, A. J. (1990). A model for rates of disequilibrium melting during metamorphism, *in* "High-Temperature Metamorphism and Anatexis" (J. R. Ashworth and M. Brown, eds.), pp. 57–86, Unwin Hyman, London.

Rudnick, R. L. (1990). Growing from below, *Nature* **347,** 711–712.

Rudnick, R. L. (1992). Restites, Eu anomalies, and the lower continental crust, *Geochem. Cosmo.* **56,** 963–970.

Rudnick, R. L., and Taylor, S. R. (1987). The composition and petrogenesis of the lower crust: A xenolith study, *J. Geophys. Res.* **92,** 13981–14005.

Rushmer, T. (1991). Partial melting of two amphibolites: Contrasting experimental results under fluid-absent conditions, *Contrib. Mineral. Petrol.* **107,** 41–59.

Rutter, M. J., and Wyllie, P. J. (1988). Melting of vapor-absent tonalite at 10 kbar to simulate dehydration-melting in the deep crust, *Nature* **331,** 159–160.

Ryan, M. P., and Blevins, J. Y. K. (1987). The Viscosity of Synthetic and Natural Silicate Melts and Glasses at High Temperatures and 1 Bar (10^5 Pascals) Pressure and Higher Pressures," U.S. Geological Survey Bulletin 1764.

Sawyer, E. W. (1991). Disequilibrium melting and the rate of melt-residuum separation during migmatization of mafic rocks from the Grenville Front, Quebec, *J. Petrol.* **32,** 701–738.

Scott, D. R., and Stevenson, D. J. (1984). Magma solitons, *Geophys. Res. Lett.* **11,** 1161–1164.

Scott, D. R., and Stevenson, D. J. (1986). Magma ascent by porous flow, *J. Geophys. Res.* **91,** 2973–2988.

Sengun, M. Z., and Probstein, R. F. (1989). High-shear-limit viscosity and the maximum packing fraction in concentrated monomodal suspensions, *Physicochem. Hydrodyn.* **11,** 229–241.

Shaw, H. R. (1972). Viscosities of magmatic liquids: An empirical method of prediction, *Am. J. Sci.* **272,** 870–893.

Sherrod, D. R., and Smith, J. G. (1990). Quaternary extrusion rates of the Cascade Range, Northwestern United States and Southern British Columbia, *J. Geophys. Res.* **95,** 19465– 19474.

Shirley, D. N. (1986). Compaction of igneous cumulates, *J. Geol.* **94,** 795–809.

Shirley, D. N. (1987). Differentiation and compaction in the Palisades Sill, New Jersey, *J. Petrol.* **28,** 835–865.

Singer, B. S., Myers, J. D., Linneman, S. R., and Angevine, C. L. (1989). The thermal history of ascending magma diapirs and the thermal and physical evolution of magmatic conduits, *J. Volc. Geotherm. Res.* **37,** 273–289.

Sleep, N. H. (1974). Segregation of magma from a mostly crystalline mush, *Geol. Soc. Am. Bull.* **85,** 1225–1232.

Smith, M. K. (1988). Thermal convection during the directional solidification of a pure liquid with variable viscosity, *J. Fluid Mech.* **188,** 547–570.

Stevenson, D. J., and Scott, D. R. (1987). Melt migration in deformable media, *in* "Structure and Dynamics of Partially Solidified Systems" (D. E. Loper, ed.), pp. 401–416, Martinus Nijhoff, Dordrecht.

Tepper, J. H., Nelson, B. K., Bergantz, G. W., and Irving, A. J. (1993). Petrology of the Chilliwack Batholith, North Cascades, Washington: Generation of calc-alkaline granitoids by melting of mafic lower crust with variable water fugacity, *Contrib. Mineral. Petrol.* **113,** 333–351.

Thompson, A. B. (1982). Dehydration melting of pelitic rocks and the generation of H_2O undersaturated granitic liquids, *Am. J. Sci.* **282,** 1567–1595.

Thompson, A. B. (1983). Fluid-absent metamorphism, *J. Geol. Soc. London* **140,** 533–547.

Thompson, A. B. (1988). Dehydration melting of crustal rocks, *Rend. Soc. Ital. Mineral. Petrol.* **43,** 41–60.

Thompson, A. B., and Algor, J. R. (1977). Model systems for anatexis of pelitic rocks. I. Theory of melting reactions in the system $KAlO_2$–$NaAlO_2$–SiO_2–H_2O, *Contrib. Mineral. Petrol.* **63,** 247–269.

Thompson, A. B., and Tracy, R. J. (1979). Model systems for the anatexis of pelitic rocks. II. Facies series melting reactions in the system CaO–$KAlO_2$–$NaAlO_2$–Al_2O_3–H_2O, *Contrib. Mineral. Petrol.* **70,** 429–438.

Tikoff, B., and Teyssier, C. (1992). Crustal-scale, en echelon "P-shear" tensional bridges: A possible solution to the batholithic room problem, *Geology* **20,** 927–930.

Touloukian, Y. S., Judd, W. R., and Roy, R. F. (1981). "Physical Properties of Rocks and Minerals," McGraw–Hill, New York.

Trent, D. S., and Eyler, E. E. (1991). "TEMPEST: A Computer Program for Three-Dimensional Time-Dependent Hydrothermal Analysis. Vol. 1. Users Manual," Battelle, Pacific Northwest Laboratory publication PNL-4348, Vol. 1, Rev. 3.

Turcotte, D. L., and Schubert, G. (1982). "Geodynamics: Applications of Continuum Physics to Geologic Problems," Wiley, New York.

Tuttle, O. F., and Bowen, N. L. (1958). Origin of granite in the light of experimental studies in the system $NaAlSi_3O_8$–$KAlSi_3O_8$–SiO_2–H_2O. *Geol. Soc. Am. Memoir* **74.**

Valasek, P. A., Hawman, R. B., Johnson, R. A., and Smithson, S. B. (1987). Nature of the lower crust and moho in eastern Nevada from "wide angle" reflection measurements, *Geophys. Res. Lett.* **14,** 1111–1114.

Vielzeuf, D., and Holloway, J. R. (1988). Experimental determination of the fluid-absent melting relations in the pelitic system, *Contrib. Mineral. Petrol.* **98,** 257–276.

Vielzeuf, D., and Vidal, P. (1990). "Granulites and Crustal Evolution," NATO ASI, Series C, Vol. 311, Kluwer, Dordrecht.

Vigneresse, J.-L. (1994). Granite emplacement and the regional deformation field, *Tectonophysics*, in press.

Voshage, H., Hofmann, A. W., Mazzucchelli, M., Rivalenti, G., Sinigoi, S., Raczek, I., and Demarchi, G. (1990). Isotopic evidence from the Ivrea Zone for a hybrid lower crust formed by magmatic underplating, *Nature* **347,** 731–736.

Weinberg, R. F. (1992). Internal circulation in a buoyant two-fluid Newtonian sphere: Implications for composed magmatic diapirs, *Tectonophysics* **110,** 77–94.

Wells, P. R. A. (1980). Thermal models for the magmatic accretion and subsequent metamorphism of continental crust, *Earth Planet. Sci. Lett.* **46,** 253–265.

Wickham, S. M. (1987). The segregation and emplacement of granitic magmas, *J. Geol. Soc. London* **144,** 281–297.

Wickham, S. M., and Peters, M. T. (1992). Oxygen and carbon isotope profiles from Lizzies Basin, East Humboldt Range, Nevada: Constraints on mid-crustal metamorphic and magmatic volatile fluxes, *Contrib. Mineral. Petrol.* **112,** 46–65.

Wildemuth, C. R., and Williams, M. C. (1984). Viscosity of suspensions modeled with a shear-dependent maximum packing fraction, *Rheol. Acta* **23,** 627–635.

Wolf, M. B., and Wyllie, P. J. (1989). The formation of tonalitic liquids during the vapor-absent partial melting of amphibolite at 10 kbar, *EOS Trans. Am. Geophys. Union* **70,** 506.

Wolf, M. B., and Wyllie, P. J. (1991). Dehydration-melting of solid amphibolite at 10 kbar: Textural development, liquid interconnectivity and applications to the segregation of magmas, *Mineral. Petrol.* **44,** 151–179.

Wyllie, P. J. (1977). Crustal anatexis: An experimental review, *Tectonophysics* **43,** 41–71.

Yardley, B. W. D. (1986). Is there water in the deep continental crust? *Nature* **323,** 111.

Yoder, H. S. (1990). Heat transfer during partial melting: An experimental study of a simple binary silicate system, *J. Volc. Geotherm. Res.* **43,** 1–36.

Younker, L. W., and Vogel, T. A. (1976). Plutonism and plate tectonics: The origin of circum-Pacific batholiths, *Can. Mineral.* **14,** 238–244.

Zingg, A. (1990). The Ivrea crustal cross-section (northern Italy and southern Switzerland), *in* "Exposed Cross Sections of the Continental Crust" (M. H. Salisbury and D. M. Fountain, eds.), pp. 1–20, Kluwer, Dordrecht.

Chapter 14 Two-Component Magma Transport and the Origin of Composite Intrusions and Lava Flows

Charles R. Carrigan

Overview

Observations of layering and other evidence for chemical variations in intrusions and in lava flows are presented to evaluate several fluid dynamical models for two-component effusive volcanism. The newest model involves the hydrodynamic process of encapsulation or, more generally, viscous segregation. Viscous segregation is described as the migration of the low viscosity component of a two-component magma to the high shear regions of a flow that are normally adjacent to the walls of a dike or conduit. The process is explained in terms of the tendency for a two-component flow to have its energy losses to viscous dissipation minimized. In effect, the lower viscosity component becomes a lubricant for the passage of the higher viscosity component. Both laboratory studies of the encapsulation of a higher viscosity layer by a lower viscosity layer during flow in a pipe or slot and numerical simulations show that the viscosity ratio of the two components is the major factor in determining whether a flow evolves to become a zoned, lubricated flow. The lubrication equations for two-component flow in a rectangular dike geometry are derived assuming a power-law rheology. In addition to the substantial reduction in equivalent viscosity resulting from the presence of thin, lower viscosity wall layers, it is found that, for certain power-law magmas, the equivalent viscosity is less for two-component flow than for either of the components flowing singly in the dike. With fracture propagation models, it is shown that lubrication can substantially increase propagation rates for viscous magmas and decrease the probability that solidification will occur in transit. Viscous heating resulting from the concentration of shear into the narrow lubricating layers may explain how such layers of basalt can remain fluid in the presence of both cooler rhyolitic magmas and wall rock. The effect of inlet conditions involving the distribution of the two magmatic components at the entry of the dike is also considered. In this context, a generalized and considerably more realistic model for the simultaneous withdrawal of two magmas through a dike is proposed as an alternative to the often used box and tube withdrawal model. This new model has the advantage that two or more layers can be tapped simultaneously without concern for the draw-up-depth parameter that is conventionally associated with layered withdrawal models. The model is then used to develop a scenario for the contemporaneous effusion of three composite domes in the Inyo volcanic chain of Long Valley, California. Finally, self-lubrication is considered as a means for enhancing the probability that highly silicic magmas in the presence of lower viscosity components will reach Earth's surface. The bias of magma transport toward lubricated silicic magmas reaching the Earth's surface over unlubricated silicic magmas suggests that the frequency of association of two end-member magmas, such as rhyolite and basalt in crustal reservoirs, may be somewhat lower than would be indicated by their occurrence in zoned lava flows and in near surface intrusions.

Notation

		Units
$C_{1,2}$	constants, Eq. (36)	$W \cdot m^{-2}$, $W \cdot m^{-1}$
C_1^c	constant, Eq. (10)	Pa, MPa
C_1^w	constant, Eq. (10)	Pa, MPa
D	constant, Eq. (23), or diameter of extrusion die	$Pa^{-1/n} \cdot s^{-1} \cdot m^{(1+2n)/n}$ or m
E	constant, Eq. (23)	$Pa^{-1/n} \cdot s^{-1} \cdot m^{(1+2n)/n}$
F	constant, Eq. (23)	$Pa^{-1/n} \cdot s^{-1} \cdot m^{(1+2n)/n}$
G	negative of driving pressure gradient, Eq. (9)	$Pa \cdot m^{-1}$
G_{cross}	value of G at which viscosities of two different power-law magmas are equal at interface between magmas	$Pa \cdot m^{-1}$
J	constant coefficient, Eq. (36)	$Pa \cdot s^{-1} \cdot m^{-(1+n)/n}$

Magmatic Systems
Edited by M. P. Ryan

		Units
L	length of tube or extrusion die	m
M_s	elastic shear modulus, Eqs. (27)–(30)	Pa
Q	volumetric flux of magma per unit horizontal length of dike	$m^3 \cdot s^{-1} \cdot m^{-1}$
Re	Reynolds' number ($wd\rho/\mu$)	dimensionless
S	volumetric viscous heating rate	$W \cdot m^{-3}$
T	temperature of magma	°C
T_r	temperature at which magma is rigid	°C
a	width of lubricating layer at dike wall	m
c_p	specific heat at constant pressure	$J \cdot kg^{-1} \cdot °C^{-1}$
d	width of dike	m
$\mathbf{i}$	unit vector across dike	dimensionless
k	thermal conductivity	$W \cdot m^{-1} \cdot °C^{-1}$
$\mathbf{k}$	unit vector along dike (vertical)	dimensionless
l	fracture or dike length, Eq. (27)	m
m	power-law coefficient	$Pa \cdot s^n$
n	power-law exponent	dimensionless
p	fluid pressure	Pa
p_c	fluid pressure (chamber)	Pa
p_o	fluid pressure (outlet)	Pa
q	heat flux	$W \cdot m^{-2}$
t	time	s
u	horizontal velocity component	$m \cdot s^{-1}$
$\mathbf{v}$	flow velocity vector	$m \cdot s^{-1}$
w	vertical velocity component	$m \cdot s^{-1}$
x	horizontal coordinate	m
z	vertical coordinate	m
ϕ_v	dissipation function	s^{-2}
$\dot{\epsilon}$	strain rate in magma	s^{-1}
$\dot{\epsilon}_{cross}$	strain rate at which two power-law magmas have same viscosity	s^{-1}
η	generalized Newtonian viscosity	$Pa \cdot s$
μ	molecular viscosity	$Pa \cdot s$
μ_{loc}	local viscosity, Eq. (19)	$Pa \cdot s$
μ_{eqv}	equivalent viscosity, Eq. (20)	$Pa \cdot s$
ν	Poisson's ratio	dimensionless
ρ	magma density	$kg \cdot m^{-3}$
$\boldsymbol{\tau}$	viscous stress tensor	Pa, MPa
τ_{ij}	*ij*th component of viscous stress tensor	Pa, MPa
c	as sub/superscript, indicates core layer	
w	as sub/superscript, indicates wall layer	
$\lvert\ \rvert$	absolute value	
∇	del or nabla, the gradient operator	m^{-1}

Introduction

Petrologic studies of lava flows and magmatic intrusions tend to stress chemical variations and their spatial relationships, since it is believed that such information can be used to constrain partially the physical and chemical state of the magmatic source region as well as the petrogenic nature of the local processes involved in magmatism. Thus, a full analysis of how multicomponent magmas flow should involve more than just treating the efficacy of transport between the magma reservoir and the surface. It is also necessary to understand how transport along a magmatic pathway can affect and/or produce the observed chemical and physical variations in flows and in shallow intrusions. This is a requirement if one is to distinguish between the characteristics of the flow that are contributed by the source region and those contributions imposed by the transport process itself.

At one end of the range of possible volcanic activity, large and violent eruptions, such as are responsible for the thick ash flow deposits of Long Valley Caldera, California, and Yellowstone, Wyoming, probably are the result of significant overpressuring of magmatic reservoirs that are rich in volatiles. In these eruptions, flow velocities may be great (300–500 $m \cdot s^{-1}$; Wohletz and Valentine, 1990) and the volume of erupted products may be 10% of the total magma reservoir volume (Smith, 1979). Because of the large magmatic driving pressures and the presence of large quantities of gas associated with the exsolution of volatiles that ultimately lead to the complete disruption of the continuous molten phase, the magmatic viscosity is likely to play, at most, a secondary role in the dynamics of this class of eruptions. In particular, magmatic viscosity probably will not significantly affect the ordering of the layered eruption products and the layering on the surface will tend to be a simple inversion of the layering of the contents in the magmatic reservoir (Hildreth, 1981). Toward the other end of the range, eruptions of an effusive nature probably involve much lower magmatic driving pressures and possibly lower magmatic volatile contents than those that are typical of large pyroclastic eruptions. Magmatic viscosity may now exert a much stronger influence in determining what is observed at the surface. In this chapter, the focus is on effusive eruptions involving two different magmatic components. A general model is synthesized for the fluid dynamics of two-component magmas that can explain why the effusive eruption of very viscous magmas is often associated with the occurrence of discrete chemical zoning and why this zoning at the surface is often a reversal of that found in much larger and more energetic eruptions. It is presumed in the model that silicic magmas that are associated with a component having a significantly lower viscosity can be "self-lubricating." In certain cases, the low viscosity component (usually the more mafic component) will encapsulate the higher viscosity component during flow in a dike or conduit, giving rise to the common mafic-to-silicic zoning. It is further argued that silicic magma lubricated by a less viscous component has a greater probability of ascending to the Earth's surface than does the silicic magma by itself. The model also considers those two-component magmas flowing in a dike that do not as readily produce encapsulation of the higher viscosity component by the lower viscosity one. In such flows little zoning may be present. Rather, the mixture, at the surface, takes the form of a magma "emulsion" containing blobs of one component dispersed throughout the continuous "matrix" component.

To provide an observational basis for the fluid dynamical models discussed here, two of the many extensive field investigations of the variations in chemical zoning that are manifested in surface flows and intrusions are summarized. In one example of two-component flows and intrusions, the ranges of chemical and physical differences between the two magmatic components (rhyolite and basalt) are large, whereas in the other example (rhyolite and rhyodacite), chemical analyses are required for the differences to be apparent. The discussion is limited to features that are thought to be archetypal of the petrology of such systems. Some of the earlier models developed to explain the origin of the discrete zoning of rock types are then reviewed. Such models have the common characteristic that the two different magmas involved erupt sequentially from a vent. A more rational model that involves encapsulating and self-lubricating two-component flow is presented and describes how different magmas can erupt *simultaneously* from a vent to form a composite flow in a single-step process. As a reason for the occurrence of viscous segregation, the hydrodynamic implications of the minimization of dissipative shear forces in two-component flows are considered. To support the argument that encapsulation is an important mechanism for the formation of composite intrusions and flows, a collection of results is presented from laboratory-scale experiments and from numerical models of the two-component mode of transport in pipes and channels. Then, a general derivation of the lubrication equations is used to investigate the degree to which encapsulation can mitigate the effects of high viscosity on driving pressure and dike propagation. In addition, the concentration of shear into lubricating layers coupled with viscous dissipation is investigated as a means of offsetting the heat losses that can lead to solidification. On the basis of our understanding of the flow of multicomponent fluids in pipes and channels, we outline the general types of inlet conditions that would be required to produce the kinds of zoning that occur at the surface. We also suggest how different magmatic compositions can be sampled simultaneously by a single dike that breaches the nonhorizontal roof of a magma reservoir to produce a variety of layering in near-surface intrusions and in surface flows. Having now developed a preferred model for two-component magma transport, we apply it to the formation of three contemporaneously extruded composite flows that are part of the much studied Inyo volcanic chain of Long Valley, California.

Observations of Composite Flows, Dikes, and Conduits

Basalt and Rhyolite Systems

An enduring mystery is how basaltic and rhyolitic magmas, with their strongly contrasting chemical and physical properties, must be spatially distributed at depth to produce the contemporaneous extrusion of layered lavas. Linking the observed distribution of the two components at the surface to their initial distribution in a magma chamber is one goal of the modeling of two-component transport that may help resolve a mystery that has petrogenic implications. Another goal of the modeling is to better understand how the thermal and flow regimes may interact to permit the more refractory basalt to flow large distances in a thin layer that loses heat to the wall rock on one side and to the somewhat cooler rhyolite on the other.

Some of the best examples of end-member rocks in composite intrusions and rarer composite lava flows are found in Iceland. A plausible model to explain both the thermal and mafic magma input into the diverging Icelandic crust via dike swarms and sheets requires the emplacement of shallow mafic magma chambers residing several kilometers beneath the surface in the thinnest part of the newly accreting crust (Gudmundsson, 1990; Ryan, 1990). Possibly by the melting of the hydrated crust or by extreme crystal fractionation of mafic magma, or a combination of both these processes (Yoder, 1973; Vogel & Wilband, 1978; Huppert and Sparks, 1980; Vogel, 1982; Furman and Spera, 1985), significant quantities of silicic magmas have been produced and have flowed within the crust and to the surface.

In a study of the Tertiary volcanics of eastern Iceland by Gibson and Walker (1963), five composite lava flows with associated intrusions were mapped (Fig. 1). The intrusions are dikes of the classic or normally zoned type in which a silicic core is sandwiched between layers of more mafic rock (Walker and Skelhorn, 1966). In such intrusions, the mafic margins tend to be thinner than the more silicic core. According to Gibson and

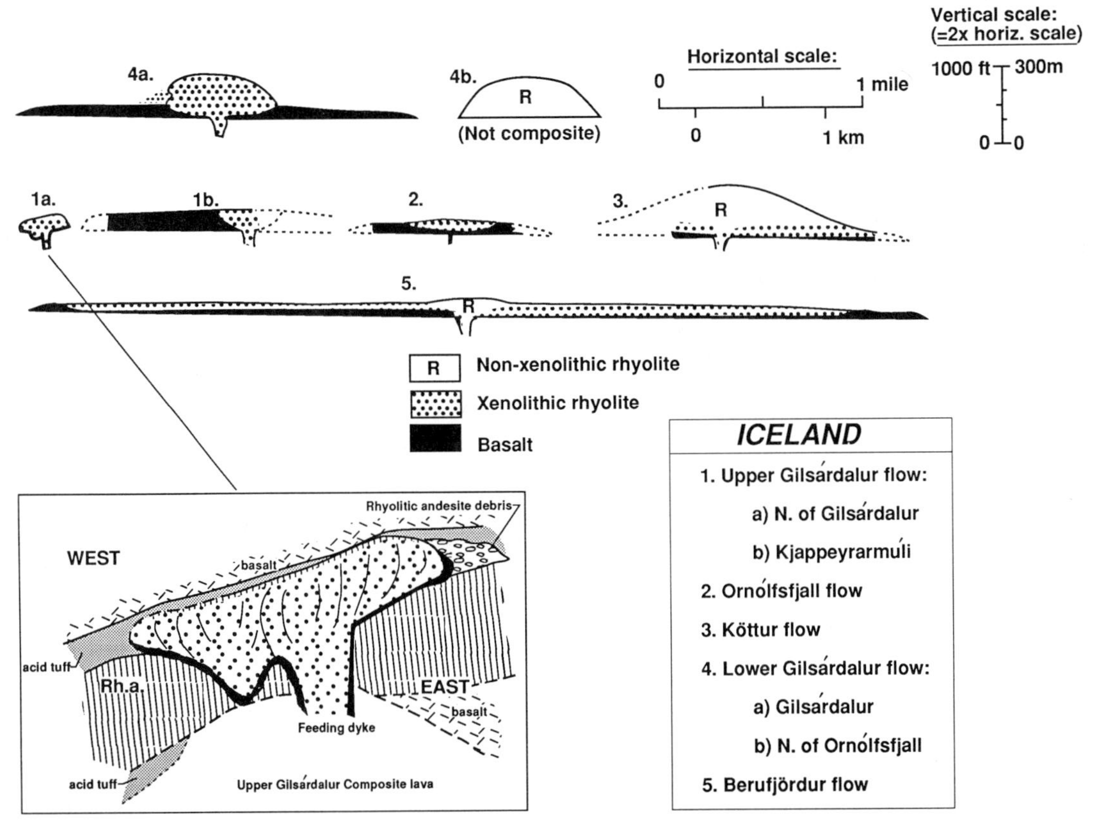

Figure 1 Mapped profiles of five composite rhyolite/basalt lava flows and intrusions located in the Tertiary volcanic region of eastern Iceland. As discussed in the text, several of the profiles clearly indicate that basalt and rhyolite lava effused simultaneously from feeder dikes to form these flows. The inset shows a detailed profile of a composite flow that filled a basin-like depression. The thin layer of basalt following the contours of the basin could only have been produced by the simultaneous effusion of the two lavas. Adapted from Gibson and Walker (1963, Fig. 6) with permission of the Geologists' Association, London.

Walker, all the field evidence suggests that the basaltic and rhyolitic magmas flowed at the same time in feeder dikes and effused simultaneously from these dikes to form layered lava flows. The form and distribution of the basalt and rhyolite, such as the wrapping of the basalt margin around the toe (left side) of the Upper Gilsárdalur flow (Fig. 1a and inset), require the simultaneous flow of these two components. If the two components had effused sequentially, with the basalt coming from the vent first, basalt would have had to completely fill the depression on the west side of the feeding dike before rising any further along the toe. Instead, the basalt forms a thin layer that follows the contours of the basin, a feature that can be explained only by the simultaneous filling of the basin from a *composite layered* feeder dike. Furthermore, from examination of Fig. 1, the relatively close correspondence in the horizontal dimension of the two components of the Köttur and Berufjördur flows (3 and 5) and the complementary nature of variations in the thickness of the basalt and rhyolite layers (1b, 2 and 5) indicate that both components were in the fluid state at the same time and that the effusions of both the upper and lower layers occurred simultaneously. If the magmas had flowed at separate times, there is no reason other than chance to explain the correspondence of the horizontal dimensions of the upper and lower flows or the complementary nature of thickness variations in the two layers. More detailed observations also support the simultaneous effusion of the flows. For example, the rhyolite veins basalt in the Ornólfsfjall flow while the basalt is chilled against the rhyolite. In the Upper Gilsárdalur flow (Fig. 4 of Gibson and Walker, 1963), rhyolite is chilled to a pitchstone at the contact with the host rock only where the thin, insulating basaltic layer is missing. At other locations where the 0.3- to 0.5-m insulating basaltic layer exists, hot basalt has protected the rhyolite from chilling. These observations are consistent with basalt and rhyolite both being molten in each other's presence.

The simultaneous flow of basalt and rhyolite might also be inferred from observations that the rhyolite of composite flows tends to be rich in xenoliths of basalt that are invariably similar to the basaltic component of the flow (Gibson and Walker, 1963). Furthermore, it was found that xenoliths of basalt tend to be more abundant nearer the basaltic layers of the flow and that the billowy nature of these inclusions or enclaves suggests that they were molten at the time of inclusion in the rhyolitic component. Numerical models of two-component flow in a channel (Stockman *et al.*, 1990) show that parcels of one fluid can be ripped off by the shear and entrained in the flow of the other component. Another example of this apparent tendency for mafic xenoliths to detach from mafic margins and be captured by the rhyolitic core flow is well documented by Skelhorn *et al.* (1969) in a map (see their Fig. 12a) of a composite intrusion at Scallastle Bay, Isle of Mull, Scotland.

Rhyodacite and Rhyolite Systems

More subtle chemical variations characterize the zoning in a volcanic conduit that has effused a layered flow, which, along with its feeder conduit, has been the subject of a recent scientific drilling investigation (Younker *et al.*, 1987). Obsidian Dome is one of seven rhyolitic flows making up the 12-km-long Inyo chain that also includes phreatic explosion craters and normal faults that cut across the northwestern boundary of the Long Valley caldera, California. The fact that three of the domes including Obsidian Dome and Inyo Craters, and a collection of explosion craters at the south end of the chain, erupted along a line about 600 yr ago (Miller, 1985; Vogel *et al.*, 1989, and references therein) suggests that these eruptions were the surface manifestations of a dike emplaced near the surface over a 10-km horizontal distance. In fact, a dike having a thickness of 7 m was intersected during another drilling study at a depth of 600 m just south of Obsidian Dome at the northern end of the Inyo chain (Eichelberger *et al.*, 1985; Younker *et al.*, 1987). Although the dike itself does not break the surface, the three contemporaneously erupted domes are associated with conduits that evidently "budded" from the dike sheet (Delaney and Pollard, 1981; Bruce and Huppert, 1989, 1990), causing a significant volume of magma to reach the surface. Since cylindrical feeder conduits are more efficient for magma transport than dikes because the pressure gradient required to drive a given magma flux is lower and also because the rate of heat loss for a given flow rate is lower, it is reasonable to expect that widened parts of the dike, which eventually become conduits, are more likely to permit

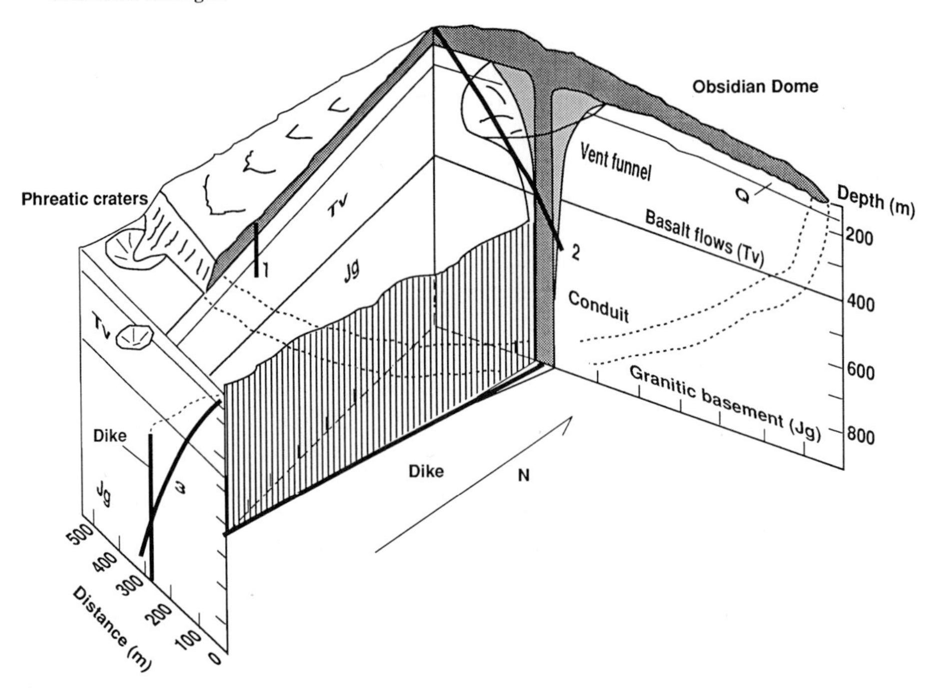

Figure 2 Cross section of the Obsidian Dome composite flow located in the Inyo volcanic chain, Long Valley, California. One core hole sampled the flow near its margin (1) and another slant hole (2) sampled the composite flow near its vent, a basal zone of breccia shed before the advancing margin of the flow and also the composite feeder conduit. The data on compositional zoning presented in Fig. 3 were obtained from hole 2. To test the hypothesis that a subsurface dike connects Obsidian Dome with two others, a third hole was drilled between Obsidian Dome and Glass Creek Dome (see Fig. 22). The hole intersected a shallow, unzoned rhyolitic dike. Adapted from original figure provided courtesy of J. C. Eichelberger.

magma to reach the surface and are favored for long-term ascent.

Obsidian Dome is a kidney-shaped flow with a horizontal dimension that varies from 1 to 2 km and a thickness that varies from 100 m over the more or less central conduit to about 50 m at the edge of the flow (see Vogel *et al.*, 1989, and accompanying references). Figure 2 illustrates the relationships between the dome, its feeder conduit, and the associated dike. This diagram attempts to synthesize the surface geology and observations obtained from three core holes. One hole was drilled about 150 m from the dome margin to sample the distal part. Another hole was slanted to penetrate both a central part of the dome and the feeder conduit. As already mentioned, the third hole was drilled a short distance away from the dome and slanted to intersect the hypothesized subsurface dike that was encountered.

The slant hole through the central dome and conduit (2 in Fig. 2) provided cores of the flow, basal breccia, and the feeder that strongly support a model for the contemporaneous effusion of rhyolitic and less viscous rhyodacitic lavas that compose Obsidian Dome (Carrigan and Eichelberger, 1990). The arrangement of a more viscous magma overlying one of lower viscosity is shared with the two other domes erupted at about the same time in the Inyo Chain (Sampson and Cameron, 1987) and with a third lava flow on the Medicine Lake Highland in California (Eichelberger, 1975).

The connection between zoning in the Obsidian Dome lava flow and zoning in the feeder conduit is shown in the chemical profiles (Fig. 3) of the drill cores obtained from the flow/conduit hole. The cores clearly show that the more silicic rhyolite flowing up the center of the conduit formed the overlying layer and the lower viscosity rhyodacite flowed along the margin of the conduit and vented to form the layer at the bottom of the flow. Another dome near Obsidian Dome that was exposed in cross section by an explosion exhibits its flowlines in its texture that support both this

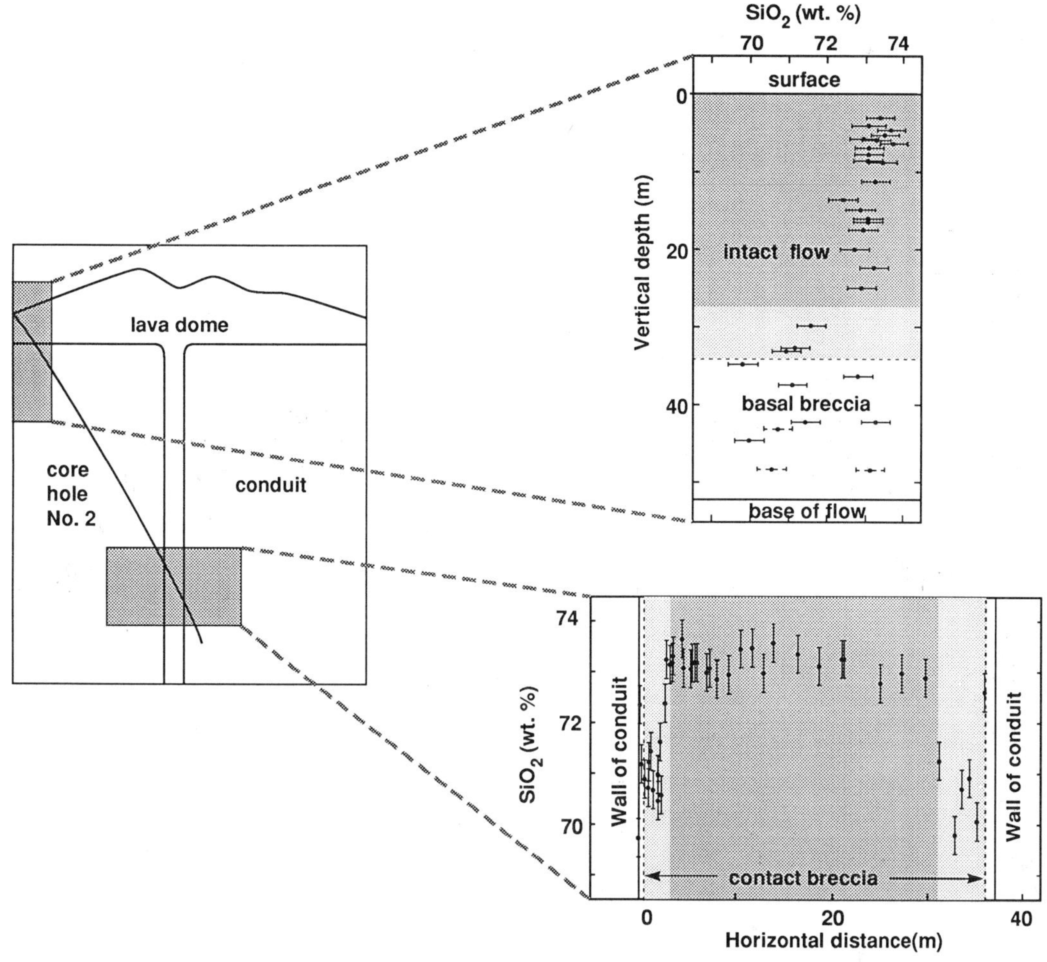

Figure 3 The silica content of cores from hole 2 of Obsidian Dome, Long Valley, California, as a function of location in the flow and in the feeder conduit. The less silicic and presumably the less viscous component occupies regions at the walls of the conduit and at the base of the intact flow that rests on a zone of basal breccia. The more silicic and more viscous component composes the center of the conduit and the top layer of the flow. This arrangement indicates that distinct mafic and silicic layers of magma flowed simultaneously up the conduit and effused onto the surface, with the marginal mafic layer forming the base of the intact flow and the central silicic layer forming the topmost layer of the flow. Adapted from Carrigan and Eichelberger (1990) with permission from *Nature* and Macmillan Magazines Limited.

model of effusion and the simultaneous venting of the two chemically distinct components of the lava (Fig. 4). The simultaneous transport of the two components is also consistent with the observation that both layers of the flow have margins that are everywhere nearly coincident (Carrigan *et al.*, 1992). This would not be expected if both flows arrived at the surface separately. Several finer-scale observations also support the conclusion that both lavas vented together (Carrigan and Eichelberger, 1990). The chemical boundary between the two lavas is physically continuous in the flow and in the conduit with fine interbanding present at the boundaries of the two components. In addition, both rhyolite and rhyodacite are present in the basal breccia of the flow on which the lower layer consisting of rhyodacite lies. This brecciated layer, which underlies the flow, has been created from the slaggy crust falling off the margin of the flow just ahead of the advancing flow front. If the lower rhyodacitic layer had effused from the vent alone and was followed later

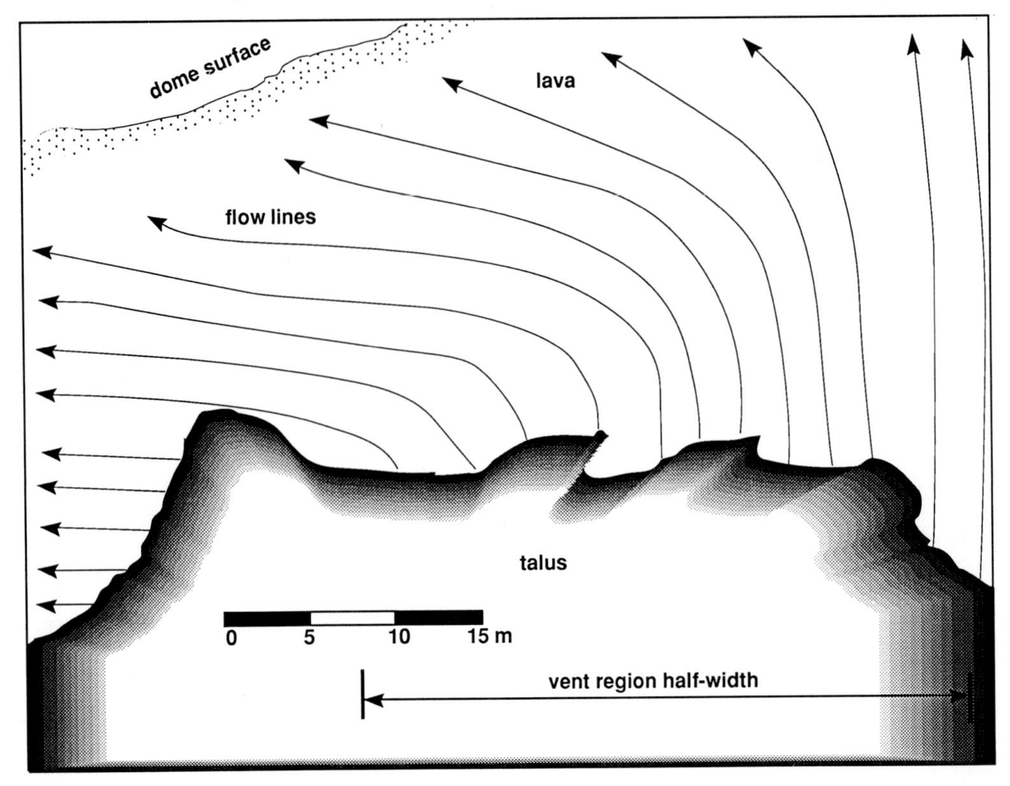

Figure 4 Photomosaic-based illustration of explosively sectioned flow near Obsidian Dome. Flow lines are observable as variations in texture and support the interpretation of drill cores (hole 2) that the dome was emplaced by the simultaneous flow of a marginal zone of magma that flowed out horizontally to form a basal layer, and a central zone that flowed out on top of the marginal layer to form the top of the flow. Adapted from Carrigan and Eichelberger (1990) with permission from *Nature* and Macmillan Magazines Limited.

by the rhyolitic layer, the underlying breccia would not contain remnants of the crust of the upper rhyolitic layer.

Models of the Origin of Mafic-to-Silicic Zoning

Sequential Magma Transport

A traditional view of the origin of composite intrusions and flows with mafic-to-silicic zoning might be referred to as the sequential emplacement model. In the basic scenario, the mafic and silicic magmas may flow contemporaneously but also sequentially (one behind the other in the same dike). Initially, the mafic magma somehow fills the dike or conduit first and may vent at the surface, forming the lowest layer of a flow. Immediately following the rise of the mafic component, the silicic component penetrates and flows along the center of the molten mafic dike or conduit, which gives rise to the composite nature of the intrusion. At the surface, the silicic lava flows out upon the more mafic layer to form the composite flow. Blake *et al.* (1965) suggest a version of this model for the origin of the composite basalt and rhyolite flows illustrated in Fig. 1. Such a sequential scenario may aid the rhyolite in rising to the surface in two ways. The initial flow of basalt into the dike insulates the silicic magma from the cool wall rock. The hotter basalt may also heat the rhyolite by a few tens of degrees and decrease its viscosity, allowing it to flow in the dike more easily.

Basalt and rhyolite lavas appear to have extruded sequentially to produce the Kjappeyrarmúli or Ornólfsfjall flows of Iceland that exhibit mainly *horizontal variations* in lava type (Fig. 1, flows 1b and 2). According to the sequential model, the basalt would have preceded the rhyolite to the surface, forming a layer that was pushed outward from the feeder upon the arrival of the rhyolitic magma. How the denser basalt, which tends to underlie rhyolite in the typical model of a stratified chamber, gets ahead of the rhyolite in the dike poses an interesting problem that will be considered further. An alternative model proposed here is that the basalt and rhyolite were simultaneously withdrawn and simultaneously flowed in the dike or conduit—side-by-side—in the normally zoned arrangement, which is shown here to be dynamically stable. Because of its lower viscosity, the basalt could, in certain circumstances, flow out of the vent from its margins at a higher rate, covering the surface ahead of the more slowly advancing rhyolite flow effusing from the central part of the vent.

The sequential model cannot explain flows that are characterized by predominantly *vertical variations* in lava type involving more or less coincidental margins and finer-scale features, indicating simultaneous flow such as the Berufjördur, Köttur, and Upper Gilsárdalur flows (Fig. 1, flows 5, 3, and 1a). *To obtain such layered flows requires both lava types to issue from the vent simultaneously.* The sequential model is also inappropriate in the case of Obsidian Dome, where both lava types must have arrived simultaneously at the surface in a core–annular arrangement with a more mafic rhyodacite in the annular part of the flow adjacent to the conduit wall and a more silicic rhyolite occupying the core region of the flow.

Simultaneous Transport of Different Magma Types

Various numerical and laboratory experiments have shown that withdrawing the contents of a gravitationally stratified magma reservoir through a tube or slot centered in the lid of the reservoir can extract magma from different layers simultaneously (Spera, 1984; Blake and Fink, 1987; Blake & Ivey, 1986; Spera *et al.,* 1986; Trial *et al.,* 1992). A parameter that determines whether a particular layer will be tapped during withdrawal is the so-called draw-up-depth. For a given reservoir geometry and set of fluid properties, the draw-up-depth is a function of the withdrawal rate through the tube. Layers shallower than the draw-up-depth will be tapped whereas the contents of deeper layers will not be entrained as magma flows into the tube. In principle, the withdrawal process can explain how magmas from different layers in a stratified reservoir can be tapped simultaneously through a conduit even if the deeper layers are separated from the outlet by intervening magma layers. Such a draw-up process may be particularly applicable to the very large eruptions that are characterized by exceedingly large magma withdrawal rates through the volcanic feeder dike/conduit. The mass flow rates that are thought to be typical of such eruptions fall between 10^5 and 10^9 kg $\cdot$ s^{-1} (Trial *et al.,* 1992).

For the conventional model of a stratified magma chamber with a small diameter cylindrical conduit centered on top, the draw-up effect is critical to explaining the origin of composite layering in the intrusions and in the flows of effusive volcanism even though the effect is likely to be much weaker owing to the much lower magma withdrawal rates of effusive eruptions (10^{-1}–10^2 kg $\cdot$ s^{-1}; Trial *et al.,* 1992). In a later section, we examine the implications of the conventional model for transporting different magmas and we present a more general model for a magma reservoir + transport system as an alternative to the symmetrical reservoir and pipe arrangement. The new model is more realistic in terms of our understanding of both how eruptions are initiated and how magmatic pathways evolve, and it permits withdrawal of two or more layers simultaneously without regard to magma withdrawal rates and the draw-up effect. This new model offers a better fit to the extensive observations of the Inyo Domes system than do conventional models.

The withdrawal of two layers simultaneously from a stratified magma chamber will tend to produce a flow regime in which the lower (more dense) layer occupies the central region of a conduit while the upper (less dense) layer forms a surrounding annular regime (Fig. 5). Since the mafic layer in a two-component system is usually the more dense, mafic magma will initially tend to occupy the central region of flow during withdrawal through a cylindrical conduit and the silicic magma will make up the outer annular region. How the transport of two magmas might

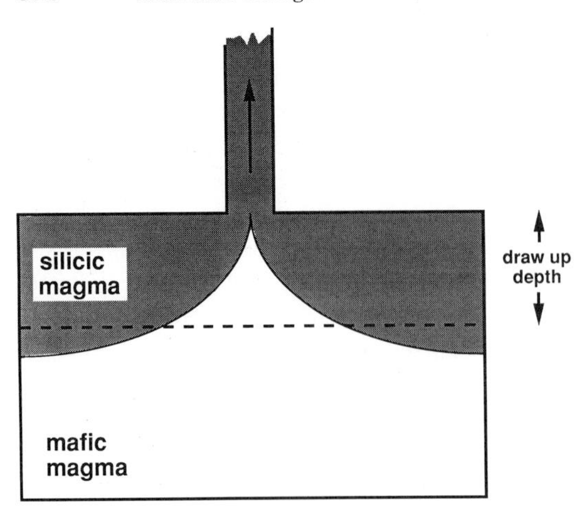

Figure 5 Cross section of a conventional model of withdrawal from a stratified magma chamber consisting of a flat-lidded box with a cylindrical tube centered in the top. If the withdrawal rate is sufficient, i.e., the draw-up-depth is large enough, the lower mafic layer will be drawn up simultaneously with the overlying silicic layer. In this case, the *initial* configuration at the inlet of the conduit will be a core–annular flow with more viscous silicic magma adjacent to the walls and less viscous mafic magma occupying the core. Experiments and theory indicate that such an arrangement of magmas with the higher viscosity magma adjacent to the wall cannot persist in the conduit. Such a box and tube model for magma withdrawal is likely to be too restrictive regarding the simultaneous withdrawal of both the upper and the lower layers since there is no reason why the magma chamber top or lid must be perfectly parallel to the layering such that only the top layer is in contact with the lid in the absence of withdrawal. Furthermore, a good case can be made that withdrawal will usually take place through a slot-like orifice attached to a tabular structure such as a dike rather than through the localized entry of a cylindrical conduit. This model is not geologically realistic and conveys false illusions about how real magma reservoirs work.

allow this initially reversed orientation of zoning to be switched to the more normally observed mafic- (annular) to-silicic (core) zoning has been the object of laboratory experiments (Koyaguchi, 1985; Blake and Campbell, 1986).

Model for Simultaneous Inlet Flow Transformed to Sequential Flow at the Outlet

In an experiment intended to model the dynamics of pressure-driven, two-component magma transport in a long cylindrical conduit, Blake and Campbell (1986) used the gravity-induced flow of two fluids having nearly the same density but differing viscosities in a vertical cylindrical tube. To simulate the inlet arrangement expected for withdrawal from a reservoir with a silicic layer overlying a mafic layer, they produced a core–annular flow at the inlet of the tube with a central lower viscosity region and an annular higher viscosity region next to the wall of the tube. This was accomplished by metering lower viscosity fluid along the central axis of the flow of higher viscosity fluid. By adjusting the rate at which the lower viscosity component was metered into the flow of the higher viscosity component, they found that the stability of the flow was affected. For the highest rates of flow of the central low viscosity component relative to the flow rate of the higher viscosity component, the central zone became wavy and appeared to shed streamers of the central lower viscosity component into the higher viscosity flow. Blake and Campbell (1986) used these observations to argue correctly that the arrangement of high viscosity magma in an annular zone and low viscosity magma in the core zone of a volcanic conduit, as an inlet condition of their hypothesized withdrawal model, would lead to mixing (breakup of the low viscosity core/high viscosity annular flow) of the two magmas at Reynolds' numbers far less than the critical value required for the onset of turbulence in pipe flow. However, they also argued that composite intrusions such as those of Fig. 1 could not be treated as "snapshots" of an entirely liquid structure. It was suggested, using the results of Koyaguchi (1985), that during the vertical propagation of a dike, the lower viscosity mafic magma could ascend in the center of a propagating, silicic magma-filled dike and overtake the more viscous component at the propagating tip of the dike. As the dike propagated upward beyond this point, the mafic component would advance ahead of the silicic component, filling the dike and plating its walls. The silicic magma would then follow along, flushing the dike of the mafic component except at the walls where the more refractory mafic magma had already chilled. According to Blake and Campbell (1986), the result of this overtaking process is to transform simultaneous, core–annular flow at depth to the sequential emplacement of the two magmatic components at shallower levels in the crust (see Fig. 9 of their paper). As such, this model cannot explain those observations requiring the simultaneous emplacement and venting of lavas. In addition, the Blake– Campbell and Koyaguchi models also require that the simultaneous withdrawal of the two components be coincident with the vertical propagation of the dike. Such a restriction on timing is not necessary for producing composite intrusions and flows in the model developed here and discussed later.

Model for Encapsulating and Self-Lubricating Flow

A contrasting view for the origin of most normally zoned intrusions is presented in this chapter. It is argued here and elsewhere (Carrigan and Eichelberger, 1990; Carrigan *et al.*, 1992) that in many, but possibly not all, cases, composite intrusions *do* represent snapshots in time of zoned structures that were formed by two magmas flowing together in adjacent layers. *We argue here that the mafic-to-silicic zoning observed in dikes and conduits is often a natural consequence of the fluid dynamics of the simultaneous transport of two magmas with differing viscosities.* Pressure differences arising between the magmas will cause the encapsulation of the more viscous component by the second, less viscous component. Thus, the lower viscosity component isolates the higher viscosity magma from zones of strong shear at the walls and acts as a lubricant for the passage of the higher viscosity component (Fig. 6). This lubrication enhances the probability that a viscous magma will reach the surface by reducing the required pressure gradient. In the case of the Deadman Dome flow in the Inyo chain, the lubrication of crystal-rich, high silica rhyolite by a less viscous rhyolite reduced the required pressure gradient to 1% of that needed by the crystal-rich magma to reach the surface alone (Carrigan *et al.*, 1992). Even if the withdrawal of magma layers dictates in some cases a silicic- (wall) to-mafic (central) reversed ordering at the inlet, we argue that such a state is unstable and will break down quickly into the simultaneous flow of the two magmas with a mafic- (wall) to-silicic (central) ordering without invoking Koyaguchi's overtaking model. In considering inlet conditions that give rise to lubricated, core–annular flow, we also propose a considerably more realistic alternative model for simulating magma pathways and chambers than the conventional withdrawal model employing a symmetric, boxlike geometry. In addition to providing a basis for understanding the origin of *simultaneous* side-by-side, two-component flow, it is further argued that this alternative model can be used to explain observations where the lower vis-

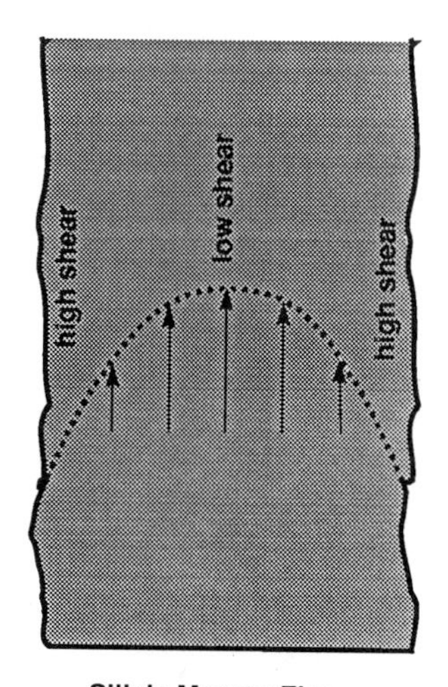

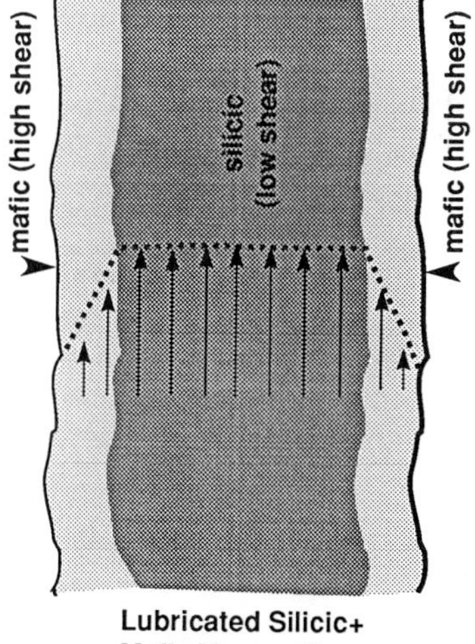

Figure 6 The flow of a very viscous, single-component, silicic magma will include zones of high shear and high dissipation near the walls of a dike (left). Encapsulation of the silicic magma component by the mafic magma component during flow in a dike leads to isolation of the higher viscosity component from the zones of high shear at the dike walls (right). Most of the shear (velocity gradient) will be concentrated in the mafic layers while the central, higher viscosity, silicic magma will tend to be characterized by little-to-vanishing shear; i.e., it will tend to be transported as a solid body carried between the lubricating layers. Lubrication of the more viscous component can substantially decrease the pressure gradient required to maintain a given flow rate in the dike. Alternatively, lubrication can greatly increase the flow rate of magma in the dike for a given pressure gradient. This can be important for determining whether a dike can propagate to the surface before solidifying.

cosity magma could have flowed ahead of the higher viscosity component in a conduit to effuse, for at least the initial part of the eruption, *sequentially* from a vent, e.g., Fig. 1, flows 1b and 2.

Fluid Dynamics of the Simultaneous Flow of Two Magmatic Components

When two fluid layers of differing viscosity are driven by a pressure gradient in a channel or tube, an interesting process is initiated. As the two layers flow along the channel, the lower viscosity layer will gradually flow around the higher viscosity component encapsulating it (Fig. 7). This encapsulation of the higher viscosity component also results in self-lubrication of the flow since the lower viscosity component now occupies all zones of strong shear near the walls and the higher viscosity component is relegated to a zone of weaker to vanishing shear in the center of the flow. This encapsulation/self-lubrication process is robust and has been long used to practical advantage in industry for transporting viscous oil in pipelines by the addition of water (Charles and Redberger, 1962; Oliemans and Ooms, 1986) and for forming composite plastic extrusions (Southern and Ballman, 1973; Han, 1975, 1976, 1981; Minagawa and White, 1975; White and Lee, 1975a; Karagiannis *et al.*, 1988).

Why encapsulation occurs has often been explained in terms of the minimization of the energy dissipated by viscosity in the process of pumping a given mass flux of two different liquids in a channel or pipe (MacLean, 1973; Everage, 1973; Williams, 1975). A higher viscosity core and a lower viscosity annular (near wall) zone produce lower energy dissipation for a given mass flux than either the opposite arrangement of higher viscosity at the wall and lower viscosity in the core (Joseph *et al.*, 1984a) or a side-by-side arrangement in which two different viscosity layers are fed together (one on top of or on the side of the other) into a pipe or channel (Karagiannis *et al.*, 1988). Another way of determining the preferred arrangement of high and low viscosity layers in a flow is to perform a stability analysis. Generally, flows that are not in a preferred state for a given mass flux will be unstable to perturbations that grow with time. Ultimately, the flows evolve to a preferred state that is stable. The case of a lower viscosity core flow surrounded by a higher viscosity annular flow in a circular pipe was shown to always be unstable at any value of the Reynolds' number, i.e., at any flow rate (Hickox, 1971). On the other hand, a tube with a higher viscosity liquid occupying the central zone and a lower viscosity liquid flowing in the annular zone was found to be stable to applied perturbations over a wide range of Reynolds' numbers (Joseph *et al.*, 1984b). However, it was also found that if the zone of lower viscosity at the wall of the tube becomes too thick (>30% of the tube radius), the core–annular flow regime becomes unstable. The authors speculated that the layering might break up to form an emulsion in such cases. It is interesting to note that all the intrusions illustrated in Fig. 1 have basaltic margins characterized by thicknesses that fall well within the stability criterion obtained by Joseph *et al.* (1984b); i.e., none are too thick for the flow to be regarded as unstable. Furthermore, the generalized description of a composite intrusion with its thinner mafic margins (Walker and Skelhorn, 1966) is also con-

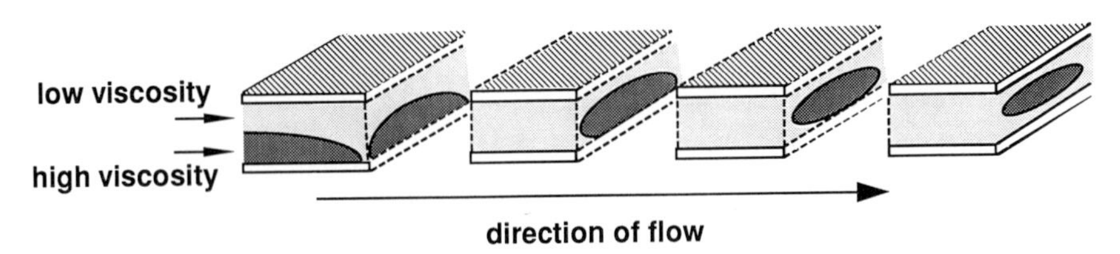

Figure 7 A channel sectioned at different locations along the direction of flow of two liquid layers with different viscosities that are injected side-by-side illustrates the progressive nature of the encapsulation or self-lubrication process. After entering the channel the initially flat interface between the layers begins to be distorted. The interface between the two liquids increasingly bends back on itself at each successive station downstream as the lower viscosity component (light shading) of the flow encroaches on the contact between the wall and the higher viscosity component (dark shading). Eventually, the higher viscosity component is completely isolated from the high shear wall region by the lower viscosity liquid.

sistent with emplacement by a core–annular regime that is stable according to the criterion of Joseph *et al.* It is possible that magma "emulsions" consisting of a silicic component with blobs or pillows of mafic magma with little or no mafic layering (see Gautneb, 1988, Fig. 5) may be produced when there is so much mafic magma that the layers are too thick to be stable.

Polymer Coextrusion Experimental Analogues

Of all the encapsulation studies, coextrusion experiments involving molten polymers have dynamics that are probably most like those characterizing the transport of magmas. The dynamics of flowing magmas are dominated by their high viscosities. Depending on parameters such as temperature, volatile content, and crystal content, the rheology of magma is typically found to vary from purely Newtonian (vanishing dependence of viscosity on shear rate) to a shear thinning power-law behavior (Shaw, 1969; Ryan and Blevins, 1987; Ryerson *et al.*, 1988; Spera *et al.*, 1988). The flow of molten polymers in capillary extrusion dies is also dominated by viscosity. As in the case of magmas, polymers can exhibit both Newtonian and power-law behavior (Han, 1981). In both two-component polymers and magmas, the surface tension can be neglected compared with the viscous effects. Polymers may also have elastic behavior, but this additional aspect is not responsible for the encapsulation process (White and Lee, 1975b). Encapsulation is also not limited to either purely Newtonian or power-law rheology. The process is observed for a variety of different liquids having a wide range of material properties and seems to depend only on viscosity contrasts between the liquids involved.

Another feature of polymer coextrusion experiments common to magma flow in a dike or conduit is that any inertial forces in the flow can be neglected compared with the viscous forces. Thus, the Reynolds' number, which represents the ratio of the inertial to the viscous forces, is much less than unity for both the coextrusive magmatic and polymeric flows of interest. The similarities in the physical properties of the molten materials and the gross dynamics of the respective flow systems suggest strongly an excellent correspondence between the coextrusion experiments and the simultaneous flow of two magmas in a dike.

The results of a coextrusion experiment (Fig. 8) performed by Han (1973) illustrate the effects of viscosity differences on the side-by-side flow of two polymers having different viscosities. Using cylindrical dies of different aspect ratios, it was found that the lower viscosity component (low density polyethylene) gradually flowed around the higher viscosity component (polystyrene) to completely encapsulate it. Although the higher viscosity component never became centered in the lower viscosity encapsulant for dies with length-to-diameter (L/D) ratios of up to 18, it was fully encapsulated (separated from the die wall) within only four diameters of the die inlet. In polymer experiments the exact details of the encapsulation/self-lubrication process, such as the interface shape, may vary in ways that depend on the rheology of a particular melt system. (In flowing polymeric systems, interfacial effects between two components will in general be a function of the viscosity and elasticity differences in the two layers.) However, available experimental evidence supports the conclusion (Han, 1981) that encapsulation in two-polymer systems is determined by the differences in viscosity just as it is in purely Newtonian systems.

Another set of experiments by Minagawa and White (1975) have produced similar results for the side-by-side extrusion mode in both a tubular die and a rectangular (dike-like) die. The coextrusion experiment using tubular geometry (Fig. 9a) illustrates the gradual bending around of the interface between the melts as the higher viscosity (dark) component is encapsulated by the lower viscosity component with increasing distance down the tube. As a result, the high shear, contact area between the higher viscosity component and the tube wall gradually decreases with distance down the tube. Encapsulation would eventually isolate the higher viscosity layer from the wall, as in the case of Han's (1973) experiment, if a longer tubular die had been used. The rectangular slot experiments (Fig. 9b) varied both the viscosity ratio of the layers and their initial orientation at the inlet of the slot. In the cases where both components had the same viscosity (viscosity ratio equal to 1), no encapsulation occurred so that the interfaces remained essentially flat at the outlet for both the horizontal and the vertical initial orientations. When both layers had different viscosities (viscosity ratio different from 1), the interface be-

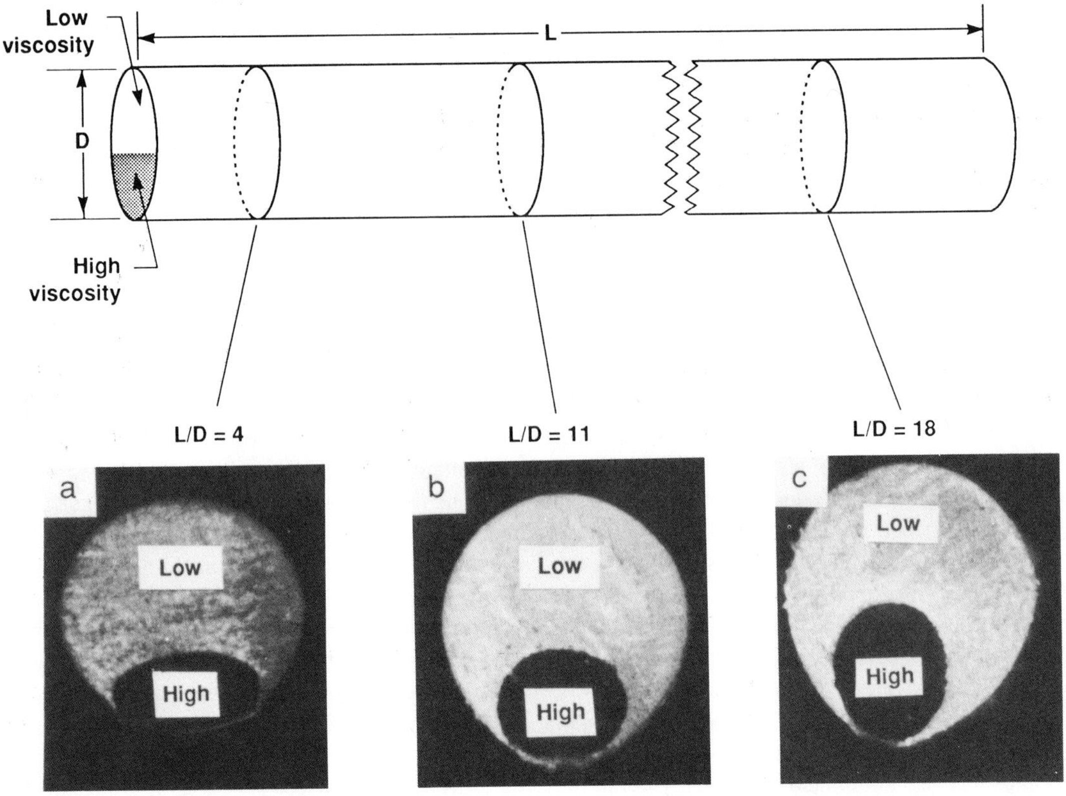

Figure 8 A coextrusion experiment involving two molten polymer layers injected side-by-side into tubular capillary dies having different length to diameter (L/D) ratios illustrates the encapsulation of the higher viscosity component. The ratio of the viscosities indicated for the two layers varies from 0.14 to 0.34 over the range of extrusion rates studied. Careful examination of the extrusion at $L/D = 4$ shows that the lower viscosity component has already almost completely flowed around the higher viscosity component. With greater distances downstream the higher viscosity component migrates away from the wall toward the center of the flow. Adapted from C.D. Han (1973) with permission of John Wiley and Sons, Inc.

tween the higher and the lower viscosity components curved around so as to isolate the higher viscosity melt from the wall of the slot by interposing an intervening layer of low viscosity melt—by the time the layers reached the outlet 15 slot thicknesses downstream from the inlet. The slot experiments suggest that only relatively small deviations of the viscosity ratio from unity (<1.56) are required for encapsulation to occur over distances (L) measured in terms of slot thickness (D) that are small ($L/D < 100$) compared to the typical vertical lengths of dikes ($L/D > 1000$). For comparison, the viscosity ratio is estimated to fall between 2 and 200 for the extruded lavas of the Inyo Domes (Carrigan and Eichelberger, 1990).

An alternative inlet condition to side-by-side, two-component flow into conduits or dikes is core–annular flow with the higher viscosity melt adjacent to the wall and the lower viscosity melt in the center of the flow. This configuration, of course, is a reversal of what is preferred from energy minimization considerations. Such an inlet arrangement is anticipated if the simultaneous withdrawal of an upper silicic and lower mafic layer occurs from a stratified magma chamber (Blake and Campbell, 1986). In the engineering and polymer physics fields, this ordering of layering at the inlet has been the subject of both theoretical (Hickox, 1971; Joseph *et al.*, 1984b) and laboratory investigations (Han, 1975). The stability analyses of Hickox and Joseph *et al.* show that this particular inlet arrangement is unstable. A coextrusion experiment by Han employing this reversed core–annular inlet condition produced additional details regarding the instability of reversed

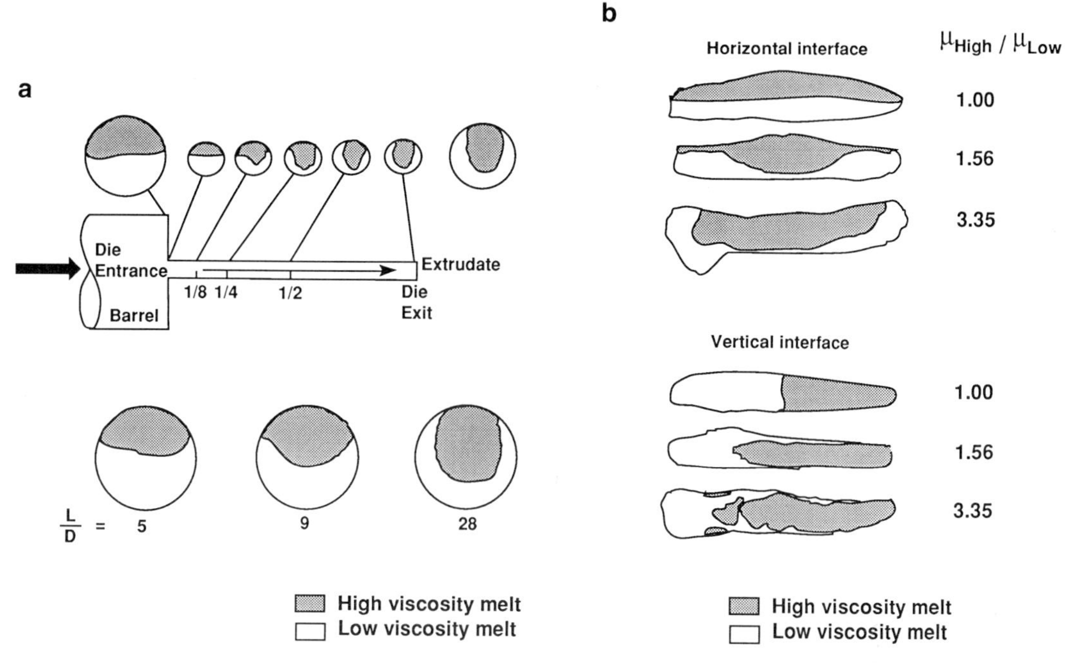

Figure 9 (a) Another coextrusion experiment with side-by-side initial conditions illustrates the bending of the interface and the gradual isolation of the higher viscosity component from the wall with increasing distance downstream (normalized by the pipe diameter D). The viscosity ratio is 1.56. (b) In a rectangular, dikelike channel, side-by-side injection is studied for two different orientations of the interface between the two layers and for different viscosity ratios. The aspect ratio (downstream length : thickness) is 15. For viscosity ratios of unity, significant interfacial distortion does not occur. The maximum encapsulation for given length of the channel takes place when the interface is parallel to the shortest dimension (thickness) of the channel. This is the anticipated orientation that will give rise to encapsulation in the generalized model for magma withdrawal through a dike. Adapted from Minagawa and White (1975, Figs. 6–9) with permission from the Society of Plastics Engineers.

zoning. For perfectly concentric inlet layering, Han found that the lower viscosity core tended to drift toward the wall of the tube, producing a slightly eccentric interface at a distance of $L = 18D$ from the inlet. (It is anticipated that the lower viscosity core eventually contacts the wall in a longer die given the results of the next experiment.) Han also produced a slightly eccentric inlet layering for both normal and reversed layering by injecting the core off center of the die inlet. At the end of the die ($L = 18D$), the normal layering (Fig. 10a) continued to exist with the higher viscosity melt along the center and the lower viscosity melt on the outside. For the reversed case (Fig. 10b), the concentric layering had broken down by the end of the die and the lower viscosity component that originally made up the core now contacted the wall of the tube along more than 50% of its circumference. Both the concentric and eccentric experiments indicate that the reversed core–annular inlet arrangement is unstable. Within a few tens of diameters downstream from the inlet, the results suggest that even the concentrically fed reversed arrangement would have completely broken down and started to form the energetically preferred, normally zoned core–annular arrangement.

In their investigation of two-liquid transport in a cylindrical tube, Blake and Campbell (1986) also found the reversed core–annular arrangement to be unstable for higher flow rates. However, it is curious that at the low flux rates of the inner, lower viscosity layer, they obtained a seemingly stable flow in contrast to the analyses of Hickox and Joseph *et al.* and the experiments of Han. The experiment of Blake and Campbell differs from these analyses and experiments in its use of the gravitational body force to help drive the flow. Considering that it is the unequal pressure gradients arising in the two different components that cause encapsulation (White and Lee, 1975b), it is possible that gravity-driven flow may be stabiliz-

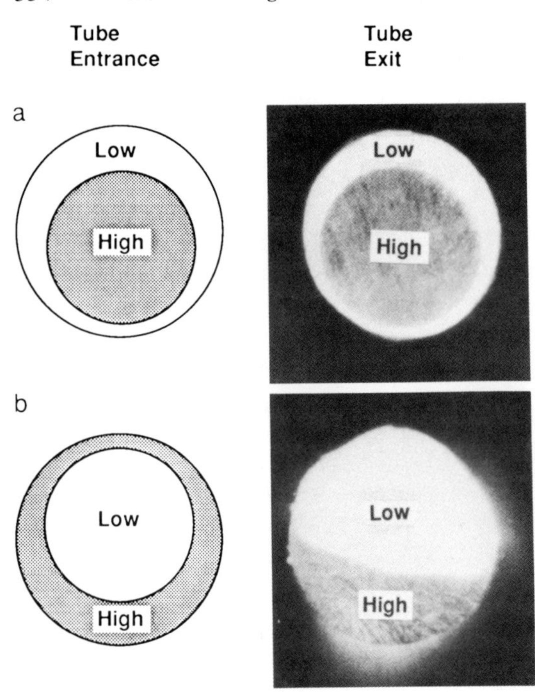

Figure 10 (a) Normal core–annular zoning (higher viscosity melt in the core) is stable when eccentrically introduced into long cylindrical die. (b) Reversed core–annular zoning (lower viscosity melt in the core) is unstable and rapidly migrates to wall when eccentrically introduced into long cylindrical die. This suggests that any reversed core–annular flow produced at a conduit inlet will change into a normally zoned flow at the surface. Adapted from C. D. Han (1975) with permission from John Wiley and Sons, Inc.

ing the reversed core–annular regime by minimizing pressure differences between the components. On the basis of Han's experiment, it is expected that any reversed core–annular flow of magma will rapidly become unstable over distances downstream from the inlet of a dike corresponding to tens of dike thicknesses. The lower viscosity magma will increase its contact with the wall until the higher viscosity component is isolated and the normal core–annular arrangement is established. Thus, extrapolating the results of Han's experiment, any reversed core–annular arrangement produced at the inlet of a dike or conduit by the simultaneous withdrawal of two layers from a magma chamber should be transformed to the normal core–annular arrangement by the time it arrives at the surface. Such a model is in contrast with the models of Blake and Campbell (1986) and Koyaguchi (1985), in which different magmas will arrive at the surface sequentially rather than simultaneously in a normal core–annular arrangement.

Numerical Models of Encapsulation and Self-Lubrication

Recent numerical models of two-component flow support arguments that encapsulation involves a process that tends to rearrange a two-component flow to minimize viscous dissipation and the driving pressure gradient. Using a finite element approach, Karagiannis *et al.* (1988) showed that the encapsulation of viscous fluids in the absence of surface tension is predicted by the minimum energy dissipation principle. Figure 7 is consistent with the results of their model for side-by-side, two-component flow in a channel. More recently, a novel numerical model based upon the lattice-gas approach was used to simulate side-by-side and dispersed two-component flow in a pipe (Stockman *et al.,* 1990). The lattice-gas method simulates the equations of motion for a liquid in an averaged sense using collections of discrete particles that are constrained to move on fixed triangular lattices (Frisch *et al.,* 1986; Rothman and Keller, 1988). As the particles move along the links of the lattice network during a time step, they collide with other particles conserving both momentum and particle number. By averaging a number of time steps, the continuum behavior of the many-particle system is obtained.

Both the reversed core–annular and the side-by-side, two-component inlet conditions were used in the pipe flow simulations (Figs. 11B and 11C). It is not surprising that in both cases, the flows break up to eventually form the normal core–

Figure 11 (A) Two-dimensional lattice gas model of a 50% mixture of high (black) and low (white) viscosity fluid flowing in a channel. Each frame from the top (initial conditions) downward represents a successively later time in the development of the flow with the final frame representing the evolved two-component flow morphology. (Numbers on the frame sides indicate the number of site updates or time-steps required to reach that point in the flow's evolution.) Flow is to the right and the viscosity ratio is 8 in cases A–C. Model predicts that viscous segregation will be more or less complete several thousand channel thicknesses downstream. (B) Reverse zoned flow is shown to be unstable as it evolves from the top frame downward to a normally zoned flow. Higher viscosity layers at the walls are exchanged for layers having lower viscosity. (C) As in the case of polymer

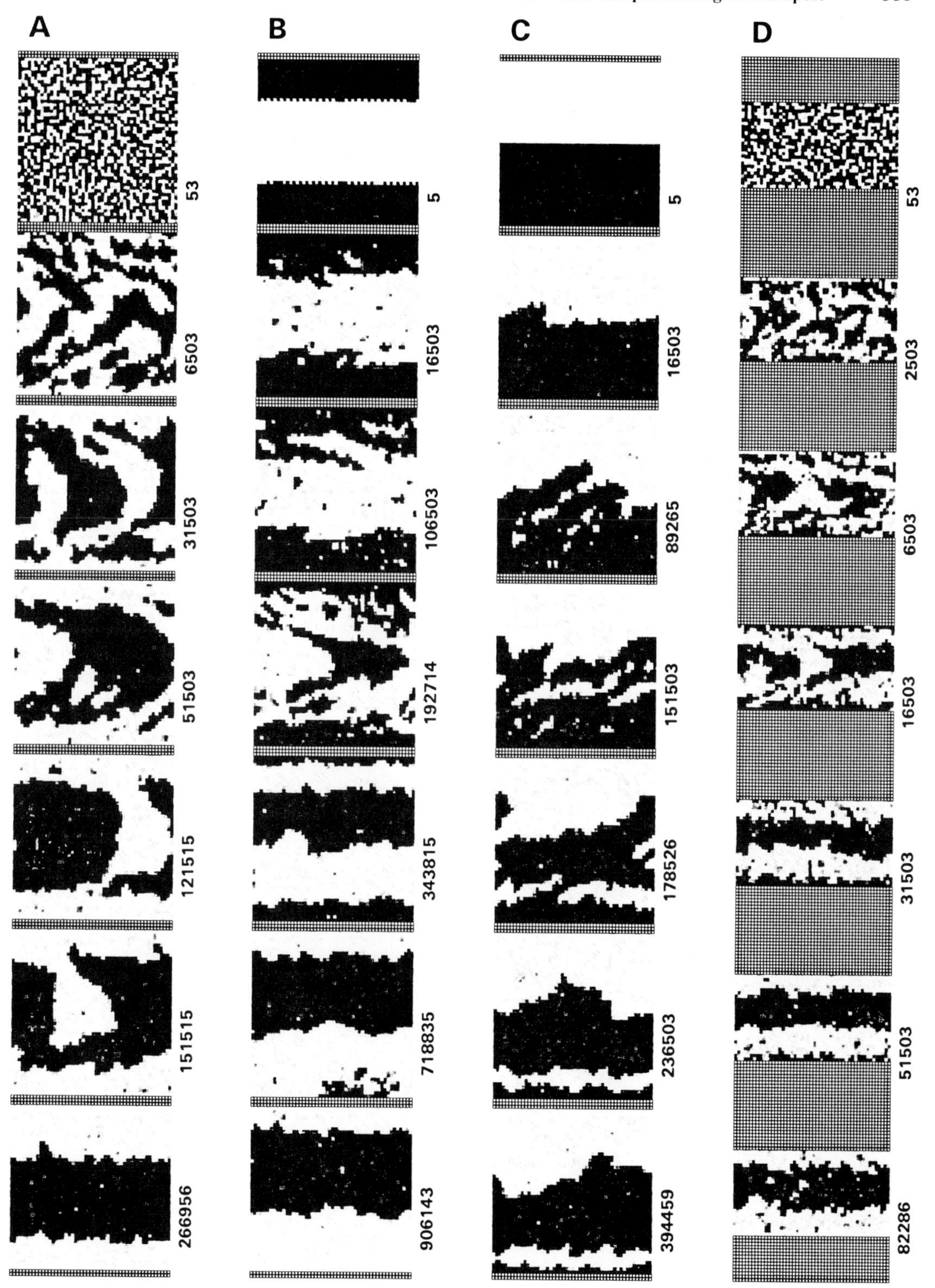

experiments, side-by-side flow is shown to change to a normal core–annular distribution. Note the thin high viscosity layer along the bottom wall of the channel. Such a layer does not destabilize the normal core–annular regime as demonstrated by Hu and Joseph (1989). (D) Same as (A) except that a viscosity ratio of 2.5 is used. Reproduced from Stockman *et al.* (1990) with permission from *Nature* and Macmillan Magazines Limited.

annular configuration. In one case (Fig. 11C), a thin margin of higher viscosity material (black) remains on the bottom boundary while the rest of the flow has changed to the normal core–annular arrangement. Hu and Joseph (1989) found that such an arrangement involving a thin layer of higher viscosity liquid on the boundary was in fact stable as long as the layer remained thin. This has an important implication for magma transport. Magma flowing in a dike or conduit loses heat to the host rock. As heat diffuses across the boundary of the dike, a boundary layer or narrow zone of cooler magma develops. This cooler layer also becomes a more viscous layer at the boundary, due to the temperature dependence of viscosity in silicate melts. Boundary roughness, rising vapor bubbles, and other effects responsible for deviations from laminar flow will tend to work against thermal diffusion to keep this layer thin and the boundary temperatures high (Carrigan *et al.*, 1992). The results of Hu and Joseph (1989) and Stockman *et al.* (1990) suggest that such a thin, thermal boundary layer of magma will not impede the formation of the core–annular regime.

Another inlet condition considered by Stockman *et al.* (1990) consisted of a 50% mixture of dispersed higher and lower viscosity components. Such a mixture might be anticipated with the formation of mafic pillows or inclusions (enclaves) in the chamber of a silicic system (Blake *et al.*, 1965; Eichelberger, 1980; Vogel, 1982; Huppert *et al.*, 1982; Bacon, 1986). In the numerical experiment of Stockman *et al.* (1990), the mixture also segregated into wall and core layers with flow along the tube for Newtonian viscosity ratios of 8:1 and 2.5:1 (Figs. 11A and 11D). However, the segregation was never complete and blobs of lower viscosity fluid continued to remain suspended in the core flow for both the dispersed and the side-by-side cases. This observation is probably explained by a large reduction in shear in the central region of the flow as the core–annular regime is set up. (In the absence of shear, viscous segregation should vanish.) Polymer blends of higher and lower viscosity components also show a tendency to segregate with flow along a tube (Lee and White, 1975), although the segregation rate is not nearly as dramatic as for side-by-side coextrusion. It is not surprising that blends do not form a core annular regime as rapidly as side-by-side, two-component flow since the pressure differences needed for the lower viscosity component to migrate to the wall would seem to form much less effectively in a well-dispersed system. There is also some evidence that the evolution of a two-component dispersed system is rather sensitive to both the type of flow and the rheology of the suspending component (Han, 1981). The process of segregation in dispersed, two-component systems is clearly more complex than in side-by-side flow and is a good candidate for further investigation using both laboratory and numerical methods.

One conclusion can be made about the application of viscous segregation to systems where one component is dispersed in the other. If the dispersed low viscosity component must remain warmer than the matrix component to retain its fluidity, e.g., a basalt in rhyolite, then the characteristic size of the dispersion must be large enough that the diffusion time scale for cooling is large compared to the total time that the more fluid component exists in the dispersed mode. A crude estimate based upon the thermal diffusion of heat from a spherical body suggests that a total time of 3 h in the dispersed mode before segregation is complete would require the suspended component to have a parcel size of 0.5 m. If the dispersed mode in a two-component system has greater fluidity because of chemical differences, e.g., a higher water content, then the dispersion size can be much smaller since chemical diffusivities are much smaller than the thermal diffusivity of magma.

Analytical Modeling of Lubricated Magma Transport

Derivation of Lubrication Equations

The vector forms of the equations governing the flow of magma in a dike are

$$\frac{\partial \mathbf{v}}{\partial t} + \mathbf{v} \cdot \nabla \mathbf{v} = \frac{-1}{\rho} \nabla p - \frac{1}{\rho} \nabla \cdot \boldsymbol{\tau} \quad (1)$$

(motion)

$$\nabla \cdot \mathbf{v} = 0 \quad (2)$$

(continuity)

$$\frac{\partial T}{\partial t} + \mathbf{v} \cdot \nabla T = \frac{k}{\rho c_p} \nabla^2 T + \frac{\eta}{\rho c_p} \phi_v \quad (3)$$

(energy),

where the temperature, pressure and two-dimensional, local flow velocity are T, p, and $\mathbf{v} = u\mathbf{i} + w\mathbf{k}$, respectively, with unit vectors $\mathbf{i}$ and $\mathbf{k}$ oriented across and vertically along the dike. The $\boldsymbol{\tau}$ in the equation of motion is the viscous stress tensor. The form of $\boldsymbol{\tau}$ is determined by the nature of the magma rheology, e.g., Newtonian or power law. Since a magma may exhibit both rheologies over different ranges of temperature and/or volatile contents, the viscous term is left in its most general form. Also, note that the pressure gradient term involving ∇p is assumed to be much larger than the body force resulting from thermal buoyancy so that the latter can be neglected. If a volcanic conduit becomes plugged and the forced flow of magma is reduced to a small or vanishing amount, this assumption is no longer valid and the thermal buoyancy term may provide the dominant driving force resulting in convective circulations (Carrigan, 1983). The quantity η in Eq. (3) is a generalized Newtonian viscosity and ϕ_v is the dissipation function (Bird *et al.*, 1960). Vector and scalar terms may be expressed in rectangular coordinates as

$$\mathbf{v} \cdot \nabla \mathbf{v} = \left(u\frac{\partial u}{\partial x} + w\frac{\partial u}{\partial z} \right)\mathbf{i} + \left(u\frac{\partial w}{\partial x} + w\frac{\partial w}{\partial z} \right)\mathbf{k}, \tag{4}$$

$$\nabla \cdot \boldsymbol{\tau} = \left(\frac{\partial \tau_{xx}}{\partial x} + \frac{\partial \tau_{zx}}{\partial z} \right)\mathbf{i} + \left(\frac{\partial \tau_{xz}}{\partial x} + \frac{\partial \tau_{zz}}{\partial z} \right)\mathbf{k}, \tag{5}$$

and

$$\phi_v = \left(\frac{\partial w}{\partial x} + \frac{\partial u}{\partial z} \right)^2 + 2\left[\left(\frac{\partial u}{\partial x} \right)^2 + \left(\frac{\partial w}{\partial z} \right)^2 \right]. \tag{6}$$

Steady solutions of these equations have been obtained for the flow of a single-component magma with temperature-dependent, Newtonian viscosity in high aspect ratio (10^3:1 or greater) dikes that lose heat across their boundaries into the host regime (Carrigan *et al.*, 1992). For typical magma properties and rates of flow, thermal boundary layers of the order of 0.1 m in thickness develop in the magma adjacent to the boundaries of the dike. However, Carrigan *et al.* found that even weak deviations from laminar flow such as caused by the buoyant rise of gas or vapor bubbles or boundary roughness will suppress the rate of growth of a thermal boundary layer and produce near core-flow magmatic temperatures in the boundary layer 1–2 km downstream from the dike inlet. What cooling of magma at the wall does occur, with its attendant increase in viscosity, should not immediately affect the dynamics of lubricated flow according to Hu and Joseph (1989). Thus, in most of the derivations we do not consider the energy equation although we do consider the thermal effects of viscous dissipation on the lubricating layers in a later subsection.

The velocity of flow in a dike is assumed to have the form $\mathbf{v} = w(x)\,\mathbf{k}$, which automatically satisfies Eq. (2). The magnitude of flow is slow enough for a given dike and range of magma vis7151cosities that the Reynolds' number is appropriate for the limit of creeping flow, i.e., $Re = wd\rho/\mu \ll 1$ where d, ρ and μ are the characteristic values of the dike width, magma density, and magma dynamic viscosity, respectively. The Reynolds' number may be interpreted as the ratio of inertial terms (left side of Eq. (1)) to the viscous term in Eq. (1). In the creeping flow limit, only the pressure gradient term is left to balance the viscous term so that Eq. (1) in rectangular coordinates reduces to

$$0 = \frac{dp}{dz} + \frac{d\tau_{xz}}{dx}, \tag{7}$$

where the tensor component τ_{xz} is the shear stress in the z direction acting on a surface whose normal is in the x direction. The magnitude of the vertical component of velocity can be related to the stress tensor component for both Newtonian and power-law rheologies by

$$\tau_{xz} = -m\left|\frac{dw}{dx}\right|^{n-1}\frac{dw}{dx}. \tag{8}$$

When $n = 1$, this equation results in Newtonian rheology in which $m = \mu$. For $n > 1$, the equation describes a shear thickening or dilatant fluid, whereas for $n < 1$ a shear thinning or pseudoplastic fluid is modeled.

To investigate the lubricative effects of two-component flow on the flux of magma in a dike for a given pressure gradient or, alternatively, on the pressure gradient reduction for a given magma flux, we can easily solve Eq. (7) for two-component magma systems with sidewall lubrication. We consider a dike of half-thickness $d/2$ with a lubricating layer of thickness a on each sidewall. Consistent with the lubrication approximation, we assume that fluid pressure variations in the cross-stream direction are small to vanishing compared to variations along the vertical direction of flow. In our model, the negative of the pressure gradient is given by

$$G = -\frac{dp}{dz} = \frac{p_c - p_o}{L}, \tag{9}$$

where subscripts c and o denote the chamber and outlet (atmospheric) pressures, respectively, and L is the length of the dike. Because the problem is symmetrical, we need only to solve for the velocity or volume flux in the layers falling between a wall and the centerline of the dike. Integrating Eq. (7) with respect to x in the wall and central layers gives

$$\tau_{xz}^{w} = Gx + C_1^{w}, \qquad 0 \le x \le a \tag{10a}$$

$$\tau_{xz}^{c} = Gx + C_1^{c}, \qquad a \le x \le d/2, \tag{10b}$$

with the superscripts w and c denoting the wall and central layers, respectively. The fact that $\tau_{xz}^{w} = \tau_{xz}^{c}$ at $x = a$ means that the integration constants C_1^{w} and C_1^{c} are equal and will be referred to as C_1. Rewriting Eqs. (10a) and (10b) as

$$\tau_{xz}^{w} = G(x - C_1), \qquad 0 \le x \le a \tag{11a}$$

$$\tau_{xz}^{c} = G(x - C_1), \qquad a \le x \le d/2, \tag{11b}$$

permits the constant C_1 to be interpreted as that x value in the dike where the shear stress vanishes. Since this can happen only at one location in the dike, i.e., $x = C_1$, and the dike is symmetric about the centerline, C_1 must be equal to $d/2$. The absolute value sign appearing in the generalized shear stress + rate-of-strain relationship (Eq. (8)) can be eliminated if only one-half of the dike channel is considered, where dw/dx is already greater than or equal to zero. Thus, from Eq. (8) we can write the shear stress as a function of strain rate in the wall and central layers as

$$\tau_{xz}^{w} = -m_w \dot{\varepsilon}^{n_w}, \qquad 0 \le x \le a \tag{12a}$$

$$\tau_{xz}^{c} = -m_c \dot{\varepsilon}^{n_c}, \qquad a \le x \le d/2, \tag{12b}$$

where m_w, n_w, m_c, and n_c are the coefficients and exponents needed to define the power-law viscosity for the wall layer and for the central layer, respectively, and where

$$\dot{\varepsilon} = \frac{dw}{dx} \quad \text{or} \quad dw = \dot{\varepsilon}\, dx. \tag{13}$$

Eliminating the shear stresses τ_{xz}^{w} and τ_{xz}^{c} in Eqs. (12a) and (12b) using Eqs. (11a) and (11b), and then taking the derivative with respect to x produces

$$dx = -\frac{n_w m_w \dot{\varepsilon}^{n_w - 1}}{G}\, d\dot{\varepsilon}, \qquad 0 \le x \le a \tag{14a}$$

$$dx = -\frac{n_c m_c \dot{\varepsilon}^{n_c - 1}}{G}\, d\dot{\varepsilon}, \qquad a \le x \le \frac{d}{2}. \tag{14b}$$

Using Eq. (13), we can now eliminate dx in Eqs. (14a) and (14b) to obtain

$$dw = -\frac{n_w m_w \dot{\varepsilon}^{n_w}}{G}\, d\dot{\varepsilon}, \qquad 0 \le x \le a \tag{15a}$$

$$dw = -\frac{n_c m_c \dot{\varepsilon}^{n_c - 1}}{G}\, d\dot{\varepsilon}, \qquad a \le x \le \frac{d}{2}. \tag{15b}$$

These last two equations may be readily integrated to obtain expressions for $w(x)$ over the intervals $0 \le x \le a$ and $a \le x \le d/2$. Integration of Eq. (15a) between 0 and x utilizes $w(0) = 0$ and $\dot{\varepsilon}(0)$ as given by Eq. (21) as boundary conditions. Integration of Eq. (15b) between a and x requires the evaluation of $w(a)$ from the expression for w in the interval $0 \le x \le a$ and $\dot{\varepsilon}(a)$ from an equation for $\dot{\varepsilon}(x)$ obtained from Eq. (14b). These integrations finally yield the desired result for the cross-stream dependence of two-layer flow in the dike over the half-width $0 \le x \le d/2$:

$$w(x) = \frac{n_w}{n_w + 1}\left(\frac{G}{m_w}\right)^{1/n_w} \cdot \left[\left(\frac{d}{2}\right)^{(n_w+1)/n_w} - \left(\frac{d}{2} - x\right)^{(n_w+1)/n_w}\right], \qquad 0 \le x \le a \tag{16a}$$

$$w(x) = \frac{n_c}{n_c + 1}\left(\frac{G}{m_c}\right)^{1/n_c} \cdot \left[\left(\frac{d}{2} - a\right)^{(n_c+1)/n_c} - \left(\frac{d}{2} - x\right)^{(n_c+1)/n_c}\right] + \frac{n_w}{n_w + 1}\left(\frac{G}{m_w}\right)^{1/n_w} \cdot \left[\left(\frac{d}{2}\right)^{(n_w+1)/n_w} - \left(\frac{d}{2} - a\right)^{(n_w+1)/n_w}\right], \qquad a \le x \le \frac{d}{2}. \tag{16b}$$

The total volume flux of magma (volume per unit horizontal length of dike per unit time), Q, in the dike can be determined by summing the contributions from the wall and core layers,

$$Q = Q_w + Q_c = \int_0^a w\, dx + \int_a^{d/2} w\, dx, \quad (17)$$

where the integration of the expressions for $w(x)$ in Eqs. (16a) and (16b) over the limits of each layer yields

$$Q_w = \frac{n_w}{n_w + 1}\left(\frac{G}{m_w}\right)^{1/n_w}\left[a\left(\frac{d}{2}\right)^{(n_w+1)/n_w} + \frac{(d/2 - a)^{(2n_w+1)/n_w} - (d/2)^{(2n_w+1)/n_w}}{(2n_w + 1)/n_w}\right] \quad (18a)$$

$$Q_c = \frac{n_c}{n_c + 1}\left(\frac{G}{m_c}\right)^{1/n_c} \cdot \left(1 - \frac{1}{(2n_c + 1)/n_c}\right)\left(\frac{d}{2} - a\right)^{(2n_c+1)/n_c} + \frac{n_w}{n_w + 1}\left(\frac{G}{m_w}\right)^{1/n_w} \cdot \left[\left(\frac{d}{2}\right)^{(n_w+1)/n_w} - \left(\frac{d}{2} - a\right)^{(n_w+1)/n_w}\right] \cdot \left(\frac{d}{2} - a\right). \quad (18b)$$

Other useful expressions for power-law, single- and two-component flow may be developed or inferred from these equations. The local viscosity of a power-law magma at an arbitrary point in the flow can be written as

$$\mu_{\text{loc}} = m\, \dot{\varepsilon}^{n-1}. \quad (19)$$

according to Eq. (8) over the half-width, where $\dot{\varepsilon} \geq 0$. This viscosity is just a constant for a Newtonian magma ($n = 1$), but it is a function of shear rate and therefore location in the dike for non-Newtonian magmas ($n \neq 1$). Another kind of viscosity might be called the equivalent viscosity, μ_{eqv}, corresponding to the viscosity of a Newtonian magma that would result in an equivalent flux of magma through a dike for the same values of the pressure gradient, $-G$, and dike width, d. Equating the volume flux Q of magma for any given type of flow in a dike, e.g., single-layer non-Newtonian, multilayer non-Newtonian, or multilayer Newtonian, to the expression for the flux of a single-component, Newtonian flow yields

$$\mu_{\text{eqv}} = \frac{Gd^3}{12Q}. \quad (20)$$

This expression is particularly useful in a later section for evaluating the effect of viscosity on dike propagation rates. For time-independent flow, the velocity gradient or rate of strain at the boundary is given by

$$\dot{\varepsilon}(0) = \left(\frac{Gd}{2m_w}\right)^{1/n_w} \quad (21)$$

and the velocity gradient in the wall layer of magma at the interface between the wall and central layers is

$$\dot{\varepsilon}(\text{a}) = \left[\frac{G}{m_w}\left(\frac{d}{2} - a\right)\right]^{1/n_w}. \quad (22)$$

In the common case of two magmas having the same power-law exponent, $n_w = n_c = n$, such as when both magmas are treated as Newtonian fluids, the pressure gradient $-G$ required to produce a given total volume flux of magma Q is readily obtained from (18a) and (18b),

$$G = \left(\frac{Q}{D + E + F}\right)^n, \quad (23)$$

where the terms in the denominator are written as

$$D = \frac{n}{n + 1}\left(\frac{1}{m_w}\right)^{1/n}\left[a\left(\frac{d}{2}\right)^{(n+1)/n} + \frac{(d/2 - a)^{(2n+1)/n} - (d/2)^{(2n+1)/n}}{(2n+1)/n}\right],$$

$$E = \frac{n}{n + 1}\left(\frac{1}{m_c}\right)^{1/n} \cdot \left(1 - \frac{1}{(2n + 1)/n}\right)\left(\frac{d}{2} - a\right)^{(2n+1)/n},$$

and

$$F = \frac{n}{n + 1}\left(\frac{1}{m_w}\right)^{1/n} \cdot \left[\left(\frac{d}{2}\right)^{(n+1)/n} - \left(\frac{d}{2} - a\right)^{(n+1)/n}\right] \cdot \left(\frac{d}{2} - a\right).$$

When the power-law exponents are different for the wall and core layers of magma, the value of G required for a given total flow rate may be

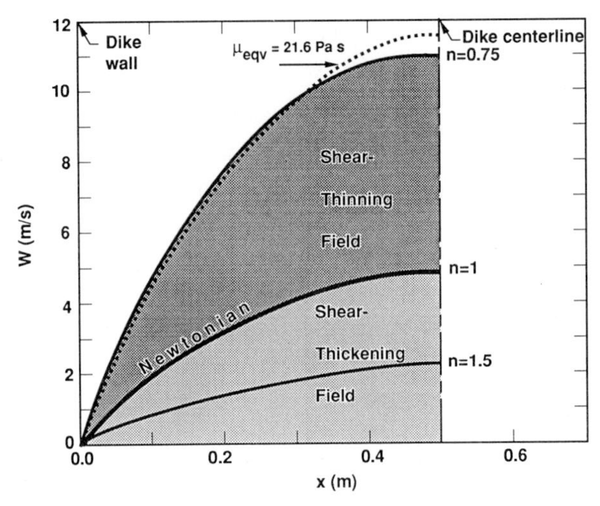

Figure 12 Velocity profiles are plotted over the dike half-width for single-component basaltic flows having different values of the power-law index n. The zero shear viscosity is 52 Pa · s for the three cases $n = 0.75, 1, 1.5$. For comparison with the $n = 0.75$ case, a Newtonian profile that would yield the same volumetric flux is plotted. In terms of the volumetric flux, the $n = 0.75$ case behaves as a Newtonian magma ($n = 1$) of 21.6 Pa · s. The calculations assume a dike width (d) of 1 m and a pressure gradient ($-G$) of 2 MPa · km^{-1}.

obtained from Eqs. (17) and (18) using iterative techniques.

Effect of Lubrication by Newtonian and Non-Newtonian Wall Layers

Given the occurrence of viscous segregation, the derived equations permit the evaluation of the pressure reduction and flow enhancement associated with the presence of lower viscosity Newtonian or power-law wall layers. For comparison, single-component velocity profiles for a given nominal pressure gradient (2×10^3 Pa · m^{-1}) obtained from Eqs. (16a) and (16b) are illustrated in Fig. 12 for several different values of the power-law exponent n. The same value of the power-law coefficient m is used for each case and corresponds at $n = 1$ to a viscosity of 52 Pa · s, which is characteristic of basaltic compositions (Shaw *et al.*, 1968; Ryan and Blevins, 1987; Ryerson *et al.*, 1988). Not surprisingly, a shear-thinning magma ($n < 1$) has lower viscosity behavior in regions of high shear where the majority of viscous dissipation occurs with the result that maximum flow velocity and volumetric flux (Fig. 13) are greater than for the Newtonian ($n = 1$) or shear-thickened cases ($n > 1$). The shear-thinned case considered here ($n = 0.75$) has an equivalent viscosity, as defined by Eq. (20), of about 21.6 Pa · s according to Fig. 13 and the Newto-

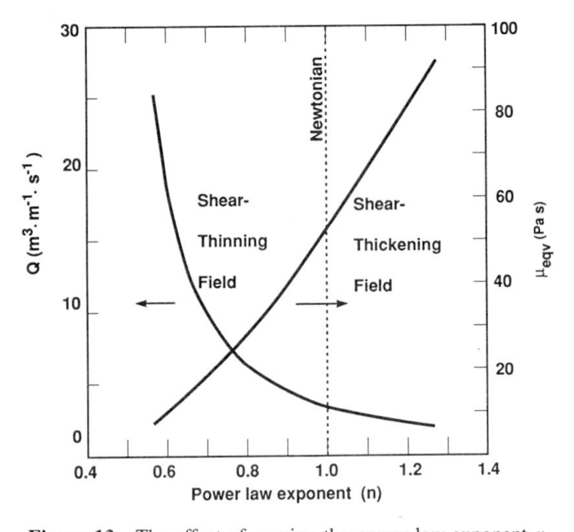

Figure 13 The effect of varying the power-law exponent n on the volumetric flux Q, defined by Eqs. (17) and (18), and the equivalent viscosity μ_{eqv}, defined by Eq. (20), is plotted for a single-component magma, with a power-law coefficient $m = 52$, flowing in a 1-m dike and driven by a 2 MPa · km^{-1} pressure gradient. At $n = 1$, the viscosity has no dependence on the rate of strain and the flow is Newtonian with molecular viscosity equal to the equivalent viscosity. For $n < 1$, the magma is shear thinning and is characterized by an equivalent viscosity that is less than that for the Newtonian case (52 Pa · s). For $n > 1$, the magma is shear thickening and in terms of the volumetric flux behaves as a Newtonian magma with viscosity exceeding 52 Pa · s. The quantity Q is the volume of magma flowing per unit time and per unit horizontal length of the dike. (Arrows indicate appropriate axis for each plot.)

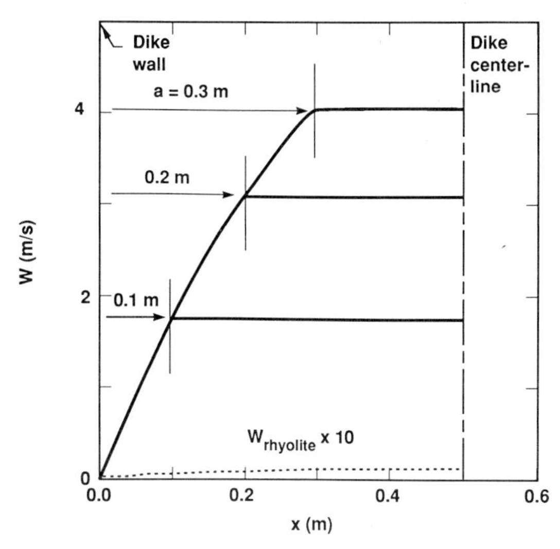

Figure 14 The effect of varying the width, a, of the lubricating layers on a two-component velocity profile is plotted for three values of a. *For a basaltic viscosity of 52 Pa · s in the lubricating layer and a rhyolitic viscosity of 2.2 × 10⁴ Pa · s in the core layer, it is found that a lubricating layer of almost any thickness strongly affects the profile and, hence, the flow rate of the viscous component through the dike.* For comparison, the single-component velocity profile (multiplied by 10) for rhyolite is shown as a dashed curve. As a approaches 30% of the dike width (1 m), the ascent rate of the rhyolite core becomes comparable to a purely basaltic flow ($n = 1$, Fig. 12) driven by the same pressure gradient (2 MPa · km^{-1}).

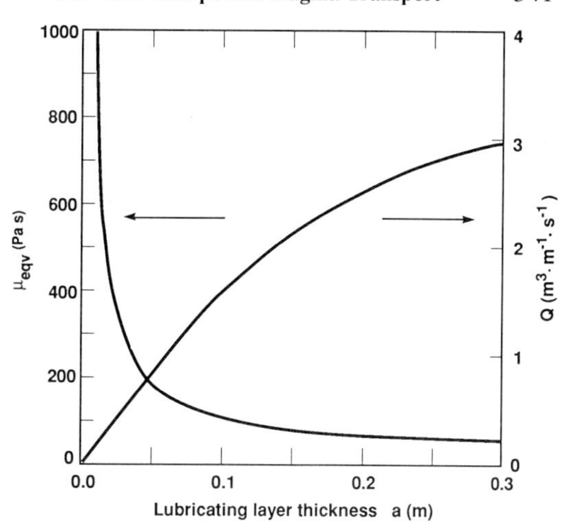

Figure 15 The volumetric flux Q and the equivalent viscosity μ_{eqv} are plotted as a function of the lubricating layer thickness, a, assuming the same parameter values given in Fig. 14. The equivalent viscosity rapidly decreases for $0 < a < 0.05$ m. Beyond a width of $a = 0.1$ m, the equivalent viscosity approaches the molecular viscosity (52 Pa · s) of the lubricating layers much more slowly.

nian profile for a magma with this viscosity is plotted in Fig. 12 for comparison with the shear-thinned case.

The effects of lubrication on the velocity profile, volumetric flux, and equivalent viscosity are shown in Figs. 14 and 15 for Newtonian, rhyolitic magmas with Newtonian, basaltic lubricating layers. As is apparent from Fig. 14, the shear is virtually eliminated in the central rhyolitic core (2.2×10^4 Pa · s; Carrigan and Eichelberger, 1990), even with thin lubricating layers (52 Pa · s). From Fig. 15, the equivalent viscosity falls from rhyolitic values to less than 200 Pa · s when the lubricating layer on each wall is about 0.05 m thick in a 1-m-thick dike and to about 75 Pa · s with 0.15 m-lubricating layers. Also plotted in Fig. 15 is the volumetric flux through the 1-m-wide dike as a function of lubricating layer thickness. For a pressure gradient of 2 MPa · km^{-1}, a single-component rhyolitic flow is characterized by a volume flux of only 8×10^{-3} m^3 · m^{-1} · s^{-1} whereas a rhyolitic core layer with 0.10- to 0.15-m lubricating wall layers produces volumetric fluxes in the 2 m^3 · m^{-1} · s^{-1} range. In this case, lubrication produces a 250-fold increase in the volumetric flux for the same driving pressure gradient, an effect that is particularly important for the production of dikes by magma-driven fractures. The role of lubrication in magma-driven fracturing is discussed in more detail later.

So far we have looked only at two-layer systems with purely Newtonian rheologies ($n = 1$). When each viscosity is characterized by a different value of the power-law exponent, Eqs. (18a) and (18b) predict a curious result. Two-component flows in a dike can produce volumetric fluxes for a given pressure gradient that exceed the fluxes produced at the same pressure gradient by single-component flows of either of the magma constituents. Alternatively expressed, the equivalent viscosity, as given by Eq. (20), can be smaller for two-component flow than the equivalent viscosity of either component flowing in the dike by itself. Models for the coextrusion of power-law materials (Han and Chin, 1979) predict that this behavior can occur when both magmas have the same local viscosity (Eq. (20)) at the interface, $x = a$, between the wall and core layers. Equating the shear stresses in Eqs. (12a) and (12b) and solving for the value of strain rate at which the viscosity relationships cross give

$$\dot{\varepsilon}_{\text{cross}} = \left(\frac{m_c}{m_w}\right)^{1/(n_w - n_c)}. \quad (24)$$

The value of $\dot{\varepsilon}_{\text{cross}}$ is the value of shear strain rate where both magmas have the same viscosity. If this value of the shear strain rate also occurs at the interface, Eq. (22) can be used to write

$$\dot{\varepsilon}_{\text{cross}} = \left[\frac{G}{m_w}\left(\frac{d}{2} - a\right)\right]^{1/n_w}. \quad (25)$$

The value of G where this occurs is obtained by eliminating $\dot{\varepsilon}_{\text{cross}}$ between Eqs. (24) and (25):

$$G_{\text{cross}} = \frac{m_w}{(d/2 - a)}\left[\left(\frac{m_c}{m_w}\right)^{n_w/(n_w - n_c)}\right]. \quad (26)$$

Thus, for given magma viscosities, dike widths, and wall layer thicknesses, we can now calculate the pressure gradient at which a two-magma system will have a lower equivalent viscosity than either of its components.

Figure 16 shows the dependence of the volumetric flux Q on the parameter G for a 1-m-wide dike with 0.1-m wall layers. Equations (17), (18a), and (18b) were evaluated for a range of G about G_{cross} as given by Eq. (26). The power-law exponents ($n_1 = 0.5$ and $n_2 = 0.8$) and coefficients ($m_1 = 100$ and $m_2 = 50$) assumed in this example appear to be characteristic of values that are appropriate for a picritic magma at different temperatures and crystal contents (Ryerson *et al.*, 1988) with magma 1 representing the lower temperature component. In the vicinity of $G_{\text{cross}} = 794$ Pa · m^{-1} ($\approx$0.8 MPa · km^{-1}), the maximum value of Q is achieved for magma 1 in the lubricating wall layer and magma 2 in the core layer. The minimum value results from the reversal of this arrangement. In between fall the values of Q for the two magmas flowing in the dike singly. In terms of the equivalent viscosity, the two-magma example giving the highest flow rate at G_{cross} has a μ_{eqv} of 30.6 Pa · s while the two components flowing alone in the dike have values of μ_{eqv} of 32.3 and 33.6 Pa · s. One implication of such a prediction is that two-component, power-law magmatic flows are potentially better for minimizing viscous losses during dike propagation than either component alone propagating the dike with the same pressure gradient. Whether this behavior can really occur for a particular two-component system will depend critically on the validity of empirically determined power-law relationships of each component in the vicinity of $\dot{\varepsilon}_{\text{cross}}$.

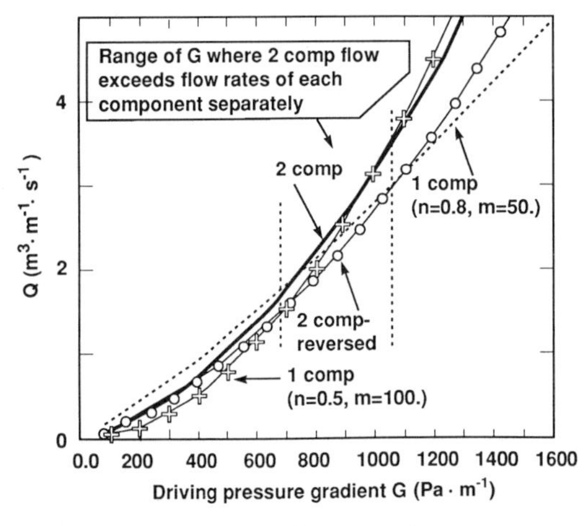

Figure 16 For two power-law magmas, the two-component volumetric flux can exceed the single-component volumetric flux for the same value of the pressure gradient. The power-law exponents ($n_1 = 0.5$ and $n_2 = 0.8$) and coefficients ($m_1 = 100$ and $m_2 = 50$) are within the parameter range observed by Ryerson *et al.* (1988). The maximum flux with component 1 in the lubricating layer and component 2 in the core is obtained in the range of G between the vertical dashed lines. Reversal of this order reduces the flux to values less than would be obtained for either component flowing in the dike individually.

Effect of Lubricated Flows on Dike Propagation

Analyses indicate that viscous losses associated with magma flow in narrow fractures may be far more important than rock fracture toughness (often designated K_{I_c}) in determining if the pressure in a magma chamber is adequate to propagate a dike significant distances (e.g., Spence and Turcotte, 1985; Lister and Kerr, 1991). Thus, it is expected that lubrication may play an important role in determining whether a particularly viscous magma, such as rhyolite, is able to penetrate to the surface through the crust overlying a magma chamber or significant distances along the horizon of neutral buoyancy (Ryan, 1987). To evaluate the effect of lubrication on dike propagation, any of the analyses by Geertsma and Haafkens (1979), Spence and Turcotte (1985), or Nilson and Griffiths (1986) could be used. The differences are minor and we use the Geertsma and Haafkens (1979) derivation for the propagation of parallel-sided fractures that is consistent with the plane-parallel models derived here. It should be noted that the Geertsma and Haafkens analysis does not take into account the decrease in the confining stress as the fracture tip propagates vertically upward. On the other hand, the magmatic pressure-head loss associated with the weight of the magma in the vertical fracture is also not considered. Both effects approximately cancel each other. (Depending on the initial crustal stress state, other, more detailed fracture propagation models may be required to adequately predict fracture propagation rates.) The solutions used also assume that the energy associated with fracturing at the crack tip is negligible compared to viscous dissipation losses. This is not a good assumption at the time just after a fracture is initiated if the crustal rock is *competent* at the point of initiation, but it becomes a progressively better assumption as the magma-filled fracture lengthens. However, for the purposes of estimating dike propagation rates between a magma reservoir and the surface, it probably is a good assumption even near the time of fracture initiation at the magma reservoir boundary since the wall rock there is likely to be at least partially molten so that the fracture toughness will be substantially reduced.

The model assumes a constant flux Q ($m^3 \cdot s^{-1} \cdot m^{-1}$) of fluid into the fracture with time so that the driving pressure in the magma chamber can therefore vary as the propagating fracture both widens and lengthens with time. The fracture length as a function of time t is written as

$$1 = 0.68\ Q^{1/2} \left(\frac{M_s}{\mu(1-\nu)} \right)^{1/6} t^{2/3}, \quad (27)$$

where M_s (2×10^{10} Pa; Spence and Turcotte, 1985) and ν (0.25) are the elastic shear modulus and Poisson's ratio, respectively. Taking the derivative of Eq. (27) with respect to time yields the instantaneous velocity of the fracture tip:

$$\frac{dl}{dt} = 0.453 Q^{1/2} \left(\frac{M_s}{\mu(1-\nu)} \right)^{1/6} t^{-1/3}. \quad (28)$$

The width of the fracture d is given as

$$d = 1.87 Q^{1/2} \left(\frac{\mu(1-\nu)}{M_s} \right)^{1/6} t^{1/3} \quad (29)$$

and the pressure is

$$p = 1.13 (Q\mu)^{1/4}\ l^{-1/2} \left(\frac{M_s}{\mu(1-\nu)} \right)^{3/4}. \quad (30)$$

Replacing μ with μ_{eqv} allows the effect of lubrication to be evaluated for two-component flows that exhibit either Newtonian or power-law behavior. As an example, consider the injection of a rhyolitic magma into a fracture. Once the fracture is established and magma is flowing in the dike, the required driving pressure (not necessarily the actual chamber pressure) would fall off from about 22 MPa to only a few MPa if the dike were to reach 5 km in length (Fig. 17). During the propagation of the dike out to a distance 5 km from the chamber, its width would increase from about 0.3 m to about 1.5 m (Fig. 17). However, the fracture tip velocity as given by Eq. (28) and plotted in Fig. 18 would fall from about 0.04 $m \cdot s^{-1}$ to about 0.01 $m \cdot s^{-1}$, which is probably much too slow to prevent solidification over the 5-km distance. Now consider the same driving pressure history of Fig. 17 for a rhyolitic magma lubricated by 100 Pa · s basaltic wall layers that are 16% as thick as the dike width. The distribution of dike width (Fig. 17) with fracture length is the same as predicted for the rhyolitic flow assuming that it somehow traversed the 5 km. But now, rather than having a viscosity of 2.2×10^4 Pa · s as in the case of the pure rhyolitic flow, the equivalent viscosity of the lubricated flow is only 146 Pa · s,

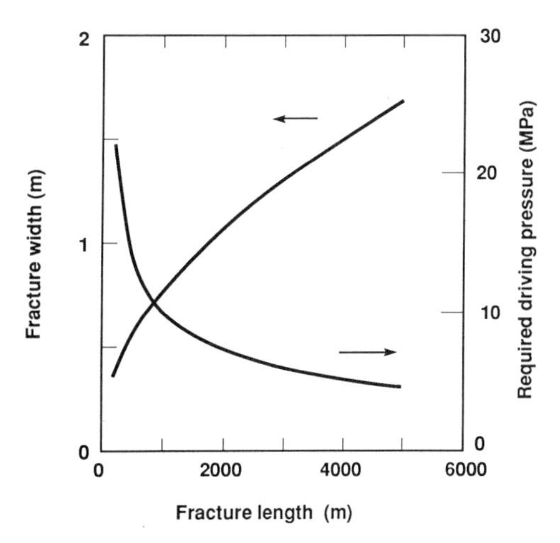

Figure 17 In the formulation for fracture propagation presented here, a constant volumetric flux of magma is injected into the fracture as it propagates vertically toward the surface. As injection occurs, the fracture grows in length and it also widens. Since the required driving pressure is more sensitive to changes in the fracture width than in its length, the required driving pressure for a given rate of injection of magma falls off with length. In this case the fracture grows from a width of about 0.3 m to about 1.5 m as it lengthens from about 0.2 to 5 km.

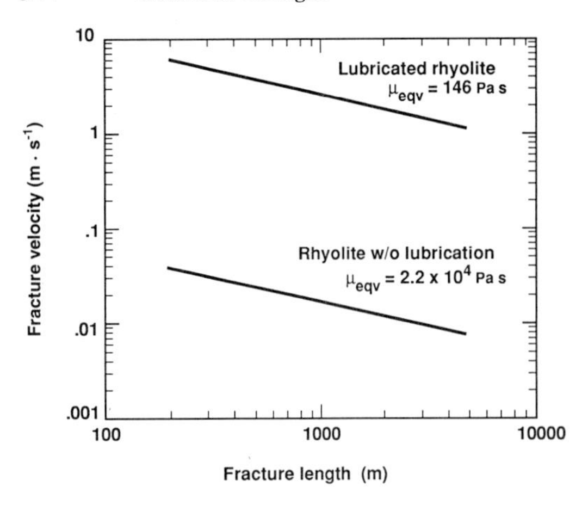

Figure 18 Comparison of fracture propagation velocities produced by a rhyolite (2.2×10^4 Pa · s) without lubrication and a rhyolite lubricated by a basalt ($a = 0.16d$, $\mu_w = 100$ Pa · s, $\mu_{eqv} = 146$ Pa · s) subjected to the same magma chamber pressures (Fig. 17). The lubricated rhyolite drives the fracture at about 150 times the rate of the unlubricated rhyolite for the same pressures.

which is about 150 times smaller. The resulting crack-tip velocity of the lubricated dike is more than 100 times larger, varying between 6 and 1 m · s^{-1} as the fracture lengthens and widens (Fig. 18). Thus, given the same pressure history as the single-component rhyolitic flow, the lubricated flow has a much greater chance of reaching the surface or extending significant distances in horizontal (lateral) propagation modes. (Lateral rift zone propagation (≈40–60 km) can be some 10– 30 times the fracture length required to move magma from shallow crustal reservoirs to the surface.)

Effect of Viscous Heating in Lubricated Flows

In the two-component flow of a rhyolite and basalt, virtually all the shear is concentrated in the thin wall layers (Fig. 14) and, as a result, virtually all the viscous dissipation will occur in these layers too. This presents yet another way that lubrication by thin wall layers can promote dike propagation, since it has been shown already that the distribution of viscous dissipation in single-component flow may actually *increase* the amount of heat in the magma as it flows along the dike but still allow solidification of magma at the dike walls (Carrigan *et al.*, 1992).

A model for the effect of viscous heating on the lubricating layers can be derived using the last term involving the dissipation function ϕ_v in Eq. (3). According to the energy equation, this term represents the amount of heat produced per unit volume per unit time by the dissipation of flow energy. For an incompressible plane parallel flow, the energy S dissipated per unit volume per unit time assuming a general power-law rheology is

$$S = \eta\phi_v = \eta\left(\frac{dw}{dx}\right)^2 = m_w\left(\frac{dw}{dx}\right)^{n_w - 1}\left(\frac{dw}{dx}\right)^2. \tag{31}$$

In terms of the gradient of the heat flux q, we have

$$\frac{dq}{dx} = S = m_w\,\dot{\varepsilon}^{n_w+1}, \tag{32}$$

where

$$\dot{\varepsilon}(x) = \left[\frac{G}{m_w}\left(\frac{d}{2} - x\right)\right]^{1/n_w} \tag{33}$$

$$\text{for } 0 \le x \le a.$$

Integration with respect to x of Eq. (32) gives

$$q = -m_w\left(\frac{n_w}{2n_w + 1}\right)\left(\frac{G}{m_w}\right)^{(n_w+1)/n_w} \cdot \left(\frac{d}{2} - x\right)^{(2n_w+1)/n_w} + C_1, \tag{34}$$

where C_1 is a constant of integration. By Fourier's law, we have

$$-k\frac{dT}{dx} = q. \tag{35}$$

To solve for temperature, the right side of Eq. (34) is substituted for q in Eq. (35) and the equation is integrated with respect to x. Another constant of integration is introduced. These are evaluated by applying boundary conditions on the temperatures at $x = 0$ and at $x = a$. In selecting temperatures for each endpoint, it is assumed that there is some temperature below which the lubricating magma component is simply too viscous to flow. Because of cooling on both the wall and core layer sides of the lubricating layer, it is expected that thin, more or less rigid zones of basalt will form when the temperature falls below that required to sustain

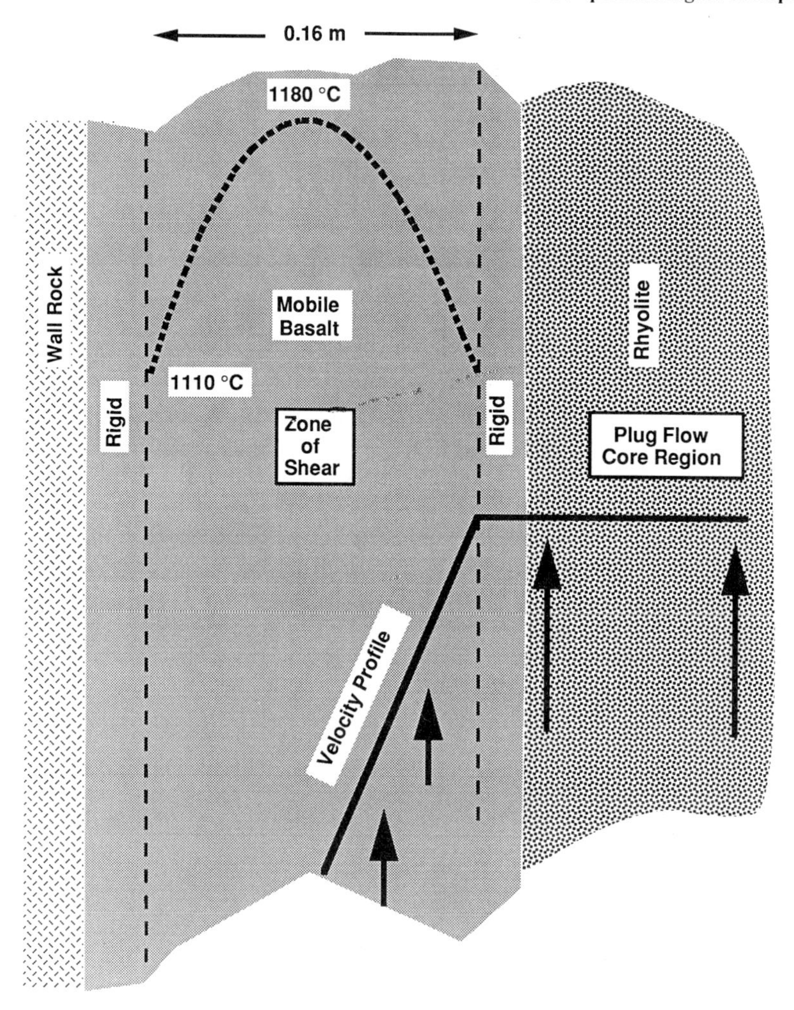

Figure 19 The thermal effect of concentrating shear in the lubricating layer (a = 0.16 m) is examined assuming that the boundaries of a basaltic layer are chilled against the cooler wall rock and rhyolite core and become too rigid to flow. The temperature at which this occurs in the basalt is taken to be 1110°C. For the case with lubrication in Fig. 18, it is found that within the range of predicted fracture velocities (6–1 $m \cdot s^{-1}$), substantial viscous heating of the lubricating layer would occur. The center of the mobile basalt layer would reach a temperature of about 1180°C. Enough heat would be produced in this case to heat both margins of the mobile zone at the rate of about 2 $kW \cdot m^{-2}$. By comparison, a single-component flow of rhyolite could creep only at 0.04–0.01 $m \cdot s^{-1}$ and produce a negligible amount of heating for offsetting heat loss to the wall rock.

the zone of mobility (Fig. 19). In principle, such a model would allow chilling of basalt at the wall rock contact and at the interface between the basalt and rhyolite with a zone lying in between that remains mobile. The stability, i.e., the transverse growth or shrinkage, of such a mobile lubrication zone depends on the balance between the rate of heat generation in the zone and the rate of heat loss across both faces of the lubricating layer into the wall rock and central rhyolite layer. In this particular model T_r represents the temperature at which the transition from mobile to effectively rigid behavior occurs. After solving for the constants of integration, one may then write a time-independent solution for the temperature in a power-law mobile zone as

$$T = -\frac{1}{k}\left[J\left(\frac{d}{2} - x\right)^{(3n_w+1)/n_w} + C_1 x + C_2\right], \tag{36}$$

where

$$J = m_w \left(\frac{n_w}{2n_w + 1}\right)\left(\frac{n_w}{3n_w + 1}\right)\left(\frac{G}{m_w}\right)^{(n_w+1)/n_w},$$

$$C_1 = \frac{J}{a}\left[\left(\frac{d}{2}\right)^{(3n_w+1)/n_w} - \left(\frac{d}{2} - a\right)^{(3n_w+1)/n_w}\right],$$

and

$$C_2 = -\left[kT_r + J\left(\frac{d}{2}\right)^{(3n_w+1)/n_w}\right].$$

Figure 19 illustrates the temperature distribution in the mobile zone for the lubricated flow considered in the previous section ($d = 1$ m, $a = 0.16$ m, $\mu_w = 100$ Pa · s, $\mu_c = 2.2 \times 10^4$ Pa · s and $k = 1.3$ W · m^{-1} · °C^{-1}). Using the rheological data of Shaw *et al.* (1968) to estimate the temperature below which a basaltic magma behaves rigidly, T_r has been set to 1110°C, and it is readily apparent from Fig. 19 that the average temperature of this zone is well above this transition temperature. The assumption of a viscosity of 100 Pa · s for this mobile zone is reasonable and may even be a little large for the temperatures indicated. This model can be used to argue that localized viscous heating can promote the transport of exceptionally viscous magmas even in thermally unfavorable environments.

The temperature gradient in the mobile zone is given by

$$\frac{dT}{dx} = -\frac{1}{k}\left[C_1 - J\left(\frac{3n_w + 1}{n_w}\right) \cdot \left(\frac{d}{2} - x\right)^{(2n_w+1)/n_w}\right], \tag{37}$$

which can be used to evaluate the rate of heat loss into the wall rock and the core layer. For the example considered, viscous dissipation supplies well over 2 kw · m^{-2} to the boundaries of the mobile zone. This is a significant rate of heating that may more than balance losses to the rhyolite and the wall rock (Carrigan *et al.*, 1992). Such a mobile zone may actually increase in mean temperature and grow in width with time. We have modeled several important aspects of two-component flow in a dike. It is now worth looking in greater detail at the models for supplying two-component magmas to a dike.

A Rational Model for Magma Chamber Withdrawal

A cylindrical tube attached to the center of a lid on a density-stratified reservoir (Fig. 5) has become a frequently used model for magma chamber geometry in considering the withdrawal and effusion of multiple layers of magma. It is likely that such a model is so often employed because it is easier to analyze numerically or to construct for laboratory experiments. Unfortunately, this storage withdrawal system is likely to be very unrealistic in at least two important ways that are critical for an understanding of the transport of two magmas and the timing of their arrival at the surface.

First, the lid of the conventional model is precisely parallel to surfaces of constant density in its stratified interior. This represents a singular geometry since there are no physical constraints that require the top and constant density surfaces to be perfectly parallel. When the lid is perfectly horizontal, it will be in contact only with the uppermost layer in a density-stratified system. The only way that lower layers can reach the lid and, thus, be tapped by the tube is to require a sufficiently high withdrawal rate that the draw-up-depth is great enough to sample these lower layers. Even if the lid is domed or pitched, the symmetric placement of the tube on the lid still requires a sufficient withdrawal rate for lower layers to be sampled. Within the context of the conventional model, it is difficult to understand, for example, how the relatively low velocities of transport required for the effusion of a lava dome can produce the simultaneous withdrawal of layers, particularly when the magmas may have significant density differences as in the case of simultaneous basalt and rhyolite venting in Iceland.

Second, cylindrical conduits, as used in the standard model, represent, in general, an unrealistic pathway for the flow of magma from a reservoir. Models of fracture growth near magma bodies produce sheetlike channels (Spence and Turcotte, 1985) and not conduits. Surface studies suggest that conduits are only secondary structures that "bud" from sheetlike dikes and gradually evolve during eruption by mechanical erosion (Delaney and Pollard, 1981) or by melting of the adjacent wall rock (Bruce and Huppert, 1989, 1990) as magma is channeled from the dike into the conduit. Furthermore, a conduit-like structure is likely to be a feature of only the top part of a dike where the cooler host regime aids the conduit-forming process by plugging narrower portions of the dike through solidification. However, in the roof of the magma chamber, solidification at the dike inlet during the flow of magma is unlikely—especially considering numerical estimates of dike boundary temperatures near the inlet (Carrigan *et al.,* 1992). Thus, it is argued that a linear or dike-like breach in the roof of a magma chamber is a much more realistic exit in both the early and later stages of the flow of magma from its reservoir than is a circular hole that is a characteristic feature of many withdrawal models. Both thermal and mechanical considerations suggest that a mature magmatic pathway should have a dike-like inlet at the end penetrating the roof of the magma reservoir and a more localized, conduit-like vent at or near the surface.

Figure 20 illustrates a more general and considerably more realistic alternative model for a reservoir + withdrawal system that avoids the two concerns associated with the conventional model. In this schematic model, the reservoir roof is inclined at an arbitrary angle relative to the surfaces of constant density, and the dike inlet for withdrawal of magma is fracture-like rather than circular. The presence of a dike on the sloping boundary permits withdrawal from more than one layer in the system simultaneously—irrespective of the withdrawal rates or magma viscosities and densities. In this model, the magmatic components can arrive at the inlet as two or more discrete layers. Such inlet flow conditions exactly correspond to the so called two-layer, side-by-side flow

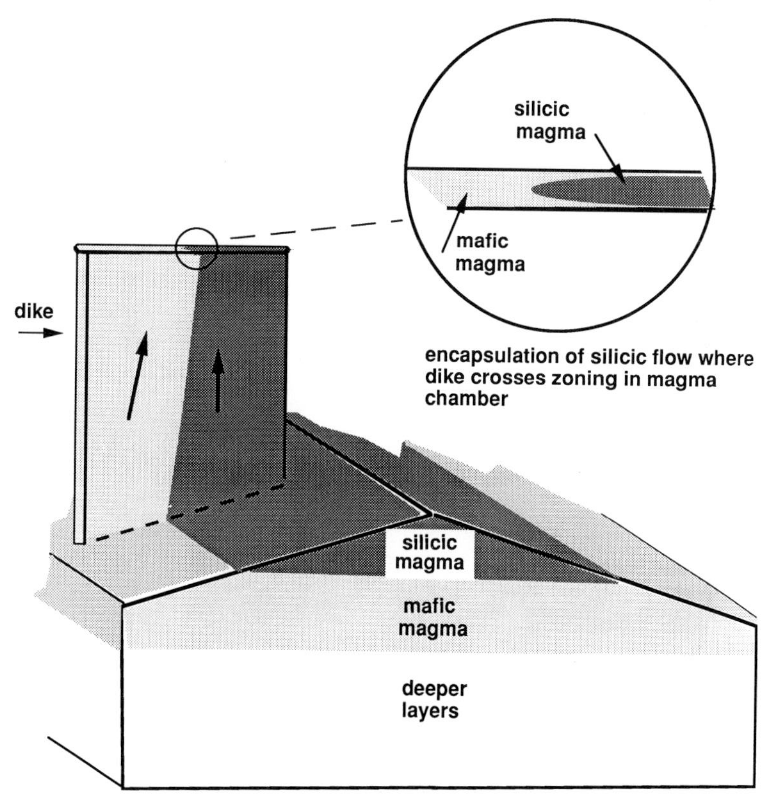

Figure 20 Generalized model of magma chamber and dike system in which the roof is not parallel to the compositional layering in the chamber and withdrawal is through a dike positioned asymmetrically on the top. This geometry allows the dike to withdraw more than one layer simultaneously at any flow velocity in contrast to the conventional model in which the draw-up-depth parameter is a function of withdrawal rate and magma properties. Encapsulation, as illustrated by the inset, is a natural consequence of the side-by-side withdrawal of the two magma layers.

conditions used in polymer coextrusion experiments to produce encapsulation (zoning) in slots with the low viscosity component at the outside of the flow adjacent to the wall (see Fig. 9b, bottom).

Some potential implications of this model deserve further discussion. One is that a dike breaching a magma reservoir roof in the direction of the dip or tilt of the roof can sample—along its length—different layers of the magma body simultaneously. Such a dike will produce at the surface lavas that have a compositional gradient along horizontal segments of the dike length (Fig. 20 inset). Such compositional variations are observed in the lavas of the Inyo Domes chain, which are thought to have contemporaneously ascended in the same dike (Sampson and Cameron, 1987). Encapsulation of the higher viscosity component by the lower viscosity fluid should result in chemical zoning of flows over locations where a horizontally propagating dike crosses between compositional zones in the magma as it breaches the sloping roof of the magma chamber. It is suggested in the next section that at least three of the Inyo Domes represent a manifestation of this encapsulation process as a common (master) dike intersected different compositional boundaries in the underlying magma chamber.

Another implication is that dikes propagating normal to the dip of the magma chamber roof tend to "see" smaller compositional gradients since they will breach the roof parallel to surfaces of constant density and hence similar composition (Fig. 21). Thus, one dike, propagating downward along the slope of the chamber roof, may have significant compositional gradients occurring along its length on the surface whereas a second dike, emplaced perpendicular to the first, may only sample magma of one composition. A corollary to this is that selecting dike orientations and the shape of the magma chamber roof in this model permits any layer in a reservoir to be tapped without regard to a restricted consideration of the draw-up-depth associated with a particular effusion rate as well as to the potentially restrictive requirements on flow rates.

A third implication is that lower viscosity magmas may determine both the order of reservoir withdrawal and whether the higher viscosity magmas will ultimately reach the surface (or, in the case of volcanic rift zones, the lateral extent of dike propagation from the source reservoir). A breach in a magma chamber roof initiated where a layer of very high viscosity magma resides may not result in the propagation of a dike to the sur-

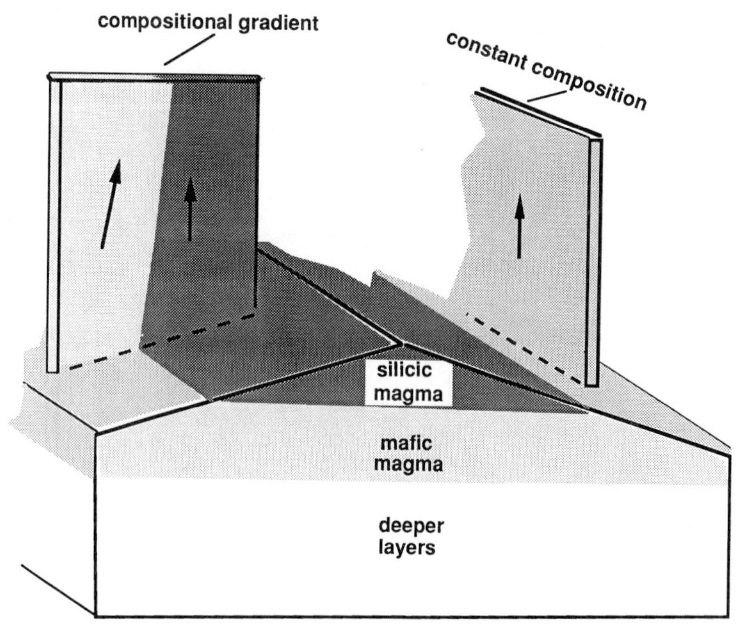

Figure 21 An implication of the generalized model for magma withdrawal is that dike orientation and location alone can determine what magma is tapped from the chamber. Here two differently oriented dikes produce lavas with either a compositional gradient or a constant composition depending on how they breach the chamber roof.

face from that location on the roof owing to the excessive driving pressures required for magma flow. In this case, magma pressure in the chamber is not relieved by vertical propagation and, therefore, the dike can propagate along a slope that breaches the roof until it begins to sample and withdraw another layer of magma having a lower viscosity. At this point encapsulation of the higher viscosity component by the lower viscosity component occurs. Lubrication of the higher viscosity component in this newly forming composite intrusion may then be adequate to permit magma to reach the surface to form a composite flow. Such a scenario is suggested by the relationship between the rhyolitic dike that did not breach the surface and the composite feeder for Obsidian Dome consisting of a rhyolitic core lubricated by a less viscous rhyodacitic annular zone.

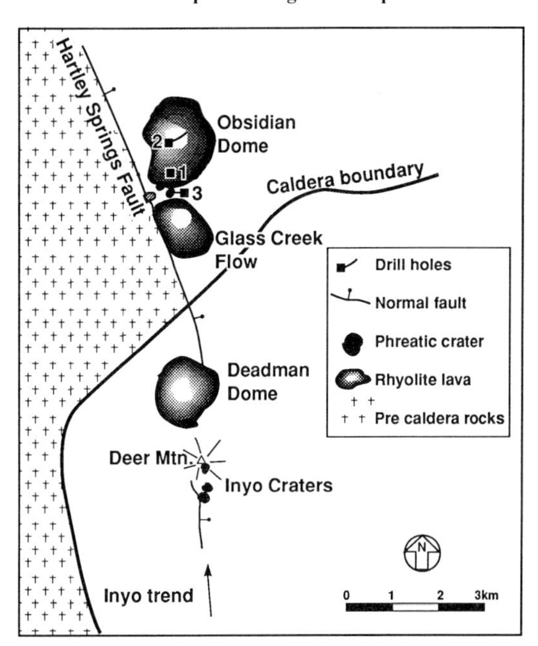

Figure 22 Map view of Inyo domes with associated volcanic and geologic features. The numbered squares on and by Obsidian Dome indicate the locations of scientific drill holes. Adapted from Eichelberger *et al.* (1988) with permission from the American Geophysical Union.

A Physical Model for the Contemporaneous Eruption of Three Domes in the Inyo Volcanic Chain, Long Valley, California

As mentioned previously Obsidian Dome is one of three composite units (Fig. 22) erupted more or less contemporaneously about 600 yr ago in Long Valley along a north–south trending line that crosses the northwestern boundary of the caldera (Miller, 1985; Sampson and Cameron, 1987). The hypothesis that a common dike connects all three domes has been partially validated by the drill hole intersection of a shallow, rhyolitic dike between Obsidian Dome and Glass Creek Dome. At least three physically distinct magmas have contributed to the formation of the composite domes according to Sampson and Cameron (1987). Deadman Dome, the southern-most flow in the chain, consists of a coarse-grained, crystal-rich rhyolite (70–74% SiO_2) that overlies a finer-grained rhyolite of lower crystal content, which also tends to have a slightly lower silica content (71.5%). It is possible that the latter lava is a mixture of the coarser-grained, crystal-rich rhyolite and a finer-grained rhyodacite that is lower in crystal content. The middle flow in the chain, the Glass Creek Dome, lies on the topographic rim of the caldera. Like Deadman Dome, it consists of a coarse, crystal-rich rhyolite overlying a finer-grained rhyolite with lower crystal content. Obsidian Dome lies beyond the rim of the caldera and consists of a fine-grained, low crystal content rhyolite overlying a fine, low crystal content and low silica rhyolite or rhyodacite (Vogel *et al.*, 1989; Carrigan and Eichelberger, 1990).

A model for dike propagation and composite flow formation that includes the breaching of a sloping magma chamber roof and the encapsulation of different viscosity magmas can explain the gross zoning of these three domes by the eruption of magma from a common, chemically zoned chamber. Figure 23 illustrates how the progressive breaching of the sloping roof of a zoned chamber could produce the observed compositional variations during a sustained period of increased pressure in the chamber. The simplest, but not necessarily the only, chronology for the formation of the domes is used. The formation of a common dike may have started by a breach in that portion of the roof in contact with a layer of coarse, crystal-rich, high silica rhyolite. Owing to its very high viscosity, the magma does not allow a dike to reach the surface. At the same time, the dike also propagates horizontally, breaching the chamber roof along the line indicated in Fig. 23. This horizontal breaching continues until the lower

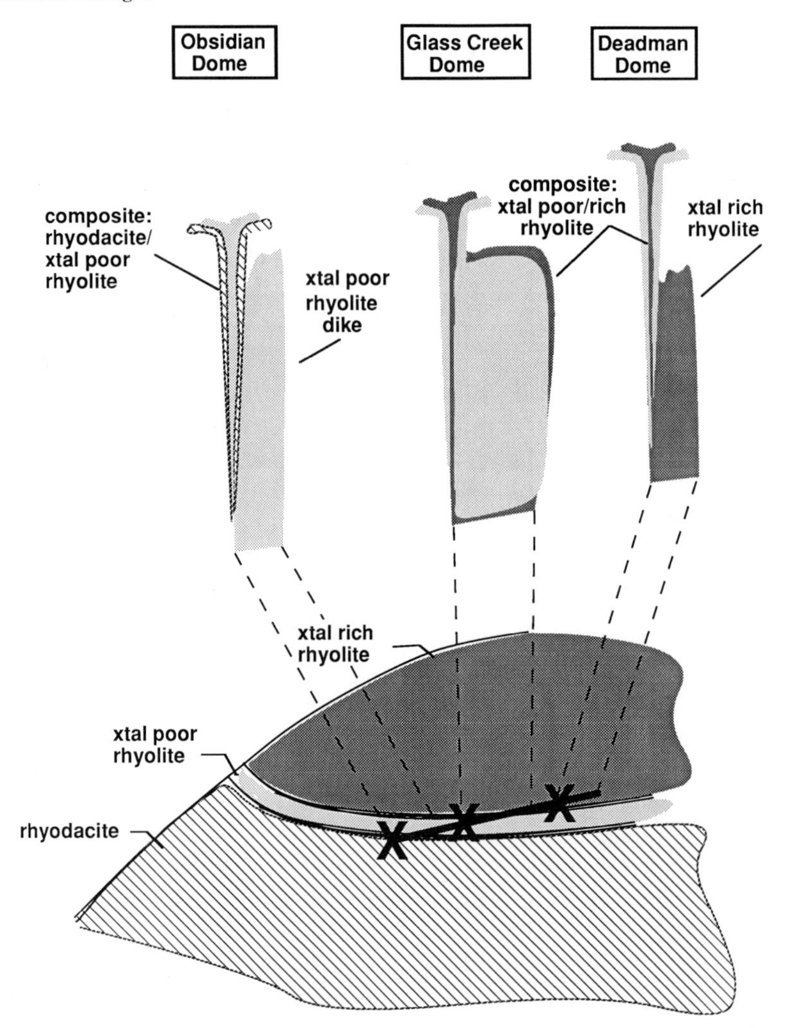

Figure 23 Using an encapsulation model for two-component magma transport and the generalized withdrawal model of Fig. 20, a new model is suggested for the contemporaneous formation of three 600 yr-old composite lava flows in the Inyo volcanic chain. If a breach in the magma chamber roof is initiated over the crystal-rich rhyolite, the high viscosity would tend to prevent this lava from reaching the surface. As the breach progresses down slope, it crosses into a zone of lower viscosity, crystal-poor rhyolite. This magma lubricates the rise of the higher viscosity, crystal-rich component and results in the composite Deadman Dome lava flow (X's indicating the points over which lava flows have formed). The continued breaching of the roof causes the composite dike to finally "bud" a conduit that reaches the surface to form the Glass Creek Dome, which has a similar composition to Deadman Dome. The breach continues its descent across the crystal-poor rhyolite zone and produces a rhyolite dike that does not reach the surface. A rhyodacitic layer is finally sampled by the dike as it crosses a compositional boundary and provides lubrication for the rhyolite to erupt to the surface to form Obsidian Dome as the magma chamber pressure wanes.

silica and lower crystal content rhyolitic layer begins to be withdrawn in addition. Because of its lower viscosity, this magma encapsulates the crystal-rich rhyolite, successfully lubricating its rise to the surface to form the composite Deadman Dome lava flow by the simultaneous effusion of both crystal-rich and crystal-poor lavas. Pressures in the magma chamber evidently remain sufficiently high to permit the continued breaching of the chamber roof along the line. Because of the dike's proximity to both the crystal-poor and crystal-rich rhyolite layers as determined by the path of breaching, both are withdrawn simultaneously and encapsulation ensues to form a composite dike that once again reaches the surface to form the composite Glass Creek Dome consisting of the same types of lavas as Deadman Dome. Continuation of the breach takes it across the zone of crystal-poor rhyolite away from the zone of crystal-rich rhyolite so that the dike now consists of a uniformly crystal-poor rhyolite. This is consistent with the cores obtained from drilling the dike near Obsidian Dome. Finally, the breach in the roof approaches and begins to sample the contents of the less viscous rhyodacitic zone. Encapsulation and self-lubrication of the crystal-poor rhyolite by the rhyodacite results in the budding of a conduit to form the Obsidian Dome flow.

Summary and Conclusions

Voluminous and violent eruptions, such as the one that resulted in the formation of the Long Valley caldera and that vented silicic magma followed by more mafic material, usually empty the underlying magma chamber from the top layer downward so that the sequence of layering on the surface is just an inversion of the chamber's layering prior to the eruption (Hildreth, 1981). In such cases the driving pressure is evidently so great that the viscosity of magma, within a very wide range, plays an insignificant role in determining whether an eruption is initiated. Whatever magma is in contact with the roof at the moment that it is breached by a dike is the first magma to arrive at the surface.

In contrast, effusive eruptions appear to be characterized by much smaller magmatic driving pressures, and the effective viscosity of a flow may be the pivotal influence in determining whether a magma can reach the surface. As long as the dike taps only the most viscous top layer through the roof of the chamber, the propagation velocity of the dike simply may be too low for it to reach the surface before cooling causes plugging. If the new dike also happens to sample a lower viscosity component across a compositional interface, then encapsulation and lubrication of the more viscous component by the less viscous component may reduce pressure losses enough to allow the flow to reach the surface. This chapter has considered the surface observations, laboratory experiments, numerical models, and analytical development needed to evaluate the processes pertinent to this scenario.

Other, more conventional, models are also evaluated and critiqued on the basis of the evidence presented. It was found that sequential-flow models, in which one magma follows another up a dike or conduit to be extruded one at a time, cannot explain several important features of the composite flows discussed here. It is possible, however, that the sequential scenario provides an explanation for that less common class of dikes that are reverse-zoned with the mafic magma in the center and the silicic magma at the walls (Gautneb, 1988, Fig. 4). In such a case, silicic magma withdrawn from the top layer of a magma chamber could have been followed by a more mafic, underlying component.

Given that encapsulation of a high viscosity component by a lower viscosity component occurs, we have derived time-independent equations for the lubrication of a magma of high viscosity by a second one of lower viscosity. The derivation is general enough to permit consideration of both Newtonian and power-law magmas. It is shown that only relatively thin lubricating layers ($<10\%$ of the dike width) are needed to substantially reduce the pressure gradient required to drive a magma of very high viscosity through a dike. Furthermore, the addition of thin lubricating layers to a flow of rhyolite can increase fracture propagation rates several orders of magnitude for a given value of the chamber pressure. In a basalt–rhyolite system, the core layer of rhyolite will flow as a solid body with virtually all the shear concentrated in the basalt lubricating layers. Viscous dissipation will then be concentrated in the lubrication layers. As a result, the thin basaltic layers can become significant heat producers during the ascent of a composite magma through a dike and offset wall heat losses ($>2\ \mathrm{kW \cdot m^{-2}}$) for reasonable dike propagation rates ($\geq 1\ \mathrm{m \cdot s^{-1}}$). If the two components of a flow are characterized by different power-law rheologies, a range of shear rate can be shown to exist, over which the pressure gradient for two-component flow is less for a given flux of magma than the pressure gradient required to drive the same flux of each component flowing by itself in the dike.

The important effects of different dike inlet conditions (in the magma reservoir roof or walls) on the encapsulation process have been discussed. In this context, a new and considerably more general model for the simultaneous withdrawal of two magmas through a dike has been presented. It is a characteristic of this model that a generalized geometry and not special—and ad hoc—appeals to magma withdrawal rates or to physical properties permits the simultaneous flow of two magmatic components in a dike. This inlet model in conjunction with the encapsulation process provides the basis for explaining the contemporaneous origin of three of the Inyo volcanic domes in Long Valley, California.

The process of self-lubrication of two-component magmas needs to be considered as a potentially important factor in any explanation for the origin of intrusions and lava flows. In some cases it seems likely that we observe at the surface a certain high silica and high crystal content lava, such as the coarse-grained, high crystal and silica content rhyolite of Deadman Dome, because of that lava's association with a more fluid component that provided lubrication. In fact, the statistics for the occurrence of extreme compositional contrasts in lava flows, e.g., rhyolites and basalts, may not remotely reflect the degree of coexistence of end-member magmas at depth because of a bias of reservoir-to-surface transport toward such two-component associations. In this regard, the observed products of a large eruption may be more indicative of the original content of the source-magma chamber than are the composite lavas from a somewhat later effusion driven by a much reduced magmatic pressure.

(Author's Note: A utility written in Fortran for either Macintosh or MS-DOS operating systems is available from the author for evaluating most of the mathematical expressions and models developed in this chapter.)

Acknowledgments

This chapter is dedicated to the memory of my father, Charles F. Carrigan. I thank Charles Bacon, John Eichelberger, Wes Hildreth, Richard Knapp, Bob Nilson, Michael Ryan, Rick Ryerson, Tom Vogel, Lee Younker, and George Zandt for both enlightening comments and stimulating discussions that have improved this manuscript. Many of the figures were redrawn by Priscilla Proctor of the Technical Information Department. This work was performed under the auspices of the U.S. Department of Energy by the Lawrence Livermore National Laboratory under Contract W-7405-ENG-48. Support was provided through a Lawrence Livermore LDRD grant and the Institute of Geophysics and Planetary Physics.

References

Bacon, C. R. (1986). Magmatic inclusions in silicic and intermediate volcanic rocks, *J. Geophys. Res.* **91,** 6091–6112.

Bird, R. B., Stewart, W. E., and Lightfoot, E. N. (1960). "Transport Phenomena," Wiley, New York.

Blake, D. H., Elwell, R. W. D., Gibson, I. L., Skelhorn, R. R., and Walker, G. P. L. (1965). Some relationships resulting from the intimate association of acid and basic magmas, *Quart. J. Geol. Soc. London* **121,** 31–49.

Blake, S., and Campbell, I. H. (1986). The dynamics of magma-mixing during flow in volcanic conduits, *Contrib. Mineral. Petrol.* **94,** 72–81.

Blake, S., and Fink, J. H. (1987). The dynamics of magma withdrawal from a density stratified dyke, *Earth Planet. Sci. Lett.* **85,** 516–524.

Blake, S., and Ivey, G. N. (1986). Magma mixing and the dynamics of withdrawal from stratified reservoirs, *J. Volcanol. Geotherm. Res.* **27,** 153–178.

Bruce, P. M., and Huppert, H. E. (1989). Thermal control of basaltic fissure eruptions, *Nature* **342,** 665–667.

Bruce, P. M., and Huppert, H. E. (1990). Solidification and melting along dykes by the laminar flow of basaltic magma, *in* "Magma Transport and Storage" (M. P. Ryan, ed.), Wiley, Chichester/Sussex, England.

Carrigan, C. R. (1983). A heat-pipe model for vertical, magma-filled conduits, *J. Volcanol. Geotherm. Res.* **16,** 279–298.

Carrigan, C. R., and Eichelberger, J. C. (1990). Zoning of magmas by viscosity in volcanic conduits, *Nature* **343,** 248–251.

Carrigan, C. R., Schubert, G., and Eichelberger, J. C. (1992). Thermal and dynamical regimes of single- and two-phase magmatic flow in dikes, *J. Geophys. Res.* **97,** 17377–17392.

Charles, M. E., and Redberger, P. J. (1962). The reduction of pressure gradients in oil pipelines by the addition of water: Numerical analysis of stratified flow, *Can. J. Chem. Eng.* **40,** 70–75.

Delaney P. T., and Pollard, D. D. (1981). "Deformation of Host Rock and Flow of Mafic Magma during Growth of Minette Dikes and Breccia-Bearing Intrusions near Ship Rock, New Mexico," U.S. Geological Surv. Prof. Paper 1202.

Eichelberger, J. C. (1975). Origin of andesite and dacite: Evidence of mixing at Glass Mountain, California, and at other circum-Pacific volcanoes, *Geol. Soc. Am. Bull.* **86,** 1381–1391.

Eichelberger, J. C. (1980). Vesiculation of mafic magma during replenishment of silicic magma reservoirs, *Nature* **288,** 446–450.

Eichelberger, J. C., Lysne, P. C., Miller, C. D., and Younker, L. W. (1985). Research drilling at Inyo Domes, California: 1984 results, *EOS Trans. AGU* **66,** 186–187.

Eichelberger, J. C., Vogel, T. A., Younker, L. W., Miller, C. D., Heiken, G. H., and Wohletz, K. H. (1988). Structure and stratigraphy beneath a young phreatic vent: South Inyo Crater, Long Valley Caldera, California, *J. Geophys. Res.* **93,** 13208—13220.

Everage, A.E. (1973). Theory of stratified bicomponent flow

of polymer melts. I. Equilibrium Newtonian tube flow, *Trans. Soc. Rheol.* **17,** 629–646.

Frisch, U., Hasslacher, B., and Pomeau, Y. (1986). Lattice-gas automata for the Navier–Stokes equation, *Phys. Rev. Lett.* **56,** 1505–1508.

Furman, T., and Spera, F. J. (1985). Co-mingling of acid and basic magma with implications for the origin of mafic I-type xenoliths: Field and petrochemical relations of an unusual dike complex at Eagle Lake, Sequoia National Park, California, U.S.A., *J. Volcanol. Geotherm. Res.* **24,** 151–178.

Gautneb, H. (1988). Structure, age and formation of dykes on the island of Smøla, Central Norway, *Norsk Geol. Tidsskrift* **68,** 275–288.

Geertsma, J., and Haafkens, R. (1979). A comparison of theories for predicting width and extent of vertical hydraulically induced fractures, *Trans. ASME J. Energy Res. Technol.,* **101,** 8–19.

Gibson, I. L., and Walker, G. P. L. (1963). Some composite rhyolite/basalt lavas and related composite dykes in eastern Iceland, *Proc. Geol. Ass.* **74,** 301–318.

Gudmundsson, A. (1990). Mafic dykes and emplacement mechanisms, *in* "Proceedings, 2nd Intern. Dyke Conf., Adelaide, 1990," pp. 47–62.

Han, C. D. (1973). A study of bicomponent coextrusion of molten polymers, *J. Appl. Polym. Sci.* **17,** 1289.

Han, C. D. (1975). A study of coextrusion in a circular die, *J. Appl. Polym. Sci.* **19,** 1875–1883.

Han, C. D. (1976). "Rheology in Polymer Processing," Academic Press, New York.

Han, C. D. (1981). "Multiphase Flow in Polymer Processing," Academic Press, New York.

Han, C. D., and Chin, H. B. (1979). Theoretical prediction of pressure gradients in coextrusions of non-Newtonian fluids, *Polym. Eng. Sci.* **19,** 1156.

Hickox, C. E. (1971). Instability due to viscosity and density stratification in axisymmetric pipe flow, *Phys. Fluids* **14,** 251–262.

Hildreth, W. (1981). Gradients in silicic magma chambers: Implications for lithospheric magmatism, *J. Geophys. Res.* **86,** 10133–10192.

Hu, H., and Joseph, D. D. (1989). Lubricated pipelining: Stability of core–annular flow, Part 2, *J. Fluid Mech.* **205,** 359–396.

Huppert, H. E., and Sparks, R. S. J. (1980). The fluid dynamics of a basaltic magma chamber replenished by influx of hot dense ultrabasic magma, *Contrib. Mineral. Petrol.* **75,** 279–289.

Huppert, H. E., Sparks, R. S. J., and Turner, J. S. (1982). Effects of volatiles on mixing in calc-alkaline magma systems, *Nature* **297,** 554–557.

Joseph, D. D., Nguyen, K., and Beavers, G. S. (1984a). Non-uniqueness and stability of the configuration of flow of immiscible fluids with different viscosities, *J. Fluid Mech.* **141,** 319–345.

Joseph, D. D., Renardy, M., and Renardy, Y. (1984b). Instability of the flow of two immiscible liquids with different viscosities, *J. Fluid Mech.* **141,** 309–317.

Karagiannis, A., Mavridis, H., Hrymak, A. N., and Vlachopoulos, J. (1988). Interface determination in bicomponent extrusion, *Polym. Eng. Sci.* **28,** 982–988.

Koyaguchi, T. (1985). Magma mixing in a conduit, *J. Volcanol. Geotherm. Res.* **25,** 365–369.

Lee, B.-L., and White, J. L. (1975). Experimental studies of disperse two-phase flow of molten polymers through dies, *Trans. Soc. Rheol.* **19,** 481–492.

Lister, J. R., and Kerr, R. C. (1991). Fluid-mechanical models of crack propagation and their application to magma transport in dykes, *J. Geophys. Res.* **96,** 10049–10077.

MacLean, D. L. (1973). A theoretical analysis of bicomponent flow and the problem of interface shape, *Trans. Soc. Rheol.* **17,** 385–399.

Miller, C. D. (1985). Holocene eruptions at the Inyo volcanic chain, California—Implications for possible eruptions in the Long Valley Caldera, *Geology* **13,** 14–17.

Minagawa, N., and White, J. L. (1975). Co-extrusion of unfilled and TiO_2-filled polyethylene: Influence of viscosity and die cross-section on interface shape, *Polym. Eng. Sci.* **15,** 825–830.

Nilson, R. H., and Griffiths, S. K. (1986). "Self-Similar Analysis of Planar Hydraulic Fractures Driven by a Constant Flow Rate," *Sandia National Laboratories Report,* SAND86–8694.

Oliemans, R. V., and Ooms, G. (1986). Core–annular flow of oil and water through a pipeline, *in* "Multiphase Science and Technology 2" (G. F. Hewitt, J. M. Delhaye, and N. Zuber, eds.), pp. 427–472, Hemisphere, Washington, DC.

Rothman, D. H., and Keller, J. M. (1988). Immiscible cellular-automaton fluids, *J. Stat. Phys.* **52,** 1119–1127.

Ryan, M. P. (1987) Neutral buoyancy and the mechanical evolution of magmatic systems, *in* "Magmatic Processes: Physicochemical Principles" (B. O. Mysen, ed.), pp. 259–287, The Geochemical Society, University Park, PA.

Ryan, M. P. (1990). The physical nature of the Icelandic magma transport system, *in* "Magma Transport and Storage" (M. P. Ryan, ed.), pp. 175–224, Wiley, Chichester/Sussex, England.

Ryan, M. P., and Blevins, J. Y. K. (1987). "The Viscosity of Synthetic and Natural Silicate Melts and Glasses at High Temperatures and 1 Bar (10^5 Pascals) Pressure and at Higher Pressures," U.S. Geol. Surv. Bull. 1764.

Ryerson, F. J., Weed, H. C., and Piwinskii, A. J. (1988). Rheology of subliquidus magmas. 1. Picritic compositions, *J. Geophys. Res.* **93,** 3421–3436.

Sampson, D. E., and Cameron, K. L. (1987). The geochemistry of the Inyo volcanic chain: Multiple magma systems in the Long Valley region, eastern California, *J. Geophys. Res.* **92,** 10403–10421.

Shaw, H. R., Peck, D. L., Wright, T. L., and Okamura, R. (1968). The viscosity of basaltic magma: An analysis of field measurements in Makaopuhi Lava Lake, Hawaii, *Am. J. Sci.* **266,** 225–264.

Shaw, H. (1969). Rheology of basalt in the melting range, *J. Petrol.* **10,** 510–535.

Skelhorn, R. R., MacDougall, J. D. S., and Longland, P. J. N. (1969). "The Tertiary Igneous Geology of the Isle of Mull," Geol. Ass. Guides, No. 20, Geologists' Association, London, Benham and Co., Ltd.

Smith, R. L. (1979). "Ash Flow Magmatism." Geol. Soc. Amer. Special Paper 180, pp. 5–27.

Southern, J. H., and Ballman, R. L. (1973). Stratified bicomponent flow of polymer melts in a tube, *Appl. Polym. Sci.* **20,** 175–189.

Spence, D. A., and Turcotte, D. L. (1985). Magma-driven propagation of cracks, *J. Geophys. Res.* **90,** 575–580.

Spera, F. J. (1984). Some numerical experiments on the withdrawal of magma from crustal reservoirs, *J. Geophys. Res.* **89,** 8222–8236.

Spera, F. J., Yuen, D. A., Greer, J. C., and Sewell, G. (1986). Dynamics of magma withdrawal from stratified magma chambers, *Geology* **14,** 723–726.

Spera, F. J., Borgia, A., Strimple, J., and Feigenson, M. (1988). Rheology of melts and magmatic suspensions. 1. Design and calibration of concentric cylinder viscometer with application to rhyolitic magma, *J. Geophys. Res.* **93,** 10273–10294.

Stockman, H. W., Stockman, C. T., and Carrigan, C. R. (1990). Modelling viscous segregation in immiscible fluids using lattice-gas automata, *Nature* **348,** 523–525.

Trial, A. F., Spera, F. J., Greer, J., and Yuen, D. A. (1992). Simulations of magma withdrawal from compositionally zoned bodies, *J. Geophys. Res.* **97,** 6713–6733.

Vogel, T. A. (1982). Magma mixing in the acidic–basic complex of Ardnamurchan: Implications on the evolution of shallow magma chambers, *Contrib. Mineral Petrol.* **79,** 411–423.

Vogel, T. A., and Wilband, J. T. (1978). Coexisting acidic and basic melts: Geochemistry of a composite dike, *J. Geol.* **86,** 353–371.

Vogel, T. A., Eichelberger, J. C., Younker, L. W., Schuraytz, B. C., Horkowitz, J. P., Stockman, H. W., and Westrich, H. R. (1989). Petrology and emplacement dynamics of intrusive and extrusive rhyolites of Obsidian Dome, Inyo craters volcanic chain, eastern California, *J. Geophys. Res.* **94,** 17937–17956.

Walker, G. P. L., and Skelhorn, R. R. (1966). Some associations of acid and basic igneous rocks, *Earth Sci. Rev.* **2,** 93–109.

White, J. L., and Lee, B.-L. (1975a). An experimental study of sandwich injection molding of two polymer melts using simultaneous injection, *Polym. Eng. Sci.* **15,** 481–485.

White, J. L., and Lee, B.-L. (1975b). Theory of interface distortion in stratified two-phase flow, *Trans. Soc. Rheol.* **19,** 457–479.

Williams, M. C. (1975). Migration of two liquid phases in capillary extrusion: An energy interpretation, *AIChE J.* **21,** 1204–1207.

Wohletz, K. H., and Valentine, G. A. (1990). Computer simulations of explosive volcanic eruptions, *in* "Magma Transport and Storage" (M. P. Ryan, ed.), pp. 113–136, Wiley, Chichester/Sussex, England.

Yoder, H. S. (1973). Contemporaneous basaltic and rhyolitic magmas, *Am. Mineral.* **58,** 153–171.

Younker, L. W., Eichelberger, J. C., Kasameyer, P. W., Newmark, R. L., and Vogel, T. A. (1987). Results from shallow research drilling at Inyo Domes, Long Valley caldera, California, and the Salton Sea geothermal field, Salton Trough, California, *in* "Proceedings 3rd Intern. Symp. on Observation of the Continental Crust through Drilling, Mora and Orsa, Sweden, 1987," vol. 2, pp. 172–187.

Chapter 15 | Fluid and Thermal Dissolution Instabilities in Magmatic Systems

J. A. Whitehead and Peter Kelemen

Overview

Magma source regions at great depths in the mantle are now thought to have melt dynamics much more potentially complex than the common perception of laminar grain boundary flows with a spatially smooth melt migration front. As the melt rises by porous flow along the edges of grains in partially melted material, it can develop a hydrodynamic instability whereby fingers of faster moving melt will dissolve parts of the porous network and lead to yet higher velocity flows. Criteria for this instability are discussed. Laboratory experiments with water flowing through salt crystals demonstrate this instability. The results follow earlier studies of thermal erosion that have been demonstrated in a laboratory experiment with syrup flowing through a chilled tube (Whitehead and Helfrich, 1991). Flow is either steady or periodic depending on the temperature of the liquid and the flow rate into the reservoir. An analytic theory indicates that the transition from steady to periodic flows depends on the nonlinearities in the steady-state relationship between pressure and flow rate. A general stability criterion that states that the Peclet number must be within a certain range for an instability to develop is then advanced. Parameters governing the oscillation period are determined. The thermal theory has also been extended to flow through a conduit, and finger development is predicted. The instability is very similar to the corrosive instability of salt + water permeation experiments. Qualitative laboratory experiments with paraffin that spreads radially over a cold plate also reveal the fingering and demonstrate features that mimic flow regimes in advancing lava flows. The instability is very similar to the corrosive instability of water flowing through salt.

Notation

		Units
A	scale of viscosity variation	dimensionless
A_c	constant magma chamber cross-section area	cm², m²
$A'(y,t)$	small area variation from pressure change	cm², m²
D	solid-state diffusion coefficient	$cm^2 \cdot s^{-1}$
E	Coefficient of elasticity times geometric terms	$dyn \cdot cm^{-4}$
F	scaled force accumulated in elastic region	dimensionless
L	length of slot	cm, m
H_h	thickness of gravity current $(g'q^3t_d/\nu)^{1/5}$	m
P	scaled pressure $(pd^4/96\nu_H\rho\kappa L^2)$	dimensionless
P_e	Peclet number (wd/κ)	dimensionless
Pe_D	diffusive Peclet number (ud^2/DL)	dimensionless
Q	uniform volumetric flux per unit length	$cm^2 \cdot s^{-1}$
Q'	scaled volume flux $(Q/W_r d)$	dimensionless
T	temperature of the fluid in slot	°C
T_H	temperature at which viscosity is ν_H	°C
W	scaled velocity $(=w/W_r)$	dimensionless
W_c	velocity perturbation amplitude	dimensionless
W_r	velocity scale $(8\kappa L/d^2)$	$cm \cdot s^{-1}$
d	thickness of slot	cm, m
d_c	crystal size	cm
g'	(gravity) × (normalized density difference) reduced gravity	$cm \cdot s^{-2}$
$f(W)$	scaled flow resistance in slot	dimensionless
l_c	cross-section length scale of magma chamber	cm, m
m	measure of mass	dimensionless
p	pressure	$dyn \cdot cm^{-2}$, MPa
q	volume flux per unit length (paraffin experiment)	$m^2 \cdot s^{-1}$
t	time	s
t'	scaled time $(12\nu_H t/d^2)$	dimensionless

Magmatic Systems
Edited by M. P. Ryan

		Units
t''	scaled time for earthquake problem	dimensionless
t_d	duration time of paraffin experiment	s
u'	scaled earthquake sliding velocity	dimensionless
v	velocity along the axis of magma chamber	$cm \cdot s^{-1}$
w	vertical velocity leaving magma chamber	$cm \cdot s^{-1}$
w_0	steady scaled velocity	dimensionless
w'	spatial and temporal perturbation to W	dimensionless
x	direction across the slot	cm
y	direction along chamber axis	cm
y'	scaled direction along chamber axis y/L	dimensionless
z	direction up the slot	cm, m
ΔT	temperature difference between top and bottom of slot	°C
Γ	measure of change of resistance $(4\gamma \partial f/\partial w_0)$	dimensionless
Σ	paraffin surface roughness $(\sigma_x/H_h L)$	m^{-1}
Ψ	ratio of cooling time to advection time scale t_s	dimensionless
α	viscosity–temperature coefficient	$cm^2 \cdot s^{-1} \cdot °C^{-1}$
β	measure of friction coefficient	dimensionless
δ	measure of aspect ratio $(12A_c l_c^2/d^3 L)$	dimensionless
γ	measure of pressure expansion $[(12\nu_H/d^2)^2 (\rho L/Ed)]$	dimensionless
σ_x	standard deviation of thickness of paraffin	m
κ	thermal diffusivity	$cm^2 \cdot s^{-1}$
ρ	constant density of fluid	$g \cdot cm^{-3}$
$\nu(T)$	kinematic fluid viscosity	$cm^2 \cdot s^{-1}$
ν_H	kinematic fluid viscosity in magma chamber	$cm^2 \cdot s^{-1}$
ν_p	kinematic viscosity of paraffin	$cm^2 \cdot s^{-1}$

Introduction

Volcanism occurs where hot silicate liquids (magma) from deep in the Earth flow to the surface, where they cool and solidify. Before solidification is complete, the flow resistance in both deep and shallow conduits can change. This resistance to flow can either *decrease* due to the failure of the surrounding rock, the thermal or chemical corrosion of the surrounding rock, or the formation of gases (which have vastly lower viscosity), or instead *increase* due to the action of numerous processes that retard the flow upon cooling of the magma or upon ascent from great depth. Some processes that increase the resistance of magma to flow are constrictions due to crystallization along conduit walls, or an increase in the fluid viscosity due to cooling. The increase in viscosity through cooling (Hughes, 1982) can result from the inherent properties of the material's viscosity–temperature relationship (Ryan and Blevins, 1987), bulk composition changes through crystal fractionation, or the addition of suspended crystals to the fluid upon cooling, with a consequent dramatic increase in bulk viscosity due to two-phase effects.

Changes in the aggregate flow resistance through cooling lead to changes in the dynamic pressure, which leads, in turn, to a number of interesting effects. In this chapter, we describe recent studies on the dynamics of flows that develop decreased resistances to flow due to chemical dissolution, and we show that the results are similar to flows that develop an increased resistance as they flow into cooler regions. In particular, we describe the mechanics of a number of intriguing fluid dynamic instabilities that may develop. We have concentrated our effort developing an analytic theory of fluid flow instability in magmatic conduits and have combined this with revealing laboratory experiments that illustrate some possible thermal, velocity, and pressure instabilities.

In general, these fluid instabilities are characterized by the development of fingers of melt and/or time-dependent effects. Volcanic features such as time-dependent surges, for example, complicated free-surface shapes such as pahoehoe texture development, and lava tubes and restricted lava fountains are thought to be the result of this change in flow resistance with temperature. The intriguing possibility that these processes happen at great depth also exists, since an increase of flow resistance with cooling is one of the most prevalent processes in magmatic systems. Our approach has been to study simple problems that capture the basic features that develop and then to suggest possible applications. Duplication of the full geophysical and geological complexities for a given application is not our present objective. Instead, the aim is to capture and illustrate—in experiment and in theory—the essential physical process.

Numerous geological systems—beyond those mentioned previously—also involve a fluid that develops a change in resistance upon cooling. Many aquifers, for example, dissolve minerals under high pore fluid pressures and elevated temperatures. Such minerals in solution may be redeposited along fractures and veins in cooler portions of the system. In both terrestrial and deep-sea hydrothermal systems, numerous instances occur in which the systems fluctuate in their flow behavior, become restricted to a few localized springs, and may ultimately have fluid permeabilities that diminish due to the deposited minerals.

A well-known instability in fluid dynamics that involves a low viscosity fluid flowing into regions containing a more viscous resident fluid is called the "Saffman–Taylor" (Saffman and Taylor, 1958) instability. A fluid intrudes, for example, into a porous region or a Hele–Shaw cell composed of two plane walls separated by a small gap that contains a second more viscous fluid. Under suitable conditions, the interface between the two fluids will develop finger-like protrusions that contain the lower viscosity fluid and periodically interpenetrate into the viscous fluid. The lower viscosity fluid possesses less hydraulic resistance to the large-scale pressure field and moves rapidly into the finger. This forces the finger tip to advance yet further into the viscous fluid. The examples of Saffman–Taylor instabilities that have been studied to date, whether with mathematical analysis or with laboratory experiments, are inherently time dependent, and the tips continue to move indefinitely. After a long time the region is filled with veins of the low viscosity fluid; each vein is surrounded by islands of viscous fluid that is slowly moving away from the source. The final state is never truly steady or even periodic with time.

Although numerous thoughts about structure generation in flow below volcanic systems have been expressed, actual attempts to analyze the mechanical behavior that lead to structure generation or time-dependent flow are rare. One of the first such studies was by Marsh (1979), who suggested that a Rayleigh–Taylor gravitational instability might produce the roughly periodic spacing of island–arc volcanos. Whitehead *et al.* (1984) and Whitehead and Helfrich (1990) followed this lead to suggest that Rayleigh–Taylor diapiric structures may form under mid-ocean ridge spreading centers. An analysis similar to the Saffman–Taylor (1958) instability but with more direct geochemical applications was conducted by Chadham *et al.* (1986) and Ortoleva *et al.* (1987b). In their study an advancing front reacted with host material to produce a scallop-shaped interface. They suggested that numerous geological features may be generated by this geochemical self-organization (Ortoleva *et al.*, 1987a). Although they showed, in general, that such wavy interfacial disturbances may grow with time, no clear stability criteria that included a prediction of fundamental length scales or fundamental time scales, for the predicted changes were developed. A second physical process for the thermal control of basaltic fissure eruptions was developed by Bruce and Huppert (1989, 1990). Melt flowing through a dike is shown to be fundamentally unsteady and the flowing magma will either gradually solidify on the conduit walls until it completely blocks the flow or melt back the walls of the conduit, eroding the adjacent country rock. It was suggested that these results show that such a process leads to flow localization, as earlier suggested by Delaney and Pollard (1982). A recent third model by Stevenson (1989) showed that a partial melt undergoing deformation is unstable to the small-scale redistribution of melt, so that melt migrates parallel to the minimum compressive stress and accumulates in veins, where the bulk viscosity is lowest and becomes even lower as more melt accumulates. Although only a very small length scale was indicated by the instability analysis, the important mechanism in the accumulation of melt in slip planes has long been hypothesized by geologists.

Here, a thermal approach that involves a general decrease in resistance with higher velocity flow like that of Bruce and Huppert (1989, 1990) is combined with instability considerations to produce a thermal and chemical equivalent to either the Saffman–Taylor instability or the geochemical self-organization cases of Chadham *et al.* (1986) or Ortoleva *et al.* (1987a, 1987b). It is completely analogous to a melt flowing through a matrix in which the matrix resistance to flow is altered by the rate of flow of the fluid. For simplicity, instead of two fluids differing materially, we have one fluid with a temperature- or chemically dependent viscosity. The fluid flows from the source as an initially hot fluid and is then cooled through thermal conduction to the the cold sidewalls of a Hele–Shaw cell or, by analogy, altered by chemical interaction with the porous medium. Unlike previously studied problems, the final state may become truly steady (although periodic or chaotic states are also possibilities).

In this chapter, we lay the foundations for the chemical modulation of a flow problem by first reviewing a laboratory experiment conducted by Whitehead and Helfrich (1991) illustrating one possible situation: hot fluid with a temperature-dependent viscosity flowing through a cold slot from an elastic reservoir. If along-slot perturbations are neglected, the flow is found to be either steady or periodic, depending on the degree of viscosity increase produced in the relatively cool region and the rate of reservoir replenishment. If along-slot perturbations are included, it is found that small wavelength instabilities are the most unstable and that instability is encountered when the Peclet number (ud/κ) is of order L/d, where u is the velocity scale, κ is the thermal diffusivity, L is the length of the conduit and d is the width of the conduit. This appears to be the first quantitative criterion found for the onset of this class of instabilities. The flow in the limit of infinite wavelength (zero wavenumber) reduces to the laboratory experiment described in the section Transition of One-Dimensional Flows to Oscillations: An Experimental Study.

Another laboratory experiment is then described in the section Stability of Thermally Eroded Flows, which illustrates the spatial instability. To perform this experiment, hot paraffin was made to spread radially through a slot from a point source over a cold plate. When the Peclet number is of the order of the conduit length divided by its width, the radially symmetric flow experiences a transition to a fingering flow. At first a number of fingers are visible at the outer front of the expanding circular pool of paraffin. In the intermediate stage, these fingers advance substantially. Each finger is fed by a tube of flowing melted paraffin. The rest of the paraffin stops advancing and ultimately solidifies. At later times, all but one of the tubes slow down and stop, and the melted paraffin flows in one final tube for as long as the experiment continues. The paraffin everywhere else gradually solidifies. This experiment is simpler than experiments by Fink and Griffiths (1990) where paraffin was fed from a central tube over a flat plate in water as a model of viscous gravity currents with a solidifying crust. In this experiment the free surface of the paraffin varied in depth and the structure became considerably more complicated. No axisymmetric flow was ever found. As in the slot experiments, the structure of the solidified flow was more complicated for Peclet numbers of order 1. Recent experiments with paraffin flows into water with a Cartesian geometry (Ito and Whitehead, in progress) find that sheet flows dominate when lava solidifies late in the flow sequence whereas pillow morphologies occur when solidification begins earlier during the eruption.

Thermal Erosion

Transition of One-Dimensional Flows to Oscillations: An Experimental Study

Figure 1 shows the schematic of the experimental apparatus consisting of a vertical glass tube 3.8 cm i.d. and 1 m long, located below a reservoir containing Karo brand corn syrup. A variable control valve leading from the reservoir allowed the syrup to flow into the glass tube at a controlled rate. Projecting out from the bottom of the glass tube was a stopper with a hole and a 0.383-cm i.d. copper tube. Varying tube lengths were used. The copper tube projected downward and was shaped like the letter "J." The lowest part of the J-tube was immersed in a refrigerated thermostatic bath with a plastic flexible tube extending from the copper tube to a point outside the bath over a beaker placed to catch outflow. In a typical experiment, syrup flows from the reservoir into the tube. Syrup in the glass tube builds up to a height h that can be easily measured and flows out through the bottom copper tube. The frictional resistance to flow takes place principally within the copper tube because it is much smaller in internal

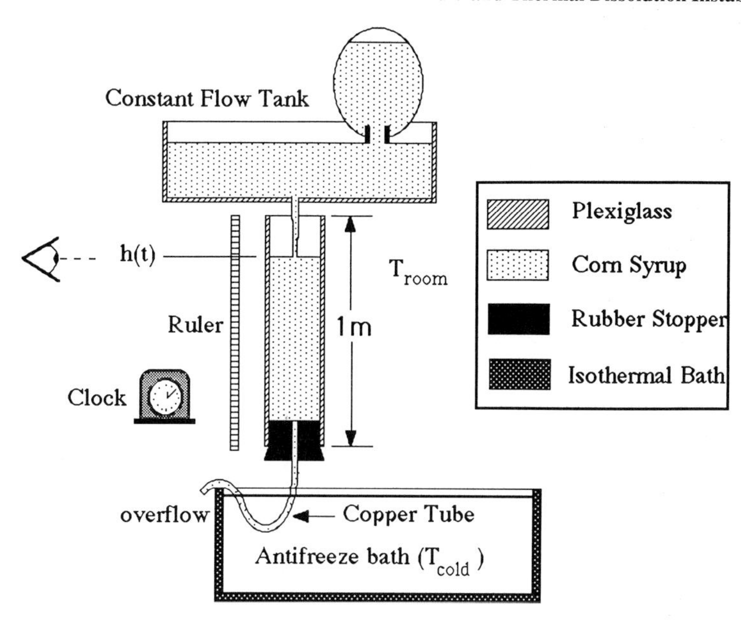

Figure 1 Schematic of the apparatus for generating flow instability modulations from fluids with temperature-dependent viscosity. Corn syrup is fed into a vertical tube at a constant rate. As the syrup accumulates, it increases pressure across a small outlet "J" tube that is in contact with a cold bath. When the syrup flows rapidly, it remains hot with low viscosity, but when the flow rate slows, it becomes relatively cool and viscous.

diameter than the glass tube. As the syrup flows out, it is cooled to some extent by thermal conduction through the copper tube in contact with the liquid bath maintained below room temperature.

The apparatus is a simple (and *highly* idealized) upside-down model of portions of a magmatic system subjected to either thermal gradients and/or chemical corrosion. The glass tube represents either a compressible magma chamber or a matrix that can hold some liquid. The height of the free surface in the glass tube conceptually represents either pressure in the chamber or percentage melt in the matrix. The copper tube in the refrigerated bath represents the tube through which magma flows to the surface of the earth or the matrix through which melt flows upward. The cool temperature of the bath represents either relatively cool country rock or the fluid flow resistance of the matrix in the presence of corrosion by the magma. If the experimental temperature bath is sufficiently cool and if the syrup has a sufficiently great viscosity increase upon cooling, an unsteady flow develops even if the flow from the reservoir is steady. Some estimate of how much viscosity change was necessary was obtained from theoretical considerations.

Run-down experiments were first conducted to estimate the resistance as a function of flow rate. Theoretical considerations indicate that a time dependence is not to be expected unless the fluid resistance is inversely proportional to the flow rate. Figure 2 shows data from two runs. Experiments in Figs. 2a, 2c, and 2e had a bath temperature set to 0°C, a room temperature of 24.1°C, and a copper tube 30 cm long. Experiments in Figs. 2b, 2d, and 2f had a bath temperature of −11.0°C, a room temperature of 24.0°C, and a copper tube 14.5 cm long. Figures 2a and 2b show height versus time from which velocity was found as a time derivative to produce Figs. 2c and 2d. For Figs. 2a, 2c, and 2e the run-down is close to exponential, which would be expected for constant viscosity conditions. In contrast, the run shown in Figs. 2b, 2d, and 2f with the bath at −11.0°C is similar to the first run only for the first thousand seconds or so; then there is a transition to a much slower run-down. Presumably at this latter stage the viscosity of the syrup is very large due to cooling in the copper tube. The difference between the two states is particularly clear in the height versus velocity plot and even more so in the log–log plot. For the run on the right, the transition region from

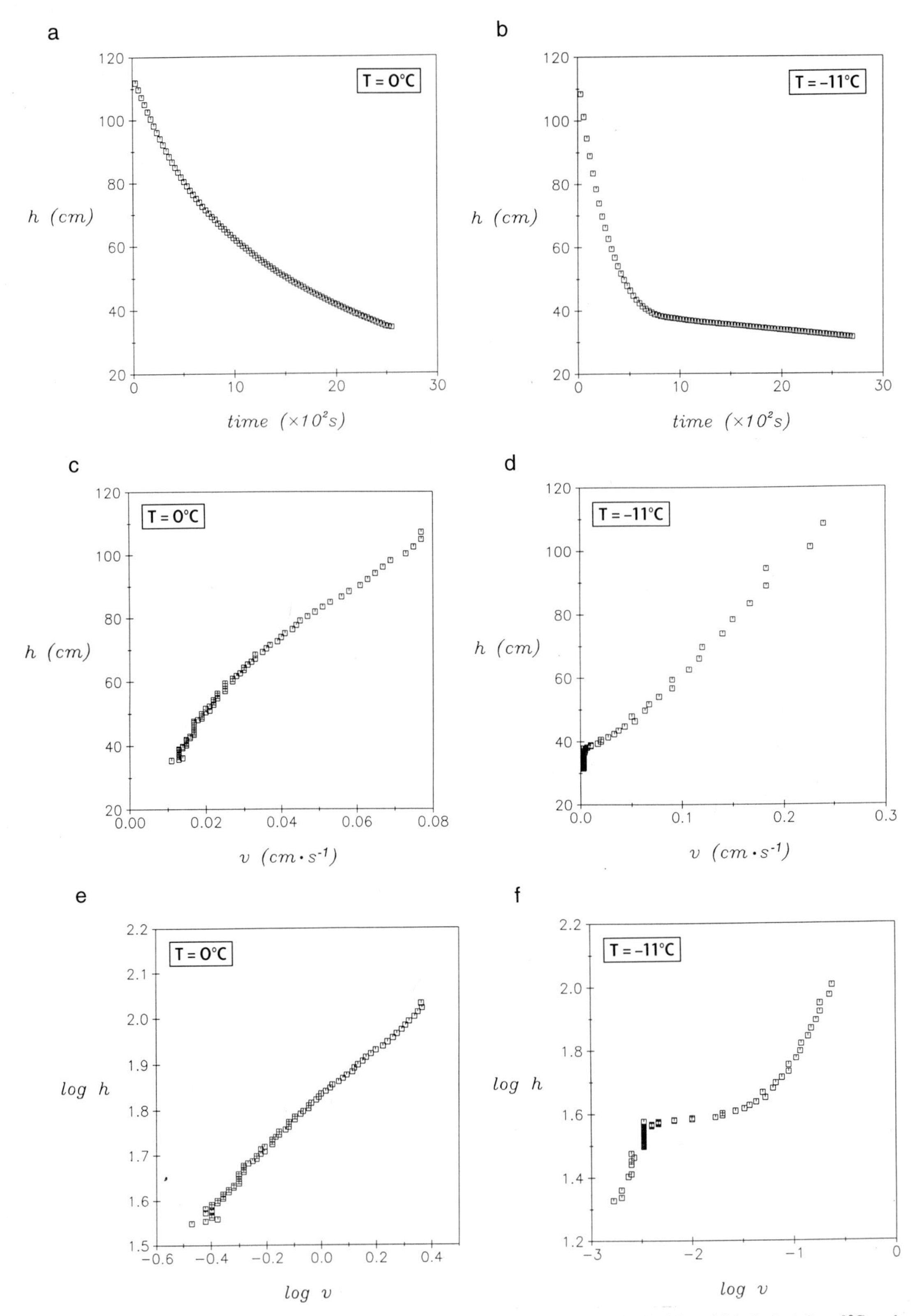

Figure 2 Results from two run-down experiments with different bath temperatures. In (a), (c), and (e), the bath is at 0°C, and in figures (b), (d), and (f), the bath is at -11.0°C. Height versus time is plotted in (a) and (b), height versus velocity is plotted in (c) and (d), and log height versus log velocity is plotted for (e) and (f). In (a), (c), and (e), the run-down is very similar to exponential decay. In (b), (d), and (f), the run-down is characterized first by a rapid decay when hot syrup fills the tube and then by slower decay when cold syrup fills the tube. Transition from rapid to slow is characterized by a plateau in v, h space as shown in (f).

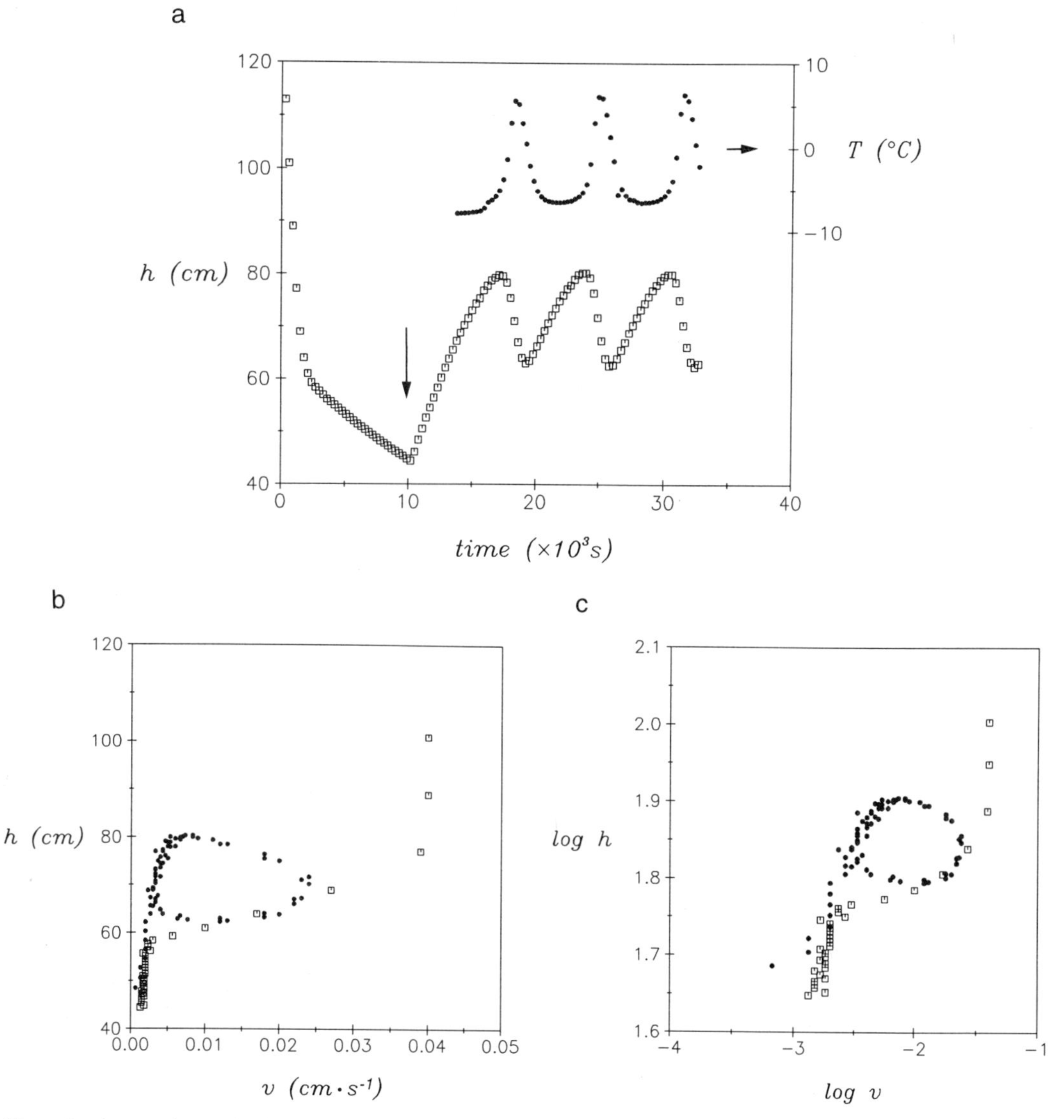

Figure 3 An experiment that first had a run-down and then had a steady source turned on at the time shown by the arrow in (a). Plotted are (a) the height (squares) and temperature (dots) versus time, (b) the height versus velocity, and (c) log height versus log velocity. In (b) and (c) run-down is shown by squares and steady source by dots.

fast to slow flow is characterized by a plateau in the height–velocity logarithmic curve. Theoretical considerations will illustrate the significance of the plateau in the section Stability of Thermally Eroded Flows.

When the volume flux of the source was set within a certain range of values, the fluid height oscillated with time. Figure 3 shows fluid height versus time, fluid height versus velocity, and log fluid height versus log time for one example. This run was started as a run-down with the glass tube filled with syrup but with no inflow of syrup to the top of the tube. The sequence illustrated in Figs. 2b, 2d, and 2f with a rapid run-down followed by a slow run-down is clearly visible. After this, the volume flux from the reservoir was started at the time denoted by the arrow in Fig. 3a. Three complete and nearly identical oscillations were seen thereafter. The plot of h and velocity was made from the data in Fig. 3a by subtracting the constant fluid velocity from the source as measured by the abrupt change in slope immediately before and after the arrow. The oscillations produce a closed curve in fluid height–velocity space that lies on the top of the plateau in the run-down. The plots of the logarithms (Fig. 3c) more clearly show the limit cycle oscillation on the plateau.

The experiments exhibited a transition from steady flow to a more complicated flow pattern when the velocity ranged between 0.003 and 0.025 cm · s^{-1}. Peclet numbers (ur/κ, see next section), based on the above velocities, a tube inner radius r of 0.2 cm, and a thermal diffusivity of $\kappa = 1.4 \times 10^{-3}$ $cm^2 \cdot s^{-1}$, range from 0.4 to 3.5. The parameter $ur^2/\kappa L$ ranged from 0.005 to 0.048.

Stability of Thermally Eroded Flows

General Equations

A theory is developed for a magma chamber that is two-dimensional and in Cartesian coordinates rather than one-dimensional and in cylindrical coordinates as in the preceding experiment. In this way, spatial as well as temporal instabilities are considered. The term magma chamber is used in a very general sense. It really means the storage area within which and from which magma can easily flow and which is cooled slowly if at all. Pressure in the chamber changes as the volume of the chamber becomes larger or smaller. The fluid leaves the chamber through a region where the flow resistance increases by sidewall cooling on the conduit. The fluid ultimately leaves that cooling-resistance area and passes into the outside world, which is at constant ambient environmental pressure.

A very simplified system with the minimal essential features is sketched in Fig. 4. A narrow slot with, for example, hydrothermally cooled walls, is fed from below by fluid from the magma chamber. The conduit width in the x direction is d and the depth of the slot in the vertical z direction is L. The y direction is along the axis of the magma chamber. The bottom of the magma chamber is fed by a uniform volumetric flux per unit length Q. The magma chamber itself has a cross-sectional area $A_c + A'$, where A_c is considered constant and $A'(y, t)$ is a small variation in area due to inflation or deflation. The chamber also has linearly elastic walls so that the pressure p in the magma chamber is related to A' by the formula $p = EA'$, where the coefficient E is a coefficient

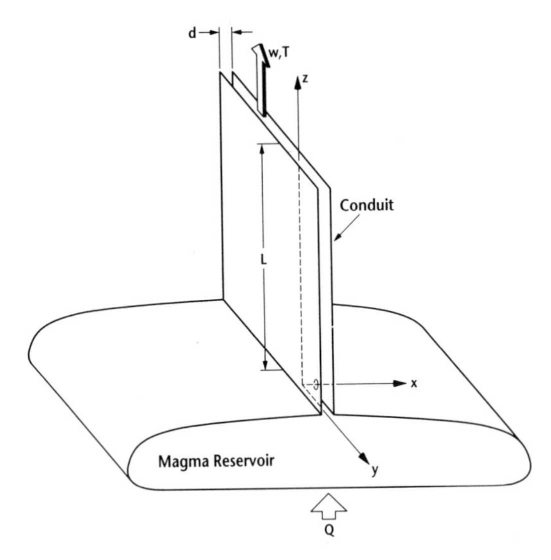

Figure 4 Sketch of the idealized system. A constant magma flux Q enters at the bottom of the elastic-walled magma chamber. Magma can flow up the slot with a local velocity w and cool through the side walls. It can also flow along the chamber axis.

of elasticity times some geometric terms that depend on the shape of the magma chamber.

The expression of conservation of volumetric flux is then

$$E^{-1}\frac{\partial p}{\partial t} = Q - dw(y,t) - A_c\frac{\partial v}{\partial y}. \tag{1}$$

Velocity w is the velocity of the magma leaving the magma chamber in the vertical ($+z$) direction and v is the (y component) velocity in the magma chamber along the axis of the magma chamber.

In the laboratory experiment described in the previous section, there is no y direction, but otherwise the preceding equation describes the experiment to a good first approximation. For such a description, Q in the equations represents the volume flux, d represents the area of the exit tube "J," the term E^{-1} represents the glass tube area, and p represents fluid height in the glass tube H.

Within the magma chamber, we also assume there is a balance between the viscous resistance from the velocity v and the pressure drop, so that

$$\frac{1}{\rho}\frac{\partial p}{\partial y} = \frac{\nu_H}{l_c^2}v, \tag{2}$$

where ν_H is the viscosity of the hot fluid in the magma chamber and l_c is a cross-sectional length scale of the magma chamber. For a circular magma chamber, l_c^2 will be of the order of the cross-sectional area, A_c, but other chamber cross sections may have smaller values of l_c^2.

Combining (1) and (2),

$$E^{-1}\frac{\partial p}{\partial t} = Q - wd + \frac{A\ l_c^2}{\nu_H \rho}\frac{\partial^2 p}{\partial y^2}. \quad (3)$$

We must now relate the vertical velocity w to the pressure drop along the slot from the magma chamber to the exit. Steady viscous flow is assumed,

$$-\frac{1}{\rho}\frac{\partial p}{\partial z} = 12\nu(T)\frac{w}{d^2} + \frac{\partial w}{\partial t}, \quad (4)$$

where the viscosity is given as

$$\nu = \nu_H + \alpha(T_H - T). \quad (5)$$

Here T is the temperature of the fluid in the conduit, T_H is the temperature at which the viscosity is ν_H, and α is the viscosity–temperature coefficient. To find p as a function of w, it is necessary to determine the temperature of the fluid in the conduit. Assume that the material is flowing with uniform slab flow (w = constant across the slot) in the z direction, so that

$$w\frac{\partial T}{\partial z} = \kappa\frac{\partial^2 T}{\partial x^2}. \quad (6)$$

Let the temperature of the boundaries decrease linearly in the z direction at a rate of $\Delta T/L$ so that along the conduit walls

$$T = T_H - \frac{\Delta T z}{L} \quad \text{at } x \pm \frac{d}{2}. \quad (7)$$

A solution to (7) of the form

$$T = T_H - \frac{\Delta T z}{L} + \frac{w\Delta T}{2\kappa L}\left(\frac{d^2}{4} - x^2\right) \quad (8)$$

exists. In particular, the temperature of the center plane ($x = 0$) is higher than the wall temperature by an amount $w\Delta T d^2/8\kappa L$. The temperature of the sidewall and centerline from Eq. (8) is sketched in Fig. 5.

For $z < wd^2/4\kappa$, it will be assumed that a thermal boundary layer is developing and T will be set

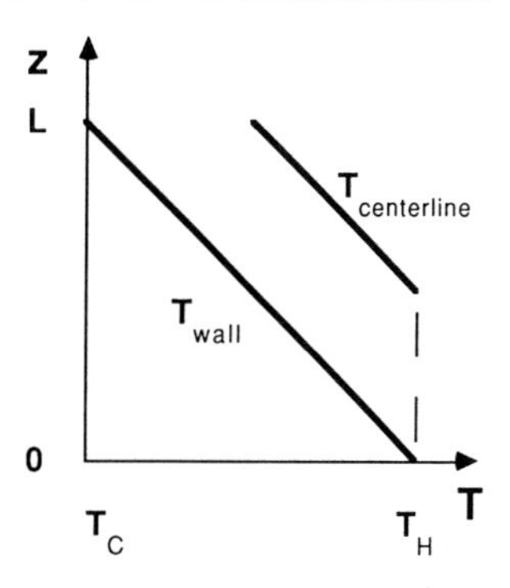

Figure 5 Sketch of the temperature of the conduit wall and the temperature of the fluid centerline for this idealized problem whose solution is Eq. (8).

to T_H. For $wd^2/4\kappa < z < L$, the temperature at the centerline will be calculated using Eq. (8) evaluated at $x = 0$. In that case, Eq. (4) assuming steady flow is combined with Eqs. (5) and (8) and integrated as

$$\frac{p}{\rho} = \frac{12w}{d^2}\left(\int_0^L \nu_h\, dz + \int_{wd^2/8\kappa}^{L} \alpha\left(T_H - \left(T_H - \frac{w\Delta T d^2}{8\kappa L} - \frac{\Delta T z}{L}\right)\right) dz\right), \quad (9)$$

so

$$\frac{p}{\rho} = \frac{12w}{d^2}\left(\nu_H L + \frac{\alpha\Delta T}{2L}\left(L - \frac{wd^2}{4\kappa}\right)^2\right), \quad (10)$$

which simplifies to

$$P = \frac{pd^4}{96\nu_H \rho\kappa L^2} = \frac{w}{W_r}\left(1 + \frac{A}{2}\left(1 - \frac{w}{W_r}\right)^2\right), \quad (11a)$$

$$w < W_r$$

$$P = \frac{w}{W_r}, \qquad w > W_r, \quad (11b)$$

where

$$W_r = \frac{8\kappa L}{d^2} \quad (12)$$

and

$$A = \frac{\alpha\Delta T}{\nu_H}. \quad (13)$$

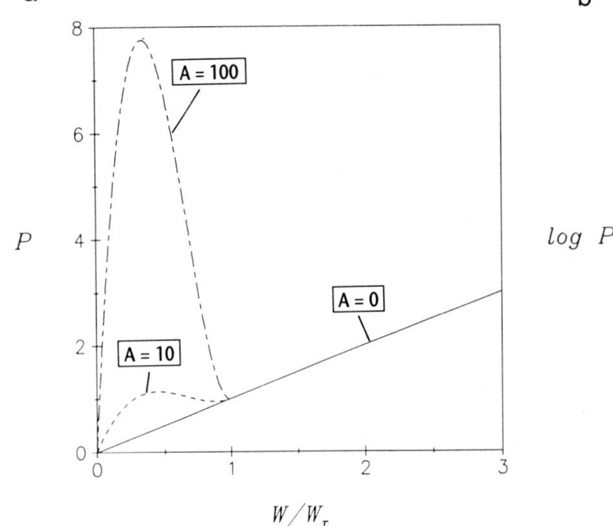

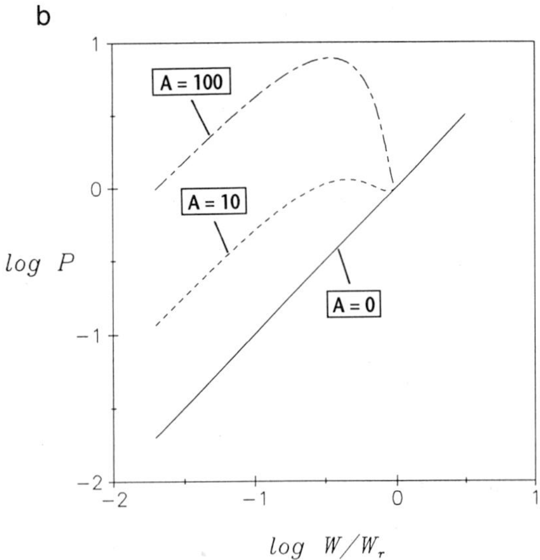

Figure 6 (a) Pressure drop across the slot calculated from Eq. (11) as a function of velocity of the fluid for the three values $A = 0$ (straight line), 10, and 100. (b) The same data plotted as log–log.

For $w > W_r$, the fluid stays at T_H during its entire traverse of the slot.

The crucial result for possible hydrodynamic instabilities is that the pressure can drop off with an increase in fluid velocity within a certain range of w/W_r if $A > 6$. To illustrate this, the function P is plotted as a function of w/W_r for some fixed values of A in Fig. 6.

It is useful to nondimensionalize the final equations using $W = w/W_r$ for the scaled velocity, $Q' = Q/W_r d$ for the scaled volume flux, $t' = 12\nu_H t/d^2$ for the scaled time, P for the scaled pressure, and $y' = y/L$ for the scaled length so that the equations now read

$$\gamma \frac{\partial P}{\partial t'} = Q' - W + \delta \frac{\partial^2 P}{\partial y'^2} \tag{14a}$$

and

$$\frac{\partial W}{\partial t'} + f(W) = P, \tag{14b}$$

where the newly scaled flow resistance in the slot

$$f(W) = W\left(1 + \frac{A}{2}(1 - W)^2\right), \quad W < 1 \tag{15a}$$

$$f(W) = W, \qquad W > 1. \tag{15b}$$

The governing dimensionless numbers are

$$A = \left(\frac{12\nu_H}{d^2}\right)^2, \qquad \gamma = \left(\frac{\rho L}{Ed}\right), \tag{16}$$

$$\delta = \left(\frac{12A_c l_c^2}{d^3 L}\right).$$

The parameter γ is a measure of the ability of the magma chamber to expand with pressure and δ is a measure of the aspect ratio of the slot.

Oscillatory Instability for One-Dimensional Flow

If y variations are neglected, Eqs. (14b), (15a), and (15b) resemble a simple set of earthquake equations (Whitehead and Gans, 1974) that have the dimensionless form

$$\frac{dF}{dt''} = 1 - u' \tag{17}$$

$$m\frac{du'}{dt''} + \frac{u'}{\beta^2 + u'^2} = F. \tag{18}$$

Here F is the scaled force accumulated in an elastic region, u' is the scaled sliding velocity, m and β are dimensionless measures of mass and friction coefficient, respectively, and t'' is a scaled time. The friction law $u'/(\beta^2 + u'^2)$ is similar to that in Eq. (15a) in the sense that the sliding resistance

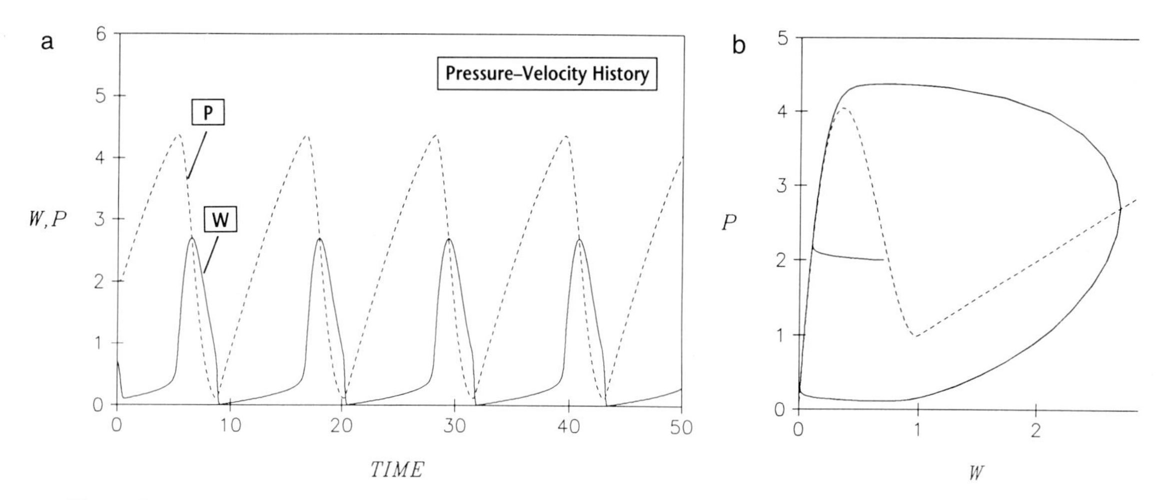

Figure 7 Dimensionless pressure and velocity history during flow in the conduit after exiting the magma reservoir. (a) Numerical solution of Eqs. (14a) and (14b) with $\partial/\partial y = 0$ for P (dashed curve) and W (solid curve) versus time. The governing parameters are $(A, Q, \gamma) = (50.0, 0.7, 1.0)$. (b) P–W phase plane of the preceding solution. The steady-state relation $P = f(W)$ from Eqs. (15a) and (15b) is also shown as a dashed curve.

increases linearly as u' increases from zero, but it then falls off as the velocity increases above a set value. The earthquake equations have a steady solution $u' = 1$. When $\beta > 1$ this solution is stable. When $\beta < 1$ this solution is unstable and there are growing oscillations. The oscillations become increasingly episodic when the inertia term m and the friction term β get very small. It is necessary to add only one term to the earthquake equations to make them similar to our present equations. Equation (18) possesses a friction term that falls off to zero as u gets large, whereas the present problem has a friction term that stays proportional to W for large W where the constant of proportionality corresponds to the viscosity of the hot fluid. This feature could be incorporated into the earthquake equations by adding a term λu, where $\lambda \ll 1$. Oscillations would then grow when $\alpha^2 + \lambda < 1$.

For no spatial dependence, $\partial/\partial y = 0$, and Eqs. (14a) and (14b) can be combined to give

$$\frac{d^2W}{dt'^2} + \frac{\partial f}{\partial W}\frac{dW}{dt'} + \frac{(W - Q')}{\gamma} = 0. \quad (19)$$

This is the form of an equation describing a mass-spring system with nonlinear friction as given by $\partial f/\partial W$. The topology is similar to the classic Van der Pohl equation with a region of negative friction bordered by a region of positive friction. For Q in the linearly unstable region $(df/dW < 0)$ a limit cycle oscillation will always develop. When A is large and $f(W)$ has a large region of negative slope, W will oscillate between long periods of nearly constant low flow and an eruptive phase in which W increases rapidly and then returns to low flow—a behavior like that of the earthquake equations. Behaviors for two sets of parameters in these limits are shown in Figs. 7 and 8. When the parameter A has a value close to the critical value of 6, the oscillations are close to sinuosoidal, as shown in Fig. 9.

Stability of Two-Dimensional Flows

The equations are also unstable to spatial disturbances. We now linearize (14a) and (14b) around a steady flow so that $W = w_0 + \epsilon w'$. One equation can be found,

$$\gamma\frac{\partial^2 w'}{\partial t^2} + \gamma\frac{\partial f}{\partial w_0}\frac{\partial w'}{\partial t} - \delta\frac{\partial^3 w'}{\partial t\partial y^2} - \delta\frac{\partial f}{\partial w_0}\frac{\partial^2 w'}{\partial y^2} + w' = 0. \quad (20)$$

Take $w' = W_c\, e^{iky+\sigma t}$, where W_c is a constant. The roots are

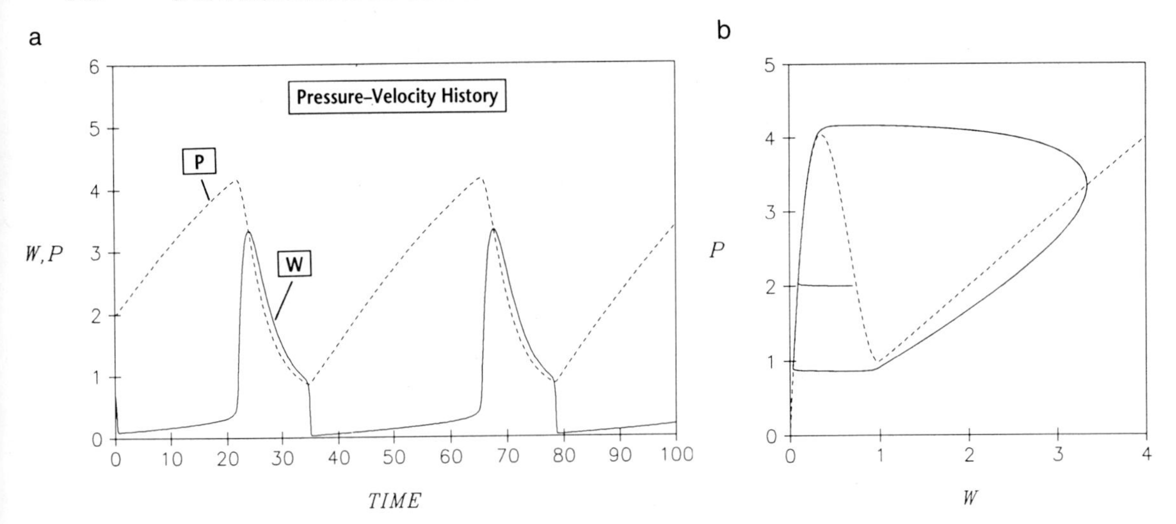

Figure 8 Dimensionless pressure and velocity history during flow in the conduit after exiting the magma reservoir. Variables same as in Figs. 7a and 7b, except that governing parameters are $(A, Q, \gamma) = (50.0, 0.7, 5.0)$ so the magma chamber is more easily inflated.

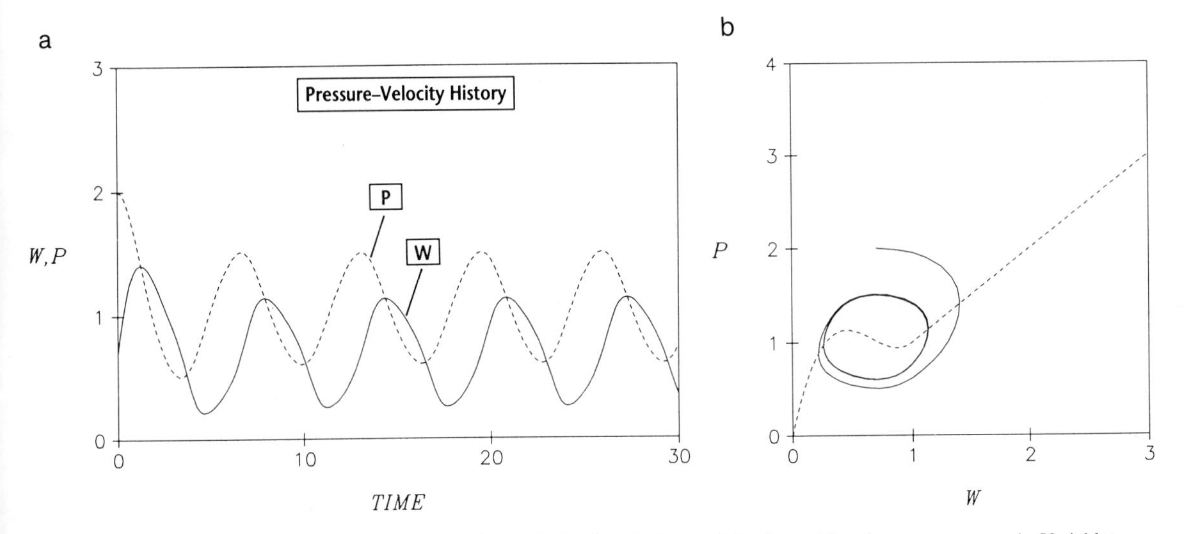

Figure 9 Dimensionless pressure and velocity history during flow in the conduit after exiting the magma reservoir. Variables same as in Figs. 7a and 7b except that the governing parameters are $(A, Q, \gamma) = (10.0, 0.7, 1.0)$ so the system is close to neutral stability.

$$\sigma = -\frac{1}{2}\left(\frac{\partial f}{\partial w_0} + \frac{\delta k^2}{\gamma}\right) \pm \left(\frac{1}{4}\left(\frac{\partial f}{\partial w_0} + \frac{\delta k^2}{\gamma}\right)^2 - \frac{1}{\gamma}\left(1 + \delta k^2 \frac{\partial f}{\partial w_0}\right)\right)^{1/2}. \tag{21}$$

For $k = 0$ the result for the time-dependent solution is recovered, namely, that $\partial f/\partial w_0 < 0$ is required for instability. For $k = 0$ and when $\gamma \gg 1$, the growth rate is $\partial f/\partial w_0$. When $\gamma \ll 1$ the growth rate is one-half as fast and the instability is in the form of growing oscillations with frequency of $\gamma^{-1/2}$. Figure 10 illustrates the behavior of the real part of the growth rate. A number of curves are shown corresponding to different values of $\Gamma = 4\gamma(\partial f/\partial w_0)$. For $\Gamma < 1$, only steady-growing modes are found. For $\Gamma > 1$ the steady root intersects an oscillating root that is shown as a straight downward-sloping line that intersects

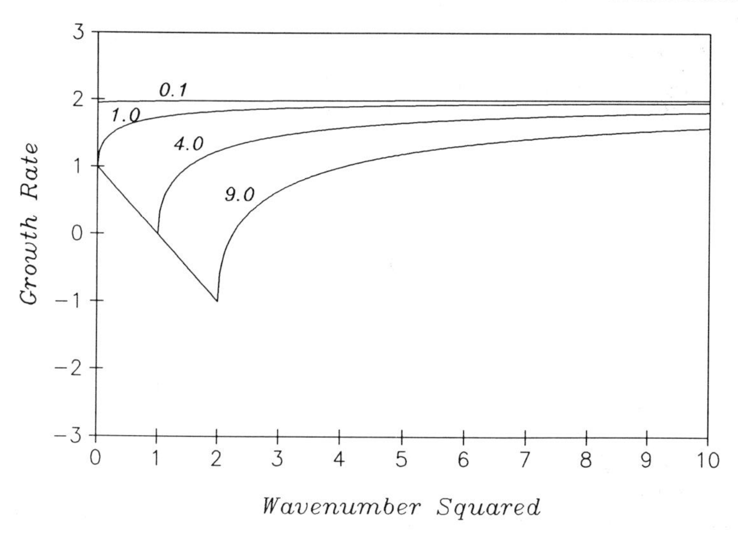

Figure 10 Plot of the growth rate from Eq. (21) as a function of wavenumber squared. The growth rate is the real part of σ divided by $(\frac{1}{2})$ $(\partial f/\partial w_0)$ and the wavenumber squared is k^2 divided by $(\gamma/\delta)(\partial f/\partial w_0)$. Four values of the parameter $(\gamma/4)(\partial f/\partial w_0)$ are shown. The curves asymptote to 2 for large wavenumber.

$k = 0$. For some values of k the growth rate is negative so that the perturbations decay. However, for large enough k, all growth rates are positive and all curves asymptotically approach the limiting growth rate of 2. Therefore, the shortest wavelengths grow most rapidly in this theory.

These results lead to some physical insights into the nature of instabilities that are produced by an increased resistance to fluid flow in conduits upon cooling. First, for an instability of the uniform flow, the parameter A must be above the value 6 so that the resistance must actually fall off with increasing fluid velocity over some range of w/W_r. As a dynamical comparison, this feature was also found in the earthquake theory, and most earth scientists probably would have agreed that this feature was necessary even before the theories were developed.

Second, w/W_r must be of order of some fixed number less than 1 and/or approaching 1 for a growing instability. The parameter group w/W_r may be interpreted as Pe d/L, where Pe is Peclet number $= wd/\kappa$. This is a measure of the ability of a flow to advect heat in competition with the dissipative effects of diffusion. Thus, according to the theory, Pe must be between $8L/3d$ and $8L/d$ (for $\alpha\Delta T \gg 6$) for an instability. For values of Pe outside this range (either smaller or larger), no instability will occur. In the former case, the fluid will be so slow that all fluid in the slot will have a high viscosity and no pressure can become sufficiently large to make the fluid in the slot flow rapidly enough to reach the limit where hot fluid fills the slot. In the latter case, fluid in the slot is hot and nothing can make the fluid flow slowly enough to become cold. These first two features were borne out by the laboratory results in the preceding experimental section. These dynamical features differ from the earthquake theory analog where an instability always happened above a critical slip rate.

Third, in this theory, the most rapid growth is at large k, so small wavelengths are to be expected as the preferred mode of growth. This third prediction was not verified in the preceding experiments because the experiments correspond to $k = 0$. Experiments with paraffin described in the next section correspond to other values of k.

Since the linear analysis predicts the fastest instability growth at infinite wavenumber, some modification of the theory is needed to produce explicit predictions of the expected wavelength. Since fluid velocities in the conduit in the direction of the axis of the magma chamber were neglected, the preceding theory can be thought of as a long wavelength theory. A theory that includes flows in two directions in the slot has not yet been developed. Secondly, no finite-amplitude theory has yet been produced. Such a theory might lead to long wavelength cutoffs as well.

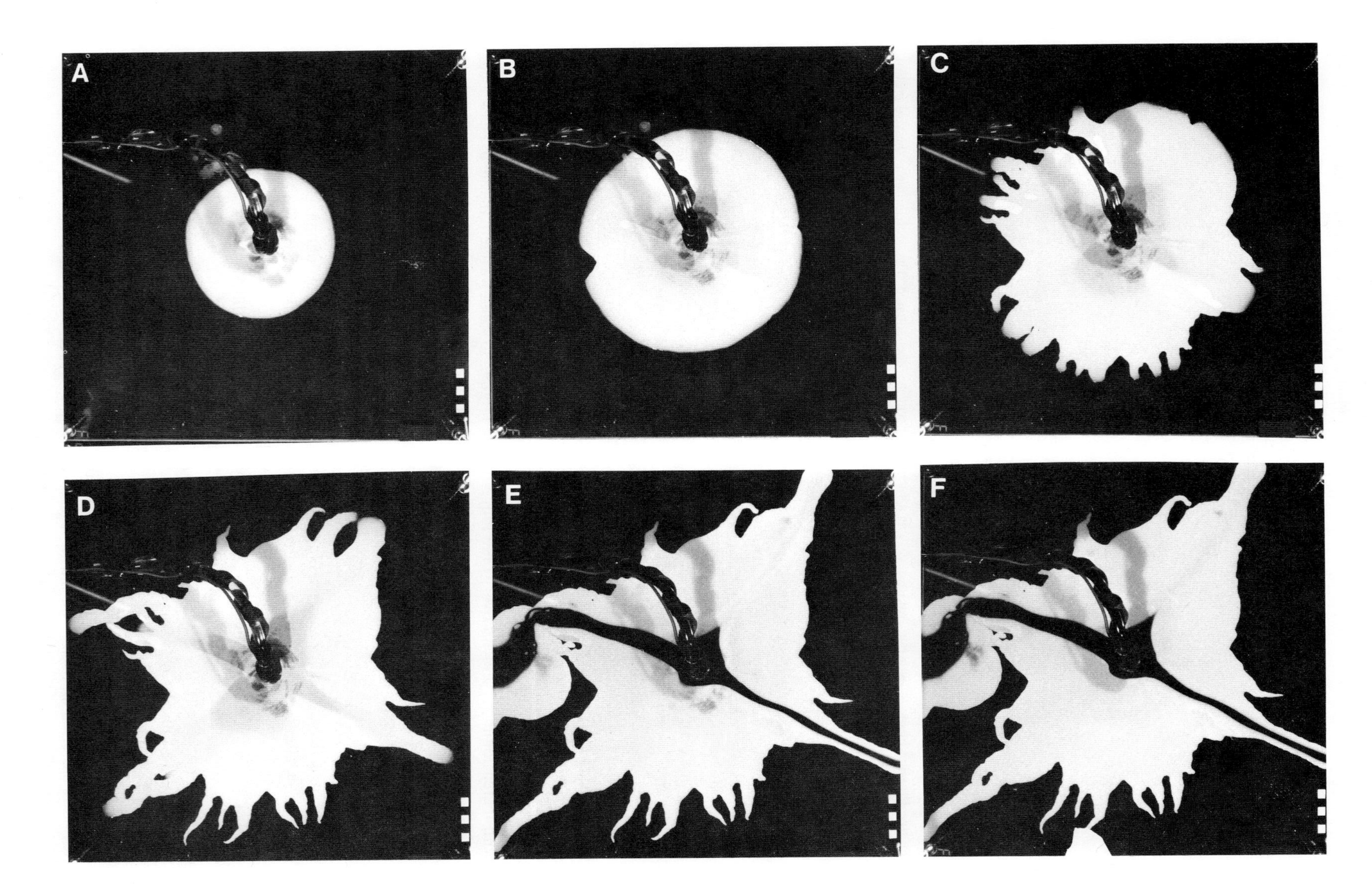
A
B
C
D
E
F

Experiments with Paraffin

Two-Dimensional Flow through a Slot

Experiments with liquid paraffin demonstrate a transition from uniform flow to fingering-instability-dominated flow as time progresses. The apparatus consisted of a 1.2-cm-thick square aluminum plate 61 cm on a side in an ice water bath. A 1.1-cm-thick square Plexiglas plate 46 cm on a side was clamped over the aluminum with spacers between the aluminum and the Plexiglas so that a narrow gap of 0.25 cm remained. A hole in the center of the Plexiglas was connected to a reservoir containing melted paraffin. A camera was positioned above the apparatus to record the progress of the experiments.

As a run commenced, paraffin was delivered to the hole at a rate of 5.5 cc $\cdot$ s^{-1}. For approximately the first 16 s, the paraffin spread out in a radially growing pattern that was close to perfectly circular (Fig. 11A). Small deviations from perfect circles appeared to be produced from the surface tension effects arising from slight irregularities in the texture of the black painted aluminum, but these deviations produced less than a 10% deviation in the radius of the circle. After 16 s, the circular front rapidly developed small radial undulations (Fig. 11B) that signified a sudden decrease in velocity at select points on the circle. Between these undulations, there was a concomitant rapid growth of radial finger-like bulges (Fig. 11C) with round tips. Ten or twelve such fingers grew within 4 s but many stopped growing during the next 4 s (Fig. 11D). The only subsequent change in the pattern was that four fingers reached the edge of the tank, and the rest froze. Oil-soluble dye was then injected into the paraffin source, and it was observed that when the dye had arrived in the cell, most of the dye fluid flux was moving into the two largest fingers (Fig. 11E). Melt tubes feeding the other fingers seem to have stopped and solidified. The dye had begun to intrude onto a third finger tube but it apparently stopped shortly after arriving in the cell. Forty-eight seconds later, dye of another color was injected, and by that time flow was going out of only one finger through one tube (Fig. 11F). For 40 more seconds, flow out through that finger continued in a clearly defined channel with little apparent change.

Distances were measured from the photographs with dividers and then tabulated. For photographs illustrating an almost circular intrusion, the extremes of the radius were measured. For the finger cases, the radial distance to the tip of each finger from the feeding tube exit was measured and one measurement was also taken of the distance between each pair of fingers. The results are shown in Fig. 12. While the front was circular, it advanced according to simple conservation of mass laws; a line with the formula $r = (Qt/\pi h)^{1/2}$ is shown for comparison, where the volume flux Q is 5.5 cc $\cdot$ s^{-1}, t is time, and the gap width h is 0.241 cm. As the fingers developed, the fronts of the fingers accelerated and the front between fingers decelerated, finally stopping. At this time the paraffin appeared to begin to solidify in the quiescent regions. As time passed, many of the smaller finger fronts also halted their advance, but those fingers remaining reaccelerated. The width of the final active channel as marked by the colored dye varied along the channel axis from 1.4 to 1.7 cm.

A simple explanation of why the instability must happen is that the paraffin would solidify if it remained in a wholly circular flow pattern because then it would stay in the gap for more than a thermal time constant. However, it was clear that the paraffin had never completely solidified anywhere during this run (it appears to experience a dramatic increase in bulk viscosity as it cools). Assuming that both the lid and the aluminum plate cool the paraffin as it flows along the conduit, the value of the thermal time constant for the paraffin in a gap of 0.241 cm can be estimated from the formula $h^2/4\kappa$. Using the above value of h and $k = 0.0004$ cm$^2 \cdot$ s^{-1}, the time constant is 36 s. We saw not only that the interface became un-

Figure 11 Photographs of the evolution of finger flow instabilities in paraffin flowing with constant volume flux through a cooled slot from a point source. (A) A circular intrusion 12 s after start of the experiment. (B) Two small notches have grown at 16 s. (C) Numerous fingers break out by 20 s. (D) The fingers have grown considerably by 24 s. (E) At 52 s, dye reveals there is flow through only two channels. There is also evidence that a third channel stopped just as the dye entered the tank. (F) At 92 s, darker dye reveals that there is flow in only one channel. The scale bar shows alternating white and black rectangles 1 cm long.

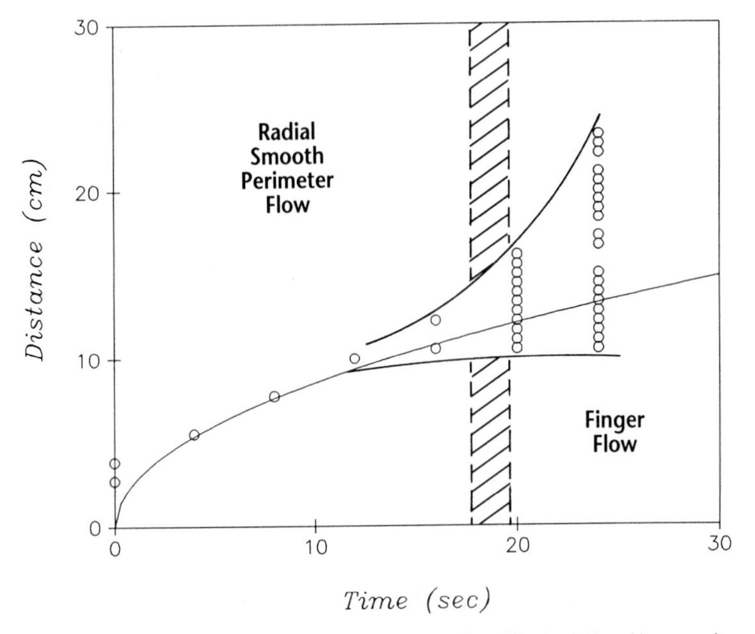

Figure 12 Transition from radial smooth perimeter flow regime to finger flow. Hatchured band is approximate transition region. Data (dots) for the advance of the front of the paraffin with time. As the unstable front breaks into fingers, the fingers advance much more rapidly than the neighboring regions. In other places the front may actually stop. The smooth curve is for a solution to the simple conservation of volume for a circular front, $r = 2.7t^{1/2}$. Diverging lines are a hand-fit envelope for the rapidly expanding radial distance spread after finger development.

stable after 16 s but also that the final channel of approximately 1.5 cm width admitted a flow from the hole to the edge of the plate of around 15 cm · s^{-1}. Therefore, fluid leaves the region after being in the slot for less than 2 s, a time that is short compared to the time constant estimated earlier. Of course, these numbers have considerable uncertainties and are meant to be only approximate and suggestive, but the simple concept that a flow pattern is formed that would allow fluid to escape before it cools is consistent with the experiment.

The parameter group $wd/W_rL = wd^2/8\kappa L$ at which the fingers were first seen can also be estimated. Using $w = 3Q/4\pi rd$, with $r = L =$ 13 cm, we get $wd/W_rL = 0.6$, which is in agreement with the notion that this number must be in the range of 1 for instability.

A detailed application to a specific geological system has not been made. It would be premature at this time and would have to be tailored specifically to the geological case being considered. Numerous magmatic and hydrothermal systems, however, are known to possess an increase in fluid flow resistance upon cooling. The hypothesis by Delaney and Pollard (1982) and further developed by Bruce and Huppert (1989, 1990) that flow through a dike is fundamentally unsteady and either is gradually blocked or melts back the country wall rock could be combined with these constraints to produce yet more realistic stability studies. Lowell (1990) suggests that the spatial focusing of black smokers may result from the thermal contraction upon cooling of the vent walls, although mineral precipitation along veins and fractures is the primary cause of permeability reduction in hydrothermal systems. Kent *et al.* (1992) have pointed out the presence of lamprophyre intrusions in the Damodar Valley of northeast India whose cylindrical melt tubes have apparently been produced from low viscosity melt penetration into adjacent carbonaceous sediments.

Two-Dimensional Flows with a Free Surface: Lava Flow Modeling

The paraffin experiments may be considered the magmatic counterparts to experiments conducted on model lava flows by Fink and Griffiths (1990). They injected polyethylene glycol at the base of a water bath and correlated the developing morphology of the solidified material with the model temperature and flow rate. Their experiments had the additional feature of possessing a large exposed free surface so that the resulting structures had additional degrees of freedom (such as the reduced gravity g') compared with the preceding experiments. They observed that progressively cooler fluid experiments had surface morphologies that resembled features on cooling lava flows and lava lakes. For flows with a crust that cooled very slowly, marginal levees that contained and channeled the main portion of the current developed. Yet cooler flows with more rapid crustal growth rates had regularly spaced folds and multiarmed rift structures complete with shear offsets and bulbous lobate forms similar to pillow lavas.

Ito and Whitehead (work in progress) are investigating the cooling of paraffin from a line source. By using paraffin, the final morphology of the solid can be quantified with thickness measurements. Photographs of two typical samples are shown in Fig. 13. Preliminary results are that the parameter $\Sigma = \sigma_x/L$ becomes progressively smaller as the parameter Ψ becomes larger, as shown in Fig. 14. In this case, σ_x is the standard deviation of the thickness of the flowing fluid, and L is the flow length. Here, L is volume of paraffin added divided by H_h times tank width. H_h is the thickness of a gravity current. In Huppert (1982), $H_h = (g' q^3 t_d/\nu_p)^{1/5}$, where g' is the reduced gravity $= \Delta\rho g/\rho$, q is the volume flux per unit length, ν_p is the viscosity of paraffin, and t_d is the duration of the flow (eruption). The parameter Ψ is the ratio of time required to cool to the solidification temperature to the advection time scale t_s.

The conclusion is that lava morphology depends on the lava solidification rate during the flow process. The primary controlling factors are the effusion rate and the lava temperature. More rugged morphology results if the lava cools early during the flow process and smoother textures form if solidification rates are slow. Bonatti and Harrison (1988) found a larger abundance of pillow basalts at slower spreading ridges and a proportionally larger number of sheet flows at faster spreading ridges. This suggests that slow spreading ridges tend to erupt cooler lavas at lower eruption rates than fast spreading ridges.

Magma Flow Instabilities from Chemical Corrosion

The final set of studies are attempts to understand the effects of chemical dissolution on melts ascending by porous flow in the Earth's mantle. Porous flow accompanied by compaction is now a well-accepted mechanism for separating melt from a mantle plume or diapir that is ascending and undergoing decompression melting (Sleep, 1974; Turcotte and Ahern, 1978; Turcotte, 1981, 1982; Scott and Stevenson, 1984; McKenzie, 1984; Fowler, 1985, 1990a, 1990b) but must be understood as representing a single extraction mechansim in a large heirarchy of transport mechanisms that include ascent by magma fracture as well as the rise of relatively viscous melt by melt-dominated diapirism in island arcs and in the silicic systems of contential interiors. However, the most successful models of compositional variation assume "closed system melting" where a mantle source is melted and the liquid is removed without further intimate chemical contact with the matrix as it ascends to the surface. Such a rapid removal of melt from its source seems necessary in understanding some magmatic systems, for example, at mid-ocean ridges where liquids apparently record equilibrium with high pressure phases, most notably garnet (Salters and Hart, 1989). However, perhaps chemical reaction during ascent of liquid through the upper mantle does play a role in melt extraction in other types of settings (Kelemen *et al.*, 1990, 1992).

It is not difficult to visualize that the same types of instabilities described previously could be produced from chemical dissolution rather than from strictly thermal effects. We assume that melt is rising by compaction through a matrix of mantle residua and that as the liquid ascends to lower pressures it departs from the melting point along an adiabatic decompression path. The matrix at the higher level must stay in equilibrium with this liquid, which is now slightly superheated. The result will be that the liquid, which is still in intimate contact with the matrix, dissolves some of the matrix such that the local region

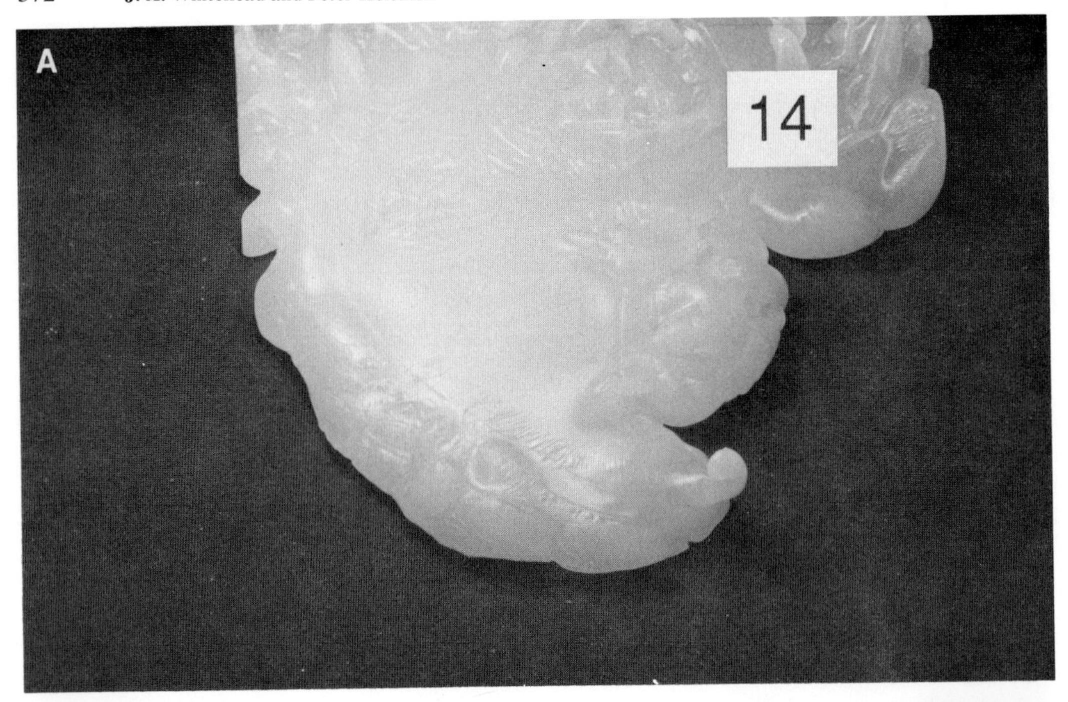

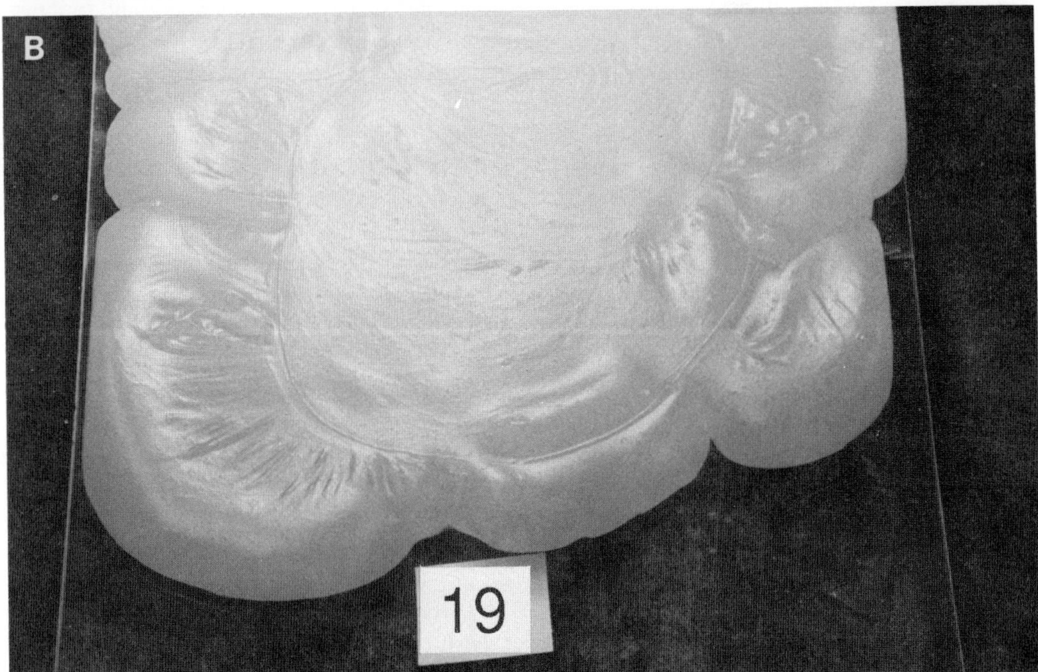

Figure 13 Photographs of representative wax flows labeled with run number. The flow direction is from top to bottom in each frame. (A) Pillows and deep clefts. (B) Lobed flow fronts. Run numbers 14(a) and 19(b).

evolves to a new thermal equilibrium. This evolution is sketched in Fig. 15. The dissolution therefore increases the melt fraction (the percentage of melt) slightly and also increases the permeability (reduces the resistance to fluid flow) of the melt, since for a fixed geometry, permeability is a function of melt fraction porosity (e.g., Turcotte and Schubert, 1982). Dissolution cannot

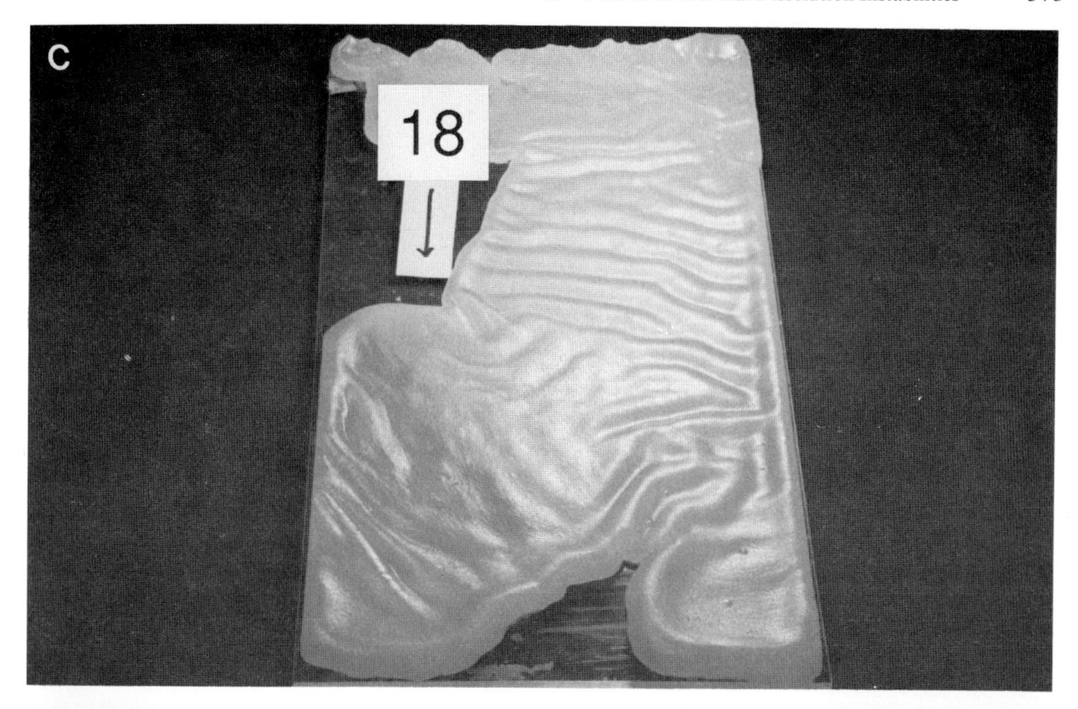

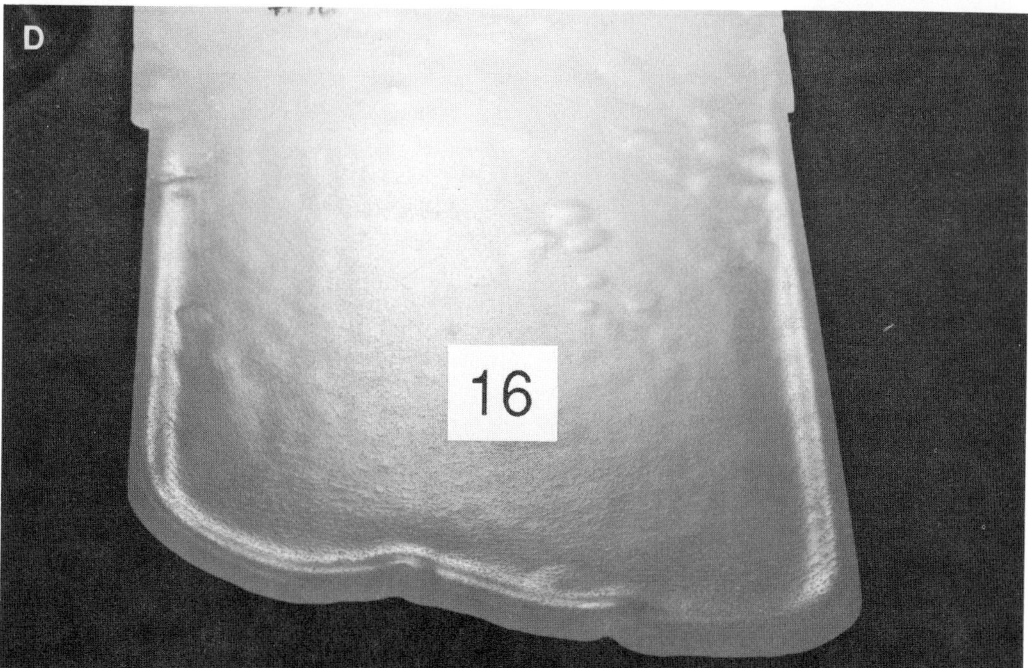

Figure 13 Continued. Photographs of representative wax flows labeled with run number. The flow direction is from top to bottom in each frame. (C) A rippled sheet; (D) Smooth sheet flow. Run numbers 18(c) and 16(d).

immediately return the melt/matrix system to thermodynamic equilibrium since the rate of dissolution is governed by the rate of diffusion across a boundary layer in the liquid. Dissolution will be slightly slower (on the order of 0.5 to <0.01 cm · day^{-1} at 1300 to 1050° C; Kutolin and Agafanov, 1978; Scarfe *et al.*, 1980; Donaldson, 1985; Thornber and Heubner, 1985; Kuo and Kirk-

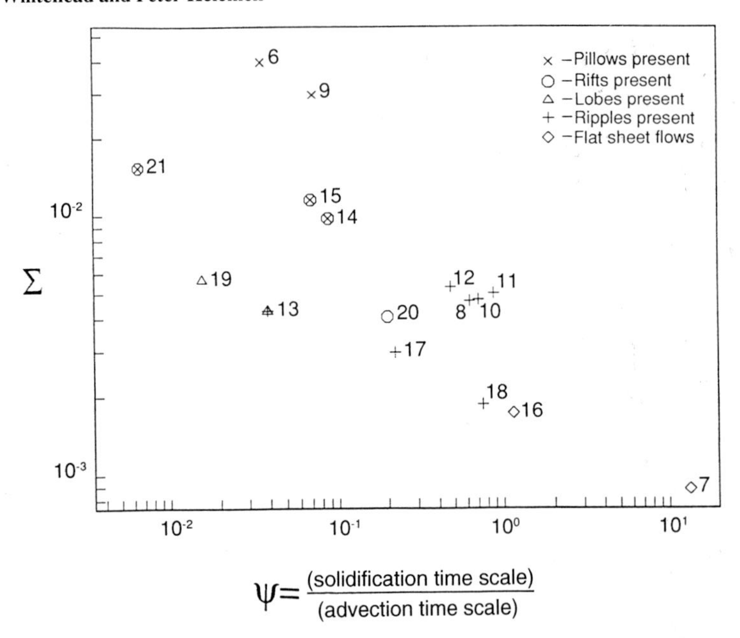

Figure 14 Scaled roughness Σ versus Ψ. Roughness decreases with increasing Ψ. Morphology types are denoted by symbols. There is a general progression from pillows to flows with deep clefts, to lobes, to folds and then finally to flat sheets.

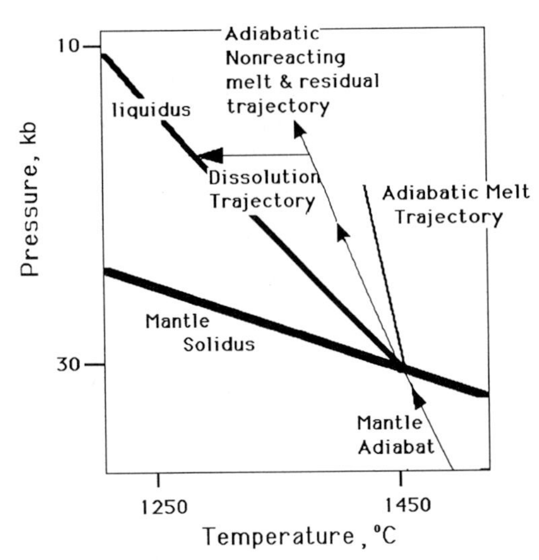

Figure 15 Sketch of the pressure–temperature relation for mantle material. The mantle material rises to the solidus. Liquid forms and flows upward by compaction. If the liquid were rising without thermal contact with the residual mantle material, it would rise at the adiabatic melt trajectory (right curve). However, the melt and residual material are in thermal contact, so if they both rise at the same temperature without dissolution, they rise at the adiabatic nonreacting melt and residual trajectory (middle curve). However, as the liquid rises to lower pressures, the liquid is above its liquidus (left curve), so it is superheated. It would dissolve residual material and follow the dissolution trajectory over to the liquidus.

patrick, 1985a, 1985b; Brearley and Scarfe, 1986; Tsuchiyama, 1986a, 1986b) than the rate of porous flow due to compaction (on the order of 2 to 0.1 cm · day^{-1} at 1300 to 1050° C; Watson, 1982; Riley and Kohlstedt, 1990, 1991; Riley *et al.*, 1990). This will be especially true as chemical equilibrium is approached, since dissolution rates for a given solid phase decrease markedly as saturation of the liquid with that phase is approached (Kuo and Kirkpatrick 1985a, 1985b; Brearley and Scarfe, 1986).

This can be quantified by assuming a liquid diffusion constant of $D = 10^{-7}$ to 10^{-10} $cm^2 \cdot s^{-1}$ for components such as Al and Si at temperatures of 1300 to 1050° C (Hofmann, 1980; Chekmir and Epel'baum, 1991; Shimizu and Kushiro, 1991; Watson and Baker, 1991) and an effective boundary layer thickness d_b ranging from 10 to 1000 μm for the case of a well-stirred, moving liquid (Kuo and Kirkpatrick, 1985b). The e-folding time to return to thermodynamic equilibrium d_b^2/D is about 10^1 to 10^8 s. A "diffusive Peclet number" Pe_D is generated by dividing this time by the time it takes melt to rise through a mantle region of interest L/w, so $Pe_D = wd_b^2/DL$. Recall that the thermal studies discussed earlier in this paper showed that the parameter group $wd^2/\kappa L$ must be of order 1 for the instability to arise. It is probable that the rise time for melt will be in the above range and that instability will arise in regions with small L (of order a few meters) where melt suddenly encounters a matrix it can dissolve.

Two exploratory laboratory experiments have been conducted to attempt to produce this instability. In the first, a thin tank of size 58.3 cm high × 63.5 cm long × 1.0 cm wide was filled with dry table salt. Water was slowly fed in from above through a pipe with many small holes to even the flow. The water level was increased to 10–20 cm above the salt so that the flow evened itself out through a natural stratification. The water slowly descended through and dissolved the salt. At the bottom, the water filtrate escaped through a 3-cm layer of foam rubber that lay above approximately 50 small holes in the base of the tank.

For the 10 runs to date, flow rates have ranged from 1 to 5 cc · s^{-1}, and in all cases the free surface of the salt became unstable. Figure 16 shows a typical evolution of the salt surface. At first, the salt surface began to move downward uniformly as the descending water dissolved the salt. However, within 5 min, undulations with a wavelength of roughly 2 to 4 cm were visible on the surface of the salt. As time progressed, the amplitude of the undulations became greater and the wavelength appeared to increase. When the slope of the salt-free surface reached a fixed angle, probably the angle of repose, the salt would tumble down the slope and the interface thereafter resembled a sawtooth pattern of peaks and valleys. As time progressed further, the distance between the peaks of the sawtooth became larger as small valleys were eliminated and large ones persisted. The small valleys appeared to be worn down on one side or the other by the larger valleys and by grain saltation down the slopes.

The experiment, although suggestive, had some serious drawbacks. Air dissolved in the water was driven out of solution as the salt dissolved and the water became saturated. This happened even though the air had been degassed as much as was practical by using very hot water over long time periods. Air bubbles would accumulate in the upper few centimeters of the salt and then percolate upward. Additionally, the experimental re-

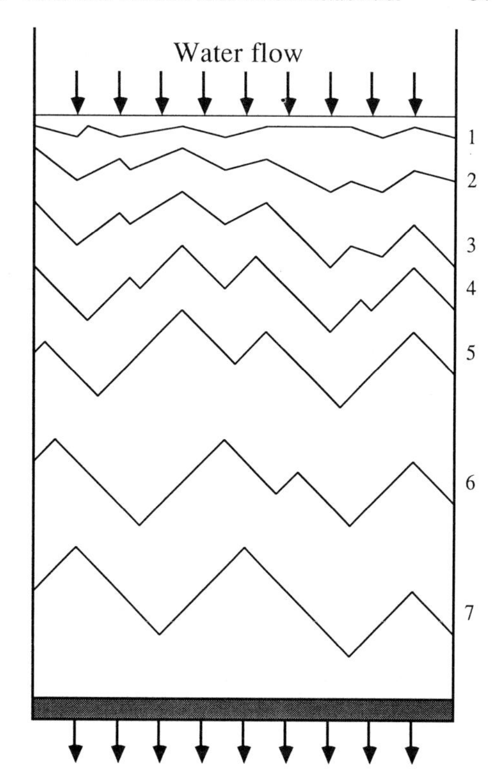

Figure 16 Sketch of the evolution of the upper surface of salt produced by the downward percolation of water and exit flow through a porous base. The numbers correspond to time progression.

sults were complicated by the absence of a rigid network in the porous media. The shape of the top salt surface was constantly modified by "saltation" of the salt grains, which rolled down from high points, where relatively little dissolution had occurred, into valleys where more dissolution had occurred. Thus the morphology of the dissolution channels was obscured.

A second experiment duplicated the first geometry but included glass spheres of 0.1 mm diameter mixed with roughly 30% salt (with grains 0.03 mm). The glass and salt grains were intended to duplicate relatively soluble (pyroxene) and insoluble (olivine) components in the mantle matrix. The presence of glass balls provided a rigid network that impeded "saltation" of the salt grains, allowing observation of the morphology of the dissolution channels. A thin layer of salt was placed over the glass bead/salt mixture so that the descending water would be close to saturated before it percolated down into the glass/salt mixture.

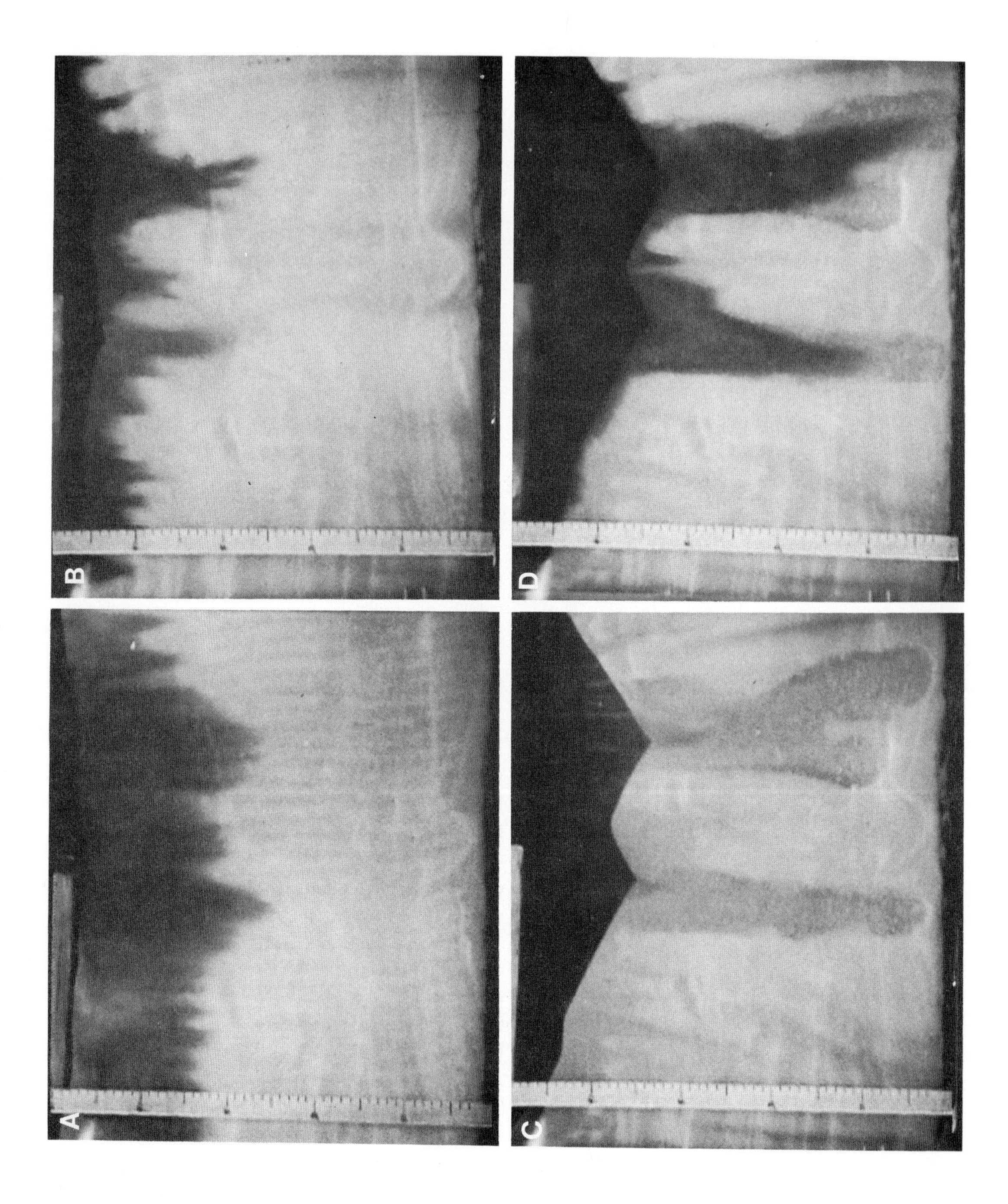

Figure 18 The glass sphere and salt mixture experiment at an advanced stage of development. After about 17 min, one region is open. Vertical fingers of dissolution are seen in other regions.

This experiment also exhibited the surface instability, but as the instability grew, the peaks and valleys along the top surface were much more subdued since the glass ball network underwent relatively minor compaction due to salt dissolution. Instead a series of finger-like vertical veins of a salt-free glass matrix were formed, alternating with regions still retaining a salt-glass mixture. Figures 17 and 18 show photographs of the tracing dye descending into the model matrix. The water flow is channeled into the salt-free fingers where the permeability is higher. The qualitative aspects of the instability, particularly channelization of the water flow into the salt-free zones with a lower flow resistance, and the emergence of these channels as a hydrodynamic instability seem clearly confirmed.

Acknowledgments

Particular thanks are due to Robert E. Frazel who assisted with numerous laboratory experiments and produced many of the the photographs. We also thank Peter Olson and Frank Richter for constructive suggestions. Supported by the Experimental and Theoretical Geophysics Division, National Science Foundation, under Grants EAR 87 08033 and EAR 89–16857.

References

Bonatti, E., and Harrison, C. G. A. (1988). Eruption styles of basalt in oceanic spreading ridges and seamounts: Effects of magma temperature and viscosity, *J. Geophys. Res.* **93,** 2967–2980.

Brearley, M., and Scarfe, C. M. (1986). Dissolution rates of upper mantle minerals in an alkali basalt melt at high pres-

Figure 17 Photographs of an experiment with a mixture of about 80% glass spheres and 20% salt. Water flows from the top to bottom. (A) After about 2 min, injected blue dye reveals vertical fingers of flow. The dissolution front at the top is becoming uneven. (B) After about 9 min, injected blue dye shows that water flows at higher velocity and penetrates preferentially in some regions. (C) After about 13 min, dissolution fingers are clearly visible. (D) After about 13.5 min the injected blue dye shows that almost all flow occurs within the largest fingers.

sure: An experimental study and implications for ultramafic xenolith survival, *J. Petrol.* **27,** 1157–1182.

Bruce, P. M., and Huppert, H. E. (1989). Thermal control of basaltic fissure eruptions, *Nature* **342,** 665–667.

Bruce, P. M., and Huppert, H. E. (1990). Solidification and melting along dykes by the laminar flow of basaltic magma, *in* "Magma Transport and Storage" (M. P. Ryan, ed.), Wiley, Chichester/Sussex, England.

Chadam, J., Hoff, D., Ortoleva, P., and Sen., A. (1986). Reactive-infiltration instability, *J. Appl. Math.* **36,** 207–238.

Chekmir, A. S., and Epel'baum, M. B. (1991). Diffusion in magmatic melts: New study, *in* "Physical Chemistry of Magmas" (L. L. Perchuk and I. Kushiro, eds.), Applications in Physical Chemistry, Vol. 9, pp. 99–119, Springer-Verlag, New York.

Delaney, P. T., and Pollard, D. D. (1982). Solidification of basaltic magma during flow in a dike, *Am. J. Sci.* **282,** 856–885.

Donaldson, C. H. (1985). The rates of dissolution of olivine, plagioclase, and quartz in a basalt melt, *Min. Mag.* **49,** 683–693.

Fink, J. H., and Griffiths, R. W. (1990). Radial spreading of viscous-gravity currents with solidifying crust, *J. Fluid Mech.* **227,** 485–509.

Fowler, A. C. (1985). A mathematical model of magma transport in the asthenosphere, *Geophys. Astrophys. Fluid Dyn.* **33,** 63–96.

Fowler, A. C. (1990a). A compaction model for melt transport in the Earth's asthenosphere. Part I. The basic model, *in* "Magma Transport and Storage" (M. P. Ryan, ed.), pp. 3–14, Wiley, Chichester/Sussex, England.

Fowler, A. C. (1990b). A compaction model for melt transport in the Earth's asthenosphere. Part II. Applications, *in* "Magma Transport and Storage" (M. P. Ryan, ed.), pp. 15–32, Wiley, Chichester/Sussex, England.

Hofmann, A. W. (1980). Diffusion in natural silicate melts: A critical review, *in* "Physics of Magmatic Processes" (R. B. Hargraves, ed.), pp. 385–418, Princeton Univ. Press, Princeton, NJ.

Hughes, C. J. (1982). "Igneous Petrology," Elsevier Scientific, New York.

Huppert, H. E. (1982). The propagation of two-dimensional and axisymmetric viscous gravity currents over a rigid horizontal surface, *J. Fluid Mech.* **121,** 43–58.

Kelemen, P. B., Johnson, K. T. M., Kinzler, R., and Irving, A. J. (1990). High field strength depletions in arc basalts due to mantle–magma interaction, *Nature* **345,** 521–524.

Kelemen, P. B., Dick, H. J. B., and Quick, J. E. (1992). Production of harzburgite by pervasive melt-rock reaction in the upper mantle, *Nature* **358,** 635–641.

Kent, R. W., Ghose, N. C., Paul, P. R., Hassan, M. J., and Saunders, A. D. (1992). Coal–magma interaction: An integrated model for the emplacement of cylindrical intrusions, *Geol. Mag.* **129**(6), 753–762.

Kuo, L. C., and Kirkpatrick, R. J. (1985a). Dissolution of mafic minerals and its implications for the ascent velocities of peridotite-bearing basaltic magmas, *J. Geol.* **93,** 691–700.

Kuo, L. C., and Kirkpatrick, R. J. (1985b). Kinetics of crystal dissolution in the system diopside–forsterite–silica, *Am. J. Sci.* **285,** 51–90.

Kutolin, V. A., and Agafanov, L. V. (1978). Composition of the upper mantle in light of the relative stability of ultramafic nodules, *Geol. Geofiz.* **19,** 3–13.

Lowell, R. P. (1990). Thermoelasticity and the formation of black smokers, *Geophys. Res. Lett.* **17,** 709–712.

Marsh, B. D. (1979). Island arc development: Some observations, experiments and speculations, *J. Geol.* **87,** 687–713.

McKenzie, D. P. (1984). The generation and compaction of partial melts, *J. Petrol.* **25,** 713–765.

Ortoleva, P., Merino, E., Moore, C., and Chadam, J. (1987a). Geochemical self-organization. I. Reaction-transport feedbacks and modeling approach, *Am. J. Sci.* **287,** 979–1007.

Ortoleva, P., Chadam, J., Merino, E., and Sen, A. (1987b). Geochemical self-organization. II. The reactive-infiltration instability, *Am. J. Sci.* **287,** 1008–1040.

Riley, G. N., Jr., and Kohlstedt, D. L. (1990). An experimental study of melt migration in an olivine–melt system, *in* "Magma Transport and Storage" (M. P. Ryan, ed.), pp. 57–92, Wiley, Chichester/Sussex, England.

Riley, G. N., Jr., and Kohlstedt, D. L. (1991). Kinetics of melt migration in upper mantle type rocks, *Earth Planet. Sci. Lett.* **105,** 500–521.

Riley, G. N., Jr., Kohlstedt, D. L., and Richter, F. M. (1990). Melt migration in a silicate liquid–olivine system: An experimental test of compaction theory, *Geophys. Res. Lett.* **17,** 2101–2104.

Ryan, M. P., and Blevins, J. Y. K. (1987). "The Viscosity of Synthetic and Natural Silicate Melts and Glasses at High Temperatures at 1 Bar (10^5 Pascals) Pressure and at Higher Pressures," U.S. Geol. Surv. Bull. **764,** p. 563.

Saffman, P. G., and Taylor, G. I. (1958). The penetration of a fluid into a porous medium or Hele–Shaw cell containing a more viscous liquid, *Proc. Roy. Soc. London A* **245,** 312–29.

Salters, V. J. M., and Hart S. R. (1989). The Hf-paradox, and the role of garnet in the MORB source, *Nature* **342,** 420–422.

Scarfe, C. M., Takahashi, E., and Yoder, H. (1980). Rates of dissolution of upper mantle minerals in an alkali–olivine basalt melt at high pressures, *Carnegie Inst. Washington Yearbook* **79,** 290–296.

Scott, D. T., and Stevenson, D. J. (1984). Magma solitons, *Geophys. Res Lett.* **11,** 1161–1164.

Shimizu, N., and Kushiro, I. (1991). The mobility of Mg, Ca and Si in diopside-jadeite liquids at high pressures, *in* "Physical Chemistry" (L. L. Perchuk and I. Kushiro, eds.), Vol. 9, pp. 192–212, Springer-Verlag, New York.

Sleep, N. H. (1974). Segregation of magma from a mostly crystalline mush, *Geol. Soc. Am. Bull.* **85,** 1225–1232.

Stevenson, D. J. (1989). Spontaneous small-scale melt segregation in partial melts undergoing deformation *Geophys. Res. Lett.* **16,** 1067–1070.

Thornber, C., and Heubner, J. S. (1985). Dissolution of olivine in basaltic liquids: Experimental observations and applications, *Am. Mineral.* **70,** 934–945.

Tsuchiyama, A. (1986a). Experimental study of olivine-melt reaction and its petrological implications, *J. Volcanol. Geotherm. Res.* **29,** 245–264.

Tsuchiyama, A. (1986b). Melting and dissolution kinetics: Application to partial melting and dissolution of xenoliths, *J. Geophys. Res.* **91,** 9395–9406.

Turcotte, D. L. (1981). Some thermal problems associated with magma migration, *J. Volcanol. Geotherm. Res.* **10,** 267–278.

Turcotte, D. L. (1982). Magma migration, *Annu. Rev. Earth Planet. Sci.* **10,** 397–408.

Turcotte, D. L., and J. L. Ahern. (1978). A porous flow model for magma migration in the asthenosphere, *J. Geophys. Res.* **83,** 767–772.

Turcotte, D. L., and G. Schubert, (1982). "Geodynamics, Applications of Continuum Physics to Geological Problems," Wiley, New York.

Watson, E. B. (1982). Melt infiltration and magma evolution, *Geology* **10,** 236–240.

Watson, E. B., and Baker, D. R. (1991). Chemical diffusion in magmas: An overview of experimental results and geochemical applications, *in* "Physical Chemistry of Magmas" (L. L. Perchuk and I. Kushiro, eds.), Applications in Physical Chemistry Vol. 9, pp. 120–151, Springer-Verlag, New York.

Whitehead, John A., and Gans, Roger F. (1974). A new, theoretically tractable earthquake model, *Geophys. J. Roy. Astron. Soc.* **39,** 11–28.

Whitehead, J. A., Jr., Dick, H. J. B., and Schouten, Hans. (1984). A mechanism for magmatic accretion under spreading centers, *Nature* **312,** 146–148.

Whitehead, J. A., and Helfrich, K. R. (1990). Magma waves and diapiric dynamics *in* "Magma Transport and Storage" (M. P. Ryan, ed.), pp. 53–76, Wiley, Chichester/Sussex, England.

Whitehead, J. A., and Helfrich, K. R. (1991). Instability of flow with temperature-dependent viscosity: A model of magma dynamics, *J. Geophys. Res.* **96**(B3), 4145–55.

Author Index

Geographical Index

Subject Index

International Geophysics Series

EDITED BY

RENATA DMOWSKA
Division of Applied Sciences
Harvard University
Cambridge, Massachusetts

JAMES R. HOLTON
Department of Atmospheric Sciences
University of Washington
Seattle, Washington

Volume 1 BENO GUTENBERG. Physics of the Earth's Interior. 1959*

Volume 2 JOSEPH W. CHAMBERLAIN. Physics of the Aurora and Airglow. 1961*

Volume 3 S. K. RUNCORN (ed.). Continental Drift. 1962*

Volume 4 C. E. JUNGE. Air Chemistry and Radioactivity. 1963*

Volume 5 ROBERT G. FLEAGLE AND JOOST A. BUSINGER. An Introduction to Atmospheric Physics. 1963*

Volume 6 L. DUFOUR AND R. DEFAY. Thermodynamics of Clouds. 1963*

Volume 7 H. U. ROLL. Physics of the Marine Atmosphere. 1965*

Volume 8 RICHARD A. CRAIG. The Upper Atmosphere: Meteorology and Physics. 1965*

Volume 9 WILLIS L. WEBB. Structure of the Stratosphere and Mesosphere. 1966*

Volume 10 MICHELE CAPUTO. The Gravity Field of the Earth from Classical and Modern Methods. 1967*

Volume 11 S. MATSUSHITA AND WALLACE H. CAMPBELL (eds.). Physics of Geomagnetic Phenomena. (In two volumes.) 1967*

Volume 12 K. YA. KONDRATYEV. Radiation in the Atmosphere. 1969*

Volume 13 E. PALMÉN AND C. W. NEWTON. Atmospheric Circulation Systems: Their Structure and Physical Interpretation. 1969*

Volume 14 HENRY RISHBETH AND OWEN K. GARRIOTT. Introduction to Ionospheric Physics. 1969*

Volume 15 C. S. RAMAGE. Monsoon Meteorology. 1971*

*Out of Print.

Volume 16 James R. Holton. An Introduction to Dynamic Meteorology. 1972*

Volume 17 K. C. Yeh and C. H. Liu. Theory of Ionospheric Waves. 1972*

Volume 18 M. I. Budyko. Climate and Life. 1974*

Volume 19 Melvin E. Stern. Ocean Circulation Physics. 1975

Volume 20 J. A. Jacobs. The Earth's Core. 1975*

Volume 21 David H. Miller. Water at the Surface of the Earth: An Introduction to Ecosystem Hydrodynamics. 1977

Volume 22 Joseph W. Chamberlain. Theory of Planetary Atmospheres: An Introduction to Their Physics and Chemistry. 1978*

Volume 23 James R. Holton. An Introduction to Dynamic Meteorology, Second Edition. 1979*

Volume 24 Arnett S. Dennis. Weather Modification by Cloud Seeding. 1980

Volume 25 Robert G. Fleagle and Joost A. Businger. An Introduction to Atmospheric Physics, Second Edition. 1980

Volume 26 Kuo-Nan Liou. An Introduction to Atmospheric Radiation. 1980

Volume 27 David H. Miller. Energy at the Surface of the Earth: An Introduction to the Energetics of Ecosystems. 1981

Volume 28 Helmut E. Landsberg. The Urban Climate. 1981

Volume 29 M. I. Budyko. The Earth's Climate: Past and Future. 1982

Volume 30 Adrian E. Gill. Atmosphere–Ocean Dynamics. 1982

Volume 31 Paolo Lanzano. Deformations of an Elastic Earth. 1982*

Volume 32 Ronald T. Merrill and Michael W. McElhinny. The Earth's Magnetic Field: Its History, Origin, and Planetary Perspective. 1983

Volume 33 John S. Lewis and Ronald G. Prinn. Planets and Their Atmospheres: Origin and Evolution. 1983

Volume 34 Rolf Meissner. The Continental Crust: A Geophysical Approach. 1986

Volume 35 M. U. Sagitov, B. Bodri, V. S. Nazarenko, and Kh. G. Tadzhidinov. Lunar Gravimetry. 1986

Volume 36 Joseph W. Chamberlain and Donald M. Hunten. Theory of Planetary Atmospheres: An Introduction to Their Physics and Chemistry, Second Edition. 1987

Volume 37 J. A. Jacobs. The Earth's Core, Second Edition. 1987

Volume 38 J. R. Apel. Principles of Ocean Physics. 1987

Volume 39 Martin A. Uman. The Lightning Discharge. 1987

Volume 40 David G. Andrews, James R. Holton, and Conway B. Leovy. Middle Atmosphere Dynamics. 1987

Volume 41 Peter Warneck. Chemistry of the Natural Atmosphere. 1988

Volume 42 S. Pal Arya. Introduction to Micrometeorology. 1988

Volume 43 Michael C. Kelley. The Earth's Ionosphere. 1989

Volume 44 William R. Cotton and Richard A. Anthes. Storm and Cloud Dynamics. 1989

Volume 45 William Menke. Geophysical Data Analysis: Discrete Inverse Theory, Revised Edition. 1989

Volume 46 S. George Philander. El Niño, La Niña, and the Southern Oscillation. 1990

Volume 47 Robert A. Brown. Fluid Mechanics of the Atmosphere. 1991

Volume 48 JAMES R. HOLTON. An Introduction to Dynamic Meteorology, Third Edition. 1992

Volume 49 ALEXANDER A. KAUFMAN. Geophysical Field Theory and Method. Part A: Gravitational, Electric, and Magnetic Fields. 1992. Part B: Electromagnetic Fields I. 1994. Part C: Electromagnetic Fields II. 1994

Volume 50 SAMUEL S. BUTCHER, GORDON H. ORIANS, ROBERT J. CHARLSON, AND GORDON V. WOLFE. Global Biogeochemical Cycles. 1992

Volume 51 BRIAN EVANS AND TENG-FONG WONG. Fault Mechanics and Transport Properties in Rock. 1992

Volume 52 ROBERT E. HUFFMAN. Atmospheric Ultraviolet Remote Sensing. 1992

Volume 53 ROBERT A. HOUZE, JR. Cloud Dynamics, 1993

Volume 54 PETER V. HOBBS. Aerosol–Cloud–Climate Interactions. 1993

Volume 55 S. J. GIBOWICZ AND A. KIJCO. An Introduction to Mining Seismology. 1993

Volume 56 DENNIS L. HARTMANN. Global Physical Climatology. 1994